AF664178

CROP PROTECTION:

Management Strategies

CROP PROTECTION:
Management Strategies

Editor
Dr. D. Prasad
Division of Nematology,
Indian Agricultural Research Institute, Pusa,
New Delhi - 110 002

2024
DAYA PUBLISHING HOUSE®
A Division of
Astral International Pvt. Ltd.
New Delhi - 110 002

First Impression, 2005
Reprinted, 2024

ISBN 978-81-7035-372-0 (Hardbound)
ISBN 978-93-5124-373-1 (International Edition)

Published by : **Daya Publishing House®**
A Division of
Astral International Pvt. Ltd.
– ISO 9001:2015 Certified Company –
4736/23, Ansari Road, Darya Ganj
New Delhi-110 002
Ph. 011-43549197, 23278134
E-mail: info@astralint.com
Website: www.astralint.com

Laser Typesetting : **Classic Computer Services**, Delhi - 110 035

Printed at : **Replika Press Pvt. Ltd.**

PRINTED IN INDIA

Preface

Crop Protection plays a pivotal role in agricultural production. During the past few decades, the science of crop protection has developed rapidly resulting in many diverse techniques to manage a variety of pests, pathogens and nematodes. Currently the prospects of chemical control do not appear good because of accidental and incidental trauma encountered on the field of chemical pesticides. Therefore, vigorous research efforts for exploring newer but progamatic tools for the control of insect pests, diseases and nematodes, have become imperative.

A number of cultural, biological and physical methods, are known to reduce pest population, many of them are modifications of the agro-practices being followed by the farmers, since ages, were also developed either for economic or environmental reasons. But very soon thought and realized that none of these techniques individually could serve as a panacea against target pests and pathogens including nematodes. Often invoking them resulting in resurgence of the target species of suppression of useful enemies. Therefore, our approach should be multifaceted to combat the disease complex caused by nematode-fungus, insect-fungus or insect-virus but the management tactics have not reached to the farm communities.

The book "*Crop Protection: Management Strategies*" is an attempt to analyses the problems and purpose, suitable measures for the effective management of the insect pests, plant diseases and plant parasitic nematodes. This book is an effort to bring together available information on the various plant protection subjects like entomology plant pathology, nematology, agrochemicals and a chapter on weed. The experts of the relevant fields have given a critical appraisal of the various aspects of crop protection and suggested suitable management tactics based on locally available management options.

There are 4 parts of the book. *Part I* deals with the insect problems as a whole covering microbial pesticides, *Hexamermis* for insect control, recent advances on *Ocimum* spp. for IPM, perspective in biocontrol in pest management, *Helicoverpa armigera* NPV, ovicides of lepidopterous pests, and termite management: past, present and future. *Part II* discusses Karnal bunt disease, foliar blight of wheat, diseases of crops in NEH region, ecofriendly management wilt of chickpea, powdery mildew disease, viral diseases of fruits, vegetables and spices, tomato leaf curl virus, IDM of linseed, induced resistance for disease management, rice blast epidemiology chitinases in defense against fungi, rice blast in J&K, insect pest and diseases of cold arid region, pest and disease management in bulbous crops, pests and diseases of nut fruits and application of molecular markers in disease resistance genes in plant. *Part III* brings out interactions of pollutants on plant parasites, nematode problems of vegetables and pulses, precision farming in nematode management, acarines as biocontrol agents and regulatory and phytosanitary measures for nematode management and *Part IV* highlights decontamination of pesticide residues, microbial metabolites, plant protection management through seeds and weed management for sustainable agricultural production.

The book "*Crop Protection: Management Strategies,*" not only provides references but also serves as guide and inspiration for future research into the realm of biological, chemical, physical and quarantine aspects and onslaughts of modern agriculture on it.

The scientists, teachers, students, scholars, administrators and policy makers dealing with pest and disease management in particular and plant protection in general expected this book very useful and informative.

I am grateful to all the contributors for their valuable articles and kind cooperation. We are also thankful to *Shri Anil Mittal, Daya Publishing House*, New Delhi for his technical guidance and help in bringing out this book.

New Delhi **D. PRASAD**

Contents

PART II: PLANT PATHOLOGY

Part I
ENTOMOLOGY

1

Microbial Pesticides

V.M. Pawar, P.S. Borikar and U.T. Thombre
Marathwada Agricultural University, Parbhani – 431 402, M.S.

The term 'Biopesticides' is taken to mean the use of fungi, bacteria, viruses, protozoa and nematodes through inundative or incoculative releases for the biological control of insects pests, diseases, weeds and nematodes in agricultural, medical, veterinary, forestry and horticultural systems. The use of pathogens as biological control agents has a long history with the first nematode-trapping fungus *Arthrobotrys obligospora* Fres. described in 1852. The commercial use of *Bacillus thuringiensis* Berliner (*Bt*) as a bio-insecticide dates back nearly 60 years when the first commerical formulation "Sporeine" was made available in France. Since that time there has been a gradual but steady increase in the interest and exploitation of pathogens as biopesticides.

Bt underwent substantial development during the 1970s, largely under the umbrella of IPM, where the advantage of *Bt* (specificity, non-toxic to the environment or beneficial organisms) could be more fully exploited in tandem with other pest control measures. Between 1985 and 1996, pharmaceutical and agrochemical companies attempted to research and develop other insect control pathogens such as *Beauveria bassiana* (Balsamo) Vuillemin, *Metarhizium anisopliae* (Metschnikoff) Sorokin and baculo virus. In general, there was an increase in the amount of R and D undertaken with pathogens in relation to their use as biopesticides throughout the 1980s, with the viruses receiving some consideration attention, followed by a range of fungi and nematodes.

Issues in Microbial Control

Consideration of the strength and weakness of microbial insecticides is an essential prelude to seek improvement in applications to insect control, since the entomologists and molecular geneticists will need to overcome or minimize certain disadvantages while retaining or enhancing certain advantages.

A desirable feature of microbial pathogens is that they are capable of reproduction and have the potential to amplify themselves in the field. The high degree of host specificity that makes biological insecticides so ecologically and environmentally attractive also constitutes a potentially serious drawback. Their narrow host range prevents their successful use to control the multiplicity of pests that cycle through planted areas during the course of a typical growing season.

Potency is another factor that has limited the use of biological insecticides. For economic reason, the user will not apply an insecticide prophylactically in anticipation of an outbreak that might not occur or might not reach significant proportions. The insecticide must therefore be capable of controlling an insect infestation once evidence of a threat becomes clearly established. Some biological insecticides, though highly effective under laboratory conditions, have suffered in the field where they have only been effective on light infestations. Lack of potency has been ascribed, in some cases, to poor storage capability, requirements in the field for optimum temperature, relative humidity or sunlight, short residual activity or slow killing times.

Benefits of Biopesticides

Safety and Environment

There is a considerable body of evidence that attests to the relative safety of naturally occurring fungi, viruses, nematodes and bacteria used as biopesticides compared to chemical pesticides. Safety has been assessed in terms of mammalian toxicity, non-target effects and environmental impact and biopesticide pathogens have consistently been shown to be remarkably safe for use provided basic safety precautions are taken. However, this is not to say that use of biopesticides is not without risk and it is generally recognised that biopesticides need to be evaluated and regulated by National and International regulatory authorities to ensure their safety under normal conditions of use.

Additional Control

The environmental safety of biopesticides is a strong marketing message but it is only half the message. The logical second half is that the environment which is protected provides additional pest control in the form of conserved, local natural enemies. The host specificity of the pathogens used as biopesticides means that natural enemy populations are preserved to exert control in tandem with the biopesticide. In addition the specificity of biopesticide will maintain the bio-diversity of fauna within a crop unlike chemical insecticide that generally act to reduce bio- diversity.

Reproduction and Host Finding Capability

An ideal biological control product, that a farmer would want most, would have the following properties: it would apply itself to the pest, it would kill the pest and it would give the long-lasting and widespread control by its own reproduction and spread. Biological control agents meet a large number of these attributes. For instance, fungi such as *Metarhizium* have been shown to exhibit secondary cycling and thereby reinfect other individuals beyond the point of initial infection. This is a characteristic that cannot be found in chemical pesticides but it is a significant benefit of some biopesticides that, if fully exploited, can lead to less frequent application and hence reduced costs. Nematode biocontrol agents have the advantage that, to a limited extent, they can seek out their prey, *i.e.*, they have a host finding capability. Again this is an important attribute which can have a significant impact that cannot be found in chemical pesticides.

Compatibility with IPM

The environmentally friendly credentials of biopesticides have made them ideal candidates for inclusion in IPM systems and scientists have over a twenty-five years period looked at the range of interactions between entomopathogens and other control measures relating to their use in IPM. Insect pathogens can be used in pest management in four ways:

1. By utilising naturally occurring diseases.
2. By introduction of insect pathogens into insect pest population.
3. By application of insect pathogens or their products as microbial insecticides for the immediate and temporary control.
4. By novel application of insect pathogen genes, usually involving recombinant DNA technology.

In the majority of cases the interactions between biopesticides and other means of control are extremely positive, adding to and complementing control provided by host plant resistance, cultural control and biological control. In the latter case largely through conservation of natural populations of predators and parasitoids which are not killed by the biopesticides and hence provide additional levels of control.

Production Technology Appropriate for Developing Countries

A large number of fungi are easy to mass produce on natural substrates which are cheaply available and in India, the future outlook for the development of mycoinsecticide industries providing avenues of employment opportunity for rural people are bright.

Many entomo-viruses can be produced *in vivo* using labour intensive methods that have proved highly successful in developing countries, however more sophisticated and industrial scale production technologies are available. In general, the production processes available for biopesticides tend to be more amenable to lower capital, labour intensive, intermediate technology utilising readily available substrates than chemical pesticides which makes them an eminently more suitable production technology for developing countries.

Biopesticides are advantageous due to their eco-safety, target specificity, no development of resistance, reduced number of applications, yields and quality improvement, higher acceptability and value of produce for exports and suitability for rural areas. In view of the advantages of Biopesticides, there has been wide acceptance of these globally, amounting to around 10 per cent share of agrochemical market in 2000 AD, with the growth rate of 10–15 per cent per annum. There has been a total ban on use of chemicals in green houses in Europe and there is a niche market for "Green Label" (no chemicals used) vegetables and fruits, apart from clothes from organic cotton. The use of BTs in neighbouring countries are high: China–3000 T, Malaysia–250 T, Thailand–160 T and Philippines–300 T in comparison to Indian's BT market approximately at 50 T only.

Constraints

There are a lot of constraints that have come in the way of wide adoption of IPM and Biopesticides by the Indian farmers. Some of these include:

1. Awareness training and education short falls.
2. Lack of multidisciplinary approach.

3. Technology constraints–delivery systems.
4. Technology constraints–biopesticide quality.
5. Toxicology and regulatory matters.
6. Decline of national and international research support.
7. Small, medium and micro-biopesticide enterprise.
8. Lack of knowledge and suspicion in farmers.
9. Lack of simple illustrated IPM information.
10. Absence of really well executed IPM demonstrations.
11. Degradation of information during knowledge transmission.
12. Non-adoption of IPM practices in large areas.
13. In-consistent/unplanned demands of biopesticides and biocontrol agents.
14. Short shelf life of Biocontrol Agents.
15. Absence of well equipped quality testing laboratories for biopesticides and biocontrol agents.
16. Inappropriate application technologies and equipments.
17. Dealers' interests in chemical pesticides.
18. Major Indian Pesticides Companies and Multi National Companies not yet interested in biopesticides.

Table 1.1: Candidate Biopesticides Available for Pest Control

	Pathogen	*Major Target Groups*
	Bacteria	
1.	*Bt* (Cry I A–G, Cry II A–C, Cry III A–D, Cry IV A–D, C IIA)	Lepidoptera, Coleoptera, Diptera
2.	*B. sphaericus*	Diptera
	Viruses	
1.	Nuclear Polyhedrosis Viruses	Lepidoptera, Hymenoptera
2.	Granulosis Viruses	Lepidoptera
	Fungi	
1.	*Beauvaria bassiana* and *Metarhizium anisopliae*	Coleoptera, Lepidoptera, Hymenoptera, Orthoptera
2.	*Verticilium lecani*	Homoptera and Thysanoptera

Table 1.2: Availability of Biopesticides in India

Biopesticides/Bioagents	*Quantityt/Annum*	*Constraints*
BTK	Over 50,000 kg	–
NPV (Liquid)	Over 5 Lakh LE	Quality, Stability
NPV (WP)	Nil	–
Beauveria	Meagre	Quality, Stability
Trichoderma	Over 500 T	Quality, Stability

Table 1.3: Interaction Matrix for Five Commonly Used Entomopathogens

Factors	*Nuclear Polyhedrosis Virus*	*Bacillus thuringiensis*	*Beauveria bassiana*	*Verticillum lacanii*	*Metarhizium anisopliae*
Nuclear Polyhedrosis Virus		+	+		
Bacillus thuringiensis	+		+		
Verticillium lecanii					
Beauveria bassiana	+	+			
Metarhizium anisopliae		+			
Paecilomyces spp			+		
Other Natural Enemies					
Other Viruses	+				
Protozoa			+		
Nematodes		+	+		
Predators		+	+		
Parasitoids	+	+			
Chemical Pesticides					
Insecticides	+	+	+	+	+
Fungicides	+				
Herbicides			+		
Insect Growth Regulators	+	+	+		
Natural Products					
Oils			+		
Neem	+		+		
Plant Metabolites		+			
Host Plant Effects					
Species	+	+	+		
Cultivar	+	+			
Surface Factors	+				
Plant Extracts			+		
Cultural Control					
Soil Type				+	
Soil Condition			+		
Fertilizer			+		
Irrigation	+		+		
Environmental Conditions					
Temperature	+	+	+	+	+
Humidity				+	+
Irradiation	+	+	+		+
Rainfall		+	+		
Formulation	+	+	+	+	+
Adjuvants	+	+			
Application Technology	+	+	+	+	+

Table 1.4: Nuclear Polyhedrosis Viruses Recorded in India

Pest	*Crop*
Nuclear Polyhedrosis Viruses	
Helicoverpa armigera	Chickpea and others
Spodoptera litura	Tobacco and others
S. exigua	Tomato and others
S. mauritia	–
A. moorei	Pulses
Agrotis ipsilon, A. segetum	Potato and others
Anadevidia peponis	Gourds
Chrysodeixis chalcites	–
Trichoplusiani	Potato and others
Thysanoplusia orichalcea	Potato and others
Spilarctia obnliqua	–
Adisura atkinsoni	Pulses
Lymantria obfuscata	Walnut
Plutella xylostella, Hellula undalis	Cabbage
Corcyra cephalonica	Rice
Mythimna separata	Rice
Galleria mellonella	Bee wax combs
Phthorimaea operculella	Potato
Dasychira mendosa	–
Antheraea mylitta	–
Pericallia ricini	–
Granulosis Viruses	
Cnaphalocrocis medinalis	Rice
Pericallia ricini	
Achaea janata	Castor
Agrotis ipsilon	Potato
A. segetum	Potato
A. biconica	Potato
Phthorimaea operculella	Potato
Spodoptera litura	Tobacco, Cabbage
Galleria mellonella	Bee wax combs
Thysanoplusia orichalcea	Potato
Trichoplusia ni	–
Plutella xylostella	Cabbage
Corcry cephalonica	Rice
Chilo infuscafellus	Sugarcane
Other Viruses	
Virus	Pest
Non occluded virus	*Oryctes rhinoceros*
Cytoplasmic polyhedrosis virus	*Helicoverpa armigera, Spilarcita obliqua, Sphanarches anisodactylus, Maruca testulalis, Papilio demoleus, Euproctis fratema* *H. armigera, S. litura*
Asco virus	*H. armigera, Amsacta Moorei*
Pox virus	*Spodoptera litura, Achoea janata*
Rickettsiae like organism (RLO)	

Table 1.5: Engineering of Baculoviruses with Pestidal Genes to Enhance the Virulence

Pesticidal Gene	*Source*
AaIT	*Andractonus australis*
	Andractonus australis
Endotoxin of *Bt.*	*Bt.* kurstaki HD73
Tx P–1	*Pyemotus tritici*
Alpha latro insect toxin	Black widow spider

Table 1.6: Entomopathogenic Bacteria Recorded from India

Species	*No. of Hosts*
Bacillus thuringiensis	68
B. cerus	7
B. megaterium	1
B. subtilis	1
Pseudomonas aeruginosa	6
Serratia marcescens	28
Streptococcus faecalis	4

Table 1.7: *Bacillus thuringiensis*: Target Pests and Crops

Plutella xylostella, Crocidolomia binotalis, Pieris brassicae	Crucifers
Heliothis armigera	Tomato, Sunflower
Spodoptera litura, H. armigera	Tobacco
S. litura	Beet root
Athalia lugens proxima	Radish
Papilio demoleus	Citrus
Achaea janata	Castor
Helicoverpa armigera, Amsacta moorei	Pulses
Acigona steniellus	Sugarcane
Earias vittella	Okra
Spoladea (=Hymenia) recurvalis	Amaranthus
Phthorimaea operculella	Potato

Table 1.8: *Bt.* Expression in High Value Crops

Plant	*Promoter*	*Genes*	*Target Insects*
Cotton-Bollguard (R)	Monsanto	Cry I A–C	Tobacco budworm
Corn-Knock out (R)	Ciba Seeds	Cry I A	European corn borer
Corn-Nature Gard (R)	Mycogen	Cry I A b	European corn borer
Potato–Newleaf (R)	Nature Mark	Cry 3 A	Colorado Potato Beetle

Table 1.9: Transgenic Plants with Pest Resistance

Crop	Gene	Pest
Cotton	*Bt* toxin Cry 1 A (b)	Boll worm
Maize	Syn Cry 1 A (b)	*O. nubilalis*
Tomato	Cry 1 A (c)	Fruit borer
Rice (Japonica)	Modified Cry 1 A (b)	Striped stem borer + *C. medinalis*
Tobacco	Cry 1 A (a)	
Cauliflower	Cry 1 A (a)	
Rice (Indica)	Syn Cry 1 A (b)	*S. incertulas* *C. medinalis* *C. suppressalis* *M. patnalis*

Table 1.10: Registration of Commercial Formulation of *Bacillus thuringiensis* Formulation in India

Sl.No.	Trade Name	Product Name	Crop Pest	Dose
1.	Bioasp 32000 IU/mg 4–6% a.e. (Water dispersible Wettable powder)	*Bt.* var. kurstaki H- serotypes-3a 3b strian 7-a	*H. armigera* *S. litura* *Pectionophora gossypiella*	2 kg
2.	Biolep 3000 IU/mg 5–8% a.i. (WDP)	*Bt.* var. kurstaki H- serotypes-3a 3b strian Z-52	*Heliothis* sp. *Pectinophora* sp. *Earias* sp.	1–2 kg
3.	Delfin WG	*Bt* Berliner 5 sp. *kurstaki*	*Plutella maculipennis* *H. armigera* *Spodoptera* sp. *Earias insulana*	
4.	Dipel 8-L (3.5%) a.i.	*B. thuringiensis*	Boll worm Fruit borer Diamond back moth	
5.	Halt (WDP)	*B. thuringiensis*	*Plutella maculipennis*	
6.	Spicturin (25% Flowable concentrate)	*B. thuringiensis* var galleriae	*Plutella maculipennis*	10 L/h
7.	HIL *BTK* (32.000 IU/mg) water dispersible granules (WG) 6.4% a.i.	*B. thuringiensis var kurstaki*	Lepidoptera larvae	1.5 lit/ha

Table 1.11: Entomofungal Pathogens which could be Registered for Commercialisation

Pathogen	Target	Production
Aschersonia sp.	While files, scale insects liquid fermentation	
Beauveria bassiana	Borers	Semi-solid
Hirsutella spp.	Mites	Semi-solid
Metarhizium anisopliae	Rhinoceros beetle, pyrilla	Semi-solid, liquid solid
Nomouraea rileyi	Caterpillars (*Spodoptera, Helicoverpa*)	Liquid fermentation
Verticillium lecanii	Green scale, aphids	

Table 1.12: Entomofungal Pathogens Recorded from India

Total Recorded	*Important Species*	*Important Insects Attacked*
92	*Beauveria bassiana*	*Helicoverpa armigera, Spodoptera litura, Chilopartellus, Scirpophaga incertulas, Myllocerus* sp., *Lymantria obfuscata*
	Metarhizium anisopliae	*Pyrilla perpusilla, Oryctes rhinoceros, Holotrichia consanguinea*
	Verticillium lecanii	*Cocus viridis, Archips termias, Emposca* sp., *Aphis gossypii, Tetranychus* sp.
	V. aphidicola	*Aphis spiraecola, Lipaphis erysimi, Brevicoryne brassicae, Myzus persicae*
	Nomuraea rileyi	*Spodoptera litura*
	Helicoverpa armigera	*Trichoplusia* spp.
	Hirsutella spp.	*Nilaparvata lugens, Tetranychus* sp.
	Erynia neoaphidis	*Lipaphis erysimi*
	Paecilomyces	*Bemisia tabaci, Aleurocanthus woglumi*
	Farinosus	
	Entomophthora aulicae	*Lymantria obfuscata*

Table 1.13: Entomophilic Nematodes Recorded from India

Total Recorded	*Important Species*
40	*Steinernema carpocapsae, S. feltiae, S. bibionis*
	Heterorhabditis indicus, H. bacteriophora
	Ovomermis albicans
	Rhabditis sp.

Current Status

There is now considerable know-how available that is increasingly being utilised to develop commercial biopesticide products, virus, fungi and nematode products are now commercially available in developed countries which are produced and marketed by multinational companies or SMEs. A number of biopesticides are also produced and are widely available in developing countries.

An analysis of the 194 papers dealing with the entomopathogens *B. thuringiensis*, nucleopolyhedrovirus (NPV), *B. bassiana* and *Verticillium lecanii* (Zimmerman) Viegas that were published between 1989–96 (CABPESTCD ROM 1988–96) indicated that there are significant differences in the amount of research undertaken and the pathogens studied between different geographical regions throughout the world. The majority of work published on these entomopathogens is undertaken in South East Asia and North America followed by Western Europe, Eastern Europe and Russia. Generally, little research on entomopathogens originates from Africa, the Middle East, Australia or the Caribbean. Central and South American research is dominated by work on *Bt.* and *V. lecanii* whereas in North America and South East Asia the amount of work published on these fungi is second to the number of publications dealing with *Bt.* and NPVs.

Table 1.14: Entomopathogenic Nematodes Intercepted from India

Nematode	*Host*
Agamermis sp.	*S. incertulus*
Agamermis sp.	*Lymentria obfuscata*
Agamermis sp.	*Scirophaga incertulus*
Coccinellimermis sp.	*Coccinella septempunctata*
Heterotylenchus	*Musca crassirostris*
H. hydrabadensis	*Morella hortensia*
Howardula albopunctata	*Sepsis* spp.
H. aptini	*Microcephalothrips abdominalis*
H. aptini	*Megaluriothrips* spp.
Hexamermis sp.	*Nilaparvata lugens*
Hexamermis sp.	*Chrysopa* sp.
Hexamermis sp.	*Chilo infuscatellus, D. sacchariphagus, Scirpophaga nivella*
Hexamermis sp.	*Leptocorysa augur*
Hexamermis sp.	*Spodoptera* sp.
Geomermis indica	*Pempherulus affinis*
Limnomermis sp.	*Neochetina* sp.
Mermis sp.	*Amascta moorei*
Mermis sp.	*C. infuscatellus*
Ovamermis sp.	*Helicoverpa, H. assulta*
Pentatomermis sp.	*Nezara viridula*
Romanomermis iyengari	*Culex tritaenlorhynchus*
Romanomermis sp.	*C. fatigans*
Romanomermis culicivorax	*C. fatigans*
Panagrolaimus sp.	*C. partellus*
Panagrolaimus mygophilus	*Musca domestica*
Parasitorhabdus sp.	*Sesamia inferens*
Pelodera sp.	*M. domestica*
Protrellus chauhani	*P. americana*
P. travassosi	*Neostylopyaga sexpustulata*
P. phyllodromi	*Blarra orientalis*
P. murtii	*Aryllus domesticus*
Praecocilenchus feruginiphorus	*Rhynchophorus ferruginens*
Thelostoma guptai	*Periplaneta americana*
T. kherai	*Periplaneta americana*
Steinernema carpocapsae	*Scirpophaga, S. incertulus, Mythimna separata*
S. feltiae	*H. armigera, Mylabris pustulata*
	H. armigera
S. feltiae	*Pieris brassicae*
H. bacteriophora	*Amasacta albistriga*
H. indica	*Scirpophaga excerptalis*

Table 1.15: Antagonists Effective Against Plant Parasitic Nematodes

Antagonists	*Nematodes*	*Crop*
Paecilomyces lilacinus	*Meloidogyne incognita*	Tomato
P. lilacinus	*M. javanica*	Tomato
P. lilacinus	*M. javanica*	Okra
P. lilacinus	*M. javanica*	Brinjal
P. lilacinus	*M. incognita*	Chickpea
P. lilacinus	*M. incognita*	Chilli
P. lilacinus	*M. incognita*	Brinjal
P. lilacinus	*Rotylenchulus reniformis*	Pigeonpea
P. lilacinus	*M. incognita*	Tomato
P. lilacinus	*M. javanica*	Chickpea
P. lilacinus	*M. incognita*	Brinjal
P. lilacinus	*M. incognita*	Sunflower
Verticillium chlamydosporium	*Heterodera cajani*	Pigeonpea
V. chlamydosporium	*M. javanica*	Chickpea
V. chlamydosporium	*M. incognita*	Tomato
V. chlamydosporium	*M. incognita*	Tomato
V. chlamydosporium	*H. cajani*	Cowpea
Pasteuria penetrans	*M. javanica*	Tomato
Pasteuria penetrans	*M. incognita*	Tomato
Pasteuria penetrans	*M. incogita*	Black pepper

Table 1.16: Protozoan Pathogens of Insect Pests in India

Species	*Pests*
Vairimorpha sp.	*Helicoverpa armigera, Spodoptera litura,* and *S. exigua*
Farinocystis tribolii	*Tribolium castaneum*
Nosema sp.	*Anadevidia* (=*Plusia*) *peponis, Spilarctia obliqua, Adisura atkinsoni, Crocidolomia binotalis*
Mattesia dispora	*Ephestia cautella*
Bacillidium sp.	*Thrips flavus*
Microsporidium sp.	*Spodoptera litura, S. exigua, Utethesa pulchella*
Tetrahymena sp.	*Chilo partellus*

Table 1.17: Mass Production of Entomopathogens

Entomopathogen	*Production*	*Entomopathogen*	*Production*
Viruses	*In vivo*	Nematodes	*In vivo* culturing
Bacteria	*In vivo*		*In vitro* culturing
	In vitro	Protozoa	*In vivo*
Fungi	Surface		
	Submerged		

Table 1.18: Field Recommendations of Microbial Agents for Insect Pest Control

Crop	*Pest*	*Agent*	*Recommendation*
Sugarcane	Sugarcane shoot borer *Chilo infuscatellus Snellen* (Lepidoptera: Pyralidae)	Granulosis virus (GV)	250 LE/ha or 750 virosed larvae (10^6–10^7) + 0.50 per cent Sandovit mixed in 200 L of water and sprayed on 30th day of crop growth. Subsequent sprays at 15-day intervals. 0.5 kg/ha for proper distribution of spores, 2g of spore dust (Containing 200 million spores) is deposited with a spacing of 10 feet.
	White grub *Leucopholis lepidophora* Blanchard (Coleoptera : Scarabaedae)	*Bacillus popilliae* Dutky	
		Metarhizium anisopliae (Metchnikoff) Sorokin	Required quantity of spores are mixed with 0.05 per cent Sandovit and a water suspension is prepared for spraying.
Cotton	Old world bollworm *Helicoverpa armigera* (Hubner)	*Ha*NPV	450 LE/ha with 5 per cent jaggery and 0.1 per cent Ranipal at ETL 7 second instar larvae/20 plants.
	Leaf eating caterpillar *Spodoptera litura* (Fabricius) (Lepidoptera: Noctudae)		450 LE/ha with 0.025 per cent boric/tannic acid (or along with 0.5 per cent jaggery and 0.1 per cent Ranipal) at ETL of 20 second instar larvae/20 plants, 2–4 sprays needed.
Tobacco	Tobacco caterpillar	*Sl*NPV	250 LE mixed in 125 L of water, 1 per cent crude sugar and 0.1 per cent Teepol/ha for nursery, 3 times at fortnightly intervals. Subsequent sprays could be altered with 2 per cent need seed kernel sprays. On other crops sprays are given at 7–10 day intervals. 250–450 LE/ha, 200–400 L of water, 1 per cent crude sugar and 0.01 per cent Teepol.
	Old world bollworm *Helicoverpa armigera* (Hubner)	*Ha*NPV	250 LE/ha directed on the inflorescence, one or two well timed applications to protect seed crop.
Pulses			
Chickpea Pigeonpea Field bean	Old world bollworm *Helicoverpa armigera* (Hubner)	*Ha*NPV	250 LE/ha (1 LE = 6× 10^9 PIB), 3–4 sprays.
Pigeonpea Lablab	Pod borer *Adisura erysimi* (Kaltenbach)	*Aa*NPV	First spray of 250 LE at the peak of egg hatching at the tender pod stage followed by two more need-based sprays at weekly intervals.
Oilseeds			
Mustard	Mustard aphid *Lipaphis erysimi* (Kaltenbach) (Homoptera : Aphididae)	*V. lecanii*	Spores at 10 × 10^6/ml + 0.05% + Teepol in water suspension are sprayed.

Contd...

Table 1.18–Contd...

Crop	*Pest*	*Agent*	*Recommendation*
Groundnut	*Helicoverpa armigera*	*Ha*NPV	250 LE/ha (1 KE = 6 × 10^6 PIB), 3–4 sprays.
	Spodoptera litura	*Sl*NPV	Sprays are given at 7–10 day intervals, 250–450 LE/ha, 200–400 L of water, 1 per cent crude sugar and 0.01 per cent Teepol.
	White grub *Holotrichia consanguinea* Blanchard (Coleoptera : Scarabaeidae)	*Bacillus popilliae* Dutky	0.05 kg/ha.
		M. anisopliae (Metchnikoff) Sorokin	Applied as water solution.
Coconut	Rhinoceros beetle *Oryctes rhinoceros* Linnaeus (Coleoptera : Scarabaeidae)	*M. anisopliae*	Two bottles of culture (80 mL) containing 10^9–10^{11} spores/mL and fungal mycelium per m^3 after onset of monsoon every year.
		Baculovirus (Rhabdion virus) *Oryctes* Huger	Infected beetles @ 10/ha are released in the ground of centre of the plots during June–August.
Fruit Crops			
Citrus	Green scale *Coccus viridis* (Green) (Homoptera : Coccidae)	*Verticillium lecanii* (Zimmerman) Vieges	Spray *V. lecanii* spores 10 × 10^6/mL + 0.005 per cent quinalphos + 0.05 per cent Teepol during the rainy season.
Vegetable Crops			
Tomato	Fruit borer *Heliccoverpa armigera*	*Ha*NPV	250 LE/ha (1 LE = 6 × 10^9 PIB), 3–4 sprays.
		Bacillus thuringensis Berliner	Commercial powder/liquids of *Bt* var. kurstakii 0.3–0.5 kg/ha.
Crucifers	Diamond back moth *Plutella xylostella* (Linnaeus) (Lepidoptera : Yponomeutidae)	*Bt*	Commercial powder/ iquids of *Bt* var. kurstakii 0.3–0.5 kg/ha.
	Spodoptera litura	*Sl*NPV	Sprays are given at 7–10 day intervals, 250–450 LE/ha, 200–400 L of water, 1 per cent crude sugar and 0.01 per cent Teepol.
		Nomuraea nileri (Farlow) Samson	The fungus is mixed in Tween 80 to 0.04 per cent and sprayed by diluting 6 parts of the powder in 100 parts of water in the evening hours, 30 packets/ha.
Field beans	Pod borer *Adisura atkinsoni* Moore (Lepidoptera: Noctuidae)	*Aa*NPV	250 LE/ha (1 LE = 6 × 10^9 PIB), 3–4 sprays.

Table 1.19: Commercial Products of *Trichoderma* Available in India

Product/Antagonist	*Target Pathogen/Host*
Sun-derma (*T. viride*)	*R. Solani, Macrophomina phaseolina*
Ecofit (*T. viride*)	*R. Solani, Macrophomina phaseolina*
Trieco (*T. viride*)	Root rot and wilts
Antagon–TV (*T. viride*)	*R. Solani, Macrophomina phaseolina*
Trico (*T. viride*)	Several
Biohit (*Trichoderma* spp.)	Several
Biofit (*T. viride*)	Several

Table 1.20: Bacterial Antagonists of Plant Diseases in India

Antagonist	*Diseases*
Pseudomonas fluorescens	*Pyricularia oryzae* (rice blast)
	Rhizoctonia solani (rice sheath blight)
	Sclerotium rolfsii (groundnut root rot)
	Fusarium oxysporum f sp. *clceri* (chickpea wilt)
	Fusarium oxysporum f sp. *udum* (pigeonpea wilt)
	F. oxysporum f. sp. *cubense* (banana wilt)
Other *fluorescent pseudomonads*	*Sarocladium oryzae* (sheath rot of rice)
	Rice tungro virus
	Xanthomonas axonopodis (bacterial blight of cluster beans)
	Cercospora moricola (mulberry leaf spot)
Bacillus subtilis	*Colletotrichum capsici* (die-back of chilli)
	Macrophomina phaseolina (chickpea, cowpea, sesame root rots)
	Fusarial wilts
	Corticum invisum (black rot of tea)
Streptomyces spp.	*Plasmodiophdra brassicae* (club root of crucifers)
	Sclerotium rolfsii (groundnut root rot)
Serratia marcescens	*Xanthomonas campestris pv. citri* (citrus canker)
Erwinia herbicola	*Xanthomonas campestris pv. vignaeradiatae* (cowpea blight)

An analysis of the number of biopesticide companies and those with biopesticide products listed in the Agrochemical Companies Fact File 1997 indicates that there is a marked difference in distribution of these companies around the world and the value of their biopesticides sales. There are over 30 companies listed from South East Asia compared to 11 from Europe and 8 in North America. East Europe and Russia, Africa, Middle East, Australasia and the Caribbean have few commercial biopesticide companies.

Table 1.21: Major Fungal Pathogens of the Two Most Important Weed Targets in India

Weed	*Fungal Pathogens*
Parthenium hysterophorus L.	*Cercospora partheniphila* Chupp and Greene
	Colletotrichum capsici (Syd.) Butter and Bisby
	Colletootrichum gloeosporioides (Penz.) Sacc.
	Curvularia lunata (Walkar) Boedjin
	Fusarium pallidoroseum (Cooke) Sacc.
	Myrothecium roridum Tode ex Fr. *Oidium parthenii* Satyaprasad and Usharani
	Rhizoctonia solani Kuhn.
	Sclerotium rolfsii Sacc. *Sphareotheca fuliginea* (Schlecht.) Poll.
	Puccina abrupta Diet. and Holw. var. *parthenilcola* (Jackson) Parmalee
Eichhormia crassipes (Martius) Solms-Laubach	*Acremonium zonatum* (Sawada) W. Gams
	Alternaria alternata (Fr.) Keissler
	Alternaria eichhprniae Nag Raj and Ponnappa
	Bipolaris sorokiniana (Sacc.) Shoem
	Cercospora piaropi Tharp *Cercospora rodmanii* Conway *Cochliobolus bicolor* Paul and Parbery *Cochliobolus lunatus* Nelson and Haasis
	Curvularia clavata B.L. Jain
	Cylindrocladium scoparium var. *brasiliense* Batista
	Epicoccum nigrum Link
	Fusarium equiseti (Corda) Sacc.
	Fusarium solani (Mart.) App. and Wr.
	Fusarium sulphureum Schlechtendal
	Marasmiellus. inoderma (Berkeley) Singer
	Myrothecium roridum Tode ex Fries
	Pestalotiopsis palmarilm (Cooke) Steyaert
	Phoma sorghina (Sacc.) Boerema, Dorenbosch and van Kesteren *Rhizoctonia sp.*
	R. solani Kuhn
	Thanatephorus cucumeris (Frank) Donk

Microbial Insecticides: The Global Scenario

The market for microbial pesticides is growing though it represents less than 1 per cent of the total crop protection market most of which is accounted for by *Bt.* (*Bacillus thuringiensis*)–based products. The estimated market for bio insecticides in 1995 was US$ 380 million including $ 3–4 million for viruses. By 2000, the market for bioinsecticides is predicted to increase by 116–141 million. Of the current microbial insecticides available, the success story has undoubtedly been Bt which dominates the current market. Bt has seen tremendous genetic improvement through biotechnology and at the global level, Bt's are claimed to be increasingly cost-effective which has made them more competitive with chemical insecticides.

Insect viruses particularly of the family baculoviridae have great potential for development as microbial insecticides. The first commercial viral insecticide viral insecticide Elcar (R), the *Heliothis zea* nuclear polyhedrosis virus was registered by Sandoz with the EPA for use on crops like corn, cotton and soybean. Examples of other commercial products include the *Spodoptera exigua* NPV (Biosys) and the *Lymantria dispar* NPV (USDA/Cyanamid). The *Cydia pomonella* granulosis virus) (GV) has seen considerable development towards commercialisation. Madex and Granupom are two commercial products of this virus registered for commercial production in Europe. A world survey of virus control of insect pests has summarised the progress made on the use of baculo viruses mostly NPV and GV in 11 different zoogeographical regions of the world.

Entomopathogenic fungi particularly *Beauveria* and *Metarhizium* spp. are of considerable importance because of their ability to infect a wide range of insects like aphids, white flies, leaf miners, weevils, grass hoppers and cockroaches. Bayer AG developed B 10 1020 a strain of *Metarhizium anisopliae* for the control of the black vine weevil *Otiorhynchus sulcatus,* but the production was discontinued because the market value was considered inadequate. Despite this, products based on *Beauveria* and *Metarhizium* have probably the highest business potential for smaller companies due to the large number of available strains against a wide range of insect pests. *Verticillium lecanii* has been extensively used for whitefly and aphid control in glass houses but is disadvantaged by requiring a relative humidity of 95–100 per cent to be effective. *Hirsutella thompsoni* a fungus showing great promise in the control of the citrus rust mite was registered in the U.S. as early as 1981 as Mycar(R) by Abbott. *Nomuraea rileyi* is another promising fungus effective against several lepidopteran pests.

Entomopathogenic nematodes particularly of the genera *Steinernema* and heterorhabditis characterized by their symbiotic association with bacteria of the genera *Xenorhabdus* and *Photorhabdus* have been found promising particularly against soil insects. The high cost of production however limits the use of these nematodes to cash crops such as cranberry and pot plants. Although mass production can be done on insect larvae, commercial production is done using either bedding method or liquid fermentation.

Types of Pathogen for Insect Control

Bacteria

The most widely used biopesticides are those based on *B. thuringiensis* (Bt), although other species such as *Bacillus sphaericus* (Meyer and Neide) have also been used. Several important agricultural pests are controlled with *Bt*. This has often been stimulated by problems resulting from widespread resistance to chemical insecticides, *e.g.*, the Diamondback moth *Plutella xylostella* (Linn.) in Malaysia. In China, *Bt* was first introduced in 1953 and the first production plant was set up in 1966. Today, the main target crops are rice, corn, vegetables and cotton. Wider use of *Bt* in developing countries not only depends on effective control and environmental safety but also on reduced cost. Imported *Bt* products cost more than many chemical insecticides. This has resulted in many countries investigating cheap methods of local production.

Viruses

Insect viruses have not been commercially developed on the scale of *Bt* but there are several examples of their large-scale use in developing countries, the majority based on local production of viruses. Insect viruses are particularly suited to low input, local production. At present they can only be effectively produced *in vivo* which, although labour-intensive, does not need the high-cost capital equipment required for *Bt* fermentation. Labour, of course, is readily available and comparatively cheap in many developing countries. The insect factory approach has been used at least on an experimental or pilot scale in a number of countries, *e.g.*, Egypt, India, Guatemala and Nicaragua.

Fungi

Entomopathogenic fungi have been exploited on a reasonably large scale in some developing countries, predominantly through local production. There are examples of large-scale use in Brazil, Honduras, Columbia, Venuzuela, Indonesia and China. Successful control of locust and grasshoppers in Africa has also been achieved. Local production of fungus normally utilises agricultural products or by-products, for instance maize or rice seed or bran. Simple equipment is normally used and the fungus is inoculated into bottles or plastic bags containing the substrate or in some cases, *e.g.*, Venezuela and Columbia, sporulation takes place in shallow open trays.

Protozoa

No commercial use of protozoa has been reported in developing countries. Some trials have been undertaken, most notably the large-scale trials to control grasshoppers in Cape Verde. Protozoa have to be produced *in vivo* by inoculating susceptible insects, hence this technology should be suitable for developing countries.

Nematodes

Expectations for nematodes are based on the global potential for a mode of action which would in theory target the 90 per cent of insect pests that have a soil phase in their life cycle. However, although large numbers of insect species have been shown to be susceptible in the laboratory, the host range in the field is limited by dose response and the behavioural avoidance patterns in the target host. Entomopathogenic nematodes have been successful in the control of pests such as the black vine weevil in a wide range of protected and field crops. There are now more than eight species of nematode (*Heterorhabditis* and *Steinernema* spp.) produced and sold in the USA, Australia and Europe, Most of the $4–5 million sales in the USA are in high value markets in citrus, cranberry and ornamental crops.

Adoption of Microbial Pesticides in India

The use of biopesticides and biocontrol agents in India is on the increase but not up to the desired level of growth. Although presently, a decrease or decline in chemical pesticide consumption is indicated, whether it is real or due to weather abnormalities only, is a subject of analysis. Many small entrepreneurs are coming up with the biopesticides and biocontrol agent products, but many of them have little or no quality consciousness.

The concept of biopesticides and biocontrol in farming community is still in infancy. Only 1 per cent of 143 million ha. crop area and only 2500 villages out of over 6 lakh villages in the country have been covered so far under IPM. Basic IPM modules have been designed by agricultural scientists of ICAR, SAUs and Directorate of Plant Protection, but these need to be refined for local needs. It is heartening that establishment of some "organic cotton", "organic tea" and "organic vegetable" units has been initiated, but this activity really needs to be implemented in extensive areas.

Mass Production of Microbial Pesticides

Biopesticides based on insect viruses, bacteria, entomopathogenic fungi, microsporidians and entomopathogenic, nematodes offer considerable scope as crop protection agents against several insect pests, mainly of insect species belonging to the group Lepidoptera and Homopetra. With few exceptions, the potential of several insect pathogens have not been exploited in the control of insect pests mainly due to lack of knowledge on the mass production, formulation and application of these pathogens. The success of microbial control of insect pests depends not on mere isolation, characterization and pathogenicity, but on successful mass production formulation and application of microbial agents. Though insect pathogens cause tremendous epizootics under natural conditions and bring down the insect population to zero level, they fail to multiply on the commonly available growth media.

Research Under MAU

There is a need to encourage mass production and use of NPV's and GV's by the government, public sector cooperatives, private agencies and Agricultural Universities. The commonly employed methods for production of, baculoviruses are:

1. Insectary rearing of host insects on large scale,
2. Collection of host larvae from infested crop and infecting them in laboratory and field collection of virosed larvae from the infested field after application of HaNPV (Pawar and Thombre, 2001).

The authors have developed technique for rearing of host insects and for mass multiplication of baculovirus in the laboratory. The important steps involved in mass multiplication are: host insect rearing, infection of larvae through artificial diet, harvesting, purification of diseased larvae, maceration, decantation, purification, standardization and storage of HaNPV.

The extensive work carried out at Marathwada Agricultural University, Parbhani in HaNPV revealed significant findings (Thombre, 1996). The dose of 0.5 LE per 300 larvae was optimum which accounted for 1.85×10^9 PIBs/larva with productivity ratio of 1 : 1114. During September to December, 1990 a total of 87960 reared showed virosis to 74362 larvae accounting for 85 per cent of the diseased larvae @ 1 LE/300 larvae. The yield of PIBs per larvae ranged from 0.95×10^9 to 3.5×10^9 with an average of 2.00×10^9. Infection of field collected larvae revealed that an average of 69.85 per cent larvae died due to NPV with average yield of PIBs per larva ranging from 1.10×10^9 to 1.36×10^9. The field recovery of HaNPV after spraying cotton pigeonpea and chickpea showed 56.47, 61.97 and 58.00 per cent virosis with 0.98×10^9, 1.12×10^9 and 1.02×10^9 average PIB yield/larva, respectively.

Success Stories of Microbial Pesticides in India

There are many encouraging success stories of biopesticides and biocontrol agents in Indian Agriculture. These include:

1. Control of Diamand back moth by BTK (Biolep delfin, and biobit etc.) in Najafgarh (Delhi) and Ajmer.
2. Control of mango hopper and mealy bugs by *Beauveria* (Biorin and Daman etc.) in Malihabad (Lucknow).
3. Control of *Helicoverpa* on cotton, pigeon-pea, tomato by BT (Biolep delfin, and biobit etc.) in Haryana, Punjab, Gujarat and West Bengal.
4. Control of *Helicoverpa* on gram by NPV (Biovirus-H and Helicide etc.) in Maharashtra and Rajasthan.

5. Control of rots, wilts in various vegetable crops by *Trichoderma* based product like Bioderma in Haryana, Rajasthan and Maharashtra.
6. Control of coffee pod borer by Beauveria in Karnataka.

Registration and Quality Control

The registration of microbial pesticides is similar in many of the countries where the agents have been developed. The processes have evolved from the existing requirements of chemical resources out in light of the experience have been modified to take account of the biological nature of the active ingredients involved.

There are two classes of biological and biologically derived pesticides:

Microbial Pest Control Agents (MPCAs)

Pesticides referred to this category include (but are not limited) bacteria, viruses, fungi and protozoa.

Biochemical Pest Control Agents (BPCAs)

They generally fall into four functional categories. Semiochemicals, hormones, natural plant regulators and enzymes.

In India pesticides manufacturing requires registration of the product with the Central Insecticides Board Registration Committee, Department of Plant Protection and Quarantine, Ministry of Science and Technology, Government of India. Under the section 9(i) of the insecticides Act 1968 any person desiring to import or manufacture an insecticide included in the schedule annexed to the Insecticides Act, has to apply for the registration of the product to the Registration Committee. Different kinds of classifications are in practice in India for the registration of plant protection agents.

The manufacturer is required to submit the data for the following categories:

1. Chemistry/Product Characteristics
2. Bioefficacy
3. Toxicity
4. Processing, Packaging and Labeling
5. Data requirements for the registration of form
6. Bioefficacy tests
7. Field tests.

Recommended Minimum Quality Control for Fungi

Careful Culture Maintenance and Preservation

1. Accurate identification and characterization of the fungal strain by a recognized taxonomist.
2. Use of established long-term storage methods for the isolate obtained from the original isolation.
3. Use of single spore(ss) isolation to reduce genetic drift.
4. For production, use of stock culture material bred from the ss isolation and kept in long-term storage.

In-process Contamination Monitoring

1. Take sub samples at critical stages throughout the course of production for the detection of contaminants entering the system

Total Contaminating Organism Count (assuming prior screens for human pathogens are clear)

1. Measured by dilution plating onto generalist media. The acceptable level of contamination will be product specific, but should always be backed up with results of in-process contamination monitoring to show that contaminants have not entered during the production process and have not multiplied with the production fungus. Under ideal conditions, a typical limit for an un-formulated fungal product could be as low as 1×10^6 cfu g^{-1} in pure conidia powder containing 5×10^{10} conidia g^{-1}. However, the issue of detecting potential human pathogens in the final product requires further investigation

Viability

1. Greater than 85 per cent viable using a direct counting method *e.g.*, counting germinated spores under the microscope (up to advised product expiry date when kept at recommended temperature).

Virulence/Potency

1. This aspect of quality control requires further study before defied procedures and limits can be set. In principle, a standard bioassay appropriate to the agent in question should be developed and the performance of product batches compared against a fungal standard at appropriate intervals. Product performance should be compared with that of the standard and a tolerance limit set beyond which batches are re-tested to confirm loss in efficacy and discarded on second failure.

Other Checks Appropriate to Specification

1. Moisture content
2. Quantity of active ingredient per gram product
3. Particle size distribution as established by the producer and indicated on label (*e.g.*, for stable ultra low-volume (ULV) formulations, all particles should be < 100 μm, 99.9 per cent should <60 μm and 80 per cent <10 μm).

Recommended Minimum Quality Control of Baculoviruses

1. DNA restriction profiles
2. Characterisation of the isolate using at least three commonly used restriction enzymes, *e.g.*, Eco R1, Hind III, Pst1, Barn H1
3. OB microscope counts (NPV).

On the technical product (after processing prior to final formulation) or on formulated product (this may not be practicable with some powder formulations due to presence of carrier).

Total Bacterial Count

Dilution plating on generalist agar media for formulated product. Ideally, viable counts should normally not exceed 1×10^8 cfu g^{-1} for formulated products with an activity of 1×10^9 OB ml^{-1} or 1×10^8 cfu g^{-1} for formulated dry products.

Virulence (Pathogenicity)

Standardised bioassay (LD_{50} or LC_{50}) on final product using defined instar of a laboratory-reared strain of the target pest. Results should confirm that the potency ratio is within defined limits specified for the product and determined during product development or >0.5 of a standard virus preparation.

Market

The global insecticides market was estimated at 8 billion in 1995 (Wood McKenzie Report, 1995). The estimated market for biopesticides was 119 million whereas in 1991, the biopesticide market was estimated to be up to 71 million. This represents a 68 per cent growth in four years. However, due to the introduction of new technologies, the growth of the biopesticides market is not now expected to exceed 141 million in the year 2000.

Niche Markets

At present, biopesticides have a competitive advantage in small markets, where demand for biopesticides is present and competition from conventional chemicals is limited. Most currently registered products for insect control are produced in developing countries, and included (104 products on the market, mostly *Bt*), nematodes (44 products), fungi (12 products), virus (8 products) and protozoa (6 products). Commercialization in developing countries is limited but growing. Most production is associated with the plantation industry, where central production units can service the large crop area, and usually involves predators, parasitoids and fungi (*e.g.*, *M. anisopliae*). Prospects for further development of augmentation in developing countries are good: technologies are not complex, labour is available and the incentive to invest in pest control technologies is great given the foreign exchange currently expended to import chemical pesticides.

Ecological Use Strategies

Like any living control agent, the effectiveness of a biopesticide depends on two factors: its capacity to kill pests and its capacity to reproduce on pests and therefore continue to compound its killing action; in ecological terms, its functional and numerical responses. Biopesticide development has tended to the numerical response of pathogens and has concentrated only on their direct killing action. To some extent, this involves selecting pathogens that are virulent but which do not reproduce well in the crop environment. Increasingly, however, the value of making use of pathogen's power of reproduction in biological control is being realised. This involves moving from regular application of the biopesticide to strategic application early in the crop season, to establish populations that will intercept and check the growth of invading pest populations. This strategy greatly extends the utility of biopesticides and can be much cheaper for farmer.

There is enormous potential in looking for opportunities to utilise the additional properties of host finding and reproduction, which are precisely those properties that make biopesticides superior to chemicals.

Hence, there are many opportunities to exploit the ecological benefits of biopesticides that have so far been little explored. The use of biopesticides in this was open up whole new markets that chemical pesticides just cannot address.

Microbial Control of Plant Parasitic Nematodes

Various organic amendments, parasitic fungi like *Paecilomyces lilacinus* and like *Pasteuria penetrans, Bacillus* sp. have been utilised for control of nematodes like *Meloidogyne* sp., *Heterodera* sp., *Globodera* sp., *Rotylenchulus* sp., on crops like brinjal, tomato, citrus, coleus in nurseries and small scale field trials. *Bacillus subtilis* along with *Eichhornia* as green manure suppressed root knot on tomato. Combination of *Paecilomyces* with Farm Yard Manure (FYM) was useful in controlling root knot on *Coleus. Pasteuria penetrans* has so far been cultured on rootknot affected roots. Synthetic media are required to be standardised. However the parasitic fungi and bacteria offer potential against nematodes. There is a need for viable commercial formulation for field use.

Despite numerous attempts over many years to utilize nematophagous fungi for biological control, there is no current example of the widespread use of these fungi in agriculture. One of the reason that nematophagous fungi have not been commercialized is that, they are normally mass produced and delivered into soil has been limited to traditional methods such as infested grains or in aqueous spore suspensions. Formulation, method and time of application and selection of biological control agents play an important role in the successful introduction of antagonists and subsequent biocontrol of certain plant parasitic nematodes.

In spite of the pioneering work on the exploitation of pathogens for weed control done in the 1960s and 1970s by plant pathologists based at the erstwhile Indian Station of the Commonwealth Institute of Biological Control (CIBC) in Bangalore, much work on weed control by means of plant pathogens in recent years. This is mainly because increased public awareness on health and environment has paved the way for research on alternative management strategies.

Microbial Control of Weeds

In spite of the pioneering work on the exploitation of pathogens for weed control done in the 1960s and 1970s by plant pathologists based at the erstwhile Indian Station of the Commonwealth Institute of Biological Control (CIBC) in Bangalore, much work on mycoherbicides has not been done in India. However, there has been a spurt in the work on weed control by means of plant pathogens in recent year. This is mainly because increased public awareness on health and environment has paved the way for research on alternative management strategies.

Developed countries like the United States of America (USA) and Canada are far ahead of India in the field of mycoherbicides. Although 37 microbial herbicides and 8 microbial herbicide technologies have been patented the world over, none of them are from India.

Microbial Control of Plant Diseases

The use of anatagonists has got a heightened stimulus because of the fact that for many diseases there are no well-perfected control measures. As is common the world over, in India as well most of the research is carried out against soil-borne diseases. The reasons are obvious: First, only a very few fungicides are available to control soil-borne diseases and second, resistance breeding against root pathogens has not been popular. Also, the delivery systems employed for the control of rots and wilts are not complicated unlike in the case of foliar diseases of plants. A number of commercial products based on *Trichoderma* spp. have entered the Indian market in recent years.

Biotechnology and Microbial Pesticides

There are tremendous opportunities for genetic improvement of microbial pathogens for virulence, persistence, host range, speed of kill and stability in storage. A number of genes with potential insecticidal properties have been inserted into baculoviruses to enhance their activity (Table 1.1). However, recombinant baculovirses, thus eliminating *in vivo* insect production as a viable economic option. Instead, commercial production will need to be undertaken *in vitro* utilizing optimised insect cell lines grown in large-scale bioreactors. It is believed that *in vitro* production technology for baculoviruses is currently at about the 250 litre bioreactor level. In order to attain the necessary economics of scale of a successful commercial product, reliable production will ultimately need to be carried out in large reactors of 16,000 litre capacity and above.

Genetic improvement of *Bt* can be achieved by plasmid curing and conjugal transfer and recombinant DNA technology.

Culture condition can influence the characteristics of fungal spores and can be manipulated to increase mycoinsecticide efficiency.

Future Thrust

1. Biological control agents are available only for some selected insect pests and very little for other insects, diseases, nematodes and weeds. Extensive surveys for isolating and identifying naturally occurring biocontrol agents in various agro-ecosystems should be made.
2. Even the available biocontrol agents are not produced adequately for supply to farmers. Creation of awareness among the farmers and extension personnel and development of entrepreneurship are important issues.
3. The microbials used for biological control of insect pests have to be registered. Adequate data of bioefficacy, mass production technology, shelf-life, formulations, safety and quality control are not available in all cases.
4. While considering the proposals for registration of microbial pesticides, they should be less exposed to the rigors of the schedules prescribed for the chemical pesticides. In particular, biopesticides should be exempted from generating toxicology data pertaining to carcinogenicity, allergenicity, teratogenicity etc., as has been done by the Environment Protection Agency (EPA) of the USA.
5. To encourage the entrepreneurs to take up to commercial production of biopesticides large number of pilot plants should be established. The scientists involved in the programme should provide help in technology transfer, consultancy, training, quality control, trouble shooting and market development.
6. Some of the biocontrol agents are affected by various kinds of environmental stresses like UV light, temperature, moisture, pH etc. Adequate efforts have not been made to exploit biotechnology and genetic engineering concepts to develop more efficient strains of biocontrol agents.
7. In view of the labour intensive nature of the biocontrol programmes the cost of crop protection may be somewhat higher in certain cases. However, in view of the environmental safety and sustainability, the commercial ventures in biological control should receive adequate support and encouragement by the Government in view of the overall efficiency of biological control.
8. For successful biocontrol of pests, these agents should be used at the appropriate time during the cropping season alongwith the various components of IPM.
9. The cost of biological control of pests can be optimised by proper rate and timing of the release of the biological control agents. Information on pest ecology and monitoring through insect sex pheromone technology should be collected.
10. Quantification of microbial biodiversity.
11. Characterization.
12. Selection/development of superior or virulent strains.
13. Development of economically viable mass production systems.

14. Evolution of proper storage techniques.
15. Development of suitable formulations and safety testing.
16. Selection of suitable adjuvants and UV protectants.
17. Quality control of microbial products.
18. Registration of most effective and safe microbial products.
19. Standardization of packaging and transport methods.
20. Development of simple, efficient and economical methods of field application (including dosage and frequency of application) and advise for the users.
21. Standardization of utilization of microbial agents as apart of BIPM.
22. Development of cell lines for tissue culture of obligate pathogens.
23. Development of transgenic plants by incorporation of toxic *Bt* genes in plant systems.
24. Use of genetic engineering techniques for creating new combination of multipurpose pathogens.
25. Development of novel methods of transfer of technology.
26. Training and human resource development.
27. Creating village level crop protection units in which among other things microbial pathogens could be multiplied and supplied to the needy farmers on payment. This will also create productive job opportunity for unemployed youth, farm women and even under employed farmers in lean period could be engaged.

Conclusion

Recent advances in the development of biopesticides or microbial pesticides offer opportunities for the worldwide exploitation of biological agents as replacements for more hazardous and environmentally unacceptable chemical insecticides and for inclusion in Integrated Pest Management programmes. The advantages and value to be obtained through development, exploitation and use of biopesticides in developing countries is immense. The production, sale and use of biopesticides in a developing country can provide local employment opportunities, reduce health risks and costs due chemical poisoning and environmental damage, improve export earnings through reducing chemical residue levels on export commodities; and in addition to this there are the benefits obtained through the extra control achieved by preserving natural, enemies in crop systems and by maintaining indigenous bio-diversity.

REFERENCES

Rabindra, R.J., N. Sathian and J.M. Kennedy (1999). *Summer School Manual on Emerging Trends in Microbial Control of Crop Pest,* May 5–26, 1999. Department of Agricultural Entomology, Centre for Plant Protection Studies, TNAU, Coimbatore.

Singh, S.P., S.T. Murphy and C.R. Ballal (2001). *Augmentative Biocontrol,* Proceedings of the ICAR-CABI Workshop June 29th to July 1st 2000. Published by CABI, Bioscience, UK and PDBC, ICAR; India, pp. 66.

Srivastava, R.P. (2003). *Biopesticides and Bioagents in Integrated Pest Management of Agricultural Crops.* Published by International Book Publishing Company, Lucknow, pp. 163.

2

Parasitisation by *Hexamermis* sp. (Nematoda–*Mermithidae*) to *Leptocoris augur* Fabr. (Heteroptera-Coroidea-Rhopalidae)

S.C. Dhiman and Kumkum

Entomology Research Lab, Department of Zoology, M.S. (P.G.) College, Saharanpur – 247 001

Mermithid nematodes are endoparasite of many insects such as grasshoppers, beetles, Odonata and Diptera. They infest the host in juvenile stage and come out of the host body in fully developed condition as post-parasitic stage which moult in soil to become adult. The host insect dies. Thus, mermithids are potential biocontrol agent, hence, these are of paramount significance. *Hexamermis* sp. (Nematoda-Mermithoidea-Mermithidae) is an entomophilic nematode parasitizing *L. augur* Fabr. (Heteroptera-Coroidea-Rhopalidae) which in turn is a pest of kusum tree, *Schleichera oleosa* (Sapindaceae) (Dhiman, 1983; Dhiman and Gulati, 1988; Malhotra, 1958).

The bug infests severely seeds and leaves of aforesaid plant in Saharanpur and adjacent areas. Only few aspects of *Hexamermis* sp. have been worked out till date by Dhiman (1984, 1994, 2003), Dhiman and Singh (1989, 1991), Dhiman and Ghayyur (1993), Dhiman and Kumar (1996a, 1996b), Ghayyur (1996). Yet, no observations on the effect of parasitism on host bug have been made and it is an endeavour to describe the said aspects.

Parasitized bugs can be identified easily by swollen abdomen, letharginess and thin cuticle through which coils of parasitic nematode appears even by naked eyes. Hence, field collected parasitized bugs were reared in lab, in wooden wire gauze cage and hurricane lamp chimneys covered at top by fine muslin cloth. Fresh food supply was maintained daily and stale food was replaced, to maintain necessary R.H. a wet cotton swab was kept in a watch glass inside the cage and chimney. Behaviour of the bugs was noticed with the aid of hand lens. Dissections were made time to time to ascertain the anatomical affect. Biochemical tests for the measurement of glucose, amino acids, cholestrol, fatty acids were performed on healthy as well as parasitized bugs.

Affect of parasitization of *Hexamermis* sp. on *Leptocoris augur* has been studied using following criteria:

Morphological Effect

Influence of parasitism of this nematode on physical changes of *L. augur* was studied by watching the behaviour of parasitized bugs during laboratory rearing and field also.

1. First entrance of parasitic juveniles in the host body makes the bug slightly irritated and it moves frequently but, as the development of parasitic stage proceeds inside the host, there appears letharginess in the behaviour of the bug and at last prior to the emergence becomes quite sluggish.
2. The abdomen of parasitized bug swells up greatly and attains biconvex shape.
3. Movement of antennae, legs and wings are greatly affected and the parasitized bug developed inability in flight.
4. A bluish spot (preferable towards left side) develops on the posteroventral part of the abdomen of the bug.
5. Body cutical of the hosts abdomen becomes thin and transparent so that the coils of the parasitic *Hexamermis* can be seen externally under binocular microscope or even by naked eye having good eye sight.
6. The parasitized bug usually live alone and avoid company.

Finally, after the emergence of parasitic stage of *Hexamermis* sp. the host appears much tired and die soon.

Anatomical Effect

Parasitic juvenile causes great anatomical changes of the host body which are:

1. As the parasitic juvenile of *Hexamermis* sp. resides in haemocoelomic cavity of the host *L. augur,* hence it first takes nutrients from haemocoelomic fluid, causing quantitative loss in it.
2. Secondly, the fat bodies of the host which lie around the gut, reproductive organs and in haemocoelomic fluid are first dissolved and then consumed by diffusion through body wall.
3. Thoracic muscles, including flight muscles (dorsal longitudinal, dorso ventral and oblique muscles as well as wing auxiliary muscles) and abdominal muscles are first dissolved and absorbed. In late stage, when the parasitic juvenile extends in head also, antennal intrinsic muscles and some mouth parts muscles are also devoured.

4. Highly parasitized bugs have negligible amount of haemocoelomic fluid, since, most of their body is filled up by the coiling of parasitic juveniles.
5. However, nervous system and tracheal systems are not seen much affected.

Physiological Effect

1. Mermithids stimulate catabolism and inhibits anabolism of host fat body proteins, which thus provides the necessary dietary amino nitrogen *via* the host haemolymph. Cessation of vitellogenesis in the host bug can be attributed to an inability of the oocytes to sequester available proteins from the haemolymph. Subsequent oocyte resorption probably results from a depletion of vitellogenic proteins in the haemolymph due to the nutrient requirements of the developing mermithids. This results in sterilization of the female host.
2. *Hexamermis* infections diminish the efficiency of the excretory system. A heavy infection of second and third instar nymphs increased the uric acid level of the haemolymph five times more that of the controls. Furthermore, the concentration of faecal uric acid was reduced to one quarter that of uninfected controls.
3. Amino acids are taken up in large quantity by developing mermithid larvae and incorporated in to protein at varying rates during mermithid development. Dipeptides and polypeptides are not taken up by the developing larvae. Under experimental conditions protein was synthesized most rapidly from (14C) leucine by 17 days old larvae. This timing of maximum synthesis coincides with the rapid increase in total dry weight and in protein level that occurs mostly between 17 and 21 days after infection (Gordon and Webster, 1972) as the developing nematode accumulates stored proteins and lipids in the trophosome (Chitwood and Jacobs, 1938; Gordon and Webster, 1972) prior to emergence from the host. These observations are seen true of *Hexamermis* sp. in which rounded spheres of proteins and fats are stored in large number in trophosomes.
4. Parasitic larvae take up glucose from the haemolymph of the host rather than trehalose, the host's main blood sugar. They do this faster in the mid phase (*i.e.*, about 10 days old parasitic stage) than in the closing phase (*i.e.*, after 18 days of their parasitic development). Glucose absorption through the cuticle is probably mediated by an active transport system in the trophosome which sets up a diffusion gradient across the pseudocoel, hypodermis and cuticle.
5. The fatty acid composition of the host haemolymph is not significantly changed by mermithid parasitism but the levels of cholesterol and cholestanol appears to be increased. The rapid and persistant uptake of glucose modifies the fat body haemolymph for the parasite. The depletion of the glycogen phosphorylase in the fat body prevents further glycogenolysis there and helps to maintain a constant low level of glycogen in the host fat body.
6. The parasitic mermithids may digest the protein reserves of the host fat body by secreting hydrolytic enzymes, it is more likely that mermithids indirectly utilize host fat body proteins by inducing changes in the host's metabolism. The level of haemolymph, proteins and amino acids remains relatively constant in adult host, parasitized by mermithids, but a significantly decrease in fat body proteins and amino acids occurs (Gorden and Webster, 1971).

Hexamermis not only utilize fat body proteins but also the lipid content of the fat as a result of which fat bodies disappear from the haemolymph.

On Feeding Capabilities

During the host searching, sensory organs of the pre-parasites (amphids and phasmids) play a vital role which direct the presence of host in near by vicinity and on contact with a host it searches the weak point of the host body for penetration such as wing auxillaries, coxal joints etc. After entrance in the host, the parasitoid derive its complete nourishment from the haemolymph, adipose tissue and body muscles of the host bug. It causes increased demand of food and thus the feeding requirement of host bug greatly increases and it devotes more time in feeding on *S. oleosa* seeds and sucking sap from the leaves of adjacent host plants (Dhiman and Gulati, 1986).

Super parasitism, which is a general tendency, further enhance the continuous supply of nutrition by which feeding time is increased many fold.

On Copulation and Reproductive Organs

1. No marked effect has been observed on the copulation ability of early parasitized adult bugs as the development of parasitoid occurs in the body of host bug, mating ability of the bug is reduced and finally no mating occurs in mid and late stage of parasitized bugs.
2. Testis are greatly reduced in size. Vasa deferentia are dissolved, hence, there is no connection between male gonads and ejaculatory duct. Bulbous ejaculatorius and genital muscles have also been observed damaged. Thus, 100 per cent sterility occurs in male bugs by parasitism.
3. In female, maturation of ovum in ovary is greatly atfected as fat and protein contents of the host body are consumed by the parasitic stage, hence, ovalies in parasitized bugs have been recorded smaller due to retardation in vitellogenesis. Moreover, oviducts are destroyed at several points and muscles of ovipositor reduced. Thus, like the male bug, female sex also get sterilized.

On Fecundity

1. Early parasitization of female bug has less effect on oviposition and female lay eggs as usual but number of eggs laid decreases to half. As the development of parasitic nema proceeds further in haemocoel, it causes marked effect on ovaries and oviduct and fecundity is adversely effected and no oviposition occurs in mid and late parasitized female. Thus, fecundity in female, too, is reduced to zero.

On Ecdysis

1. Parasitization in nymphal instars inhibits the ecdysis and the nymph prolonged its life time till the emergence of nematode causes its death.
2. If the pre parasitic stage enters just after moulting in host instar, then the nymph prolonged its life time till the emergence of nematode causes its death.
3. If the pre-parasitic stage enters just prior to moulting in host instar, then the nymph moults into next stage which die with the emergence of nema.

Thus, *Hexamermis* sp. of melmithid nematode has good potential to be used as a biocontrol agent for the control of Kusum pest, *Leptocoris augur.* Hence, present studies recommend the use of this nematode as a biocontrol agent.

Conclusion

Hexamermis sp. is a parasitoid of kusum bug, *Laptocoris augur* which in turn is a pest of kusum tree, *Schleichera oleosa* (*Sapindaceae*). Penetration of preparasitic juveniles into host makes it irritated. Development inside the body brings letharginess and finally at the emergence of parasitic nema from host body the bug dies within few hrs. Abdomen of the host swells, critical gets thin and inability in flight develops. Anatomically, quantitative loss of haemococlornic fluid occurs. Thoracic muscles, abdominal muscles, fat bodies and reproductive organs are lost.

Efficiency of excretory system diminishes amino acids are taken up in large quantities. Glucose level in haemolymph of the host decreases. No change in free amino acids but level of cholestrol increases. Fat bodies, proteins and amino acids decrease. Testis and ovaries are reduced. Parasitized bug turns sterile. Parasitism inhibits the ecdysis in nymphal instars and cent percent mortality occurred in parasitized bug population. Thus, *Hexamermis* sp. has a good potential as a biocontrol agent for *L. augur*.

REFERENCES

Chitwood, B.G. and L. Jacobs (1938). Stored nutritive materials in the trophosome of the nematode, *Agamermis decaudata* (mermithidae). *J. Washington Academy of Sciences*, 28(1): 12–13.

Dhiman, S.C. (1983). Studies on the oviposition and egg structure of *Serinetha augur*, Fabr. (Heteroptera: Coreidae). *U.P. J. Zool.*, 3: 166–169.

Dhiman, S.C. (1984). A new record of *Hexamermis* sp. (Nematoda : Mermathidae) parasitizing *Serinetha augur* Fabr. (Heteroptera : Coreidae). *Curr. Sci.*, 53(24): 1310.

Dhiman, S.C. (1994). Mermithid nematods as potential biocontrol agent of insect pest. *Proc. Biol. Cont. Ins. Pest*, p. 159–165.

Dhiman, S.C. (2003). Effect of host *Leptocoris augur* and parasitoid (*Hexamermis* sp.) density on percentage of parasitization. *J. Appl. Zool. Res.*, 14(1): 113–114.

Dhiman, S.C. and U. Gulati (1986). Aggregation behaviour of *Serinetha augur* (Fabr.), (Heteroptera : Coreidae). *Ind. Jour. Biol. Res.*, 2(4): 318–323.

Dhiman, S.C. and U. Ghayyur (1993). Laboratory rearing technique of *Hexamermis* sp. (Nematoda : Mermithidae). a parasitoid of *Leptocoris augur* (Fabr.). (Heteroptera : Coroidae : Rhopalidae). *Bioved*, 4(2): 171–174.

Dhiman. S.C. and A. Kumar (1996a). Ethology of *Hexamermis* sp. (Nematoda : Mermithidae), a parasitoid of *L. augur* (Fabr.). *IPM and Sustain. Agri. Anent. Appr.*, 6: 159–168.

Dhiman, S.C. and A. Kumar (1996b). Effect of host stage on the development size of parasitic juvenile of *Hexamennis* sp. (Nematoda : Mermithidae). *J. Appl. Zool. Res.*, 7(2): 173–174.

Dhiman, S.C. and R.K. Singh (1989). Studies on the parasitism of *Leptocoris augur* (Fabr.) (Heteroptera: Rhopalidae) by *Hexamermis* sp. (Nematoda : Mermithidae). *Nutri. Ecol. Ins. & Env.*, p. 296–301.

Dhiman, S.C. and R.K. Singh (1991). Preliminary observations on *Hexamermis* sp., (Nematoda : Mermithidae), a parasite of *Leptocoris augur* (Fabr.). *Bioved*, 2(2): 111–112.

Gordon, R. and J.M. Webster (1971). *Mermis nigrescence*: Physiology relationship with its host, the adult desert locusts *Schistocera gregaria*. *Exp. Parasitol.*, 29: 66–79.

Gordon, R. and J.M. Webster (1972). Nutritional requirement for protein synthesis during parasitic development of the entomophilic nematode *Mermis nigrescence*. *Parasitology*, 64: 161–172.

Ghayyur, U. and S.C. Dhiman (1996). Ethology of *Hexamermis* sp., a Mermithid Parasitoid of *L. augur*, a pest of Kusum Plant *Schleichera oleosa*. *Ann. Pl. Protec. Sci.*, 4(1): 80–84.

Gulati, U. (1988). Bionomics and ecology of *Serinetha augur* Fabr. (Heteroptera : Coreidae). Ph.D Thesis, Meerut University, pp. 113.

Malhotra, C.P. (1958). Bionomics of *Serinetha augur* Fabr. and its association with *Dysdercus singulatus* Fabr., the red cotton bug. *Ind. For.*, **81**: 669–671.

3

Perspectives in Biocontrol as Effective Tool in Pest Management

Suvasish Das, N.C. Gupta, M. Sehgal and B.B. Bhosle
School of Environmental Management, GGSIP University, Delhi – 110 006

Indian agriculture has seen glorious years of green revolution and today it is our onus to maintain and rather increase production in a sustainable manner. Being a billionaire in terms of population and with an ever increasing population, our food grain requirement will be around 262.0 million tonnes by 2020. When we look at the contribution of crop protection technology, it stands around 30–40 per cent. The mere Rs. 300 crores industries of 60s has grown today to staggering Rs. 3000 crores. Use of pesticide in India is increasing at the rate of 2–5 per cent per annum. India stands next only to China in production of pesticides. Currently there are 153 pesticides which are registered among, which 35 are most widely used. Average consumption of pesticide per hectare in 480 gm a.i./ha. Inspite of such a rate of increase in pesticide consumption agriculture production in our country is not increasing in an expected manner. Although pesticide consumption is much less in our country than that of Japan and USA, still this expensive input is not supplementary in terms of output. 30 per cent of the entire agricultural produce in field and in storage is lost to insect and pests and diseases, the estimated cost of this may well be over Rupees 60000 crores annually.

There is skepticism about the potential of crop protection industries everywhere. Looking at the conditions the foremost need of the hour is to generate new plant protection molecules with appreciable frequency, breeding of resistance varieties, and genetically modified crops. Whereas the staggering cost of research and development is additional obstacle for country like ours. A recent estimate shows that it requires US $ 300 million to produce a new molecule. As a result, agriculture in our country is still being carried out on use of conventional, highly toxic, ineffective plant protection chemicals.

Rejection of agricultural exports upto Rs. 1000 crore per annum is reported, only because of high pesticide residue content in produce. Suicide cases reported among farmers from Maharashtra, Karnataka, A.P., Punjab are matter of great concern, which is related to crop failures as a result of ineffectiveness of crop protection chemicals. Unregulated and excessive use of pesticide has rendered about 500 pests resistant, compounding our problem.

Problems of biomagnification is reportedly confirmed in case of DDT and more recently Endosulfan, in various reports. Hence associated health hazard due to pesticide are nothing newly and at this moment it is gaining much importance.

After this long exploitation phase of pesticide, the only recourse is IPM, latest buzz world in agriculture. The word control is replaced by management now-a-days, as it is futile exercise to control, rather our aim should be to manage. IPM is now looked upon as a national strategy to conserve biodiversity, comprising beneficial pollinators, parasites and predators to obtain high yield of crops through superior pest control.

Development of insecticide resistance has become well known among most organochlorine, organophosphorus, carbamates, formamidines and pyrethroid insecticides. This is solely attributed to extensive and indiscriminate use of these chemicals and others restricted in use or prohibited due to potential environmental hazards. Under prevailing scenario it becomes imperative to direct present day research towards synthesis of new molecules, which satisfies successfully the various prerequisites according to need that takes care of safety and sustainability both. Such problem aroused due to disturbed ecological balance and expression of dormant genetic trait or existing polymorphism among target pests. Answer to such problem lies in nature, which is endowed with enormous treasure in form of biodiversity. Secondary metabolites produced by both plants and microbes in situ which is found to show antagonistic effect against its counterparts and certain insect pests has been documented for long time and is continuously explored. It is estimated that to date only 10 per cent of plant species in the world have been examined chemically. Therefore, a challenging task to discover newer class of pesticide and new mode of action from natural plant source lies beforehand. There is greatly reduced likelihood of harmful bioaccumulation on soil and ground water residues in addition to resistance development.

Biological Control

It involves control of insect pest by using the biological agents.

Essential part of IPM practices due to its mode of action based on three principles:

1. Modification of host and vector population.
2. Modification of reservoirs of pathogens.
3. Direct use of microbial pathogens and predators.

The biological agents which are used in pest control are Bacteria, Viruses, Fungi, Nematodes, Protozoans, and Rickettsiae.

List of important micro-organisms, known to control the pests.

Viruses

Insect Pest, Against which Used	*Scientific Name*	*Family of Virus*
Gypsy moth	*Lymantria dispar*	NPV
Red headed leaf roller	*Argyortaenia velutinana*	GV
Cabbage worm	*Pieris rapae*	GV
Cotton boll worm	*Helicoverpa* sp.	NPV
Cabbage looper	*Trichoplusia* sp.	NPV, GV
Pine caterpiller	*Denotrolimus spectabilis*	CPV

Bacteria

In bacteria, *Bacillus thuringiensis* is most important and widely used sp. for pest control. Other bacteria like *Serratia entomorphila* is also used for pest control.

Fungi

Fungal Pathogens	*Pest, Against which Used*
Beauveria bassiana	Cotton bollworms
	Coffee berryborer
Metarhizium anisopliae	Sugarcane pyrilla
	Rhinoceros beetle
Nomouraea rileyi	*Helicoverpa armigera*
	Spodoptera litura
	Achaea janeta

Entomogenous Nematodes: Nematodes Feeding on Insects

Nematodes	*Pests Group*
G. operculella	Lepidopterous, Coleopterous
A. crtalaunalis	Hemipterous
Neoaplectana	Dipterous

Parasitoids and Predators

Parasitoids	*Target Pests*
Egg parasitoids *e.g., Trichogramma* sp.	Lepidopterous pests of different crops
Egg-larval parasitoids *e.g., Chelonus blackburni*	*Earias* spp. And *Pectinophora gossypiella* *Phthorimoea operculella*

Parasitoids	*Target Pests*
Larval parasitoids	
e.g., Bracon brevicornis	*Opisina arenosella*
B. kirkpatricki	*Pectinophora gossypiella*
Cotesia marginiventris	*Helicoverpa armigera*
C. plutellae	*Plutella xylostella*
Nymphal parasitoids	
e.g., Anagyrus dactylopii	*Maconellicoccus lirsutus*
Epiricama melanoleuca	*Pyrilla perpusilla*
Predator/parasite	
Tricogramma minutum	Sugarcane and cotton borers
Lixophoga diatraceae	Sugarcane borers
Opius vandenboschi	Fruit fly
Aphelinus mali	Wolly aphid
Rodolia cardinalis	Cottony cushion scale
Cyrtorhinus mudulus	Sugarcane leaf hopper

Among numerous biosuppression agents, most widely used one is egg parasitoids, *Trichogramma chilosis,* and *T. japonicum* against tissue borers in sugarcane and rice ecosystem. Recently PDBC, (Project Directorate of Biocontrol). Banglore, has developed superior strain of *T. chilosis* with physiological tolerance against upto 0.07 per cent endosulfan for the first time in world. Such biocontrol agents effectively fits into concept of IPM, in tune with recent needs.

Varietal Resistance

Conventional breeding strategies have not been successful in developing insect pest resistant varieties. In recent years, the area of genetic engineering has provided a altogether new approach for circumventing the use of chemical insecticides for controlling the insect pest. This approach employs the techniques of molecular biology whereby genes coding for insecticidal proteins can be identified, isolated, mobilized and expressed in crop plants. Such transgenic plants thus acquire the ability, which is heritable, to protect themselves from insect pest attack. The most common gene candidates are used for engineering plants resistant to insects pests are:

1. Delta-endotoxin coding gene from *Bacillus thuringensis* strain.
2. Protease inhibitor genes from cowpea, soyabean.etc.
3. Amylase inhibitors from cereals and legumes
4. Lectins, primarily from wheat germ, snowdrop plant etc.

Bacillus thuringiensis Crystal Proteins

B. thuringiensis, a gram positive soil bacterium is omnipresent. It produces intracellular crystal proteins or delta endotoxin during its sporulation.

The crystal proteins characterized by their insecticidal activity and spectrum are grouped in four classes, which are specific to specific order of insects. These are:

Cry I → for Leipidoptera – specific

Cry II → Leipidoptera and Diptera specific

Cry III → Coleoptera specific

Cry IV → Diptera specific

In terms of toxicity, the target organ for cry toxin is the larval midgut.

The effectiveness of the transgenic plants in insect pest management, however, turned out to be very limited, primarily because of the sub-optimal expression of the *Bt* crystal protein gene in plant cells. However, in the last few years, spectacular progress has been made in enhancing the cry gene expression in the transgenic plants which include important monocot crops like rice and maize. Expression of proteins in specific tissues (using tissue specific promoter) during certain stages of crop growth and development *viz.* seed development will reduce the selection pressure on insects allowing limited damage to less important plants parts. Agronomic practices like raising a crop of transgenic plants interspersed with normal plants is yet another approach which could be investigated. In India Monsanto and Mahyco are about to introduce *Bt* cotton.

Protease Inhibitors

Many protease inhibitors have been identified from many plant species, which are active against certain insect species. The mechanism of antimetabolic action of protease inhibitors is not yet fully elucidated. Direct inhibition of digestive enzymes is unlikely to be the only effect and possibly the proteases vital in insect development, such as those involved in moulting are also affected. This protease inhibitors has been shown to be an effective antimetabolite against a range of field and storage pest including members of the Lepidoptera, Coleoptera and Orthoptera. *e.g.*, Gene from cowpea inhibits trypsin of first instar larvae of tobacco bud worm. (*Heliothis virescenes*) and *H. zea, Spodeptera littoralis* and *Manduca sexta*.

α-amylase Inhibitors

α-amylase inhibitors in cereal and legume seeds have been shown to be inhibitory to insect gut α-amylases. α-amylase inhibitor is responsible for the protection of *Phaseolus vulgaris* seeds against attack by *Callosobruchus chinesis*, a storage pest of certain legumes.

Lectins

Lectins are carbohydrate binding proteins found mainly in storage tissues and seeds. Most of the lectins are toxic to mammals, birds and insects. Like Bean and winged been lectins, Soyabean lectin is toxic to larvae of the lepidopteran, *Manduca sexta*, Wheat grain aggultinin has inhibitory effect on Europian Corn borer (*Ostrinia. nubilalis*) and corn root worm (*Diabrotica. undecimpuncetella*). Among all the above methods, most successfully exploited gene is that of *Bacillum thuringiensis*. In India, *Bt.* cotton is about to be commercialized within few months. Other *Bt* crops, produced are tomato, soyabean etc.

Insect Growth Regulators

Although insect have an ability to fight back attack of invading microbes, they could be made more susceptible to the infection by suppressing various immune mechanisms.

The main neuroendocrine growth regualtors are: Juvenile hormone, Ecdysteroids Azadirachtin, Plumbagin

Juvenile Hormone

These are the most effective chemicals that interfere with the morphogenesis in insects (growth and development). These are found as useful insecticides against dipterans, hemipterans and recently against orthopterans like cockroaches. Synthetic Juvenile. Hormone have short residual life, and low toxicity to biological agents. *e.g.*,.Hydropene, Mehtoprene, Kinoprene, Phenoxycarb etc.

Ecdysterods

Ecdysones promote the moulting process and are sterodidal in nature. They are also found in plant kingdom. They are more polar and hence penetrate very less into insects via cuticle and causes the destruction of granulocytes, plasmatocytes and prohaemocytes. Mainly used for the control of grasshopper.

Azadiractin and Neem

Pure azadirachtin derived from neem seed kernels is both a strong antifeedant and growth regulators. More than 200 species of insects are sensitive to this molecule. Many formulations of neem are commercially available in India.

Lead for New Generation Pesticides

Botanical pesticides can play an important role in controlling pests. Screening of plant preparations, followed by bioassay-guided fractionation, leading to isolation and identification of pure active plant constituents, is considered one of the most successful methodologies for the discovery of new products.

Pesticidal Properties of Parthenin and Related Compounds

The sesquiterpene lactone parthenin is the main secondary metabolite of *Parthenium hysterophorus* L., a noxious weed known for its environmental and health hazard properties. It exhibits a wide spectrum of biological activities, which include cytotoxic, antilumour, allergic, antimicrobial, antifeedant, phytotoxic and insecticidal properties (Datta and Saxena, 2000) Parthenin along with its chemically and photochemically derived derivatives was tried for the antifeedant activity against *Spodoptera litura,* for insecticidal activity against *Callosobruchus maculates,* for herbicidal effect against *Cassia tora* and for nematicidal activity against *Meloidogyne incognita.* Antifeedant bioassay revealed that parthenin is moderately antifeedant. Other analogues were found to be most effective inhibitor of germination and seedling growth of *Cassia tora* further evaluation has confirmed its effectiveness in terms of nematicidal activity. These compounds were therefore evaluated as lead molecule for developing products in pest control.

Ailianthone from *Alianthus altissima* as Herbicide

The phytotoxic compound isolated from *Ailanthus altissima* root bark, was identified to be ailanthone, a quassinoid compound. It exhibited potent pre and post emergence herbicidal activity in green house trials. Post emergence activity was specially striking event at the lowest application rate (0.5 kg/ha). Ailanthone is rapidly detoxified in field as a result of microbial activity (Heisey, 1996). Efficacy of Ailanthone is comparable to glyphosate and paraquat most widely used for weed control (Table 3.1).

Table 3.1: Efficacy of Ailanthone, Glyphosate and Paraquat Application Rate of Various Active Ingredients

Active Ingredient	*Application Rate (Kg/ha)*
Glyphosate	0.6–4.5
Paraquat	0.3–1.7
Ailanthone	0.5–1.0

The high level of post emergence herbicidal activity in conjunction with rapid biodegradation in soil suggest ailanthone may have potential for development as a natural product herbicide with a broad spectrum.

Allelopathic Potential of *Castila tortusa*

Chaparrinone, another quassinoid closely related to ailanthone, was identified to be active constituent of Castela tortusa. This compound is also phytotoxic and its activity is quite similar to that of ailanthone in *Brassica juncea* and *Eragrostes* tef. (Table 3.2.) (Lin., 1995).

Table 3.2: Effective Phytotoxic Dosage of Ailanthone and Chaparrinone on Various Plant Species

	Lemna minor	*Brassicajuncea Eragrostes tef.*	
	I_{50} *(μM)*	I_{50} *(μM)*	I_{50} *(μM)*
Ailanphone	2.1	0.9	2.6
Chaparrinone	4.2	0.9	2.1

Allelopethic Potential of *Heleanthus annus*

A number of sesqueterpenoids have been extracted from sunflower. Heliannuol-A and Guainolides both are excellent candidate as pre-emergent herbicides on monocotyledons at low dosage (Macias *et al.*, 1996).

Insecticidal Activity of Allelochemicals

Studies showed that plant flavonoids, other phenols and α, β unsaturated carbonyl compounds acted as potent inhibition of Glutathion S-transferase a group of multifunctional enzymes that carry out detoxification of insecticides in lepidopterous insects. Thus these allelochemicals found in certain foliage may serve as synergists by interfering with GST-mediated detoxification in phytophagous insects (Yu and Abo, 2000).

Among flavours apigenin, among phenols Tannic acid and among α, β unsaturated carbonyl compounds Ethacynic acid showed good inhibitory activity.

Antifeedant Activity of Cystisine

Cystisine a secondary metabolite compound belonging to quinolizidine alkaloid family and is extracted from *Sophra alspecuroides* (L)–a leguminous plant. Cystisine acts as a feeding deterrent against several insect species (Table 3.3.) Cystisine was much more toxic than anabasine and nicotine to apterous adults of the mustard aphid (Lee, 2001).

Table 3.3: Antifeedant Activity of Cystisine and Nicotine on Mustard Aphid Mortality (%)

Alkaloid	*(mg/l)*	*24 hrs*	*48 hrs*
Cystisine	1000	65.00	96.67
	2000	95.00	100.00
Nicotine	1000	10.00	23.33
	2000	96.77	98.23.

Fumigant Activity of Essential Oil from *Mentha arvensis L. var. piperescens*

Essential oil from *M. arvensis* had potent activity against the rice weevil (LC_{50} = 45.5 ml litre^{-1} air). The menthone present in essential oil showed some promise as an alternative to those fumigants currently used to control stored grain insect pests.

Natural and Semisynthetic Ceveratum Alkaloids

Bioactive principles of sabadilla, a botanical insecticide prepared from the seeds of *Schoenocaulon officinale,* and related semisynthetic ceveratrum alkaloids revealed antimicrobial effects against *Erwinia, Corynebacterium, Fusarium, Glomerella* and *Plasmopara* species. Significant activities were displayed against the sunflower downy mildew fungus, *Plasmopara halstedii.* The ceveratrum alkaloids are thought to interact with the cell membrane, disrupting its integrity and/or altering its function. In contrast to commercial fungicides affecting either the parasite thallus or the asexual spores of *P. halstedii, i.e.,* only a single developmental stage, the alkaloids tested disrupt its life cycle at several points in host-dependent and host-independent stages. Characteristically most of the ester alkaloids and cevagenine exhibited remarkable but developmental stage-dependent anti peronospora activity disrupting the life cycle of *P. halstedii* in both host dependent and host independent stages. Among the natural alkaloids, cevacine and cevadine were the most effective against sunflower downy mildew and the activity of the semi-synthetic oleoyl vercevine and cevagenine was comparable to that of metalaxyl. The results indicates that the modest anti-oomycete activity of the natural sabadilla components can act as potential antifungal agents with a novel mode of action (Oros and Ujavary, 1999).

Insecticidal Activities of Daphnone Diterpenes

Methanol extract of roots of *Lasiosiphon Kraussianns* (Meisn) (Thymelaeaceal) showed potent insecticidal activity against *Aphis gossypii* (G Lov) and *Drosophila melanogaster* (Meig). Bioassay driven fractionation of this extract led to the isolation and characterization of two known daphne diterpenoids Exoercaria toxin and Wikstrotoxin D. The two natural products were inferior to methomy in activity against *Aphis gossypii* and *Myzus persicae* in contact. assays were superior in ingestion assay against *D. melanogaster.* Thus the two natural daphne diterpenoid Exoercaria toxin and Wikstrotoxin D showed potent insecticidal activity against *Aphis gossypi* and *Drosophila melanogatser* and in addition, these two have been reported as antitumour compound (Bala *et al.*, 1999).

Activity of Botanical Aphicides

1,5-Diphenyl-1-pentanone and 1,5-diphenyl-2-pentene-1-one the natural products extracted from *Stellera chamaejasme* exhibited potent aphicidal activity and very good antifeedant activity against *Aphis gossypi* and *Schizaphis graminum.* Thus these are potentially valuable agents for use as insecticides with high effect, reduced toxicity and increased safety for non-target organisms (Ping *et al.*, 2001).

Fungicidal Activity of β-Thujaplican Analogues

β-thujaplicin is a naturally occurring compound responsible for the resistance to fungal decay and insect attack on heartwood of several cupressaceous trees such as *Thuja plicata* Donn, *Thujopsis dolobrata* Sieb and Zucc and *Thuja standishii* Carr. In addition to these insectiidal and fungicidal activities, β-thujaplicin also shows antibacterial and plant-growth regulating activity, cytotoxic effects on mammalian cells, scavenging activity against active oxygen species and inhibitory activity against various enzymes (Baya *et al.*, 2001).

New Rocaglamide Derivatives from *Aglaia roxburghiana*

In course of screening for novel naturally occurring pesticides from the plant family Meliacea, an extract of the stem bark of *Aglaria roxburghiana* was found to exhibit significant insecticidal activity. In addition to rocaglamide, a known insecticide isolated from several species of genus *Aglaia,* 15 new natural products were isolated from this plant. These compounds show potent insecticidal activity, rocaglamide being the most active (Molleyers, 1999).

Molluscicidal and Antibacterial Compounds from *Petunia hybrida*

Molluscicidal and antibacterial compounds from three novel and six known petuniasterones and sucrose esters were isolated from the aerial parts *Petunia hybrida* (*Solanaceae*) by (Moser *et al.*, 1999). Some sucrose esters showed molluscidal activities at concentrations as low as 15mg ml^{-1} and were active against *Bacillus subtilis* and *Pseudomonas stutzeri*

Plant Derived Napthoquinone Pesticides

A series of compounds with structures based on insecticides/acaricidal napthoquinenes isolated from *Calceolaria andina* were synthesized (Khambay *et al.*, 1999). A feature of the series was the lack of resistance shown by strains resistant to established class of pesticides.

Allumin spp. Thiosulfinates as Substitute Fumigants for Methyl Bromide

Methyl bromide, the most widely used fumigant, is considered to be one of the major factors causing depletion of the ozone layer, and this is likely to lead to it being banned in the near future. *Allium* sulfur volatiles known to be nematicides was evaluated as insecticides by (Auger *et al.*, 1997) against insect pests in stored products, in comparison with their degradation compounds (disulfides) which have already been tested. Methyl and allyl thiosulfinates, with 24-h LD_{50} values of 0.02–0.25 mg litre^{-1} were more active than disulfides against six test insects and were superior to methyl bromide; it is suggested that they could be used as alternative to methyl bromide in stored product control.

Lead for Fumigant

Allium sp. provides excellent lead for active ingredients against stored grain pest (Table 3.4).

Microbes as Potential Source of Pesticides

Activity of Ilicicolins against Plant Pathogenic Fungi

Ilicicolins and dechloroilicicolin D, ascofuranone and arthrichifin were isolated from a fungicidally active the fenuentation broth of *Nectria sp.* The ilicicolins and related metabolites ascofurainone and arthichitin all contributed to the fungicidal activity of the broth (Bal Tembe *et al.*, 1999).

The ilicolins showed good activity at 125 mg/litre while ascofuranone was active only at 500 mg litre^{-1}. The ilicicolins were mainly active against the two oonycetes *Phytophthora infestans* and

Phytophthora viticola. Thus this work confirmed that ilicicolins show good fungicidal activity particularly against the oomycestes and indicates that they may provide useful leads for further investigations.

Table 3.4: Constituent of *Allium* Sulfur Volatile and its Activity on Stored Grain Pest

Fumigants LD_{50} (mg litre^{-1})	*Sitiphilus oryzae*	*Calosobruchus maculatua*
Methyl thiosulfinate	0.19	0.25
Allyl thiosulfinate	0.43	0.16
Methyl disulfide	1.23	1.06
Allyl disulfide	1.98	0.58
Dibromoethane	2.66	0.72
Dichloroethane	4.13	3.71
Methyl bromide	1.05	–

Source: Auger *et al.* (1999)

β-methoxy Acrylates for Synthesis of New Types of Strob Ilurin Fungicides

The strobilurin family of natural products are derivatives of β-methoxy acrylic acid and are found in several genera of small fungi such as *Strobilurus tenacellus.* The fungicidal activity of the strobilurins arises from their ability to inhibit mitochondrial respiration by blocking electron transfer between cytochromes b and c. (Aspinall *et al.*, 1999) Work in Zeneca led to the discovery of azoxy strobin which is sold as broad spectrum systemic fungicide. Strobilurins of other type gave rise to a new generation of active fungicides in which the oxime ether bond provided a mimic for the pyrimidine. A further class of active strobilurins prepared were oxyamidines.Good levels of activity amongst all three series of strobilurins was found against a broad spectrum of fungal pathogens in glass house tests, examples being *Erysiphae granunis* DC, *Plasmopora viticoha* Beri and de Toni and Septoria Modurum Berk. The oxamidine series also showed excellent activity against *Puccinia recondia* Rob. ex *Desm* and *Venturia inequalis* (Cooke) wint. Thus working upon the lead from natural source *i.e., Strobilurus tenaculus* newer and better fungicidal molecules can be prepared.

Natural Cyclopeptides as Lead Molecule

Cyclopeptides are produced by pathogenic fungi. Destruxins are class of cyclopeptides first isolated by Kodaira from the entomopathogenic fungus *Metarhizium anisopliae* (*Metsch.*) *Sorokin,* and recently from the entomogenous fungus *Aschersonia sp.* Other destruxins, such as destruxin B, are produced by phytopathogenic fungi like *Ophiosphaerella herpotricha* and *Alternaria brassicae* (*Berk*) *Sacc.* Tentoxin is produced by a phytopathogenic fungus of the same family, *Alternaria alternata* (*Fr.*) *Keissler* (*A. tenuis Nees*) These toxins have different pathogenic effects. Destruxins isolated from *M. anisopliae* have an insecticidal activity. Some insecticidal properties could be related to the inhibition of immune reactions. Destruxins from entomogenous fungi also possess antitumour and antiviral properties. Those of the destruxins extracted from *A. brassicae,* particularly destruxin B. have shown phytotoxic properties mainly directed against Brassicae and other crucifers. Tentoxin exhibit phytotoxicity by inducing chlorosis of many plants (Cavalier, 1998).

Herbicidal Metabolites from *Streptomyces viridochromogens* Tii 6105

Two fatty acids (2E, 4Z)-decadienoic acid and (2E, 4Z, 7Z)-decatrienoic acid were detected the latter being described for the first time as a natural product, in the culture filterate of *Streptomyces viridochromogens* Tii 6105 (Armin Maier *et al.*, 1999).

Both metabolites showed strong herbicidal activity against *Lemma minor* and *Lepidium sativum*. It is assumed that these compound effect the cytoplasmic membrane of the plant cells because of their lipophilic nature and free acidic group.

Conclusion

The active principles screened by various workers in plants and microbes are of much relevance in terms of today's agriculture. Our answer to various ills associated chemical pesticides. Natural pesticide as an alternative is need of the hour. However the problems faced in large-scale production of such crop protection agents act as disincentive to research. Biodiversity legislation, ownership rights and finally low margin market and activity attributed to volume renders it very difficult to adapt this approach.

However respite is in form of leads for synthesis of novel pesticides. Plant and microbes produces various bioactive metabolites, analogues and intermediate produced through manipulation of biochemical pathways provides excellent opportunity opening up new vista in form of endeavours to develop elegant routes and methodology to construct complex framework, functionality, with correct stereochemical orientation of a natural product. In addition, these new products also provide suitable tools both in the identification of the molecular target and in elucidation of its mechanism of action.

REFERENCES

Aspinall, I. and P.A. Northington (1999). β-methoxyarylates; synthesis of new types of strobllurin fungicides with extended side chains. *Pesticide Science.* 55: 197–198.

Auger, J., F. Cadoux, and E. Thibour (1999). *Alluim* spp. thiosulfinates as substitute fumigants for methyl bromide. *Pesticide Science,* 55: 200–202.

Bala, A. and R. Delorme (1999). Insecticidal activity of Daphnone diterpenes from *Lasiosiphon kraussianus* (Meisn) (*Thymelaceae*) roots. *Pesticide Science,* 55: 745–750.

Bat Tembe, S. and S. Kundu (1999). Activity of the ilicicolins against plant pathogenic fungi. *Pesticide Science,* 55: 645–647.

Baya, M. and P. Soulounganga (2001). Fungicidal activity of β-Thujaplicin analogues. *Pest Management Science,* 57: 833–838.

Cavalier, F. and I. Verducc (1998). Natural Cyclopeptides as leads for novel pesticides: Tentoxins and Destruxin. *Pesticide Sciences,* 52: 81–89.

Datta S. and D.B. Saxena (2000). Pesticidal properties of parthenin (from *Parthenium hyseero trophus*) and related compounds. *Pest Manag. Sci.,* 57: 95–101.

Heisey, R.M. (1996). Identification of an Aclelopathic compound from *Aclanthus altessima* and characterization of its herbicidal activity. *American Journal of Botany,* 83(2): 192–200.

Khambay, B.P.S. and D. Batty (1999). A new group of plant derived napthoquinone pesticides. *Pesticide Science,* 55: 645–647.

Lee, S.E. *et al.* (2001). Fumigant toxicity of volatile natural products from Korean species and medicinal plants towards rice weevil, *Sitophilus oryzae. Pest Manag. Sci.*, 59: 548–553.

Lin L.J. (1995). Identification of Plant growth inhibitory principles in *Acilanthus attessunia* and *Castela tortusa. J. Agric. Food Chem.*, 43: 1708–1711.

Macias F.A., A. Torres, J.M.G. Melinillo, R.M. Verala and D. Castellano (1996). Potential allelopathic sesquiterpene lactones from sunflower leaves. *Phytochemistry*, 43: 1205–1215.

Maier, A. (1999) (2E, 4Z)–Decadienoic acid and (2E, 4Z, 7Z)–decatrienoic acid two herbicidal metabolites from *Streptomyces viridochromogens* T ü 6105. *Pesticide Science*, 55: 733–739.

Molleyers, L. (1999). Insecticidal natural products: new rocaglamide derivatives from Aglaia roxburghiana. *Pesticide Science*, 55: 494–497.

Moser, D. and I. Klaiber (1999). Molluscicidal and antibacterial compounds from *Petunia hybrida. Pesticide Science*, 55: 336–339.

Oros, G. and J. Ujavary (1999). Botanical fungicides: Natural and semisynthetic cerveratrum alkaloids. *Pesticide Science*, 55: 253–264.

Ping, G., and H. Taiping (2001). Activity of the botanical aphicides 1, 5-diphenyl-1-pentanone and 1,5-diphenyl-2- pentene-1-one on two species of Aphididnae. *Pest Management Science*, 57: 307–310.

Rice, M.J., M. Legg and K.A. Powell (1998). Natural products in agriculture: A view from the industry. *Pesticide Science*, 52: 184–188.

Yu, S.J. and G.E. Abo-Elghar (2000). Allelochemicals as inhibitors of Glutathione S. transferases in the fall amyosm. *Pesticide Biochemistry and Physiology*, 68: 173–185.

4

Helicoverpa armigera Nuclear Polyhedrosis Virus and its Effectiveness

Saqia F. Chandel R. Ahmad** and P.K. Singh****

**Lecturer, ***Head, Department of Agriculture Zoology and Entomology, RBS College, Agra*

*** Principal Scientist, Division of Crop Protection, Indian Institute of Pulses Research, Kanpur – 208 024*

Native pathogens of *Heliothis armigera* (Hubner) reported from India include nuclear polyhedrosis virus, cytoplasmic polyhedrosis virus, granulosis virus, *Bacillus thuringiensis* Berliner, *Beauveria bassiana* (Balsamo). *B. brongniartie,* (Sacc.) *Metarhizium anisopliae* (Metschnikoff), *Nosema* sp. *Vairimorpha* sp, and *Ovomermis albicans,* (Sieb). Of these, nuclear polyhedrosis virus (NPV) has been most extensively studied for its morphology, dissolution and haemagglutination characteristic, pathogenicity. persistence in soil and foliage, mass production, safety and field efficiency in the control of pest on chickpea, pigeonpea, field bean, cotton, sunflower and tomato.

Of the different species of *Heliothis* occurring in India, *Heliothis armigera* (Hubner) is the most widely prevalent and devastating pest, with an extensive host range, causing sever damage to many food and fiber crops. The damage due to *H. armigera* is particularly severe on chickpea and pigeonpea in the Indian sub-continent, where 75 per cent of the world's chickpea and 90 per cent of the world's pigoenpea crops are grown. An average infestation of one larva per plant of pigeopea can cause an yield loss 1015 kg/ha. In sorghum, yield losses of 18 per cent to 26 per cent have been reported, in the cotton 41 per cent to 50 per cent. In tomato the damage caused may range from 40 per cent to 50 per cent. Though the damage caused by this pest in U.P. is high but the sustained researches have shown the potential of *Ha* NPV in the management of *H. armigera*. This wonder substances can be applied alone or in combination with other substances like insecticide, biorationals or chemicals and good results have been achieved.

Pathogenic Microbes Reported from India

Of the native pathogens of *H. armigera*, the nuclear polyhedrosis virus (NPV) has been the most extensively studied for its pathogenecity, mass production safety and full efficacy in controlling the pest of chickpea, pigeonpea, field beans, sunflower, cotton and tomato etc. Besides viruses, *H. armigera* also gets infected by other microbes. A list of microbes found parasitizing *H. armigera* is given in Table 4.1.

Table 4.1: Microbial Pathogens of *Heliothis* spp. reported from India

Pathogens	*References*
Virus	
Nuclear polyhedrosis virus (NPV)	Patel *et al.* (1968)
Cytoplasmic polyhedrosis virus (CPV)	Rabindra and Subramanian (1973a)
Granulosis virus (GV)	Narayanan (1985a)
Bacteria	
Bacillus thuringiensis (Berliner)	Majumdar *et al.* (1955)
Fungi	
Beauveria bassiana (Balsamo)	Jayaramaiah (1981)
Beauveria brongniartii (Sacc)	Jayaramaiah (1981)
Metarhizium anisopliae (Metschnikoff)	Urs and Govindu (1971)
Protozoa	
Nosema sp.	Anon (1985)
Vairimorpha sp.	Narayana (1985b)
Nematode	
Ovomermis albicans (Seib)	Bhatnagar *et al.* (1982)

Nuclear Polyhedrosis Virus

The occurrence of nuclear polyhedrosis virus in a laboratory culture of *H. armigera* in India was first reported by Patel *et al.* (1986) in Gujarat.

It is a baculovirus. These are DNA virus occluded in silicaceous and many sided proteinaceous bodies called polyhedron. Each polyhedron encloses 4–5 rod shaped virus, therefore these are called polyhedral inclusion bodies (PIBs). The polyhedron is about one micron in size which can be seen by a magnification of X 400. When the larva along with the food ingests the polyhedral inclusion bodies, the silicaceous frame work gets dissolves, the protein is digested by proteolytic enzymes and virus rods are released in gut. They invade the epithelial cells of the gut, haemolymph, fat cells and other cells of the viscera. They multiply in the nucleus. The food intake of larvae gets reduced. The body becomes soft and they die in 3–6 days.

The symptoms of the disease are visible 2 to 3 days after ingestion of the virus. The cuticle of the infected larvae in the first instars becomes milky white in appearance. Those infected in the later instars turns slightly pinkish or change in colour. The diseased larvae sometimes discharge a yellow gray fluid from their mouth. At death, which occurs normally between 3 to 7 days depending upon the stage of insect, the integument becomes fragile and burst liberating the liquefied body contents. The larvae generally die on the sides and on the lid of the rearing container. Some of the larvae die showing the classical position of hanging head downwards with posterior prolegs fixed on to the top of the plant twig in the form of an inverted "V".

Since the discovery of nuclear polyhedrosis virus studies have been taken up on various aspects, both fundamental and applied, studies have revealed a number of fruitful information, which are of immense importance. Some of the works relating to the present studies are discussed in this chapter.

The literature on the control of *Heliothis* species through nuclear polyhedrosis virus (NPV) has been reviewed extensively by Ignofo and Couch (1980). Three inclusion viruses from *Heliothis* are reported from India: two inclusion viruses (granulosis and NPV) in the genus Baculoviruses (Jacob and Subramanian 1972, Narayan,1979) and a Cytoplasmic polyhedrosis in the Reoviridae (Rabindra and Subramanian, 1973 a). Besides these, an Iridovirus and an unidentified spindle shaped virus has also been reported (Smith, 1967; Stadelbacher *et al.*, 1978; Carner and Hudson, 1983).Since the virion of the NPV of *H.armigera* is rod shaped, it is a grouped baculovirus. In India, the incidence of NPV in a laboratory culture of *Heliothis armigera* was first reported by Patel *et al.* (1968) in Gujarat. However the NPVs are the only virus that have shown promise for use as microbial pesticides.

In Tamil Nadu, Jacob and Subramanian (1972) described the polyhedra of *H. armigera* NPV (*Ha*NPV) to be irregular in shape, the diameter ranging from 0.5 to 1.4 µ with an average of 0.89±0.012µ. The virions were occluded singly. Some of the physico-chemical properties of PIB of NPV were studied by Narayana (1979). Rabindra (1973) found that the diameter of the polyhedra varied from 0.17 to 2.21µ with an average of 1.31±0.03µ. The carbon replicas did not show any surface patterns (Jacob and Subramanian, 1972). Characters of *Ha*NPV varied from that of *Agrotis stolonifera* NPV or *Pseudaletia separata* granulosis virus the polyhedrin exists in the form of aggregates (Sun-GuoXun *et al.*, 2002)

The factors responsible for the dissolution of polyhedral inclusion bodies of *H. armigera* NPV were studied by Narayana (1979), who examined the role of silicon, magnesium and iron in the skeletal lattice frame work of polyhedra and the influence of alkaline compounds, pH and proteolitic enzymes of the insect host on the dissolution of the polyhedra.

Narayana (1985c) has studied the invasive mechanism of NPV of *H. armigera* through haemagglutination technique and found it has more or less the same mechanism of adsorption of viruses to erythrocytes mediated to the same receptors as vertebrates viruses. Thus, it is possible by haemagglutination technique to detect and monitor the viruses in the environment.

Studies on the susceptibility of the different larval instars revealed that the early instars were very highly susceptible to the viruses (Rabindra and Subramanian, 1974) and the LT_{50} was shorter in younger larvae than in older ones (Narayana 1979).These findings indicate that for successful control of pest, the application of the virus should coincide with the occurrence of early stage larvae in the field. The virus was found to infect the nuclei of the fat body, hypodermis tracheal matrix, haemocytes, nerve cells of ganglion and brain, foregut, connective tissues surrounding the midgut, silk gland and epithelium of the gonads (Rabindra and Subramanian, 1974). Growth parameters such as consumption index, growth rate, approximate digestibility gross efficiency of ingested food to the body substance and net efficiency of the digested food to the body substance were lower in NPV infected larvae than the healthy ones.

Mass Production of NPV

Narayanan (1979) working on mass production of NPV, found that the optimum dose of inoculum required for obtaining the maximum harvest of virus from fourth instar larvae was 1.1 × 10^4 PIB/cup/larva by the diet surface contamination method. The larvae inoculated with this dose yielded the maximum virus of 2.5 × 10^9 PIB/larva. There was a significant positive correlation between the larval weight and the number of polyhedra in NPV infected larvae. The virus needed to treat 0.4 ha at 250 LE/ha can be obtained from 24 diseased final instar larvae. Similar methods of mass production of the virus have been described subsequently by Rabindra and Jayaraj (1985d). Gopali and Lingappa 2001 worked on mass production of the virus *Helicoverpa armigera* nuclear polyhedrosis virus (*Ha*NPV) to increase its production and productivity and found that NPV can be multiplied at all stages of a pest but recovery varied with the stage. Fourth instar larvae registered higher larval mortality and higher quantity of viral yield per larva (2.81 × 10^4 polyhedral occluded bodies; POB). The optimum viral dose required for virus mass production to both third and fourth instar was 10^8 POB/ml.

Comparison of *Ha*NPV Isolates from Different Locations

Odak *et al.* (1982) found considerable difference in infectivity of *Ha*NPV isolated from different locations. The culture isolated locally from Narasinghpur (M.P.) gave a higher mortality (93.3 per cent) than cultures from other centers. Sachan and Ahmad (1994) also reported better virulence of Kanpur isolate over Bangalore, Coimbatore and ICRISAT isolate in bringing about mortality of *H. armigera* larvae. Gopali and Lingappa (2001) studied various isolates of *Ha*NPV and reported Gulbarga, and Coimbatore isolate to cause higher larval mortality compared to others on fifth, seventh and tenth day after viral inoculation. Chandel *et al.* (2004) besides reports of differential mortality of the pest by using *Ha*NPV suggesting existence of races, experiment were conducted to evaluate the efficacy of five isolates of *Ha*NPV against second and third instar larvae of *H. armigera* under lab condition. Kanpur isolate and PCI, Bangalore isolate of *Ha*NPV was the most virulent and efficacious as it brought about highest mortality in second and third instar larvae of *H. armigera*.

Persistence of Virus in Soil and on Foliage

Studies on some viruses have shown that inclusion viruses may persist for long periods in soil. *T.ni* NPV retained about 15 per cent of its original activity six years after application to surface soil in field plots. Viral inclusion bodies appear to be absorbed quite firmly to soil particles and are not leached through the soil by rain. Maximum viral activity remained in the top of 0–2.5 cm of the undisturbed soil during the 223 week of observation. Little viral activity was detected at a depth 10–12.5 cm. The pH of the soil may affect persistence of viruses. *T.ni* NPV was inactivated more quickly in soils of low P than in soils near neutral. It is apparent, therefore, that virus persisting in soil following treatments of natural epizootics in succeeding generations of host insects and is a mechanism by which a long term effect of virus introductions may be attained. Polyhedra remain inactive in the soil for years and are spread to the plants by rain splashing and cultivation of soil. As the host feeds on several wild plants, neighbouring uncultivated soil may accumulate polyhedra from natural infections and preserve them as pathogenic reservoirs.

Adsorption of polyhedra in two types of soil was studied by Narayanan *et al.* (1985). It was found that the adsorption rate was higher in Alfisols (0.88) than in Vertisols (0.69). The activity of NPV on tomato leaves steadily decreased with increasing time of field exposure (Anon 1985), four days after exposure, activity was only 36.7 per cent of the original and by the tenth day it was as low as 6.4 per cent. This findings stresses the need to use suitable UV protectants along with the virus.

Stability of Viruses

Effect of Sunlight on Potency of *Ha*NPV

Sunlight is probably the most destructive environmental factor affecting microbial insecticides (Ignoffo and Hostetter, 1977) In general the half life of unprotected insect viruses both occluded and nonoccluded is 24 hours direct sunlight. Ignoffo and Batzer (1971), tried micro-encapsulation and UV light protectant with *Ha*NPV in artificial diet and natural sunlight. Non-encapsulated virus + carbon was as effective as encapsulated preparations with carbon. Carbon was found to be having some value in protecting the virus of *Heliothis*. Various additions of blocking and absorbing agents such as carbon black, lignin sulphate, shade and keltose and water soluble polymers bound to UV screening agents were all tried. It is better to advocate the spraying of virus at sunset to avoid exposure of virus to solar radiation. Several studies have demonstrated that deposits of viral inclusion bodies are inactivated within a few days after the application on foliage of crops. Evidence indicates that exposure to sunlight is the most important factor contributing to this inactivation. Bullock (1967) observed that *Ha*NPV retained little activity two days after application to cotton. There are reports that it is the UV portion of sunlight that inactivates insect virus. The exposure of suspensions of *Ha*NPV to ultraviolet light (254nm) reduced activity more than did exposure to a similar quantity of ultraviolet light of 307.5 nm with exposure to longer wavelength (UV 365 nm) and a broad band mixture of visible and infra-red light did not affect activity. It is recommended that viruses should be applied in the evening rather than in the morning to gain time before the viral deposits would be exposed to an intensity of sunlight sufficient to cause inactivation. The rapid inactivation of foliar deposits of virus by sunlight and UV light led to search for materials that could be added to spray mixtures to protect viral deposits to prolong their activity and increase their effectiveness in control. The ultra violet fraction of the sunlight (Mc Leod *et al.*, 1977) and the leaf surface of crop plants like chickpea (Rabindra *et al.*, 1994) inactivate the *Ha*NPV, necessitating frequent repeated sprays. It is better to advocate the spraying of viruses to sunset to avoid exposure of viruses to solar radiation. The rapid inactivation of foliar deposits of viruses by sunlight and UV light led to search for materials that could be added to spray mixture to protect viral deposits to prolong their activity and increase their field effectiveness.

Extracts of mango leaves was used as UV protectant with *Ha*NPV and was found to delay sunlight inactivation of the virus for about 4 days and prevented inactivation up to 4 hours of exposure to germicidal lamp. (Kranti *et al.*, 2001)

Effect of Temperature on Potency of *Ha*NPV

Temperature in a typical agro-ecosystem range from 5–50°C. Temperature that inactivated the NPV are usually too high to be encountered under natural conditions. For example, the thermal inactivated point of virus *S. litura* is between 90 and 95° C and 93 per cent for *Ha*NPV (Stuermer and Bullock, 1968) and 70°C for purified GV for *Pieris brassicae.*

Viruses, being a crystalline type of structure, are devoid of enzymes and other life process and therefore they are affected less by low temperatures as other forms of nucleoprotein. Deep freeze has little effect on the viral activity (David and Gardiner, 1967) and this method is used for storage of stock supplies of virus as powders or suspensions. Whereas, exposure to low temperature has little effect on activity of insect viruses, but exposure to high temperatures causes inactivation.

The bioassay studies indicated that storage of *Ha*NPV under refrigerated conditions did not affect the virulence of the virus throughout the year of the experiment. Results showed that storage of

*Ha*NPV in earthenware pots surrounded by wet sand preserved the virulence of the virus by 2 months longer as compared to storage under ambient conditions (Gopali and Lingappa, 2001)

Tests showed that a ten minutes exposure to temperature has little effect on activity of insect viruses, exposure to high temperatures (70–80°C) causes inactivation. Since the temperature in the field would not reach these high levels. Studies on the effect of long exposure to moderate high temperature are of more practical interest in considering stability of microbial insecticides in the field environments. There is need to search strains which can withstand maximum temperature normally encountered in the field environment that is, air temperature of 43°C (110°F) or soil temperature of (123°F), at least for the period when it is lying idle in the environment. At high temperatures sometimes increases the resistance of host insect to viruses. For instance, the resistance in *Pieris rapae* to granulosis virus increases with temperature above 36°C, *Trichoplusia ni* and *Heliothis zea* to their NPV at 39°C. *Colias eurytheme* to CPV at 35°C and *Diprion sertifer* to NPV at 29°C. The exposure of *Pieris brassicae* GV to a slightly higher temperature for 20 days destroyed most virus activity. Exposure of natural population to extreme heat of summer may cause higher failure of control or disappearance of latent infections. The effect of heat and radiation should be considered at the time of use of virus under hot conditions.

In contrast, good infection by GV of cut worm *Euxoa segetum* on cotton was reported even at 45°C. Mostly temperature influences the host and pathogen relationship as a whole. The optimum temperature range of the host insect is normally favourable to the disease.

Effect of Humidity and Rainfall on Potency of *Ha*NPV

It is generally accepted that humidity has less effect on stability of viruses than on stability of other types of pathogens. The relative humidity has no influence on the stability of deposits of *Pieris brassicae* GV on glass slides held for 2 days at 20°C but inactivation of dried deposits at low humidity was slightly greater in 7 days. There is good evidence that certain substances on or in leaves may influence stability of viruses. The deposits of *Ha*NPV were inactivated during the night when leaves were wetted with dew. It was found that the dew on the leaves was alkaline pH (8.2–9.1) and that washings of the leaf surfaces were pH 9.7–10.1 and the immersion of NPV caused swelling of the polyhedral bodies. The chemical content of the dew differs with the plant. Cotton dew was normally more alkaline (pH–8.8) than the dew on soybean leaves (pH–7.8). Various chemicals (bicarbonates, carbonates, sodium potassium and calcium) found in dew or leaves could have been responsible for effects of the dew on deposits of the virus.

The washing of virus from leaves by rain does not appear to be major factor affecting persistence of foliar deposits. The activity of pure and crude preparation of *Pieris brassicae* GV deposits on leaves of cabbage was not reduced by exposure for 5 hours to 250 cm of simulated rain. *Ha*NPV was not washed from cotton leaves in similar tests by Bullock (1967) and Ignoffo *et al.* (1965).

Efficacy of *Ha*NPV

Laboratory Experiments

These were conducted to assess the effects of selected low cost and widely available adjuvants on the efficacy of *Ha*NPV on sunflower. Results indicated that jaggary, chickpea flour, molasses, extract of sunflower leaves and egg albumen may be used as adjuvant for nuclear polyhedrosis virus in the control of *H.armigera*. Another laboratory experiment was conducted by Chunduwar *et al.* (1990) to test the efficacy of *Ha*NPV in combination with boric acid and tannic acid on the larvae of *H.armigra* fed on treated chickpea leaves. The greatest larval mortality was obtained with NPV+ 0.5 per cent boric acid with an LT_{50} of 136.8 hours. Polyhedrin and virion were isolated and purified from *Agrotis* *segetum* nuclear polyhedrosis virus (*As*NPV) and administered to 3rd instar *Heliothis armigera* larvae with *Ha*NPV. Both polyhedrin and virion could increase the mortality of *H. armigera* larvae, shorten the time to 50 per cent mortality and decrease the concentration required for 50 per cent mortality. (Bai-XiaoDong *et al.*, 2000)

Field Trials

Studies on *Ha*NPV for the management of *H.armigera* in pigeonpea and chickpea were conducted by Gopali and Lingappa (2001). Many isolates of *Ha*NPV tested were from Gulbarga, Bangalore, Dharwad, Coimbatore and Kanpur. Chickpea shoots were dipped in viral isolates and fed to insects. Mortality of larva was recorded starting on third day. Pigeonpea was sprayed with *Ha*NPV isolate (250 LE/ha) prior to insect release on the plots. It was reported that *Ha*NPV isolates from Gulbarga and Coimbatore caused higher larval mortality compared to other isolates on the fifth, seventh and tenth days after viral inoculation in the bioassay. Gulbarga and Coimbatore isolates on pigeon pea plots resulted in higher grain yield compared to other isolates. Gulbarga isolates maintained their superiority and were highly cost effective. On chickpea the efficacy of *Ha*NPV has been demonstrated in Gujarat, where high mortality of *Helicoverpa* larvae is obtained. In the preliminary trials, 47 per cent increase in grain yield has been indicated over control (Anonymous 1987).

Pokharkar and Chaudhary (1998) studied the spray schedules of *Ha*NPV in tomato and found that six sprays of *Ha*NPV at 5 days interval, resulted in complete reduction of larval population, 8.08 per cent mean fruit damage, 271.45 q/ha mean total yield and 253.95 q/ha mean marketable yield. However, it was on a par with 5 sprays of *Ha*NPV at 7 days interval, 4 sprays of *Ha*NPV at 10 days interval and 2 sprays of endosulfan at 15 days interval.

*Ha*NPV at the rate of 1.5×10^{12} polyhedral occlusion bodies (POB)/ha sprayed three to four times in the evening hours at an interval of 7-10 days when the pest was in the very early stages effectively controled the pest on chickpea (Jayaraj *et al.*, 1987) and pigeonpea (Santharam *et al.*, 1981) and lablab bean (Jayaraj *et al.*, 1987). Application of *Ha*NPV 125 LE/ha along with 0.035 per cent endosulfan has been found to give better control of *H. armigera* on chickpea (Jayaraj 1981) and pigeonpea (Santharam *et al.*, 1981). In Maharashtra, two applications of 450 larval equivalents/ha at a 10 day interval are most effective in reducing damage and increasing yield (Dhamdhera and Kaire 1986). Similarly two sprays of *Ha*NPV at 500 LE/ha are as effective as two sprays of 0.05 per cent endosulfan in reducing infestation by *H.armigera* larvae and pod damage and in increasing seed yield (Pawar *et al.*, 1987). *Ha*NPV can also be applied with diflubenzuron against *H. armigera* on chickpea (Ethiraju 1986). Two applications of 0.07 per cent endosulfan or *Ha*NPV at 500 LE/ha alternated with endosulfan significantly reduced pod damage in pigeonpea and increased grain yield (Sanap and Deshmukh,1988).

In Rajasthan, field trials were conducted to determine the effect of trichogrmmatid egg parasitoid *Trichogramma pretiosum* and of nuclear polyhedrosis virus (NPV) for the control of noctuid *H. armigera* on chickpea. All the treatments were carried out at flowering stage. Spraying with *Ha*NPV at 250 LE/ha was carried out when the population of *H.armigera* on chickpea was mainly of early instars. The percentage reduction in *H. armigera* numbers following spraying with *Ha*NPV alone ranged from 46.2 to 51.3 per cent. Treatment with *Trichogramma pretiosum* and *Ha*NPV produced 69.7–76.4 per cent reduction in number of *H.armigera* compared to the untreated control. It is concluded that the release of *Trichogramma pretiosum* followed by spraying with *Ha*NPV is effective for controlling *H. armigera* (Balasubramanian *et al.*, 1989). Another field trials were conducted in Madhya Pradesh by Sharma *et al.* (1997) where *Ha*NPV, *Trichogramma chilonis*, monochrotophos (Nuvacron 36 EC) and endosulfan

(Thiodan 35 EC) were evaluated for the control of *H. armigera* on chickpea. It was reported that nuclear polyhedrosis virus gave the best control of the pest.

Venkatesan *et al.* (1997) used *Ha*NPV and green lacewing, *Chrysoperla carnea* Stephens for *Helicoverpa armigera* (Hubner) management on sunflower. *Ha*NPV reduced the larval population from 3 to 1/head and increased the seed yield. The larval population of the pest was reduced from 3 to 2 per head by green lacewing larvae (*Chrysoperla carnea*) released twice at weekly intervals, at the rate of 2/head.

Field efficacy of *Ha*NPV alone and in combination with delfin was studied by Srinivasan *et al.* (1994) for the control of gram pod borer, *H.armigera* (Hubner) on chickpea. The efficacy of *H. armigera* and *Ha*NPV culture in different diets individually and in combination with *Bacillus thuringiensis* (*Bt.* Delfin) were evaluated at two doses of 125 LE/ha (D1) and 250 LE/ha (D2) on chickpea following 6 sprays application between 30 and 90 days after sowing (DAS) and compared with endosulfan (350 g.a.i./ha).Larval count recorded 7 and 10 days after combined of NPV (D1) + *Bt.* (1Kg/ha) was as effective as *Ha*NPV (D2) but statistically superior to endosulfan in the first two sprays. However from the third spray onwards all the treatments were on par with each other. There was no difference in virulence of *Ha*NPV obtained from different sources of diet. Endosulfan treatment results in minimum pod damage while treatment involving *Ha*NPV either alone or in combination were equally effective in reducing pod damage where in *B.t* alone, pod damage was maximum. The cost benefit analysis indicated that *Ha*NPV (D2) was the most cost effective treatment followed by endosulfan, *B.t* either alone or in combination with *Ha*NPV proved uneconomical due to its high cost.

Field experiment was conducted by Neharka *et al.*, 1999 for the control of pigeonpea pod borer complex *H. armigera, Exelastis atomosa* and *Melanagromyza obtusa* on pigeonpea. A total of 9 insecticidal treatments were given. These were acephate, chlorpyriphos, etrofolan (isoprocarb), lindane fenvalerate, endosulfan, *Ha*NPV, *Bacillus thuringiensis* subsp. *kurstaki* (Dipel EC) and neemark (extract of *Azadirachta indica*). The per cent pod infestation due to *H.armigera* and *E.atomosa* varied from 26.51 to 54.88 per cent. Neharka *et al.* (1999) recorded the lowest pod damage (26.51 per cent) due to *Ha*NPV treatment. The infestation of pods due to *Melanagromyza obtusa* varied between 18.53 and 35.0 per cent. Endosulfan at 0.07 per cent showed less pod damage 18.53 and 18.66 per cent respectively. Treatment with *HaNPV* resulted in highest yield (18.74 gm/ha). Among all the insecticides *A. indica* extract and isoprocarb were least effective against pod borer complex.

Another field experiment was also conducted by Wanjari *et al.* (1998) for the management of *H.armigera* in which 8 treatments comprising of *Ha*NPV, a commercial formulation of *Bacillus thuringiensis* subsp. *kurstaki* (Dipel), neem seed extract and endosulfan either alone or in combination with two products. It was shown that all the treatments were effective in reducing larval population as well as pod damage due to this pest and produced greater yields compared to the untreated control. According to Wanjari *et al.* (1998) in their field experiment the treatments with *Bacillus thuringiensis* subsp. *kurstaki* alternated with endosulfan proved to be the best.

Biradar *et al.* (2002) conducted field experiment to study the efficacy *Ha*NPV, *Bacillus thuringiensis var. Kurstaki*, Neemark alone and in combination was studied against cotton bollworms in comparison with endosulfan (0.07 per cent). The results indicated that the least percent bollworm damage and maximum seed cotton yield were obtained from *Ha*NPV followed by *Ha*NPV+*B.tk* and endosulfan. Sarode and Sonalkar (2001) compared biological control agents (BCA) and conventional chemical method to control *Helicoverpa armigera* and found that the pod damage caused by *H. armigera* was lowest (26.20 per cent) in the plots sprayed with *Ha*NPV compared to 29.00 per cent in the plots treated with insecticides. An additional investment of Rs 353.75/ha on BCA-treated plots increased net output, which resulted in higher cost benefit ratio of 1:2:40 compared to the conventional method (1:2:14). The incremental cost benefit cost ratio for BCA was 1:4.65 over the conventional method.

In order to evaluate the efficacy of biopesticides, Loganathan *et al.* (2000) conducted two field experiments. In both the field trials, chlorpyrifos proved superior in recording the lowest larval population and pod damage, followed by Spicturin. The highest pod yield was recorded with chlorpyrifos followed by Spicturin + *Ha*NPV. The cost benefit ratio was higher with chlorpyrifos (1:3.50) followed by *Ha*NPV individually (1:3.32), Spicturin (1:2.18) and Spicturin + *Ha*NPV (1:2.07). Chandel *et al.* (2001) evaluated biopesticides (*B.b*, *B.t*, NSKE and *Ha*NPV) under field conditions against *H. armigera* and found that these biopesticides worked very effectively in the farmers field and considerably brought down the loss and consequently increased yield. Sprays of commmerical formulation of *B.b* (Dispel), *B.t*, (Delfin and Halt) and *Ha*NPV resulted in lower pod damage. These treatments were on par with traditional sprays of 0.07 per cent endosulfan. The interactive effect of chickpea genotypes and nuclear polyhedrosis virus on the management of *Helicoverpa armigera* (Hubner) was studied by Bhagwat (2001) and found that malic acid did not reduce the efficacy of *Ha*NPV, even did not contribute to the control of the pest. Susceptible cultivars produced higher yields than the resistant ones when sprayed with *Ha*NPV + endosulfan.

The efficacy of four integrated pest management (IPM) modules along with two controls, *i.e.*, one standard state recommendations and an untreated control, was evaluated. The standard control (two sprays of endosulfan) was more effective than module 3 (*Ha*NPV followed by Nimbecidine) and module-4 (two sprays of *Ha*NPV + endosulfan) with regard to chickpea protection and production as well as CBR (1:5.26) (Singh *et al.*, 2001) Ahmad *et al.* (1999) conducted field trials which revealed that 2 applications of *Ha*NPV either alone or *Ha*NPV followed by insecticides resulted in effective control of *H.armigera* infesting chickpea and there was significant increase in grain yield. He narrated that the amount of insecticide in the ecosystem can be drastically reduced with the use of *Ha*NPV without compromising the degree of pest control.

Efficacy of *Ha*NPV with and without adjuvant and UV protectants against *H.armigera* in chickpea crops was evaluated in the field trials by Rawat and Shukla (2001). All the treatments registered significantly lower larval population and higher seed yield as compared to untreated control. Significantly higher seed yield of 1612.5 and 155.0 kg/ha was observed in the treatment of *Ha*NPV 250 LE + Ranipal (0.5 per cent) respectively and both were on par.

Effect of Adjuvant on Efficacy of *Ha*NPV

Rabindra and Jayaraj (1985a) conducted a series of pot culture experiments to identify suitable adjuvants, which could be cheap and easily available. Fifteen 20 to 25 days old chickpea plants were sprayed with 100 ml suspensions of different treatments containing the virus and adjuvants. Spraying was done with hand atomizer, and the plots were rotated for uniform coverage. After drying, 15 to 25 early second instar larvae of *H. armigera* of uniform age were released on the plants and allowed to feed for 18 to 24 hours. They were then transferred to individual vials containing diet (lacking formaline) and the per cent mortality was recorded on the seventh day.

Rabindra and Jayaraj (1985a) conducted many field experiments. In the first experiment, addition of crude sugar 0.5 per cent was found to give increased mortality from *Ha*NPV.In the subsequent two experiments, the efficacy of crude sugar was compared with those of coconut, groundnut and cottonseed oil. All the adjuvant significantly increased *Ha*NPV caused mortality over *Ha*NPV alone, crude sugar being the most efficient. In the fourth experiment, the efficacy of different oil cakes and flours was compared with that of crude sugar. Groundnut oil cake 3 per cent and chickpea flour 1 per cent were found to be as good as crude sugar. Because of high cost of applying groundnut oil cake at 3 per cent, the efficacy of the cake at 1 per cent was compared with crude sugar 0.5 per cent and with a combination of crude sugar, oil cake, chickpea flour and cottonseed oil. Groundnut oil cake 1 per cent was found to be as effective as crude sugar 0.5 per cent but a combination of crude sugar 0.25 per cent + chickpea flour 0.5 per cent + groundnut oil cake 0.5 per cent was found to give significantly higher mortality of *H. armigera* than crude sugar 0.5 per cent. Roome (1975) has reported that addition of molasses 0.6 per cent greatly improved the performance of *Ha*NPV in the field of *H. armigera* larvae on sorghum and cotton. These encouraging results indicate clearly the possibility of using such adjuvants as crude sugar or groundnut oil cake which are cheap and easily available to Indian farmers and which possess the required properties for an ideal adjuvant as described by Bell and Kanavel (1978) to obtain better control of *H. armigera* especially on chickpea.

Field Trials

The results of field trials with *Ha*NPV on chickpea have been very promising (Jayaraj 1985), probably young larvae feed on the leaflets, scraping the epidermal tissues and the older larvae feed on the leaves, flower buds and pods of chickpea. When larvae feed on the foliage they acquire the virus better than when they feed on flower or pods.

Narayanan (1979) conducted two field experiments at doses ranging from 125 to 375 LE/ha, either alone or in combination with insecticide. The efficacy of three sprays of *Ha*NPV at 125 and 250 LE/ha was compared with that of single application of endosulfan 0.07 per cent by high volume application. The virus was applied in the evening hours along with Triton x 100 at 0.01 per cent as wetting agent. It was found that larval population was significantly reduced in virus treatments when compared with the control. The virus treatments reduced leaflets damage considerably and were as effective as endosulfan. The damage to flowers and pods were significantly lower than with endosulfan. Both total grain yield and number of pods formed were maximum in pots treated with *Ha*NPV at 250 LE/ha. The virus treatments were as effective as endosulfan. In the second set of experiments, the virus was applied in the morning hours along with adjuvants, which was a liquid flow material formulated by Sandoz–Wander, Inc. at 10 per cent spray fluid concentration with inclusion of 2.5 per cent cane sugar. It was noted that application of three rounds of *Ha*NPV at 250 LE/ha plus adjuvants effectively checked larval population, reduced damage of leaflets, flowers and pods and increased total number of pods formed and yield. Reduction of pod damage in chickpea due to application of *Ha*NPV has also been reported by Makode (1978), Bakwad (1979), Santharam and Balasubramanian, (1982), Mistry *et al.* (1984), Ahmad *et al.* (1999) and Rabindra and Jayaraj (1985c). Application of *Ha*NPV at 375 LE/ha was found to reduce pod damage and was comparable to carbaryl (Santharam and Balasubramanian 1982). Rabindra and Jayaraj (1985 c) found that application of *Ha*NPV at 250 LE/ha followed by endosulfan 262 g a.i/ha 5 days later, reduced the larval population and pod damage better than either treatment alone. Mistry *et al.* (1984) studied the field efficacy of *Ha*NPV and found the application of five rounds of *Ha*NPV at 250 LE/ha at weekly intervals could reduce the larval population as well as damage to pods.

Application of a mixture of virus and insecticide has also been reported to give better control than either applied alone. *Ha*NPV at 125 LE/ha and endosulfan 0.035 per cent spray with high volume knapsack sprayer in the evening hours reduced pod damage significantly. The pod damage in plots receiving NPV spray @ 250 LE/ha was 14.0, where as it was only 3.7 per cent in plots treated with a mixture of *Ha*NPV and endosulfan 0.35 per cent. Untreated plots recorded 21.4 per cent pod damage. Praveen and Dhandapani (2001) evaluated biopesticides applied with adjuvants and predator *Chrysoperla carnea* against *H. armigera* infesting tomato and observed *Bt.* + *Ha*NPV + Neem 0.3 per cent at 0.5 litre/ha was the most effective treatment against the leaf borer, *H. armigera,* and resulted in increased fruit yield and high cost: benefit ratio, followed by *Bt.* alone and *Bt.* + *Ha*NPV. Several adjuvants like crude sugar (Rabindra and Jayaraj, 1988b), crude larval extract of certain lepidopteron including *H. armigera,* whole milk, whole egg homogenate, egg white and tender coconut milk were reported to be effective in increasing the efficacy of the virus against *H. armigera* (Rabindra and Jayaraj, 1988c). These adjuvants were also found to be effective in the field control of pest (Rabindra *et al.,* 1989).

The efficacy of *Ha*NPV was reported to be increased by adding phagostimulant (Jaggary 0.5 per cent) and UV protectants (Tinopal 0.1 per cent) (Muthuswami *et al.,* 1994). Addition of 0.25 to 0.5 per cent boric acid increased the efficacy of *Ha*NPVs. Cotton seed flour has been reported as the most preferred adjuvant for *H. virescens* (F.). Bell and Kanavel (1978) and "Coax" a commercial adjuvant (Traders oil mill Co., Texas, USA), consisting mainly 62.3 per cent of cottonseed flour improved the field efficacy of NPV against *H. viriescens* on cotton (Bell and Romine, 1980). The performance of these adjuvants can be enhanced by the addition of whitening agents like Ranipal or Robin Blue (commercial whitening agents), which were reported to increase the persistence of *H. armigera* nuclear polyhedrosis inclusion bodies (Rabindra and Jayaraj, 1988c; Sachan and Ahmad, 1994). Adjuvants were screened for improving the efficacy of *Ha*NPV against *H.armigera* on sunflower and pigeonpea in Karnataka. Application of virus plus 0.5 per cent jaggary significantly reduced larval populations compared to virus alone. The addition of boric acid or jaggary also reduced head damage in sunflower and pod seed damage in pigeonpea and increased yield (Bijjur *et al.,* 1991).

Production of PIBs

Studies were conducted to find out the ideal stage of the pest and optimum dose of pathogen to get maximum harvest of polyhedral inclusion bodies with high virulence. There was a significant interaction between the age of larva and dose of the virus with the recovery of PIBs. A maximum yield of 250, 33.33 × 10 PIBs was obtained when the virus at 1.1×10^6 PIBs/cup/larva was used at late fourth instar, which could cause 90–100 per cent mortality (Narayanan, 1979).

Gopali and Lingappa (2001) worked on mass production of the virus *Helicoverpa armigera* nuclear polyhedrosis virus (*Ha*NPV) to increase its production and productivity and found that *Ha*NPV can be multiplied at all stages of a pest but recovery varied with the stage. Fourth instar larvae registered higher larval mortality and higher quantity of viral yield per larva (2.81×10^9 polyhedral occluded bodies; POB). The optimum viral dose required for virus mass production to both third and fourth instar was 10^8 POB/ml.

Safety of *Ha*NPV to Non-target Organism

The *Ha*NPV of *H. armigera* was tested against 12 species of parasitoids, 4 species of predatory insect and one species of predatory mite and was found to be safe to all of them. The virus was also found to be non-pathogenic to silkworms *Bombyx mori* L and *Philsomia ricini* Boisduval (Anon 1983). *Ha*NPV is non toxic to lady bird beetle (Biradar *et al.,* 2002)

The effects of synthetic insecticides, *i.e.,* endosulfan, chlorpyrifos, methomyl and cypermethrin, and biological insecticides, *i.e.,* neem seed kernel extract (NSKE) and *Ha*NPV was studied on the predatory behaviour of black drongo, Gopali and Lingappa (2001). In plots treated with *Ha*NPV and NSKE, the predatory activity of black drongo was not interrupted. The predatory efficiency of black drongo in *Ha*NPV-treated plots increased on the 4th and 5th day after spraying. Bird droppings contained virulent virus particles (7×10^{-8} POB/ml) and therefore, served as virus-disseminating agents in nature. Srinivas (1985) found the *Ha*NPV to be safe to the parasitoids *Apenteles* sp., *Bracon greeni* Ashmead, *B.hebetor* Say., *Chelonus blackburni* (Cameron), *Eriborus* sp., C. *chloridae* and *Eucelatoria bryani* Sabrosky.

In preliminary studies, oral administration of NPV of *H. armigera* produced no deleterious effect on albino rats, white broiler, chick or fish (common *Carp, Cryrinus Carpio* L and *Tilapia, Tilapia mozambica* P) (Narayanan, 1979). There is no significant difference between the NPV treated and control organisms in weight gain, food consumption, feed efficiency, weight of different organs and blood and urine chemistry. *Ha*NPV did not effect the development of parasitoid *E. bryani* larviposition and behaviour of adult flies emerging from diseased host larvae was normal (Narayanan 1980). Insect viruses are apparently harmless to other life forms. The NPV of *Heliothis zea* using white mice and guinea pigs by various methods were tested and showed that all the test animals were healthy and made normal weight gains. Similar tests with rats and dogs also showed the safety of the virus. NPV of *Heliothis zea* when introduced into the culture system of several established lines of mammalian tissue culture had not brought about any cytological changes or viral replication. The virus was safe to oyster, shrimps, fish, minnows, trout, etc. Human feeding tests with volunteers (ten men and women were each fed 5.82 billion polyhedra over period of 5 days) showed them as healthy. *Baculovirus amsacta* is also safe to poultry bird and white mice. The noninclusion virus in the nuclei of cells of *Galleria mellonella* was tested in adult mice, rabbits and newborn mice. None of the treated animal died or showed pathological symptoms. Endosulfan resulted in the highest percentage of natural enemy reduction (45 per cent) over all the treatments including *Ha*NPV, which were safe to the natural enemy fauna living in the chickpea ecosystem (Suganthy *et al.*, 2002).

Studies on the cross infectivity of seven nuclear polyhedrosis viruses (NPV) and one granulosis virus (GV) to second instar larvae of *Helicoverpa armigera* showed that the NPV of *H. armigera* is host specific. NPV of other insects did not infect *H. armigera* larvae. Larvae inoculated with NPV of *Galleria mellonella* (*Gm*NPV) died showing typical virus symptoms. Restriction endonuclease analysis with PstI, HindIII, BamHI and EcoRI, however, revealed that the progeny virus was only *H. armigera* NPV (*Ha*NPV). The *Gm*NPV had obviously triggered expression of a latent virus infection in *H. armigera* larvae and a comparison of the DNA fragments revealed that this latent virus was a strain different from the Coimbatore strain of *Ha*NPV.

Effect of NPV on the Biochemical Changes in the Infected Larvae

The carbohydrate, glycogen and fat contents decreased in diseased larvae and the reduction was highly pronounced at 120 and 144 hour after inoculation over comparable healthy ones. In the case of minerals lower levels of calcium, potassium, copper and iron and higher levels of magnesium, sodium, phosphorus, manganese and zinc were noticed.

Transmission of *Ha*NPV by Parasitoids

Campoletis chlorideae transmitted the virus to *H.armigera* larvae directly by oviposition causing 100 per cent mortality (Odak *et al.*, 1982). *E bryani* transmits *Ha*NPV during larviposition (Narayanan 1980). Studies of the ability of *Eucelatoria* sp.,nr. *armigera* to transmit the virus during larviposition revealed the successful transmission of the pathogen (Narayanan, 1980). Surveys of natural enemies of *H.armigera* in South-Central India have so far revealed 27 spp. of insect parasitoids (Bhatnagar *et al.*, 1982) and it would be worthwhile to study the promising ones for their ability to transmit the virus to *H.armigera* larvae.

NPV Infection and Insecticide Susceptibility

Nuclear polyhedrosis virus infection in *H.armigera* increased the larval susceptibility to insecticide. When fifth instar larvae were exposed to a dose of 315 PIB/mm^2 of diet surface, the susceptibility to endosulfan, fenvalerate, cypermethrin and chlorpyriphos increased several fold (Srinivas, 1985). It is suspected that *H.armigera* populations may have developed resistance to some of the pesticides used against them. Entomopathogens when properly introduced into the population can break such insecticide resistance (Listov and Nesterov 1976, Rabindra and Jayaraj 1985b, Rud and Bellonick 1984). Larvae of *Helicoverpa armigera* resistant and susceptible to fenvalerate and endosulfan were tested for susceptibility to a nuclear polyhedrosis virus. The LC_{50} values were less in resistant strains than in susceptible strains. The resistance ratio to *Ha*NPV was greatest in fenvalerate resistant strain (0.88) followed by the endosulfan resistant (0.64) strain (Goud *et al.*, 1997).

NPV and Parasitoid Interaction

Studies on the interaction between the *Ha*NPV infection and parasitism by *Eucelatoria* sp nr. *armigera* Coq. have shown that NPV infection in *H.armigera* larvae did not affect the development of the parasitoid. Maggot development, adult emergence, sex ratio, longevity and larviposition behaviour of adult flies emerging from diseased host larvae were normal (Narayanan, 1980). But NPV infection in *H.armigera* larvae suppressed development of *Campoletis chlorideae* Uchida as reported by Odak *et al.* (1982). However, preliminary field observations on the percentage parasitism by *Campoletis chlorideae* on *H.armigera* in NPV treated plots suggested that there was no adverse effect on virus application on the development of parasitoid either in the larval or in pupal stage (Anon., 1983).

Effect of Virus on Egg

NPV of *H.armigera* when applied at different concentrations on the eggs, the larval mortality was noticed from the fourth day onwards at 2×10^6 and 1.5×10^6 PIB/ml doses respectively.

Conclusion

The gram pod borer *Helicoverpa armigera* (Hubner) is an important pest of agriculture and horticulture crops. Besides chickpea and pigeonpea it also damages cotton, tomato, lady's finger and sunflower. It is a serious pest of pulses in India and damages both chickpea and pigeonpea. On an average 30 per cent crop loss is reported. Due to absence of true resistant varieties in both the crops, resorting to chemicals had been the only alternative. But this has resulted to environment disaster in the form of many fold against insecticide and other social problems. Among various alternatives to chemicals microbial control with nuclear polyhedrosis virus (NPV) is favoured for the management of *H. armigera* (Hubner) on many crops including chickpea (Jayaraj *et al.*, 1985) and cotton (Rabindra and Jayaraj, 1995). It is mainly due to its host specificity, safety to ecosystem and having no effect on non-target species so that all can work together. Indiscriminate use of pesticides against *H. armegera* has remitted into many fold resistance against common insecticide: use of *H. armigera* nuclear polyhedrosis virus is the most efficient tool to control it. Mass production of *Ha*NPV, maximum harvest of PIBs, persistence in soil and on foliage, the effect of sunlight, temperature, humidity on potency of *Ha*NPV, increasing the effect of *Ha*NPV by adjuvants, safety of *Ha*NPV to parasitoids and non-target organisms have been discussed in detail.

REFERENCES

Ahmad, R., C.P. Yadava and S.S. Lal (1999). *Indian J. Pulses Res.*, 12(1): 92–96.

Anonymous (1983). *Proc. 5th All India Workshop on Biocontrol*, 22–24 February, 1983, Bangalore, India.

Anonymous (1985). *Proc. 5th All India Workshop on Biocontrol,* 13–16 September, 1985, Coimbatore, India.

Anonymous (1987). Annual Report, All India co-coordinated Research Project on Biocontrol of Crop Pest and Weeds. Indian Council of Agricultural Research, IIHR, Bangalore, India *Technical Document* No. 20: 1–236

Bai, XiaoDong, Cui Ding, X.D. Bai, C. Ding (2000). *Chinese J. App. Env. Biol.,* 6(1): 52–55.

Bakwad, D.G. (1979). M.Sc. thesis, Marathwad, Agric. Univ., Parbahani, India.

Balasubramanian, S., R.S. Arora and A.D. Pawar (1989). *Plant Prot. Bull.,* Faridabad, 41(3–4): 1–3.

Bell, M.R. and R.F. Kanavel (1978). *J. Econ. Entomol.,* 71, 350–352.

Bhatnagar, V.S., S.S. Lateef, S. Sithanantham, C.S. Pawar and W. Reed (1982). Research on *Heliothis* at ICRISAT, Proc. International Workshop on *Heliothis* Management, 15–20 Nov., 1981, ICRISAT, Patancheru, A.P., India.

Bhagwat,V.R. (2001). *Indian J. Plant Prot.,* 29: (1–2), 8–16.

Bijjur, S., K.A. Kulkarni and S. Lingappa (1991). *Indian J. Ent.,* 53(13): 479–483.

Biradar, V.K., P.R. Shivpuje, B.N. Rawale, R.S., Bansod, A.T. Munde, M.P. Badgujar (2002). *J. Soil & Crops,* 12 (1): 66–67.

Biradar, V.K., P.R. Shivpuje, B.N. Rawale, R.S. Bansod (2002). *J. Soils & Crops,* 12(1): 151–152.

Bullock, H.R. (1967). *J. Inv. Patho.,* 9: 434–436.

Bull, D.L., R.L. Ridgway, V.S. House and N.W. Pryor (1976). *J. Econ Ent.,* 69: 731–736.

Carner, G.R., and J.S. Hudson (1993). *J. Inv. Patho.,* 41.

Chandel, S.F., R. Ahmad, P.K. Singh (2004). *Archives of Phyto-Phatology and Plant Protection,* 37(2): 129–132

Chandel, S.F., R. Ahmad and Hem Saxena, (2001). Proceedings of National Symposium on biocontrol based pest management for quality crop protection in the current millennium, 18–19 July, pp. 200–201, P.A.U. Ludhiana, India

Chundurwar, R.M., V.M. Pawar and M.R. More (1990). *International Chickpea Newsletter,* 23: 17–18.

Dhamdhere, S.G. and V.M. Khaire (1986). Current Research Reporter, Mahatma Phule Agriculture University, Rahauri 2 (2): 221–226.

Ethiraju (1986). M.Sc. (Agri.) Thesis. Tamil Nadu Agricultural University, Coimbatore.

Gopali, J.B. and S. Lingappa (2001). *Karnataka J. Agric. Sci.,* 14(4): 1067–1071.

Gopali, J.B. and S. Lingappa (2001). *Karnataka J. Agric. Sci.,* 14(2): 336–341.

Gopali, J.B. and S. Lingappa (2001). *Karnataka J. Agric. Sci.,* 14:(2): 332–335.

Gopali, J.B. and S. Lingappa (2001). *Karnataka J. Agric. Sci.,* 14:(4): 947–954.

Gopali, J.B. and S. Lingappa (2001). *Karnataka J. Agric. Sci.,* 14:(4): 1072–1074.

Goud, K.B., S. Lingappa and K.A. Kulkarni (1998). *J. of Bio. Cont.,* 11(1–2): 1–4.

Ignoffo, C.M. and D.L. Hostetter (1977). *Misc. Publ. Ento. Soc. Am.* 10, 117–119.

Ignoffo, C.M., A.L. Chapman and D.F. Martin (1965). *J. Invertbr. Pathol.*, 7: 227–235

Jacob, A., and T.R. Subramanian (1972). Nuclear polyhedrosis on some Lepidoptera. *Curr. Sci.*, 41:536.

Jayaraj, S. (1981). *Proceedings of the International Workshop on Heliothis Management*, 15–20,November, 1981, ICRISAT.

Jayaramaiah, M. (1981). *Ph.D Thesis*, Univ.Agric. Sci. Bangalore, India, pp. 112.

Jayaraj, S., R.J. Rabindra and G. Santharam (1987). *Indian J. Agric. Sci.*, 58(10): 738–741.

Kaushik, U. K., V.S., Rathore and N. K. Sood (1969). *Indian J. Ent.*, 31: 175–177.

Kranthi, K.R., Kranthi Sandhya, K.S., S. Mohan Kranthi (2001). *J. Ent., Res,.* 25(2): 155–160.

Listov, M.V. and V.A. Nesterov (1976). *Zaschita Rastenii* No. 6, 48(Ru) VN II *Prikladnoi Microbiologii*, U.S.S.R.

Loganathan, M., P.C.S., Babu and G. Balasubramanian (2000). *Indian J. Ent.*, 62(1): 53–59

Mc Leod, P.J., W.C. Yearian and S.V. Young (1977). *J. Inverter. Pathol.*, 30: 237–2241.

Majumdar, S.K., M. Mathu and S.V. Pingale (1955). *Curr. Sci.*, 24: 122–123.

Makode, D.L. (1978). *M.Sc. Thesis*, Marathwad Agric. Univ., Parbhanai, India.

Mistry, A., D.A. Yadava and R.C. Patel (1984). *Indian J. Plant Prot.*, 12: 31–33.

Muthuswami, M., R.J. Rabindra, S. Jayaraj (1994). *J. Biol Control*, 8 (1): 27–32.

Narayanan, K. (1979). *Ph.D., Thesis*, Tamil Nadu Agricultural University, Coimbatore.

Narayanan; K. (1980). *In Proc. 3rd Workshop of AICRR on Biological Control of Crop Pests and Weeds*, 27–30.October, (1980. Punjab Agri. Univ., Ludhiana, India.

Narayanan, K. (1985c). *Curr. Sci.*, 54: 816–817.

Narayanan, K., S. Jayaraj and T.R. Subramanian (1985). *Legume Res.*, 5: 13–17.

Neharkar, P.S., N.R. Patange, M.S. Kendre and A.T. Munde (1999). *J. Soils & Crops.*, 9(2): 178–181.

Odak, S.L., D.K. Shrivastav, V.K. Misra and K.K. Nema (1982). *Legume Res.*, 5: 13–17.

Patel, R.C., R. Singh and P.B. Patel (1968). *J. Econ. Entomol.*, 61: 191–193.

Pawar, V.M., M. Aleumuddin and B.B. Bhosle (1987). *International Chickpea Newsletter*,16: 4–6.

Pokharkar, D.S., S.D. Chaudhary (1998). *J. Ins. Sci.*, 11(2): 117–121.

Praveen, P.M., N. Dhandapani (2001). *Proceedings qf a National Seminar,* Coimbatore, Tamil Nadu, India, 28–30 August South Indian Horticulture, pp. 261–264.

Rabindra, R.J., and S. Jayaraj (1985b). Tamil Nadu Agricultural University, Press, Coimbatore, India.

Rabindra, R.J. and S. Jayaraj (1985d). *Paper presented at group meeting on Plant Protection.* 27 May–1 June (1985, Tamil Nadu Agricultural University, Coimbatore, India.

Rabindra, R.J. and S. Jayaraj (1988b). *Indian J. Exp. Biol.*, 26: 60–62.

Rabindra, R.J. and S. Jayaraj (1988c). *Indian J. Biol. Cont.*, 2 (2): 102–105

Rabindra, R.J. and S. Jayaraj (1995). *J. Biol. Cont.*, 9(1): 34–36.

Rabindra, R.J., N. Sathian, C. Muthiah and S. Jayaraj (1989). *J. Biol. Control*, 3(1): 37–39.

Rabindra, R.J. and T.R. Subramanium (1973a). *Madras Agri., J.* 60:642–643.

Rabindra, R.J. and T.R. Subramanium (1974). *Madras Agric., J.* 61 (7): 217–220.

Rawat, R.L. and A. Shukla (2001). *J. of Biol. Cont.,* 15(2): 171–175.

Roome, R.E. (1975). *Bull.Ent. Res.,* 65: 507–514.

Rud, E.W., and S. Bellonick (1984). *J.Econ. Entomol.,* 77: 989–994.

Sachan, J.N. and R. Ahmad (1994). *Final Report Indian Institute of Pulses Research, Kanpur,* pp. 98.

Santharam, G., and M. Balasubramanian (1982). *J. Ent. Res.* 6, 179–181.

Santharam, G., M. Balasubramanian and S. Chelliah (1981). *Madras Agri., J.* 68: 417–420.

Sanap,M.M. and R.B. Deshmukh (1988). *Indian J. Pulses Res.,* 11: 80–82.

Sarode,S.V., V.U. Sonalkar (2001). *Shashpa,* 8(1): 85–87.

Sharma,M.L., H.S. Rai and M.L. Verma (1997). *Int. Chickpea & Pigeonpea Newsletter,* 4: 26–27.

Singh, R.B. and A.R. Reddy (1976). *Presented at all India Coordinated Improvement Project, Rabi Workshop,* BHU, Varanasi, India.

Srinivas, P.M. (1985). *Ph.D Thesis,* Tamil Nadu Agricultural University, Coimbatore, India.

Srinivas, G., P.C. Sundara Babu, N. Sathiah and G. Balasubramanian (1994). *Pest Mgt. & Eco. Zool.,* 2(1): 45–48.

Suganthy, M., G.V.R. Rao and S.T. Kumar (2002). *Legume Res.,* 25(1): 66–68.

Sun, GuoXun., QiLian Qin, XiuYu Cai, Cui Ding, G.X. Sun, Q.L. Qin, X.Y. Cai and C. Ding (2002). *Acta Entomologica Sinica,* 45(1): 24–29.

Summers, M.D. and G.E. Smith (1975). *J. Virol.,* 16:1108–1116.

Urs, N.V.R.R.and H.C. Govindu (1971). *Mycopathol. Mycol. Appl.,*44:317–320.

Vaishampayan, S.M and R. Verma (1987). *Indian J. Agril. Sci.,* 57(3): 200–205.

Vikram, Singh., N.M. Mathur, R.K. Kalyan, Akhter Hussain, G.K. Sharma, V. Singh and A. Hussain (2000). *Indian J. Ent.,* 62(1): 24–27.

Wanjari, R.R., G.D. More, N.R. Supare, K.S. Turkar, and V.K. Aquarkar (1998). *J. Soils & Crops,* 8(1): 34–37.

5

Ovicides for Management of Lepidopterous Pests

Vanit Kathuria, R.K. Saini and Pala Ram

Department of Entomology, CCS Haryana Agricultural University, Hisar–125 004, India

Young stages of insects are generally more susceptible to the action of insecticides than the older ones. Such stages not only get killed at lower doses, which in turn, results in the reduction in cost of plant protection but the probable damage to the crops is also prevented. Egg stage of a pest offers a unique opportunity to nip the evil in the bud. This stage, in most cases, remains exposed to the outside environment. Against lepidopterous pests, the present day integrated pest management (IPM) still relies heavily on chemicals intended for hitting the damaging stage *i.e.*, the larva. However, under field conditions the observed control is a combination of the result of toxic action of chemicals on all the stages. Thus, the contribution of ovicidal action to the total effect usually remains undetermined or least studied. In fact, the ovicidal activity of some chemicals has generally been discovered accidentally while using them for other purposes. Smith and Salkeld (1966) comprehensively reviewed the earlier work on ovicidal action of the available insecticides. However, regular updating of the strategy and tactics of pest management with the introduction of new concepts and approaches together with the introduction of new insecticides (or ovicides) is required in order to have better management of pests. To achieve this objective a clear understanding of the factors affecting mortality of eggs in response to chemicals, mechanisms of action of different ovicides, their scope and limitations etc. is quite necessary so that ovicides may be employed more efficiently.

What matters most is the actual proportion of the egg population coming into contact with the ovicides under field conditions. Here, the egg laying behaviour of the pest becomes very important. For exposing the maximum egg population to pesticides, modified pesticide application equipment or techniques may sometimes be required or even changes in plant type may become imperative if good ovicides are available. For example, with high volume power operated sprayers, the foliage may be thoroughly covered with the pesticidal sprays. Similarly, plant type with lesser foliage or reduced leaf size (*e.g.*, okra-leaf in cotton) may allow better penetration of insecticides through the crop canopy which may enhance the chances of eggs being exposed to the chemicals.

Timing of application of ovicides is also important. It would be desirable if ovicides are applied early in the season when the target pest has started laying eggs. Though the control achieved at this stage may be modest, yet it may help in holding the population below economic injury levels and in containing the size of population in the following generations. Therefore, while employing different components of IPM, the possibilities of accommodating ovicides need to be explored which may further add to the pest management successes.

Among the different families of Lepidoptera, Noctuide includes some serious pests of agricultural crops. Therefore, much of the available information relates to these pests. The prominent genera include *Helicoverpa, Heliothis, Spodoptera, Earias* etc. and these have been prominently covered in the present review. Further, support from species other than those belonging to Lepidoptera has also been taken to explain certain aspects.

Factors Affecting Susceptibility of Eggs to Ovicides

Under field conditions, there are a number of potential variables that may influence the effectiveness of ovicides. Some of the important variables are: egg location, their structure and development, age, the resistance level of the species to ovicides, and nature of ovicide.

Egg Location

The insects may deposit their eggs in, on or near the host plant. Further, these may also be placed within the tissue of the host plants which make them inaccessible to the direct application of insecticides, though these may become accessible to the insecticides with systemic or fumigant action. Similarly, eggs deposited in soil may be treated by adding chemicals to the soil. Therefore, it is necessary to have knowledge of egg locations in order to select the appropriate ovicide. For an insecticide to be more effective as an ovicide, the eggs should preferably be in an exposed condition. For example, in the case of *Helicoverpa armigera* (Hubner), quite a large proportion of eggs is laid on the upper surface of foliage in the upper canopy of the plant making them more vulnerable to insecticidal sprays. On the other hand, females of *Spodoptera litura* (Fabricius) lay their eggs in masses which are covered and are generally on the undersurface of leaves which may make them less vulnerable to the action of ovicides. Apart from their accessibility to ovicides, the surface on which eggs are laid is also important in terms of concentration of the toxicant. Deposits of the same chemical may vary greatly on different surfaces which may influence the effectiveness of an ovicide. Reidl *et al.* (1995) observed that on an inert surface such as waxed paper, young eggs of *Cydia pomonella* (Linnaeus) were three times as susceptible to horticultural mineral oils as eggs laid on apples.

Egg Structure and Development

Eggs of different species may have inherent structural and/or physiological differences to affect their susceptibility to different insecticides. Further, their susceptibility to an insecticide may change during the course of embryonic development. The embryonic material is surrounded by a number of layers and their physical and chemical properties differ in different species which affect their permeability to ovicides. In *Pieris brassicae* L. the embryonic membrane becomes tough and thickened which makes it resistant to chemicals; while shortly before eclosion this membrane is softened and partially dissolved by the embryo (Beament and Lal, 1957). Salkeld and Potter (1953) observed that in Lepidoptera the period of maximum susceptibility to ovicides corresponded closely to the time during incubation before the epembryonic membranes were formed and the time just prior to eclosion. Further, the penetration of ovicides in Lepidoptera occurred through solid portion of the chorion (Salkeld and Potter, 1953; Beament and Lal, 1957). In some insects, water proofing of the inner surface of chorion is provided by a very thin layer of wax of lipid. In fact, chorion serves as a protective covering against various chemical and other environmental stresses.

Egg Age

In Lepidoptera, egg age has been reported to affect its susceptibility to different insecticides. In general, the susceptibility of eggs to insecticides decreased with the increase in egg age. Kathuria *et al.* (2000) observed the highest egg mortality in young (0–24h old) as compared to older (24–72h old) eggs demonstrating that mortality had a direct correlation with the egg age. Similarly, Ascher and Nemmy (1974) noted that older eggs of *Spodoptera littoralis* (Boisduval) were less susceptible to an ovicide, pH 60–40. On the other hand, some studies revealed that older eggs were more sensitive to ovicides as compared to eggs of early age (Allen *et al.*, 1997). However, irrespective of egg age a direct relationship between egg hatch inhibition and dose of an insecticide has been well established (Hargreaves and Cooper, 1984; Ahmed *et al.*, 1990; Ramesh and Khan, 1996; Sharma *et al.*, 1996; Pachori and Sharma, 1996).

Resistance Level

Insects have the ability to develop resistance to any chemical to which they are exposed repeatedly. Eggs from different areas may respond differently to the insecticides under test possibly due to development of resistance depending upon the history of pesticides used earlier in that area. This may necessitate the identification of ovicides on area-wide basis.

Nature of Ovicide

Different chemical classes of insecticides and their doses may have varying effect on egg mortality. By virtue of the chemical properties, some insecticides prove better ovicides over others. Similarly, insecticides with systemic or fumigant action may kill the eggs deposited into the plant tissues while such eggs may remain unaffected by the contact insecticides. Therefore, identification of selective ovicides needs attention.

Mode of Action of Major Insecticide Classes

Insecticides are primarily meant for killing the insects. However, control under field conditions may result from either killing of the insects or preventing the insect from engaging in behaviours deemed destructive. Insecticides may be natural or man-made and act on the pests in different ways. How an insecticide works is called its mode of action. Although most insecticides have several biological effects, toxicity is usually attributed to a single major effect. As far as ovicides are concerned, insecticides with ovicidal activity are available in different categories of compounds, but only a few have emerged as good ovicides. Nevertheless, it is important to discuss major insecticide classes that have been or are used today to cope with insect pests. Most insecticides affect on of five biological systems in insects. These include:

1. The nervous system,
2. The production of energy,

3. The production of cuticle,
4. The endocrine system, and
5. The water balance.

Insecticides Affecting Nervous System

Chlorinated hydrocarbons, cyclodiene, organophosphorous (OP), carbamate and pyrethroid insecticides fit into this category. Ovicidal activity of chlorinated hydrocarbons against lepidopterous species was reported by Samy (1964). Eggs treated at any stage in their development with these compounds continued to develop to eclosion, at which point death occurred. Not much information is available with respect to their toxicity to embryos. OP compounds have been used as contact, fumigant, and systemic poisons against many species of insects and mites. Although ovicidal activity of these compounds against several species of insects attacking field crops is well established, they have rarely been used exclusively for this purpose. A large volume of literature exists on the mode of action of OP compounds and other chemical classes of insecticides in insects and mammals. However, the question is whether the mode of action of poisoning in the embryo differs from that in the later stages and in the mammals. In several of the instances the insecticides did not act during embryogenesis but rather against the young larvae shortly before or during hatching after the larvae had eaten the contaminated egg membrane as reported by Otto *et al.* (1984).

Organophosphorus and carbamate insecticides are synaptic poisons. They bind to an enzyme found in the synapse called acetylcholinesterase. This enzyme stops a nerve impulse after it has crossed the synapse. These insecticides bind to and prevent the enzyme from working. Acetylcholine then builds up and proper nerve function is disturbed in the nerve synapse *i.e.*, continued stimulation of the nerve occurs. A number of currently available good ovicides belong to OP and carbamate groups. Pyrethroids are axonic poisons (*i.e.*, they poison the nerve fibre) which bind to a protein in the nerves called the voltage-gated sodium channel. Normally, this protein opens causing stimulation of the nerve and closes to terminate the nerve signal. Pyrethroids bind to this gate and prevent it from closing normally which results in continuous nerve stimulation.

Ovicidal effects of synthetic pyrethroids have been reported by several workers (El-Guindy *et al.*, 1981; Allen *et al.*, 1997; Kathuria *et al.*, 2000; Patel and Patel, 2000; Panickar *et al.*, 2003).

It is important to mention here that symptoms of poisoning such as hyperactivity, convulsions, and paralysis which are visible in other life stages cannot be seen in the egg stage. However, studies on respiratory rate of treated eggs showed that treatment in early or late stages of development caused no marked changes in oxygen consumption until near the normal time of hatching (Smith, 1955). It implied that the system affecting respiration is attacked but played no vital role until late stage of development or the system attacked is not present in the early stages, the toxicant being retained until the appearance of the vulnerable system. It was important to note that, in contrast to eggs, later stages showed a sharp rise in respiration following treatment (Lord, 1950).

Among the recently introduced insecticides nicotinoids (or chloronicotinyls) (*e.g.*, imidacloprid, acetamiprid, thiamethoxam, nitenpyram, clothianidin, dinotefuran and thiacloprid), spinosyns (*e.g.*, spinosad) and fiproles (or phenylpyrazoles) (*e.g.*, fipronil) are important (Watkinson, 1994; Ware and Whitacre, 2004). The nicotinoids act on central nervous system of insects causing irreversible blockage of post-synaptic nicotinic acetylcholine receptors. Imidacloprid mimics the action of a neurotransmitter called acetylcholine. Acetylcholine normally turns on a nerve impulse at the synapse but its effects are terminated very quickly. Imidacloprid turns on the nerve impulse but can not terminate it. Therefore, the nervous system is overexcited resulting in tremors and uncoordinated movement. Spinosad is a fermentation metabolite of a soil actinomycete bacterium *Saccharopolyspora spinosa* and is a mixture of spinosyn A and D (Saldaga *et al.*, 1997). It acts by disrupting binding of acetylcholine in nicotinic acetylcholine receptors at the post-synaptic cell. Pineda *et al.* (2000) reported ovicidal action of spinosad against *S. littoralis*. Fipronil is similar to cyclodiene insecticides with respect to mode of action. It causes blocking of GABA-gated chloride channels (Hope, 1994). In poisoned insects excessive release of acetylcholine takes place but the enzyme cholinesterase is not inhibited. However, ovicidal activity of this compound is not reported.

Insecticides Affecting Insect Endocrine System

These chemicals are typically called as insect growth regulators or IGRs. They act on the endocrine or hormone system of insects and have low mammalian toxicity. Most of the IGRs mimic the juvenile hormone produced in the insect brain and keep the insect in immature state. Insects treated with these chemicals are unable to molt successfully to the adult stage, and cannot reproduce normally. IGRs have also been reported to have ovicidal effect. Pyriproxyfen suppressed over 90 per cent egg hatch of whitefly, *Bemisia tabaci* Genn. (Ishaaya and Horowitz, 1992; Ishaaya *et al.*, 1994). Likewise, eggs laid by adults produced from *Grapholita funebrana* Tr. [*Cydia funebrana*] larvae sprayed in plum orchard with fenoxycarb, an insect growth regulator, did not hatch (Charmillot *et al.*, 1986). Gardner (1991) reported that per cent eclosion in fenoxycarb sprayed egg masses of *Spodoptera frugiperda* (J.R. Smith) was significantly lower in one day old than in 2 to 3 days old eggs. Trisyono and Chippendale (1997) observed that more than 90 per cent eggs of European corn borer died when egg masses were dipped in a solution of 100 ppm methoxyfenozide or tebufenozide. Further, IGR-pesticide mixtures against pests have also been advocated (Ravi and Verma, 1999).

Insecticides Inhibiting Cuticle Production

These chemicals are known as chitin synthesis inhibitors or CSIs. They are often grouped with the IGRs. They inhibit the production of chitin which is a major component of insect exoskeleton. The treated insects are unable to synthesize new cuticle, thereby preventing them from molting successfully to the next stage. However, no information is available with respect to their role, if any, as ovicides.

Insecticides Inhibiting Energy Production

The well known energy inhibiting insecticide is hydramethylnon belonging to the chemical class amidinohydrazone. It binds to a protein called cytocrome in the electron transport system of mitochondrion and blocks the production of ATP. Another such insecticide is sulfluramid belonging to the chemical class halogenated alkyl sulphonamide. Other compounds developed as energy production inhibitors belong to class dinitrophenols, pyrrole, and quinazoline. Ovicidal activity of all these chemical classes is reported.

The dinitrophenol compounds were earlier extensively used as ovicides. The use of nitrophenols in the form of dormant sprays against fruit pests was first suggested by Tattersfield *et al.* (1925), and Gimingham *et al.* (1926); and their use along with petroleum oil provided control of all the major fruit pests with a single application. However, their combined use decreased mainly because of the phytotoxic hazards, unpleasant handling properties and application problems. Similarly, availability of more effective insecticides these days have further prevented their wide spread use. Nevertheless, the dinitro compounds have a broad spectrum of biological activity as insecticides, fungicides, herbicides, etc. The susceptibility of eggs to dinitro poisoning is due to changes in permeability of the egg shell rather than to changes in embryonic susceptibility (Potter and Tattersfield, 1943). These compounds appeared to be adsorbed onto the protein of the shell. The fumigant action may then prolong their toxic effect (Beament and Lal, 1957). Dinitrophenols penetrate the chorion of the lepidopteran *Diataraxia oleracea* (L.) very quickly and cause a marked initial increase in oxygen consumption (Potter and Tattersfield, 1943). Apart from accelerating oxygen uptake and uncoupling oxidative phosphorylation the dinitros also affect the amino acid content of the poisoned insects in some species.

Insecticides Affecting Water Balance

Insecticides under this category include boric acid, diatomaceous earth, and sorptive dusts. A thin waxy layer on the insect cuticle helps in preventing water loss. When an insect comes into contact with these chemicals it absorbs the protective waxy covering resulting in rapid water loss from the cuticle and the insect dies of desiccation. However, little is known about their use as ovicides.

Other Insecticides

Formamidines, unlike conventional insecticides, have very unique action against phytophagous mites, ticks and certain insects (primarily Lepidoptera and Hemiptera) by acting as agonists of octapamine receptors. The expression of formamidine (such as chlordimeform, amitraj) actions depends on the species, the life stage, chemical, and the test and use conditions. In Lepidoptera, usually the eclosion from the egg is the critical stage and the hatched larvae remain within the egg shell and die due to the action of formamidines.

Nereistoxin is a poison from a marine annelid, *Lumbrineris brevicirra* (= *Lumbriconereis heteropoda* Marenz). Nereistoxin analogues have broad spectra, particularly effective on Coleoptera and Lepidoptera, and are active as contact and stomach poisons with ovicidal, larvicidal and adulticidal effects against sucking and leaf-biting insects. Nereistoxin is an antagonist, blocking cholinergic transmission, and resulting in paralysis and death of insects.

Avermectins (commonly called as abamectin) belonging to a novel class of macrocyclic lactones (or macrolactones) are isolated from the fermentation products of a fungus, *Streptomyces avermitilis*, a member of actinomycete family. Emamectin benzoate is an analogue of abamectin and was first registered in 1999. Avermectins block the neurotransmitter, γ-aminobutyric acid (GABA), at the neuromuscular junction in insects and mites. Feeding and egg laying stops shortly after exposure, though death may not occur for several days (Dybas, 1983; Ware and Whitacre, 2004).

Petroleum Oils

The petroleum oils as insecticides have been used in three ways: (a) as principal toxicant, (b) as surface active agents and (c) as solvents and carriers of other toxicants. They may act as ovicides in different ways interfering with the normal gaseous exchange of respiration in a manner which caused death. For example, they may (a) prevent the normal exchange of gases through the outer covering; (b) harden the outer covering so as to prevent hatching; (c) interfere with the water balance; (d) soften or dissolve the outer covering of eggs; (e) penetrate the egg and cause coagulation of the protoplasm; (f) penetrate the egg and interfere with enzyme or hormone activity; (g) come in contact with the emerging insect and exert toxic effect on the delicate integument (Smith and Salkeld, 1966).

Neem possesses a number of properties affecting insects. Azadirachtin, a principal constituent bearing insecticidal properties, is structurally similar to insect hormones, ecdysones, and seems to be an ecdysone blocker so that insect will not molt. It also interferes with the function of taste, so that many leaf eating insects are deterred by plants sprayed with azadirachtin. Neem may affect the insect in several ways: it may cause: partial reduction or complete inhibition of fecundity and/or sometimes egg hatchability; reduction of the life span of adults; antifeedant effects against larvae, nymphs or adults; formation of permanent larvae; and growth regulating effects at molting between larval (or nymphal) instars (Ascher, 1993; Perry *et al.*, 1998).

Promising Ovicides

A perusal of the available literature reveals that several insecticides have shown good ovicidal activity against a number of insect-pests. Stoeva (1979) considered methomyl as an excellent ovicide giving complete mortality of *Heliothis* (=*Helicoverpa*) *armigera* eggs at 0.15 per cent concentration. Hargreaves and Cooper (1984) using methomyl sprays against eggs of *H. armigera* on tomato also confirmed its ovicidal toxicity wherein the highest ovicidal action was recorded at 0.0125–0.05 per cent concentration. Under laboratory conditions Lannate LV (methomyl) at 0.225 lb/acre significantly reduced the number of hatched live larvae of *H. armigera* (Allen *et al.*, 1996). Similarly, Kathuria *et al.* (2000) observed 77 per cent mortality of *H. armigera* eggs when these were dipped for five seconds in 0.025 per cent active ingredient of methomyl. Other workers have also reported ovicidal activity of methomyl against *H. armigera* (Gunning *et al.*, 1992; Mala *et al.*, 1993; Ramesh and Khan, 1996; Pachori and Sharma, 1996; Sharma *et al.*, 1996). A high ovicidal activity of methomyl has been reported against other lepidopterous pests also such as *Helicoverpa zea* (Boddie) (Bradley and Agnello, 1988), *Heliothis virescens* (Fabricius) (Bradley and Agnello, 1988; Leonard *et al.*, 1991), and *S. litura* (Reddy and Divakar, 1998).

Another important ovicide is chlorpyriphos. Its ovicidal activity has been observed against *H. armigera* (Patel and Patel, 2000), *S. litura* (Reddy and Divakar, 1998) and *S. littoralis* (El-Guindy *et al.*, 1981; Ascher and Nemmy, 1990; Rizk *et al.*, 1990). Similarly, chlordimeform has been found effective against eggs of *Heliothis virescens, Helicoverpa zea* and *S. littoralis* (El-Guindy *et al.*, 1981; Bradley and Agnello, 1988; Leonard *et al.*, 1991). Susceptibility of lepidopteran eggs to synthetic pyrethroids has been highlighted by several workers (Patel and Patel, 1989; Vekaria and Vyas, 1985). Cypermethrin and fenvalerate, each at 0.006 per cent concentration, gave more than 80 per cent mortality of *H. armigera* eggs within 24 hours of their exposure to these insecticides (Kathuria *et al.*, 2000). Ovicidal activity of cypermethrin against eggs of *S. littoralis* (El-Guindy *et al.*, 1981) and that of lambdacyhalothrin against *H. armigera* (Allen *et al.*, 1997) has also been reported. Apart from the above insecticides, endosulfan also had ovicidal effect against eggs of *H. armigera, S. litura, Earias vittella* (Fabricius), *Spilosoma obliqua* Walker and *Chilo partellus* (Swinhoe) (Patel and Patel, 2000; Singh, 2002; Srinivas and Panwar, 2002; Panickar *et al.*, 2003). Similarly, quinalphos (Patel and Patel, 1989) and triazophos (Mala *et al.*, 1993; Kathuria *et al.*, 2000) were found to have ovicidal effect against *H. armigera* eggs. Among the recently introduced insecticides, thiodicarb (Bradley and Agnello, 1988; Ascher and Nemmy, 1990; Mala *et al.*, 1993; Reddy and Divakar, 1998) and profenophos (Ascher and Nemmy, 1990; Sharma *et al.*, 1996; Pachori and Sharma, 1996; Panickar *et al.*, 2003: Reddy and Divakar, 1998) have been found to have good ovicidal action.

Insecticides of plant origin, for example, nicotine, pyrethrins and derries products etc., have been reported to possess some ovicidal properties against eggs of Lepidoptera (Potter and Tattersfield, 1943). However, the potential contribution of these insecticides as ovicides has not been fully explored, except in the case of neem. Neem leaf extracts at 0.5–2.0 per cent concentrations gave 20–74 per cent mortality of *H. armigera* eggs (Sharma *et al.*, 1997). Other workers have also reported the ovicidal properties of neem (Srinivas and Panwar, 2002; Panickar *et al.*, 2003). Neem, apart from having insecticidal properties against different stages of a number of pests species, is also known for its oviposition deterrent properties. Jeyakumar and Gupta (1999) observed that neem seed kernel extract at 10 per cent concentration reduced oviposition of *H. armigera* by 60 per cent. Huang *et al.* (2000) reported that components of garlic (*Allium sativum* L.) oil reduced egg hatching of *Tribolium castaneum* (Herbst). Tripathi *et al.* (2003) obtained 99.5 per cent reduction in egg hatching of *Spilarctia obliqua* Walker with essential oils of *Aegle marmelos* leaf.

The aqueous extracts of the plants possessing insecticidal properties are generally poor in controlling pests and have only limited efficacy against various pests, even at the higher concentrations. On the other hand, their extracts in alcohol or other solvents may provide greater proportion of the insecticidal compounds. There is much scope for their exploitation, particularly against the young stages (egg and early instars) of the pests. Greater concerns these days regarding environment and health demand that these compounds may be evaluated against young stages of the important pests. It would be worth investing in research on bioactive compounds obtained from plant material.

Case Study

Susceptibility of *H. armigera* eggs to different insecticides was tested in the laboratory using egg-dip method (Kathuria *et al.*, 2000). The insect eggs were obtained on Whatman filter paper No. 1 by placing it over a nylon netting covering the oviposition glass jars having mated females of the insect. Such papers with eggs were designated as 'egg cards'. Egg cards bearing eggs of different ages were cut into small pieces in a way that each piece had 10 eggs on it. Such pieces were dipped for 5 seconds in field recommended concentrations of different insecticides and egg mortality was recorded after 24, 48 and 72 h of treatment. The insecticide tested were endosulfan[1] (0.07 per cent), quinalphos[2] (0.05 per cent), monocrotophos[3] (0.04 per cent), triazophos[4] (0.04 per cent), methomyl[5] (0.025 per cent), malathion[6] (0.05 per cent), carbaryl[7] (0.05 per cent), fenvalerate[8] (0.006 per cent) and cypermethrin[9] (0.006 per cent).

Though all the insecticides had ovicidal effect giving at least 50 per cent kill of *H. armigera* eggs, yet there were considerable differences in egg mortalities among different insecticides (Figure 5.1). One-day-old eggs exhibited greater susceptibility to all the insecticides, followed by two-days and three-days old eggs. Synthetic pyrethroids *viz.*, fenvalerate and cypermethrin recorded the highest egg mortality (>80 per cent), followed by triazophos and methomyl (>75 per cent each). Quinalphos, monocrotophos and endosulfan exhibited comparatively less egg mortality (60–70 per cent). Much of

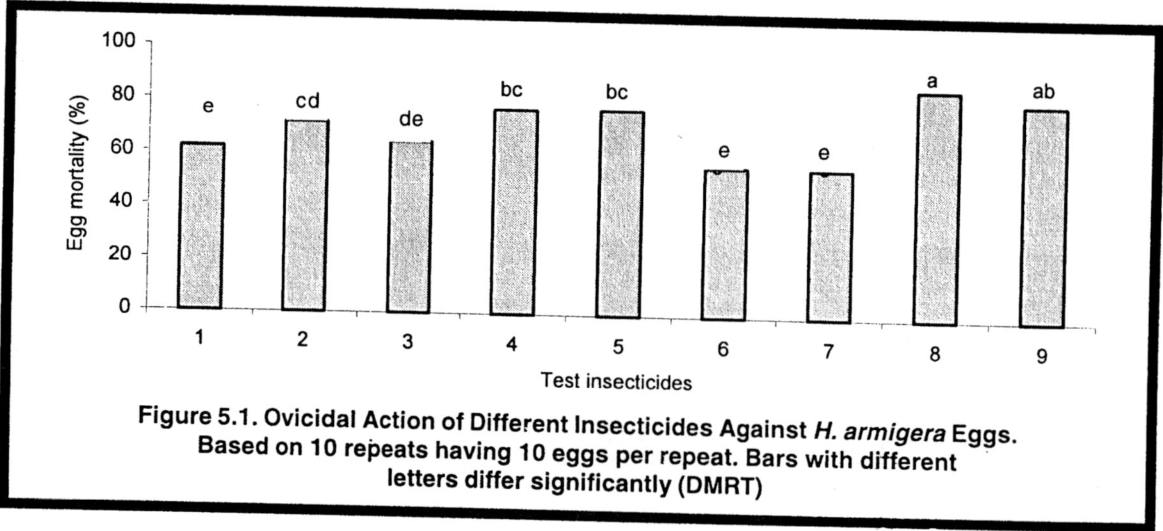

Figure 5.1. Ovicidal Action of Different Insecticides Against *H. armigera* Eggs. Based on 10 repeats having 10 eggs per repeat. Bars with different letters differ significantly (DMRT)

the egg mortality in eggs of all ages occurred within 12 h of their treatment with different insecticides. No insecticide under study provided additional egg mortality after 24 h of treatment.

Insecticides are expected to continue as a major component of IPM programme during the near future. With the discovery of new compounds with novel modes of action, there is a significant potential for the expansion of the global arsenal of new insecticide products. At least some of the new products will find great utility, particularly in managing resistant strains of insect pests. This will add new opportunities for the development of more effective IPM. The success of IPM depends on a number of control methods. Therefore, each and every component is to be employed in the most efficient manner. For this, the existing management practices need to be reassessed and new ones added by carefully taking into consideration the most vulnerable stage. Knowledge of pests' life history, their occurrence, location and susceptibility of the eggs to various environmental stresses is pre-requisite to their successful management. In an effort to identify good ovicides, screening of new compounds belonging to different chemical classes against eggs of different species of insects will have to be continued. Some success has already been achieved and a number of good ovicides are currently available. Further progress in the understanding of modes of action of various compounds will help in selecting the compounds with greater ovicidal action.

Conclusion

There is a great concern these days about the environmental pollution and harmful residues of insecticides in foods. Therefore, the need of the hour is to manage pests with the lowest possible amounts of environment friendly chemicals. Insecticides of plant origin may find suitable place in this scheme. Depending upon the egg laying behaviour of the pest, selective use of ovicides will help in sparing the non-target pests and natural enemies. Since there may be overlapping generations of a pest, the use of ovicides early in the season may help in minimizing the need for later treatments which may cause residue problems at harvest. However, sound knowledge of pests' population dynamics in relation to the environmental factors is necessary in deciding timing of application of ovicides so that their application synchronizes with the period of maximum oviposition in the field. In view of several potential benefits from the use of ovicides mentioned above, it is desirable to have their better integration in the pest management programmes. Further, ovicides may be successfully used in combination with other control agents like *Bacillus thuringiensis* (Klein *et al.*, 1995; Klein *et al.*, 1996). Good ovicides with favourable attributes may play greater role in the future pest management programmes.

REFERENCES

Ahmed, K., N.H.P. Rao and A.S. Reddy (1990). Ovicidal action of insecticides on eggs of gram pod borer (*Heliothis armigera*). *Indian J. Agri. Sci.*, 60(2): 154–156.

Allen, C.T., S. Firizzell and A.C. Riddle (1997). Bollworm ovicides and how they work. Special Report, Agricultural Experiment Station, Division of Agriculture, University of Arkansas, 183: 173–176.

Ascher, K.R.S. (1993). Non–conventional insecticidal effects of pesticides available from neem tree *Azadirachta indica*. *Arch. Insect Biochem. Physiol.*, 22: 433–449.

Ascher, K.R.S. and N.E. Nemmy (1974). The ovicidal effect of pH 60–40 (1–(4–chlorophenyl)–3–(2,6–diflurobenzoyl)–urea) in *Spodoptera littoralis* Boisd. *Phytoparasitica*, 2(2): 131–133.

Ascher, K.R.S. and N.E. Nemmy (1990). Ovicides for *Spodoptera littoralis* (Boisd.). *Int. Pest Control*, 32(5): 124–128.

Beament, J.W.L. and R. Lal (1957). Penetration through the egg shell of *Pieris brassicae* (L.). *Bull. Ent. Res.*, 48: 109–125.

Bradley, J.R. Jr. and A.M. Agnello (1988). Comparative persistence of the ovicidal activity of thiodicarb, chlordimeform and methomyl against *Heliothis* spp. (Lepidoptera: Noctuidae) on cotton. *J. Econ. Ent.*, 81(2): 705–708.

Charmillot, P.J., B. Bloesch, M.I. Frischknecht, E. Gingins and H. Hohn (1986). Control of the plum fruit moth *Grapholita funebrana* Tr. by means of an insect growth regulator (IGR) used for its ovicidal action. *Mitteilungende Schweizerischen Entomologischen Gesellschaft*, 59(1–2): 39–46.

Dybas, R.A. (1983). Avermectins: their chemistry and pesticidal activities. In: IUPAC Pesticide Chemistry, J. Miyamoto and P.C. Kearney (eds.), Pergamon Press, Oxford, 1983, Vol. 1, p. 83.

El-Guindy, M.A., S.M. Madi and M.M. El-Sayed (1981). The ovicidal action of insecticides and insect growth regulators of eggs of susceptible and resistant strains of the Egyptian leafworm, *Spodoptera littoralis* (Boisd.). *Bull. Ent. Soc. Egypt, Economic Series*, 1996–97, 10: 285–292.

Gardner, W.A. (1991). Ovicidal properties of fenoxycarb against fall armyworm (Lepidoptera: Noctuidae). *Florida Ent.*, 74(2): 257–261.

Gimingham, C.T., A.M. Massee and F. Tattersfield (1926). A quantitative examination of the toxicity of 3: 5-dinitro-o-cresol and other compounds to insect eggs, under laboratory and field conditions. *Ann. Appl. Biol.*, 13: 446–465.

Gunning, R.V., M.E. Balfe and C.S. Easton (1992). Carbamate resistance in *Heliothis armigera* (Hubner) (Lepidoptera: Noctuidae) in Australia. *J. Australian Ent. Soc.*, 31(2): 97–103.

Hargreaves, J.R. and I.P. Cooper (1984). Ovicidal tests with insecticides against tomato grub, *Heliothis armigera* (Hubner). *Queensland J. Agric. Animal Sci.*, 39(2): 105–108.

Hope, J.H. (1994). Fipronil: A new soil and foliar insecticide for cotton. In: *Challenging the Future: Proceedings of the World Cotton Research Conference–I*, (Eds.) G.A. Constable and N.W. Forrester, Brisbane, Australia, February 14–17, CSIRO, Melbourne, pp. 534–536.

Huang, Y., S.X. Chen and S.H. Ho (2000). Bioactivities of methyl allyl disulfide and diallyl trisulfide from essential oil of garlic to two species of stored pests, *Sitophilus zeamais* (Coleoptera: Curculionidae) and *Tribolium castaneum* (Coleoptera: Tenebrionidae). *J. Econ. Ent.*, 93(2): 537–543.

Ishaaya, I. and A.R. Horowitz (1992). Novel phenoxy juvenile hormone analog (pyriproxyfen) suppresses embryogenesis and adult formation of sweet potato whitefly (Homoptera: Aleyrodidae). *J. Econ. Ent.*, 85: 2113–2117.

Ishaaya, I., A. DeCock and D. Degheele (1994). Pyriproxyfen, a potent suppressor of egg hatch and adult formation of the green house whitefly (Homoptera: Aleyrodidae). *J. Econ. Ent.*, 87: 1185–1189.

Jeyakumar, P. and G.P. Gupta (1999). Effect of neem seed kernel extract (NSKE) on *Helicoverpa armigera*. *Pestic. Res. J.*, 11(1): 32–36.

Kathuria, V., P. Ram and R.K. Saini (2000). Susceptibility of *Heliothis armigera* (Hubner) eggs to different insecticides. *J. Ent. Res.*, 24(3): 213–217.

Klein, C.D., D.R. Johnson and A.M. Jordan (1995). The role of *Bacillus thuringiensis* plus ovicides in management of the Heliothine complex. *Proc. Beltwide Cotton Conferences, San Antonio*, Tx, USA, Jan. 4–7, 1995, Vol. 2: 880–881.

Klein, C.D., D.R. Johnson, H.B. Myres and L.M. Page (1996). Evaluation of *Bacillus thuringiensis* and ovicide combinations for bollworm and tobacco budworm. *Proc. Beltwide Cotton Conferences, Nashville, TN*, USA, Jan. 9–12, 1996, Vol. 2: 798–799.

Leonard, B.R., J.B. Graves and T.C. Sparks (1991). Ovicidal effects of selected insecticides against pyrethroid–susceptible and resistant tobacco budworm. *Southwestern Entomologist*, 15 (Supplement): 163–172.

Lord, K.A. (1950). The effect of insecticides on respiration. II. The effect of a number of insecticides on the oxygen uptake of adult *Tribolium castaneum* Hbst. at 25°C. *Ann. Appl. Biol.*, 37: 105–122.

Mala, S.R., C. Peter and B.V. David (1993). Ovipositional behaviour and eclosion of eggs of *Heliothis armigera* as affected by insecticides. *Entomon.*, 17(3,4): 177–181.

Otto, D., C.D. Edlich and G. Casperson (1984). The ovicidal effect of insecticides on the eggs of Colorado beetle *Leptinotarsa decemlineata* Say and of the turnip moth *Scotia segetum* Schiff. *Archiv. Fur. Phytopathologie-und-Pflanzenschutz*, 20(6): 487–502.

Pachori, R. and D. Sharma (1996). Ovicidal effects of various concentrations of profenophos and methomyl on the eggs of *Heliothis armigera* Hubner. *Pestology*, 20(3): 5–7.

Panickar, B.K., T.M. Bharpoda, J.J. Patel and J.R. Patel (2003). Ovicidal effects of botanical and synthetic insecticides on bollworms. *Indian J. Ent.*, 65(2): 292–293.

Patel, P.S. and J.R. Patel (1989). Ovicidal action of some insecticides against the eggs of *Heliothis armigera* (Hubner). *Pestology*, 13(8): 36.

Patel, S.R. and I.S. Patel (2000). Ovicidal effect of some synthetic insecticides and neem formulations on *Helicoverpa armigera* (Hub.). *Pest Mgmt. and Eco. Zool.*, 8(2): 201–203.

Perry, A.S., I. Yamamoto, I. Ishaaya and R. Perry (1998). *Insecticides in Agriculture and Environment: Retrospects and Prospects*. Spring-Verlag Berlin Heidelberg (Applied Agriculture), 261 pp.

Pineda, S., F. Budia, M.I. Schneider, A. Gobbi, E. Vinuela and P. Estal (2000). Biological effectiveness of spinosad and methoxyfenozide (RH-2485) (IGR) on *Spodoptera littoralis* (Boisduval, 1813) (Lepidoptera: Noctuidae) eggs. *Boletin de Sanidad Vegetal, Plagas*, 26(4): 483–491.

Potter, C. and F. Tattersfield (1943). Ovicidal properties of certain insecticides of plant origin (nicotine, pyrethrins and derris products). *Bull. Ent. Res.*, 34: 225–244.

Ramesh, P.C. and H.K. Khan (1996). Ovicidal action of methomyl 40SP Dunet against eggs of pigeonpea pod borer *Helicoverpa armigera* (Hub.). *Pestology*, 20(1): 35–36.

Ravi, G. and S. Verma (1999). Pesticide–IGR mixtures: an effective ecofriendly approach against *Heliothis armigera* (Hubner). *Pestology*, 23(2): 285–294, (Special issue).

Reddy, A.R. and B.J. Divakar (1998). Ovicidal action of certain insecticides on *Spodoptera litura*. *Indian J. Plant Prot.*, 26(1): 78–79.

Riedl, H., J. Halaj, W.B. Kreowski, R.J. Hilton and P.H. Westigard (1995). Laboratory evaluation of mineral oils for control of codling moth (Lepidoptera: Tortricidae). *J. Econ. Ent.*, 88(1): 140–147.

Rizk, G.A., M.A. Soliman and H.M. Ismael (1990). Semi–field and field evaluation of acyl–ureas on the cotton leafworm, *Spodoptera littoralis* (Boisd.) in Middle Egypt. *Assiut J. Agric. Sci.*, 21(3): 44–57.

Saldaga, V.L., G.B. Watson and J.J. Sheets (1997). Studies on mode of action of spinosad, the active ingredient in Tracer insect control. *Proc. Beltwide Cotton Conf.*, 2: 1082–1086.

Salkeld, E.H. and Potter, C. (1953). The effect of the age and stage of development of insect eggs on their resistance to insecticides. *Bull. Ent. Res.*, 44: 527–580.

Samy, O. (1964). Effect of insecticides on the eggs of certain cotton insects. *Agr. Res. Rev.*, 42: 18–28.

Sharma, P.D., M.S. Jaglan and S.P. Singh (1996). Ovicidal action of insecticides as a component of *Helicoverpa armigera* Hubner management. *Pestology*, 20(8): 5–7.

Sharma, P.L., S. Chander and C. Sheikher (1997). Toxic and morphogenic effects of neem leaf extracts on *Helicoverpa armigera* (Hubner). *Pest Mgmt. Eco. Zool.*, 5(2): 95–100.

Singh, A.N. (2002). Preliminary studies on the ovicidal actions of some ecofriendly insecticides against *Spilosoma obliqua* Walker. *Insect Env.*, 8(20): 65–66.

Smith, E.H. (1955). Further studies on the ovicidal action of parathion to eggs of the peach tree borer. *J. Econ. Ent.*, 48: 727–731.

Smith, E.H. and E.H. Salkeld (1966). The use and action of ovicides. *Ann. Rev. Ent.*, 11: 331–368.

Srinivas, P.S. and V.P.S. Panwar (2002). Ovicidal effect of endosulfan and neem formulations against *Chilo partellus* eggs. *Ann. Plant Protec. Sci.*, 10(1): 173–174.

Stoeva, R. (1979). Studies on the biology of the cotton bollworm *Heliothis armigera* (Hubner) (Lep., Noctuidae) and the determination of suitable chemicals for its control. *Rasteniev dni Nauki*, 16(8): 92–104.

Tattersfield, F., C.T. Gimingham and H.M. Morris (1925). Studies on contact insecticides. III. A quantitative examination of the insecticidal action of the chloro–, nitro–, and hydroxyl derivatives of benzene and naphthalene. *Ann. Appl. Biol.*, 12: 218–262.

Tripathi, A.K., V. Prajapati, A.A. Naqvi and S.P.S. Khanuja (2003). Feeding deterrent, growth inhibitory and toxic effects of essential oil of *Aegle marmelos* leaf against three lepidopteran insects. *J. Medicinal & Aromatic Plant Sci.*, 25(2): 466–472.

Trisyono, A. and G.M. Chippendale (1997). Effect of the nonsteroidal ecdysone agonists, methoxyfenozide and tebufenozide, on the European corn borer (Lepidoptera: Pyralidae). *J. Econ. Ent.*, 90(6): 1486–1492.

Vekaria, M.V. and H.N. Vyas (1985). Studies on ovicidal toxicity of certain insecticides against the eggs of *Heliothis armigera* Hubner. *Pesticides*, 19(10): 43–45.

Ware, G.W. and D.M. Whitacre (2004). An introduction to insecticides. In: *The Pesticide Book, 6th ed.* Published by MeisterPro Information Resources, A Division of Meister Media Worldwide Willoughby, Ohio.

Watkinson, I.A. (1994). Novel biological and conventional insecticides: what's new? In: *Challenging the Future: Proceedings of the World Cotton Research Conference–I*, (Eds.) G.A. Constable and N.W. Forrester, Brisbane, Australia, February 14-17, CSIRO, Melbourne, pp. 569–578.

6

Recent Researches on *Ocimum* Species for Insect Pest Management

Vanit Kathuria and Nutan Kaushik

TERI-School of Advanced Studies, Darbari Seth Block, Lodhi Road, New Delhi – 110 003, India

The plants of genus *Ocimum* (Lamiacea formely Labiatae) are popular herbs found in the tropical regions of Asia, Africa, and Central and South America. Nearly 50 to 150 annual or perennial species are known (Bailey, 1924; Hortus III, 1976; Darrah, 1980; Wister, 1991). About nine species have been reported from India and three are exotic (Anonymous, 1982). In India, basil (*Ocimum sanctum*) plant is known for its religious significance and medicinal properties. When fully grown, the plant may reach a height of two feet. The plant has square stems, green shiny 1–2 inch long fragrant opposite leaves and spikes of flowers (Darrah, 1980). The green (*Sri tulsi*) type and purple type (*Krishna tulsi*) are the commonest grown plants. The diversity within basil species has been highlighted by the existence of chemotypes or chemical races that do not differ morphologically (Anonymous, 1982). Therefore, the taxonomic nomenclature is complicated and confused (Tucker, 1986). The predominant classification is based on essential oil constituents such as camphor, eugenol, linalool, methyl chavicol and cinnamate. These properties are being used for identification of basil oil in international commerce.

Commercial Utility

The plants are valued for having cooking and medicinal (Thakur *et al.*, 1989) uses, and for variety of other purposes (Anonymous, 1982). In cooking, the herb is often used either fresh or dried to impart its distinctive flavour into various preparations. As an herbal medicine, basil is believed to have a soothing effect on the digestive system. The essential oils of the plant are treasured in perfumery industry because of its pleasant odour.

Several scientists have assessed the insecticidal properties of *Ocimum* spp against various groups of insect- pests and also for various reasons. An overview of the work done in the last decade for insecticidal properties of different plant parts of *Ocimum* species is given below.

Leaf Powder

Laboratory study revealed that powdered leaf of *Ocimum sanctum* when mixed in diet has growth regulating effects against *Spilosoma obliqua* (Ahmed and Bhattacharya, 1991). Weaver *et al.* (1994) reported that the powdered leaves of *O. canum* at 1 per cent concentration caused 100 per cent mortality of *Zabrotes subfasciatus* adults in 48 hrs. Another study revealed that the grains of maize and sorghum when treated with dried ground leaves and essential oils of *O. kilimandscharicum* at concentration of 25.0 g and 0.3 g per 250 g of grain, respectively, resulted in no adult survival or progeny production of *Sitophilus zeamais, Rhyzopertha dominica* and *Sitotroga cerealella.* Both powder and oil were also found to show repellent activity (Jembere *et al.*, 1995). Further, grains treated with 30 g of ground leaves and 750 mg of essential oil caused significant reduction in the progeny produced by *R. dominica* and *S. cerealella*. All the grain moths were killed after 24 h of treatment with essential oil. The same treatment evoked repellent action to *S. zeamais*. The same group also reported the potential of *O. suave* (wild) as grain protectant to similar insects in maize and sorghum (Bekele *et al.*, 1996).Ignatowitz (1997) also reported the repellent action of *O. basilicum* powder against *Sitophilus granarius* in wheat grain. Recently, Kathuria and Kaushik (2003) have observed the negative effects of *O. sanctum* leaf powder on the growth indices of *H. armigera* larvae following chronic feeding bioassays at 5 per cent concentration. The larval weight index on 7th day reduced to 0.104 with pupal and larval-pupal index values of 0.447 and 0.582 with no survival index.

Leaf Extracts

Satpathi and Ghatak (1990) observed exposure of 1 per cent leaf extract of *Ocimum sanctum* on leaves of aubergine in the laboraratory caused 90 per cent mortality of *Henosepilachna vigintioctopunctata* larvae in 12–24 h treatment periods. Report showed that addition of methanolic fractions of *O. sanctum* increased the efficacy of NPV preparations when tested against the 3rd instar larvae of *Spodoptera litura* (Prasad *et al.*, 1993). Another study revealed that ether extract of *O. sanctum* leaves exhibit growth inhibiting effect when second instar nymphs of *Dysdercus koenigii* treated at 1.0 per cent and 2.0 per cent concentrations under laboratory conditions (Guddewar *et al.*, 1994).

Sharma *et al.* (1992) reported the larvicidal action of *O. americanum* against *Anopheles stephensi.* The other study revealed that acetone extract of *O. sanctum* leaves when tested against fourth instar larvae of *A. stephensi* resulted in dose dependent larval mortality, growth regulatory effects and morphogenetic aberrations (Bhagat, 1992). Studies with the acetone crude extract of *O. sanctum* at 1400 ppm under laboratory conditions affected the pupae formation and emergence of *Aedes aegypti* larva with mortality as high as 90 per cent Morphological aberrations and growth disruptions were also observed (Kumari *et al.*, 1994).

Nadarajan *et al.* (1996) reported that spraying the rice plants with oil of *O. gratissimum* reduce feeding damage by rice leaf folder up to 64.5 per cent at 0.6 per cent concentration. Another report of laboratory studies revealed that the nymphs of brown plant hopper, *Nilaparvata lugens* when fed on rice plants treated with extracts from roots, leaves and flowers of *O. basilicum* exhibited growth inhibitory effects. Also *O. canum (O. americanum)* showed the similar effects (Subramanium and Nachiappan,

1996a and b). Gakuru and Foua (1996) reported that the ether extracts of *O. basilicum* at 1 per cent w/w dose as seed coat gave LD_{50} value after 24 hrs against the bruchid, *Callosobruchus maculatus* and curculionid, *Sitophilus oryzae*.

The study conducted in 1994 at Suwan, South Korea on paddy brown plant hopper, *Nilaparvata lugens* indicated that the extract of *O. americanum* gave 59.1 per cent mortality of adult female hoppers (Hiremath and Ahn, 1997).

Results of field experiment with *O. sanctum* extract showed decrease in okra jassid and fruit borer population (Adiroubane and Letchoumanane, 1998). Larval mortality up to 63.3 per cent at 1.0 per cent concentration was observed against ber fruit borer *Meridarchis scyrodes* when fed on ber fruit treated with methanolic extract of *O. sanctum* dissolved in acetone (Dhembare *et al.*, 1998).

Umerie *et al.* (1998) reported the insecticidal potential of *O. basilicum* leaf extract against adult mosquitoes when used in aerosol and mosquito coil formulation. Kelm and Nair (1998) also reported that fractions of hexane extract of *O. sanctum* leaves and stems when tested against fourth instar larvae of mosquito, *Aedes aegypti* showed activity. Pitasawat *et al.* (1998) reported larval mortality against fourth instar larvae of *Culex quinquefasciatus* with ethanolic extracts of *O. gratissimum*. The other report revealed that extracts of *O. sanctum* when tested against mosquito, *Anopheles stephensi* gave 84 per cent and 78 per cent larval mortality and adult repellency, respectively at 4.0 per cent concentration. In addition, ovipositional deterrency, reduction in pupation and adult emergence were also observed (Murugan and Jeyabalan, 1999). There are reports on the efficacy of *O. sanctum* leaf extracts at 10 per cent on the population count of sorghum earhead bugs during field experiments (Ramamurthy and Rajaram, 2001). Effects of powdered leaves of *O. basilicum* against *C. chinensis* on pea seeds have also been reported in terms of damage and weight loss (Singh *et al.*, 2001).

Leaf Oil

Essential oil of *Ocimum* species has been reported to possess insecticidal, particularly as grain protectant for storage pests, (Chavan and Nikam, 1981), nematicidal (Chatterjee *et al.*, 1982), fungistatic (Reuveni *et al.*, 1984) and antimicrobial properties (Ntezurubanza *et al.*, 1984). In addition, minor components of oil such as juvocimene I and II have been classified as juvenile hormone analogs (Nishida *et al.*, 1984). The greenish yellow fixed seed oil has good drying properties (Anoymous, 1982).

Laboratory studies showed that the essential oil of *O. basilicum* leaves and its major constituent, methyl chavicol, were more effective than *O. sanctum*. Further, they were more toxic to *Anopheles stephensi* as compared to *Aedes aegypti* and *Culex quinquefasciatus* (Bhatnagar *et al.*, 1993). Regnault and Hamroui (1994) observed the fumigant toxic effect of essential oils from *O. basilicum* on the adults of kidney bean bruchid, *Acanthoscelides obtectus* in addition to the larvicidal and ovicidal effects. The other report suggested that the oleanic acid from the leaves of *O. gratissimum* showed activity to *Aedes aegypti* (Njoku *et al.*, 1997). The laboratory experiment showed that the grain treatment of wheat and maize grains with 1,8-cineole extracted from the oil of *O. kenyense* completely inhibits the development of eggs and immature stages within grain kernels as well as progeny emergence of *Sitophilus granarius*, *S. zeamais*, *Tribolium castaneum* and *Prestephanus truneatus*. It also exhibits the repellent action against all the four beetles (Obeng *et al.*, 1997).

An overview of anti-insect properties of different *Ocimum* species against a range of insect-pests is presented in Table 6.1.

Table 6.1: Activity of *Ocimum* Species Against Different Insect Pests

Sl.No.	Plant Species	Test Insect	Activity*	Test Material	Results/Remarks	Reference
1.	*O. canum*	Bean bruchids	Tox and Rpl	Leaves, flowers and fruits	Leaves proved toxic but no repellency	Kayitare and Ntezurubanza, 1991
2.	*O. sanctum*	*Spilosoma obliqua*	GI	Leaf powder and extract	Significant results with diet incorporation method	Ahmed and Bhattacharya, 1991
3.	*O. sanctum*	*Spodoptera litura*	—	Methanolic extract fractions	Synergized the activity of NPV	Prasad *et al.*, 1993
4.	*O. basilicum*	Kidney bean bruchid	Rep	Essential oil	Inhibited biological development	Regnalut-Roger and Hamraoui, 1994
5.	*O. canum*	*Zabrotes subfasciatus*	Fum	Leaf powder	Oviposition EC50 value of 0.45 per cent (w/w) of powdered leaves and fumigant toxicity of 24g/lt had an LT50 value of 8.15h	Weaver *et al.*, 1994
6.	*O. kilimand-scharicum*	Storage pests	Tox and Rpl	Leaves and essential oil	Essential oil protected the maize and sorghum grains against feeding by all pests	Jemebere *et al.*, 1995
7.	*O. suave*	Storage pests	Tox and Rpl	Leaves and succulent stems	Dose dependent activity and essential oil provided the greatest protection	Bekel *et al.*, 1996
8.	*O. gratissimum*	Rice leaf folder	Fd	Essential oil	Reduced feeding damage by 64 per cent at 0.6 per cent oil spray solution	Nadarajan *et al.*, 1996
9.	*O. canum*	*Nilaparvata lugens*	GI	Extract	Laboratory studies showed the effect	Subramanian and Nachiappan, 1996
10.	*O. basilicum*	*Nilaparvata lugens*	GI	Roots, leaves and flower extract	Effect of nymphs fed on treated rice plants	Subramanian and Nachiappan, 1996
11.	*O. kenyense*	Storage pests	Tox and Rpl	Leaves and succulent stems	Significant protection of stored grains of maize and sorghum	Bekele *et al.*, 1997
12.	*O. basilicum*	*S. granarius*	Rpl	Leaf powder	Dosage dependent action 1 and at 5 per cent concentration, 70–80 per cent weevils repelled after 7 days of treatment	Ignatowicz, 1997

Contd...

Table 6.1–Contd...

Sl.No.	*Plant Species*	*Test Insect*	*Activity**	*Test Material*	*Results/Remarks*	*Reference*
13.	*Ocimum* spp.	Storage pests	Tox and Rpl	Cineole, eugenol and camphor; all essential oil component	100 per cent mortality at 0.5µl/kg grain after 24 h of treatment, and also inhibited development of eggs and juveniles	Obeng-ofori *et al.*, 1997a and b; Obeng-ofori and Reichmuth, 1997
14.	*O. suave*	Storage coleopterans	Tox and Rpl	Eugenol	Dose-dependent mortality and 100 per cent mortality at 1µl/kg grain, also caused significant repellency (80–100 per cent) and inhibited development of eggs and immature stages	Obeng-ofori and Reichmuth, 1997
15.	*O. basilicum*	*Earias vittella*	Fec	Leaf odours	Reduced egg output and egg hatching	Shukla *et al.*, 1997
16.	*O. basilicum*	*Opogona sacchari*	—	Aqueous and acetone extract	High larval mortality (second instar) and interference with development cycle	Benedicto *et al.*, 1998
17.	*O. sanctum*	Safflower aphid	LTS	Methanol extract	Proved effective at 0.5 and 1.0 per cent concentration	Dhembare *et al.*, 1998
18.	*O. sanctum*	Ber fruit borer	Tox	Methanol extract in acetone along with emulsifier	63.3 per cent larval mortality after 48 h at 1.0 per cent concentration	Dhembare *et al.*, 1998
19.	*O. kilimand-scharicum*	Stored product beetles	Tox and Rpl	Camphor, an essential oil component	Dosage dependent mortality and high repellency (80-100%) against the beetles	Obeng-ofori *et al.*, 1998
20.	*O. basilicum* & *O. sanctum*	*Nephotettix virescens*	Rpl	Ethanol crude extract	Effect on feeding behaviour	Chittihunsa and Samngamnim, 1999
21.	*Ocimum* spp.	Mosquito	Rpl	Essential oil	Lotion of different formulations proved effective	Bindra *et al.*, 2000
22.	*O. gratissimum* & *O. basilicum*	*Aedes aegypti* adult mosquitoes	Rpl	Essential oil	Activity at par with commercial repellent	Gbolade *et al.*, 2000
23.	*O. canum* & *O. basilicum*	*Callosobruchus maculatus*	Fum	Essential oil	Significant impact on oviposition, hatching and adult emergence	Keita *et al.*, 2000

Contd...

Table 6.1–Contd...

Sl.No.	*Plant Species*	*Test Insect*	*Activity**	*Test Material*	*Results/Remarks*	*Reference*
24.	*O. basilicum*	*Tribolium castaneum*	GD	Leaf extract	Reduced survival rate and prolonged developmental period	Lohra and Singhvi, 2000
25.	*Ocimum* spp.	Silkworm fly (*Exorista bombycis*)	Ovi	Extract	Significant egg mortality	Narayanaswamy *et al.*, 2000
26.	*Ocimum* spp.	Blumea leaf beetle	Ovi	Leaf extract	Reduced hatchability of eggs	Oudhia, 2000
27.	*O. viride*	*Tribolium castaneum* & *Sitophilus oryzae*	Rpl and FD	Hexane + isopropyl alcohol leaf extract	Reduced survival (< 25 per cent) at 0.1mg/ml after 10 days of treatment	Owusu, 2000
28.	*Ocimum* spp.	*Helicoverpa armigera*	AT	Essential oil	Reduced feeding of second instar larvae	Rao *et al.*, 2000
29.	*O. gratissimum*	*Sitophilus zeamais*	—	Leaf powder	Significant protection of maize grains	Banjo *et al.*, 2001
30.	*O. kenyense* & *O. kilimand-scharicum*	*Sitophilus zeamais* & *Rhizopertha dominica*	Tox	Essential oil	Toxicity due to combined action of different components of oil	Bekele and Hassanali, 2001
31.	*O. gratissimum* & *O. basilicum*	*Callosobruchus maculatus*	Tox	Essential oil	More than 70 per cent mortality at 25ml/vial during 12h fumigation	Keita *et al.*, 2001
32.	*O. basilicum*	*Tribolium castaneum*	OD	Plant extract	Reduced oviposition on jowar seeds	Lohra *et al.*, 2001
33.	*O. sanctum*	*Callosobruchus chinensis*	—	Leaf powder	Reduced fecundity at 3 per cent on stored black gram	Mishra, 2000
34.	*O. gratissimum*	*Callosobruchus maculatus*	OD	Leaf powder	Reduced oviposition and adult emergence in cowpea seeds	Odutayo *et al.*, 2001
35.	*O. basilicum*	*Callosobruchus maculatus*	Rep	Aqueous leaf extract	Reduced egg laying potential during storage of grains in gunny bags	Raja *et al.*, 2001

Contd...

Table 6.1–Contd...

Sl.No.	*Plant Species*	*Test Insect*	*Activity**	*Test Material*	*Results/Remarks*	*Reference*
36.	*O. basilicum* & *O. sanctum*	*Spodoptera litura*	I, GI and AT	Leaf essential oil	100 per cent deterrence of feeding of third instar larvae at 10 per cent and *O. sanctum* also caused mortality up to 20 per cent	Sharma *et al.*, 2001
37.	*O. americanum*	Mosquito vector	Rpl	Volatile oil	Addition of 5 per cent vanillin proved effective	Tawatsin *et al.*, 2001
38.	*O. basilicum*	*Callosobruchus maculatus* and its parasitoid (*Dinarmus basalis*)	Tox	Essential oil (Estragol)	Significant affect on parasitoid than bruchid	Ketoh *et al.*, 2002
39.	*O. basilicum*	*Acanthoscelides obtectus*	Rpl, Tox & RI	Essential oil	Oil vapours evaluated in choice and no choice tests	Papachristos and Stampolous, 2002
40.	*O. selloi*	Mosquito (*Anopheles braziliensis*)	Rpl	Oil in ethanol (10 per cent v/v)	Reduced number of mosquito bites on volunteers skin and showed no risk to human skin	Paula *et al.*, 2003
41.	*O. basilicum*	*Callosobruchus maculatus*	AI	Essential oils of 18 accessions	Presence Methyl chavicol and linalool in essential oil exhibited oviposition repellency and toxicity	Villalobos and Acosta, 2003
42.	*O. sanctum*	*Helicoverpa armigera*	GD	Leaf powder	Negative effects on the growth indices	Kathuria and Kaushik, 2003
43.	*O. basilicum*	*Bemisia tabaci*	Tox	Essential oil vapours	Significant activity in greenhouse conditions	Aslan *et al.*, 2004

* Activity keywords: AI: anti-insect, AT: antifeedant, Fec: fecundity, Fd: feeding, FD: feeding deterrent, Fum: fumigant, GD: growth and development; GI: growth inhibition, I: insecticidal; LTS: life table studies; Ovi: ovicidal, OD: ovipositional deterrent; Rep: reproduction, RI: reproduction inhibitory; Rpl: repellent, Tox: Toxic.

Seed Extract

Very little work has been done on seed oil extraction. There are reports of insecticidal properties of *O. sanctum* seed extract against *Culex fatigans* (*C. quinquefasciatus*). In water, the mosquito larvae attach to the seed exude by their mouthparts and die. Nearly 3–4 million larvae can be trapped with the use of I kg seed (Hasan and Deo, 1994).

Conclusion

The studies conducted with *Ocimum* species against a wide range of insect pests reveal its potential as repellent, insect growth inhibitor, pupicidal, feeding inhibitor, synergist as well as inhibitor of the hatching of the eggs. In spite of all these properties more research is required on the applied aspects in order to utilize the results effectively in insect pest management programmes.

References

Adiroubane, D. and S. Letchoumanane (1998). Field efficacy of botanical extracts for controlling major insect pests of okra (*Abelmoschus esculentus*). *Indian J. Agri. Sci.*, 68(3): 168–170.

Ahmed, S.R. and A.K. Bhattacharya (1991). Growth inhibitory effect of some plants for *Spilosoma oblique* Walker. *Indian J. Ent.*, 53(3): 453–474.

Anonymous (1982). The Wealth of India, Raw Materials. Vol III, CSIR, New Delhi, pp. 60.

Aslan, I., H. Özbek, Ö. Calmasur and F. Sahin, (2004). Toxicity of essential oil vapours on two greenhouse pests, *Tetranychus urticae* Koch and *Bemisis tabaci* Genn. *Industrial Crops and Products*, 19(2): 167–173.

Bailey, L.H. (1924). Manual of cultivated plants. MacMillan Co. New York.

Banjo, A.D., O.I. Odutayo and T.O. Ojirinde (2001). The use of some locally available plant parts as protectants of maize (*Zea mays*) grains against infestation of *Sitophilus zeamais*. *Crop Res.* (Hisar), 21(2): 208–213.

Bekele, A.J., D. Obeng-Ofori and A. Hassanali (1996). Evaluation of *Ocimum suave* (wild) as a source as source of repellents, toxicants and protectants in storage against three major stored product insect pests. *Int. J. Pest Mgt.*, 42(2): 139–142.

Bekele, A.J., D. Obeng-Ofori and A. Hassanali (1997). Evaluation of *Ocimum kenyense* (Ayobangira) as source of repellents, toxicants and protectants in storage against three major stored product insect pests. *J. Appl. Ent.*, 121(3): 169–173.

Bekele, J. and A. Hassanali (2001). Blend effects in the toxicity of essential oil constituents of *Ocimum kilimandscharicum* and *Ocimum kenyense* (Labiatae) on two post-harvest insect pests. *Phytochemistry*, 57(3): 385–391.

Benedicto, L.H., E.C. Bergmann, M.R. Potenza, M.E. Sato and R.K. Nishimori (1998). Evaluation of the control of *Opogona sacchari* (Bojer, 1856) (Lepidoptera: Tineidae) using plant extracts in laboratory conditions. *Arquivos do Instituto Biologico* (Sao Paulo), 65(1): 21–27.

Bhagat, O.P. (1992) Morphogenetic abberration and growth regulatory effect of *Ocimum sanctum* against *Anopheles stephensi L. Bioved*, 3(2): 153–156.

Bhatnagar, M., K.K. Kapur, S. Jalees and S.K. Sharma (1993). Laboratory evaluation of insecticidal properties of *Ocimum basilicum* Linnaeus and *O. sanctum* Linnaeus plant's essential oils and their major constituents against vector mosquito species. *J. Ent. Res.*, 17(1): 21–26.

Bindra, R.L., A.K. Singh, R. Gupta, Y.N. Shukla and S. Kumar (2000). Use of essential oils containing preparation for human protection against mosquitoes. *J. Med. Arom. Plant Sci.*, 22(1/3): 707–709.

Chatterje, A., N.C. Sukul, S. Laskal and S. Ghoshmajumdar (1982). Nematicidal principles from two species of Lamiaceae. *J. Nemat.*, 14:118–120.

Chavan, S.R and S.T. Nikam (1982). Mosquito larvicidal activity of *Ocimum basilicum* Linn. *Indian J. Med. Res.*, 75:220–222.

Chittihunsa, T. and N. Samngamnim (1999). Studies on the effects of some crude extracts to the mortality and feeding behaviour of *Nephotettix virescens*. In 37th *Kasetsart University Annual Conference*, 3–5 Februray, pp. 70–77.

Darrah, H.H. (1980) The cultivated basils. Buckeye Printing Company, Independence, MO.

Dhembara, A.J., G.M. Pondhe, R.D. Kharde and J.A. Dhumal (1998a). Life table studies and control of safflower aphid by plant product. *J. Maharashtra Agril. Univ.*, 23(1): 1–3.

Dhembare, A.J., G.M. Pondhe, M. Padol and S. Gulave (1998b). Efficacy of some plant extracts against ber fruit borer *Meridarchis scyrodes* (May). *Adv. Plant Sci.*, 11(7): 299–301.

Gakuru, S. and B.K. Foua (1996). Effects of plant extracts on the cowpea weevil (*Callosobruchus maculutus* Fab.) and the rice weevil (*Sitophilus oryzae* L.). *Cahiers Agricultures* 5(1): 39–42.

Gbolade, A.A., A.O. Oyedele, M.B. Sosan, F.B. Adewoyin, and O.L. Soyelu (2000). Mosquito repellent activities of essential oils from two Nigerian *Ocimum* species. *J. Trop. Med. Plants*, 1(1/2): 146–148.

Guddewar, M.B., R. Chandra, B. Kumar, R. Chandra, and B. Kumar (1994). Evaluation of insecticidal and growth inhibiting properties of some indigenous plants against *Dysdercus koenigii* (Fab.). *Plant Prot. Bull.*, Faridabad, 46(2–3): 1–3.

Hasa, S.B. and P.G. Deo (1994). *Ocimum sanctum* seeds for mosquito control. *Int. Pest Control* 36(1): 20–21.

Hiremath, I.G. and Y.J. Ahn (1997). Parthenium as a source of pesticide. *First International Conference on Parthenium Management*, Dharwad, India, 6–8 October, pp. 86–89.

Hortus III (1976). Staff of L.H. Bailey Hortorium. 1976. Hortus III. MacMillan, New York, pp. 166.

Ignatowicz, S. (1997). Powdered herbs of the mint family (Lamiaceae) as insect repellent for protection of stored wheat grain. *Polskie Pismo Entomologiczne*, 66(1/2): 135–149.

Jembere, B., D. Obeng-Ofori, A. Hassanali and G.N.N. Nyamasyo (1995). Products derived from the leaves of *Ocimum kilimandscharicum* (Labiatae) as post-harvest grain protectants against the infestation of three major stored product insect pests. *Bull. Ent. Res.*, 85(3): 361–367.

Kathuria, V. and N. Kaushik (2003). Studies on the impact of certain plants on the growth indices of *Helicoverpa armigera* (Hübner). In: *5th National Symposium on Bio-Control Agents for Sustainable Management of Pests*, 18–20, December 2003, GBPUAT, Uttaranchal, India.

Kayitare, J. and L. Ntezurubanza (1991). Evaluation of the toxicity and repellent effect of certain plants from Rwanda against the bean bruchids: *Acanthosclides obtectus* Say and *Zabrotes subfasciatus* Boheman. *Insect Science & its Appl.*, 12(5–6): 695–697.

Keita, S.M., C. Vincent, J.P. Schmit, J.T. Arnason and A. Belanger (2001). Efficacy of essential oil of *Ocimum basilicum* L. and *O. gratissimum* L. applied as an insecticidal fumigant and powder to control *Callosobruchus maculatus* (Fab.) (Coleoptera: Bruchidae). *J. Stored Products Res.*, 37(4): 339–349.

Keita, S.M., C. Vincent, J.P. Schmit, S. Ramaswamy and A. Belanger (2000). Effect of various essential oils on *Callosobruchus maculatus* (F.) (Coleoptera: Bruchidae). *J. Stored Prod. Res.*, 36(4): 355–364.

Kelm, M.A. and M.G. Nair (1998). Mosquitocidal compounds and a triglyceride, 1,3–dilinoleneoyl–2–palmitin, from *Ocimum sanctum*. *J. Agril. & Food Chem.*, 46(8): 3092–3094.

Ketoh, G.K., A.I. Glithe and J. Huignard (2002). Susceptibility of the bruchid *Callosobruchus maculatus* (Coleoptera: Bruchidae) and its parasitoid, *Dinarmus basalis* (Hymenoptera: Pteromalidae) to three essential oils. *J. Eco. Ent.*, 95(1): 174–182.

Kumari, C.P., C.L. Sharma and R.C. Saxena (1994). Pupicidal effect of *Ocimum sanctum* on the vector *Aedes aegypti* (Diptera: Culicidae). *J. Ecob.*, 6(1): 69–70.

Lohra, Y. and P.M. Singhvi (2000). Effectiveness of biopesticides on the development of *Tribolium castaneum* Herbst infesting stored sorghum. *J. Appl. Zool. Res.*, 11 (2/3): 128–131.

Lohra, Y., P.M. Singhvi and M. Panwar (2001). Effect of certain plant extracts on oviposition of rust–red flour beetle, *Tribolium castaneum* Herbst infesting stored jowar. *J. Appl. Zool. Res.*, 12(1): 67–70.

Misra, H.P. (2000). Effectiveness of indigenous plant products against the pulse beetle, *Callosobruchus chinensis* on stored black gram. *Indian J. Ent.*, 62(2): 218–220.

Murugan, K. and D. Jeyabalan (1999). Effect of certain plant extracts against the mosquito, *Anopheles stephensi* Liston. *Curr. Sci.*, 76(5): 631–633.

Nadarajan, L., J. Thomas and S. Mathew (1996). Essential oils against rice leaf folder. *Insect Env.*, 2(3): 99–100.

Narayanaswamy, K.C., M. Gowda, Y.A.N. Reddy and T.N.V. Gowda (2000). Ovicidal action of certain plant extracts against silkworm uzi fly, *Exorista bombycis* (Louis). *Insect Env.*, 6(2): 80.

Nishida, R., W.S. Bowers and P.H. Evans (1984). Synthesis of highly active juvenile hormone analogs, juvocimene I and II, from the oil of sweet basil *Ocimum basilicum* L. *J. Chem. Ecol.*, 10:1435–1450.

Njoku, C.J., Lu Zeng, I.U. Asuzu, N.H. Oberlies, J.L. McLaughlin and L. Zeng (1997). Oleanolic acid, a bioactive component of the leaves of *Ocimum gratissimum* (Lamiaceae). *Int. J. Pharm.*, 35(2): 134–137.

Ntezurubanza, L, J.J.C. Scheffer, A. Looman and S. A. Baerhiem (1984). Composition of essential oil of *Ocimum kilimandscharicum* grown in Rwanda. *Planta Medica*, 385–388.

Obeng-Ofori, D. and C. Reichmuth (1997). Bioactivity of eugenol, a major component of essential oil of *Ocimum suave* (wild) against four species of stored product Coleoptera. *Int. J. Pest Mgt.*, 43(1): 89–94.

Obeng-Ofori, D., C. Alder and C. Reichmuth (1997a). Toxicity and repellency of 1,8 cineole, eugenol and camphor against stored product insects. *Mitteilungen der Deutschem Gesellschaft fur Allgemeine and Angewandte Entomologie*, 11(1/6): 259–264.

Obeng-Ofori, D., C.H. Reichmuth, A.J. Bekele and A. Hassanali (1998). Toxicity and protectant potential of camphor, a major component of essential oil of *Ocimum kilimandscharicum* against four–stored product beetles. *Int. J. Pest Mgt.*, 44(4): 203–209.

Obeng-Ofori, D., C.H. Reichmuth, J. Bekele and A. Hassanali (1997b). Biological activity of 1,8 cineole, a major component of essential oil of *Ocimum kenyense* (Ayobangira) against stored product beetles. *J. Appl. Ent.*, 121(4): 237–243.

Odutayo, O.I., A.D. Banjo and A.T. Adegoke (2001). Protection of stored cowpea from *Callosobruchus maculatus* F. using plant products. *Crop Res.*, (Hisar) 21(2): 214–218.

Oudhia, P. (2000). Effect of some leaf leachates in hatchability of Blumea leaf beetle (Chrysolina madrasae Jackoby) eggs. *Indian J. Weed Sci.*, 32(3/4): 206–207.

Owusus, E.O. (2000). Effect of some Ghanian plant components on control of two stored–product insect pests of cereals. *J. Stored Prod. Res.*, 37(1): 85–91.

Papachristos, D.P. and D.C. Stampolous (2002). Repellant, toxic and reproduction inhibitory effects of essential oil vapours on *Acanthoscelides obtectus* (Say) (Coleoptera: Bruchidae). *J. Stored Prod. Res.*, 38(2): 117–128.

Paton, A. (1991). A synopsis of Ocimum (Labiatae) in Africa. *Kew Bull.*, 47: 403–435.

Paula, J.P., M.R. Gomes-Carneiro and F.J.R. Paumgartten (2003). Chemical composition, toxicity and mosquito repellency of *Ocimum selloi* oil. *J. Ethnopharm.*, 88(2–3): 253–260.

Pitasawat, B., W. Choochote, D. Kanjanapothi, A. Panthong, A. Jitpakdi and U. Chaithong (1998). Screening for larvicidal activity of ten carminative plants. *Southeast Asian J. Trop. Med. & Pub. Hlth.*, 29(3): 660–662.

Prasad, V.D., S. Jayaraj, R.J. Rabindra and G.P.V. Reddy (1993). Studies on the interaction of certain botanicals and nuclear polyhedrosis virus against tobacco caterpillar, *Spodoptera litura*. *Bot. Pest. Int. Pest Mgt.*, 190–196.

Raja, N., A. Babu, S. Dorn and S. Ignacimuthu (2001). Potential of plants for protecting stored pulses from *Callosobruchus maculatus* (Coleoptera: Bruchidae) infestation. *Biol. Agril. Hort.*, 19(1): 19–27.

Ramamurthy, R. and V. Rajaram (2001). Effect of plant products on sorghum earhead bug. *Ann. Plant Prot. Sci.*, 9(2): 326–327.

Rao, M.S., G. Prathibha and G.R. Korwar (2000). Evaluation of aromatic oils against *Helicoverpa armigera*. *Ann. Plant Prot. Sci.*, 8(2): 236–238.

Reganault-Roger, C. and A. Hamraoui (1994) Inhibition of reproduction of *Acanthoscelides obtectus* Say (Coleoptera), a kidney bean (Phaseolis vulgaris) bruchid, by aromatic essential oils. *Crop Protection* 13(8): 624–628.

Reuveni, R., A. Fleisher and E. Putievsky (1984) Fungistatic activity of essential oils from *Ocimum basilicum* chemotypes. *Phytopath.* Z. 110:20–22.

Satpathi, C.R. and S.S. Ghatak (1990) Evaluation on the efficacy of some indigenous plant extracts against *Henosepilachna vigintioctopunctata* (F) (Coccinellidae: Coleoptera) a pest of brinjal. *Env. & Ecol.*, 8(4): 1293–1295.

Sharma, R.N., A.S. Gupta, S.A. Patwardhan, D.S. Hebbalkar, V. Tare and S.B. Bhonde (1992) Bioactivity of Lamiaceae plants against insects. *Indian J. Exp. Biol.*, 30(3): 244–246.

Sharma, S.S., K. Gill, M.S. Malik and O.P. Malik (2001) Insecticidal, antifeedant and growth inhibitory activities of some medicinal plants. *J. Med. & Arom. Pl. Sci.*, 22/23(4A/1A): 373–377.

Shukla, A., S.C. Pathak and R.K. Agarwal (1997) Effect of some plant odours in the breeding environment on fecundity and hatching of okra shoot and fruit borer, Earias vittella Fab. under laboratory conditions. *Crop Research* (Hisar) 13(1): 157–161.

Singh, R., B. Singh and R.A. Verma (2001). Efficacy of different indigenous plant products as grain protectant against *Callosobruchus chinensis* Linn on pea. *Indian J. Ento.*, 63(2): 179–181.

Subramanian, S. and R. Nachiappan (1996a). Insect growth regulatory activity of aqueous extracts of *Ocimum sanctum* on rice brown plant hopper *Nilaparvata lugens*. *Madras Agri. J.*, 83(3): 200–201.

Subramanian, S. and R. Nachiappan (1996b). Growth retardation of *Nilaparvata lugens* by extracts of *Ocimum canum*. *Madras Agril. J.*, 83(3): 202–203.

Tawatsin, A., S.D. Wratten, R.R. Scott, U. Thavara and Y. Techadamrongsin (2001). Repellency of volatile oils from plants against three mosquito vectors. *J. Vector Zool.*, 26(1): 76–82.

Thakur, R.S., H.S. Puri and A. Hussain (1989). *Major Medicinal Plants of India*. Central Institute of Medicinal and Aromatic Plants, Lucknow, India, pp. 56

Tucker, A. O. (1986). Botanical nomenclature of culinary herbs and potherbs. In: *Herbs, Spices, and Medicinal Plants: Recent Advances in Botany, Horticulture, and Pharmacology, Vol. 1*, (Eds.) L.E. Craker and J.E. Simon, Oryx Press, Phoenix, pp. 33–60.

Umerie, S.C., H.U. Anaso and L.J.C. Anyasoro (1998). Insecticidal potentials of *Ocimum basilicum* leaf extract. *Biores. Tech.*, 64(3): 237–239.

Villalobos, M.J.P. and M.C.B. Acosta (2003). Chemical variation in an *Ocimum basilicum* germplasm collection and activity of essential oils on *Callosobruchus maculatus*. *Biochem. System & Ecol.*, 31(7): 673–679.

Weaver, D.K., F.V. Dunkel, R.C. Potter and L. Ntezurubanza (1994). Contact and fumigant efficacy of powdered and intact *Ocimum canum* Sims (Lamiales: Lamiaceae) against *Zabrotes subfasciatus* (Boheman) adults (Coleoptera: Bruchidae). *J. Stored Products Res.*, 30(3): 243–252.

Wister, J.C. (1991). Basil. In *Collier's Encyclopedia*, Vol. 3, Macmillan Educational Company, New York, pp. 676.

7

Termite Management: Past, Present and Future

Berin Pathrose and Swaran Dhingra
Division of Entomology, Indian Agricultural Research Institute, New Delhi – 110 012

Termite attack on field and tree crops and on forestry trees, especially in the semi-arid and sub-humid tropics causes significant yield losses. Annual termite control expenses of United States are in the tune of $1.5 billion (Su and Scheffrahn, 1990). Of the approximately 2500 species in the order Isoptera, about 300, spread across all families, have been recorded as pests (including those damaging buildings). However, measures to control them frequently ignore the diversity in ecology and behaviour manifested by the group. A basic understanding of termite diversity and biology is a prerequisite for adequate pest management.

Termites are social insects living in colonies usually comprising a reproductive pair (king and queen) and numerous sterile workers and soldiers whose tasks include foraging, nest building and maintenance, care of eggs and youngs and defence. All species maintain symbiotic relationships with micro-organisms which are essential for digestion. Few species build the often spectacular mounds for which these insects are renowned. There are major differences in distribution, biology and pest status among the seven conventionally recognized families; these are outlined below. The seven families are divided into fifteen subfamilies as follows:

Mastotermitidae

Kalotermitidae

Termopsidae
  Termopsinae
  Porotermitinae
  Stolotermitinae
Hodotermitidae
  Cretatermitinae (fossil)
  Hodotermitinae
Rhinotermitidae
  Psammotermitinae
  Heterotermitinae
  Stylotermitinae
  Coptotermitinae
  Termitogetoninae
  Rhinotermitinae
Serritermitidae
Termitidae
  Termitinae
  Apicotermitinae
  Nasutitermitinae
  Macrotemitinae

The Mastotermitinae contain only a single species, *Mastotermes darwiniensis* confined to Northern Australia and New Guinea. It lives in subterranean colonies, often associated with wood, and is locally a pest in sugarcane and forestry (Sands, 1973). More seriously it can cause dramatic damage to buildings.

The dry-wood termites, belonging to the family Kalotermitidae, are generally scarce but widely distributed especially in coastal and other humid regions in the tropics. They live in small colonies in wood maintaining no contact with soil as a moisture source or dispersal medium. One species, *Cryptotermes brevis* cause serious damage to buildings, but in agriculture and forestry they are relatively minor pests. It also attacks certain crops like tea (Sands, 1973).

Termopsidae is a small family of termites living in damp, rotten wood, partially or wholly buried underground.

The Hodotermitidae; harvester termites, are locally distributed in dry and arid parts. They build subterranean nests and forage in the open, predominantly on grass and grass litter.

The Rhinotermitidae include some widespread and common termites. They generally nest underground, often associated with dead wood. Some species build mounds. Two genera, *Reticulitermes* and *Coptotermes*, include important building damaging species. *Coptotermes* spp. are major pests of tree crops, particularly in Australia and South-east Asia, often attacking trees through the roots deep in the ground and eventually excavating the trunk (Cowie *et al.*, 1989).

The Serritermitidae contains a single South American species which is not a pest.

The above six families, known as lower termites, are largely wood feeders and their symbiotic intestinal microbiota is dominated by cellulolytic protozoa and considerable number of facultative and obligate anaerobic bacteria. The remaining family, Termitidae, known as higher termites, lacks the cellulolytic protozoa and, with the exception of the Macrotermitinae, bacteria are the dominant gut microbes. The Termitidae include about 75 per cent of termite species, divided among four subfamilies. Macrotermitinae are fungus-growing termites with obligate exosymbiosis with the fungus *Termitomyces* spp. Most members of this subfamily feed on living plants, causing serious and widespread damage, particularly in semi-arid Africa and India (Logan *et al.*, 1990). *Microtermes* and *Odontotermes* are the most predominant one in the Indian subcontinent. *Odontotermes* spp. build both subterranean and especially in India, epigeal nests.

Termites generally attack plants when they are not in a vigorous state of growth and when their vitality is low either due to internal growth factors or due to environmental causes such as drought or poor soil. Long lived plants such as trees and plantation crops are more susceptible to attack in the seedling stage or as transplants, and annual crops near harvest time (Roonwal, 1979).

Control measures can be divided broadly into those which attempt to:

1. Prevent termites gaining access to plants
2. Reduce termite numbers in the vicinity of the plants and
3. Render the plants themselves less susceptible to attack by the termites (Logan *et al.*, 1990).

Detection of Termites

The first and foremost step in the management of termites is the detection of termites as the damage may be visualized at an advanced stage of infestation. The sudden emergence of winged alates are the first obvious sign of a nearby termite colony. Swarming occurs in mature colonies that typically contain several thousand termites. Another sign of termite presence is mud tubes and fecal pellets. Acoustic emission devices are also used for the detection of termites infesting wood (Scheffrahn *et al.*, 1993; Mankin *et al.*, 2002).

Physical Barriers

Ebeling and Pence (1957) suggested the use of fine mineral particles (sand or crushed volcanic cinders) as a barrier to subterranean termites. The barrier is impenetrable because the particles are too large and heavy for the termites to carry away yet are small enough so that in packing they create no continuous passages through which the termites can move. Particle size is the crucial factor and must be between about 1.7 and 2.4 mm (Logan *et al.*, 1990). The technique may have potential as a preventive measure in control of subterranean termites in buildings but its applications in agriculture are clearly limited. Germinating coconuts are said to be protected by this method or by covering them with sand rather than soil (Harris, 1971).

The efficacy of excessive heat or cold, electrocution and microwaves were evaluated against dry wood termites attacking wood (Lewis and Haverty, 1996). Except electrocution all the other methods gave mortality ranging from 90–100 per cent. These methods also have limited application in agriculture.

Cultural Methods

Healthy plants may sometimes be damaged by termites, but unhealthy and stressed plants are generally more susceptible (Cowie *et al.*, 1989). Drought, disease, weeds, lack of fertilization, transplant shock, mechanical and fire damage and general neglect may all increase the likelihood of attack

(Logan *et al.*, 1990). Cultural control therefore includes those procedures which help to maintain or enhance plant vigour and which are generally good agricultural practice. Good quality seeds, healthy seedlings and appropriate transplanting procedures help to maintain plant vigour especially in early stages of crop growth.

Water Management

Levels of rainfall and irrigation may be related to termite attack on plants because:

1. Levels of termite foraging activity are frequently related to soil moisture levels, and, more importantly.
2. Deficiency or excess of water may stress plants and encourage attack.

In India, termite attack on tea is greater during the rainy season (Srivastava and Butani, 1987) and attack on maize and groundnuts was greater in parts of experimental fields with higher soil moisture levels where termite foraging activity was greater (Reddy and Sammaiah, 1988). In sugarcane, *Odontotermes obesus* commonly occurs in dry and loamy soils where irrigation is not practiced; attack by *Odontotermes assmuthi* is severe in light soils in spite of abundant moisture; *Microtermes obesi* prefers damp but not waterlogged soil and its colonies mostly occur on raised embankments and near irrigation channels (Roonwal, 1979). Generally, termite attack is low in waterlogged areas or where soil aeration is poor (Agarwal, 1972).

In irrigated areas regular irrigation will limit periods of water stress. More frequent irrigation reduced attack by *Microtermes obesi* in wheat (Verma, 1980). Water logging may weaken plants, enhancing their susceptibility. Even distribution of irrigation throughout a crop ensuring that no areas remain dry or waterlogged is important and may necessitate careful leveling of fields prior to planting (Krishnamoorthy and Ramasubbiah, 1962).

Crop Rotation

Intensive cropping, especially monocropping, for long periods reduces soil fertility and structure. Under such conditions crops will be less vigorous and therefore more susceptible to attack. Crop rotation, especially including fallow periods, to reduce detrimental effects on soil fertility and structure is to be recommended (Logan *et al.*, 1990).

Damage to Plants

Both dry-wood and subterranean termites frequently gain access to living tissue in plantation crops through wounds. In plantation crops dead wood be removed and cut back to the living tissue and pruning be carried out carefully with clean cuts and in a manner to avoid dieback. Pruning cuts and accidental wounds should be treated with paint or tar (Kashyap *et al.*, 1977).

Intercropping

Certain grasses are grown in West Africa to repel termites (Sands, 1961). Intercropping maize with soyabean, groundnut and common beans caused a significant reduction in termite attack, reduced loss in grain, yield of maize and increased the nesting of predatory ants in maize fields (Sekamatte *et al.*, 2003).

Sivapalan *et al.* (1980) found that the presence of shade trees around highly susceptible clones of tea in Sri Lanka was correlated with reduced attack by the dry-wood termite *Glyptotermes dilatatus*, which initiated colonies in the shade trees but could not survive subsequently.

Crop Residues

The addition and removal of organic matter from the field have both been suggested as methods of reducing levels of attack. Essentially, the conflicting principles are:

1. That removal of residues and other debris from the field will reduce potential termite food supplies and hence lead to a reduction in termite numbers and subsequent attack; and
2. That leaving residues in the field or adding further organic matter will provide alternative food to which the termites will be attracted, thereby reducing levels of attack.

The selection of residue management practice depends on the termite species and its behaviour. Clearing and destruction of debris before planting should be effective for the control of dry-wood termites and others such as *Coptotermes* spp., but may be less successful for the control of Macrotermitinae. The former use the wood both as nest sites and as food so clearing debris will remove existing colonies and potential nest sites. Macrotermitinae, on the other hand, build their nests in soil and are not dependent on wood for nest sites. Existing colonies and nest sites are not then directly affected by removal of debris and since the termites can survive on the food stored in their fungus combs when other food is not available (Logan *et al.*, 1990).

Weeding

Weeds competing with crops, especially young ones for nutrients, light and water may lead to stress and hence increased susceptibility to termite attack (Logan *et al.*, 1990).

Breaking Up Termite Galleries

Termite damage is not generally a problem in crops planted on deep cracking soils because the frequent cracking prevents adequate building and maintenance of mounds, runways and galleries (Lee and Wood, 1971). Artificial breaking up of soil by deep ploughing may have a similar effect (Kakde, 1985).

Seed Rate

Many farmers sow seeds at a higher rate than recommended in order to offset anticipated losses due to termites; surviving plants may then thinned to the appropriate number (Wood and Cowie, 1988).

Varietal Resistance

In general crops showing resistance or tolerance are indigenous while the susceptible crops are exotic. Presumably, indigenous crops have evolved defence mechanisms against the local termite species. Where there are major losses to termites, resistant crops could provide options by way of rotations or intercropping. (Logan *et al.*, 1990). Grafting susceptible fruit trees on to rootstocks of resistant species has been practiced successfully (Fuller, 1912). The American and Egyptian varieties of cotton are more severely damaged than the *desi* (indigenous) ones (Roonwal, 1979). The damage to sugarcane setts by termite *Odontotermes* was lower in varieties S 2315–82 and Co J 64. The quick germinating varieties of sugarcane suffered less damage to the cane setts as compared to the slow germinating ones (Singla *et al.*, 1988).

Biological Control

DeBach (1964) defined biological control as the action of parasites, predators, or pathogens (disease-causing organisms) in maintaining another organism's population density at a lower average than would occur in their absence. Termites are eaten by a wide range of vertebrates and invertebrate predators whose natural influence on termite numbers and population dynamics is partially understood in a small number of cases (Wood and Johnson, 1986).

Predators

The predation hazards confronting termites are enormous. They are preyed upon by a wide range of predators, specialist or opportunistic. Attack is mainly upon individuals outside the nest, foraging worker castes or on alate reproductives. Predation of alate reproductives is opportunistic as the emergence of alates is seasonal. Arthropod predators include scorpions, spiders, centipedes, dragonflies, cockroaches, mantids, crickets, beetles, flies, ants and wasps (Wood and Sands, 1978). It is likely that ants account for a significant proportion of the mortality of swarming alates. Among the vertebrate predators birds are the best documented ones (Logan *et al.*, 1990). Opportunistic predation of foraging sterile castes are mainly by ants. Transplanting *Solenopsis* sp. to nurseries of tree seedlings controlled termites (Beeson, 1941). Burying 'gur' (unrefined sugar) near trees to attract ants which, prey on termites has also been suggested (Anon., 1898). Specialized invertebrate predators of foraging termites seem to be confined to a few species of ponerine and myrmicine ants like *Termitopone commutata* and *Megaponera foetens*. Burrowing doryline ants predates on colony brood centers. Several vertebrate predators of the hive are also been reported from different parts of the world (Logan *et al.*, 1990).

Pathogens

Termites have been considered as good candidates for pathogenic control because they live in a conducive environment humid, crowded and intimate. But the well-known termite behaviour of isolating infected colony members in walled off parts of the nests leads to reduced effectiveness of the pathogens. *Bacteria* -Khan *et al.* (1977a) isolated a naturally occurring termite strain of *Bacillus thuringiensis* from *Bifiditermes beesoni* and demonstrated its pathogenecity to *Heterotermes indicola* and *Microcerotermes championi* but in the field *Bt* has not been that effective (Logan *et al.*, 1990). Various other bacteria, notably *Serratia macescens*, are pathogenic to a number of termite species and can be transferred by trophallaxis (Khan *et al.*, 1977b). Chemicals are tried for suppressing the immune systems of termites like eicosanoid biosynthesis inhibitors (eicosanoids protect insects from bacterial infection). *Serratia marcescens* along with eicosanoid biosynthesis inhibitors like dexamethasone and ibuprofen increased the mortality of *Coptotermes fromosanus* (Connick *et al.*, 2001).

Fungi

It is theorized that termite social behavioiur, trophallaxis and grooming, will aid the transfer of pathogenic fungi within termite colony. Therefore, spores picked up by foraging workers could be transferred from termite to termite (Wright *et al.*, 2002). The insect pathogenic fungus *Metarrhizium anisopliae* caused heavy mortality of the Australian termite *Nasutitermes exitiosus*. Conidia of *M. anisopliae* started an epizootic when introduced into field colonies of *N exitiosus*, irrespective of whether the termites were treated by dusting or spraying in the field itself (Hanel and Watson, 1983). Bapatla isolate of *Beauveria bassiana* was found to be effective against agroforestry termite *Odontotermes wallonensis* (Khan *et al.*, 1992). Different strains of *Beavaria bassiana* and *M anisopliae* showed difference in pathogenecity to *Coptotermes formosanus* (Delate *et al.*, 1995; Wright *et al.*, 2002). Entomopathogenic fungi holds promise for further development as baits for termite control (Delate *et al.*, 1995a). Bioblast™ a product containing *M. anisopliae* as the active ingredient is available in the American markets (Lax and Osbrink, 2003). Imidacloprid showed ability to increase termite susceptibility to *M. anisopliae* and *B. bassiana* (Ramakrishnan *et al.*, 1999).

Hanel and Watson (1983) reported the inhibition of germination of fungal spores inside the termite colony. Antifungal volatiles like naphthalene and fenchone have also been identified from termite nests (Milner *et al.*, 1997; Chen *et al.*, 1998). Improved strains of pathogens and providing a continuous source of the inoculum, as in baits, may help to overcome these problems.

Virus

A nuclear polyhedrosis virus isolated from *Spodoptera littoralis* was found to infect termite castes of *Kalotermes flavicollis.* But for the use of NPV to control termites periodic applications would be more efficacious than one treatment (Fazairy and Hassan, 1988).

Nematodes

Termites live and forage in habitats that are moist, cool and without direct sunlight such as soil or wood materials. These environmental conditions are ideal for the survival and movement of steinernematid and heterorhabditid nematodes. *Heterorhabditis* spp. were more effective than *Steinernema* spp. (Wang *et al.*, 2002). In a field study on *Heterrhabditis* sp., effective control was achieved against a dry-wood termite, *Glyptotermes dilatatus*, to protect tea bushes in Sri Lanka (Danthanarayana and Vitarana, 1987).

Although nematodes can reproduce in termites, they were rarely seen emerging from dead termites because mites usually consumed the nematode-killed termites. The high numbers of termites in a colony and the wide foraging range are obstacles for nematodes to eliminate termite colonies. The limited mobility of nematodes and low rate of reproduction in dead termites make it unlikely that nematodes will reach and maintain a large enough density to eliminate a termite colony in the field (Wang *et al.*, 2002).

Chemical Control

A plethora of compounds have been tried against termites for the past hundred years. Before the discovery of organochlorines compounds like naphthalene emulsion, lead arsenate, crude oil emulsion, rosin compounds, creosote, Bordeaux mixture, carbon bisulphide, potassium-sulpho-carbonate, tar, mordant, Paris green, mercuric chloride, Clensel and, phenyl were used in India against termites. After 1940 organochlorines like DDT, BHC, chlordane, toxaphane, heptachlor, endrin, aldrin were the major insecticides in use upto the 1980s (Narayanan and Rattanlal, 1952; Butani, 1967; Roonwal, 1979).

As the organochlorines were phased out due to environmental concerns organophosphates, synthetic pyrethroids, neonecotinoids and several other newer groups are available the control of termites. Low mammalian toxicity and low environmental hazard are added advantages of the current group of soil barrier termiticides. Pyethroids are highly repellent to termites, which invokes low mortality within a termite population, whereas organophosphates are non-repellent and cause high, localized mortality and secondary repellency caused by the presence of the dead termites within the treated zone. Pyrethroids like permethrin, cypermethrin, fenvalerate, bifenthrin, cyhalothrin, esfenvalerate, silaneophane, tralomethrin and deltamethrin prevented the penetration of *Reticulitermes flavipes* through the treated soil due to their repellency (Su and Scheffrahn, 1990b). Pyrethroid-based soil barrier termiticides are much more useful in high temperature, low moisture environments than organophosphate-based soil barrier termiticides because they do not degrade as rapidly in this environment. The chloronicotinyls, of which imidacloprid is the most active compound, are not readily detected by termite (Lax and Osbrink, 2003).

Different state agricultural universities have given varied recommendations for the control of termites in different crops (Table 7.1). Sugarcane setts treated with monocrotophos was effective than diazinon and carbofuran while the latter two insecticides protected the growing canes better (Kolo *et al.*, 2000). Sett treatment with 0.20 per cent solution of Gaucho 70 WS, soil treatment with phorate 10 G 2.50 kg a.i./ha, chlorpyriphos 15 G 2.50 kg ai/ha and chlorpyriphos 20 EC @ 1.00 kg ai/ha were the best in controlling termite infestation of sugarcane (Singh and Singh, 2001). Sharma *et al.* (2003) reported seed treatment (3.0 g a.i./kg) followed by spraying (0.5 per cent on 15 days old crop) of endosulfan to be the most effective treatment against termites infesting maize.

Table 7.1: Recommendations of Different SAU's for Termite Management

Crop	*State*	*Management*
Paddy	Assam	Endosulfan 4 per cent 25 kg/ha
	Tamil Nadu	Apply chopped paddy straw treated with quinalphos 1.5D 50 kg/ha
Wheat	Haryana	Treat one quintal of grains with chlorpyriphos 20 EC @ 150 ml or formothion 25 EC @ 250 ml or endosulfan 35 EC @ 350 ml and dilute with water to make 5 litre emulsion
	IGKVVV, Raipur	Destroy termitaria to eradicate colony, practice deep ploughing, frequent interculture and irrigation
	Orissa	Methyl parathion 2 per cent
	Punjab	Seed treatment with chlorpyriphos 20EC @ 4ml/kg seed or spray 40 kg seed with 160ml chlorpyriphos (in 1 litre water)
	Rajasthan	Spraying chlorpyriphos 20 EC @ 4 litre or endosulfan 1 35 EC 2.5 litre/ha during irrigation
	Uttar Pradesh	Seed treatment with chlorpyriphos 200ml diluted with 3–4 litres of water/ quintal seed
Maize	Orissa	Methyl parathion 2 per cent dust @ 25 kg incorporated in soil
Sugarcane	Himachal Pradesh	Apply well rotten FYM, remove the stubbles and debris of the previous crop at the time of planting
	Rajasthan	Soil treatment with quinalphos 1.5 per cent or endosulfan 4 per cent dust @ 25 kg/ha or use chlorpyriphos 20 EC @ 4–5 litre on the sugarcane seed setts and apply the insecticides before covering the seeds with soil or endosulfan 35 EC 2.5 litre/ha with irrigation water.

Source: Puri *et al.* (1999).

Scheffrahn *et al.* (1997); reported spinosad to be effective against dry-wood termites. Though exposure to an insect growth regulator fenoxycarb resulted mortality of termites the responses of different termite colonies were extremely variable (Jones and Lenz, 1996). Microencapsulation of permethrin helps to reduce its repellency and the delay in toxic effect ensures the distribution of insecticide through the entire population (Schoknecht *et al.*, 1994). Use of antibiotic tetracycline eliminated protozoa and subsequently killed *Reticulitermes* flavipes (Mauldin and Rich, 1980).

Control of termites using baits has shown encouraging results, but this novel technology has not yet been tried in India. A major factor affecting the efficacy of termite baits is the amount of the bait consumed. The earliest bait matrices consisted of wood blocks dusted with arsenic (Esenther, 1961) and mirex (Esenther and Beal, 1978), but more recent commercially available bait products consist of cellulose matrix, cardboard or paper impregnated with a slow acting termiticide. Termite workers feed upon the bait and transfer it by grooming or trophallaxis to other colony members, eventually reducing or eliminating the entire colony. Such baits could be improved by making them more palatable to the termites based upon termite food preferences (Rojas and Morales-Ramos, 2001). The toxicant should be slow acting because termites tend to avoid sites where sick and dead termites accumulate. Slow acting poisons like hydramethylnon (Logan and Abood, 1990), avermectin B_1 (Su *et al.*, 1987), sulfuramid (Su and Scheffrahn, 1991) and chitin synthesis inhibitor hexaflumuron (Su, 1994; Su and Scheffrahn, 1996; Rojas and Morales-Ramos, 2001) have been identified as potential candidates for use in baits. Insect growth regulators, because of their apparent slow action, can also be incorporated in baits (Jones, 1984; Su *et al.*, 1985; Haverty *et al.*, 1989; Su and Scheffrahn, 1989). Feeding preference for bait matrix masked any feeding deterrence caused by the inhibitors and can be used as an improved delivery vehicle for slow-acting toxicants employed in a baiting control strategy. Increased preference and consumption of the matrix helps to overcome any innate repellency of the active ingredient. This also allows the use of lower concentrations of active ingredient (Lax and Osbrink, 2003). Rojas and Morales- Ramos (2001) has given chemical composition of a bait matrix for delivery of chitin synthesis inhibitors to *Coptotermes formosanus* (Table 7.2).

Table 7.2: Bait Matrix for Delivery of Chitin Synthesis Inhibitors

Ingredient	*Amount, g*
Cellulose	250.00
Distilled water	5.62
Drinking water	750.00
Ethyl alcohol	3.75
Ergosterol	0.45
Lecithin	1.25
Water-storing polymer	0.187
Yeast hydrolysate	1.50

Fumigation is also an important tool for controlling termites, especially drywood termites and wood boring termites. Carbon dioxide (Delate *et al.*, 1995b), methyl bromide and sulfuryl fluoride (Su and Scheffrahn, 1986) has been found useful for termite management. In Sri Lanka, Danthanarayana and Fernando (1970) found fumigation of termite galleries inside tea bushes with phosphine as an effective method of control.

Resistance of Termites to Chemical Control

The selection of resistant individuals in a colony that can moult into secondary reproductives could lead to the development of colonies resistant to some insecticides (Osbrink *et al.*, 2001). Different colonies of Coptotermes formosanus showed different tolerance levels to several commonly used termiticides. While there were substantial differences in the levels of various detoxification enzymes such as cytochrome P450 oxidase, cytosolic esterase and microsomal esterases, there was no direct correlation between the levels of enzyme activity observed and the levels of insecticide tolerance noted (Valles and Woodson, 2002).

Plant Products

A wide range of plants insects or has antifeedant properties (Table 7.3). Waste saw dust or wood chips from trees containing repellent chemicals may provide some protection if incorporated into soil or used as around crops or trees. Several mulches using leaves or berries of the plant or cake are reported to be effective (Logan *et al.*, 1990). Extracts of neem, *Acorus calamus* (Roonwal, 1979), *Acacia catechu*, *Gardenia gummifera*, *Ferula jaeschukeana*, *Aloe* spp.; oil from *Canarium strictum*, *Hopea* sp, *Shorea* sp. and castor oil cake gave protection against termites (Beeson, 1941). Water extracts of *Calotropis procera* leaves as soil treatment (Parihar, 1994) and as set treatment of sugarcane (Singh *et al.*, 2002) are effective in controlling *Odontotermes*. Wood ash heaped around the base of the trunk has been recorded as preventing termite infestation of coffee bushes (Kashyap *et al.*, 1984).

Table 7.3: Plant Species Reported to be Toxic or Repellent to Termites

Plant	*Method/Use*
Acanthaceaceae	
Adathoda vasica	Infusion used against termites
Agavaceae	
Agave americana	Water extract (leaves) to protect paper
A. angustifolia	Water extract as paint for tree wounds
A. vera	Water extract as paint for tree wounds
A. cantela	Water extract as paint for tree wounds
Sansevieria libericum	Water extract to protect crops
Anacardiaceae	
Anacardium occidantale	Seed oil/stem gum to protect timber
Araliaceae	
Hedera helix	Saponins in leaves inhibit feeding
Asclepiadaceae	
Calotropis procera	Latex from stem to protect timber
Sarcostemma acidum	Decoction with salt as termite repellent
S. brevistigma	Plants/salt bag hung in irrigation water, decoction with salt as termite repellent
Burseraceae	
Boswellia dalzielli	Gum to protect wood
Canarium strictum	Gum from leaves to paint for tree wounds
Commiphora africana	Water extract in tree nursery irrigation, resin used as termite repellent
Caricaceae	
Carica papaya	Immature fruit juice toxic
Caryophyllaceae	
Gypsophila paniculata	Saponins in roots inhibit feeding
Compositae	
Chrysocoma tenuifolia	Oil from leaves/flowers toxic/repellent
Parthenium hysterophorus	Extract of leaves toxic
Tagetes minuta	Water extract in tree nursery irrigation
Tagetes sp.	Water extract in tree nursery irrigation

Contd...

Table 7.3–Contd...

Plant	Method/Use
Convolvulaceae	
Ipomoeafistulosa	Mulch to protect drying groundnuts
Cucurbitaceae	
Cissus quadrangularis	Planted astennite repellent
Cupressaceae	
Chamaecypariy formosensis	?
Juniperus sp.	Extract of leaves toxic
Diptocarpaceae	
Hopea sp.	Oil used as paint for tree wounds
Shorea sp.	Oil used as paint for tree wounds
Euphorbiaceae	
Euphorbia tirucalli	Planting hole to protect *Eucalyptus*
Euphorbia continifolia	Planted as repellent
Ricinus communis	Oil cake mulch in tree nursery, oil cake as paint for tree wounds, oil cake mulch in sugarcane fields
Fabaceae	
Daniellia oliveri	Resin as termite repellent
Graminae	
Cymbopogon citrates	Planted to protect crops
C. schoenanthus	Planted to protect crops
Digitaria sp.	Planted to protect crops
Fennisetum purpureum	Planted to protect crops
Hippocastanaceae	
Aesculus hippocastanum	Saponins in fruit toxic, inhibit feeding
Labiateae	
Hypis spicigera	Whole plant laid below stored millet
Ocimum basilicum	Water extract as seed dressing for yam
O. canum	Water extract in tree nursery irrigation
O. urticifolium	Water extract in tree nursery irrigation
Leguminosae	
Acacia catechu	Leaves/gum as paint for tree wounds
Albizzia anthelmintica	Saponins in roots toxic, inhibit feeding
Cassia siamea	Leaf mulch in tree nursery
Glycyrrhiza glabra	Saponins in roots toxic, inhibit feeding
Hardwickia manii	Stem and branches toxic
Leucaena leucocephala	Leaf mulch in tree nursery
Parkia clappertoniana	Water extract to protect crops
Pterocarpus erinaceus	Leaves used to protect millet granaries
Robinia pseudacacia	Wood shavings repellent
Swartzia madagascarienis	Powdered fruit toxic, powdered pods to protect stored grains

Contd...

Table 7.3–Contd...

Plant	Method/Use
Liliaceae	
Aloe graminicola	Water extract for *Eucalyptus* irrigation
Meliaceae	
Azadirachta indica	Oil cake mulch in *Eucalyptus* nursery, leaf mulch in tree nursery, mulch to protect drying groundnuts, seed extract (methanol)
Melia azaderach	Leaf/belry mulch/extract in tree nursery, soil treatment with leaves protects wheat, bark, seed, leaf and fruit extracts
Moraceae	
Ficus carica	Sap used to destroy termite mounds
Naucleaceae	
Uncaria gambir	Leaves/gum as paint for tree wounds
Palmae	
Elaeis guineensis	Oil from fruit as seed dressing, yam
Papaveraceae	
Argemone mexicana	Oil from seed, juice from plant
Pinaceae	
Pinus roxburgii	Needle extract repellent
Primulaceae	
Cyclamen europaeum	Saponins in tuber toxic, inhibit feeding
Primula elatior	Saponins in tuber toxic, inhibit feeding
Rosaceae	
Quillaja saponaria	Saponins in bark inhibit feeding
Rubiaceae	
Gardenia gummifera	Water extract to protect trees
Gardenia lucidia	Water extract as paint for tree wounds
Sapotaceae	
Butyrospermum parkii	Water extract of kernels repellant
Chrysophyllum albidum	Fruit exocarp to protect crops
Scrophulariaceae	
Digitalis purpurea	Saponins in seeds toxic, inhibit feeding
D. lanata	Saponins in seeds toxic
Simaroubaceae	
Quassia indica	Water extract of leaves
Smilacaceae	
Smilax aristolochiaefolia	Saponins in seeds toxic, inhibit feeding
Solanaceae	
Nicotiana tabacum	Water extract in tree nursery irrigation, Water extract in crop irrigation, Water extract in tree irrigation, dust to protect crops
Solanum spp.	Saponins inhibit feeding, some toxicity

Contd...

Table 7.3–Contd...

Plant	Method/Use
Theaceae	
Thea sinensis	Saponins in fruit toxic
Umbelliferae	
Ferula laeschukeana	Water extract-tree band
F erula asafoetida	Reduced attack on wheat
Verbenaceae	
Lippiajavanica	Water extract in tree nursery
Acorus calamus	Water extract to protect crops, Water extract to dip sugarcane setts

Modified after Logan *et al.*, 1990

Terpenoids like linalool, citronellal and carvone showed toxicity and limonene showed repellency against *Odontotermes brunneus* (Sharma and Raina, 1998). Treatment with vetiver oil and nootkatone affected tunneling activity, wood consumption, survival and falgellate numbers of *Coptotermes formosanus* (Maistrello *et al.*, 2002).

Quarantine

A number of termite species have been accidentally introduced to various countries where they have subsequently become pests, but there are only a few instances in which introduced termites have caused major problems in agriculture (Logan *et al.*, 1990).

Future Management Strategies

With the increasing awareness about environmental and health hazards posed by synthetic insecticides, the current strategy of using chemicals as the sole/major method for controlling termites will have to undergo a radical change in the coming years. Correct identification of the species and knowledge of its behaviour will be a key component in any of the management strategy to be adopted. Often there is a tendency among plant protection workers to treat different species of termites as one while choosing a control method. This will not yield much dividend, as we have seen in the case of crop residue management, where the behaviour of the termite species will determine whether to remove the debris or to leave it in the field.

Keeping the crops in a healthy condition will be the most simplest and cheapest method of avoiding/reducing loss due to termites. Stress resulting from drought, disease, weeds, lack off fertilization, mechanical and fire damage should be avoided. Though very less studies have been done to assess the effect of intercropping on termite infestation, there seems to be a great potential (Sekamatte *et al.*, 2002) in using intercropping as a simple and inexpensive tool for the resource poor farmers. Use of indigenous plants with any termite repellent/toxic activity as mulches in the crops may provide some protection (Logan *et al.*, 1990). Development/identification of varieties/cultivars showing resistance to termites is another area of promise as a management strategy to counter termite infestation from the early stages of the crop itself, with a least input cost.

The use of naturally occurring pathogens of termites, principally fungi, bacteria, viruses and nematodes, offers unique advantages over chemically based termiticides. Chief among them are ease of registration, environmental responsibility and safety to non-target organisms. Use of biocontrol agents against termites has not been that succesful due to their avoidance behaviour of infected individuals, though their social activities (grooming, trophallaxis, etc) helps in the spread of inoculum. As there is lot of variation in the pathogenicity of different strains (Khan *et al.*, 1992), isolation of improved strains of entomopathogenic fungi will be key in the use of fungi as a control agent against termites. Presence of antifungal volatiles in the colonies of termites may be another reason for the inability of fungi to propagate within the colony (Chen *et al.*, 1998). This can be overcome by selecting fungi resistant to antifungal compounds (Lax and Osbrink, 2003). The limited mobility, low rate of reproduction in dead termites (Wang *et al.*, 2002) and tendency of termites to avoid areas with large number of entomogenous nematodes (Epsky and Capinera, 1988) are the reasons for low success of entomopathogenic nematodes against termites. A combination of nematodes/fungi with chemicals may help to improve their control over termites (Ramakrishnan *et al.*, 1999; Wang, *et al.*, 2002).

The use of natural products will be one of the environmentally safe termite management tactic of the future. There is a lot of potential in using the locally available plant materials like *Calotropis procera* (Singh and Singh, 2002) as water extracts mixed with irrigation water, sprayed on to plants or mixed with soil to protect crops (Logan *et al.*, 1990).

Environmentally safe and less persistent insecticides will continue to be the major tool for termite management. Making it available for longer periods without polluting the environment can further enhance the effectiveness of the current formulations. This can be done by preparing controlled release or microencapsulated formulations of various compounds. Baiting technology with slow acting chemicals like insect growth regulators and other slow acting poisons is another method of improving the delivery of chemicals. But the success of bait depends on the bait matrix formulation and it should closely resemble the chemical composition of highly preferred wood species of the termite (Rojas and Morales-Ramos, 2001). Use of bait matrix will also help to overcome any innate repellency of the compound.

Area-wide management will be the key to the success of any termite management strategy in the future. Termite colonies have been demonstrated to have foraging territories that may range upto 300 feet from the primary nest (Lax and Osbrink, 2003). This large territory of foraging will help them to avoid any treated area. So devising a management strategy to a larger area rather than to smaller one will help in the long run.

Conclusion

An integrated approach on a wide-scale will be a viable option for the management of termites in crop fields as well as in buildings. Use of tolerant varieties/cultivars and crop rotation can act as preventive measures in termite management. Proper irrigation scheduling will help in containing the infestation in fields. Biological control options like use of entomopathogenic fungi, viruses, parasites and predators will augment other management practices. Better understanding of chemicals involved in communication among termites can open a new avenue in termite management.

Termite infestation, being a perennial problem, will need the use of crop protection agents, which can provide control for longer periods. As persistent chemicals are known to pollute the environment, the focus should be on improved delivery systems, which are able to release the toxicant only at the required levels for extended periods. Identification of repellants/feeding deterrents of natural origin will be a better option to ward off termite threat from uninfested areas.

REFERENCES

Agarwal, R.A. (1972). Problems of termites of sugarcane in India. In: *Termite Problems in India* (Ed.) Roonwal, M.L. Council of Scientific and Industrial Research, New Delhi.

Al Fazairy, A.A. and F.A. Hassan (1988). Infection of termites by *Spodoptera littoralis* nuclear polyhedrosis virus. *Insect Science and its Application*, 9: 37–39.

Anon. (1898). White ants as a pest of agriculture: A note by the settlement officer of Balaghat, Central Provinces, prefixed by certain passages reprinted from the Indian Museum Notes, on termites in relation to crops etc. *Entomological Series No. 7, The Agricultural Ledger* 1897 No. 18, 434. Office of the Superintendent, Government Printing, Calcutta.

Beeson, C.F.C. (1941). A guide to the control of termites for forest officers. *Indian Forest Records* (New Series), *Ent.*, 4: 44–90.

Butani, D.K., (1967). Sugarcane termites: A resume. *Indian Sugar*, 17: 543–549.

Chen, J., G. Hederson, C.C. Grimm, S.W. Lloyd and R.A. Laine (1998). Naphthalene in Formosan subterranean carton nests. *J. Agri. & Food Chem.*, 46: 2337–2339.

Connick, W.J., W.L.A. Osbrink, M.S. Wright, K.S. Williams, D.J. Daiglc, D J Boykin and A R Lax (2001). Increased mortality of *Coptotermes formosanus* (Isoptera: Rhinotermitidae) exposed to eicosanoid biosynthesis inhibitors and *Serratia marcescens* (Eubacteriales: Enterobacteriaceae). *Environ. Entomology*, 30: 449–455.

Cowie, R.H. J.W.M. Logan and T.G. Wood (1989). Termite (Isoptera) damage and control in tropical forestry with special reference to Africa and Indo-Malaysia: A review. *Bulletin Entomological Res.*, 79: 173–184.

Danthanarayana, W. and S.I. Vitarana (1987). Control of the live-wood tea termite *Glyptotermes dilatatus* using *Heterhabditis* sp. (Nemat.). *Agriculture, Ecosystems and Environment*, 19: 333–342.

Danthanarayana, W. and S. Fernando (1970). A method of controlling termite colonies that live within plants. *Int. Pest Control*, 12(1): 10–14.

Delate, K.M., J.K. Grace and C.H.M. Tome (1995a). Potential use of pathogenic fungi in baits to control the Formosan subterranean termite (Isoptera : Rhinotermitidae). *J. Applied Ent.*, 119: 429–433.

Delate, K.M., J.K. Grace J.W. Armstrong and C.H.M. Tome (1995b). Carbon dioxide as a potential fumigant for termite control. *Pesticide Science*, 44: 357–361.

Ebeling, W. and R.J. Pence (1957). Relation of particle size to the penetration of subterranean termites through barriers of sand or cinders. *J. Economic Ent.*, 50: 690–692.

Epsky, N.D. and I.J. Capinera 1988). Efficiency of the entomogenous nematode *Steinernema feltiae* against a subterranean termite *Reticulitermes tibialis* (Isoptera : Rhinotermitidae). *J. Economic Ent.*, 81: 1313–1317.

Esenther, G.R. (1961). Termite attractant from fungus-infected wood. *Science*, 154: 50.

Esenther, G.R. and R. Beal (1978). Insecticidal baits on field plot perimeters suppress *Reticulitermes*. *J. Economic Ent.*, 71: 604–607.

Hanel, H. and J.A.L. Watson (1983). Preliminary field tests on the use of *Metarhizium anisopliae* for the control of *Nasutitermes exitiosus* (Hill) (Isoptera : Termitidae). *Bull. Ent. Res.*, 73: 305–313.

Harris, W.V. (1971). *Termites: Their Recognition and Control.* Longmans, London.

Haverty, M.I., N.Y. Su, M. Tamashiro and R. Yamamoto (1989). Concentration dependent presoldier induction and feeding deterrency: Potential of two insect growth regulators for remedial control of the Formosan subterranean termite (Isoptera : Rhinotermitidae). *J. Economic Ent.*, 82: 1370–1374.

Jones, S.C. (1984). Evaluation of two insect growth regulators for the bait-block method of subterranean termite (Isoptera : Rhinotermitidae) control. *J. Economic Ent.*, 77: 1086–1091.

Jones, S.C. and M. Lenz (1996). Fenoxycarb-induced caste differentiation and mortality in *Coptotermes formosanus* (Isoptera : Rhinotermitidae). *J. Economic Ent.*, 89: 906–914.

Kakde, J.R. (1985). *Sugarcane Production.* Metropolitan, New Delhi.

Kashyap, R.K., A.N. Verma and J.P. Bhanot (1984). Termites of plantation crops: Their damage and control. *J. Plantation Crops*, 12: 1–10.

Khan, H.K., S. Jayaraj and M. Gopalan (1992). Testing of entomopathogenic fungi against agro-forestry termite, *Odontotermes wallonensis* (Wasmann). *Indian J. For.*, 15: 342–345.

Khan, K.I., Q. Fazal and R.H. Jafri (1977a). Pathogenicity of locally discovered *Bacillus thuringiensis* strain to the termites: *Heterotermes indicola* (Wassman) and *Microcerotermes championi* (Snyder). *Pakistan J. Sci. Res.*, 29: 12–13.

Khan, K.I., Q. Fazal, R.H. Jafri and M. Ahmad (1977b). Susceptibility of various species of termites to a pathogen, *Serratia marcescens. Pakistan J. Sci. Res.*, 29: 46–47.

Kolo, I.N., K.I. Nwosu and L.D. Busai (2000). Control of termites in sugarcane using insecticides. *Sugar Tech.*, 2(4): 17–20.

Krishnamoorthy, C. and K. Ramasubbiah (1962). Termites affecting cultivated crops in Andhra Pradesh and their control: Retrospect and prospect. In: *Termites in the Humid Tropics*, UNESCO, Paris.

Lax, A.R. and W.L.A. Osbrink (2003). United States Department of Agriculture: Agriculture Research Service research on targeted management of the Formosan subterranean termite *Coptotermes formosanus* Shiraki (Isoptera : Rhinotermitidae). *Pest Management Sci.*, 59: 788–800.

Lee, K.E. and T.G. Wood (1971). *Termite and Soil.* Academic Press, London.

Lewis, V.R. and M.I. Haverty (1990). Evaluation of six techniques for control of the Western drywood termite (Isoptera : Kalotermitidae) in structures. *J. Economic Ent.*, 89: 922–934.

Logan, J.W.M. and F. Abood (1990). Laboratory trials on the toxicity of hydramethylnon (Amdro; AC 217,300) to *Reticulitermes santonensis* Feytaud (Isoptera : Rhinotermitidae) and *Microtermes lepidus* Sjostedt (Isoptera : Termitidae). *Bull. Ent. Res.*, 80: 19–26.

Logan, J.W.M., R.H. Cowie and T.G. Wood (1990). Termite (Isoptera) control in agriculture and forestry by non-chemical methods: A review. *Bull. Ent. Res.*, 80: 309–330.

Maistrello, L., G. Henderson and R.A. Laine (2002). Comparative effects of vetiver oil, nootkatone and disodium octaborate tetrahydrate on *Coptotermes formosanus* and its symbiotic fauna. *Pest Management Sci.*, 59: 58–68.

Mankin, R.W., W.I. Osbrink and F.M. Oi and J.A. Anderson (2002). Acoustic detection of termite infestations in urban trees. *J. Economic Ent.*, 95: 981–988.

Mauldin, J.K. and N.M. Rich (1980). Effect of chlortetracycline and other antibiotics on protozoan numbers in the Eastern subterranean termite. *J. Economic Ent.*, 73: 123–128.

Milner, R.J., R.A. Staples and G.G. Lutton (1997). The effect of humidity on germination and infection of termites by the hyphomycete, *Metarhizium anisopliae. J. Invertebrate Path.*, 69: 64–69.

Narayanan, E.S. and Rattan Lal (1952). A short review of recorded information on the control of termites damaging crops in India, along with results of recent control experiments at Delhi. *Indian J. Ent.*, 24(1): 21–30.

Osbrink, W.L.A., A.R. Lax and R.J. Brenner (2001). Insecticide susceptibility in *Coptotermes formosanus* and *Reticuliotermes virginicus* (Isoptera : Rhinotennitidae). *J. Economic Ent.*, 94: 1217–1228.

Parihar, D.R. (1994). Termite management in arid zone of Rajasthan, India. *Pest Management and Economic Zool.*, 2: 81–84.

Puri, S.N., Surender Kumar, O.P. Sharma and T. Banerjee (1999). *Database on Plant Protection Recommendations for Agricultural Crops in India, Part I.* National Center for Integrated Pest Management, New Delhi.

Ramakrishnan, R., D.R. Suiter, C.H. Nakatsu and R.A. Humber (1999). Imidacloprid-enhanced *Reticulitermes flavipes* (Isoptera : Rhinotermitidae) susceptibility to the entomopathogen *Metarhizium anisopliae. J. Economic Ent.*, 92: 1125–1132.

Reddy, M.V. and C. Sammaiah (1988). *Odontotermes brunneus* (Hagen) (Termitidae : Isoptera) as a new pest of maize and groundnut. *Entomon.*, 13: 47–50.

Rojas, M.G. and J.A. Morales-Ramos (2001). Bait matrix for delivery of chitin synthesis inhibitors to the Formosan subterranean termite (Isoptera : Rhinotennitidae). *J. Economic Ent.*, 94: 506–510.

Roonwal, M.L. (1979). *Termite Life and Termite Control in Tropical South Asia.* Scientific publishers, Jodhpur.

Sands, W.A. (1961). Foraging behaviour and feeding habits in five species of *Trinervitermes* in West Africa. *Entomologia Experimentalis et Applicata*, 4: 277–288.

Sands, W.A. (1973). Termites as tree, and crop pests. *Mededelingen Fakulteit Landbouwwetenschappen, Gent.*, 38: 817–830.

Scheffrahn, R.H., W.P. Robbins, P. Busey, N.Y. Su and R.K. Mueller (1993). Evaluation of a novel, hand-held, acoustic emissions detector to monitor termites (Isoptera : Kalotermitidae, Rhinotermitidae) in wood. *J. Economic Ent.*, 86: 1720–1729.

Scheffrahn, R.H., N.Y. Su, and P. Bussey (1997). Laboratory and field evaluations of selected chemical treatments for control of drywood termites (Isoptera : Kalotemlitidae). *J. Economic Ent.*, 90: 492–502.

Schoknecht, U., D. Rudolph and H. Hertel (1994). Termite control with microencapsulated permethrin. *Pesticide Sci.*, 40: 49–55.

Sekamatte, B.M., M. Ogenga Latigo and A. Russell-Smith (2003). Effects of maize-legume intercrops on termite damage to maize, activity of predatory ants and maize yields in Uganda. *Crop Prot.*, 22: 87–93.

Sharma, R.K., K. Sharma and J.C. Sekhar (2003). Evaluation of plant protectants on damage and yield of rainfed maize by termites. *Pesticide Res. J.,*. 15: 36–39.

Sharma, R.N. and R.M. Raina (1998). Evaluating chemicals for eco-friendly pest management-I: Terpenoids and fatty acids for building termites. *J. Scientific and Industrial Res.*, 57: 306–309.

Singh, M. and N.B. Singh (2001). Application of insecticide for termite control and its effect on yield contributing characters in sugarcane. *Sugar Tech.*, 3(4): 146–153.

Singh, M., K. Lal and S.B. Singh (2002). Effect of calotropis (*Calotropis procera*) extract on infestation of termite (*Odontotermes obesus*) in sugarcane hybrid. *Indian J. Agri. Sci.*, 72: 439–441.

Singla, M.L., M.S. Duhra and M.A. Aulakh (1988). Varietal resistance in sugarcane to *Scirpophaga excerptalis* Wlk. and *Odontotermes* spp. *J. Insect Sci.*, 1: 99–101.

Sivapalan, P., A.A.C. Karunaratne and D.G.S. Jayatilleke (1980). Clonal susceptibility and the influence of shade trees on the incidence of *Glyptotermes dilatatus* (Bugnion and Popoff) (Isoptera : Kalotermitidae) in tea. *Bull. Entomological Res.*, 70: 145–149.

Srivastava, K.P. and D.K. Butani (1987). Insect pests of tea in India and their control. *Pesticides*, 21(2): 16–21.

Su, N.Y. (1994). Field evaluation of a hexaflumuron bait for population suppression of subterranean termites (Isoptera : Rhinotermitidae). *J. Economic Ent.*, 87: 389–397.

Su, N.Y. and R.H. Scheffrahn (1996). Comparative effects of two chitin synthesis inhibitors, hexaflumuron and lufenuron, in a bait matrix against subterranean termites (Isoptera : Rhinotermitidae). *J. Economic Ent.*, 89: 1156–1160.

Su, N.Y. and R.H. Scheffrahn (1986). Field comparison of sulfuryl fluoride susceptibility among three termite species (Isoptera : Kalotermitidae, Rhinotermitidae) during structural fumigation. *J. Economic Ent.*, 79: 903–908.

Su, N.Y. and R.H. Scheffrahn (1989). Comparative effects of an insect growth regulator, S-31183, against the Formosan subterranean termite and eastern subterranean termite (Isoptera : Rhinotermitidae). *J. Economic Ent.*, 82: 1125–1129.

Su, N.Y. and R.H. Scheffrahn (1990a). Economically important termites in the United States and their control. *Sociobiology*, 17: 77–94.

Su, N.Y. and R.H. Scheffrahn (1990b). Comparison of eleven soil termiticides against the Formosan subterranean termite and eastern subterranean termite (Isoptera : Rhinotermitidae). *J. Economic Ent.*, 83: 1918–1924.

Su, N.Y. and R.H. Scheffrahn (1991). Laboratory evaluation of two slow acting toxicants against Formosan and eastern subterranean termite (Isoptera : Rhinotermitidae). *J. Economic Ent.*, 84: 170–175.

Su, N.Y., M. Tamashiro and M.I. Haverty (1985). Effects of three insect growth regulators, feeding substrates, and colony origin on survival and presoldier production of the Formosan subterranean termite (Isoptera : Rhinotermitidae). *J. Economic Ent.*, 78: 1259–1263.

Su, N.Y., M. Tamashiro and M.I. Haverty (1987). Characterization of slow-acting insecticides for the remedial control of the Formosan subterranean termite (Isoptera : Rhinotermitidae). *J. Economic Ent.*, 80: 1–4.

Valles, S.M. and W. D. Woodson (2002). Insecticide susceptibility and detoxication enzyme activities among *Coptotermes formosanus* Shiraki workers sampled from different locations in New Orleans. *Comparative Biochemistry & Physiology*, 131: 469–476.

Verma, A.N. (1980). Effect of number of irrigations on termite damage in wheat crop. *Haryana Agricultural University J. Res.*, 10: 564–565.

Wang, C., J.E. Powell and K. Nguyen (2002). Laboratory evaluations of four entomopathogenic nematodes for control of subterranean termites (Isoptera : Rhinotermitidae). *Environmental Entomology*, 31: 381–387.

Wood, T.G. and R.H. Cowie (1988). Assessment of on-farm losses in cereals in Africa due to soil insects. *Insect Science & its Application*, 9: 709–716.

Wood, T.G. and R.A. Johnson (1986). The biology, physiology and ecology of termites. In: Vinson, S.B. (Ed.) *Economic Impact and Control of Social Insects.* Praeger, New York.

Wood, T.G. and W.A. Sands (1978). The role of termites in ecosystems. In: *Production Ecology of Ant and Termite*, Brian, M. V. (Ed.). Cambridge University Press, pp. 230.

Wright, M.S., W.L.A. Osbrink, and A.R. Lax (2002). Transfer of entomopathogenic fungi among Formosan subterranean termites and subsequent mortality. *J. Applied Ent.*, 126: 20–23.

8

Status and Distribution of Insect Pests on Crops and their Management by Organic Methods in North Eastern Hill Region of India

K. Rajasekhara Rao, A. N. Shylesha and N. S. Azad Thakur

Division of Entomology, ICAR Research Complex for North Eastern Hill Region, Umroi Road, Umiam (Barapani) – 793 103, Meghalaya, India

The North Eastern Hill (NEH) Region consisting the states of Arunachal Pradesh, Meghalaya, Mizoram, Manipur, Nagaland, Tripura and Sikkim lies between 20° 5′ to 29° 30′ latitude and 89° 49′ E to 97° 3′ E longitude. The region possesses a great diversity in soil and climate as influenced by varied altitudes ranging from 18m amsl to more than 3000m amsl. The region is characterized by rainfall (2000 mm to 4000 mm/year) humidity is very high (60-100 per cent). The region is very rich in natural resources with varied types of vegetation.

Almost all agricultural crops grown in India are grown in the region. Organic farming is introduced recently in Meghalaya and already the states of Sikkim are declared Organic. In view of these policy changes in the agriculture sector, this paper on the distribtution of insect pests and their natural bioagents widens the scope of organic farming. The insects dominate animals in both diversity and magnitude in the region. Insect pest problems in the region are very wide and more than six hundred species of insects have assumed pest status in various crops in the last half century. Sixty per cent of predatory coccinellid beetles existing in India were reported from NEH Region.

Rice

Rice is the major cereal crop grown extensively in plains, valleys, terraces and uplands in hill slopes and jhum upto an altitude of more than 2000 meters. Under rainfed condition only one crop is taken which is sown in the month of June-July and harvested in October-November. In wet land conditions three crops are taken which are locally called Ahu (March-April), Aman and Sali (July-August) and Boro or winter crops (December–January). The major insect pests of rice in the north eastern hill region are the stem borer *(Scirpophaga incertulas* (Wlk), leaf roller, slug caterpillar, the white rice leaf hopper and hairy caterpillar among the foliage feeders, root aphid and white grub as soil borne pests and gundhibug as panicle feeder. Rice insect pests (Table 8.1) continue to be one of the major constraints to rice production which causes about 20 per cent yield losses and complete failure of the crop was observed during epidemics (Sachan and Pathak, 1981).The natural enemies recorded from the region on stem borer include mermithid nematodes, entomogenous fungi *Beauveria velata, B. bassiana* and hymenopteran parasites *Tetrasticus* sp.

The leaf roller (*Cnaphalocrocis medinalis* (Guen) is a common pest on upland rice. The availability of alternate grass hosts during the off-season helps the pest to thrive well in the region. The case worm (*Nymphula depunctalis*) is a serious pest in Barapani and many other areas of the region. Army worm (*Spodoptera mauritia* Boisdual) is sporadic pest on rice and severely occurs in Umraiang valley of Jowai in Meghalaya. Caterpillars feed on leaves and in severe infestation entire seed beds and fields are destroyed and the fields appear as if it has been grazed by animals/cattle. The larvae feed voraciously in the night and hide in the cracks and crevices of the soil during the day time. Thrips (*Thrips oryzae* Will) nymphs and adults feed on the plant sap from tender leaves. The leaf tips of the affected plants fade and rolled up, the thrips remain within the folds.This is serious pest in uplands paddy in Meghalaya. The Horned caterpillar (*Malantis ismene* Guen.) adult is dark brown butterfly which lays white eggs on rice leaves. Caterpillar green with a roughened skin, head bearing a pair of red horned like processes and anal end with two slender processes. It feeds on the leaves at night and remains stationary during day time. Pupation takes place on leaf. The Skipper (*Pelopidas mathias)* caterpillar is yellowish green with four white dorsal stripes. It has a large head and tapering body. The caterpillars defoliate the plant. The adult is a fast moving skipper.

The slug caterpillar *Parasa lepida* C. (Cochlididae: Lepidoptera) is a sporadic pest on rice. The larvae feed on the leaves leaving only the midrib. This pest occurred in large scale feeding the leaves during the kharif 2003 in upland rice in Meghalaya (Figures 8.1 and 8.2).

Gallmidge, *Orseolia* (= *Pachydiplosis*) *oryzae* (Wood-Mason) (Diptera Cecidomyiidae) is found in lowland or upland conditions during the tillering stage of the rice plant. It has also been reported in upland and deepwater rice. Its population density is favored mainly by cloudy or rainy weather, cultivation of high-tillering varieties, intensive management practices, and low parasitization.

Rice hispa *Dicladispa armigera* (Olivier) is a serious pest in the areas lying in 500-600 m and >1500-2000 meters altitude.Severity of damage is related to the crop growth stage of attack. The larvae move down between the leaf sheaths until they reach the apical bud or one of the lateral buds. They lacerate the tissues of the bud and feed until pupation. The feeding causes formation of a gall called "Silver" or "Onion" shoot. Hispa occurs severely in Ri-Bhoi district and other bordering areas of Meghalaya.

Table 8.1: Distribution of Insects Pests on Rice in NEH Region

Rice	*ML*	*MZ*	*Ar. P.*	*NL*	*SK*	*TR*	*MN*
Scirpophaga incertulus	+	+	+	++	++	++	+++
Scirpophaga innotata	+	+	+	+	+	++	++
Chilo suppressalis	+++	+++	++	++	++	+	+
Sesamia inferens	++	+	+	++	+	++	+++
Chilo polychrysus	++	+	+	++	+	++	+++
Leptocorisa oratorious	+++	+++	++	+++	++	+++	+++
Cletus signatus	++	++	++	++	++	++	++
Cnaphalocrosis medinalis	+++	+++	++	++	++	+++	+++
Marasmia trapezalis	+++	+++	++	++	++	+++	+++
Dicladispa armigera	++++	++	+++	++	+	+++	+++
Leptispa pigmoea	++	+	++	+	+	++	++
Nilaparvata lugens	+		++			++	++
Sogatella furcifera	++	++	++	++	++	++	++
Melanitis ismene	+	+	++	+	+	++	++
Nephotettix varicornis	+++	++	++	++	++	++	+++
Nephotettix virescens	+++	++	++	++	++	++	+++
Recelia dorsalis	+	+	+	+	+	+	++
Tettigella spectra	++	++	++	++	++	++	++
Nymphula depunctalis	+++	++	+++	++	++	++	+++
Orseolia oryzae	+	+	+			+	++++
Mythimna separata	+++	+	+	+	+	+	+++
Hydrellia philippina	+++	++	++	++	++	++	+++
Tetraneura-nigriabdominalis	++++						
Baliothrips biformis	++						++
Donasia provosti	++						
Odontotermes sp.	+	+	+	++	+	++	+
Pelopidas methias	+	+	+	+	+	+	+
Chrotogonus chinensis	+	+	+	+	+	++	+
Attractomorpha crenulata	+	+	+	++	+	++	+
Oxya velox	+	+	+	++	+	++	+
Gryllotalpa africana	+	+	+	+	+	+	+
Chaetocnema basalis	++	+	+	+	+	+	+
Monolepta signata	++	+	+	+	+	+	+
Nisaga simplex	++	++					
Nezara viridula	+	+	+	+	+	+	+
Latoea lepida	+	+	+	+	+	+	+

Note: ++++ and +++: Major; ++: Moderate; +: Minor

ML-Meghalaya; MZ-Mizoram; Ar.P.-Arunachal Pradesh; NL-Nagaland; SK-Sikkim; TR-Tripura; MN-Manipur

Few minor insect pests become major pests in some years. Outbreaks of rice ear-cutting caterpillar *Mythimna separata* Walk was recorded during 1982 initially in Assam and then spread to Manipur, Arunachal Pradesh, Meghalaya and Tripura (Barwal, 1983). During 1977, an outbreak of black hairy caterpillar, *Nsaga simplex* (Figure 8.3) was first reported in Meghalaya (Sachan and Gangwar, 1979) and occurred regularly from 1999 in Umiam. Rice hispa outbreaks were recorded during 1987 in Meghalaya and Arunachal Pradesh.

Two hemipteran shield bugs *viz. Eusarocoris guttieger* (Jhunl.) and *Nezara viridula* L. (Pentatomidae) were found for the first time infesting paddy during milky soft dough and hard dough stages of growth during 1986 to 1990 in Manipur. Rice green semilooper, *Naranga aenescens* Moore was recorded as a major pest on 30–40 days old upland crop and continued to damage the crop upto 65 days. The larvae were parasitized by *Apanteles* sp. upto 80 per cent in field conditions. Leaf folder (*Cnaphalocrocis medinalis*), stink bug (*Cletus signatus*), gundhi bug (*Leptocorisa oratorius*), root aphids are major pests in upland conditions.

Two species of root aphids were found severely damaging upland rice crop *i.e.*, black root aphid (*Rhopalosiphum rufiabdominalis*), and brown root aphid (*Tetraneura nigriabdomi-nalis*) (Figure 8.4). The brown root aphid infestation was severe. The population fluctuation was studied in upland paddy (var. IRAT 141). The infestation starts from 48 days after sowing. The aphids were found feeding on roots in colonies around the upper and middle parts of roots by sucking the sap. The leaves of infested plants became yellowish, wilted and stunted in growth. The infestation reduced the number of tillers and finally the yield.

The other minor pests include flea beetles *Chaetocnema basalis, monolepta signata* stemfly, mole and field crickets, crab, black bug, blue beetles, black aphids, rats and birds. *Altica* sps. a metallic blue coloured chrysomelid beetle (Figure 8.5) occur on rice in Jowai district and other adjoining areas surround the district.

Management of Insect Pests in Rice

Varietal Screening

During 1984, 341 rice accessions were studied against natural infestation by leaf folder. Among the characters studied, leaf broadness has maximum effect on infestation followed by plant height and leaf length, 44 germplasm accessions were found tolerant to leaf folder infestation *viz.* Maibiring, Khonorullo, Ngoba, Ryllo white, Ryllo Red-4, Local Black A, Lyngsi, Kuki, Tongla, Themru red, Kalisatiya, Sonamukhi (A), Hnapdai, Mantai, Tura 80, Sonamukhi (B), Botama paddy, Konoldhan, IC 25693, IC 25705, Phonlem, Dziikero, Phaichang, Ponbu, Immo inkhe, Pyre, Pyat kogya, Mayatsuk, Manin, Yimon tsuk, Kiyir tsuk, Yaribat tsuk, Chinairi, Buhte, Leitetbuhban, Beti, Birat, Binni, AR 2-87, Horipuri, Longkuni, Tsuk, Kupru, Choksu, IC 25685, IC 25686 (Majumdar *et al.*, 1984). For upland paddy IRAT-141 was found resistant to stem borer, leaf roller and gundhi bug. In lowland paddy RCPL 1-3, RCPL 1-87-8 recorded low levels of green leafhoppers, stem borer, leaf roller and gundhi bug infestation. Manipuri local and TRC-10 varieties were free from thrips (Anonymous, 1998).

Under screening National Nursery (Hills), 62 advanced yield trial entries were evaluated during the year 2000 for resistance against leaf folder and gundhibug. Six entries were found to be highly resistant and 21 were found to be moderately resistant to gundhibug infestation. Out of 62 entries, 15 were identified as susceptible and VL-93-2767 was found to be highly susceptible to gundhibug. Three entries were found to be resistant and 14 were found to be susceptible to leaf folders.

Figure 8.1: Male and Female Moths of *P. lepida*

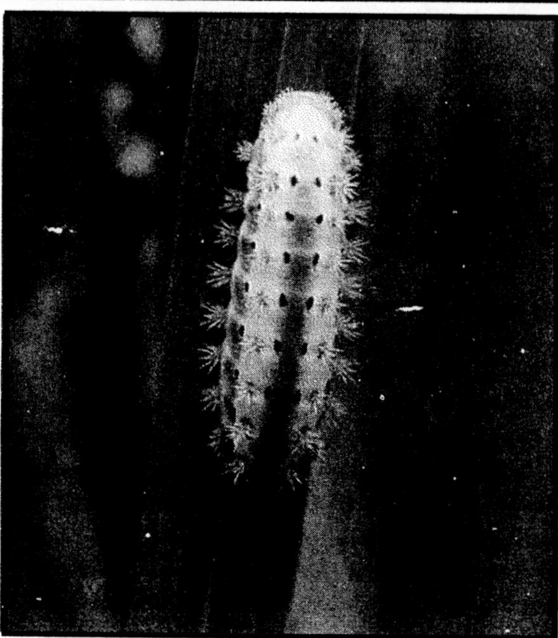

Figure 8.2: Larva of *P. lepida* Feeding on Rice Leaf

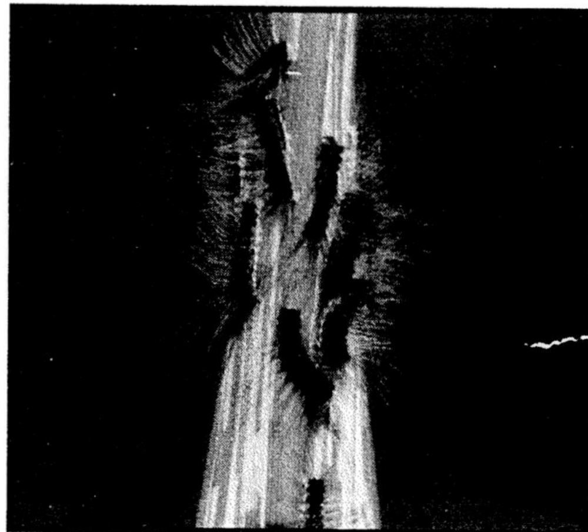

Figure 8.3: Black Hairy Caterpillar *N. simplex* Feeding Rice Leaves

Figure 8.4: Brown Root Aphid *Tetraneura nigriabdominalis* on Rice Roots

Figure 8.5: *Altica* spp. Sitting on the Tip of Rice Leaf

During 2001, the entries *viz.* VSR-8, 15, 31, 33, VL 4041, BL 245, CB 96001, 98013, 98076, JGL 3827, 3918, MTU 1044, RNR 19994, RYT 2662, TTB 281-47-2 and PHB 71 showed zero damage (immunity) to both leaf folder and stem borer at 30 and 50 days after sowing (DAS). Another 7 entries *viz.* CB 97033, 98006, BR 8, BPT 1788, JGL 420, 3858, MTU 1036 showed less than 10 per cent damage by leaf folder and stem borer at 30 and 50 DAS.

Natural Control

The favourable climatic conditions for build up of insect pests and very low quantity use of pesticides by the farmers of NEH region also responsible for the higher activity of natural enemies which kept the pests at a natural equilibrium and thus the pests never became epidemic.

Rice gall midge, *Orseolia oryzae* has been parasitized by 6 hymenopterous parasites like *Tetrastichus* sp., *Ceratosolen* sp., *Aneristus ceroplastate*, *Coccophagus* sp., *C. tschirchi*, *Metaphyars* sp. (Anonymous, 1982). Twelve species of the insect bioagents which parasitized 30 per cent of the caterpillar pests of rice, dipterous tachnids, *Servillia sobria* and *Blepharella* spp. were most active followed by braconid, *Apanteles ruficrus* and eulophid *Tetrastichus ayyari*. Braconid was hyperparasitized by *Mesochours* spp. (Barwal, 1989).

The high humidity and high rainfall conditions of the region are also congenial for build up of the beneficial insect pathogens. Th microbes spread through rainwater flowing from one terrace to another. The first report of infection of the pupal stages of three lepidopteran rice pests *viz.* the rice skipper (*Pelopidas mathias* F.), the horned caterpillar (*Melantis ismene* Cramer) and the leaf folder (*Cnaphalocrosis medinalis* Gn.) by a pathogenic fungus *Beauveria velata* Samson and Evans in Arunachal Pradesh (Padmanabhan *et al.*, 1990). The infected pupae are hard and brittle. The entire body was covered with a whitish fungal mat with fluffy to powdery appearance. The percentage of infected pupae was highest (27.7–60.0 per cent) for *C. medinalis* followed by *M. ismene* (20.0–43.6 per cent) and *P. mathias* (12.5–51.1 per cent). *Entomopthora* fungal infection was severe on white leaf hopper *Cicadellina spectra* (Figure 8.6) in the entire region. The entomopathogens infection was noticed consequently for the three years only in the low land rice ecosystem during 1987 to 1989. The pathogen outbreak was recorded when the crop was in booting stage (Padmanabhan, 1993). Mermethid nematodes were reported infecting the leaf folder larvae to the extent of 5 to 6 per cent (Anonymous, 1987). The publications confirm occurrence of 23 species of insect parasites and 15 species of predators on different insect species in NEH Region (Anonymous, 1981 and 1984).

Indigenous Methods of Trapping Gundhbug in Upland Rice

Gundhbug (Figure 8.7) was the most serious pest of upland rice for last 10-15 years in many states of the region. When the population of gundhibug is high, *L. oratorius* was found to follow negative binomial distribution. When the population was low, it tends to follow Poisson distribution as evidenced by very high values of 'k' and negative values of I_{DM} (Rajasekhara Rao and Shylesha, 2002 a) *C. signatus* was found to follow the positive binomial distribution (Rajasekhara Rao and Shylesha, 2002 b).

Farmers in the region especially those in Meghalaya and Nagaland use dead crabs for trapping the gundhbugs. In this practice, dead crab is inserted into a bamboo stick and the stick is kept in upland paddy. Each trap attracts 30–40 gundhibugs/day and these bugs are collected and killed mechanically. The method was a very successful control measure for the gundhbugs.

Maize

Maize cob borer *Stenachroia elongella* Walker and stem borer *Chilo partellus* Swinhoe (Figure 8.9) are the major pests of maize both in jhum and terraced cultivation in the North East Hill region. Termites, ear cutting caterpillars and field crickets are also found in the early crop growth stage. The incidence of pink borer *Sesamia inferens* was very low in the region. The other pests on maize are Bihar hairy caterpillar (*Diacrisia obliqua* Walk), aphids (*Rhopalosiphum maidis* Fitch), Semilooper (*Trichoplusia orichalcia*), tussel caterpillars, grasshoppers (*Chrotogonus roberstoni* B., *Oxya chinensis* Thunb. and armyworm (*Cirphis unipunctata*) and termites are of minor importance. The incidence of stalk borer *Sesamia inferens* was very low during the past 5 years, the other pests of importance include the *Xylotrupes giddeon*, which feeds on the cobs during May-August, Swarming grasshopper, *Aularchis meliaris*, Aphids, leafhoppers and white backed plant hoppers were all either occurred for a short period or were minor pests.

Varietal Screening

At silking stage varieties Yayung early, SW DMR-91, Aswini, Prabha, Agetti-76, NLD-composite and Palmira-8529 had more than 40 per cent damage due to cob borer compared to other varieties. At harvest varieties like VL-88, Kiran, Yayung early, Gujarat Makka-1, SW-DMR 91(145-1) F-2 and Monong 8531 were found to be less susceptible compared to Dharwad. NLD composite, which was used as control, had a mean cob damage of 34.25 per cent at harvest.

Cultural Practices

Studies on different dates of sowing with the variety local yellow, a fairly susceptible line revealed that early sown crop completely escaped cob borer attack. Crop sown during 12th May onwards suffered heavily and the late sown crop during 12th June had the highest cob damage of 86.6 per cent. The crops sown during 27th April, 12th May and 27th May had 2 per cent, 22 per cent and 57 per cent cob damage, respectively. It was also concluded from the experiments that crop should be planted early *i.e.*, in the month of March/April to overcome cob borer damage.

Different trap crops were tested for attracting the cob borer. The studies revealed that sorghum being the preferred host of the maize cob borer (*S. elongella*) also known as sorghum web worm, can be successfully used for the management of cob borer in maize (Anonymous, 1997). The sorghum varieties IS-2312, DJ-6514 and DSV-1 had compact earhead and the plant height was less (2 m) compared to local sorghum, which was a tall variety (4 m). In case of the three Dharwad varieties, earhead formation coincided with the cob formation whereas in the local sorghum it was delayed till all the cobs were matured. Cob borer incidence was very high on all three sorghum lines except the local sorghum. DSV-1 attracted maximum number of cob borers. Average larval population on DSV-1 was 34 and on IS-2312 it was 29. In case of DJ 6514 larval population ranged from 25–54. RCM 1-1 escaped from borer damage with only 2 per cent of the cob damage. RCM 1-2 (pop corn) even though a preferred host for cob borer did not suffer from cob borer when planted adjacent to sorghum.

Pulses

The important pulses are pigeonpea, cowpea, rice beans, rajmash, bengalgram, mungbean, urd bean, lentil, fava bean, and lathyrus etc. Because of the favourable climatic conditions, pulse crops are infested with a variety of pests throughout the cropping season.

Pigeon Pea

Apion clavipes called as pod boring weevil is a new record and was observed a serious pest on this crop in Naya bunglow, Burnihat and Barapani areas of Meghalaya, Arunachal Pradesh, Assam, Manipur, Nagaland and Mizoram. In Tripura, termites (*Odontotermis sp.*) were recorded as serious pest. In Imphal and Basar pod-boring weevil was a major pest of arhar, which caused 10-15 per cent damage to the pods in the states. *A. clavipes* alone causes 77.8 per cent damage to the pods and 43.1 per cent damage to the grains (Figure 8.10). It was reported that early maturing varieties are susceptible than late maturing varieties. It is a shiny black coloured beetle with a prominent snout (Figure 8.11). *A. clavipes* followed positive binomial distribution at higher incidence and random when the pest incidence was low (Rajasekhara Rao *et al.*, 2002 a).

Varietal Screening

The pod infestation by *Apion clavipes* varied from 22.75 to 38.15 per cent in the early maturing varieties and 15.6 to 26 per cent in the late maturing varieties. Early maturing varieties like Pant 1-A, Pant-A-3, Pant-A-4 were most susceptible than the late maturing varieties like JNAM-374, 456, 243, 183, Pant-A-2 (Sachan and Gangwar, 1984).

Natural Enemies

Two hymenopteran parasites (Figure 8.12 and 8.13) have been recorded on this pest feeding on the grubs and one egg parasite has also been observed. The Eulophid parasite lays eggs on the grubs by piercing ovipositor through the pod and or from the entry hole of the grub. The parasite grub feeds as an external parasite and emerge on the 8th day.

Entomogenous fungus *Beauveria bassiana* was tried in laboratory for the management of the adult weevils (Figure 8.14). Spores of *B. bassiana* when treated @ 10^9 spores/ml was found to be highly effective killing 90 per cent of the test insects with in 3–4 days of treatment and the conidial development was seen on 8th day.

The mealy bugs on red gram are reported to be parasitized to the extent of 38.3 per cent by 4 types of hymenopterous parasites *viz. Tetrastichus* sp. *vr. ceroplastae* (Girault) (Eulophidae), *Aneristus ceroplastae* Howard (Aphelinidae), *Coccophagus tschirchi* Mahdihasan (Aphelinidae) and *Coccophagus* sp. (Aphelinidae) (Annonymous, 1982).

Field Pea

The cut worms (*Agrotis ipsilon, A. flammata*), pea semilooper (*Plusia orichalsia*), pod borer (*Etiella zinckenella*), lycaenid butterfly (*Catochrysops cnejus*), aphid (*Acyrthosiphon pisum*), leaf miner (*Phytomyza atricornis*), thrips (*Megaleurothrips usitatus*), Bihar hairy caterpillar (*Spilosoma obliqua*) and stem fly (*Ophiomya phaseoli*). Pea aphid followed negative binomial (contagious) and its coccinellid predator *C. transversalis* followed positive binomial (regular) distribution (Rajasekhara Rao *et al.*, 2002 b). The spatial distribution of stem fly *M. phaseoli* followed negative binomial distribution. The pupae of *M. phaseoli* in each infested plant followed negative binomial at moderate infestation and Poisson distribution when the infestation was higher (Rajasekhara Rao *et al.*, 2002 c).

Chalcid wasps parasitize pea aphids upto 40–60 per cent during December and showed negative correlation with minimum temperature. The parasitized larvae turned to white colour from green and later to light brown and abdomen enlarge to 5–10 times. The chalcid wasps emerged from the abdomen

Figure 8.6: Entomopthora Fungus on White Leaf Hopper *Cicadellina spectra*

Figure 8.7: Rice Gundhbug

Figure 8.8: Gundhbugs Attracted to Dead Crabs in Rice

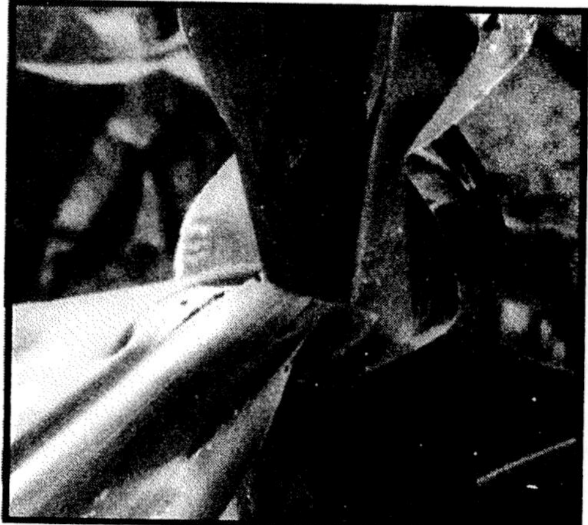

Figure 8.9: Stem Borer *Chilo partellus* Damage on Maize

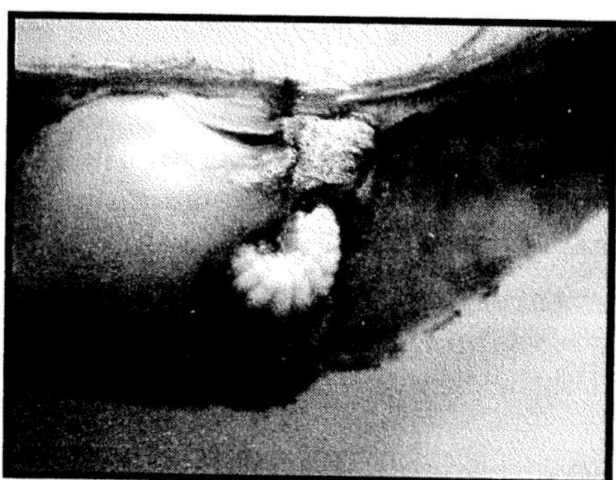

Figure 8.10: *A. clavipes* Grub Feeding on Pigeonpea Pod

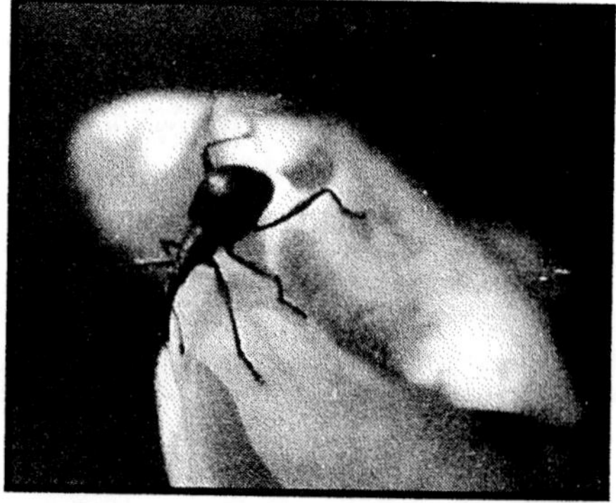

Figure 8.11: Adult *A. clavipes* Pigeonpea

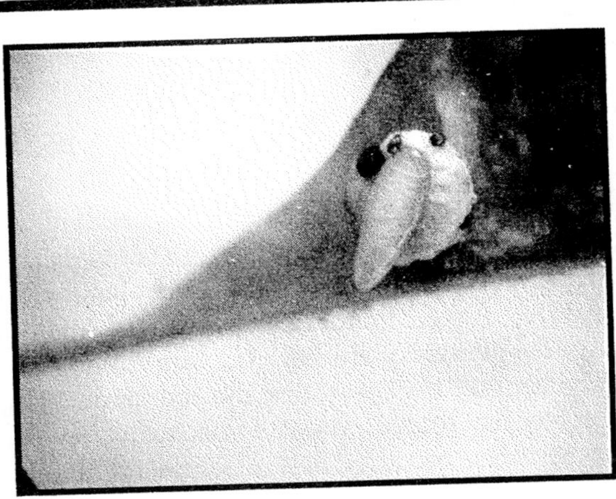

Figure 8.12: Parasite Grub on *A. clavipes* Grub

Figure 8.13: Adult Parasite of *A. clavipes*

Figure 8.14: *B. bassiana* on *A. clavipes*

Figure 8.15: Pea Aphid Infected by *E. neoaphidis*

of the aphid. These wasps are usually quite small, metallic blue or green in colour.The infection of pea aphid by fungus *Erinium neoaphidis* increased with the increase in the bright sunshine hours during winter season and dry cool weather (Figure 8.15). The parasitization started from last week of November and their activity was static till the end of the crop period.

Fava Beans, Lathyrus, Lentils, Mung Been, Rajmash, Rice Bean and Urd Bean

Cut worm (*Agrotis ipsilon*), stem fly (*Ophiomyia phaseoli*), flea beetle (*Monolepta signata*), aphids (*Aphis craccivora, Aphis gossipii*), jassid (*Empoasca devastans*), leaf folder/roller (*Necolea vulgalis, N. dieminalis*), white fly (*Bemissia tabaci*) semilooper (*Trichoplusia ni*), green shield bug (*Nezara viridula*), leaf footed bug (*Anoplocnemus virisis*), grass hopper (*Attractomorpha crenulata*), black cricket (*Gryllus sp), blister beetle (*Mylabris pustulata, M. phelarata* and *Epicauta* sp.), Bihar hairy caterpillar (*Spilosoma (Diacrisia) obliqua*) and tobacco caterpillar (*Spodoptera litura*). The incidence of pod boring weevil (*A. clavipes*) was also recorded on rice bean and cowpea. The damage to pods varied from 10–15 per cent.

Oilseeds

Many insect pests infest soybean, *Glycine max* (L.) among which leaf folder, *Nacoleia vulgalis* Guen. and *N. diemenalis* Guen. were the most destructive at medium altitudes of Khasi hills of Meghalaya (Sachan and Gangwar, 1980). These folders are voracious feeders and cause loss in soybean yield upto 9.3 per cent during kharif season (Azad Thakur, 1985). Azad Thakur (1985) recorded soybean stemfly, *Ophiomyia phaseoli* (Tryon) on soybean causing 15.4 per cent yield loss. The infestation starts from 15–20 days after sowing. Gangwar and Azad Thakur (1991) determined the economic injury level (EIL) for leaf folder, *N. vulgalis* and *N. diemenalis* as 8–9-leaf folders/plant. Mexican bean beetle, *Epilachna vigintioctopunctata, E. varivestis* and *Cerotoma* sp. were recorded as a major pest in the first week of July and continued to damage the crop upto last week of August.

Varietal Screening

The soybean cultivars were screened for resistance against *Nacoleia* sp. and all the varieties were found susceptible. However, the varieties are arbitrarily grouped into three categories (highly susceptible, medium and least preferred) based on average infestation of leaf folders. The scale was taken as 30-40 per cent as least preferred, 40-50 per cent damage as medium and 50 per cent as highly susceptible (Azad Thakur, 1988) (Table 8.2).

Table 8.2: Screening of Soybean Cultivars Against Leaf Folder, *Nacoleia* sp. Resistance

Sl.No.	Category	Cultivar
1.	Highly susceptible	JS-2, MGSB-77, PK-310, Manipur-4, NL-81-3, Manipur-1. PK-73-84, PK-73- 203, Davis, PK-73-94, PK-74-257, PK-74-279, PK-74-295, PK-74-293 MGSM-75, UPSM-960 and Bragg.
2.	Medium susceptible	NLC-8-2, PK-73-75, Sikkim local, DS073-8, PK-73-213, UPSM-12.
3.	Least susceptible	PK-74-294, PK-71-21, NL-81-1, MGSM-77-3, PK-74-261; DS-73-66, PK-73-109.

Source: (Azad Thakur, 1988)

Intercropping

Spiders were more in rice intercropped soybean and coccinellids were higher in maize intercropped soybean. In soybean the spider activity was very high and as high as 20 spiders/10 plants were observed in rice inter cropped soybean.

Biocontrol

Azad Thakur (1985) reported 8 species of parasites and one species of predator (Table 8.3) parasitizing/predating on larvae/pupae of soybean leaf folders in Meghalaya for the first time. The extent of natural parasitization ranged from 18–21.5 per cent. During 1988 another 2 parasites *viz.* chalcid fly, *Brachymeria lasus* (Walker) (Chalcididae: Hymenoptera) and a braconid *Ascogaster* sp. (Braconidae: Hymenoptera) were found parasitizing leaf folder (Azad Thakur, 1990). The coccinellid predators like *Coccinella septumpunctata*, *C. transversalis*, *Menochilus sexmaculatus* were recorded @ 2–3 per plant. These beetles found predating on the aphids (*Aphis gossypii*).

Table 8.3: Natural Enemies of Soybean Leaf Folders *N. vulgalis* and *N. dimenalis*

Sl. No.	Parasite	Order	Family
1.	*Goniozus* sp.	Hymenoptera	Bethylidae
2.	*Trichomalopsis apanteloctena* (Crawford)	Hymenoptera	Pteromalidae
3.	*Apanteles* sp.	Hymenoptera	Braconidae
4.	*Apanteles* sp.	Hymenoptera	Braconidae
5.	*Phanerotoma* sp.	Hymenoptera	
6.	*Bracon* sp.	Hymenoptera	Braconidae
7.	*Orgilus* sp.		
8.	*Sarcophaga* sp.	Diptera	Sarcophagidae
	Predator		
1.	*Cantheconidea furcellata* (Wolf.)	Heteroptera	Pentatomidae

Source: (Azad Thakur, 1985)

Blister beetles *Mylabris pustulalta* were found infected with *Beauvria bassiana* (Figure 8.16) from August to October and a mean number of 2 infected beetles per plant was recorded during 1 week of September of 1997 (Anonymous, 1998).

A high population of parasites was found on the puparium of pea stem fly. Coccinellid beetles like *Menochilus sexmaculatus* and *Scymnus* sp. was active in fields. Hover flies upto 3 per plant were recorded (Anonymous, 1998). The larvae of leaf folders were infected by mermethid nematode, to the extent of 5-6 per cent. The fungus, *Aspergillus flavus* infection on *Mylabris phalerata* feeding on soybean, pulses and beans was also observed in field conditions (Anonymous, 1987).

Groundnut

Many insect pests damage groundnut of which hairy caterpillar, *Amsacta moori*, *A. albistriga*, tobacoo cutworm *Spodoptera litura*, Bihar hairy caterpillar, *Spilosoma obliqua*, leafminer, *Aproaerema modicella*, leaf webber, *Necolea vulgalis*, *N. dieminalis*, aphids *Aphis craccivora*, flower beetles, *Mylabris pustulata*, *M. phaelarata* and leaf feeding beetles like *Monolepta signata* and *Chaetochnema* sp, *Epicauta* sp. and soil borne pests like white grubs (*Holotrichia consanguinea*, *H. serrata*) and termites are important.

Of the several species of white grubs attacking groundnut, *Holotrichia serrata* and *H. consanguinea* and *Leucopholis lepidophora* (Figure 8.17) were predominant. The adults emerge from soil during the first monsoon showers and feed on host plants like neem and other forest plants. After mating they and lay eggs in soil. The grubs feed on the roots of groundnut and also on the developing pods

Varieties MS-11, GN 1024, 191, 271, 362, 450 and USA-61 and No.243 are found to be resistant to leaf miner. ICG-7758 and ICG-8322 were found to be most promising with higher yields. 100 groundnut lines were screened against insect pests during 1999-2001. Twelve varieties were found to be susceptible to jassids with > 8 jassids/leaflet and ICG-4083, ICG-4070 was found tolerant. ICG-3674, ICG-3259, ICG-7030, ICG-1785, ICG-5185 and ICG 6988 were found to be susceptible to wire worm *Gonocephalum halfmansgi* with >15 per cent pod damage whereas ICG-1822, ICG-3273, ICG-4075 were found tolerant. Crop rotation with non-legume crops, collection and destruction of infested leaves were found effective in management of the leaf miner.

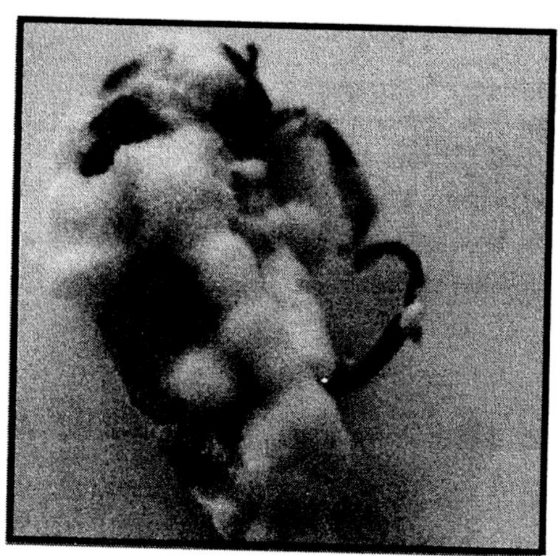

Figure 8.16: ***B. bassiana*** **on** ***M. pustulata***

Figure 8.17: White Grub ***Leucopholis lepidophora*** **Adult**

Early sown crop *i.e.,* upto June is less infested by white grubs. Deep ploughing, manual collections during first showers of monsoon in June–July are found to be the best cultural practices.

Mustard

Rapeseed mustard is the second important source of edible oil in North Eastern region. Over three dozen insect species were found infesting the cruciferous oil seed crops *viz.* saw fly *Athalia lugens proxima*, flea beetle *Phyllotreta cruciferae*, leaf feeders like DBM *Plutella xylostella*, pod borer *Crocidolomia binotalis*, cabbage butterfly *Pieris brassicae,* aphids *Lipaphis erysimi, Brevicoryne brassicae* and *Myzus persicae*. Mustard aphid *Lipaphis erysimi* (Kalt) (Aphididae: Hemiptera) is a key pest of rapeseed mustard causing 35-73 per cent reduction in yield and 5-6 per cent reduction in oil content.

Varieties of mustard RCM-198, RCM-84 and RCM-202 were found to be resistant to aphids. Early sown crop (25th Oct.-15th Nov.) escaped the aphid incidence and yielded very high. Coccinellids are highly predaceous on the aphids and many species were found active along with to syrphid predators and parasites which were found to control aphids.

Fruit Crops

Citrus

Mandarin orange (*Citrus reticulata* Blanco) is one of the major fruit crops of NEH Region of India. The citrus production in North East Hill Region had suffered from different production constraints. The environmental conditions of the region are favourable for occurrence and buildup of pests like leaf miner (Figure 8.18), *Sebaethe* leaf mining beetle (Figure 8.19) psylla (Figure 8.20), aphids, mealybugs (Figure 8.21), fruit bugs (Figure 8.22), citrus butterfly, blackfly (Figure 8.23) and scales (Figure 8.24) and trunk borer (Table 8.4). Every year, approximately 20 per cent of the trees are dying in Meghalaya because of several factors and insects are the major ones. The trunk borer causes very severe damage to the plants and is the major pest of citrus in this region. Citrus psylla and aphids suck the sap from

Table 8.4: Distribution of Insects Pests on Fruits in NEH Region

Citrus	ML	MZ	Ar. P.	NL	SK	TR	MN
Leaf Feeding Caterpillars							
Aleurocanthus woglumi Ashby.	+	+++		+++	+	+++	+++
Aonidiella aurantii	+++	+++	++	++		+++	
Aphids							
Aphis citricola F	+						
Chrysomophalus aonidium	+++	++	++		++		
Coccus viridis	+++	++			++	++	
Dialeurodes citri Ashmed		++			++	++	
Diaphorina citri (Kuw.)	++	+++	++	+++	++		+++
Drosicha stebbingi	+++	+++	+++				++
Icerya purchasi M.	+	+			++		
Latoea lepida	+++	+	—		+		
Lepidosapes beckii	+	++		+++			+
P. helenus. L.	+++	++	+++	+	+++		+
P. memnon L.	++	++	+++	+	++		++
P. polytus. L.	+++	+	+++	+	++		
P. polymnester Cramer	++	++	+++	++	++		
P. protenon Cramer	+	++	+++	+	++		
Papilio demoleus L.	+++	++	+++	++	++	++	++
Phyllocnistis citrella Stan	+++	+++	+++	+++	+++	++	+++
Planococcus citri R.	+++	++	++	++	++	++	++
Pseudococcus filamentosus	+++	+++	++	+++	++	++	
Sebaethe fulvipennis	++	+	++	+	+	+	++
Sessetia nigra	+++	+					++
T.dorsalis kirkaldy	++						
Toxoptera aurantii	++	+++	+++	++	++	++	
Toxoptera citricidus	+++	++					++
Ulodemis trigrapha Merik	—	—	++		++++		++
Bark Feeders							
Indarbela quadrinotata	+++	+++	+++	++	+++	+	+
Indarbela tetraonis	+++	++			+++		
Monohammus versteegi	+++	+++	++++	+++	++	++	++
Xylotrupes giddeon	++++				++		
Stem Feeders							
Batocera rufomaculata	++	+++			+++	+	+
Chilodonium cinctum	—	—			+		
Monohammus versteegi	+++++	+++	++++	+++	++		++
Oberia sp.	+++	++					

Contd...

Table 8.4–Contd...

Citrus	*ML*	*MZ*	*Ar. P.*	*NL*	*SK*	*TR*	*MN*
Oberia latepicalis	+++	++	++		++		
Stromatium barbatum	+	++	+++	++	+++		
Fruit Feeders							
D. cucurbitae	+	++					
Dacus dorsalis	+	++		++	++	++	
Nezara viridula	+	+++	++				
Rhynchocoris humeralis	+++	++++	+++	+++	++		++
Fruit Sucking Moths							
Achoea janata	++	++	++	+	+		
Achorentia styx	++	+	++	+	+	+	++
O. fullonica	+	++		+	+		++
O. materna	+	+	++	++	+		
Apple							
Dyscerus clatheratus	+		+++				
Eriosoma lanegerum	+		++				
Linda nigoscutata	+		++				
Malacosoma indica, Trabla vishnov, Matanastria latipennis	+		++				
Myllocrus viridanus	+		+				
Pandema sp.	+		+				
Quadrspidiotus permiciosus	+		+++				
Zeuzera sp.	+		+				
Litchi							
Cryptophleba carpophaga	++	++	++	++	++	+++	+++
Cacoecia sp.	++	++	++	++	++	+++	+++
Argyroploce leucaspis	+++	++	++	++	++	++	++
Lymantria mathuri	++	++	++	++	++	++	++
Inderbela tetraonis	+	+	+	+	+	++	+++
Eriophyes sp.	+++	+			++	+++	+++
Guava							
Inderbela tetraonis	+++	++	++	++	++	++	++
Myllocrus sp.	++	++	++	++	++	++	++
Dacus sp.	++	++	++	++	++	++	+++
Pulvinaria sp.	++	++	++	++	++	++	+++
Aristobia testudo	+++	++	+++	++	++	++	+++
Planococcus citri	++	++	++	++	++	++	++
Drosicha stebengi	++	++	++	++	++	++	++
Pineapple							
Pseudococcus bromeliae	++	++	++	++	++	++	++

Figure 8.18: *P. citrella* Damage

Figure 8.19: *S. fulvipennis* Damage

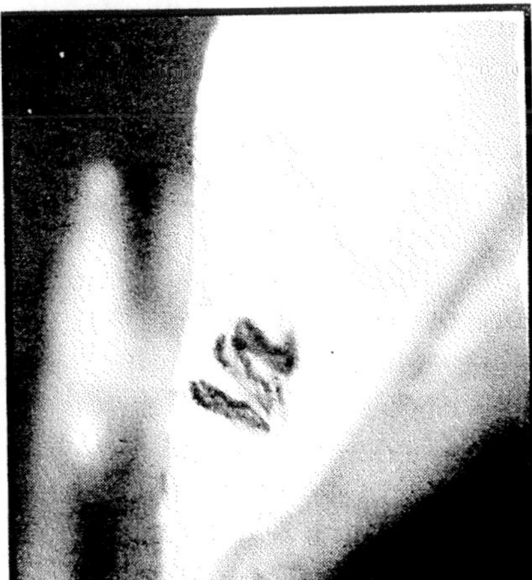

Figure 8.20: *D. citri* Adults

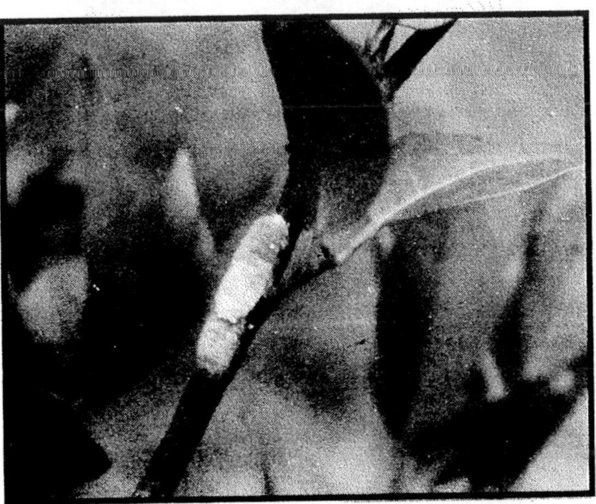

Figure 8.21: Giant Mealybug on Citrus

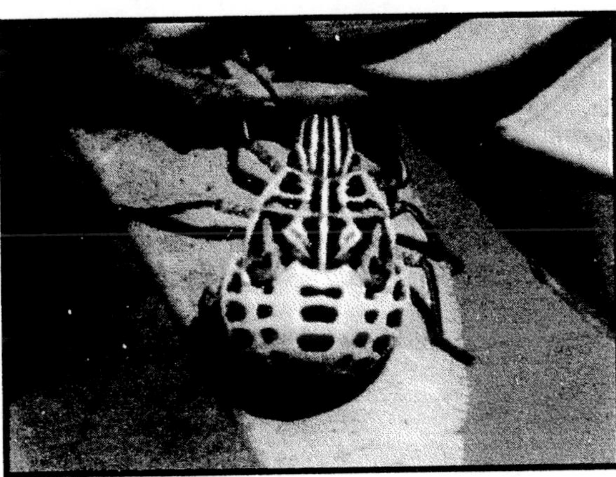

Figure 8.22: Citrus Fruit Bug Nymph

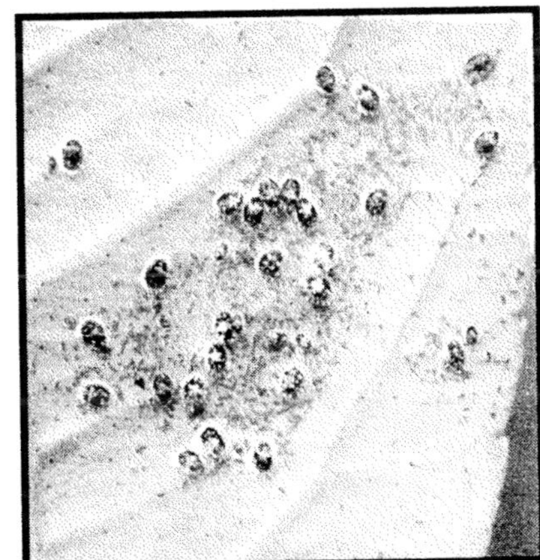

Figure 8.23: Blackfly Nymphs

Figure 8.24: Citrus Scales

the leaves and cause severe damage to the plants. The psylla is also a vector of citrus greening bacterium and the black aphid is the vector of citrus tristeza virus (CTV). Leafminer is also equally important and damage the plants from seedling stage during each new flush. The fruit sucking bug *Rhynchocoris humeralis* is another serious pest and starts sucking sap from fruits from the very beginning till harvest. In severe infestation dropping of immature fruits was observed.

Spatial Distribution

The *P. citrella* infestation followed Negative binomial (contagious) distribution with low to medium damage during the first two flushes and Poisson distribution at higher infestation levels during the third flush. The best fitted multiple regression equations which were different during the three flushes and all the weather parameters together contributing to the differential distribution pattern of the leafminer infestation (Rajasekhara Rao *et al.*, 2002 e). In all the three flushes of citrus, the *T. aurantii* was found to follow negative binomial series with contagious distribution with an exponent '*k*' ranging between 2 to 5. Correlation coefficients between pest and weather indicated positive correlation with maximum and minimum temperature during the first flush period in both the years. (Rajasekhara Rao *et al.*, 2002 d). *P. citri* followed a negative binomial model with a Iwao's regression equation $X^* = 2.6731 + 1.2916\ \delta (R^2 = 0.9794)$. These derivations are useful for analyzing the spatio-temporal changes in population including the density dependent and density independent processes under certain conditions. (Rajasekhara Rao *et al.*, 2001). *Sebaethe fulvipennis* Illiger followed negative binomial distribution within the leaf. During the initial and later phases of incidence *S. fulvipennis* followed positive binomial distribution due to low pest density (Rajasekhara Rao *et al.*, 2002 f).

Leaf mining beetle *S. fulvipennis* (Alticinae: Chrysomellidae: Coleoptera) grubs were found feeding on the foliage of Assam lemon (*Citrus limon* Burm) during new flush period. The brick red coloured beetles measured 2.5mm length and 2mm wide fed on the young leaves. Each female beetle laid 150–200 eggs in groups covered by some abdominal secretions on young leaves, which hatched in 8–10 days, and the emerging grubs entered between two-epidermal layer and fed on the leaf forming a blotch. 25–40 grubs were found within a single leaf causing complete damage to leaf. After completion of feeding they came out and entered into other leaf and like this whole twig is damaged by the time they reach full growth of 4.5–6mm length. They drop down to soil and pupate and emerge during next new flush period.

Varietal Screening

Twelve species of micropropagated citrus plants were screened for their resistance against leaf miner, *Phyllocnistis citrella* and mealybug, *Planococcus citri*. About eight species *viz.* Satkara, Indian wild orange, Assam lemon, Volkamer lemon, Khasi papeda, Sweet lime, Sohbitara and Adajamir showed moderately resistant (25–39 per cent foliage damage). The citrus psylla was recorded during

June on citrus germplams and the species Sohjara and Khasi mandarin-recorded very high population 16.66 and 19.33 adults/leaf respectively. The leafminer incidence was maximum on Gandharaj citron (75.8 per cent foliage damage) and the least in Karna khatta (11.66 per cent damage) (Rajasekhara Rao and Pathak, 2001a &b).

Figure 8.25: Citrus Trunk Borer *M. versteegi*

Citrus trunk borer, *Monohammus versteegi* Rits. (Figure 8.25) was the most destructive pest of citrus in entire region. It is found to cause 60–80 per cent damage to citrus orchards. Both adults and grubs cause damage, grubs tunnel into the trunk making zigzag feeding galleries throwing the frass out preventing nutrient uptake. Adults feed on the bark of the tree which ultimately dry up. This pest has been reported as the main cause of citrus decline in the region. By 2004 almost 50–60 per cent of citrus orchards in the state of Meghalaya are destroyed by this borer.

The trunk borer was controlled in an organic mode for the last few years. The mode includes pasting the stem with 1 per cent Bordeaux misture during May-June redued the egg laying by females. Pasting of spore suspension of a fungal pathogen Beauveria bassiana (10^{20} spores/ml) reduced the adult emerge. Mechanical shaking of the trunk makes the beetles fall back to the ground can be easily collected and killed. The pheromone extracted from the beetles was found to be very effective in trapping makes during peak emergence (Rajasekhara Rao *et al.*, 2003) We discovered the first long range crude sex pheromone for this beetle which was the first report from India and the world.

Guava

Guava is infested by more than a dozen pests in NEH Region and among them guava trunk borer *Aristobia testudo* Voet is the most serious pest causing a very serious losses to the trees. Losses due to this pest in the guava orchard at Barapani ranged from 15–100 per cent on different varieties. Allahabad safeda and L-49 were damaged to the maximum extent. Other pests recorded were aphids, leaf eating caterpillars, *Myllocerus* weevils, *Attacus* sp., *Inderbela* sp., *Dacus* sp. and mealy bugs of minor importance. Collection and destruction of adults during August to November from pigeonpea, hollyhock and *Hibiscus* plants..

Pineapple

It is another important fruit crop of this region and rated next to the citrus from commercial point of view. It is extensively grown on hill slopes. The incidence of mealy bug *Pseudococcus bromiliae* and mite *Stigmacus floridanus* has been reported from Tripura, Manipur and Mizoram. In case of mealy bug, nymphs and adults suck juices from leaves and tender shoots.

Litchi

Indigenous varieties of litchi are common and grown in the lower altitudes of the region. Tripura has the maximum area under litchi plantation. Fruit and stone borer *Cryptophleba carpophaga*, leaf roller *Cacoecia* sp., *Argyploce leucaspis, Lymantria mathuri*, bark eating caterpillar *Indarbela* sp. and aphid *Toxoptera aurantii* attack the trees. The major pests of litchi were the litchi mite *Eriophyes* sp. and the trunk borer *Aristobia testudo* Voet. The fruit and stone borer damage was recorded upto 40 per cent in the orchards at Meghalaya, Manipur, Mizoram and Nagaland.

Apple

More than 23 species of apple pests have been identified in Meghalaya and Arunachal Pradesh and their seasonal incidence, nature and extent of damage have been studies in detail. Among them, wooly aphid (*Eriosoma lanigerum*), shoot borer (*Linda nigroscutata*), fruit boring weevil (*Dyscerus clathratus*), apple stem borer (*Zeuzera* sp.) and leaf defoliating weevils (*Myllocerus* sp.) have been recorded as major pests. Other pests like slug caterpillars, tent caterpillars belonging to family Lasiocampidae, Limacodidae and Lymantriidae respectively have been recorded causing minor to moderate damage to the foliage by defoliating leaves at different stages of growth. Fruit boring weevil (*Dyscerus clathratus*) has been recorded for the first time in India and this seems to be the major pest in Meghalaya and zero in Arunachal Pradesh.

Wooly aphid has also been recorded as one of the very serious pest of apple from Sikkim (Lachan and Lachan), Arunachal Pradesh (Dirrang, Tawang, Zire) and other apple growing regions. The fruit sucking moths *Pandesma* sp., *Schoea* sp., *Ophideris* sp. and few other have been identified as the serious pest of apple fruit in Arunachal Pradesh. More than 40 per cent damage was recorded during the year 1979 to the maturing fruits. Severe attack of fruit sucking moths was also recorded during 1980 at Government orchard, Dirrang in Arunachal Pradesh.

Vegetables

Cole Crops

Insect pests recorded on cole crops (cabbage, cauliflower, radish, knol-khol and gobhi sarson) were cabbage butterfly *Pieris brassicae, P. nepalensis, P. rapae montana*, cabbage semilooper, *Plusia orichalcea, Trichoplusia ni*, diamond back moth *Plutella xylostella*, Bihar hairy catter pillar *Spilosoma obliqua*, cabbage head borer *Hellula undalis, Crosidolomia binotalis*, cut worms *Agrotis ipsilon*, aphids *Brevicoryne brassicae, Lipaphis erysimi*, flea beetle *Phyllotreta cruciferae* and painted bug *Bagrada cruciferarum*. Saw fly *Athalia proxima* was recorded as an important pest of radish and mustard in the region.

Cabbage

The temperatures ranging from 15.2°C to 30°C was found ideal for the pest multiplication during May- June. Ecological studies revealed that *P. brassicae* (L.) has 4 generations with two overlapping generations in low and mid altitude areas and 2–3 generations at higher altitudes (Azad Thakur and Deka, 1997). *P. brassicae* is one of thee most serious pests of cruiciferous crops causing 20 to 100 per cent damage (Sachan and Gangwar, 1980).

The predominant bioagents active during spring summer and monsoon were *Apanteles glomeratus* (L.) a larval endoparasite and *Pteromalus puparum* (L.) a pupal endoparasite of *P. brassicae* (Azad Thakur, 1995). Hand picking of *Pieris brassicae* eggs, use of potential natural enemies like *Apanteles glomeratus*, spraying with *Bacillus thuringiensis* formulation @ 0.2 per cent conc. or spraying with Azadiractin 0.03 per cent @ 2–3ml/lit of water at 15 days interval reduced the pests.

Brinjal

Insect pests are the major constrains in the production of brinjal and the key pest responsible for quality and quantity deterioration is the fruit and shoot borer *Leucinodes orbonalis* Guenee in the region. It attacks shoots in the early stages affecting plant growth where as in later stages larvae prefer fruits to shoots. Shoot damage continues but decreases in the later stage of the crop

Natural Enemies

Three parasites were recorded on the larvae of *L. orbonalis* they are *Trathala flavo-orbitalis* Cameron (Hymenoptera Ichneumonidae) having 5.79 per cent parasitization. Unidentified Dipteran parasite (Tachinidae:Diptera) having 1.65-2.35 per cent parasitization. The mermithid nematode was found to the extent of 5.94-9.39 per cent. It was found that the larvae of *L. orbonalis* Guen were found to be parasitized on an average of 14.74- 14.57 per cent by these three parasites.

Tuber Crops

Sweet Potato

The tribals of the region largely grow it and are one of the important crops on the hill slopes. Insect pests recorded are hawk moth *Acherontia styx, Spodoptera litura* and the sweet potato weevil *Cylas formicarius*. Most of the varieties tested for resistance against the weevil were found susceptible.

Colocasia

The major pests recorded were leaf roller, shoot and rhizome borer *Haplosonyx chalybaeus* Hope and *Aulacophora almora*. The adult beetles of borer feed on the leaves by making circles of different sizes. The female lays eggs in the shoot leaf sheath above the ground in clusters of 80-200. After hatching, the grub bore in the shoot and then finally eats on the developing rhizome. Upto 70-80 larvae per plant were recorded feeding on rhizome. The entomopathogenic nematodes like *Steinernema* and *Heterorhabditis* and the the fungi *B. bassiana* are found infecting the corm borer in the natural conditions.

Spice Crops

Ginger and Turmeric

Ginger and turmeric is cultivated largely in the hills and plains of Meghalaya, Mizoram and other states. The major insect pests reported are the rhizome borers *Libnotes punctipennis* on turmeric (Figure 8.26). Other pests include leaf feeders chrysomelid beetles (unidentified), thrips, skipper and lacewing bugs. Stem borer *Dichochrosis punctiferalis* and shoot boring weevil *Prodioctes haematicus* Chev were found to cause damage upto 35 per cent to shoots in ginger. *D. punctiferalis* starts egg laying in the last week of June and after hatching the larvae feeds inside the shoot till they pupate in the last week of July. Adult emergence was recorded in the first week of August. *P. haematicus* damage was recorded after the stem borer *D. punctiferalis* during August. Other pests include the ginger rhizome maggots, which are also associated with the rhizome rot of ginger.

Figure 8.26: ***L. punctipennis*** **Damage on Turmeric**

Heavy rains in the region cause these damaged culms to start rotting and attract secondary insect pests including the rhizome fly *Mimegralla coerulifrons*. In turmeric only leaf damage is observed and no oviposition or stem boring is observed but on *Amomum* Sp.both leaf feeding and stem boring is recorded.

Alternate Hosts

During the emergence period of the adult weevils only self-sown ginger, *Amomum sp., curcuma zedoaria* are found hence some weevils complete their life cycle on these alternate hosts. Maximum adult scraping is seen on the alternate hosts and oviposition is observed on ginger plants planted.

Natural Enemies

A larval pupal parasite (Unidentified) belonging to Ichneumonidae, Hymenoptera has been recorded on the final instar grubs. Parasitization upto 30 per cent has been observed.

The ginger culm borer can be effectively managed by adopting the integrated management practice. Before taking up sowing of ginger clean the plots by removing all the alternate hosts like *curcuma zedoaria* and *Amomum* sp. Collect all the adult beetles emerged which are found congregating and feeding on the alternate hosts and destroy. In the infested field collection of dead hearts and destruction of the same will help in reduction of the pest. Burning of the thrash and ploughing of the fields just after harvesting will kill most of the hibernating pupae and adults.

Large Cardamom

It is widely grown in Sikkim state in the forest regions and is one of the most important cash crops of the state. About 20 insect pests have been recorded causing damage to the crop. Important among them are hairy caterpillar *Cyane effracta, Pericallia gallectinia, Eupterote fabia,* stem borer *Glyphopterix sp.,* leaf eating caterpillar *Clelea plumbiola,* cardamom aphid *Pentalonia nigronervosa* and green aphid *Neomyzus circumflexes*. The aphids and the stem borers caused maximum damage. Seasonal occurrence of hairy caterpillars caused heavy defoliation. Aphids were also responsible for causing foorki and chirky disease of cardamom. Removal and burning of virus infected culm/plants reduce the spread of diseases and the incidence of foliage pests.

Conclusion

The North Eastern Hill (NEH) Region of India possesses great biodiversity in terms of insect fauna and their biocontrol agents. Insect pest problems in the region are very wide and more than six hundred species of insects have assumed pest status in various crops in the last half century. Sixty per cent of predatory coccinellid beetles existing in India were reported from NEH Region. Many insects reported on main agricultural crops are present in NEH Region. Apart from this, some insect pests like gundhi bug *Leptocorisa oratorius, L. acuta* and *Tetraneura nigriabdominalis* are serious pests on rice; cob borer *Stenachroia elongella* on maize; pod boring weevil *Apion clavipes* on pigeonpea; aphid *Acrythosiphum pisum* on pea; diamond back moth *Plutella xylostella* on cole crops; trunk borer *Anoplophora versteegi* on citrus are very severe and became regular pests. These pests were kept below their damaging levels by natural enemies in all most all the years and also because of low usage of pesticides and fertilizers by the farmers of the region. The recent introduction of organic farming in the north eastern states restricts the use of pesticides on these crops. The changes in plant protection system from conventional to organic necessitate the search for alternative pest control strategies which includes expoloration for indigenous natural enemies like entomopathogens. The present paper focuses on the pest distribution and the existence of their natural enemies distributed over the entire region.

Acknowledgements

The authors are thankful to Dr. K. M. Bujarbaruah, Director, ICAR Research Complex for NEH Region, for providing facilities for carryout the study.

REFERENCES

Anonymous (1976–99). Annual Reports of ICAR Research Complex for NEH Region, Umiam, Meghalaya.

Azad Thakur, N.S. and T.C. Deka (1997). Ecological studies on the cabbage butterfly, *Pieris brassicae* (L.) in North Eastern India. *Pest Management in Horticultural Ecosystem*, 3(1): 13–15.

Azad Thakur, N.S (1985). Yield losses caused by major soybean insect pests. *Agriculture Science Digest*, 5(4): 213–214.

Azad Thakur, N.S. (1988). Field screening of soybean cultivars against leaf rollers, *Nacoleia* sp. in Khasi Hills of Meghalaya. *Legume Res.*, 11(1): 44–46.

Azad Thakur, N.S (1985). New records of bioagents on leaf rollers *Nacoleia vulgalis* Guen. and *N. diemenalis* Guen in Khasi Hills of Meghalaya. *Entomon.*, 19(2): 183–185.

Azad Thakur, N.S (1990). Additional record of parasitoids on soybean leaf folders, *Nacoleia* sp. in Khasi Hills of Meghalaya (India). *Indian J. Hill Farming*, 3(2): 83–84.

Azad Thakur, N.S. (1995). Management of cabbage butterfly, *Pieris brassicae* (L.) in Meghalaya p.32. Paper presented at *National Symposium on IPM and Entomological Approach to Sustainable Agriculture*, September 22–24, Muzaffarnagar – 251 001.

Barwal, R.N. (1989). Natural enemies of rice earcutting caterpillar during its outbreak. Oryza 26: 204–205.

Barwal, R.N. (1983). Outbreak of rice ear–cutting caterpillar, *Mythimna separata* (Walker) in Manipur, India. *International Rice Research Newsletter*, 4(1):11–12.

Padmanabhan, B., R.G. Chaudhary and S.K. Gangwar (1990). Occurrence of *Beauveria velata* Sampson and Evans on some lepidopteran rice pets. *Oryza*, 27: 501–502.

Padmanabhan, B. (1993). Studies on the natural infection of the Entomogenous fungus, *Beauveria velata* Sams and Evans on some pests of low land rice ecosystem. *J. Biol. Cont.*, 7(2): 109–111.

Rajasekhara Rao, K. and K.A. Pathak (2001a). Field evaluation of indigenous germplasm of citrus against insect pests. *Indian J. Hill Farming*, 14 (2): 117–119.

Rajasekhara Rao, K. and K.A. Pathak (2001b). Incidence of citrus psylla, *Diaphorina citri* Kuw. on different Indigenous species of citrus in Meghalaya. *Insect Environment*, 6(4): 156–157.

Rajasekhara Rao, K., A.N. Shylesha and K.A. Pathak (2001). Spatial dynamics of citrus mealy bug *Planococus citri* Risso at medium altitudes of Meghalaya. *Indian J. Hill Farming*, 14 (2): 48–52.

Rajasekhara Rao, K., K.A. Pathak and A.N. Shylesha (2002a). Spatial distribution of pod boring weevil, *Apion clavipes* Gerst. (Apionidae: Coleoptera) on pigeonpea. *Indian J. Ent.*, 64 (1): 88–91.

Rajasekhara Rao, K., K.A. Pathak and A.N. Shylesha (2002b). Spatial distribution of green aphid *Acyrthosiphum pisum* (Harris) and its predator *Coccinella transversalis* Fab. on field pea. *Indian J. Ent.*, 64 (4): 418–423.

Rajasekhara Rao, K., K.A. Pathak and A.N. Shylesha (2002c). Spatial distribution of stemfly, Melanagromyza phaseolin Tryor on Field Pea. *Annals of Plant Protection Sciences*, 10(1): 59–61.

Rajasekhara Rao, K., K.A. Pathak and A.N. Shylesha (2002d). Spatial dynamics of black aphid, *Toxoptera aurantii* Fon. on citrus at midhill altitudes of Meghalaya. *Indian J. Citriculture*, 1(1): 72–78.

Rajasekhara Rao, K., K.A. Pathak and A.N. Shylesha (2002e). Spatio-temporal changes in the infestation of citrus leafminer, *Phyllocnistis citrella* Stantion in Meghalaya. *Entomon.*, 27(2): 169–178.

Rajasekhara Rao, K. and A.N. Shylesha (2002a). Spatial distribution of rice gundhi bug *Leptocorisa oratorius* Fab. in Meghalaya. *Insect Env.*, 7(4): 148–149

Rajasekhara Rao, K. and A.N. Shylesha (2002b). Spatial distribution of Gundhi bug *Cletus signatus* on rice in Meghalaya. *Insect Env.*, 7(4): 147.

Rajasekhara Rao, K., A.N. Shylesha and K.A. Pathak (2002 f). Spatial distribution of leafminer, *Sebaethe fulvipennis* Illiger (Coleoptera: Chrysomelidae) on Assam Lemon. *Pest Management in Horticultural Ecosystems*,8(2):91–96.

Rajasekhara Rao, K., N.S. Azad Thakur and A.N. Shylesha (2003). Citrus trunk borer *Monohammus (Anoplophora) versteegi* Rits: A major pest of Khasi mandarin and its management in North eastern Hill region. Pp. 90–91. Proc. Nat. Sym. On Frontier Areas of Ento. Res. 5–7, Nov. 2003. IARI, New Delhi.

Sachan, J.N. and S.K. Gangwar (1980). Insect pest of soybean in Khasi Hills of Meghalaya and their control. Bulletin of Entomology. 21: 105–112.

Sachan, J.N. and S.K. Gangwar (1984). Pod boring weevil, *Apion clavipes*. A serious pest of pigeonpea in Meghalaya. *Bull. Ent.*, 25(2): 186–189.

Sachan, J.N. and S.K. Gangwar (1979). *Nisaga simplex*, Black hairy caterpillar–A new threat to paddy crop in Meghalaya. *Indian J. Ent.*, 41(3): 278–280.

Sachan, J.N. and K.A. Pathak (1981). *Insect Pests of Crops in Manipur.* Division of Entomology, ICAR Research Complex, Umiam, Meghalaya, 12: 113.

9

Use of Safer Insecticides for Natural Enemy Protection

P.B. Singh

Indian Agricultural Research Institute, Regional Station, Karnal – 132 001

Management of insect pests without causing considerable loss to the environment and maintaining the ecological balance between natural enemy and its injurious host is the desired way of pest management. In this context, use of safer insecticides to the natural enemies is of paramount significance to obtain the desired extent of pest control exerted by the natural enemies on one hand and avoiding the pest resurgence, secondary pest outbreak on the other. In the past few decades, hundreds of synthetic insecticides were developed to control wide range of insect pests. Although, these chemicals were targeted to plant pests but as many of them were broad spectrum biocides had profound effects on their natural enemies. Even the biorational pesticides like, microbial insecticides, insect growth regulators, chitin inhibitors etc. derived from natural sources also exhibited their side effects to the natural enemies. Insecticides may directly affect natural enemies by killing them or indirectly by altering their biology including, fecundity, fertility, rate of development, longevity, host or prey feeding behaviour and general mobility. With the result of these, resurgence of primary and outbreak of secondary pests are often seen.

As far as our awareness to such harmful effects are concerned, as early as 1880, J.H. Comstock noted that pyrethrum insecticide sprays applied for scale insect control were counter productive due to their destructive effects on its natural enemies. In another observation, Koebele in 1883 noted that insecticidal spraying against California red scale also killed parasitic wasps, which attacked the scale and in few months the trees were again re-infested with red scale but natural enemies were absent (Croft, 1990).

Use of synthetic insecticides increased tremendously after the second world war which are still continuing but when we look critically, we find that it only provided temporary suppression of pests and at the same time killed natural enemies, which resulted the resurgence of primary pests or some minor pests assumed major status resulting the total pest problem either at the same level or at the increased level. This situation clearly convinced majority of the entomologists that pests cannot be eradicated and the problem is unending one, therefore, the best course of action is to manage them below economic threshold level, adopting integrated pest management, which envisages the use of all management measures including the need based use of selective insecticides which are relatively less toxic to natural enemies. As an important step in this direction, many countries including India have fixed mandatory norms to test the safety of pesticides to important natural enemies and mammalian fauna before their registration. In India, Paul and Thyagarajan (1992) followed by Singh (1995) compiled available information on this aspect. In foregoing text, methods for evaluation of safety of insecticides including microbials, to parasitoids, predators and compatibility of microbials with synthetic pesticides besides, measures for safe use of insecticides to natural enemies have been discussed.

Methods for Testing Toxicity of Insecticide on Natural Enemies

Earlier, standard method for testing the effect of an insecticide on a natural enemy was not available but now a uniform standard procedures has been suggested by a working group of the International Organization for Biological Control (IOBC) and West Palaerctic Regional Station, which involves a three step test procedures *viz.*, initial laboratory test, semi-field test and field test (Hassan *et al.*, 1985).

Initial Laboratory Test

In this, adult/immature stages of uniform age (0-24 hr old) are exposed to different concentrations including the recommended one to control the pest, are tested by fresh dry film of insecticide applied on a glass plate or plant leaf or soil. Potter's tower or sprayer may be used for making insecticidal film. Parasites/predators are exposed for certain period like 6, 12 or 24 hr and mortality is recorded after 24 hr. Moribund insects are also regarded as the killed ones. Control mortality should not exceed 10 per cent. If there is mortality in control, it should be corrected by Abbott's formula (Abbot, 1925). Reduction in beneficial capacity is measured as compared with water treated control. If reduction in beneficial capacity is less than 50 per cent, insecticide is regarded as harmless those causing 50-79 per cent reduction are regarded as slightly harmful, 80-99 per cent as moderately harmful and more than 99 per cent as harmful.

Semi-field Test

If the insecticide is harmless in laboratory test, no further testing in semi-field condition is necessary. Otherwise, semi-field test is conducted. Persistence studies are conducted by taking susceptible laboratory population of natural enemies, which are exposed daily to the leaves of sprayed and weathered plants up to 30 days and residues are determined. Those causing toxic effects to less than 5 days are regarded as short lived, 5-15 days as slightly persistent, 16-30 days as moderately persistent and more than 30 days as persistent.

Field Test

Effect of direct spray of insecticide on crops or soil inhabited by beneficial insects is determined with recommended dosage of insecticides. The living/dead individuals are sampled before and after the treatment and at fixed intervals. Those insecticides causing less than 25 per cent mortality are regarded as harmless, 25-50 per cent as slightly harmful, 51-75 per cent as moderately harmful and more than 75 per cent as harmful.

Indian Protocol for Assessment of Toxicity

Pesticide registration committee of the Central Insecticides Board, Directorate of Plant Protection, Quarantine and Storage, Govt. of India has approved following protocol for testing the toxicity of pesticides to parasites and predators (Paul and Thyagarajan, 1992).

(i)	Bio-assay method	Dry film topical application.
(ii)	Species	Any two important insect parasites of major crop pests and one insect predator of crop pests.
(iii)	Treatment dosages	A minimum of 5 treatment dosages, giving mortality between 20-80 per cent.
(iv)	Number of test insects	10-20 laboratory reared one day, old adult females per replicate in each case of parasites and 10-20 field collected predators per replicate.
(v)	Number of replications	Three or more and one untreated control. For the purpose of handling, insects may be chilled or anaesthetized by carbon dioxide.
(vi)	Exposure period	6-8 hr. in case of dry film method.
(vii)	Post treatment period	1 day in case of parasites and 8-12 hr. in case of predators.
(viii)	Test conditions	Test insect should be conditioned for 6-12 hr. at a temperature of 28$\pm$2°C. Same temperature should be maintained for exposure and post treatment period.
(ix)	Mortality counts	Moribund insects should be treated as dead. If mortality is observed in control population, Abbott's formula should be employed to correct mortalities in treatments.
(x)	Calculation	LC_{50} should be calculated statistically following Finney (1952) and regression equation should be provided.

Toxicity/Safety of Insecticides to Natural Enemies

In recent years, a number of studies have been carried out with a view to find out relatively safe insecticides to natural enemies (Table 9.1). In majority of the cases, endosulfan, phosalone and fenvalerate were identified as relatively safer. Most of the botanical formulations have been found to be safer to the natural enemies, though, in many instances toxicity levels of botanicals were quite high but when we look to the standards fixed by Hassan *et al.* (1985) these levels are still safer to certain extent. However, it needs testing against the individual natural enemies of importance. Toxicity/ safety of insecticides are generally found to be species specific and hence, there is scope to identify safer insecticides to the important natural enemies, which at the same time are quite effective against the insect pests. From the work of Croft (1990), it is apparent that many insecticides of microbial origin are toxic to one or the other natural enemies and it is more so for those of bacterial and fungal origin (Table 9.2).

Works carried out in India showed that NPV of *Helicoverpa armigera* (Hubner) did not show any harmful effect against natural enemies *viz.*, *Trichogramma chilonis* Ishii, *Chelonus blackburni* Cameron, *Cheilomenes sexmaculatus* (Fabr.), GV of sugarcane internode borer *Chilo sacchariphagus indicus* (Kapur) *Sturmiopsis inferens* (Townsend), *Trichogramma* spp. *Cotesia flavipes* Cameron, *Adelencyrtus mayurai* (Subba Rao), *Chilocorus nigritus* Fabr., *Pharoscymnus horni* Weise. (Jayaraj *et al.*, 1992). But entomogenous

Table 9.1: Effect of Insecticides on Parasitoids and Predators

Sl.No.	Entomophagous Insect	Stage Tested	Insecticide Tested	Safe Insecticide Identified	References
	Egg Parasitoid				
1.	*Trichogramma chilonis* Ishii	Egg, larva, Pupa, adult	Endrin, ethyl parathion, endosulfan, malathion, γ-BHC, chlorpyriphos, endrin+ thiometon, leptophos, phosalone, DDT, DDT+BHC, diazinon, carbaryl, cypermethrin, fenvalerate, deltamethrin, dimethoate, monocrotophos, oxydemeton methyl, permethrin, phosphamidon, acephate, plant extracts of *Vitex negundo* Linn., *Callotropis gigantea*, *Callistemon lanceolatus* Curtis, *Tridax procumbens* Linn., Nomento (combination of neem seed kernel extract, leaf extract of mentha (*Mentha spicata* Linn.) and tobacco (*Nicotiana tabacum* Linn.)	Endosulfan, γ-BHC, acephate, diazinon, DDT, deltamethrin, oxydemetron methyl, phosphamidon, phosalone, fenvalerate and all the tested plant extracts	Paul and Thyagarajan (1992), Chand *et al.* (2001), Solayappan *et al.* (2001)
2.	*Trichogramma japonicum* Ashm.	Egg, larva, pupa, adult	Endrin, ethyl parathion, endosulfan, malathion, γ-BHC, monocrotophos, quinalphos econeem, neem azal, nimbecidine, neemgold.	Endosulfan, γ-BHC, monocrotophos, quinalphos, malathion, econeem, neem azal	Singh (1995), Jhanshi Lakshmi *et al.* (1997)
3.	*Trichogramma achaeae* Nagaraja and Nagarkatti	Immature stage, adult	BHC, carbaryl, cypermethrin, DDT, DDT+BHC, deltamethrin, dimethoate, endosulfan, fenvalerate, malathion, monocrotophos, oxydemeton methyl, permethrin, phosalone, phosphamidon	Fenvalerate, permthrin, oxydemeton methyl, phosalone, monocrotophos	Varma *et al.* (1988) Singh (1995)
4.	*Trichogramma brasiliensis* Ashm.	Immature stage, adult	Malathion, methyl parathion, carbaryl, endosulfan, phosalone, monocrotophos, triazophos, diamethoate, deltamethrin, fenpropathrin, flucythrinate, diflubenzuron, lambda cyhalothrin, quinalphos, phenthoate, fenitrothion, permethrin, fenthion, neem oil	Phosalone, monocrotophos, fenpropathrin, flucythrinate, fenvalerate, endosulfan, cypermethrin, deltamethrin, dimethoate, permethrin, neem oil	Singh (1995)
5.	*Trichogramma perkinsi* Girault	Immature stage, adult	DDT, BHC, endosulfan, endrin, parathion, malathion, diazinon, carbaryl	Diazinon, DDT, endosulfan, malathion, endrin	Singh (1995)

Contd...

Table 9.1–Contd...

Sl.No.	Entomophagous Insect	Stage Tested	Insecticide Tested	Safe Insecticide Identified	References
6.	*Trichogramma pretiosum* Riley	—	Dimethoate, oxydemeton methyl, monocrotophos, fluvalinate, deltamethrin, fenvalerate, carbaryl, methyl parathion, dichlorvos, endosulfan cypermethrin, phosalone, acephate, chlorpyriphos, cartap hydrochloride	None	Singh (1995)
7.	*Trichogrammatoidea armigera* Nagr.	Immature stage, adult	Dimethoate, fenitrothion, monocrotophos, phosphamidon, cypermethrin, phosalone, deltamethrin, fenvalerate, fluvalinate	Fluvalinate	Singh (1995)
8.	*Trichogrammatoidea eldanae* Vigg.	Immature stage, adult	BHC, carbaryl, DDT, DDT+ BHC, endosulfan, phosalone, malathion, dimethoate, monocrotophos, phosphamidon, oxydemeton methyl, cypermethrin, fenvalerate, deltamethrin, permethrin	Oxydemeton methyl, deltamethrin	Paul and Thyagarajan, (1992), Singh, (1995)
9.	*Telenomus remus* Nixon	Adult	Monocrotophos, phosalone, endosulfan, methomyl, oxydemetron methyl, dimethoate, chlorpyriphos, carbaryl, quinalphos	Phosalone, monocrotophos	Paul and Thyagarajan, (1992
10.	*Tetrastichus pyrillae* Crawf.	Immature stages	Prophenphos, methomyl, triazophos, alenycarb, endosulfan, cypermethrin + triazophos, chlorpyriphos, triazophos + deltamethrin, quinalphos, triazophos + endosulfan, dimethoate, deltamethrin + endosulfan, cypermethrin, ethion, monocrotophos, fenvalerate, alphamethrin, deltamethrin, phosphamidon, azadirachtin	Azadirachtin, cypermethrin, alphamethrin, fenvalerate, quinalphos	Deepak and Choudhary (1998)
	Egg-larval parasitoid				
11.	*Chelonus blackburni* Cameron	Adult	Cypermethrin, fenvalerate, deltamethrin, flucythrinate, fenpropathrin, tirazophos, carbaryl, phosalone, monocrotophos, endosulfan, dimethoate, diflubenzuron, lambda cyhalothrin, permethrin, quinalphos, methomyl, bromophos ethyl, EPN	Dimethaote, diflubenzuron, phosalone, fenpropathrin, deltamethrin, permethrin, fenvalerate.	Singh (1995), Manisegaran and Kumaraswamy (1988)

Contd...

Table 9.1–Contd...

Sl.No.	Entomophagous Insect	Stage Tested	Insecticide Tested	Safe Insecticide Identified	References
	Larval parasitoid				
12.	*Allorhogas pyralophagus* Marsh	Adult	Dimethoate, phosphamidon, endosulfan, fenitrothion, cypermethrin, deltamethrin, fenvalerate, monocrotophos, oxydemeton methyl, phosalone	Phosalone	Paul and Thyagarajan, (1992)
13.	*Apanteles* sp. (*vitripennis* group)	Adult	Carbaryl, endosulfan, malathion, phosalone	Phosalone	Sheikh and Kushwaha (1987)
14.	*Apanteles angaleti* Mues	Immature stage, adult	Carbaryl, dichlorvos, dimethoate, chlorpyriphos, endosulfan, fenitrothion, methomyl, monocrotophos, oxydemeton methyl, phosalone, phosphamidon, quinalphos, thiometon, deltamethrin, permethrin, felvalerate, cypermethrin	Phosphamidon, phosalone, fenvalerate, permethrin	Paul and Thyagarajan, (1992, Singh (1995)
15.	*Apanteles papilionis* Viereck	Adult	Dimethoate, monocrotphos, phosalone, carbaryl, fenvalerate, permethrin, cypermethrin, endosulfan, quinalphos, dichlorvos	Phosalone, permethrin, fenvalerate, cypermethrin, deltamethrin, monocrotophos, endosulfan	Singh (1995)
16.	*Apanteles plutellae* Kurdj.	Immature stages, adult	Deltamethrin, permethrin, fenvalerate, cypermethrin, monocrotophos, endosulfan, phosalone, quinalphos	Phosalone, permethrin, fenvalerate	Paul and Thyagarajan (1992)
17.	*Bracon kirkpatricki* (Wilk.)	Immature stages, adult	Carbaryl, chlorpyriphos, dimethoate, endosulfan, malathion, methyl, demeton, monocrotophos, phosphamidon, phosalone, quinalphos, deltamethrin, permethrin, fenvalerate, cypermethrin	Phosalone, permethrin, cypermethrin, deltamethrin	Paul and Thyagarajan (1992)
18.	*Bracon brevicornis* Wesm.	Adult	Carbaryl, chlorpyriphos, dichlorvos, dimethoate, endosulfan, monocrotophos, phosalone, phosphamidon, parathion, imidan, malathion, endrin, diazinon, carbophenothion, trichlorphon, BHC, fenitrothion, fenthion, methomyl, fenvalerate, triazophos, deltamethrin	Endosulfan, phosalone, dichlorvos, chorpyriphos, methomyl	Singh (1995)

Contd...

Table 9.1–Contd...

Sl.No.	Entomophagous Insect	Stage Tested	Insecticide Tested	Safe Insecticide Identified	References
19.	*Campoletis chloridae* Uch.	Cocoon, adult	Endosulfan, carbaryl, monocrotophos, methomyl, oxydemeton methyl, dimethoate, malathion, quinalphos, methyl parathion, fenthion, chlorpyriphos, dichlorvos, acephate, phosphamidon, phosalone, fenvalerate, deltamethrin, cypermethrin, fluvalinate, aldrin, neem oil, karanj oil	Acephate, phosalone, endosulfan, carbaryl, monocrotophos, oxydemeton methyl, fenthion, chlorpyriphos	Singh (1995)
20.	*Carcelia illota* Curran	Adult	Carbaryl, quinalphos, monocrotophos, dimethoate, malathion, phosalone, endosulfan, chlorpyriphos, oxydemeton methyl, dichlorvos, phosphamidon, fenthion	Phosalone	Singh (1995)
21.	*Euceletoria bryani* Sabrosky	Adult	Deltamethrin, cypermethrin, fenvalerate, permethrin, quinalphos, endosulfan, carbaryl, phosalone, monocrotophos, chlorpyriphos, dimethoate, oxydemeton methyl, fenthion, malathion, dichlorvos, phosphamidon	Phosalone, cypermethrin, monocrotophos	Singh (1995)
22.	*Goniozus nephantidis* Mues.	Immature stages, adult	Carbaryl, chlorpyriphos, dichlorvos, dimethoate, endosulfan, phosalone, monocrotophos, phosphamidon, repellin, neem gaurd, margosol, diflubenzuron	Phosalone, endosulfan, neem gaurd, margosol, diflubenzuron	Radhika (1996), Radhika and Chitra (1995), Singh (1995)
23.	*Sturmiopsis inferens* Townsend	Adult	Lindane, endosulfan, monocrotophos, malathion, dimethoate, quinalphos, cypermethrin, fenvalerate, deltamethrin, carbofuran, chlorpyriphos, sevidol	Fenvalerate, soil and whorl applications of granular insecticides	Paul and Thyagarajan (1992)
	Parasitoids of Sucking Pests				
24.	*Adelencyrtus mayurai* (Subba Rao)	Adult	Carbosulfan, monocrotophos, deltamethrin, cypermethrin, fenvalerate, malathion, dimethoate	Fenvalerate, deltamethrin	Easwaramoorthy *et al.* (1994)
25.	*Aenasius advena* Compere	Adult	Monocrotophos, dichlorvos, fenthion, carbaryl, malathion, methyl parathion, endosulfan, phosalone, diazinon, dimethoate, quinalphos, chlorpyriphos, oxydemeton methyl	Diazinon	Singh (1995)

Contd...

Table 9.1–Contd...

Sl.No.	Entomophagous Insect	Stage Tested	Insecticide Tested	Safe Insecticide Identified	References
26.	*Allotropa japonica* (Mani)	Adult	Monocrotophos, dichlorvos, dimethoate, oxydemeton methyl, methyl parathion, diazinon, chlorpyriphos, phosalone, quinalphos, fenthion, carbaryl	Phosalone	Singh (1995)
27.	*Anagyrus dactylopii* (Fabr.)	Adult	Monocrotophos, dichlorvos, dimethoate, oxydemeton, methyl, methyl parathion, chlorpyriphos, diazinon, phosalone, quinalphos, endosulfan, malathion, carbaryl, fenthion	Dichlorvos, diazinon, phosalone	Singh (1995)
28.	*Aphelinus gossypii* Timb.	—	Neem oil, oxydemeton methyl, monocrotophos, phosalone, endosulfan, triazophos, fenvalerate	Fenvalerate, phosalone	Singh (1995)
29.	*Aphelinus mali* (Hald.)	—	Oxydemeton methyl, endosulfan, phosphamidon, phosalone	Endosulfan, phosalone	Singh (1995)
30.	*Aphytis proclia* Walker	Adult	Monocrotophos, ethion, acephate, deltamethrin, cypermethrin, fenvalerate, dimethoate, phosphamidon	Fenvalerate	Paul and Thyagarajan (1992)
31.	*Diaeretiella rapae* (M'Intosh)	—	Phosalone, methamidophos, vamidothion, monocrotophos, malathion, fenvalerate, quinalphos, endosulfan, oxydemeton methyl, dimethoate, phoxim, dicrotophos, chlorpyriphos, NSKE, neem oil, cotton seed oil	Pjhosalone, methamidophos, chlorpyripohos, NSKE, neem oil, cotton seed oil	Singh and Rawat (1981), Awchar *et al.* (1995), Singh (1995)
32.	*Encarsia perniciosi* Tower	Adult	Monocrotophos, malathion, acephate, deltamethrin, cypermethrin, fenvalerate, triazophos, endosulfan, phosalone, diflubenzuron	Phosalone, diflubenzuron	Singh (1995)
33.	*Encarsia shafeei* Hayat	Adult	Phosalone, triazophos, monocrotophos, endosulfan, buprofezin, diflubenzuron	Phosalone, buprofezin, diflubenzuron	Beevi & Balasubramanian (1992)
34.	*Epiricania melanoleuca* (Fletcher)	Immature stages	Malathion, dimethoate, sevidol G, vamidothion, endosulfan, BHC, quinalphos	All safe	Paul and Thyagarajan (1992)
35.	*Platygaster* sp.	Adult	Endosulfan, phosphamidon, monocrotophos, phosalone, chlorpyriphos	Phosalone	Paul and Thyagarajan (1992)

Contd...

Table 9.1–Contd...

Sl.No.	Entomophagous Insect	Stage Tested	Insecticide Tested	Safe Insecticide Identified	References
	Predators				
36.	*Amblyseius tetranychivorus*	Adult	Endosulfan, dicofol, monocrotophos, fenitrothion, chlorpyriphos, quinalphos, dimethoate, malathion, oxydemeton methyl, phosalone, methomyl, carbaryl, sulphur, zineb	Sulphur, zineb	Singh (1995)
37.	*Andrallus spinidens* (Fab.)	Egg, adult	Quinalphos, monocrotophos, fenitrothion, chlorpyriphos, dimethoate, malathion, oxydemeton methyl, phosalone, methomyl, carbaryl, sulphur, zineb	Phosalone, malathion, cypermethrin	Singh (1995)
38.	*Cheilomenes sexmaculata* (Fabr.)	Immature stages, adult	Monocrotophos, endosulfan, carbaryl, deltamethrin, oxydemeton methyl, dimethoate, fenvalerate, cypermethrin, phosalone, neem oil, thiometon, phosphamidon, malathion, formothion, repelin, chlorpyriphos, fluvalinate, nicotine sulphate, lindane	Monocrotophos, carbaryl, dimethoate, deltamethrin, oxydemeton methyl, endosulfan, phosalone, neem oil, thiometon, quinalphos, fenvalerate, repelin, lindane, nicotine sulphate	Singh (1995), Dhingra *et al.* (1995)
39.	*Brumoides suturalis* (Fabr.)	Adult	Endosulfan, deltamethrin, fenvalerate, ethofenprox, neem seed kernel extract	Endosulfan, neem seed kernel extract	Chandrababu *et al.* (1997)
40.	*Cyrtorhinus lividipennis* Reuter	Adult	Deltamethrin, methyl parathion, cypermethrin, fenvalerate, phosalone, quinalphos, permethrin, fenthion, phosphamidan, azadirachtin, econeem, neem azal, neem gold, NG-4, neemax, achook, chlorpyriphos, monocrotophos, thiamethoxam, imidacloprid, fipronil, acephate	Phosalone, phosphamidon, neem gold, neemax, acephate, thiamethoxam, fipronil	Singh (1995), Jhanshi Lakshmi *et al.* (1998, 2001)
41.	*Tytthus parviceps* (Reut.)	Nymphs, adults	Thiamethoxam 100ppm, imidacloprid 50ppm, fipronil 100ppm, acephate 1200ppm	Acephate less toxic	Jhanshi Lakshmi *et al.* (2001)
42.	*Rhinocoris fuscipes* Fabr.	Egg, adult	Quinalphos, monocrotophos, endosulfan, methyl –o- demeton, phosphamidan, chlorpyriphos, dimethoate, nicotine sulphate, nimbecidine, niconeem, neemark	Endosulfan, nimbecidine, niconeem, neemark, nicotine sulphate	Ghelani *et al.* (2000)

Contd...

Table 9.1–Contd...

Sl.No.	Entomophagous Insect	Stage Tested	Insecticide Tested	Safe Insecticide Identified	References
43.	*Microvelia douglasi atrolineata* Bergroth	Adult	Neemax, achook, fortune aza, neem azal, econeem, neem gold, rakshak, NG-4, neem oil, neem seed exctract, quinalphos, monocrotophos, chlorpyriphos, carbofuran and phorate G, thiamethoxam, imidacloprid, fipronil and acephate	Neemax, rakshak, phorate-G, carbofuran-G, quinalphos, fipronil, acephate	Jhanshi Lakshmi *et al.* (1997), Krishnaiah *et al.* (2001)
			Chlorpyriphos 50% + Cypermethrin 5% at 344ppm, betacyfluthrin 1.25% + chlorpyriphos 25% at 393ppm, acephate 45% + cypermethrin 5% at 500ppm, imidacloprid 5% + betacyfluthrin 5%	Acephate, Monocrotophos	Jhanshi Lakshmi *et al.* (2003)
44.	*Propylaea japonica* (Thunberg)	—	Methomyl, fenvalerate, cyhalothrin, deltamethrin, endosulfan, dimethoate, sulprofos, monocrotophos	Endosulfan	Zhu *et al.* (1997).
45.	*Coccinella septempunctata* Linn.	Adult	Endosulfan, oxydemeton methyl, monocrotophos, phosphamidon, thiodicarb, vamidothion, carbofuran, disulfoton, phorate, dimethoate, ethion, fenthion, phosalone, fenitrothion, DDT, quinalphos, carbaryl, malathion, formothion, parathion, endrin, diazinon, mevinphos, thiometon, alcoholic extract of sweet flag rhizome, dharek drupes, dichlorvos, carbophenothion, rotenone, heptachlor, fenvalerate, cyhalothrin, methomyl, methamidophos, cypermethrin, achook, deltamethrin, Bt, neemark	Endosulfan, endrin, oxydemeton methyl, alcoholic extract of dharek drupes and sweet flag rhizomes, aldicarb, mevinphos, carbaryl, heptachlor, thiometon, phosalone, Bt, achook, neemark	Singh (1995), Zhu *et al.* (1998), Sharma and Kashyap (2002)
46.	*Cryptolaemus montrouzieri* Muls.	Adult	Monocrotophos, dichlorvos, dimethoate, oxydemeton methyl, methyl parathion, diazinon, chlorpyriphos, phosalone, fish oil, rosin soap quinalphos, endosulfan, malthion, carbaryl, fenthion, fenvalerate, fenpropathrin, deltamethrin, flucythrinate.	Dichlorvos, chlorpyriphos, fish oil, rosin soap	Singh (1995)
47.	*Chrysoperla carnea* (Stephens)	Egg	Carbaryl, endosulfan, quinalphos, monocrotophos, abamectin	Carbaryl, endosulfan, quinalphos, monocrotophos	Srinivasan and Sundara Babu (2000)

Contd...

Table 9.1–Contd...

Sl.No.	Entomophagous Insect	Stage Tested	Insecticide Tested	Safe Insecticide Identified	References
		Larva	Endosulfan, phosalone, quinalphos, carbaryl, monocrotophos, phenthoate, fenitrothion, cypermethrin, chlorpyriphos, oxydemeton methyl, phosphamidan, malathion, dimethoate, dichlorvos, methomyl, deltamethrin neem oil, abamectin, alphamethrin,	Phosalone, endosulfan, monocrotophos, phosalone, oxydemeton methyl, phosphamidon, cypermethrin, carbaryl, fenvalerate, deltamethrin, neem oil, abamectin	Singh (1995), Srinivasan and Sundara Babu (2000), Balikai and Lingappa (2003)
		Adult	Carbaryl, endosulfan, quinalphos, monocrotophos, abamectin, methyl demeton, chlorpyriphos, malathion, alphamethrin, dimethoate	Endosulfan, abamectin,	Srinivasan and Sundara Babu (2000), Balikai and Lingappa (2003)
48.	*Chrysopa scelestis* Banks	Larva	Plant extracts of *Vitex negundo* Linn., *Callotropis gigantea*, *Callistemon lanceolatus* Curtis, *Tridax procumbens* Linn., Nomento (combination of neem seed kernel extract, leaf extract of mentha (*Mentha spicata* Linn.) and tobacco (*Nicotiana tabacum* Linn.)	All the tested plant extracts	Solayappan *et al.* (2001)
49.	*Mallada boninensis* (Okamoato)	Larva	Endosulfan, oxydemeton methyl, dimethoate, monocotophos, dichlorvos, methyl parathion, carbaryl,fenthion, malathion, quinalphos, fenvalerate, cypermethrin, phosalone, phosphamidon, acephate, deltamethrin, chrorpyriphos, fluvalinate	Dichlorvos, fluvalinate, cypermethrin, fenvalerate, endosulfan, oxydemeton methyl	Singh (1995)
50.	*Scymnus coccivora* Ayyar	Grub, adult	Monocrotophos, dichlorvos, dimethoate, oxydemeton methyl, methyl parathion, chorpyriphos, diazinon, fish oil rosin soap	None	Singh (1995)
51.	*Ishiodon scutellarius* (Fabr.)	Maggots	Chlorpyriphos, malathion, alphamethrin, endosulfan, dimethoate, methyl demeton	Endosulfan	Balikai and Lingappa (2003)
	Predatory Spiders				
52.	*Lycosa* sp	Adult	Deltamethrin, methyl parathion, cypermethrin, fenvalerate, phosalone, quinalphos, permethrin, fenthion, phosphamidan	Phosphamidan	Singh (1995)

Contd...

Table 9.1–Contd...

Sl.No.	Entomophagous Insect	Stage Tested	Insecticide Tested	Safe Insecticide Identified	References
53.	*Lycosa pseudoannulata* (Boesenberg & Strand), *Oxyopes javanus* Thorell	Adult	Acephate, BPMC, ethofenprox, quinalphos, chlorpyriphos monocrotophos, carbaryl	Acephate, BPMC, ethofenprox	Kumar and Velusamy (1996)
54.	*Tetragnatha javana* Thorell	Adult	Acephate, BPMC, ethofenprox, quinalphos, chlorpyriphos monocrorophos, carbaryl	Ethofenprox, BPMC	Kumar and Velusamy (1996)

Table 9.2: Effect of Microbial Pesticides on Natural Enemies

Sl.No.	Natural Enemy	Pathogen	Treated insect host	Dosage	Exposure Method/Toxicity
Virus					
1.	*Chelonus blackburni*	NPV	*Heliothis* sp	Min. recom. dose	Contact by adults*
2.	*Cotesia glomeratus* (Linn.)	GV	*Pieris rapae*	1.0×10^5 granula/ml	Ingested by host larvae after parasitisation****
3.	*Trichogramma evanescens* Westwood	NPV	*Xestia c. nigrum*	-	Egg before parasitisation*
4.	*Chrysoperla carnea*	NPV	*Mamestra brassicae*	1,5,10 × LC_{100} for *M. brassicae*	Ingested and contacted by larvae*
Bacteria					
5.	*Bracon brevicornis*	Bt.var. *thuringiensis* Weise	*Sesamia cretica*	3×10^{10} spores/ml	Ingested by larvae in host**
6.	*Cotesia glomeratus*	Bactospeine (Bt)	*Pieris brassicae*	0.05% a.i.	Ingested by larvae in host**
7.	*Campoletis sonorensis* (Cam.)	Thuricide (Bt)	-	Recom. field rate	Contact by adults*
8.	*Trichogramma evanescens*	Entobacterin	-	1.0% a.i.	Contact by adults*
9.	*Adonia variegata* (Goeze)	Entobacterin	*H. armigera*	2.0% a.i.	Ingested by adults*
10.	*Coccinella septempunctata*	Bitoxibacillin 202	*C. pomonella*	70 millard/g	****
11.	*Chrysoperla carnea*	B.t. var. *galleriae*	-	-	Ingested by larvae**
Fungi					
12.	*Cotesia glomeratus*	*Beauveria bassiana* (Boverin)	*P. brassicae*	-	Infects host after female oviposits*
13.	*Trichogramma evanescens*	Boverin	-	0.5 a.i.	Contact by adults*
14.	*C. septempunctata*	Boverin	-	1%	Ingested by adults and larvae*
15.	*Cryptolaemus montrouzieri*	Boverin	*Sitotroga cerealella*	1%	Ingested by larvae***
16.	*Chrysoperla carnea*	Boverin	*Sitotroga cerealella*	1%	Ingested by larvae and adult*
Protozoans					
17.	*Cotesia glomeratus*	*Nosema polyvora*	*P. brassicae*	-	Ingested by larvae***
18.	*Campoletis sonorensis*	*Nosema campoletidis*	*H. zea*	-	Exposed to host larvae*
19.	*Trichogramma evanescens*	*N. pyrausta*	*Ostrinia nubilalis*	-	Ingested by larvae***

*: No or slight toxicity (0 to 10%), **- Low toxicity (10 to 40%), *** Moderate toxicity (40 to 80%)

****: High toxicity (80-100%) mortality.

fungus *Beauveria bassiana* (Bals.) was found to infect *Coccinella septempunctata* Linn., *Brumoides suturalis* (Fabr.) and syrphids (Haseeb and Murad, 1997). It has been generally observed that majority of insect viruses are more host specific than bacterial and fungal entomopathogens. As far as compatibility of the entomopathogens is concerned, *Bacillus thuringiensis* was found compatible with cypermethrin, endosulfan, phosalone and fenvalerate (Gaffar and Kushwaha, 1994; Patel and Vyas, 1999), NPV of *H. armigera* with neem seed kernel extract and endosulfan whereas, GV of *Chilo infuscatellus* Snellen was found compatible with endosulfan (Jayaraj *et al.*, 1992). From the toxicity/safety studies carried out against natural enemies, its is evident that most of the synthetic insecticides, many botanical and microbial insecticides adversely affect natural enemies, only their level of toxicity varies hence, for safety, these should invariably be screened out against important natural enemies of a crop ecosystem. In plant protection, it is difficult to manage insect pests without using insecticides however if certain measures (Figure 9.1) are taken, harmful effects of insecticides on natural enemies may be brought down thereby resurgence of primary and outbreak of secondary insect pests may be minimized.

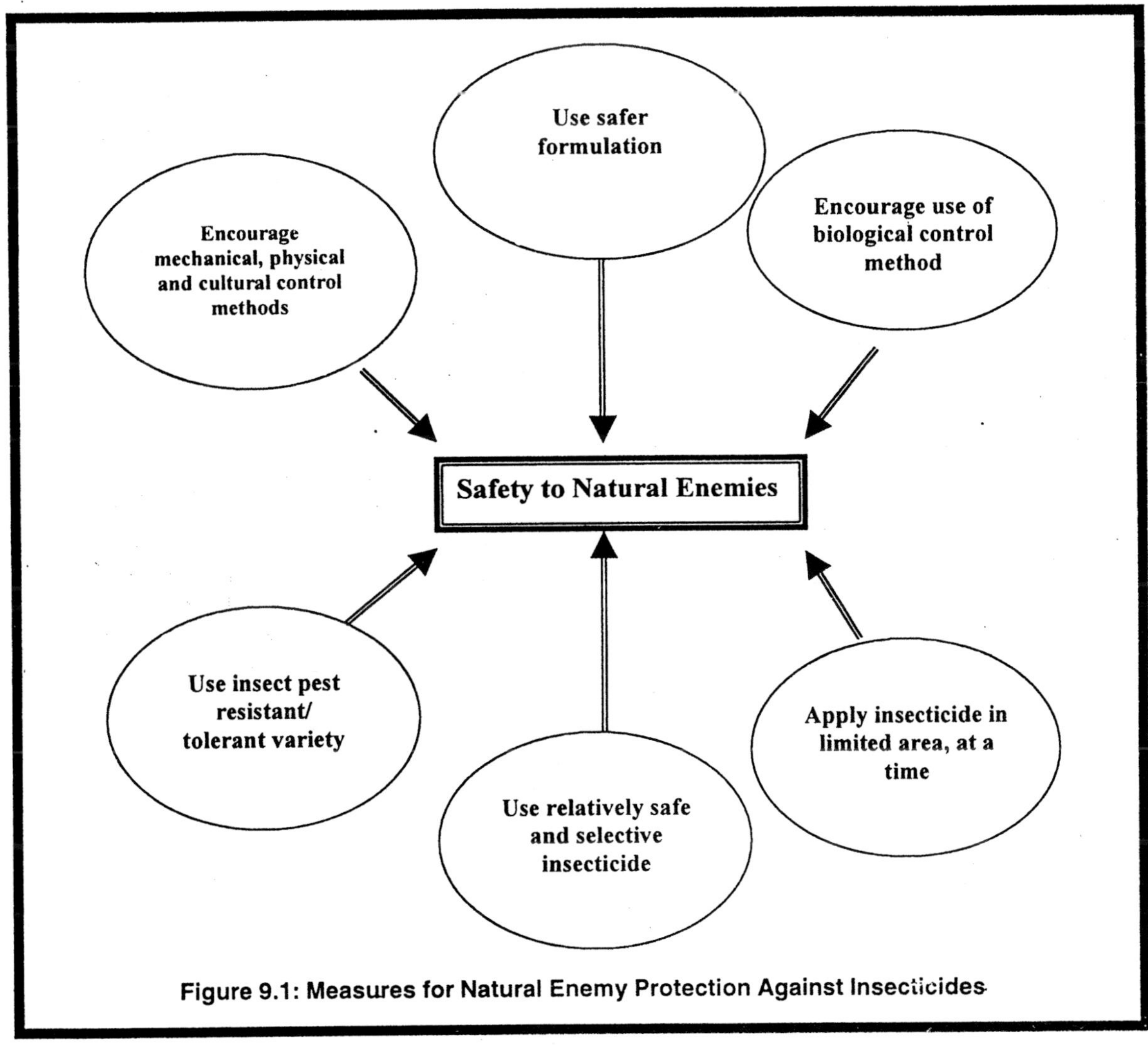

Figure 9.1: Measures for Natural Enemy Protection Against Insecticides

Plate 9.1: *Brumoides suturalis*

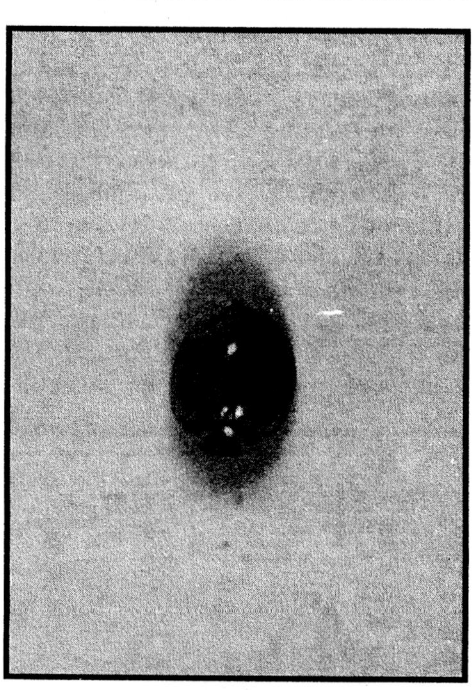

Plate 9.2: *Cheilomenes sexmaculatus*

Plate 9.3: *Chrysoperla carnea*

Plate 9.4: *Coccinella septempunctata*

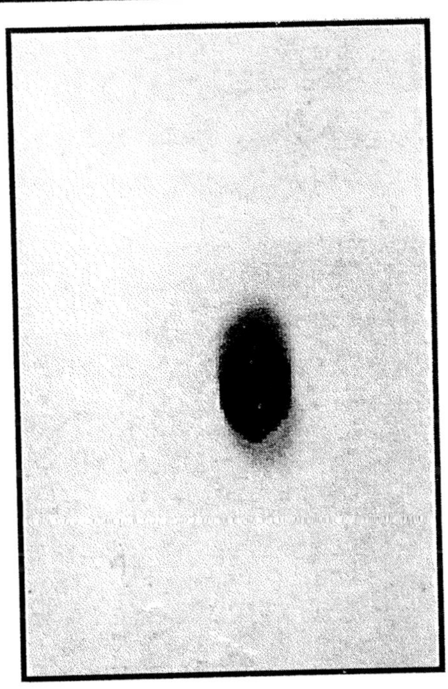

Plate 9.5: *Cryptolaemus montrouzieri*

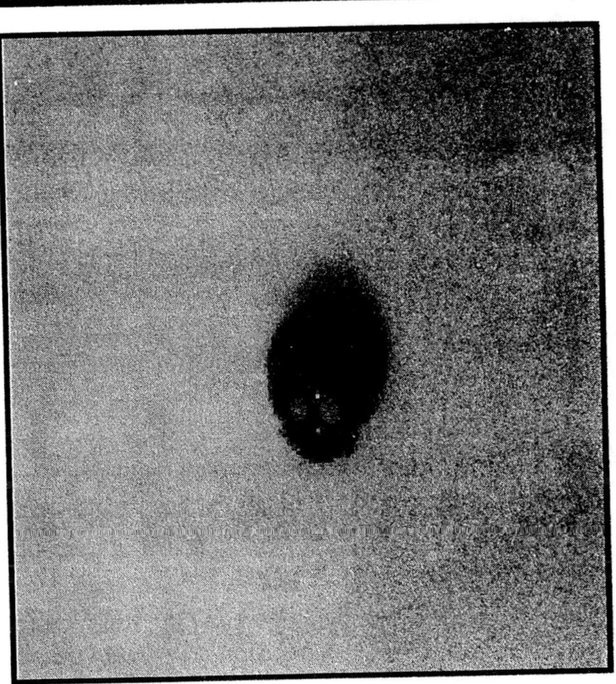

Plate 9.6: *Propylea dissecta*

Plate 9.7: *Trichogramma* sp.

Guidelines for Safe Use of Insecticides to Natural Enemies

For minimum harm to the natural enemies, while management of insect pests of crops, following measures (Figure 9.1) should be adopted:

1. Mechanical, physical, cultural, biological and varietal means of pest management should be utilized to their optimum extent so as need for insecticide application is minimized.

2. Use of insecticides should be minimum and need based. For this, relatively safe and selective insecticides should be used, taking into account economic injury level (ETL) of the pest. Application of insecticide should be at proper time with proper dosage and suitable appliance.
3. At a time, insecticide application should not be done in large area. This will help in fast build up of the natural enemies, in the area where they were killed as a result of insecticide application.
4. Safer formulations like granular ones may be encouraged for use.

References

Abbot, W.S. (1925). A method of computing the effectiveness of an insecticide. *J. Econ. Entomol.*, 18: 265-267.

Awchar, S.L., Satpute, U.S., Sarnaik, D.N. and Sarode, D.B. (1995). Effect of certain chemical and botanical insecticides on some beneficial insects associated with mustard. *J. Biol. Control* 9 (1): 13–15.

Balikai, R.A. and Lingappa, S. (2003). Effect of insecticides on aphid predators and sorghum plants. *Ann. Pl. Protec. Sci.* 11 (2): 371–373.

Beevi, S.P. and Balasubramanian, M. (1992). Relative safety of buprofezin and insecticides to *Encarsia shafeei* a parasite of cotton whitefly *Bemisia tabaci* Genn. *J. Biol. Control* 6 (2): 59–63.

Chand, P., Kashyap, N.P. and Sharma D.C. (2001). Laboratory evaluation of toxicity of commercially used insecticides against egg parasitoid, *Trichogramma chilonis* Ishii (Hymenoptera: Trichogrammatidae). *J. Biol. Control* 15 (1): 39–42.

Chandrababu, A., Gautam, R.D. and Garg, A.K. (1997). Residual toxicity of some insecticides against the predatory beetle, *Brumoides suturalis* (Fabricius). *Ann. Plant Protec. Sci.* 5 (2): 131–136.

Croft, B.A. (1990). *Arthropod biological control agents and pesticides.* John Wiley and Sons, New York. 723p.

Deepak, K.D. and Chaudhary, A.K. (1998). Toxicity of some insecticides to *Tetrastichus pyrillae* Crowford (Hymenoptera: Eulophidae), an egg parasite of sugarcane leaf hopper, *Pyrilla perpusilla* Walker. *Ann. Pl. Protec. Sci.* 6 (2): 214–216.

Dhingra, Swaran; Murugesan, K. and Sridevi, D. (1995). Insecticidal safety limits for the coccinellid, *Menochilus sexmaculatus* Fabricius predating on different aphid species. *J. Ent. Res.* 19 (1): 43–47.

Easwaramoorthy, S., David, H. and Bai, K.S. (1994). Relative toxicity of certain insecticides to *Adelencyrtus mayurai* (Subba Rao), a parasitoid of sugarcane scale insect, *Melanaspis glomerata* (Green). *J. Biol. Control* 8 (1): 14–17.

Finney, D.J. (1952). *Probit analysis.* Cambridge Univ. Press, 318p.

Gaffar, S.A. and Kushwaha, K.S. (1994). Synergistic interaction of *Bacillus thuringiensis* Berliner with some insecticides against the tobacco caterpillar *Spodoptera litura* Farbricus on cauliflower. *J. Biol. Control* 8 (2): 98–101

Ghelani, Y.H., Vyas, H.N. and Jhala, R.C. (2000). Toxicological impact of certain insecticides to *Rhinocoris fuscipes* Fabricius (Heteroptera: Reduviidae): A potential predator. *J. Appl. Zool. Res.* 11 (1): 54–57.

Haseeb. M and Murad, H. (1997). Pathogenecity of the entomogenous fungus *Beauveria bassiana* (Bals.) Vuill. to insect predators. *Int. Pest Control* 40 (2): 50–51.

Hassan, S.A. Bigler, F., Blaisinger, P., Bongenschutz, H., Brun, J.,Chiverton, P., Dickler,E., Easterbrook, M.A., Edwards, P.J., Englert, W.D., Firth, S.I., Huang, P. Inglesfield, C., Lingauf, F., Kuhner, C., Ledieu, M.S.Naton, E., Oomen, P.A., Overmeer, W.P.J., Plvoets,P., Reboulet,J.N. Rickmann, W., Somesol–Petersen,L., Shires, W.S., Staublim, A., KStevenson, J. Tuset, J.J., Vanwtscoinkel, G.and Van Zon A.Q. (1985). Standard methods to test the side effect of pesticides on natural enemies of insects and mites developed by the IOBC/WPRS Working Group Pesticides and beneficial organisms. *OEPP/EPPO Bulletin* 15: 214–255.

Jhansi Lakshmi, V., Katti, G. and Krishnaiah, N.V. (1997). Safety of neem formulations and insecticides to *Microvelia douglasi atrolineata* Bergroth (Heteroptera: Veliidae), a predator of plant hoppers in rice ecosystem. *J. Biol. Control* 11: 33–36.

Jhansi Lakshmi, V., Katti, G., Krishnaiah, N.V. and Lingaiah, T. (1997). Laboratory evaluation of commercial neem formualtions *vis–a–vis* insecticides against egg parasitoid, *Trichogramma japonicum* Ashmead (Hymenoptera: Trichogrammatidae). *J. Biol. Control.* 11:29–32.

Jhansi Lakshmi, V., Katti, G., Krishnaiah, N.V. and Kumar, K.M. (1998). Safety of neem formulations *vis–a–vis* insecticides to *Cyrtorhinus lividipennis* Reuter, a predator of brown plant hoppers *Nilaparvata lugens* (Stal.) in rice crop. *J. Biol. Control* 12(2):119–122.

Jhansi Lakshmi, V., Krishnaiah, N.V., Pasalu, I.C., Lingaiah, T. and Krishnaiah, K. (2001). Safety of some insecticides to brown mired bug, *Tytthus parviceps* (Reut.) (Hemiptera: Miridae), a predator of brown plant hopper, *Nilaparvata lugens* (Stal) (Homoptera: Delphacidae) in rice. *J. Biol. Control* 15 (1): 43–48.

Jhansi Lakshmi, V., Krishnaiah, N.V., Pasalu, I.C. Lingaiah, T. and Krishnaiah, K. (2001). Safety of thiamethoxam to *Cyrtorhinus lividipennis* Reuter Hemiptera: Miridae), a predator of brown plant hopper, *Nilaparvata lugens* (Stal) in rice. *J. Biol. Control.* 15 (1): 53–58.

Jhansi Lakshmi, V., Pasalu, I.C., Krishnaiah, N.V. and Lingaiah, T. (2003). Safety of combination products and single compound insecticides to *Microvelia douglasi atrolineata* Bergroth, a predator of plant hoppers in rice. *J. Biol. Control.* 17 (2): 121–124.

Jayaraj, S. Sathiah, N. and Sundarababu, P.C. (1992). Biopesticides research in India: The present and future status. In: *Pest Management and pesticides: Indian Scenario.* pp. 144–157 (David, B.V., ed.), Namrutha Pub., Madras.

Krishnaiah, N.V., Jhansi Lakshmi, V., Pasalu, I.C., Lingaiah, T. and Krishnaiah, K. (2001). Relative safety of some insecticides to *Microvelia douglasi atrolineata* Bergroth, an aquatic predator of hoppers in the rice ecosystem. *J. Biol. Control* 15 (1): 49–52.

Kumar, M.G.and Velusamy, R. (1996). Safety of insecticides to spiders in rice fields. *Madras Agril. J.* 83(6): 371–375

Mani, M. and Thontadarya, T.S. (1988). Response of *Cryptolaemus montrouzieri* Muls. (Coccinellidae: Coleoptera) to commonly used pesticides in vineyards. *J. Biol. Control* 2 (1): 17–20.

Manisegaran, S. and Kumaraswamy, T. (1988). Assessment of contact toxicity of insecticides to *Chelonus blackburni* Cam. (Hymenoptera: Braconidae). *J. Biol. Control* 2 (2): 126–127.

Patel, M.C. and Vyas, R.V. (1999). Compatibility of *Bacillus thuringiensis* var. *kurstaki* formulation Cutlass with cypermethrin against *Earias vitella* and *Spodoptera litura. Ann. Pl. Protec. Sci.* 7(2): 336–339.

Paul, A.V.N. and Thyagarajan, K.S. (1992). Toxicity of pesticides to natural enemies of crop pests in India. pp. 158–176. *In: Pest Management and Pesticides: Indian Scenario* (David, B.V. ed.) Numrutha Pub., Madras.

Radhika, P. (1996). Effect of diflubenzuron on larval parasitoid, *Goniozus nephantidis* M. under laboratory conditions. *J. Appl. Zool. Res.* 7 (2):138–139.

Radhika, P. and Chitra, K.C. (1995). Effect of botanicals with conventional insecticides on a larval parasitoid under laboratory conditions. *J. Appl. Zool. Res.* 6 (2): 105–108.

Sharma, D.C. and Kashyap, N.P. (2002). Impact of pesticidal spray on seasonal availability of natural predators and parasitoids in the tea ecosystem. *J. Biol. Control* 16 (1): 31–35.

Sheikh, A.G. and Kushwaha, K.S. (1987). Relative contact toxicity of four common insecticides to *Apanteles* sp. (*vitripennis* sp. group) and its host *Spodoptera litura* (F.). *J. Biol. Control* 1 (1): 16–19.

Singh, S.P. (1995). Compatible natural enemies and synthtic pesticides for use in integrated pest management in India. *Pesticide Res. J.* 7 (1): 8–28.

Singh, O.P. and Rawat, R.R. (1981). Note on safety of some insecticides for *Diaeretiella rapae* McIntosh. a parasite of mustard aphid. *Indian J. Agric. Res.* 54: 204–205.

Solayappan, A.R., Radham, M., Sujatha, T. and Jesudasan, R.W.A. (2001). Toxicity of biopesticides to *Trichogramma chilonis* Ishii and *Chrysopa scelestis* Banks, the natural enemies of sugarcane pests. *Indian J. Plant Prot.* 29 (1&2): 85–87.

Srinivasan, G. and Sundara Babu, P.C. (2000). Toxicity of certain insecticides to predatory green lacewing, *Chrysoperla carnea* Stephens (Neuroptera: Chrysopidae). *J. Biol. Control* 14 (1): 5–7.

Varma, G.C., Shenhmar, M. and Brar, K.S. (1988). Studies of the effect of insecticides on *Trichogramma achaeae* Nagaraja and Nagarkatti (Hymenoptera: Trichogrammatidae). *J. Biol. Control* 2 (1): 5–8.

Zhu, F.X., Wang, J.X. Liu, F., Yu, J.F., Mu, W., Zhang, X and Mu, L. (1997). Evaluation of insecticide susceptibility I. *Propylaea japonica* (Thunb.). *Scientia Agricultura Sinica*, 30(6): 78–80.

Zhu, F.X., Waing, J.X., Liu, F., Mu, W.and Zhang, X. (1998). Studies on insecticide susceptibility of ladybirds. *Acta Entomologica Sinica* 41(4): 359–365.

— Part II —

PLANT PATHOLOGY

10

Karnal Bunt Disease of Wheat: A Threat to Grain Trade

Anuja Gupta and V.K. Maheshwari

Indian Agricultural Research Institute, Regional Station, Karnal – 132 001, Haryana

Karnal bunt disease also known as kernel smut or partial bunt is caused by *Tilletia indica* Mitra, Syn *Neovossia indica* (Mitra) Mundkur. It is an important disease of wheat adversely affecting the yield and trade. In India, Karnal bunt of wheat was first reported from Karnal by Mitra in 1931. Karnal bunt is recorded from all wheat growing states of India maximum being in Punjab, Haryana, HP and Uttar Pradesh. The disease is particularly prevalent in the tarai and high fertility irrigated areas of the Northwest plains of India. The disease is also reported from Afghanistan, Iraq, Mexico, Sweden, Syria, Turkey, Russia and Nepal.

Green revolution during the late 1960's has revolutionized the production of wheat, increasing it manifold from around 12 million tonnes in 1964–65 to 70 million tonnes in 2000–01. The introduction of semi-dwarf varieties having uniform flowering were exploited for high yields through excessive use of agricultural inputs as nitrogenous fertilizers, irrigation water and pesticides, which helped in attaining a quantum jump in wheat production. Prior to 1970's Karnal bunt disease occurred sporadically and remained confined to a few pockets only. But in early 1970's, as a consequence of changed cropping systems and enhanced inputs, it assumed epiphytotic proportions, causing substantial losses to both quality and quantity of wheat. The problem of Karnal bunt thus got re-established in Northern States and caused losses between 0.2 to 0.5 per cent of total production (Munjal, 1975; Joshi *et al.*, 1983) which increased to 20 per cent in the year of epidemic occurrence (Shukla *et al.*, 1982; Singh, 1986). Freshly collected infected grains emit a foul smell, like rotten fish, due to the presence of trimethylamine, and the wheat products from severely infected grains are unpalatable (Singh and Bedi, 1985). The disease apart from causing direct quantitative and qualitative losses in terms of yield and seed vigour is also responsible for rejection of grains in international trade. Flour made from bunted grains is discoloured and has an unpleasant odour and taste. Even one to four per cent Karnal bunt infection renders the grains unfit for human consumption (Mehdi *et al.*, 1973). An importing country insists on zero tolerance levels for fear of introducing the disease in areas where the disease is not known to occur. Strict quarantine regulations have been imposed on the import of bunted grain. In 1996, the Governments of Turkey, Morocco and Poland detained some export consignments of wheat from India. A cargo worth US $ 5 million was not allowed to be unloaded in Poland due to Karnal Bunt infection. Thus, the disease greatly affects the potential export markets for India. All round increase in wheat yield, comfortable food stocks and freedom from blemishes like black tip and Karnal bunt may facilitate greater grain trade in India (Nagarajan, 2000) of both bread and durum wheats and their value added products as well.

Detection of Karnal Bunt (KB)

The incidence of KB disease is difficult to record in standing wheat field since the disease does not produce symptoms, which are easily visible on the plant. However, critical examination of mature ear heads in the field could lead to detection of spikelets with wide-open glumes having bunted grain (Figure 10.1). This is possible where high incidence of bunt takes place. In a field, earheads and grains are infected at random *i.e.*, neither all the earheads nor all the grains in a spike get infected. Once the pathogen infects the spikelet, the mycelium moves within the earhead infecting the adjoining spikelets and produce partially infected grains (Dhaliwal *et al.*, 1983).

The infection of *T. indica* in wheat grains is visible by naked eye. As the infection proceed, the affected portion of wheat grain is transformed into teliospores of the fungus and looks black powder on examination. Severely damaged grains (Figure 10.2) are identified very easily since the tissues along the suture and adjacent endosperm are replaced with spores but minute infection on the germ tip may be confused with black tip symptoms. The infection of KB can also be detected by seed wash tests or sodium hydroxide seed soak technique (Agarwal and Verma, 1983). In this technique, the wheat seeds are soaked overnight in 0.2 per cent solution of NaOH. Next morning the seeds are taken out of solution and spread on a blotter sheet to remove excess moisture. On examination, the KB infected seeds/portions of seeds give appearance of shiny jet-black colour. The infection is confirmed as stream of jet-black spores ooze out when bunted seed are pricked under a microscope. The seeds discoloured due to other reasons turn brown to dull black in colour. The detection of KB can also be done effectively by visual observation that does not involve any cost and little time.

The Pathogen

The causal organism, *Tilletia indica*, infects the ovary and converts the grain into a sorus of dark colour. The black powdery mass inside the seed is the teliospores. These are oval to ellipsoidal to spherical in shape and olive brown to dark brown in colour having reticulations on the epispores with covered spines (Figure 10.3). The yellow coloured sterile cells are mixed with mature teliospores. On germination teliospores produce promycelium that bear secondary allantoid sporidia, which multiply and initiate infection in the ear head.

Development of KB disease

The pathogen is seed, soil and air-borne. The soil borne inoculum is the primary source for annual recurrence of the disease. During harvesting/threshing/winnowing the KB infected grains

Figure 10.2: Healthy and Infected Wheat Seed

Figure 10.1: Bunt Affected Wheat Spike

Figure 10.3: Teliospores and Sterile Cells of *Tilletia indica*

release teliospores that contaminate both soil and seed. The teliospore density reported from Punjab fields in India range from 5×10^3 to 16×10^3 per 250g of soil. These teliospores have dormancy and retain their viability for more than 18 months when buried in soil at a depth of 5cm (Nagarajan *et al.*, 1997). Nagarajan (1991) opined that the teliospores lose their viability at extremely low (below freezing) and high temperatures. Vivek *et al.* (1995) also demonstrated the effect of snowing and thawing and found that chilling reduced the viability of teliospores.

The teliospores lying at the soil surface germinate and produce promycelium bearing a whorl of about 65-185 primary sporidia (Joshi *et al.*, 1983). Smilanick *et al.* (1986) reported that the spores lying upto 2-mm depth of the soil were capable of germinating and releasing sporidia. Primary sporidial stage is short lived and is splash dispersed. These on germination would either form the mycelium mat or secondary sporidia (allantoid shaped). The secondary or micro sporidia may multiply and subsequently produce several crops of allantoid sporidia by repetitive multiplication or germinate to produce the mycelium. As most of the area under wheat cultivation in North-West India is irrigated, it provides conducive conditions for teliospore germination and sporidial production. Thus, teliospore germination is not a limitation but its liberation, dispersal and deposition onto the leaf surface is a limitation. As a result of sporidial multiplication, the atmosphere over the crop canopy gets changed with the inoculum (allantoid sporidia).

Primary infection occurs between heading and anthesis when the sporidia germinate on the glumes and the fungal hyphae enter the epidermal cells of the glumes to penetrate the ovaries directly (Munjal and Chatrath, 1976). The microconidia get deposited on the lowermost leaves by wind and splash (Prescott, 1986). The maximum trapping of sporidia occurs in air during early morning hours when 100 per cent RH prevails. The microsporidia multiply to produce enormous amounts and travel upwards to higher leaves through monkey jumps and reach flag leaf by splashing. If at the time of earhead emergence there is mild rain these sporidia get washed down into the sheath (Nagarajan, 1998) and infect the emerging spikelets. If the weather remains favourable more crops of microsporidia are produced causing more severe KB infection. Heterothallism has been reported in this pathogen. During this process, fusion between compatible allantoid sporidia occurs, resulting in infective propagules, since the sporidia cause infection only after fusion (Duran and Cromarty, 1977). The disease progresses systemically to other florets within the spikelet initially infected and then to adjacent spikelets, including those on the alternate side of the rachis.

The maximum sporidial release occurred in second fortnight of February when the crop is at pre-anthesis stage, which is the most susceptible stage and weather conditions are most favourable. The process of infection in the field depends very much on weather conditions and host susceptibility. Host resistance influences the systemic movement of the pathogen and further spread is arrested around early dough stage when hardening of the grain takes place. The post anthesis weather condition for atleast three weeks are very important in deciding the extent of bunt damage. Climatic conditions play an important role in the sporidial multiplication, dispersal and lodging of the sporidia on the host surface. Aujla (1995) reported that soil temperature favoured sporidial production and last week of February was most conducive, as peak sporidial release occurred in 8 days when minimum and maximum soil temperature ranged between 11 to 18°C (av. 13.7°C) at PAU Ludhiana.

Dhaliwal and Bains (1992) reported that period of high humidity and somewhat cooler air temperatures favoured the release of sporidia. Relative humidity and temperature greatly affected the sporidial viability, germination and multiplication. Sporidia produced at low humidity levels (80 per cent) recorded low germination. High temperature up to 30°C at 100 per cent relative humidity enhanced sporidial germination. Aerobiological studies revealed maximum catch of the spore around 2AM in the midnight when 95–100 per cent RH prevails. The allantoid spores retain satisfactory level of viability up to 10 hours. They germinate and cause complete infection between wide temperature range of 10–30°C and in 3–4 hours if there is leaf wetness.

Thus, the disease cycle (Figure 10.4) can be summarized as:

1. Under ideal weather conditions the allantoid spores multiply on the leaf surface, and increase the inoculum potential substantially.
2. The spores multiply and serially jump from lower leaves to up till they reach the boot leaf.

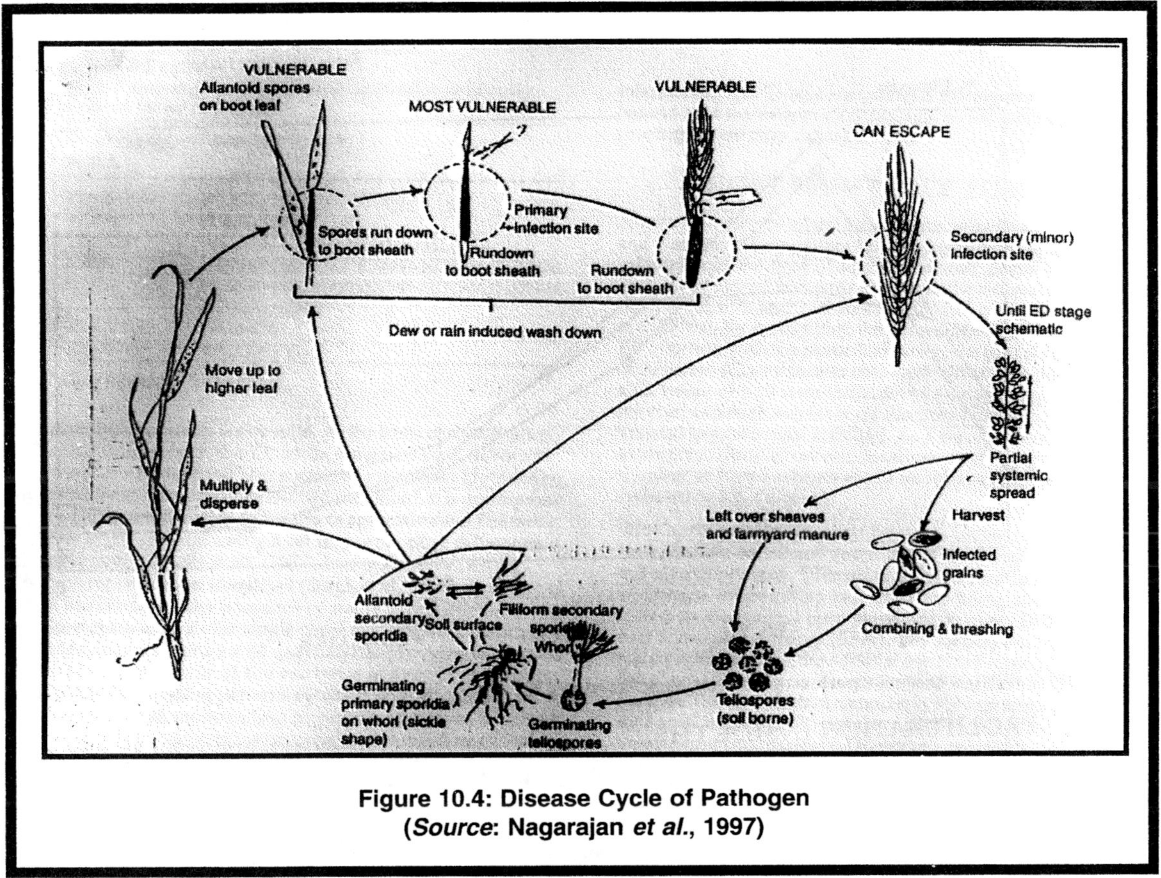

Figure 10.4: Disease Cycle of Pathogen
(*Source*: Nagarajan *et al.*, 1997)

3. From the boot leaf the allantoid spores run down into the boot sheath, when the ear head is just emerging. Frequent drizzle or heavy dew or both favour such run down into the sheath.
4. The number of primary infection sites are proportional to the number of run downs and infection happens before complete emergence.
5. Infection at anthesis time is minimal, if at all, but congenial weather through early dough stage permits the pathogen to become systemic inside the rachis and hence results in severe infection.
6. 2–3 weeks of time is needed from teliospore germination stage to the reaching of allantoid spores into the boot leaf sheath.
7. The period from flowering to early dough stage (2–3weeks) is critical for the systemic invasion of the rachis and delaying of kernel maturity by few days.
8. In a situation when both conditions coincide disease epidemic can occur. Hence, for effective prediction of the disease these two conditions are considered the most crucial as other stages in the life cycle do not pose any real constraint.

Factors Affecting Disease Development

The time of sowing influenced the infection of KB disease in wheat crop (Singh, 1980). The incidence of Karnal bunt was 0.20, 2.04 and 6.00 per cent in crop sown on 15th November, 30th November and 15th December, respectively (Table 10.1).

Table 10.1: Effect of Sowing Time and Weather on the Appearance of Karnal Bunt Disease at Karnal.

Date of Sowing	*Karnal Bunt (per cent)*	*Temperature (°C)*		*Rainfall (mm)*
		Maximum	*Minimum*	
15th November	0.20	19.9	7.3	31.8
30th November	2.04	19.8	8.2	15.4
15th December	6.00	22.9	11.4	149.2

The wheat crop sown between 25th November to 15th December at Ludhiana had more disease as compared to early or late sown crop (Gill *et al.*, 1993). The incidence of the disease also varied with the varieties. Continuous increase in the disease has also been reported with increase in nitrogen levels up to 180 Kg. per hectare and under irrigated and rainfed conditions (Aujla *et al.*, 1981).

The analysis of different grades of infection of KB (Singh, 1980) on the seed of wheat cultivar CC 464 revealed that 33.17 per cent loss in weight was caused by grade 4 infection, and with increase in disease intensity on the seed there was a gradual decrease in seed weight, seedling length and germination percentage (Table 10.2).

Table 10.2: Effect of Karnal bunt Infection on Seed Weight, Seedling Length and Germination

*Grade**	*1000 Seed Weight (g)*	*Per cent Loss in Seed Weight Over Healthy Seeds*	*Per cent Loss in Seedling Length Over Healthy*	*Per cent Germination*
0	40.7	—	—	82.00
1	40.5	0.49	6.74	73.7
2	36.3	10.81	12.06	71.50
3	33.0	18.92	26.21	69.25
4	27.2	33.17	33.89	44.00

* Grade 0= no infection; 1= infection at germ tip; 2= infection advancing from germ to 25–30 per cent length of groove; 3= ventral scutellum completely damaged, entire length of groove and both sides of endosperm tissues transformed into smut sori; 4= entire ventral surface along the groove transformed into black mass which is normally found detached thus making the seed hollow, dorsal surface remains unaffected.

Beniwal *et al.* (2000) and Bansal *et al.* (1984) also reported reduction in 1000 grain weight ranging from 4.5 to 52.3 percent depending on severity of infection. Decrease in disease severity resulted in proportional decrease in seed weight, germination and vigour.

Status of Disease in Wheat Varieties Grown at IARI-Regional Station, Karnal

During 1980's, the prevalence of KB disease in wheat varieties like HD- 2009 and CC 464 was quite high and ranged from 1.72 to 5.20 per cent at this station. But with the passage of time the infection got reduced (Table 10.3). The meteorological data (Table 10.4) shows that during February

1996, the temperature (11.5°C–24.3°C) and humidity (57–94 per cent) were congenial for the development and spread of the disease in the fourth week of February and first week of March. This is also the critical period for the development and spread of disease, as wheat crop at this stage is highly prone to infection. The total rainfall during this period was 40mm and the soil temperature (13.4–27.1°C) was also conducive for the germination of teliospores and a higher incidence of the disease was recorded on wheat seeds of different varieties harvested during Rabi 1995-96, at this station. Though during 1998, 109mm of rainfall was recorded during February (3rd and 4th week) and March (1 and 2 week), the humidity was less than 90 per cent and so probably low incidence of disease was recorded during the year. But during the same weeks in 1999, more than 90 per cent humidity was recorded which probably led to 0.13 and 0.15 per cent disease incidence in HD 2329 and HS 295 varieties, respectively. In other varieties including HD 2009, a susceptible variety, the disease however remained absent. It has already been highlighted (Gill *et al.*, 1993) that it is not the amount of rainfall but the distribution of rainfall, which accounts for the KB outbreaks. More rainfall occurring for a short period may not be of much significance for the disease, but intermittent rainfall (may be very less) uniformly distributed for longer time from boot stage to spike emergence stage enhances the probability of disease occurrence. In subsequent years the weather conditions were not conducive for the spread of the disease and hence a downward trend was observed. Warham and Flores (1988) also opined that weather fluctuations at anthesis were most important for development of the disease. Probably the threshold level of the pathogen may have got reduced in the seed and soil, resulting in decrease in the disease incidence. Sharma and Nanda (2000) also observed a decline in KB disease in wheat in Punjab area.

Table 10.3: Status of Karnal Bunt Disease in Different Varieties of Wheat at IARI Regional Station, Karnal from Rabi 1995 to 2001

Varieties	*Per cent Karnal Bunt Infection*						
	1995-96	*1996-97*	*1997-98*	*1998-99*	*1999-00*	*2000-01*	*2001-02*
HD- 2189		0.25	0.00	0.00	0.00	0.00	0.00
HD- 2329		0.30	0.00	0.09	0.13	0.00	0.00
HD- 2285		0.00	—	0.00	0.00	0.00	0.00
HD- 2009		0.30	—	0.00	0.00	0.00	0.00
Kundan		0.35	0.00	0.00	—	0.00	—
HD- 2687		Not Done	—	—	0.00	0.00	0.00
HD-2643		—	0.00	—	0.00	0.00	0.00
HD- 2428		0.45	0.08	0.00	—	—	0.00
HS- 295		—	0.00	0.00	0.15	—	0.00
HS- 240		0.50	0.00	0.00	0.00	0.03	—
HDR- 77		0.10	0.00	0.00	—	—	0.00
HS- 365		—	0.00	—	—	0.00	0.00
HD- 2402		—	—	—	—	—	0.00
HD- 2380		—	—	—	—	—	0.00
Vaishali (DL-784)		—	—	—	—	—	0.00
HD- 2733		—	—	—	—	—	0.00

Permissible Karnal bunt infected seed by number: Certified Seed = 0.250 per cent; Foundation Seed = 0.050%

Table 10.4: Meteorological Conditions During Rabi 1995-2001

Year	*Month*	*Week*	*Temperature (°C)*		*Relative Humidity (%)*		*Soil Temperature (°C)*		*Sunshine Hours*	*Rainfall (mm)*	*No.of Rainy Days*
			Min.	*Max.*	*Min.*	*Max.*	*Min.*	*Max.*			
1995–96	Feb.	III	8.4	22.5	49	89	11.4	24.2	8.5	00	0
		IV	11.0	22.3	64	91	13.4	24.0	6.7	40	3
	March	I	12.0	26.3	57	94	14.9	27.1	7.8	00	0
		II	14.8	28.6	50	86	18.2	31.1	8.0	1.8	1
1996–97	Feb.	III	5.2	22.3	37	83	10.2	24.7	8.9	00	0
		IV	9.7	24.4	51	80	12.9	26.8	10.1	00	0
	March	I	10.0	26.3	42	83	15.1	29.3	7.6	00	0
		II	13.7	24.2	56	77	17.4	29.4	5.3	4.0	2
1997–98	Feb.	III	12.4	23.7	50	87	13.7	26.4	7.6	Drizzle	2
		IV	9.0	20.4	65	87	12.4	22.0	6.2	22.0	2
	March	I	10.1	21.7	60	92	12.1	22.6	6.3	47.4	3
		II	9.6	24.4	51	83	13.8	25.1	9.4	24.2	2
1998–99	Feb.	III	10.2	24.5	50	93	13.8	24.6	7.8	00	0
		IV	9.1	23.9	47	90	12.1	26.9	7.5	00	0
	March	I	11.3	27.7	48	89	15.1	31.0	8.4	00	0
		II	10.4	27.4	39	85	14.7	32.5	9.9	0.8	1
1999–2000	Feb.	III	5.5	21.1	48	91	9.1	23.1	9.9	00	0
		IV	5.9	22.3	46	84	9.6	21.6	9.4	00	0
	March	I	8.7	25.9	37	85	13.5	24.9	8.2	1:2	1
		II	10.1	26.0	41	78	14.7	29.7	8.8	1.4	1
2000–01	Feb.	III	9.3	24.6	51	94	11.8	26.8	6.4	2.0	1
		IV	10.4	26.1	44	87	12.9	29.7	8.8	6.8	1
	March	I	7.7	26.0	40	88	12.8	31.7	9.1	00	0
		II	11.6	28.4	41	77	14.6	33.3	8.4	3.0	1
2001–02	Feb.	III	8.4	24.3	52.9	97.4	10.7	25.9	7.7	0.0	0
		IV	10.3	23.3	55.6	89.0	12.8	24.5	6.4	3.0	3
	March	I	10.8	23.0	57.0	93.1	13.4	26.0	6.9	1.1	2
		II	12.2	27.0	44.4	82.8	14.7	32.9	9.3	0.0	0

Management of Karnal Bunt

The climatic factors, crop management and the genetic vulnerability of the plant all play a role in the susceptibility of the crop to KB. For effective management of a disease, the information on the critical requirement of weather parameters, the nature of dissemination of the pathogen, life span of the pathogen in the soil/host etc. is quite essential. The best control is possible when the weak link of the pathogen in its life cycle is disturbed either through:

1. Limiting the infection causing entity
2. Altering cultural practices
3. Suppressing the pathogen in the field/on the seed by spraying and dressings with fungicides and
4. Growing resistant varieties.

Cultural Practices

Several cultural practices such as crop rotation, intercropping, fertilizers, deep ploughing, reduced irrigation, judicious use of nitrogen fertilizers, adjustment in sowing time, soil solarization have been recommended for control of KB (Singh *et al.*, 1988) Biological mulches like chickpea, intercropped with wheat reduced KB incidence by 60 per cent and covering the interrow space with transparent polyethylene sheets from the tillering stage onwards reduced disease incidence by 75 per cent (Singh *et al.*, 1991). The soil solarization with polyethylene mulching for 45 days at 0 to 10 cm depths resulted in very high inhibition of teliospore germination (94.00–98.87 per cent). At the depths of 0 to 5 cm, soil solarization for 30 days also resulted in high inhibition of teliospores (89.88–96.2 per cent). The polyethylene mulching increased soil surface temperature between 57.42 to 59.85°C compared to 45.28 to 48.28°C in non-mulched soil (Singh, 1994), thus restricting the liberation of primary sporidia from soil and reducing the disease severity.

Root exudates of soybean, pigeonpea, bajra and jowar enhanced teliospore germination and if grown preceding wheat, it would reduce the intensity of primary inoculum by forcing the teliospores to germinate before the vulnerable wheat crop. The root exudates of urdbean and KB tolerant wheat variety HD 2310 had inhibitory effect indicating the role of crop rotation in disease management (Singh, 1994).

In storage, the teliospores remain viable for 7 years and in soil they survive upto 42 months at 8 cm and 24 months at 34-cm depth. The teliospores can survive in wheat straw and farmyard manure upto 2 and 1 year, respectively (Bedi and Dhiman, 1982). One-year-old teliospores show high germination than the older spores.

Biological Control

Bio-agents like *Trichoderma viride, T.harzianum* and *Gliocladium deliquescence* have been reported to reduce the germination of teliospores. *T. viride* when applied to soil or ear heads, reduced the incidence of Karnal bunt infected grains from 1.63 per cent in control to 0.8 per cent and 0.65 per cent, respectively under natural condition. Under artificial conditions, the bioagent applied 48 hours prior to pathogen or simultaneously reduced the number of infected grains from 13.4 per cent to 9.2 and 6.4 per cent, respectively (Singh *et al.*, 1991).

Chemical Control

Seed Treatment

The fungicides *viz.* Blitox, Thiram, Emisan and Bavistin had only fungistatic action on seed containing teliospores just after application and 4 weeks of storage (Ahuja *et al.*, 1989). Slurry method of fungicide treatment gave better control of teliospore over dry seed treatment. Singh (1994) reported Bavistin and Brassicol as most toxic to teliospore germination without any adverse effect on wheat seed germination. G-696, Thiram, Captan and Dithane M-45 were also highly toxic to teliospore germination but treated seed could not be stored for long without adverse effects on wheat seed germination. Tilt and Vitavax were found to be poor seed dressers as the former was toxic to seed germination and the latter produced poor inhibition of teliospore germination. Rai and Singh (1979) found Bavistin, Dithane M-45 and Vitavax less effective against teliospore germination compared to Mercurials, Organic Tin compounds and Thiram.

Spray Treatment

The spraying of fungicides at heading stage has been found effective in controlling KB infection. Single spraying of Triadimefon and Carbendazim @ 0.2 per cent at heading stage controlled the disease upto 82–87.5 per cent (Krishna and Singh, 1982). Tilt (Propiconazole) @ 500 ml/ha gave 71.4, 91.0 and 100 per cent disease control in one, two and three sprays, respectively at Gurdaspur and Ludhiana during 1984–85 to 1986–87. Tilt has been found very effective in controlling Karnal bunt disease and enhancing 1000-grain weight (Rattan, 1988).

Systemic fungicides like Tilt, Folicur or San 619 have been found to be very effective in the control of KB. Amongst three fungicides Folicur @ 500 ml/ha gave 82 per cent control of the disease and yield of 64.69 q/ha (Gill *et al.*, 1993).

Soil Treatment

Methyl bromide reduced teliospore germinability by 98 per cent at 5 and 10 cms depths in wet soil (Smilanick *et al.*, 1987).

Host Resistance

Many cultivars and lines have been found with varying levels of resistance. Some KB free lines have also been developed (Sharma *et al.*, 2001). However, high yielding cultivars with high degree of KB resistance are not available. The currently grown cultivars tend to show moderate to high susceptibility to the disease. The varieties, HD 2329, HD 2009 and WL 711 have been shown to be highly susceptible to Karnal Bunt (Gill *et al.*, 1993; Beniwal *et al.*, 1997). Several wheat varieties with moderate levels of resistance have been identified and recommended for cultivation in KB–prone areas (Gill *et al.*, 1993; Dhaliwal and Singh, 1997). Some of the varieties of *Triticum aestivum* group such as WL 1562, PBW 154, WH 283,WH 291, HD 2285 and HD 2281 have field tolerance, while the durum wheats like PBW 34, PDW 215, PDW 233 and DWL 5023 exhibited resistance to the disease. Triticale cv. TL 1210 is resistant and could be introduced in Karnal bunt prone areas.

Seed Certification

The seed certification practice has really helped in containing the spread of some of destructive diseases. The Central Seed Certification Board has permitted a maximum of 0.05 and 0.25 per cent infected seed by number in foundation and certified seed, respectively (Tunwar and Singh, 1988). In a study at Karnal, a sample of 300g was found to be appropriate for studying Karnal bunt infection (Singh, 1981). Rai and Singh (1985) keeping ISTA practice in mind suggested that standards for Karnal bunt could be based on spore load of the grain lots even if the infection is low; seed lots may have less than 0.5 per cent infection but more than 25 spores per grain may be rejected in certification.

Even though the disease development is entirely dependant upon the environmental factors but certification standards cannot be under mind since the disease is seed borne and the pathogen remains active in soil for many years. Aerobiological studies by using aeroscopes in wheat fields for trapping of sporidia will also help in predicting the disease. Thus, if the disease can be predicted based on the prevailing weather conditions during the season, it can be managed from spreading and getting established in the seeds. The prevalence of infection in the field of course depends very much on weather conditions and host susceptibility.

Future Prospects

1. Use of clean and disease free healthy seed.
2. Extensive surveys to be conducted to demarcate disease free areas for disease free seed production.
3. Seed treatment though not very effective can be applied with Thiram or Duter 20WP or Indar 70LC @ 2.5g/Kg seed or Panogen 15 @ 0.7ml/Kg seed.
4. Foliar application either with 0.2 per cent mancozeb/carbendazim or 0.1 per cent propiconazole, applied at early heading stage is effective.
5. Avoid late planting and high plant density.
6. Avoid high rate of fertilizer application at the time of sowing and use split dose of fertilizer.
7. All the State seed testing laboratories of wheat growing areas should have trained personnel to test the seed for the disease incidence.
8. Farmers to be advised to use only certified seed and follow 3–5 year crop rotation in infested areas.
9. Cultivate seeds of resistant/tolerant wheat varieties in disease prone areas.

Conclusion

Karnal Bunt disease of wheat caused by *Tilletia indica* has assumed high concern, globally. The disease is seed, soil and air borne. The shift in the cropping systems, excessive use of fertilizers and irrigation and also the conducive environment especially in the months of February and March play a significant role in the prevalence of the disease at different times. The disease is responsible for both qualitative and quantitative losses in terms of yield, vigour, quality deterioration and toxicity. The affected seeds are partially or completely converted into black powdery mass. The disease has become a limiting factor in the export of wheat because the importing country insists on zero tolerance levels for the fear of introducing the disease. Management of disease poses problem due to the different modes of perpetuation and long survival of teliospores (3-5 years) in the soil and seed. Quarantine measures have been imposed to restrict the spread of the disease. Crop rotation for long periods, sowing of disease free seed, avoiding highly susceptible varieties in disease prone areas, judicious use of fertilizers, irrigation and sowing dates, fungicidal treatments and sprayings and the most important, breeding for disease resistance are the measures advocated to manage the disease effectively.

REFERENCES

Agarwal, V.K. and H.S. Verma (1983). A simple technique for the detection of Karnal bunt infection in wheat samples. *Seed Res.*, 11(1): 100–102.

Ahuja, S.S., S. Kaur and I. Sharma (1989). Chemical seed treatment for control of Karnal bunt of wheat. *Plant Prot. Bull.*, 31: 20–22.

Aujla, S.S. (1995). An overview of Karnal bunt of wheat. *34th All India Wheat Research Worker's Meeting* (ICAR) held at Univ. of Agric. Sci., Dharwad, Aug. 25–28, 1995, pp. 63.

Bansal, R., D.V. Singh, L.M. Joshi (1984). Effect of Karnal bunt pathogen (*Neovossia indica* (Mitra) Mundkur) on weight and viability of wheat seed. *Indian J. Agric. Sci.*, 54: 663–666.

Bedi, P.S. and J.S. Dhiman (1982). All about Karnal bunt of Wheat. *Seeds and Farms*, 6: 95.

Beniwal, M.S., J.C. Duhan, S.S. Karwasra, A. Gupta, M.L. Chhabra and M.S. Yadav (1997). *Crop Res.*, Hissar 14(1): 155–164.

Beniwal, M.S., P. Chawla, S. Madan and R. Singh (2000). Effect of Karnal bunt on grain weight and quality of wheat. *Annals of Biology*, Ludhiana, 16(1): 101–104.

Dhaliwal, H.S., A.S. Randhawa, K. Chand and D.V. Singh (1983). Primary infection and further development of Karnal bunt of wheat. *Indian J Agric. Sci.*, 53(4): 239–244.

Dhaliwal, H.S. and S.S. Bains (1992). Production of secondary sporidia of *N. indica* under different temperature and relative humidity regimes. *Indian J. Mycol. & Pl. Pathol.*, 22: 19–23.

Dhaliwal, H. S. and H. Singh (1997). Breeding for resistance to bunts and smuts: Indian Scenario. In: *Bunts and Smuts of Wheat: An International Symposium*, Research Triangle Park, North Carolina, USA, 17–20, August, 1997, pp. 327–347.

Duran, R. and R. Comarty (1977). *Tilletia indica*: A heterothallic wheat bunt fungus with multiple alleles controlling incompatibility. *Phytopath.*, 67: 812–815.

Gill, K.S., I. Sharma and S.S. Ahuja (1993). *Karnal Bunt and Wheat Production*. Book published at PAU press, Ludhiana.

International Seed Testing Association (1996). International Rules for Seed Testing: Rules 1996. *Seed Science and Tech.*, 21, Supplement.

Joshi, L.M., D.V. Singh, K.D. Srivastava and R.D. Wilcoxson (1983). Karnal bunt a minor disease that is now a threat to wheat. *The Botanical Review*, 49: 309–330.

Kumar, J. and S. Nagarajan (1998). Role of flag leaf and spike emergence stage on the incidence of Karnal bunt in wheat. *Plant Dis.*, 82(12): 1368–1370.

Krishna, A. and R.A. Singh (1982). Evaluation of fungicides for the control for Karnal bunt of wheat. *Pesticides*, 16(8): 7.

Mehdi, V., L.M. Worhi and Y.P. Abrol (1973). Studies on chapatti making quality. (vi) Effect of wheat grains with bunts on the quality of chapaties. *Bull. Grain Tech.*, 11(3): 195–197.

Munjal, R.L. (1975). Status of Karnal bunt (*Neovossia indica*) of wheat in northern India during 1968–1969 and 1969–1970. *Indian J. Mycology. Plant Path.*, 5: 185–187.

Munjal, R.L. and M.S. Chatrath (1976). Studies on mode of infection of *Neovossia indica* incitant of Karnal bunt of wheat. *J. Nuclear Agric. & Biol.*, 5: 40–41.

Nagarajan, S. (1991). Epidemiology of Karnal bunt of wheat incited by *Neovossia indica* and an attempt to develop a disease prediction system. Technical Report–Wheat Programme, CIMMYT, Mexico, 62 pp.

Nagarajan, S., S.S. Aujla, G.S. Nanda, I. Sharma, L.B. Goel, J. Kumar and D.V. Singh (1997). Karnal bunt (*Tilletia indica*) of wheat: A review. *Rev. Pl. Pathol.*, 76(12): 1209–1214.

Nagarajan, S. (2000). *Indian Wheat Newsletter*, DWR, Karnal 6(1–2): 1.

Rai, R.C. and A. Singh (1979). Effect of chemical seed treatment on seed borne teliospores of Karnal bunt of wheat. *Seed Res.*, 7: 186–189.

Rai, R.C. and A. Singh (1985). A view on seed certification standards for Karnal bunt of wheat. *Seed Res.*, 13: 94–98.

Sharma, I and G.S. Nanda (2000). Decline in Karnal bunt of Wheat: Disease status in Punjab (1997–2000). *Indian Wheat Newsletter*, 6(2): 7–8.

Skukla, D.N., N. Singh and S.N. Bhargava (1982). Wheat varieties affected by Karnal bunt. *Vijnana Parishad Anusandhan Patrika*, 25(1): 39–40.

Singh, A. (1994). Epidemiology and management of Karnal bunt disease of Wheat. Research Bulletin No.127, p. 132–140. Directorate of Experiment Station, G.B. Pant Univ. of Agric. and Tech., Pantnagar, India.

Singh, D. (1980). A note on the effect of Karnal bunt infection on the vigour of wheat seed. *Seed Res.*, 8: 81–82.

Singh, D. (1981). Studies on Karnal bunt of Wheat. 1. Standardisation of Techniques for seed sampling and minimum working sample. *Seed Res.*, 9: 76–80.

Singh, D.V. (1986). Bunts of wheat in India. In: *Problems and Progress of Wheat Pathology in South Asia*, (Ed.) L.M. Joshi, D.V. Singh and K.D. Srivastava and Published by Malhotra Publishing House, New Delhi, India, p.125–165.

Singh, P.P. and P.S. Bedi (1985). Effect of Karnal bunt infection on gluten characteristics and protein fractionation of wheat grains. *Ann. of Appl. Biol.*, 1(2): 223–225.

Singh, P., H.S. Dhaliwal, S.S. Aujla and S.K. Mann (1991). Studies on the biological control of Karnal bunt of wheat by *Trichoderma viride*. *Proc. of Symp. on Impact of Diversification in Agriculture on Plant Disease Problems and Annual Meeting of IPS* (Northern Chapter), 6 pp.

Smilanick, J.L, M. Dupler, B.J. Goates, J.S. Hoffman, D. Clark and Dobson (1986). Germination of teliospores of Karnal dwarf and common bunt fungi after ingestion by animals. *Plant Dis.*, 70: 242–244.

Smilanick, J.L., J.A. Hofmann, L.R. Secrest and K. Wiese (1987). Evaluation of Chemical and physical treatments to prevent germination of Tilletia indica teliospore. *Plant Dis.*, 72: 46–51.

Tunwar, N.S. and S.V. Singh (1988). *Indian Minimum Seed Certification Standards*. The Central Seed Certification Board, Dept. of Agriculture and Cooperation, Ministry of Agriculture, Govt. of India, New Delhi.

Vivek,S.S., D.V. Singh, K.D. Srivastava and R. Aggarwal (1995). Some epidemiological aspects of Karnal bunt of Wheat. *Indian Phytopath.*, 48; 419–426.

Warham, E.J. and D. Flores (1988). Farmer surveys on the relation of agronomic practices to Karnal bunt disease of Wheat in Yaqui valley, Mexico. *Tropical Pest Mgt.*, 34: 373–381.

Wilcoxson, R.D. and E.E. Saari (1996). *Bunt and Smut Diseases of Wheat: Concepts and Methods of Disease Management*, Mexico, D.F.: CIMMYT.

11

Status and Management of Foliar Blight of Wheat in India

A.K. Singh, R.P. Singh and Mamta Singh

Department of Plant Pathology, Narendra Deva University of Agriculture and Technology, Kumarganj, Faizabad – 224 229, U.P.

The foliar blight of wheat is generally considered as one of the complex disease which include leaf blights and blotches caused by species of *Alternaria* and *Helminthosporium,* respectively. This complex is taken as one group because in most of the cases it is rather difficult to distinguish one from other under field condition. In recent years, the pathogen has assumed major significance in North-West and Eastern part of the country (Tandon, 1985; Nagarajan and Singh, 1998). In addition to these two pathogens, few other pathogens of minor importance have also been reported from various parts of the country (Goel *et al.*, 1976; Goel and Gupta, 1979; Joshi *et al.*, 1970, 1978, Singh *et al.*, 2001a). A number of pathogens causing foliar blights, blotches and spots on wheat crop have been recorded in India. Among the pathogens causing blight on wheat, *Helminthosporium sativum* Pammel, King and Bakke (Syn. *Bipolaris sorokiniana* Sacc. et Sorokin) and *Alternaria triticina* Prasada and Prabhu are considered important (Joshi *et al.*, 1974; Misra, 1973; Singh *et al.*, 1986; Singh and Singh, 2003). It has been shown that *Helminthosporium sativum* (Syn. *B. sorokiniana*) is most important than *Alternaria* spp. and other fungal pathogens (Joshi *et al.*, 1974; Singh *et al.*, 1980; Singh and Srivastava, 1997).

Relative preponderance of different pathogens, including that of major pathogens *B. sorokiniana* and *A. triticina* has been changing over the period. A nation wide study for the period 1988–96 (Goel *et al.*, 1999) brought outB. *sorokiniana* as the most predominant pathogen in the states of Bihar, Delhi,

Gujrat, Haryana, Karnataka, Maharashtra, Rajasthan, Uttar Pradesh and West Bengal and *A. triticina* in the states of Himachal Pradesh, Punjab, J&K and Madhya Pradesh. However, analaysis of data for the period 1989–90 to 1998–99 (Singh *et al.*, 2000a) revealed uniform predominance of *B. sorokiniana* across the country including in the states where earlier (Goel *et al.*, 1999) predominant pathogens was *A. triticina*, relative frequency of occurrence of the former and later pathogens being 59.75 per cent and 37.17 per cent. Ratio of these pathogens for the periods 1990–2000 (Singh *et al.*, 2001b) further changed to 69.2 : 20.8 clearly establishing a consistent decline in the frequency of isolation of *A. triticina*. These findings were further reinforced by the results for the period 1994–99, which established the frequency for isolation of *B. sorokiniana* and *A. triticina* at 70.5 per cent and 18.90 per cent, thus bringing out *B. sorokiniana* induced foliar disease spot blotch as the most important component of the foliar blight complex (Singh *et al.*, 2001c).

Alternaria Species

Distribution of the Disease

Alternaria leaf blight on wheat was first reported authentically in India about 79 years ago in 1924 from Bombay, Presidency, Maharashtra (Kulkarni, 1924). It attacked all the aerial parts of the plant and considerably reduced the yield. In the same year Mc Rae, 1924 reported a similar disease associated with *Alternaria* from Pusa (Bihar). Subsequently it was reported in severe form on some *durum* wheats (Bansi, Bansipali 808 etc.) at Kanpur (Mehta, 1951). Later, Mathur (1956) felt that *Alternaria tenuis* was associated with the disease and visualized that it was seed borne and the presence of excessive moisture may cause seedling rot. Similar, observations were also reported for *durum* wheat from Cuttack (Orissa) and Andhra Pradesh (Bose, 1956; Govind Rao, 1961).

Most of the earlier accounts on *Alternaria* leaf spots/blights were mainly restricted to reporting the occurrence of disease but pathogencity of the organisms was not established. The disease received much more attention due to the failure of Kenphad wheat released by the Bombay States in 1952. Prasada and Prabhu (1962, 1964) established, on the basis of morphological, physiological and pathogenic characters, that the leaf blight of wheat was caused by a new species of *Alternaria* which they named as *Alternaria triticina*. Later, Sokhi *et al.* (1972) established that like *A. triticina, A. alternata (A. tenuis)* also can cause mild infection. Based on the taxonomic studies on different species of *Alternaria* causing leaf spots/blights of wheat in India, Kumar and Arya (1973 a,b) suggested that *A. tenuissima* may be the ecological types of *A. tenuis* itself failure of Kenphad wheat in the very first year of their release focused the attention to this disease. Since then the disease has been reported from many parts of Maharashtra, Gujrat, Bihar, Bengal and even Rajasthan (Singh *et al.*, 1986). Recently, Goel *et al.* (1999) reported, based on the observation on the distribution of two major foliar blight pathogen (*B. sorokinniana* and *A. triticina*), *A. triticina* predominates in the states of Himachal Pradesh, Punjab, J&K and Madhya Pradesh.

Pathogen

The genus *Alternaria* Nees ex-Fries was established in 1817 by Nees with single species, *A. tenuis* and since then a large number of species have been added to the genus and presently it comprises of about hundred species which parasitize a large number of cultivated and wild hosts. Prasada and Prabhu (1962, 1964) established that foliar blight of wheat was caused by *A. triticina*. Later, Sokhi *et al.* (1972) established that like *A. triticina, A. alternata (A. tenuis)* also can cause very mild infection. Important species of *Alternaria* pathogenic to wheat are *A. tenuis, A. tenuissima, A. triticola* and *A. triticina* (Kumar and Arya, 1973 a and b). Among these *A. triticina* is responsible for foliar blight of wheat.

Symptoms

The pathogen induces wide range of symptoms on wheat leaves under field conditions. Mc Rae (1933) mentioned that a species of *Alternaria* occurred every year on wheat at Pusa and produces symptoms characterized by brownish spots with a pale yellow ring on the leaves. In 1951, Mehta observed a leaf spot of wheat caused by *Alternaria* sp. in severe form at Kanpur. Eventhough, Mathur (1956) and Govinda Rao (1961) has suspected that leaf blight of wheat was caused by *Alternaria tenuis.* Prabhu and Prasada (1966) established that the pathogenicity of *Alternaria* isolated from leaf blight affected leaves and identified earlier as *A. triticina* producing localized leaf spots followed by blightening which is the most pronounced symptoms under field conditions on one month old wheat plants. Nema and Joshi, 1971 b; Edward *et al.*, 1972; Frisullo, 1982; Singh, 1983 and Singh *et al.*, 1986, observed similar symptoms produced by *A. triticina.*

In the field, *A. triticina* produced symptoms mainly on the leaves, but occasionally on node, stem, sheath and ear also. At initial stages of leaf blight symptom development was observed as small, oval, discoloured lesions. Sometimes small, dotted, chlorotic, oval or elliptical lesions appeared on leaves. The chlorotic borders of the lesions were diffused later turning light to dark brown in colour. Occasionally, the initial symptoms were light yellow to light brown colour, irregular or sub-oval in shape. A bright yellow marginal zone of varying dimensions was sometimes seen around the spots. In the leaves showing some advanced symptoms, the spots were variable in shape and size measuring 0.5–15 mm or more in length and 0.5–5 mm or more in width. Fully developed lesions were of brown to grey colour, sunken and sometimes with yellow margin. As the disease progressed several lesions coalesced to cover the large area and start drying from the tips, resulting in the death of the entire leaves. The first sign of the infection always appear on the lower leaves, gradually spreading to the upper leaves.

Losses

Losses due to *A. triticina* have been estimated by Chenulu and Singh, 1964; Chenulu *et al.*, 1970; Sokhi and Joshi, 1974; Prabhu and Singh, 1974; Ram and Joshi, 1978a. Foliar blight of wheat affected yield very adversely in late sown crops of Eastern Uttar Pradesh. Disease loss figures at farmer's fields are: UP 262 suffered 29.72 per cent loss in yield and 29.01 per cent loss in grain weight with 74.28 per cent disease index, Sonalika showed 35.22 per cent loss in yield and 34.38 per cent loss in grain weight and 94.28 per cent disease index, HD 2285 suffered 27.97 per cent loss in yield and 30.76 per cent loss in grain weight with 81.50 per cent disease index (Singh *et al.*, 1995a). Parashar *et al.* (1995) also reported losses upto 29 per cent in the yield.

Factors Affecting Disease Development

Temperature plays an important role during infection and incubation period. Maximum disease development take place at the temperatures between 20–25°C (Prabhu and Prasada, 1966), the optimum temperature for infections being 25°C and minimum 19°C. The disease development in Delhi is therefore slow in December and January due to the prevailing low temperatures but infection increases as the temperature rises above 20°C. It has also been observed that the disease appears when the plants are 6–7 weeks old and assumes severe form in February when they are sufficiently matured. The leaf wetness greater than or equal to 10 hours for four consecutive days following incubation, with the coincident average temperature of 10–20°C, caused disease to high blight intensities (prabhu and Prakash, 1973). Severity of the disease in the month of February could thus be attributed to the favourable temperature, humidity as well as maturity of the crop. Fertilizer application influenced the growth of the pathogens and development of the disease. Ojha and Mehta (1970) reported that susceptibility increases with increasing dose of nitrogen but decreases with increasing doses of phosphorus and potash. Ram and Joshi (1981) found that maximum disease intensity (43.2 per cent) was recorded when 75 Kg N ha^{-1} was applied in field, whereas only 35.6 per cent disease intensity was observed when no nitrogen was given. Rashid *et al.* (1988) reported that the interaction of three fertilizers significantly favoured on increase in *Alternaria* leaf blight. Singh *et al.* (1995c) reported that increasing high level of NPK and more irrigations favoured incidence of the disease.

Pathogenic Variability

Occurrence of variation in the genus *Alternaria* has been studied by several workers from time to time. In nature as well as in artificial culture new forms constantly arise by saltation. Sokhi and Joshi (1972) studied the physiologic specialization in eight isolates of *A. triticina* and concluded that at least two races exists. Later, Garg *et al.* (1972) recorded existence of six races but Jain and Prabhu (1974) did not find any appreciable morphological differences in the cultural characteristics of the few monosporial cultures. Singh *et al.* (2003a) reported that *A. triticina* isolates from distant places differed consistently with each other in cultural, morphological and pathogenic characters. Faizabad isolate was more virulent than others.

Disease Management

Varietal Resistance

A large number of wheat varieties have been screened for their resistance to the foliar blight disease of wheat caused by *Alternaria triticina* under field and glasshouse conditions by several workers Prabhu and Prasada (1966) Garg *et al.*, (1972); Dalela *et al.* (1971); Singh *et al.* (1977); Singh and Tyagi (1979); Ahmad and Singh (1983); Ahmad (1984); Casulli *et al.* (1986) and Das *et al.*, (1989). Singh *et al.* (1990) reported that out of 121 diverse genotypes of *Triticum aestivum* and *T. durum* evaluated for resistance to *Alternaria triticina* during 1986–88 and for resistance to *Puccinia recondita* during 1986–87 at 3 Bihar locations, 13 were resistant to both diseases and 42 were resistant to *P.recondita* only. They also reported that out of 252 cultivars of both species at 3 locations during rabi 1986–87, 106 were resistant to *P. recondita*. Of these, 24 were tolerant of *A. triticina* and 11 lines, including *7:aestivum* PBW 138 and HUW 284, also possessed desirable agronomic characteristics (Singh *et al.*, 1991). Shivankar and Lokhande (1993) reported that out of two hundred twenty five varieties/lines, *Alternaria* leaf blight was noticed in 142 entries where 43 entries showed resistant reaction, 81 moderately resistant and 18 susceptible. Singh *et al.* (2003b) screened 312 genotypes under artificial epiphytotics and found 43 as resistant ones.

Chemical

Fungicidal application of the disease has been attempted for the control of *Alternaria* blight by several workers (Singh and Singh 1971; Mishra *et al.*, 1970; Ram and Joshi, 1977; 1978b; Raut *et al.*, 1983). Mishra *et al.* (1989) reported that the most effective and economical treatment was 3 sprays of Manozeb at 10 days intervals, followed by 3 sprays at 15 days intervals. The percentage increase in yield following these treatments was more than double that of only 1 spray. Vats and Chauhan (1995) reported that Dithane M-45 (0.25 and 0.30 per cent), Tilt (0.025 and 0.05 per cent) and Topsin-M (0.05 and 0.10 per cent) against. *Alternaria* and *Helminthosporium* leaf blight of wheat under conditions of *in vitro* and *in vivo*. Tilt (500 ppm) provided significant control of both the pathogens at 24 and 48 h intervals of incubation while Tilt (0.05 per cent) provided significant control as foliar application in the field after three sprays at 10 days interval with a cast benefit ratio of 1:2.69. Recently, Kumar and

Parveen (2002) reported that the efficiency of combined treatment of *Trichoderma viride* with Vitavax as foliar spray in combating ravages of leaf blight of wheat.

Botanicals

From very early time plant pathologist following the lead given by men and medicine have evinced great interest in the use of plant products for control of plant diseases. In the recent past, this line has resulted in the voluminous records. Murthy and Sirsi (1957) Shah *et al.* (1959) Sehgal (1961); Sekhawat and Prasada (1971); Khan *et al.* (1974); Chalfound and Carvalho (1987); Rekha *et al.* (1989); Ali, *et al.* (1992); Kazmi *et al.* (1993); Meena and Mariappan (1993) and Raj and Kapoor (1993) summarized the status of our knowledge of those plant constituents which have antimicrobial and antiparasitic activity.

Vir and Sharma (1985) observed that the preparation from *Azadirachta indica* inhibited the growth of *Alternaria alternata* by 61.1 per cent at a concentration of 1 per cent and by 100 per cent at 10 per cent concentration. Singh *et al.* (1999b) tested six neem products *in vitro* at different concentrations against *A. triticina*. Among these nimin gave better response in respect to inhibiting the growth of*A.triticina* followed by neemgold and nimbicidine. Recently, Amaresh *et al.* (2000); Babu *et al.* (2001) and Jitendra and Majumdar (2001) also reported similar antifungal properties of neem against several fungi including *Alternaria* spp.

Helminthosporium Species

Distribution of the Disease

Helminthosporium blight was first recorded from Ussuria, Russia in 1890 by Sorokin (Drechsler, 1923). The pathogen was identified by Pammel, King and Bakke in 1910 from the material collected in U.S.A as *Helminthosporium sativum*. The first record of the disease in India on wheat leaves was made by Mohy in 1914 from Pusa, Bihar (H.C.I.O. Association No.12508). Later, Mc Rae (1930), Mitra (1930, 1934) gave a detailed description of the disease. They observed that leaf spots of wheat was mainly caused by *H.sativum*. The leaf blight/spot blotch prevalence in India (Joshi *et al.* 1978; Nema, 1986; Sharma *et al.*, 1996); Nepal (Karki, 1982; Dahal *et al.*, 1992; Karki and Karki, 1996, Sharma, 1996; Mahto, 1999a); Bangladesh (Alam *et al.* 1995); Pakistan (Bajwa, 1985; Bhatti and Ilyas, 1986; Hafiz, 1986); China (Zhimin *et al.*, 1998) and Bolivia (Toledo and Guzman, 1998) has also been mapped.

Spot blotch (leaf spot) is an important disease of wheat incited by *Helminthosporium sativum* Pammel, King and Bakke, (Syn. Bipolaris sorokiniana Sacc. Shoem.) teleomorph: *Cochliobolus sativus* (Ito and Krubayashi) Drechs, ex. Dastur, all over India. Presently the disease is prevalent almost all wheat growing areas of the country, particularly Bihar, Haryana, Himachal Pradesh, J&K, Tamil Nadu, Kamataka, Madhya Pradesh, Orissa, Kerla, Rajasthan, Punjab, Uttar Pradesh and West Bengal (Nema and Joshi, 1971a; Misra and Singh 1972; Narain *et al.*, 1973; Joshi *et al.* 1974; Bidari, 1974; Bidari and Govindu, 1975; Singh *et al.*, 1986; Duveiller and Gilchrist, 1994; Singh and Srivastava 1997; Singh *et al.*, 1998a). *H.sativum* is most frequently isolated and more destructive among several of *Helminthoporium* species (Chattopathyay and Chakarbarti, 1968, Singh *et al.*, 1980; 1997; 1999a; Singh *et al.*). Singh *et al.* (2001 a,b and c) reported that the *B.sorokiniana* and *A. triticina* are the two major pathogens responsible for foliar blight in India. Singh *et al.* (2002a) compared the existing inoculation techniques for spot blotch induction efficiency. These included application of *B. sorokiniana* inoculum on wheat variety HP 1633 in one of the following three forms, namely, spraying conidial suspension, broadcasting fungus cultures on sorghum grains and putting such grains in the leaf whorl at maximum tillering (MT), MT and boot leaf (BL) stages, MT, BL and 15 days after boot leaf (DABL) stages and MT and 15 DABL stages. Results established significant superiority of conidial spray method over putting pathogen impregnated sorghum grains in the leaf whorl and broadcast of such grains, in that order. Likewise, inoculation at all the three growth stages was found significantly superior over inoculations at MT and BL stages, MT and 15 DABL stages and at MT, irrespective of the procedures involved. Thus, foliar application of the conidial suspension of *B. sorokiniana* at all the above three stages, at 15 days intervals with 92.58 per cent disease intensity (PDI) was found to be significantly superior spot blotch induction exercise over all others.

Pathogen

The genus *Helminthosporium* was established by Link (1809). It was validated in 1821 by S.F. Gray's (Shoemaker, 1959). Nisikado (1928) divided it into two sub genera on the basis of shape and germination of conidia, *viz.*,

1. *Cylindro-Helminthosporium,* having straight cylindrical conidia that germinate by one or more germ tubes from any cell, represented by *H.tritici* and
2. *Eu-Helminthosporium* having fusoid or curved conidia that germinate only from the end cells, represented by *H. sativum* Ito (1930), later, changed the sub genus *Cylindro-Helminthosporium* to *Drechslera.* Luttrell (1951), however, divided *Helminthosporium* in three major groups:
   (a) *H. sigmoideum* group with conidia bome on sterigmata.
   (b) *H. gramineum* group with cylindrical conidia having slightly tapering ends. These are sessile and germinate from any or all cells.
   (c) *H. sativum* group with ellipsoidal to fusoid conidia germinating by polar germ tube. Shoemaker (1959) divided the gmminicolous *Helminthosporium* into two:
      (i) Species germinating from any or all the cells were included under *Drechslera.*
      (ii) Species having fusoid phragmespore with polar germination were included under *Bipolaris with B. maydis* as lactotype.

Subramanium and Jain (1966), however, reverted all the graminicolus *Helminthosporia* to genus *Drechlera.*

Current Taxonomic Status

Alcom (1988) reviewed the taxonomy of the graminicolous hyphomycetes of the genus, *Helminthosporium* and their teleomorphs and grouped them as under:

Bipolarif

Shoemaker (1959) established *Bipolaris* to accommodate fungi previously assigned to subgenus *Eu-Helminthosporium* (Nisikado 1928). These having fusoid, straight or curved conidia germinating by one germ tube from each end with *Bipolaris maydis* (Nisik and Miyake) Shoem. as lactotype. Texa such as *B.rostrata* (Drechsler) Shoem and *B .turcica* (pass.). Shoem. having protubemnt conidial hilum were included in the genus. These species had have acigerous states in Metasphaeria (Luttrell, 1957) (later called *Trichometasphaeria*), while other species of *Bipolaris* had non-protruding hila and the teliomorphs of some were in *Cochliobolus.*

Drechslera

Ito (1930) established to accommodate fungi previously assigned to the sub genus *Cylindro-Helminthosporium.* Members of this new genus were chamcterized as having "cylindrical" not curved conidia, germinating from every cell and associated with *"pyrenophora"*. *Drechslera tritici-vulgaris* (Nisik) Ito [= *D. tritici-repentis* (Died) Shoem] was designated as the lectotype by Hughes, (1958).

Exserohilum

Leonard and Suggs (1974) established the genus *exserohilum* for *Helminthosporium* species with strongly protuberant conidial hilum with *E. turcicum* (Pass). Leonard and Suggs as its lectotype. New ascomycete genus, *Setosphaeria* was proposed to accommodate its teleomorphs.

Differentation Between *Drechslera, Bipolaris* and *Exserohilum*

Shoemaker (1966) agreeing with Luttrell's (1963, 1964) additional criteria for differentiating *Drechslera* from *Bipolaris* added another, hilum morphology. In *Drechslera* the contour of the basal cell included a flat scar cell, while in *Bipolaris* hilum is slightly to strongly protuberant (Shoemaker, 1966) and in *Exserohilum* hilum is prominently exerted (Leonard and Suggs, 1974). Generic characteristics considered significant, subsequently were conidial shape, colour, germination behaviour, septum development and hilar form (Luttrell, 1977; 1978) and are listed in Table 11.1.

Table 11.1: Differential Characters *Bipolaris, Drechslera* and *Exserohilum* (Alcorn, 1988)

Character	*Bipolaris*	*Drechslelra*	*Exserohilum*
Conidial shape	Fusoid, obclavate fusoid, rarely truly cylindrical, straight or curved	Commonly cylindrical of obclavate cylindrical, mostly straight	Fusoid obclavate fusoid or cylindrical, straight or curved
Hilum	Often slightly protruding and truncate	Non-protruding, rounded, well defined intrahilar cavity	Strongly protruding, truncate, often with enveloping 'bubble'
Germination	Commonly from one or both polar cells, rarely from intermediate cells	From intermediate and/or polar cells	Commonly from one or both polar cells, rarely from intermediate cells
Basal germ tube direction of growth, position of emergence	Semiaxial, close to hilum, rarely lateral	Lateral, midway between hilum and basal septum, rarely semiaxial from near hilum	Semiaxial, close to hilum rarely lateral
Septum ontogeny	Median to submedian	Delimits basal cell	Submedian
First	Delimits basal cell	Median	Supermedian
Second	Distal	Distal	Median or variable
Third			
Conidiogenous nodes	Smooth or verruculose	Smooth	Smooth or verruculose

Correlation with Teleomorphs

The type species for the anamorph genera *Drechslera, Bipolaris and Exserohilum*, each, are connected with ascomycete genera *Pyrenophora, Cochliobolus and Setosphaeria*, respectively, other taxa within the complex show similar correlations (Ellis, 1976; Luttrell, 1977; Sivanesan, 1984; 1987). Additionally, some species of *Curvularia* also have *Cochiliobolus* teleomorphs (Luttrell, 1977; Alcorn, 1988; Sivanesan, 1984; 1987) introducing some heterogenecity into the anamorph-teleomorph correlation described here. But because of close similarity (Chaudhuri, 1969; Ellis, 1971) it has been suggested that *Bipolaris* and *Curvularia*, compared in detail by Sivanesan, (1987), be united (Von Arx and Luttrell, 1979).

Symptoms

The symptoms due to *B. sorokiniana* have been described by Joshi *et al.*, 1970; Nema and Joshi, 1971b; Misra, 1973; Narain *et al.* 1973; Vance *et al.* 1980). Gilchrist *et al.* (1991) stated that *H. sativum* attacks seedling, roots, leaves, nodes, spikes and grains during various stages of growth and development. Small dark brown spots initially appear approximately 15–20 days after infection on resistant materials. Lesions become surrounded by chlorotic margins and spot blotch development is limited. The flag leaf remains green until the peduncle turns yellow or large dark brown spots appear on the leaves and develop quickly, causing leaf death (Minh, *et al.* 1998). The symptoms produced by *B. sorokiniana* first appear on the lower leaves, progressing upwards to other leaves, sheaths, nodes, internodes and sometimes even earheads. Initially these are watersoaked, chlorotic and oval spots, later turning into irregular, eye, boat or spindle shaped blotches with reddish brown colour, lighter in center, chlorotic or diffused yellow at margins. Severely diseased fields give a burnt out appearance recognizable from a distance (Duveiller, *et al.*, 1998).

Losses

Losses caused by this pathogen range from 2.7 to 100 per cent in various countries and varieties (Nema and Joshi 1971a; Mehta, 1985; 1998; Duveiller and Gilchrist, 1994; Raemaekers, 1988; Villareal, *et al.*, 1995, Karwasra *et al.*, 1998). The average yield loss due to *Bipolaris* spot blotch estimated at four locations in Bangladesh (Dinajpur, Jessore, Jamalpur and Ishurdi) were 13.9, 16.2, 14.8 and 14.5 per cent, respectively (Alam *et al.* 1995). Yield losses of upto 10–15 per cent in China (Zhimin, *et al.* 1998), 23.6 per cent in Bangladesh (Alam *et al.*, 1998) and 57 per cent Bolivia (Toledo and Guzman, 1998) have been reported. Bazlur Rashid *et al.* (1987) reported most yield loss on cv-Kalyanasona, less on cv MS 1 and least on CY. Inia 66. This was 81.68, 72.68 and 55–81 per cent, respectively. Yield losses were 16.3 and 28.5 per cent at Pantnagar due to foliar blight in varieties UP 262 and UP 2338, respectively (Singh *et al.*, 1996). Singh *et al.* (1999a) estimated average yield loss of 29.4 per cent and 1000 grain weight reduction of 15.2 per cent. *Helminthosporium* leaf blight caused by a combination of *H.sativum* (*D.sorokiniana*) and *pyrenophora tritici-repentis* causes 3.1 to 23.8 per cent loss in grain yield in Nepal (Mahto, 1999a, 1999b). Singh *et al.* (2002d) estimated average yield losses during two years 27.9, 12.6 and 12.6 per cent at Faizabad and Kanpur in HP 1633 and at Pantnagar in UP 262, respectively and 1000 grain weight reduced significantly from 1.2 to 26.1 per cent during 1998–99 due to leaf blight. Singh *et al.* (2002c) reported that protection by 0.05 litre propiconazol (Tilt 25 EC) ha^{-1} avoided a yield loss 33.71 per cent.

Factors Affecting Disease Development

A lot of work has been conducted on the factors affecting disease development under field and green house conditions by several workers (Dosdall, 1923, Mitra, 1930; Clark and Dickson, 1958; Morton, 1962; Nema and Joshi, 1973; Misra, 1973; Khanna and Shukla, 1981 and Pant 1982). Temperature of about 28°C and relative humidity of 92 per cent promoted maximum foliar blight development. Rainfall directly increased disease intensity. Over the last few years, 1994–95 was most favourable year for disease development with an average temperature of 27.7°C and RH of 92 per cent. Factors which increased foliar blight intensity were minimum tillage, irrigation, late planting and inappropriate fertilizer regimes (Singh *et al.*, 1998b). Disease adversely affects the crop particularly under late sown conditions. Highest seed rate (125 kg ha^{-1}), lower row spacing (15 cm), full fertilizer dose (120 : 60 : 60 N.P.K.), higher number of irrigation application (6) and late sowing (20 December) increased foliar blight intensity compared to a lower seed rate (100 Kg ha^{-1}, wider spaced row (23 cm), half fertilizer dose (60 : 30 : 30 N.P.K.), lower number of irrigation applications (3) and timely sowing (30 November) (Singh *et al.*, 1998b).

Current Indian Wheat (*Triticum aestivum* L.) cultivars are, in general, highly susceptible to the disease because their parental/early generations were over estimated for resistance under inadequate disease epiphytotics created artificially in the open field conditions. Artificial epiphytotics of foliar blight developed with higher intensity inside the polythene house than in open field. Higher disease pressure was generated inside the polythene house because the congenial temperature of 27°C and relative humidity of 90 per cent could be maintained with the help of an automatic control device. But natural spells of cool and dry weather in the open field hindered the process of disease buildup (Kumar *et al.* 1999). Chauhan *et al.* (2003) reported that increasing nitrogen levels induced apparently higher spot blotch severity which ranged from 41.02 to 45.56 per cent and 42.95 to 46.99 per cent under normal and zero till conditions, but their interact on with the varieties did not show any significant differences in blotch severity under zero till situation.

Pathogenic Variability

Anamorph form of pathogen *B.sorokiniana* showed great variability in morphology, physiology, and pathogenicity (Christensen, 1925, 1934; Ashworth *et al.*, 1960; Clark, 1957, 1962; Misra and Singh 1972; Bidari and Govindu, 1975; Pidepichke and Zrazhevskaya, 1977 and Mism, 1979). Misra *et al.* (1981) observed that isolates from Dholi (Bihar) and Bhubaneshwar (Orissa) differed in their pathogenic behaviour on eight wheat varieties. The Dholi isolate was more virulent than the Bhubaneshwar one. They later reported that similar differences also occurred in five isolates of the pathogen from wheat, barley and triticales from different localities. Hetzler *et al.* (1991) identified 15 pathotypes out of 40 isolates, isolated from different countries on the basis of resistance reactions in a 12 member differential set of genotypes.

Pathogenic variability studied by the inoculations of nine monoconidial isolates of *H. sativum* on 24 wheat genotyopes using "detached leaf culture" test, revealed significant differences in incubation period (IP) lesion length (LL) and sporulation. Among nine isolates the maximum IP (3.4 days), minimum LL (2.4 mm) and moderate sporulation (12.2 spore/mrn1 of diseases leafarea were associated with isolate PHS 1. However, IP (2.2), the maximum length LL (11.5rnrn) and highest sporulation (17.3) was observed PHS 3. The isolate PHS 3 was found highly virulent and PHS 1 was least virulent on 24 wheat genotypes. The other isolates varied between the highly virulent and least virulent isolate for their virulence. (Akram and Singh, 2001). Mahto *et al.* (2001) also reported cultural variability in isolates of *Drechslera sorokiniana.*

Disease Management

Varietal Resistance

A large number of wheat strains have been found resistant to *H. sativum* by several workers (Misra and Singh, 1969; Mohanty and Manna, 1972; Narain *et al.*, 1973; Chaurasia *et al.*, 1973; Joshi and Adlakha, 1974; Dubin and Van Grinkel, 1991; Mahto *et al.*, 1995; Singh *et al.*, 1995b; Sharma *et al.*, 1995; Shrestha *et al.*, 1998; Naitao and Yousan, 1998). Das *et al.* (1999) reported wheat genotype 3336 (Lok Bhart × DARF) showing a high level of resistance to the spot blotch pathogen (*B.sorokiniana*). Moderately resistant varieties were Bobwhite and Parula and susceptible genotypes were HD 2285 and WH 542. Singh *et al.* (1999a) noted VL 738, PBW 343, PBW 373, HP 1744 and C 306 to be promising cultivars showing comparatively lesser incidence of leaf blight. Singh *et al.* (1999a) also identified ten entries, namely Uhu, K 9204, BAU3, BAU4, MRNG-ALD/ ANS, BW 14989, HW2013, HW2041, CB (BW) 346 and CB(BW) 352 as resistant Pandeyand Tiwari (1999) at Pantnagar, failed to record any variety as immune to foliar blight, but found VL738 as resistant and several other varieties with lower resistance. Akram and Singh (2000) found that varieties UP 2121, HUW 206 and UP 262 consistently tolerant to spot blotch at Pantnagar. Singh *et al.* (2000b) evaluated a large number of wheat genotypes against leaf blight pathogens *B. sorokiniana* and *Altemaria triticina.*

Attempts were made to identify sources of resistance to *B. sorokiniana* and *A. triticina* by screening a large number of genetic stock comprising moderately resistant AVT–II year and NGSN material under controlled conditions in a polyhouse tunnel for three consecutive year (1997–98, 1998–99, 1999–2000) (Shekhawat and Shekhawat, 2001). Ragiba *et al.* (2001) noted Chirya-3 and Mayoor genotypes showing high level of resistance under field conditions at Pusa, Bihar. A number of newly evolved wheat genotypes were evaluated under artificial epiphytotics against the spot blotch caused by *C. sativus* and only three genotypes mainly NW 2043, MAS 2942 and HUWL 99003 gave resistant reaction (Singh *et al.*, 2002b).

Chemical

The leaf blight/blotch of wheat was effectively controlled by 6 applications of Dithane Z-78 and Dithane M-45 with corresponding increase in yield. Other fungicides *viz.*, Bisdithane, Cuman, Karathane and cosan also reduced the disease intensity (Singh and Singh, 1971). One prophylatic spray of 0.1 per cent Fytolan sprayed one day before inoculation, suppressed the disease appearance to a large extent. Spraying of Fytolan at 10 days intervals at the rate of 0.1 per cent reduced the disease intensity effectively (Ram and Joshi, 1978b). Lapis (1985) noted propiconazole giving best control of spot blotch resulting in a 65 per cent yield gain. In field trials conducted for 3 years to manage leaf blight/blotch of wheat it was observed that three sprays of Mancozeb at 15 day and 10 day intervals was most effective and economical in checking the disease and it helped to increase the yield (Misra *et al.*, 1989). Roberti *et al.* (1992) tested many fungicides. These were Guazatine, Mancozeb, Carbendazim + Mancozeb and Thiophanate methyl + Maneb and gave good results against *C. sativus* when used in December and February. Prochloraz + Mancozeb gave best result in January.

The disease was effectively controlled by three sprays of Propiconazol (Tilt) 25 EC @ 0.5 liter ha^{-1} at two weekly intervals and recorded 16.19 per cent loss in grain yield in unsprayed plots (Mahto and Sedhai, 1994). Alam *et al.* (1995) sprayed Prapiconazole at 0.04 per cent and reduced the disease significantly. Mehta *et al.* (1998) reported different fungicides belonging to the Dithiocarbamates such as Mancozeb and Triazoles, Propiconazole, Tebuconazole, Flutriafol, Iprodione, Procloraz and Triadimenol effective against spot blotch. Singh *et al.* (1998b) observed maximum yield and grain weight and minimum disease intensity after three sprays of fungicide Tilt 25 EC @ 0.5 liter ha^{-1} followed by its two sprays and three sprays of Indofil M-45 @ 2.5 Kg ha^{-1}. Mahto (1999b) also controlled leaf blights with three sprays of Tilt 25 EC @ 125 ml @ ha^{-1} in 800 litre of water. First spray given at the on set of initial symptoms at booting stage. According to Singh and Gupta (2000) Tilt giving best result by completely inhibiting the colony growth of the pathogen *in vitro* at all the concentrations followed by Bayleton and Dithane M 45 which inhibit 80.24 and 65.44 per cent colony growth. Chauhan *et al.* (2001) conducted a study to control spot blotch of wheat caused by *Cochliobolus sativus* using 1 to 4 sprays of propiconazole (Tilt 25 EC) @ + 1.0 and 0.51itre doses per hectare. All the treatments were effective in controlling the blotch and increasing grain yield and were significantly superior over cheek (unsprayed). Application of 4 and 3 sprays at a dose of 1.0 litre and 4 sprays at dose of 0.5 litre resulted in maximum disease control and highest yield increase followed by two sprays at a dose of 1.0 litre. Singh *et al.* (2002c) also reported that propiconazol (Tilt 25 EC) @ 0.50 litre ha^{-1} control spot blotch most economically. Akram and Singh (2003) also reported that three sprays of tilt (propiconazole) @0.05 per cent 25 EC ha^{-1} protected the wheat varieties from spot blotch and their yielding potential without any significant disease interference.

Botanicals

Several higher plants have been found to posses outstanding fungitoxicants against mycelial growth or spore germination of different fungi during *in vitro*. Spancer *et al.* (1963), Nene and Kumar

(1966) exhibited antifungal activity against test fungi *Helminthosporium* spp. Shekhawat and Prasada (1971) discussed the antifungal properties of a large number of plant species including of Chinaberry against *Helminthosporium* spp. Khan *et al.* (1974) studied the effect of water soluble fraction (WSF) of different oilcakes and bitter principles from Neem *viz.*, Nimbidin and Thionemone on some fungi including *Helminthosporium nodulosum* also. Al-Abed *et al.* (1995) investigated 40 common wild plant species against several fungal pathogen including *Helminthosporium sativum* and reported fresh shoot extract of *Inula viscosa* inhibited the growth of *H. sativum,* while extract of*Anagallis arvensis* reduced growth of *B. sativum.* Pandey *et al.* (1983), Nath and Bordoloi (1984), Qasem *et al.* (1995) and Srivastava and Lal (1997) also reported antifungal properties of botanicals against several fungi including *Helminthosporium* spp. Pandey *et al.* (2002) reported that the leaf extract of *Mangifera indica* completely inhibited the mycelial growth of *Helminthosporium sativum. In vitro* test was carried out to evaluate fungitoxicity of essential oils collected from *Artemisia nilagirica and Jaglans regia* var. *Kumaonica* against *Drechslera sorokiniana* causing spot blotch of wheat. *A. nilagirica* oil was more effective than *J.regia* var. *Kumaonica* oil causing more inhibition on spore germination and mycelial growth of fungus. EC_{50} values for spore germination inhibition and mycelial growth inhibition were 12.06×10^{-2}, 21.64×10^{-2} and 10.00×10^{-2}, 18.63×10^{-2}, for *A. nilagirica* and *J. regia* var. *Kumaonica* oils, respectively (Chowdhury *et al.*, 2003). Chaudhary and Singh (2003) obtained that Nimin inhibited the growth of pathogen 95.75 per cent followed by Neemgold (82.72 per cent), Nimbicidine (76.71 per cent) and Neemoil (57.76 per cent).

REFERENCES

Ahmad, Q. and D.K. Singh (1983). Screening of wheat varieties. *Indian Phytopath.*, 36 (3): 563–564.

Ahmad, Q. (1984). Screening of wheat cultivars against *Alternaria* leaf blight–II. *Indian Phytopath.*, 37: 557–578.

Akram, M. and A. Singh (2000). Studies on tolerance to *Helminthosporium sativum* incitant of spot blotch of wheat. National Symposium on Role of Resistance in Intensive Agriculture, February, 15–17, Karnal (Abst), pp. 26.

Akrarn, M. and A. Singh (2001). Variability and characterization of host response to *Helminthosporium sativum,* incitant of spot blotch of wheat. *Indian Phytopath.*, 54(4): 429–434.

Akram, M. and A.Singh (2003). Determination of wheat varieties tolerant to spot blotch disease, *Bipolaris sorokiniana* (Sacc.) Shoem. *Ann. Pl. Protec. Sci.*, 11(1) 75–78.

Al-Abed, A.S.J.R. Qasem, H.A. Abu-Blan (1995). Antifungal effect of some common wild plant species on certain plant pathogenic fungi. *Pure- and Applied Sciences*, 20: 149–158.

Alam, K.B., P.K, Malaker, M.A, Shaheed,. M V., Ahmad, F. Ahmad, and M.S. Haque (1995). Yield loss assessment of wheat due to *Bipolaris* leaf blight in Bangladesh. *Bangladesh J. Pl. Pathol.*, 11(1–2): 35–38.

Alam, K.B., S.P. Banu and M.A. Shaheed (1998). Occurrence and significance of spot blotch in Bangladesh. In: Duveiller, E; Dubin, H.I; Reeves I. and McNab, A. (Eds.) *Helminthosporium Blights of Wheat: Spot Blotch and Tan Spot*, Mexico, D.F.: CIMMYT, pp. 63–66.

Alcom, I.L. (1988). The taxonomy of "*Helminthosporium*" species. *Annu. Rev. Phytopathol.*, 26: 37–56.

Ali, T.E.S, M.A Nasir, and A.S. Shakin (1992). *In vitro* evaluation of certain neem products as mould inhibitions against post harvest fruit rotting fungi of tomato. *Pakistan J. Phytopath.*, 4(1–2) 58–61.

Amaresh, Y.S., V.B. Nargund and B. Sarnasekhar (2000). Use of botanicals and fungitoxicants against *Alternaria helanthi* the causal agent of sunflower, leaf blight. *Pl. Bio. Res,.,* 15(2): 140–145.

Ashworth, L.I., K.A. Lahar and R.I. Collin (1960). Pathogenic specialization of *Helminthosporium sativum.* (Abstr), *Phytopathology,* 50: 627.

Babu, S., K. Seethararnan, R. Nanda Kumar and I. Johnson (2001). Inhibitory effect of leaf extract of some medicinal plants and weeds on *Alternaria solani. Pl. Dis. Res.,* 16(1): 84–86.

Bajwa, M.A (1985). Wheat research and production in Pakistan. In: Villareal, R.L. and Klatt, A.R. (Eds.) *Wheats for More Tropical Environments,* Mexico. D.F.: CIMMYT, pp. 68–70.

Bazlur Rashid, A.Q.M., Meah Bahadur and M. Ialaluddin (1987). Effect of leaf blight caused by *Drechslera sorokiniana* (Sacc.) Subrarn. and lain on some yield components of wheat. *Crop Protection,* 6(4): 256–260.

Bhatti, M.A.R. and M.B. lyas, (1986). Wheat diseases in Pakistan. In: L.M Joshi., Singh, D.V. and Srivastava, K.D. (Eds.) *Problems and Progress of Wheat Pathology in South Asia,* New Delhi, India, Malhotra Publishing House, pp. 20–30

Bidari, V.B. (1974). Studies on blight of wheat caused by *Helminthosporium sativum* Pam. King and Bakke in Mysore State. *Mysore J. Agric. Sci.,* 8: 475–477 (Abstr.).

Bidari, V.B. and Govindu, H.C. (1975). Studies on the pathogencity of three isolates of *Helminthosporium sativum* P.K. and B. on wheat at its different stages of growth in Karnataka state. *Mysore. J. Agric. Sci.,* 9: 87–94.

Bose, S.R. (1956). *Alternaria* within the pericarp of the wheat seed. *Nature* (Lond.), 178: 640–641.

Casulli, F., A. Siniscalco and S. Frisullo (1986). Behaviour of some *duraum* wheat cultivars towards *Alternaria triticina* Pras. and Prab., *Phytopathologia Mediterranea.,* 25(1–3): 131–134.

Chalfound, S.M. and D. Carvalho (1987). Effect of industrial garlic oil extracts on fungal development. *Fitopathological Brasileria,* 12: 204–234.

Chattopadhyay, S.B. and K.B. Chakrabarti (1968). Role of different pathogens in inciting the *Helminthosporium* disease of wheat in West Bengal. *Indian J. Agric. Sci.,* 38(6): 937–940.

Chaudhuri, S. (1969). Validity of the type of conidial germination as a taxonomic criterion in grarninicolous *Helminthosporium. Bull. Indian Phytopathol. Soc.,* 5: 15–19.

Chaudhary, V.P. and A.K. Singh (2003). Efficacy of neem products in inhibition of *Bipolaris sorokiniana.* Zonal conference (East Zone), Indian Society of Mycology and Plant Pathology and seminar on Plant Diseases of National Importance with Special Reference to Guava wilt and Mango Malformation, April 4–5, IISR, Lucknow (Abstr.) p. 3.

Chowdhury, H., R.D. Singh and S.K. Biswas (2003). *In vitro* antifungal activity of two essential oils on *Drechslera sorokiniana* causing spot blotch of wheat. *Ann. Pl. Protec. Sci.,* 11(1): 67–69.

Chauhan, A., R.N. Singh and AK. Singh (2001). Doses and number of sprays of propiconazole for the economic control of spot blotch of wheat and avoidable yield losses. *Indian Phytopath.,* 54(4): 507.

Chauhan, A., R.N. Singh and A.K. Singh (2003). Effect of nitrogen doses and cultivars on wheat spot blotch severity under normal and zero till conditions of rice–wheat cropping system. Zonal Conference (East Zone), Indian Society of Mycology and Plant Pathology and Seminar on Plant Diseases of National Importance with Special Reference to Guava Wilt and Mango Malformation, April 4–5, IISR, Lucknow (Abstr.) p. 2.

Chaurasia, R.K., A.B. Mishra and S.P. Singh (1973). Response of wheat varieties to *Helminthosporium sativum* causing leaf blight of wheat. *J.N.K.V.V. Res. J.*, 7: 188–89.

Chenulu, V.V. and Amar Singh (1964). A note on estimation of losses due to leaf blight of wheat caused by *Alternaria triticina. Indian Phytopath*, 17: 254–256.

Chenulu, V.V., Amar Singh and L.M. Joshi (1970). Estimation of losses due to *Alternaria* blight of wheat caused by *Alternaria triticina.* Plant Disease Problems. Indian Phytopathological Society, IARI, New Delhi: 28–31.

Christensen, J.J. (1925). Physiologic specialization and mutation in *Helminthosporium sativum. Phytopathology.*, 15: 785–795.

Christensen, J.J. (1934). Physiologic specialization and variation in *Helminthosporium gramineum* Rab. *Tech. Bull. Minn. Agric. Exp. Stat.* 95: 40.

Clark, R.V. (1957). The evaluation of variability in pathogenicity of *Helminthosporium sativum* and the relation of temperature to disease development of barley. *Dis. Abstr.*, 17: 220–221.

Clark, R.V. and J.G. Dicson (1958). The influence of temperature on disease development in barley infected by *Helminthosporium sativum. Phytopathology*, 48: 306–310.

Clark, R.V. (1962). Pathogenicity of the spot blotch organism on cereals. Abs. In: *Proc. Canad. Phytopath. Soc.*, 29: 11–20.

Dahal, G., P. Amatya and H.K. Manandhar (1992). Plant Diseases in Nepal. *Rev. Plant Path.*,. 71: 797–807.

Dalela, G.G., H.C. Sankhla, B. Sankhla, and R.L. Mathur (1971). Reaction of wheat varieties to the Rajasthan isolate *Alternaria triticina* incited of leaf blight of wheat. *Indian J. Mycol. Pl. Path.*, 1: 121–125.

Das, P.K., Ashutosh Roy and A. Roy (1989). Varietal reaction of wheat to some diseases. *J. Res. Assam Agric. Uni.*, 10(1–2): 61–62.

Das, S.R. Aggarwal, P. Dureja, D.V. Singh (1999). Leaf surface waxes in relation to resistance against spot blotch (*Drechslera sorokiniana*) of wheat. *Acta Phytopathologica Entomologica Hungarica*, 34 (1/2): 75–84.

Dosdall, L. (1923). Factors influencing the pathogenicity of *Helmlnthosporium sativum. Minn. Agric. Exp. Sta. Tech. Bull.*, 17: 47.

Drechsler, C. (1923). Some graminicolus species of *Helminthosporium. J. Agric. Res.*, 24: 641–739.

Dubin, H.J. and M. Van Grinkel (1991). The status of wheat diseases and disease research in warmer areas. In: Saunders, D.A. (Ed.) *Wheat for the Nontraditional Warm Areas*, Mexico, D.F.: CIMMYT, pp. 125–145.

Duveiller, E. and L. Gilchrist (1994). Production constraints due to *Bipolaris sorokiniana* in wheat current situation and future prospects. In: Saunders, D.A. and Hettel, G.P. (Eds.) *Wheat in Heat–Stressed Environments: Irrigated Dry Areas and Rice Wheat Farming Systems*, Mexico, D.F.: CIMMYT, pp. 334–352.

Duveiller, E., H.J. Dubin, J. Reeves and A. McNab (1998). *Helminthosporium* Blights of Wheat: Spot Blotch and Tan Spot Mexico, D.F.: CIMMYT, p. 376.

Edward, J.C., A.N. Misra, M.K. Sinha and S.C. Tripathi (1972). Studies on identification of *Alternaria* associated with wheat blight and possible use of its crude filterates for screening wheat resistance. *Allahabad Fmr.*, 46: 105–106.

Ellis, M.B. (1971). Dematiaceous Hyphomycetes CMI, Kew Surrey, England, p. 608.

Ellis, M.B. (1976). More Dematiaceous Hyphomycetes. CMI, Kew England, p. 507.

Frisullo, S. (1982). Fungal parasites of plants in Southern Italy. I. *Alternaria triticina* Prasada and Prabhu on *durum* wheat. *Phytopathologia Mediterranea*, 21: 113–115.

Garg, V.R, H.C. Arya and U. Kant (1972). Varietal reaction and physiologic specialization in *Alternaria triticina* causing leaf blight of wheat in Rajasthan *Indian J. Mycol. Pl. Path.*, 2: 24–31.

Gilchrist, L.J., W.H.Ppfeiffer and C.Valazquez (1991). Resistance to *Helminthosporium sativum* in bread wheat: relationship among infected plant parts and agronomic traits. Seventh regional wheat workshop of Eastern. Central and Southern Africa, Nakura Kenya. 16–19 September, pp. 176–181.

Goel, L.B., L.M. Joshi, R. Nath, D.V. Singh and K.D. Srivastava (1976). A new leaf spot of wheat caused by *Chaetomium* sp. *Curr. Sci.*, 45: 277.

Goel, R.K. and A.K. Gupta (1979). A new blotch disease of wheat in India. *Plant Dis. Reporter*, 63: 620.

Goel, L.B., S. Nagarajan, R.V. Singh, V.C. Sinha and J. Kumar (1999). Foliar blights of wheat: Current status in India and identification of donor lines for resistance through multilocational evaluation. *Indian Phytopath.*, 52: 398–402.

Govinda Rao, P. (1961). Proceedings of All India Wheat Research Workers Seminar, IARI, New Delhi.

Hafiz, A (1986). *Plant Diseases*. Islamabad: Pakistan Agricultural Res. Council, 552 pp.

Hetzler, J., Z. Eyal, Y.R Mehta, L.R. Campos, U. Kushnir, J. Zekaria and L. Cohan (1991). Interactions between spot blotch (*Cochliobolus sativus*) and wheat cultivars. In: *Wheat for the Non-traditional Warm Areas* (Eds.) D.A Saunders, Mexico, D.F.: CIMMYT, pp. 146–164.

Hughes. S.J. (1958). Revisions hyphomycetum aliquot cum appendice de norninibus rejiciendis. *Can. J. Bot.*, 36: 727–836.

Ito, S. (1930). On some new ascigerous stages of the species of *Helminthosporium* parasitic on cereals. *Proc. Imp. Acad.* (Tokyo) 6: 352–355.

Jain, K.L. and A.S. Prabhu (1974). Comparative studies on variation in some monoconodial isolates of *Alternaria triticina. Indian J. Mycol Pl. Path.*, 4: 67–73.

Jitendra, S. and V.L.Majumdar (2001). Efficacy of plant extracts against *Alternaria alternata* the incitant of fruit rot of pomegranate (*punica granatum* L.). *J. Phycol. Pl. Pathol.*, 31 (3): 346–349.

Joshi, L.M., B.L. Renfro, E.E. Saari, Wilconson and S.P Raychaudhari (1970). Diseases of wheat in India other than rusts and smuts. *Plant Dis. Reptr.*, 54: 594–597.

Joshi, L.M. and K.L. Adlakha (1974). Sources of resistant to *Helminthosporium sativum. Indian J. Gent. Pl. Breed.*, 34: 368–370.

Joshi, L.M., S.D. Gera, K.L. Adlakha, K.D. Srivastava, B. Ram and L.T. Palmer (1974). Some foliar diseases of wheat during 1969–70 crop season. *Indian Phytopath.*, 27: 178–181.

Joshi, L.M., K.D. Srivastava, D.V. Singh, L.B. Goel and S. Nagarajan (1978). *Annotated Compendium of Wheat Diseases in India.* Indian Council of Agric. Research, New Delhi, 332 pp.

Karki, C.B. (1982). Tan spot and other foliar diseases of wheat in Nepal, In: RM. Hosford Jr. (Ed.) *Proc. of the Tan Spot of Wheat and Related Diseases Workshop,* 1981, Fargo,ND, USA. North Dakota State University, pp. 76–81.

Karki, C.B. and P.B. Karki (1996). Wheat disease report, 1994–95. In: *Proc. of the National Winter–Crops Technology Workshop,* 1995, Kathmandu, Nepal: NARC and CIMMYT, pp. 269–287.

Karwasra, S.S., M.S. Beniwal and Rajender Singh (1998). Occurrence, cultivar reaction and yield losses due to leaf blight of wheat. *Indian Phytopath.,* 51(4): 363–364.

Kazmi, A.R., I. Niaz and G. Jilani (1993). Evaluation of some plant extracts for antifungal proporaties. *Pakistan J. Phytopath.,* 5(102): 93–97.

Khan, M.W., M.M. Alam, A.M. Khan, and S.K. Saxena (1974). Effect of water soluble fractions of oil cakes and bitter principles of neem on some fungi and nematodes. *Acta. Bot. Indica,* 2: 120–128.

Khanna, B.M. and T.N. Shukla (1981). Epidemiological studies on leaf blight of wheat caused by *Helminthosporium* sp. *Indian J. Mycol. Pl. Path.,* 11: 21–29.

Kulkarni, G.S. (1924). Report of the work done in Plant Pathology Section during the year 1922–23. *Rep. Dept. Agric. Bombay Presidency for the year,* 1922–23, p. 167–1971.

Kumar, V.R. and H.C. Arya (1973a). Taxonomy of the fungus causing leaf blight of wheat. *Proc. 62nd Sci. Cong. Part III:* 40–41.

Kumar, V.R and H.C. Arya (1973b). Nature of disease resistance and perceptibility of leaf blight of wheat caused by *Alternaria triticina* Prasada and Prabhu *in vitro* and *in vivo* conditions. *Proc. 62nd Sci. Cong. Part III:* 349–350.

Kumar, J., A.K. Sharma, R. Kumar, S. Nagarajan (1999). A new technique to develop artificial epiphytotics of foliar blight in wheat (*Triticum aestivum* L.). *Indian J.Agric. Sci.,* 69: 48–51.

Kumar, V.R. and S. Praveen (2002). Integrated disease management of leaf blight of wheat. *Ann. Pl. Protec. Sci.,* 10(2): 302–307.

Lapis, D.B. (1985). Chemical control of wheat diseases in the Philippines. In: Villareal, R.L. and Klatt, A.R. (Eds.) *Wheat for More Tropical Environments,* Mexico, D.F.: CIMMYT, pp. 204–208.

Leonard, K.G. and E.G., Suggs (1974). *Setosphaeria prolata* the ascigerous state of *Exserohilum prolatum, Mycologia,* 66: 281–297.

Link, H.F. (1809). Observation on ordines Plantarum naturals, I. Anandroarum ordines epiphytas mucedines Gastromycos et. *Fungus Gasell. Nat. Freund: Burlin Magasin,* 3: 3–42.

Luttrell, E.S. (1951). A key to the species of *Helminthosporium* reported on grass in the United States. *Plant Dis. Reptr. Suppl.,* 201: 59–67.

Luttrell, E.S. (1957). *Leptosphaeria* (*Metasphaeria*) perfect stage for *Helminthosporium turcicum* and *H. rostratum. Phytopathology,* 47: 313.

Luttrell, E.S. (1963). Taxonomic criteria in *Helminthosporium. Mycologia,* 55: 643–674.

Luttrell, E.S. (1964). Systematics of *Helminthosporium* and related genera. *Mycologia,* 56: 119–132.

Luttrell, E.S. (1977). Correlation between conidial and ascigerous state characters in *pyrenophora, Cochlioholus and Setosphaeria. Rev. Myco.*, 41: 271–279.

Luttrell, E.S. (1978.) Biosystematics of *Helminthosporium*: impact on agriculture. In: Romberger, J.A., Foote, R.H., Knutson, L.V., Lentz, P.L. (Eds.) *Biosystematics in Agriculture, Bettsville Symposia in Agricultural Research* 2. Montclair: Allanheld, Osmon and Co., pp. 193–209.

Mahto, B.N. and G.P. Sedhai (1994). Loss assessment of wheat due to foliar blight. *Wheat Proceedings,* NWRP, Kathmandu, Nepal: NARC.

Mahto, B.N. (1995). Wheat disease screening nurseries. In: Proceeding of the 1994–95, Winter Crop Workshop, NARC, pp. 287–307.

Mahto, B.N. (1999a). Assessment of yield losses due to foliar blight of wheat in Nepal, *Ann. Agric. Res.,* 20(2): 212–215.

Mahto, B.N. (1999b). Management of *Helminthosporium* leaf blight of wheat in Nepal. *Indian Phytopath.,* 52: 408–413.

Mahto, B.N., D.V. Singh, K.D. Srivastava and R. Aggarwal (2001). Cultural variability in isolates of *Drechslera sorokiniana. Ann. Pl. Protec. Sci.,* 9: 276–280.

Mathur, R.S. (1956). *Alternaria* leaf spot of wheat in *Uttar Pradesh. Agric. Anim. Husb. India,* 6: 12–14.

Mc Rae, W. (1924). Economic Botany, Part III: Mycology. *Rep. of Board of Scientific Advice India.,* 1922–23, pp. 31–35.

Mc Rae, W. (1930). Report of the Imperial Mycologist. *Sci. Rep. Imp. Agric. Res. Inst.,* Pusa, 1930–31, pp. 73–86.

Mc Rae, W. (1933). Report of the Imperial Mycologist. *Sci. Rep. Imp. Inst. Agric. Res.,* Pusa, 1931–32, pp. 120–140.

Meena, S.S. and V. Mariappan (1993). Effect of plant products on seed borne microflora of sorghum. *Madras Agric. J.,* 80(7): 383–387.

Mehta, P.R.. (1951). Some new diseases of plants of economic importance of Uttar Pradesh. *Pl. Prot. Bull.,* 2: 50–51.

Mehta, Y.R. (1985). Breeding for wheat resistance to *Helminthosporium* spot blotch. In: Villareal, R.L. and Klatt, A.R. (Eds.), *Wheat for the Non–tropical Environments,* Mexico, D.F.:CIMMYT, pp. 135–144.

Mehta, Y.R. (1998). Constraints on the integrated management of spot blotch of wheat. In: E., Duveiller. B.I. Dubin, I. Reeves, and A. McNab (Eds.), *Helminthosporium Blights of Wheat: Spot Blotch and Tan Spot.* Mexico, D.F.: CIMMYT, pp. 18–27.

Minh, T.D., T.D. Long and T.T. Ngu (1998). Screening wheat for *Bipolaris sorokiniana* resistance in Vietnam. In: Duveiller. E., Dubin, B.I., Reeves, I. and McNab, A. (Eds.), *Helminthosporium Blights of Wheat: Spot Blotch and Tan Spot,* Mexico, D.F.: CIMMYT, pp. 213–217.

Mishra, A.B., S.P. Singh, S.M. Harne and S.V. Velenkar (1970). Chemical control of *Altemaria* sp. causing leaf blight of wheat. *PANS,* 16: 522–524.

Misra, A.P. and R.A. Singh (1969). Varietal reaction of wheat against *Helminthosporium sativum* P.K. and B. *J. Appl. Sci.,* 1: 33–37.

Misra, A.P. and R.A. Singh (1972). Comparative pathogenicity of different species of *Helminthosporium* on wheat in India, *Proc. 59th Indian Sci. Cong.*, p. 567.

Misra, A.P. and R.A. Singh (1972). Pathogenic differences among three isolates of *Helminthosporium-tritici-repentis* and performance of wheat varieties against them. *Indian Phytopath.*, 25: 350–353.

Misra, A.P. (1973). *Helminthosporium Species Occurring on Cereals and Other Gramineae*. The Catholic Press, Ranchi, p. 288.

Misra, A.P. (1979). Variability, physiologic specialization and genetics of pathogenicity in graminicolus *Helminthosporia* affecting cereals crops. *Indian Phytopath.*, 32(1) 1–22.

Misra, A.P., S.P. Pandey, AK. Misra and I. Jha (1981). Pathogenic differences in isolates of *Bipolaris sorokiniana* Shoemaker (*H. sativum* P.K. and B.) from wheat, barley and triticales from different geographical regions. Abstr. 3rd Int. Symp. Pl. Path. IARI., New Delhi, pp. 131–132.

Mishra, A., N.M. Patel, D.B. Patel, I.R. Patel, R.S. Jadon and A. Mishra (1989). Economical spray schedule for the management of leaf blight/blotch of wheat. *Pestology*, 13(8): 22–25.

Mitra, M. (1930). A comparative study of species and strains of *Helminthosporium* on certain Indian cultivated crop. *Trans. Br. Mycol. Society.*, 15: 254–293.

Mitra, M. (1934). A new leaf spot disease of wheat caused by *Helminthosporium tritici-repentis* Died. *Indian J. Agric. Sci.*, 4: 692–700.

Mohanty, N.N. and M.K. Manna (1972). Varietal resistance to wheat spot blotch disease caused by *Helminthosporium sativum* Pammel, King and Bakke, *Proc. 59th Indian Science Congr.*, p. 567.

Morton, D.J. (1962). Influence of temperature humidity and inoculums concentration on development of *Helminthosporium sativum* and *Septoria passerinii* in excised barley leaves. *Phytopathology*, 52: 703–708.

Murthy, S.P. and M. Sirsi (1957). Studies on *Melia azadirachta* Part I. Antibacterial, antifungal and antitubercular activity of neem oil and its fractions. *Symp. Utilization Indian Med. Pl.*, Lucknow, pp. 55.

Nagarajan, S. and G. Singh (1998). Sustaining wheat (*Triticum aestivum*) revolution retrospect and Prospect, *Indian J. Agric. Sci.*, 68(Special issue): 396–404.

Naitao, C. and Wu, Yousan (1998). Incidence and current management of spot blotch of wheat in China. In: Duveiller. E., Dubin, B.I. Reeves, I. and McNab, A. (Eds.) *Helminthosporium Blights of Wheat: Spot Blotch and Tan Spot*, Mexico, D.F.: CIMMYT, pp. 119–125.

Narain, A, S.K. Sinha, B.K. Mahapatra, D.T. Sechler and I.M. Poehlman (1973). Incidence of *Helminthosporium* blight of wheat in Orissa State, India. *Pl. Dis. Reptr.*, 57: 278–280.

Nath, S.C. and D.N. Bardoloi (1984). Fungitoxicity of *Narvalia zeylanica* Dc. Extracts. *Pesticides.*, 18 (4): 13–16.

Nees, Von and C.G. Eschhebeck (1817). System der Pilzic and Schwamme, Wurburg, pp. 344.

Nema, K.G. and L.M. Joshi (1971a). The spot blotch disease of wheat caused by *Helminthosporium sativum*. *Proc. Second Int. Symp. Pl. Path.* IARI, New Delhi, p. 42.

Nema, K.G. and L.M. Joshi (1971b). Symptoms and diagnosis of the 'spot blotch' and 'leaf blight' disease of wheat. *Indian Phytopath*, 24(2): 418–419.

Nema, K.G. and L.M. Joshi (1973). Spot blotch disease of wheat in relation to host age, temperature and moisture. *Indian Phytopath.*, 26: 41–48.

Nema, K.G. (1986). Foliar disease of wheat leaf spots and blights. In: Joshi, L.M, Singh D.V. and Srivastava, K.D. (Eds.) *Problems and Progress of Wheat Pathology in South Asia*, Malhotra Publishing House, New Delhi, pp. 162–175.

Nene, Y.L. and K. Kumar (1966). Antifungal properties of *Erigeron linifolius* Wild Extracts. *Natururissenschaften*, 53(14): 363.

Nisikado, Y. (1928). Studies on the *Helminthosporium* diseases of Gramineae in Japan. *Spec. Report Ohara Inst. Landwirtsch Forsch. Kurashiki*, 4: IV+384+11 p. (Japanese).

Ojha, K.L. and P.P. Mehta (1970). Effect of NPK and balanced nutrients on the incidence of leaf blight disease of wheat caused by *Alternaria triticina* Prasada and Prabhu. *J. Appl. Sci.*, 2: 41–43.

Pammel L.H, C.M. King and Bakke (1910). Two barley blights with comparison of species of *Helminthosporium* upon cereals. *Iowa Agric. Exp. Stn. Bull.*, 116: 178–190.

Pandey, D.K., H. Chandra, N.N. Tripathi and S.N. Dixit (1983). Toxicity of the essential oil of *Ageratum houstonianum* against *Fusarium lateritium f. sp. Cajani Beitr. Biol. Pllanzcn.*, 58: 115–122.

Pandey, S. and A.N. Tiwari (1999). Field evaluation of wheat varieties against leaf blight and its control. *Pl. Dis. Res.*, 14(1): 52–54.

Pandey, M.K., A.K. Singh and R.B. Singh (2002). Mycotoxic potential of some higher plants. *Plant Dis. Res.*, 17(1): 51–56.

Pant, N.M. (1982). Screening of wheat genotypes for resistance against *Helminthosporium* leaf blight under artificial epiphytotic conditions. *Biovigyanous*, 8(1): 85–87.

Pamshar, M., S. Nagarajan, L.B. Goel, and J. Kumar (1995). Report of the coordinated Exp., 1994–95. Crop protection (pathol.) AICWP, Directorate of Wheat Research, Karnal, pp. 206.

Pidoplichko, V.N. and T.G. Zrazhevskara (1977). Pathogenic future ofstrains of *Helminthosporium, sativum* P.K. and B. the pathogen of root rot of wheat *Mikologiya Fitopatologiya.*, 11(2): 144–148.

Prabhu, A.S. and R. Prasada (1966). Plant Pathological and epidemiological studies on leaf blight of wheat caused by *Alternaria triticina. Indian Phytopath.*, 19: 95–112.

Prabhu, A.S. and V. Prakash (1973). The relation of temperature and leaf wetness to the development of leaf blight of wheat. *Pl. Dis. Reptr.*, 57: 1000–1004.

Prabhu, A.S. and Amar Singh (1974). Appraisal of yield loss in wheat due to foliage diseases caused by *Alternaria triticina* and *Helminthasporium sativum. Indian Phytopath.*,. 27: 632–634.

Prasada, R. and A.S. Prabhu (1962). Leaf blight of wheat caused by a new species of *Alternaria. Indian Phytopath*, 15: 292–293.

Prasada, R. and A.S. Prabhu (1964). Outbreaks and new records. Leaf blight of wheat. *P. Prof. Bull. R.A.O.*, 12: 117.

Qasem, J.R., A.S. Al-Abed and H.A Abu-Blan (1995). Antifungal activity of *Clammy inula* (*Inula viscosa*) on *Helminthosporium sativum* and *Fusarium oxysporum f. sp. Lycopersici. Phytopathologia Mediterraca.*, 34: 1–14.

Raemaekers, R.H. (1988). *Helminthosporium sativum:* Disease complex on wheat and sources of resistance in Zambia. In: Klatt, A.R. (Ed.) *Wheat Production Constraints in Tropical Environments*, Mexico, D.F.: CIMMYT, pp. 175–186.

Ragiba, M., K.V. Prabhu and R.B. Singh (2001). Characterization of wheat germplasm against *Helminthosporium* leaf blight. *Indian Phytopath.*, 54(3): 316–318.

Raj, H. and I.J. Kapoor (1993). Effect of neem oil on *Fusarium oxysporom* Schl. in Tomato. *Pestology*, 17(9): 34–36.

Ram, B. and L.M. Joshi (1977). Evaluation of some systemic and other fungicides for control of *Alternaria* leaf blight of wheat. *Pesticides*, 11: 33–34.

Ram, B. and L.M. Joshi (1978a). Effect nitrogen levels on *Alternaria* blight of wheat and losses caused by it. *Indian Phytopath*, 31: 125.

Ram, B. and L.M. Joshi (1978b). Spray schedule of Fytolan for leaf blight of wheat and its effect on yield components. *Indian Phytopath.*, 31: 348–351.

Ram, B. and L.M. Joshi (1981). Effect of nitrogen levels on leaf blight of wheat and losses in yield components. *Zeit. Pflanzen. Pflschutz*, 88: 744–747.

Rashid, A.Q.M.B., M.B. Meah, M. Jalaluddin and A.H.M. Razzak (1988). Effect of fertilizer treatments on the incidence of leaf blight of wheat *Acta Phytopathological Entomologica-Hungaria*, 23(1–2): 61–65.

Raut, J.G., S.M. Guldhe and P.D. Wangikar (1983). Seed borne infection of *Alternaria triticina* in wheat and its control. *Indian Phytopath.*, 36: 274–277.

Rekha, K., K.K. Khanna and Sudhir Chandra (1989). Effect of homeopathic drugs on seed mycoflora of wheat. *National Academy Science Letters*, 12: 39–41.

Roberti, R., P. Flori, and L. Busi (1992). Evaluation of chemical seed treatment for the control of seed brone *Fusarium cúlmorom* and *Bipolaris sorokiniana* on wheat *Mededelingen-Van-de-Faculteil-Landb-Ouwwetenschappen-Universiteit-Gent.*, 57(28): 223–230.

Sehgal, J.M. (1961). Antimicrobial substances from flowering plants–I Antifungal substances. *Hindustan Antibiot. Bull.*, 4: 3–29.

Shah, M.P., M.J. Shah, U.K. Seth, N.K. Bhide (1959). Chinical trials with new indigenous diunetisodium nimbidinate. *J. Assoc. Physicians India*, 7: 235.

Sharma, S., S.R. Ghimire, P.M. Pradhanang and R.R. Pandey (1995). Varietal screening against major diseases of wheat. *Wheat Research Report*, 1996. Nepal, NWRP, pp. 329–337.

Sharma, S. (1996). Wheat diseases in western hills of Nepal. In: *Proc. of the National Winter Crops Technology Workshop*, 1995, Kathmandu, Nepal, NARG and CIMMYT, pp. 269–287,

Sharma, A.K., J. Kumar, and S. Nagarajan (1996). Report of the Coordinated Experiments 1995–96: *Crop Protection (Pathology and Nematology)*. All India Coordinated Wheat Improvement Project, Directorate Wheat Research, Karnal, India, p. 185.

Shekhawat, P.S. and R. Prasada (1971). Antifungal properties of some plant extract in inhibition of spore germination. *Indian Phytopath*, 24: 800–802.

Shekhawat, P.S.N. and P.S. Shekhawat (2001). Source of resistance in wheat against leaf blight *Indian Phytopath.*, 54 (4): 509.

Shivankar, S.K. and Lokhande, V.V. (1993). Screening of different entries of wheat against leaf blight of wheat (Abstr.). *Indian Phytopath.*, 46(3): 320.

Shoemaker, R.A. (1959). Nomenclature of *Drechslera* and *Bipolaris* grass parasites segregated from *Helminthosporium*. *Can. J. Bot.*, 37: 879–887.

Shoemaker, R.A (1966). A pleomorphic parasite of cereal seeds *Pyrenophora semeniperda*. *Can. J. Bot.*, 44: 1451–1456.

Shrestha, K.K., R.D. Timila, B.B. Mahto and H.P. Bimb (1998). Disease incidence and yield loss and foliar blight in Nepal. In: Duveiller, E., Dubin, H.J. Reeves, and McNab, A. (Eds.) *Helminthosporium Blight of Wheat Spot Blotch and Tan Spot*, Mexico, D.F.: CIMMYT, pp. 67–72.

Singh, A.K., A.K. Sinha and M. Taiyab (1990). Studies on varietal stability in spring wheat for responses to *Alternaria* blight and brown rust in Bihar. *Indian J. Gent. P. Breed.*, 50: 91–92.

Singh, A.K., A.K. Sinha and M. Taiyab (1991). Sources of resistance to brown rust and leaf blight in spring wheat. *Annals of Agricultural Research*, 12(1): 88–91.

Singh, A.K., R.V. Singh, D. Singh, S.P. Singh, R.I. Gupta and V.P. Chaudhary (1995a). Assessment of yield lossess due to foliar blight of wheat at farmers fields. *Indian J. of Mycol. Pl. Pathol.*, 25(1–2): 92–93 (Abstr.).

Singh, A.K., R.N. Singh and S.P. Singh (2000a). Distribution of wheat foliar blight pathogens in India. National Symposium on Role of Resistance in Intensive Agriculture, February 15–17, Held at Directorate of Wheat Research, Karnal (Abst.) p. 24.

Singh, A.K., R.N. Singh and S. Kumar (2002a). Artificial spot blotch induction in Bread wheat. *J. Mycol. Plant Path.*, 32(1): 143.

Singh, A.K., R.N. Singh, Shusheel Kumar and B.N. Singh (2002b). Bread wheat genotypes resistant to spot blotch of wheat. *Indian Phytopath.*, 55(3): 378.

Singh, A.K. Susheel Kumar and R.N. Singh (2002c). Fungicidal management of avoidable yield loss caused by spot blotch of wheat. *Indian Phytopath.*, 55(3): 378.

Singh, A., R.D. Mishra, M. Singh, K.P. Singh and S.K. Malik (1996). *Wheat research at Pantnagar.* Res. Bulletin No. 128. Directorate of Experiment Station, G.B.P.U.A T, Pantnagar, p. 25.

Singh, D.P., A.K. Sharma, J. Kumar, L.B. Goel, B. Ram, A. Singh, R.V. Singh, A.N. Tiwari, A.K. Singh, R.N., Singh, S.P. Singh, P.C. Verma, B.M. Khanna, D.S. Dodan, P.S. Bagga, N.K. Patil and I. Kalappanavar (1999a). Losses due to leaf blight in wheat in different agro-climatic zones over years. *Pl. Dis. Res.*, 14: 221.

Singh, D.P., S. Nagarajan, L.M. Goel, J. Kumar, B. Ram, RM. Singh, R.V. Singh, A. Singh, A.N. Tiwari, P.C. Verma, P.S. Bagga, S.S. Matty, A.K. Singh, S.P. Singh and Ramayana Singh (2000b). Evaluation of wheat lines for resistance of leaf blight caused by *Bipolaris sorokiniana* and *Alternaria triticina*. *Pl. Dis. Res.*, 15(1): 110–112.

Singh, D.P., A.K. Sharma, A. Singh, R.V. Singh, A.N. Tiwari, A.K. Singh, R.N. Singh, S.P. Singh, B.M. Khanna, D.S. Sodan, P.S. Bagga and I.K. Kalappanavar (2002d). Losses caused due to leaf blight of wheat in different agroclimatic zones of India. *Pl. Dis. Res.*, 17(2): 313–317.

Singh, D.V. and V.K. Singh, (1971). Control of leaf blight of wheat with fungicides. *Indian Phytopath.*, 24: 694–697.

Singh, D.V., K.D. Srivastava, L.M. Joshi and S. Nagarajan (1980). Note on the occurrence of foliar pathogens of wheat in India. *Indian J. Agric. Sci,*. 50(8): 633–634.

Singh, D.V., L.M. Joshi, and K.D. Srivastava, (1986). Foliar Blights and Spots of Wheats in India. *Indian J. Genet.*, 46 (Suppl.): 217–245.

Singh, D.V. and K.D. Srivastava (1997). Foliar blights and *Fusarium* scab of wheat: Present status and strategies for management. In: V.P. Agnihotri, A. Sarbhoy, and D.V. Singh (Eds.) *Management of Threatening Plant Diseases of National Importance*, Malhotra Publishing House, New Delhi, pp. 1–16.

Singh, M. and P.D. Tyagi (1979). Varietal reaction of wheat to leaf blight in Haryana. *Indian Phytopath.*, 32(4): 635.

Singh, R.N., A. Chauhan and A.K. Singh (2001a). Succession of pathogens and other fungi associated with foliar blights of wheat in rice-wheat system. *Indian Phytopath.*, 54(4): 507.

Singh, R.N., A.K. Singh, S.P. Singh and B.N. Singh (2001b). Occurrence and predominance of wheat foliar blight pathogens in India. *J. Mycol. Pl. Path.*, 31: 339–342.

Singh, R.N., A.K. Singh, S.P. and B.N. Singh (2001c). Prevalence and distribution of foliar blight pathogens attacking wheat in India. *Indian Phytopath.*, 54(2): 175–178.

Singh, R.P., A.K. Singh and R.N. Singh (1999b). Effect of neem products on the growth of *Alternaria triticina*. *Indian Phytopath.*, 52: 308 (Abstr.).

Singh, R.P. and A.K. Singh (2003). Occurrence of *Alternaria triticina* and *Helminthosporium sativum* at different growth stages of wheat. National Symposium on Plant Pathogens Diversity in Relation to Plant Health, Jan. 16–18, Osmania University, Hyderabad, p. 180.

Singh, R.P., A.K. Singh and R.N. Singh (2003a). Pathogenic variability in *Alternaria triticina* causing leaf blight of wheat. *Ann. Pl. Prot. Sci.* 11(2).

Singh, R.P., A.K. Singh, Mamta Singh and R. Singh (2003b). Screening of wheat genotypes against leaf blight caused by *Alternaria triticina*. *New Botanist*, 29.

Singh, R.S. (1983). *Plant Diseases*. Oxford and IBH Publishing Co., p. 56.

Singh, R.V., B.N. Singh, A.K. Singh, D. Singh and V.B. Singh (1995b). Screening of wheat genotypes, against foliar blight caused by *Helminthosporium sativum*. *Pl. Dis. Res.*, 19(1): 88–90.

Singh, R.V., A.K. Singh, R. Ram, and S.P. Singh, (1997). Prevalence of *H. sativum* causing foliar blight of wheat in eastern Uttar Pradesh. Abstract, International Group Meeting on Wheat Research Needs Beyond 2000 AD, August 12–14, *Directorate of Wheat Research* (ICAR), Karnal. p. 107.

Singh, R.V., A.K. Singh and S.P. Singh (1998a). Distribution of pathogens causing foliar blight of wheat in India and neighbouring countries. In: Duveiller. E., Dubin, H.I., Reeves, I. and McNab, A (Eds.) *Helminthosporium Blights of Wheat: Spot Blotch and Tan Spot*, Mexico, D.F.: CIMMYT, pp 59–62.

Singh, R.V., A.K. Singh, R. Ahmad and S.P. Singh (1998b). Influence of agronomic practices on foliar blight and identification of alternate hosts in the rice-wheat cropping system. In: Duveiller. E., Dubin, H.I., Reeves, I. and McNab, A (Eds.) *Helminthosporium Blights of Wheat: Spot Blotch and Tan Spot*, Mexico, D.F.: CIMMYT, pp. 346–358.

Singh, S.P., A.K. Singh, B.N. Singh, R.V. Singh and A. Ahmed (1995c). Studies to minimize the effect of *Alternaria triticina* with management of irrigation fertilizer and fungicidal use. Indian Phytopathological Society 47th Annual Meeting and Symposium, held at NDUAT, Kumarganj, Faizabad, p. 51.

Singh, S.N. and A.K. Gupta (2000). Bioassay of fungicides against *Drechslera sativum* causing foliar blight of wheat. National Symposium on Role of Resistance in Intensive Agriculture, February 15–17. Held at Directorate of Wheat Research, Karnal, Abst., p. 26.

Singh, V.K., B.M. Khanna and P.C. Verma (1977). Reaction of wheat varieties to *Alternaria* blight. *Indian J. Mycol. Pl. Path.*, 7(2): 190–191.

Sivanesan, A. (1984). The *Bitunicate Ascomycetes and their Anamorphs.* Vaduz. Liechtenstein: J. Cramer, 701 pp.

Sivanesan, A. (1987). Graminicolus Species of *Bipolaris, Curvularia Drechslera, Exserohilum* and Their Teleomorphs. CAB International Mycological Institute. Mycological Papers No. 158.

Sokhi, S.S. and L.M. Joshi (1972). Physiological specialization in *Alternaria triticina. Indian J. Microbiol.*, 12: 209–210.

Sokhi, S.S., L.M. Joshi and B. Ram (1972). Leaf blight of wheat due to *Alterania tenuis* (*A. alternata*). *Indian J. Microbiol.*, 12: 203–204.

Sokhi, S.S. and L.M. Joshi (1974). Estimation of losses in yield due to leaf blight disease of wheat caused by *Alternaria triticina. Indian J. Mycol. Pl. Path.*, 4: 29–33.

Spancer, E.Y., R.A. Luding, P.De. Mayo and R.W. White (1963). *Helminthosporac* the toxin associated with *Helminthosporium sativum* (*Bipolaris sorokiniana*) in disease development. Vth International Pesticide Congress, London 70pp. Int. Union Pure an Appl. Chem. London.

Srivastava, A.K. and B. Lal (1997). Studies on biofungicidal properties of leaf extract of some plants. *Indian Phytopath*, 50: 408–411.

Submmanian, C.V. and B.L. Jain (1966). A revision of some graminicolus *Helminthosporia. Curr. Sci.*, 35: 352–355.

Tandon, J.P. (1985). *Wheat for more Tropical Environments*, (Eds.) R.L. Villareal and A.R. Klan, CIMMYT, Mexico, pp. 354.

Toledo, B.J. and A.E. Guzman (1998). Importance of spot blotch caused by *Bipolaris sorokiniana* in Bolvia. In: Duveiller, E., Dubin. H.J. Reeves and McNab, A. (Eds.) *Helminthosporium Blights of Wheat: Spot Blotch and Tan Spot*, Mexico. D.F.: CIMMYT. pp 146–149.

Vance, C.P., T.K. Krick, and R.T. Sherwood (1980). Effect of *Helminthosporium* leaf blight on wheat. *Ann. Rev. Phytopath.*, 18: 259–288.

Vats, A.S. and S.K.S. Chauhan (1995). Efficacy of fungicides *in vitro* and *in vivo* against leaf blight of wheat. *Indian J. Mycol. Pl. Path.*, 25(1 and 2): 111 (Abst.).

Villareal, R.L., A Mujeeb Kazi, L.I. Gilchrisi and E. Del Toro (1995). Yield loss to spot blotch in spring bread wheat in warm nontraditional wheat production areas. *Plant Dis.*, 79: 893–897.

Vir, D. and P.K. Sharma (1985). Efficacy of fungicides XXXII: Evaluation of neem oil for control of plant pathogens. *Asian Farm Chem.* 1(7/8): 23–24.

Von Arx, J.A and E.S. Luttrell (1979). Discussions. In: B. Kendrick (Ed.) *The Whole Fungus*, Ottawa: National Museum of Natural Sciences, 1: 260–261.

Zhimin, Xiao, Lianfa Sun and Wenli Xin (1998). Breeding for foliar blight resistance in Heilongjiang Province China. In: *Helminthosporium Blights of Wheat: Spot Blotch and Tan Spot*, (Eds.) Duveiller, E., Dubin, H.J., Reeves, J. and McNab, A. Mexico, D.F.: CIMMYT, pp. 114–118.

12

Important Diseases of Crops in NEH Region and their Management

M. Santha Lakshmi Prasad, M. Srinivas Prasad,** Y.P. Sharma*** and Sangit Kumar*****

** Directorate of Oilseeds Research, Rajendra Nagar, Hyderabad – 500 030*
*** Directorate of Rice Research, Hyderabad – 500 030*
**** ICAR Complex for NEH Region, Meghalaya – 793 103*
***** Directorate of Maize Research, IARI, New Delhi – 110 012*

North eastern hilly region comprises of Arunachal Pradesh, Assam, Manipur, Meghalaya, Mizoram, Nagaland, Sikkim and Tripura presents an unique geographical area where altitude varies from 50 to 3000 m.s.l. and the climate ranges from subtropical plains to temperate hills coupled with mean annual rainfall ranging from 800 to 4000 mm. A number of agricultural and horticultural crops are being cultivated in NEH region. Among cereals, rice and maize are cultivated as stable food crops. A number of vegetable crops, turmeric and ginger are important cash crops. In oil seeds soybean, mustard and ground nut are cultivated. Citrus and guava are the fruit crops grown in this region. Large Cardamom is the important cash crop in Sikkim.

Crop diseases are inseparable part of any farming system and cause economic losses in terms of reduction in yield, disfigurement of the farm produce resulting in low market value. Most of the crops are cultivated during kharif season which coincides with heavy monsoon rains. Existence of high rainfall, high relative humidity and moderate temperature favours the development and spread of various diseases. Disease management strategies include integration of varietal resistance, modification of cultural practices to reduce the inoculum and inturn disease incidence and chemical control with their most effective mode of application.

Cereals

Rice

Rice, the principal food/cereal crop of the entire north eastern region of India suffers from twenty four diseases. However, few diseases are posing real threat to rice cultivation and those are blast, brown spot, leaf scald, sheath blight at all places and sheath rot at high altitudes.

Blast

This is most important, endemic and destructive disease of rice in all north eastern states causing considerable yield losses of 35–50 per cent (Srinivas Prasad *et al.*, 2000). *Pyricularia grisea,* the causal organism of rice blast affects all the plant parts at all the growth stages but nursery, maximum tillering and panicle initiation are the most vulnerable stages. Typical symptoms on the leaves appear as spindle shaped spots with ashy centre and brown margin. Many such spots unite resulting in a larger irregular necrotic patches on the leaf and in severe cases the affected fields give a blasted appearance. Dark brown or black coloured discolouration also appear on leaf sheath, culm, culm nodes, glumes and neck of the panicle.

Two weed hosts, *Digitaria ciliaris* and *D. marginata* were identified as collateral hosts of the pathogen (Srinivas Prasad *et al.*, 1998a). In the beginning of the season the inoculum builds up on these weed hosts and disease spreads to the healthy crop. Low temperature (20–23°C), high relative humidity (90–95 per cent) and more number of rainy days (19–22 days/month), average rainfall of 1247.4 mm during crop season favours the outbreak of the disease. Races IC and IB are most commonly occurring races of blast fungus in this region (Lakshmi Prasad *et al.*, 1999).

Early sowing of the crop (April–May) escapes the severe disease incidence as compared to late sown crop (June–July).

Destruction of weeds present on the bunds/terraces of the field.

Avoid high levels of nitrogenous fertilizers. Nitrogen should be applied at 40–60 Kg/ha for upland and 50–60 Kg/ha for low land in three equal split doses at:

1. Week after sowing or transplanting,
2. Tillering and
3. Panicle initiation stage (Sharma *et al.*, 1986).

Resistant cultivars should be grown to minimize the disease incidence and following varieties are recommended:

1. *Up land*: IRAT 144, IET 13459, IET 13783, RCPL 1–29 and RCPL 1–27
2. *Low land*: RCPL 1–87–8, RCPL 1–87–4, Ngoba, DR 92, Megha rice 1 and Megha rice 2

Chemical Control

1. Seeds should be soaked for 12 hours in 0.1 per cent Bavistin 50 WP solution prior to sowing in the nursery beds or in the direct sown crop.
2. Seedling root dip treatment for 12 hrs in 0.1 per cent solution of Carbendazim or Fuzi–1 before transplanting the low land crop and a single spray of the same should be given at panicle emergence stage for the control of neck blast (or)

3. Foliar sprays of Bavistin (0.1 per cent) followed by Hinosan (0.1 per cent) or Dithane M–45 (0.25 per cent) should be sprayed at fortnightly intervals both in upland and low land crops when foliar infection is observed. A sticker sandovit 0.1 per cent may be mixed in the spray solution.
4. Application of Kitazin 10 G @30Kg/ ha in two equal doses at tillering and booting stages in low land (Verma and Kumar, 1985).

The integration of varietal resistance, cultural practices and need based chemical control help in combating the disease.

Brown Spot

This disease is incited by *Drechslera oryzae* which occurs mostly in nutrient deficient soils and poorly maintained crop. This disease is more common in Tripura and Meghalaya particularly during Boro season. The disease causes blighting of seedlings in the nursery and oval dark- brown spots on leaves and leaf sheath. Infection also seen on glumes, neck of the panicle and grains as brown discolouration.

As the disease is seed borne, the seed treatment with Dithane M–45 @2g/Kg gives effective control. Collection and destruction of crop residue helps in the effective management of the disease. Application of balanced fertilizer doses *viz.*, 60 : 60 : 40 Kg/h NPK nitrogen in three equal split doses is recommended. Correction of soil deficiency like Potash, manganese and Zinc are helpful. Liming of soil to correct soil acidity reduce the disease incidence. Growing of moderately resistant varieties like Bala, Krishna, Cauvery and Padma are moderately resistant to this disease. Spraying of Mancozeb 0.3 per cent or Carbendazim @0.1 per cent mixed with 0.1 per cent Sandovit after appearance of the disease two times at 15–20 days interval helps in controlling the disease effectively.

Leaf Scald

The disease caused by *Rhynchosporium oryzae* occurs in all states of north eastern hilly region coinciding with maturity stage of the crop which causes 18 per cent yield losses. Oblong lesions with characteristic alternate zonations of the dark brown and light yellow areas appear on upper part of the leaves, they either start from the tip or along the margin. *Metasphaeria albescens* the perfect stage of the fungus was identified for the first time (Maiti *et al.*, 1979).

Growing of resistant varieties like Nagaland special, Parowhite, Pokharali, Masino nemo, Batlong, Pyat kegya, VL–8, sail Bora 56–2 is recommended (Verma and Singh, 1982).

Application of higher doses of Potash (80 Kg/h) could reduces the incidence of the disease. Spraying of Carbendazim (0.1 per cent) and thiophanate-methyl three times starting from the appearance of disease at 15 days interval reduced the disease incidence.

Sheath Blight

This disease became major disease in the recent times. The disease is destructive in highly humid and warm temperatures. *Rhizoctonia solani*, incitant of this disease causes large, oblong lesions on leaf sheath with grayish white centre and brownish margin. In severe cases, all the leaves of a plant are blighted resulting in death of the plant, thereby affecting the yield. Close planting, heavy nitrogen fertilization tend to increase the incidence of the disease.

Local cultivars found resistant to this disease are Changatgai, Mandari, Champason, Ramkelan, Yamur and Nagaland special. Draining of fields also reduced the disease incidence. Spraying of carbendazim @ 0.1 per cent or Hexaconazole (Contaf 5 EC) @1 ml/l. Application of *Psuedomonas fluorescens* as seed treatment (600 g/ha) followed by soil application (2.5 Kg/ha) is found to be highly effective in reducing the disease severity.

Udabatta Disease

This is incited by *Ephelis oryzae* which causes 5–10 per cent yield losses in varieties like Lakhet, Sawteang, Ngoba etc. It was first detected in Jowai area of Nongpoh, Meghalaya during 1981.It causes 100 per cent sterility of the panicle which is transformed into ash coloured, hard, cylindrical spike having all the spikelets glued together, so that no grain formation occurs.

Variety Sarkar longedong was found to be resistant to this disease. As the disease is seed borne, hot water treatment of seeds at 50–54°C for 10 min. before sowing gives effective control of the disease.

Avoiding the use of seeds from infected field is advisable.

Seed treatment with 0.1 per cent Carbendazim and spraying of same fungicide at the panicle initiation stage reduces the disease incidence.

Sheath Rot

This disease was considered as minor, which assumed a major form in recent years and it is caused by *Acrocylindrium oryzae*. This disease is severe in low land conditions particularly in high altitude areas. Spots develop on the upper most leaf sheaths enclosing the panicle as grey coloured irregular patches with brown margins.

Destruction of infected plant debris by burning. Spraying of Dithane M–45 @0.2 per cent reduces the disease.

Bacterial Blight

This disease is incited by *Xanthomonas campestris* pv *oryzae* mostly occurs in low altitude areas of Tripura, Meghalaya and Arunachal Pradesh during July to September. Yellow to white lesions appears along the margin which may enlarge to cover the entire leaf. The disease spreads fast under favourable weather conditions like incessant rains, strong wind and warm temperature.

High yielding varieties like PUSA 2–21, Saket–4, Mashuri, IR–20, IR–22, IET 15654, IET 16562 and IET 15426 are moderately resistant to this disease.

Seed treatment with Streptocycline (100 ppm) for 8 hrs., spraying of Blitox 0.3 per cent alternately with Streptocycline (250 ppm) after the appearance of disease is effective in controlling the disease.

Root-Knot Nematode

Meloidogyne graminicola causes formation of galls on the roots. The infected plants look stunted with orange yellow leaves and highly reduced the yield. The disease appears in sandy loam soils of Tripura.

Flooding the soils, destruction of alternate or collateral hosts, rotation with non-host crops in the rice soils reduce the nematode population. Variety CR–75(Mut) 11–4 and Akashi was tolerant to this disease. Application of Carbofuran granules @1 Kg a.i./ha in nursery, 7 and 50 days after transplanting reduced the disease.

White Tip

This is incited by *Aphelenchoides besseyi*. The leaf tip becomes white, which later turns brownish and break-off. The plants appear stunted and tillering is reduced. The panicles become chaffy with small and distorted kernels. The disease is prevalent in Tripura and causes appreciable reduction in the yield.

The disease is seed borne, hot water treatment of seeds at 47–53°C for 15 minutes followed by spraying with 0.025 per cent Parathion at 4, 40 and 60 days after planting reduced the nematode infestation.

Maize

Leaf Blight

This is the most important disease in north eastern hills caused by *Exerohilum turcicum* especially in late sown crop, it causes yield losses of 15–20 per cent. Long, elliptical, grayish green or tan lesions appear on the lower leaves, then progresses to upper leaves of the plant. Severe infection causes a premature plant death and grey appearance that resembles frost or drought injury. In damp weather, grayish black spores are produced on the lesions, often in concentric/target like zones. The disease is favoured by moderate temperature (18–27°C) and heavy dew during the growing season.

Late sowing should be avoided. Early sowing upto 30th May gives higher yield with less blight intensity.

Field sanitation *i.e.*, collection and destruction of crop residue should be maintained.

Seed treatment with Thiram @2g/Kg seed. Spraying of Mancozeb @ 0.2 per cent or Carbendazim @ 0.1 per cent starting from knee high stage at 15–20 days interval manages the blight incidence. Chemical control is profitable on pop corn, sweet corn, seed crop and preservation of precious genetic material.

Growing of resistant cultivars like Deccan 103, Suwan–1, Hemant, Ganga 11, Meghalaya local yellow, Vijay composite, NLD white, RCM 1–1, RCM 1–2 is recommended (Mahajan *et al.*, 1991, 1995).

Rust

This disease is caused by *Puccinia sorghi,* occurs in cool climate (16–23°C) and high relative humidity (100 per cent). It is more in medium and high altitude areas of Meghalaya and Sikkim. Circular to elongate brownish spots appear on both surfaces of leaves, which later turns to brownish black. In case of severe infection, death of plants and leaf sheaths may occur. The pustules become erumpent and powdery. Diseased plants remain comparatively stunted in growth and exhibit sick appearance.

Removal and destruction of common weed like *Oxalis* sp. reduces the primary inoculum and disease development.

Spraying of Mancozeb @0.2 per cent at 10–15 interval after the appearance of disease reduces the disease incidence.

Downy Mildew

This is wide spread disease in NEH region caused by *Sclerospora philippinensis, S. sacchari* and *Sclerophthora rayssiae* var *zeae.* Its severe occurrence was reported from zinc deficient upland soils of Tripura.

Long, white or light yellow stripes appear along the entire length of the leaf, which changes to dark brownish yellow in late stages after 3–4 weeks. In severe infection stripes coalesce and irregular elongated patches are formed. Puckering of the leaf surface in the striped area and downy growth on the stripes is very common symptom mostly on lower surface. In susceptible varieties, small to large leaves develop on the tassel and these leaves bear white to yellow stripes with downy growth. Proliferation of axillary buds on the stalk of tassel and cob is also common in Hybrid Ganga. Major damage is caused if infection occurs at early stage. If infected plants continue to grow, these produce deformed cobs with few grains. High rains leading to high humidity (95 per cent) and moderate temperature (25°C) favour the disease development and spread. Free moisture on the host surface is essential for the formation and germination of conidia leading to the fast secondary spread of the disease.

Diseased crop residue/debris and alternate weed hosts should be destroyed. Crop rotation for longer periods help in reducing the disease.

Seed treatment with Metalaxyl @2g/Kg seed reduces the incidence of the disease and spraying of Ridomil–MZ @0.1 per cent two times at 15 days interval reduces the incidence of the disease.

OILSEEDS

Soybean

In north eastern hilly region, soybean is a traditional crop, cultivated either as monocrop or inter cropped with maize under shifting cultivation. Soybean is affected by 18 diseases, out of which rust, frog eye leaf spot, purple seed stain are important diseases and two new diseases leaf blight and zonate leaf spot were reported from this area (Maiti and Vishwadhar, 1982).

Rust

Rust caused by *Phakopsora pachyrhizi* is distributed through out north eastern hilly region. In general, the disease appears in mild form in the foot hills and becomes more severe in medium and high altitude areas (Maiti *et al.*, 1983).

Large number of light brown pustules appear on the dorsal surface of the leaves in the initial stages, later turns to reddish brown to tan colour. Lesions are generally restricted by veins giving angular appearance. In kharif crop, the disease appears in the first week of September coinciding with flowering or pod formation. The leaves gradually become yellow and premature defoliation occurs resulting in yield loss upto 36 per cent.

The disease was found to appear in the first week of September on crops sown in June/July but it did not occur in crops sown upto middle of May. The disease spreads from two possible sources of primary inoculum *i.e.*, teleutospores, which were found to form at high altitude in October and November and *Mucna* spp., a wild leguminous creeper found to harbour *P. pachirhizi. Trichothecium roseum*, a biocontrol agent caused 90 per cent inhibition in germination of infected uredospores of *P. pachyrhizi* (Sangit Kumar and Jha, 2002) under *in vitro* conditions. White mycelial growth of *Troseum* appears on leaves of rust infected soybean plants, it spreads and covered the uredopustules on the lower surface of the leaves. Colour of the mycelium changed with age from white to buff and to light pink.

Early maturing cultivars sown in May to 1st week of June escape rust infection.

Growing of resistant varieties *i.e.*, PK73–84, PK 310, PK 74–293, Manipur local, Punjab 1, JS 80–21, Himso 107, Himso 772, Himso 773 is recommended.

Spraying of Saprol @0.05 per cent/Topsin M @ 0.2 per cent followed by Dithane M–45 @ 0.2 per cent at 10–15 days interval, beginning at the appearance of the disease gives effective control of the disease (Maiti *et al.*, 1983b).

Frog Eye Leaf Spot

This disease is most common during warm and humid weather in mid altitude areas. *Cercospora sojina,* the incitant of this disease causes minute reddish brown, circular to angular spots appear on the upper surface of the leaves. As the lesions enlarge olive grey or ash grey area develops in the center, surrounded by narrow dark reddish brown border. Heavily infected leaves may fall prematurely.

Early sowing (upto 20th June) favours disease build-up in susceptible varieties. The fungus survives on crop refuse and on seed.

Healthy seed should be used for sowing. Crop rotation with grain crops help in keeping the disease incidence low. Cultivars such as JS 75–46, JS 78–78, Ankur, JS 75–46, JS 80–21, PK 262, PK 472 have high degree of field resistance.

Seed treatment with Captafol @3g/Kg and spraying of fungicides like Saprol @0.05 per cent, Topsin M @0.2 per cent, Dithane M–45 @0.25 per cent beginning at the appearance of the disease followed by a second spray at 10–15 days interval gives effective control of the disease.

Anthracnose

The disease appear at any stage of crop growth. This disease causes considerable damage by affecting seedling stand, seed quality and yield in warmer regions. The disease appears in severe form in Nagaland, Tripura and foot hills of Meghalaya, Arunachal Pradesh, Sikkim and Mizoram. Dark brown sunken lesions often appear on cotyledons of emerging seedlings. On pods, reddish brown to dark brown areas appear and turn black, finally the pod dries out. The few seeds in the diseased pods may be shriveled and moldy. When such seeds are sown, pre and post emergence damping-off may appear. The fungus survives on plant debris and on seed.

The seeds should collected from disease free fields.

Field sanitation: Destruction of crop debris. Seed treatment improves the emergence of seedlings.

Six weekly sprayings with Zineb or mancozeb (2.5Kg/ha) minimizes the infection.

Varieties like Bragg and Hardee have high degree of field resistance.

Purple Seed Stain

This is caused by *Cercospora kikuchii* and it is wide spread in this region. This disease also called as purple blotch or spot. The fungus causes a purple colouration of seed and also attacks other plant parts. The coloured areas on seed may vary from pink to dark purple and may be small or almost cover the entire seed. Cracks often appear in the coloured area. The coloured grains have low market value and poor germination. Infected seeds often produce poor seedlings. Cotyledonary infection may spread into stem producing necrotic girdle on the stem ultimately killing the young plant. The fungus survives in the infected plant refuse and infected seed.

Removal and destruction of plant debris. Use of healthy seed for sowing. Crop rotation can reduce the disease incidence.

Seed treatment with thiram + benomyl (1 : 1) @4g/Kg seed eradicates the seed borne inoculum and improves emergence (Maiti *et al.*, 1983c).

Spraying of benomyl (0.5Kg/ha) at 30 days interval for two times during flowering is effective in controlling the disease.

Zonate Leaf Spot

This has been reported for the first time from Meghalaya and is caused by *Aristoma camarographioides.*The disease is restricted to high altitude area and appears during monsoon months (June–September) when mean minimum and maximum temperatures are 15 and 24.7°C.

The disease appears first on the lower leaves about one month after sowing. Symptoms develop initially as brown pin heads on both leaf surfaces and slowly enlarge into spots with water soaked margin. Fully grown spots are irregular, often limited by veins at the margin with clear distinction of narrow dark brown and light brown zones surrounded by yellow margin. In severe infection, overlapping spots cause a blighted appearance. Numerous pycnidia are formed on the spots.

Cultivars IC 2143, IC 18383, IC 18762, EC 15694, EC 1694, EC 16419, EC 1975 are resistant to the disease.

Setosphaeria Leaf Spot/Blight

This is a newly recorded disease from high altitude areas and favoured by cool and humid weather conditions. *Setosphaeria glycines* causes water soaked spots on the leaves and gradually enlarge into brown spots. Spots are irregular, often limited by veins at the margin and surrounded by yellow halo. In severe cases, the leaves get blighted and resulted in premature death of the plants. Numerous minute ascocarps are formed on the lesions. The fungus survives in plant debris and reinfect the host in next season.

As the fungus survives in the debris, removal and destruction of plant debris reduces the inoculum.

Cultivars DS 73–16, MGSB 75–1, 78–1, 78–2, 78–3, DS 73–8 are resistant to this disease.

The disease can be controlled by spraying of Bavistin @0.1 per cent and Dithane M–45 @0.2 per cent three times at 15 days interval.

Mustard

In mustard, white rust alone and in combination with downy mildew was quite common in all the north eastern states.

White Rust

Both foliar as well as systemic infection have been found to cause considerable damage to the crop. In case of local infection white creamy yellow pustules appear on the leaves which later coalesce to form patches. Systemic infection causes hypertrophy and hyperplasia on stem. Maximum damage was recorded when systemic infection of the stem occurred. The disease incidence was favoured by minimum temperature (7.4°C) and maximum temperature of 19°C with relative humidity (60–79 per cent).

Downy Mildew

This is incited by *Peronospora parasitica.* Downy grayish white fungal growth appears on all aerial parts usually on leaves and inflorescence. Infection of inflorescence causes hypertrophy of the peduncle or the whole inflorescence. The affected inflorescence does not produce any siliqua or seed.

Cultivars belong to *Brassica campestris* and *Bjuncea i.e.,* YSB 9, Sikkim sarson–1, Toria–VET Tik–1, PR 3, PR 8, PR 30 and PR 45 were found to be resistant.

Spraying of Blitox–50 (0.25 per cent) or Fytolan (0.25 per cent) four times at fortnightly interval reduces the disease incidence.

Alternaria Leaf Blight

Alternaria brassicae causes grey spots on leaves and siliquae surrounded by reddish brown margin. The disease was severe in sikkim.

M–27, YID 3–87 and PYS–287 were least affected by leaf blight. Sikkim sarson showed 11 per cent infection on leaf and 13 per cent on siliqua. YED–843 and YID 3–87 identified as moderately resistant to white rust and *Alternaria* blight. In sikkim, the lines resistant to major disease are SKYS24–2, 3, 18, 21, 26, 27, 14, 5, 7, 20, 210–19, 9–0, 30, 28–17 and 32.

Spray application of Bavistin @0.1 per cent after appearance of disease and repeated spray at 15 days interval reduces the incidence of disease.

Groundnut

In north eastern hilly region, ground nut crop is introduced in recent years and it is grown under a wide variety of soil and climatic conditions. Six diseases were recorded on this crop, of which early leaf spot and late leaf spot are economic important diseases.

Early and Late Leaf Spot

Losses on the basis of dry pod yield were estimated to be 56 per cent and 82 per cent in JL 24 and R–19–8 cultivars of ground nut due to early and late leaf spot in absence of rust disease in Meghalaya.

Early leaf spot is caused by *Cercospora arachidicola* and *Phaeosariopsis personata* causes late leaf spot. Circular to irregular reddish brown to brown bigger lesions appears on lower surface of leaf surrounded by yellow halo in early leaf spot whereas dark brown to black leaf spots which are smaller, almost circular with less diffuse margin appear in late leaf spot than those of early leaf spot.

Crop residue should be destroyed as they may harbour infection. Crop rotation with non-leguminous crop reduces the inoculum.

Early planting (20th March and 5th April) reduces the incidence of disease and gave good results (25q/ha).Varieties like ICGV 86590, 87337, ICGV 86020, ICGS 76, NRCG 5953, NRCG 6346, NRCG 3209 exhibited consistent low disease score in Meghalaya for early leaf spot.

Tilt (Propiconazole), Topsin–M, Bavistn and mixture of Dithane M–45 +B avistin were found effective. Single spray of a mixture of Dithane M–45 (0.25 per cent) + Carbendazim (0.1 per cent) at 40–50 days after sowing was found economical for the control of early leaf spot in Meghalaya (Chandra *et al.*, 1998).

Rust

This is caused by *Puccinia arachidis.* Small brown to chestnut brown dusty pustules appear on the under surface of leaves, the epidermis ruptures and exposes a powdery mass of uredospores. Small, brown necrotic spots appear on the upper surface of leaves. In severe cases, premature defoliation occurs. The affected plants produces small, shriveled kernels.

Spraying of Dithane M–45 @0.3 per cent effectively reduces the rust incidence.

Pulses

Rice Bean

This pulse crop is grown only in this north eastern region and affected by leaf spot, powdery mildew and rust which are economically important diseases.

Leaf Spot

Cercospora canescence causes brown to reddish brown spots on leaves which are surrounded by purplish border, the centres of which may turn grey. In severe cases, spots are also appears on fruit and floral parts. In some cases, the affected tissue is killed and the dead portion drops out leaving shot-hole symptoms on the leaves. Removal and destruction of infected plant debris from the field.

In Arunachal Pradesh, RBL–3, RBL–6, RBL–9, PDBR–1, PDBR–3 were moderately resistant to Cercospora leaf spot.

Spraying of Mancozeb @0.3 per cent after the appearance of disease and three times at 15 days interval manages the disease incidence.

Powdery Mildew

This is caused by *Erysiphe polygoni*. On leaves, white powdery growth occurs spreading to cover the stem and other plant parts. The infection generally affects the plant at flowering stage and it persists until harvest causing wilting of the plant. The plant assumes a greyish white appearance and the leaves turn yellow and are finally shed. Diseased plant debris should be destroyed in the field.

Spraying of Topsin–M @0.2 per cent or Bavistin @0.1 per cent followed by wettable sulphur was recommended for the management of this disease.

Rust

This is caused by *Uromyces appendiculatus*. Reddish brown circular spots appear on lower surface of the leaves and they are surrounded by a yellow halo. In severe infection, complete defoliation occurs resulting in reduction in the yield.

RBL 36, RCR 7–68, RC local red, BC–6, RDM–30 were found resistant to this disease.

Spraying of Bavistin @0.1 per cent followed by Captaf or Cuman–L after the appearance of disease and 2–3 times at 15 days interval was superior in the management of rust disease.

Black-gram

This pulse crop is affected by Powdery mildew (*Erysiphe polygoni*), frog eye leaf spot (*Cercosporidium* sp.), leaf blight (*Corynespora casicola*), *Alternarla* leaf blight and black gram leaf crinkle disease.

Powdery Mildew

The fungus *Erysiphe polygoni* infects at any stage of plant growth but it was severe at flowering. White powdery growth appears on leaves and other parts of plant which results in the reduction in photosynthetic leaf area there by yield.

RCU 6–2, RCU–3, RCU 6–21, RCU 6–15, RCU 6–17, TAU–1X UL 203, 99U–43 were resistant to powdery mildew.

Spraying of wettable sulphur 2–3 times at 15 days interval reduces the incidence of disease.

Frog Eye Leaf Spot and Leaf Blight

Frog eye leaf spot is caused by *Cercospora canescens*. Brown to reddish brown spots appear on leaves often bound by veins with purplish border which resembles the frog eye. In some cases, the affected tissue is killed and drops out leaving shot-hole symptoms.

Leaf blight is caused by *Corynespora cassicola* and also *Alternaria* sp. Small, discoloured lesions scattered irregularly on the leaves. These spots becomes brown to grey irregular spots surrounded by light yellow zone. Several lesions coalesce to cover large areas of the leaf causing blight.

T–9, Pant U–19, Pant U–30, WBU–104 were resistant to frog eye leaf spot. RCU6–2, BSU–1 were resistant to blight. Spraying of Mancozeb 0.3 per cent or Bavistin 0.1 per cent at 15 days interval reduces the incidence of frog eye leaf spot and blight.

Black-gram Leaf Crinkle

This disease is recorded in recent times in some varieties and caused by virus. The leaf crinkle virus causes dwarfing, puckering, severe curling of the leaves and markedly reduced in size.Mottling is also observed with prominent bigger yellowish patches on the foliage. The pods are deformed results in heavy yield losses.

The virus was sap transmissible (80 per cent) with incubation period of 21 days whereas white fly (*Bemesia tabaci*) transmitted the disease (87 per cent) with incubation period of 12 days (Srinivas Prasad *et al.*, 1998).

NDU94–6, BSU–2 were resistant to leaf crinkle, disease. Spraying of neem oil @5ml/L was found effective in checking the disease as well as white fly population.

Cash Crops

Turmeric and Ginger

These are two cash crops grown in all states of NEH region.

In turmeric, leaf blotch (*Taphrina maculans*), leaf spot (*Colletotrichum capsici*), large leaf spot (*Phaeodactylium alpinae*), weavy leaf blight, sheath and leaf rot (*Fusarium* sp.) and leaf blight (*Rhizoctonia solani*) are major diseases. In ginger soft rot, leaf spot and blight are major diseases.

Both leaf blotch and spot causes heavy reduction in yield in turmeric.

Leaf Blotch

This disease appears on turmeric where ever it is grown. On leaves, large number of spots appear covering a major portion of the leaf blade and the affected leaves appear reddish brown. In severe infection, the affected tissues wither and dry, the leaves are severely blotched and blighted.

Leaf Spot

Oblong brown spots with grey centres appear on leaves. With maturity of the leaves, black dots representing the fungal acervuli occur in concentric rings on the spots. Severely infected leaves dry and wilt.

Removal and destruction of infected plant debris. Seed should be selected from disease free areas.

PTS–10, PCT 12, PCT 18, VK 77, VK 145, PTS 11, PTS 35, RCT–1, CA 69, CL 349, RCT–1, Annov K, Janaka, Kuchipudi, Manipur local, Padem local were resistant to both diseases. In Meghalaya, a Lakadang selection RCT–1 was found highly resistant to leaf spot.

Leaf blotch and leaf spot can be controlled by spraying Bordeaux mixture or Carbendazim 0.1 per cent 2–3 times at 15 days interval.

Soft Rot of Ginger

This is also called as Rhizome rot and is the most important disease of ginger. This is a complex disease caused by *Pythium* sp. or *Fusarium* sp. alone or in combination with each other. Infestation of different insects further aggravates the disease problem.

The leaves become pale in colour with yellow tips which spreads down the leaves and whole leaf turns yellow. The basal portion of the plant becomes watery and soft.

Rotting of the rhizomes occurs which converts into a pulpy mass with a putrefying smell. Seed should be collected from healthy field. Seed rhizome of 30–50g weight is recommended with a precaution of not to remove treated mother rhizome for eating purpose which is common practice in the farmer's field.

Seed rhizome dip treatment with Dithane M–45 @0.25 per cent + Ridomil MZ @0.1 per cent + Chlorpyriphos (0.2 per cent) for 1h was recommended for the control of this disease. Infected plant should be destroyed.

Vegetables

Damping Off

It is one of the most devastating nursery disease affecting the seedlings of all vegetable crops like brinjal, chillies, crucifers, okra, pea and tomato.

This is incited by *Pythium debaryanum, P. ultimum,* and *P. aphanidermatum.* It occurs in two stages *i.e.,* pre-emergence and post-emergence phase.In pre-emergence, the young seedlings are killed before they reach the soil surface. In post-emergence, the young juvenile tissues of the collar region gets infected at the ground level, they become soft and water soaked and the seedlings collapses.

Seed treatment with Captan @0.2 per cent reduces the damping off.

Seed should be used from healthy field, crop rotation also reduces the inoculum in the field.

Drenching of raised beds with 1 per cent Bordeaux mixture @11 /sq.m. bed area before sowing helps in disinfecting the nursery bed. Spraying of 1 per cent Bordeaux mixture on seedlings of 8–10 cm height check most of the foliar diseases and before transplanting the roots of seedlings should be dipped in 1 per cent Bordeaux mixture for 10–15 min. to check the soil borne diseases.

Bacterial Wilt of Solanaceous Vegetables

In NEH region, this disease is major disease affecting all solanaceous vegetables seriously and it is caused by *Ralstonia solanacearum* which infects other crops like ginger, ground nut, sesamum and ornamental plants in the region. The disease severity is more on tomato during April where as brinjal and chillies are seriously affected in May. The disease is favored by high soil moisture of 85–90 per cent and low temperature. The bacterium survives in the soil for longer periods, once the field is infested it is difficult to irradicate the pathogen.

Crop rotation with cereal crops for longer periods (3–4 Yrs) is recommended. Brinjal cultivars BWR 54, BWR 12, ARU–2–C, PPr–91, Pusa Purple Round and BB1 were resistant to this disease in Arunachal Pradesh and Nagaland. Chilli cultivar 8/83–22 was highly resistant to this disease.

Application of *Trichoderma viride* @5 Kg mixed in 500Kg FYM/h reduces the disease. Furrow application of bleaching powder in the soil @20 Kg/h one month prior to planting controls the disease effectively.

Spot application near collar region of the infected plants with a solution prepared by mixing 30 g copper oxychloride and 3 g streptocyclin in 10 lit. of water is recommended.

Drenching with 1 per cent Bordeaux mixture/Copper oxychloride (4g/1) + Streptocycline (1g/1) is also reduces the disease incidence.

Pea

Pea crop is cultivated in rabi season as vegetable crop.

Powdery Mildew

This is severe disease in all pea growing areas and is serious threat for pea cultivation. This is caused by *Erysiphe polygoni*. On upper surface of leaves, white powdery spots appears and during flowering stage the severity increases. The leaves turn yellow and finally shed. All the above ground parts, leaves, stems, tendrils, stipules and pods are covered with white powdery mass which, eventually becomes dusty in appearance. The disease appears early in sub mountaneous and hilly regions, where as it occurs late in plains. The greater incidence of *E. polygoni* on peas under dry than wet soil conditions indicate the preference for low moisture conditions. Conidial germination is higher at humidity levels of 50 per cent and above.

Pea cultivars resistant to powdery mildew are T10, P185, P6583, 6587 and 6588.

Removal and destruction of infected plant debris. Healthy seed should be used for sowing. Seed treatment with Thiram 0.3 per cent or Bavistin 0.1 per cent is recommended.

Spraying of sulfex (wettable sulphur) @0.2 per cent or Karathane and Bavistin (carbendazim) @0.1 per cent 3–4 times at 15 days interval during flowering stage is effective in controlling the disease.

Rust

This is incited by *Uromyces fabae* and reddish brown, circular uredopustules appear on lower side of leaves which later turns to black with teleuto pustules. In severe cases, leaves falloff prematurely resulting in heavy yield losses.

Seed treatment with Thiram @0.3 per cent is recommended.

Spraying of chlorothalonil (Kavach) @0.2 per cent 2–3 sprays at 10 days interval is effective against rust.

Beans–French Bean and Broad Bean

These are leguminous vegetables grown in this region. Beans are affected by rust (*Uromyces phaseoli*), anthracnose (*Colletotrichum lindemuthianum*), pod rot (*Pythium* sp.), aerial blight (*Rhizoctonia solani*), leaf blight (*Phaeoisariopsis graseola*). Among them rust, anthracnose and aerial blight are most important diseases.

French bean is an important rabi crop of Meghalaya commonly grown for green pods after Kharif paddy in north eastern states.In this crop, three serious diseases *i.e.*, rust, angular leaf spot and anthracnose are reported. In broad bean (*Vicia faba*), powdery mildew, rust and *Alternaria* leaf spot are important diseases.

Rust

This is caused by *Uromyces phaseoli* typical *Uappendiculatus.* Characteristic rust pustules are produced on hosts which are reddish brown, circular mostly on lower surface of the leaves surrounded by yellow coalescing to cover a large area of the leaf blade. Late in the season, the telia are formed in uredosori which are dark brown and linear. In severe infections the leaf may wither resulting in considerable damage to the crop. The disease is favoured by cloudy, humid weather with heavy dew and temperatures of 21–26°C.

In beans, rust resistant varieties *i.e.*, H-14-1, Nim-11, JK-1, JK-2, 9, HUR-15-1, 15, 120, 91, KUR15, PDR3, 24, 22, 14, 15, RCR11, VL63, RCR7, JK12, 1, 3 were recommended in pro-kharif and rabi season in Sikkim.

In French bean, PI 299489, PI 247761, PI 201489, Kintucy wonder, Premier, Contender, URF 626, VL-5, IC 37153, IC 47810, Meghalaya dwarf, Canadian wonder and Arka komal were resistant to rust.

In broad bean, EC 108803 and EC 117762 were resistant to rust.

Spraying of Contaf (Hexaconazole) 0.1 per cent in combination with Mancozeb was highly effective in reducing the disease in Meghalaya Local and Manipur Local (Singh and Narayanbhat, 2002).

Folpet or Topsin M or Bavistin @0.1 per cent are also effective in reducing the disease.

Anthracnose

This is common disease of all beans and caused by *Colletotrichum lindemuthianum* (*Glomerella lindemuthiana*). The fungus produces depressed spots with dark centres with bright red, yellow or orange margins.All plant parts above ground level are prone to infection and infection may occur at any stage of plant growth. On pods also, water soaked lesions appear which later turns to brown with characteristic sunken lesions. The pathogen is seed bome. Seed treatment with carbendazim 0.1 per cent decreases the seed borne infection. Mulching the field reduces the incidence of disease at Basar, Arunachal Pradesh.

In broad bean, Dholi 5 and Dholi 12 were resistant to anthracnose. Spraying of Mancozeb (0.2 per cent) or carbendazim or Benlate @0.2 per cent with disease appearance followed by second spray at 7–10 days interval.

Potato, Tomato, Chillies and Brinjal

These are common solanaceous vegetables grown in this region and these are affected by many diseases, among them early blight, late blight of potato and tomato, die back and fruit rot of chillies, *Phomopsis* fruit rot of brinjal are economically important diseases.

Late Blight

This is one of the devastating disease of potato and tomato caused by *Phytophthora infestans*. This north eastern hilly region is considered as hot spot for this fungus as all the races of *P. infestans* are existing in this region. Some times, it causes epidemics resulting in complete failure of the crop. The disease normally occurs from February to June in north eastern hills and it spreads quickly with rains.

Water soaked, light brown lesions appears on leaf blade, under humid and cloudy weather these lesions spread fast over the entire leaf let and petiole. The lesions are roundish with concentric markings generally involve the leaf margin and they becomes dirty brown to black in later stages. The fungus shows tuft of whitish growth comprising of sporangia and sporangiphores on leaves, on close examination of infected leaves. The fungus infects the tuber causing dry rot and brown discolouration of the tissues in dry soils. In wet soils, the fungus spreads fastly and causes soft rot results in watery ooze. The disease is favoured by low temperature (18–22°C), high relative humidity (100 per cent RH) and cloudy weather with occasional rains. Most of the potato races are known to attack tomato so far T_0 and T_1 is more aggressive.

Disease-free tubers should be used for cultivation.

Burrying the infected tubers and plants in soil. Removal of ground keepers and volunteer plants, high ridging, restricted irrigation during blight period results in reduction of blight disease.

Growing of resistant varieties like Kufri kundan, Ottawa 30, 31, PI 270407 in potato is recommended. In tomato, west Virginia 63, Pierline, New Yorker, Nova, Fla–814 and Hessoline are resistant to this disease.

Dipping of sprouted tubers in 0.15–0.2 per cent Metalaxyl for 30 min. suppressed the pathogen. Regular spraying of Mancozeb @0.2 per cent at an interval of 10 days protects the plants from infection. Spraying of Ridomil MZ @0.05 per cent gives systemic protection to the plant. During rainy season, some sticker like sandovit or Triton @0.1 per cent has to be added to the fungicide.

Two sprays of Metalaxyl + Mancozeb at 15 days interval could effectively control late blight and increased the yield.

Early Blight

The disease is most common in tomato and potato, caused by *Alternaria solani.* Under congenial conditions, the disease causes considerable damage to the crop along with late blight.

The fungus produces irregular, brown to dark brown spots with concentric rings. These spots unit and form large patches, resulting in leaf blight. On stem, dark elongated spots appears at any point. But these are injurious when engirdle the juncture of stem. Some times the fungus infects the tuber near the soil surface causing brown discolouration and dry rot. The fungus survives on infected leaves and seeds. The spore dispersal continues throughout the tomato growing season, however, it is maximum in the advanced stages of disease development. The young plants show resistance while the old and the weak plants are more prone to the attack of the pathogen.

In tomato PI 134417, PI 126932, PI 112835, Hawai 79980 and EC 116050 are resistant cultivars to blight.

Crop rotation with cereal crops is advisable. Removal and destruction of infected plant debris. Spraying of copper fungicides like Zineb @0.3 per cent at 15 days interval 3–4 times starting from six weeks after planting manages the disease.

Anthracnose or Die-back and Fruit Rot of Chillies

This is most economically important disease of chillies causing considerable damage to the crop and is caused by *Colletotrichum capsici.* It causes three types of symptoms such as anthracnose, die back and fruit rot.

In die back, the disease causes necrosis of tender twigs from the tip back wards. Entire branch or entire plant dies back as the attack spreads downwards.

On fruits, small black, circular spots markedly delimited by a thick and sharp black outline enclosing a lighter black or straw coloured area. Severely infected fruits lose their normal red colour and turn straw coloured. Numerous black acervuli are found in concentric rings and black, sphericalsclerotia are found in side the diseases fruit.

The fungus survives on infected parts of the plant so destruction of infected plant debris reduces the inoculum. The seeds should be selected from disease free seeds. Seed treatment with thiram @2g/ kg seed.

Spraying of Thiram/Captan on newly emerged seedlings at 5–7 days interval or Mancozeb (2Kg/h) or Vitavax on the matured plants at fortnightly interval reduces the disease occurrence.

Phomopsis Fruit Rot

This is most important disease of brinjal caused by *Phomopsis vexans* and the fungus causes seedling blight, leaf spot and fruit rot symptoms. It attacks the seedlings in the nursery causing damping off. On matured plants, clearly defined circular, grey to brown spots with light coloured centre on leaves. The affected leaves turn yellow and die. On fruits, the disease appears as minute, sunken dull and dusky spots which later merge to form rotten areas. In severely infected fruits the entire flesh may rot. Numerous black pycnidia can be seen on affected portion of fruit surface.1n some cases, mummified fruits are formed. The fungus perpetuates on infected host debris, destruction of infected plant tissues reduces the inoculum.

Seeds from healthy fruit should be selected for sowing purpose.

Seed treatment with Carbendazim @0.1 per cent or thiram @0.2 per cent protects the seedling from infection. Spraying of Difolatan (0.2 per cent) or Captan @0.2 per cent in the nursery and in field at intervals of 7–10 days manages the disease incidence.

Fruit Crops

In north eastern hilly region, many species of citrus and guava are cultivated as main fruit crops.1n temperate fruits, apple and peach are cultivated in Arunachal Pradesh and Sikkim. The citrus species are affected by scab, anthracnose, greening, decline, gummosis, powdery mildew, pre-mature fruit drop and canker. Other important diseases includes anthracnose and wilt of guava and apple scab.

Citrus

Scab

This disease is prevalent widely in North Eastern Hilly region infecting leaves, shoots and fruits caused by *Elsinoe fawcetti*. On leaves small, semi translucent dots appear that becomes warty erupted with cracks and scabby. These spots are also found on fruits as raised spots with cracks and scabby.Newly emerged leaves are most susceptible but develop resistant on maturity. Fruits remain susceptible upto 3 months after the fruit set.

The fungus survives on infected plant parts and spreads to new flushes with incession of rains. In severe infection, the whole new shoot will be distorted. The disease is favoured by minimum temperature (20–30°C), high relative humidity and high rain fall. Maximum disease intensity (30 per cent) was observed in September whereas minimum in December (8.3 per cent). Citrus species varied in scab reaction with minimum intensity observed in *Citrus latipes, Karunjamir, Soh joh, Adajamir* and *Khasi mandarins* (Srinivas Prasad *et al.*, 2000).

In Arunachal Pradesh, Citrus species of Kamla Australia, Kinnow mandarin, Wilkin mandarin, Mandarin Thorney, Mandarin King, Theipi, Mandarin loddo, Citrus zigdardia and Satsuma. In root stock like Trifoliate orange, *Citrus volkamariana, C. latipes* and sweet oranges are resistant to scab.

Gummosis

In north east, gummosis is caused by *Phytophthora nicotiana* var *parasitica* (Dastur) Comb which is often associated with *Fusarium lateritium*. The disease becomes noticeable by profuse gumming on surface of the affected bark which turns dark brown and develops longitudinal cracks. The wood are also affected. When the infection advances further the gumming may be seen on the higher branches. The intensity of infection varies with the plant species, gum formation is more in oranges and less in grape fruit. The plant blossomes heavily but not able to maintain the fruits till maturity and can die in a year or two.

Twig Blight/Anthracnose/Wither Tip

This is a serious and widely distributed problem of citrus in NEH region. This is characterized by shedding of leaves and die back of twigs. A number of stress factors *viz.*, nutritional deficiencies, insect transmission, synchronous occurrence of root rot, virus diseases and shoot infecting fungi like *Colletotrichum*, *Botryodiplodia*, *Fusarium*, *Alternaria* etc. have been isolated from the infected twigs. But the role of *Colletotrichum* is more predominant. Leaves show light green spots which turn brown. The affected plants exhibit drying of twigs from the growing tip and die back leading to reduced plant canopy and yield. On dead twigs, black dot like acervuli appear in concentric rings. Affected plants exhibit more fruit drop and even harvested fruits show high percentage of stem end rot during post harvest storage.

Powdery Mildew

The disease is widely prevalent in Darjeeling, Sikkim, NEH region on almost all the citrus cultivars. This is caused by *Acrosporium tingitaninum* which is severe at low altitude. Characteristic whitish, powdery growth appears on young leaves and twigs. The affected leaves become distorted, the white powdery covering of foliage and twigs leads to leaf drop and die back. Affected fruits drop off prematurely and yield is reduced. The disease becomes severe with delayed rains.

Pink Disease

This is most destructive and infects tree trunks, limbs and twigs. The fungus *Botryobasidium salmonicolor* also attacks rubber, jack fruit and mango. In the early stages, branches and leaves wilt and die. The affected branches get covered with fine silvery-white mycelium. Pink or white pustules appear on the affected area. Severly infected bark gets shredded and the wood is exposed. The branches may show longitudinal cracking. The disease is endemic in some parts of Meghalaya like Kenba in Mawsynram, Pynter in Pynursla etc.

Pre-mature Fruit Drop (Stylar end and Stalk end rots)

These rots are incited by *Alternaria citri* and *Botryodiplodia theobromae*. This is an important disease causing 21–26 per cent of total losses.

In *Alternaria* stem end rot, the disease appears as small tan coloured spot on rind which gradually enlarge and become dark brown or black. Infected fruit shows browning at stem end and also at stylar end. This disease occurs among mandarin oranges, lemons, grape fruits and tangerines.

In *Botryodiplodia* stem end rot, the rot starts as a discolouration around the stem end of fruits and in advanced stage, the infected rind turns brown to black. The discolouration progresses on the rind as irregular bands covering the entire fruit within 3–7 days. The infected fruit do not shrivel but emit a strong fermentative smell. Under humid conditions, sparse dark mycelial growth occurs on the fruit surface.

Greening

This is caused by fastidious bacteria namely *Liberobacter asiaticum*. The disease was detected in certain parts of north east hilly region by indexing through chromatographic technique and it has attained alarming proportion in the mandarin belts of (a) Meghalaya: parts of Garo hills; (b) Sikkim: part of East district; (c) Nagaland: Part of Mokokchung dist. of different citrus species indexed for greening rough lemon, kagzi lime, tangelo tangerine and citrus latipes were free from the disease while mandarins, grape fruit, lemon, rangapur lime, citron, sweet lime, *C. megalaxycarpa*, *C. macroptera*, *C. assamensis* and trifoliate orange showed positive reactions. Presence of disease could be observed at different elevations ranging from 100 m to 1500 m above msl. The disease had been introduced in the area through introduction of certain budded plant material and now even some of indigenous types have been infected. The susceptibility of the widely distributed semi-wild types is posing further threat as they may act as potential source of the disease for future spread.

This is caused by fastidious positive bacteria and it causes variable symptoms. Yellowing of veins, dullness of normal green colour, chlorosis resembling zinc deficiency (sometimes dotted with green islands). Additional leaf symptoms are dwarfing, thickening of leaf blade, upright growth and leaf drop, but none of these symptoms alone is specific for greening. Twigs show shortened internodes, multiple bud formation, stunting, off-season blooming and die back. Fruits are smaller, lopsided, break colour first on sides exposed to the sun, contain curved columellas, aborted seed and drop prematurely.

The disease is spread by use of infected bud wood and by citrus psylla *Diaphorina citri*. Even a single insect is capable of transmitting the greening pathogen and the latent period is about 8–12 days. Once infective, the insect can transmit the disease to healthy plants after a minimum infection feeding of four hours and it continues to spread the disease during its life period. Nursery plants should be grown in the disease free areas such as Assam and north hilly states of India. Younger plants are extremely susceptible to the disease but the adult plants are rather resistant. The susceptible varieties should be grown in the vector-free areas and the resistant varieties in the warmer regions. Removal of unproductive trees and their replacement by nursery stock budded with scion wood of parent trees known by indexing to be free of the greening virus.

Heat therapy of infected bud woods by moist hot air at 47°C for six hours effectively controlled the disease. The only vector *D. citri* can be successfully controlled by spraying with Rogor (dimethoate). Three more systemic insecticides Metasystox (0.05 per cent), Dimecron (0.1 per cent) and Nuvacron (0.05 per cent) have recently been found to be equally effective in eradicating the vector.

Citrus Decline

This is a syndrome of various symptoms caused by a number of biotic and abiotic factors. The citrus plants after 5–6 years of excellent growth usually start declining with gradual decrease in vigour, productivity and exhibit sick look. The magnitude of decline increases with the age of plant and plants become uneconomical to maintain.

The major factors for citrus decline in north-eastern region are deficiency of zinc, copper, manganese and boron; heavy and long spells (6–7 months) of rainfall; no control against insect pest and diseases, The major disease associated with citrus decline are Tristeza, Xyloporosis, Exocortis, Psorosis, Greening, stub born and Phytophthora fungal diseases.

The symptoms starts as die-back of small branches and twigs, yellowing of leaves and heavy bearing of small fruits. As the disease progresses the symptoms intensify, resulting in severe chlorosis and mottling, reduction in leaf size and number and wilting. The feeder roots may be dead, the bark of the larger roots distorted and brittle, dry rot symptoms appear in many lateral roots. After 7 or 8 years the plant may die, by which time there may be no leaf formation, the branches may be dried up and the plants wilted almost completely.

Post Harvest Diseases

Post harvest diseases causes 20–24 per cent yield losses. The most important diseases are stem end rot, anthracnose, sour rot, green mould and *Aspergillus* rots. Stem end rot is caused by *Botryodiplodia theobromae* and *Alternaria citri.*

Botryodiplodia stem end rot is characterized by softening of the rind and the underlying pulp. The rot usually starts as a discolouration around the stem end of the fruit and in advanced stage the infected rind turns brown to black. The discolouration progresses on the rind as irregular bands covering the entire fruit within three to seven days, The infected fruits do not shrivel but emit a strong fermentive smell. Under humid conditions sparse dark mycelial growth occurs on the fruit surface. The disease is favoured by high humidity. The infection is carried from the orchards, causing decay on the senescent ripe fruits.

Table 12.1: Calendar of Operations for the Management of Diseases and Pests in Citrus Orchard

Month	*Operation*
December–January	1. Pruning of diseased, dead twigs and wound dressing with Bordeaux paste.
	2. Scarification of gum oozing wounds and dressing with Bordeaux paint
	3. Prophylactic spray with Bordeaux mixture 1 per cent or Blitox (0.2 per cent)
March–April	4. Application of Bordeaux paste on tree trunks
	5. Single spray of Bavistin (1g/l) + Monocrotophos (1 ml/l) or Endosulphan (2ml/l) on new flushes
	6. After fifteen days, spray with Bordeaux mixture (1 per cent or Blitox 0.2 per cent)
May	7. Pasting of tree trunk up to 2 m with 1 per cent Carbaryl 50 WP (20g/l)
	8. Collection and destruction of trunk borer adults by shaking the branches 2–3 times at 10 days interval.
June–July	9. Repeat the spray schedule as in Sl.No. 5 and 6
August–September	10. Application of Bordeaux paste on tree trunks (up to 60cm height) and repeat the above schedule as in Sl.No. 5 and 6
	11. To kill the trunk bore grubs, clean the bored holes of the infested trunk with iron wire and insert a cotton swab soaked in Dichlorovos/petrol or inject 5ml of Monocrotphos 0.05 per cent or Dicholorovos 0.2 per cent 2.6 ml/l and plug with mud.

Alternaria stem end rot occurs among different citrus fruits such as mandarin oranges, lemons, grape-fruit and tangerines. The disease appears as small tan coloured spots on rind, which gradually enlarge and become dark brown or black. Infected fruit shows browning at stem end and also at stylar end because the fungus grows internally along the central core of the fruit. Infection of *Alternaria* takes place in the orchards during fruit development but it remains quiescent till harvest. The pathogen enters through the button stem end, stylar end or through peel wounds.

Sour rot is caused by Geotrichum candidum causing 1.8–3.5 per cent yield losses. The fungus invades through injuries and bruising and infection starts as a water soaked discolouration around bruises. The rind becomes soft and easily succumbs to a slight pressure. The infected portion of the fruit turns into watery incoherent mass and the juice smells sour. The disease develops very fast at ambient temperature and contaminates other fruits by release of hyaline spores in infected watery mass.

Green mould and black mould are caused by *Penicillium digitatum* and *Aspergillus niger,* which invades through wounds causing soft rots. The whole fruit becomes soft and loses coherence. The affected area gets covered with green in case of green and black fungal growth in case of black mould. The pathogen spreads by contact within the crates. High temperature favours the spread of the disease.

Post harvest diseases can be managed by following measures Orchards should be kept clean, healthy and free from diseases to avoid carrying over of pathogen on fruits. Since much of the primary inoculum of many post harvest rots is carried from the field this inoculum load can effectively be minimized by pre-harvest use of fungicides.If bearing trees are sprayed three times at 15 days intervalie 45, 30 and 15 days before harvest with benzimidazole fungicides (Benomyl, Bavistin) or Bordeaux, mixture covering the whole canopy of the plants provides 70–80 per cent control of post harvest diseases upto three weeks of storage at ambient conditions (25–30°C, RH 80 per cent).

Post harvest fungicidal treatment controls 70 per cent post harvest diseases caused by *Penicillium* and *Asperigillus* upto 3 weeks period. The fruit should be treated with benzimidazole fungicides by dipping the fruit for 5 minutes in fungicidal solution. Hot water treatment of fruits at 43°C for 20 minutes reduces the post harvest diseases. Storage of fruits in containers with rice straw cushioning material will reduce the incidence of post harvest diseases and increase the shelf life of fruits (Srinivas Prasad *et al.*, 2000).

Guava and Apple

Guava is another fruit crop grown in this region affected by anthracnose, scab and wilt diseases.

Anthracnose

Anthracnose disease caused by *Glomerella psidii*. Dark brown necrotic areas are develops on the leaves. On twigs, they are in the form of black necrotic patches, while black spots appear on the older fruits. If younger fruits are infected they drop down from the trees. Anthracnose combined with Pestalotia rot attacking unripe fruits are one of the major problem in Meghalaya.

Wilt of guava is incited by *Fusarium oxysporum* f. *sp. psidii*. The symptoms includes browning and wilting of the leaves, discolouration of the stem and death of the branches along one side. The leaves may falloff and the plant may die in the course of a year or more. Warm and humid climate of lower altitude favours the disease and it is more severe in acidic soils.

Sanitation of the orchard reduces the primary inoculum and further spread of diseases. For anthracnose two to three fortnightly sprays of Dithane 2–78 (0.3 per cent) starting from flowering helps in minimizing the disease. Soil treatment with lime or gypsum also helps in reduction of the wilt disease.

Apple Scab

The apple scab disease was recorded in 1980 in Arunachal Pradesh and Sikkim and it causes 43–45 per cent yield losses in New Bomdila and Shergaon areas of Arunachal Pradesh.

The fungus *Venturia inequalis* attacks leaves, shoots, buds, blossoms and fruits. On leaves, the lesions appear on under surface in a scattered manner. The lesions appear on bud scales and on adjacent parts of the stem prior to bud burst. Early infection on the fruit are very damaging and the spots generally occur near calyx and stalk end. The fungus survives on fallen leaves so removal and destruction of infected plant debris reduces the primary inoculum. Different fungicides *viz.*, Delan, Topsin M, Difolatan Dithane M–45, Carbendazim and Benomyl were evaluated at different phonological stages of bud burst on Royal Delicious and Golden Delicious varieties. A schedule of 7 sprays of fungicides at silver-tip, pink-bud, petal-fall, fruit set, fruit development (2 sprays) and pre-maturity stages followed by a common spray of urea (5 per cent) at pre-leaf fall stage have been found effective in disease management.

Large Cardamom and Arecanut

Arecanut is grown as plantation crop in all north eastern hilly region and it is affected by bud rot/Koleroga. Large cardamom is the most important cash crop grown in sub Himalayan range of Sikkim and Darjeeling district of West Bengal. It is occupying an area of 23500 ha with annual production of 3600 MT. The productivity is low (153 Kg/ha of cured cardamom) which is due to diseases. 17 diseases are reported on this plantation crop of which chirkey, foorkey, leaf streak and anthracnose are important diseases.

Chirke or Mosaic Disease

The disease causes 85 per cent crop loss within three years after infection. The viral nature of the disease was detected by Raychaudhari and Chatterjee (1958). On unfolding leaves, the disease appears as mosaic with pale streak like spots. Gradually these spots turns to pale brown resulting in leaf drying and withering of plants. Reduction in flowering and fruit setting was observed generally in healthy panicles 16–20 flowers developed however in diseased plants only one to five flowers developed.

The virus is transmitted through infected rhizomes as well as by sap inoculation. method and it is not transmitted through seed or soil.1n nature, the disease is transmitted by aphid vectors *i.e., Rhopalosiphum maidis, R. padi, Brachycaudus helichrisi* and *Silobion aveneae* (Raychaudhuri and Chatterjee, 1965).

Infected rhizomes causes primary infection and further spread in the plantation is through aphid vectors. In host range studies. The virus infected wheat, ginger, arrow root and the perennial weed Achorus calamus. Ganguly (1966) developed a serological detection test for chirke infection using agglutination technique with anti-serum developed against the virus.

Varieties Kopringe and Sawnwy were resistant to the disease.

Foorkey Disease/Dwarf Bushy Disease

This disease also a serious viral disease extensively found in the Kalimpong, Algarh and Pedong. Foorkey means dwarf in Nepali and it causes 9–94 per cent crop losses (Varma and Capoor, 1964). As the name indicates the infected plants show dwarfing and numerous minute tillers are formed while fail to form inflorescence.

This is caused by virus which is transmitted by aphid Micromyzus kalimpogenesis (Basu and Ganguly, 1968). The virus is not transmitted by seed, soil or sap inoculation.

Regular rouging of infected plants on community basis continuously reduces frequency of both chirkey and foorkey viral diseases. Replacement with healthy seedlings at shorter intervals result in checking further spread of the disease.

Leaf Streak

This disease is caused by *Pestalotiopsis roynae* and *P. versicolor* (Srivastava and Verma, 1989a, 1989b). Elongate translucent streaks appear on leaves along the veins which turn to reddish brown. In later stages, straw coloured necrotic lesion of size 5–20 mm long and 2–4 mm wide appears surrounded by prominent dark brown margin. Emerging folded leaves and younger leaves are more susceptible.

Spraying of Blitox–50 @0.3 per cent against leaf streak is effective.

Spike Rot

This is also known as collar/root rot and this is a complex disease associated with *Rhizoctonia solani*, *Fusarium* sp., insects and chirke infection. Continuous and heavy rainfall during flowering may result in spike rot. Maximum spike rot (33 per cent) was recorded in Sawney and minimum incidence in Ramla (19.7 per cent). Karibasappa *et al.* (1991) reported that Golsey and Ramsey were moderately affected.

Removal of infected spikes reduces the inoculum. Spraying of Mancozeb @ 0.2 per cent or Carbendazim @ 0.1 per cent at the time of spike formation minimizes the spike rot infection.

Koleroga/Bud Rot of Arecanut

The disease is prevalent in the arecanut growing areas of Meghalaya especially in Pynursla and Shella area, caused by *Phytophora palmivora*. Trees of all ages get infected by this disease. Leaves become yellow and die. The disease usually appears soon after the onset of monsoon usually at the beginning of June. The fungus remains dormant in the leaf base during dry weather. The disease appears on the crown also causing crown rot. Fruit dropping is the first sign of its occurrence. Fallen nuts are discoloured and covered with a whitish felt of fungal mycelium.

In the early stages of infection, the diseases portion of crown has to be removed followed by pasting with Bordeaux paste reduces the infection.

Prophylactic spraying of 1 per cent Bordeaux mixture 1–2 times during pre-monsoon period followed by 2 to 3 sprays during monsoon season at fortnightly interval reduces the disease incidence.

REFERENCES

Basu, A.N. and B. Ganguly (1968). A note on the transmission of Foorkay disease of large cardamom by aphid Micromyzus Kalimpongensis Basu. *Indian Phytopath.*, 21(1): 127.

Chandra, S., S. Kumar and A.K. Singh (1998).Management of Cercospora leaf spot of ground nut (*Arachis hypogea*) with single fungicidal spray. *International J. Pest Mgt.*, 44(33): 135-137.

Ganguly, B. (1966). A rapid test for detecting chirke affected large cardamom (Amomum subulatum Roxb) plants in the field. *Science & Culture*, 32: 95–96.

Karibasappa, G.S., L.S. Srivastava and S.K. Ray (1991). A case study of large cardamom productivity in Sikkim and Darjeeling district of West Bengal, India. *J. Plant. Crops*, 18: 387–394.

Lakshmi Prasad, M.S., M. Srinivas Prasad and Sangit Kumar (1999). Pathogenic variation in Pyricularia grisea, rice blast fungus. *National Seminar on Strategies for Agricultural Research in North East*, at ICAR Reasearch complex for NEH region, Umiam 10–12th November, 1999, pp. 91.

Lakshmi Prasad, M.S., M. Srinivas Prasad and Sangit Kumar (2000). Management of neck blast disease of rice in Meghalaya. *Oryza*, 37(2): 77–78.

Mahajan, V., Chanara, S., R.N. Verma and S.M. Hussain (1995). Genetics of resistance to northern leaf *blight* (*Exerohilum turcicum*) in Maize (*Zea mays*). *Indian J. Agri. Sci.*, 65(1): 74–77.

Mahajan, V., R.N. Verma, S. Chandra and B.K. Sarma (1991). Evaluation of maize (*Zea mays*) varieties for resistance to *Turcicum* leaf blight (*Exerohilum turcicum*) in Meghalaya. *Indian J. Agri. Sci.*, 61(7): 488–489.

Maiti, S. and Vishwadhar (1982). A new zonate leaf spot of soybean. *Transaction of British Mycological Society*, 79(3): 547–548.

Maiti, S., S. Kumar, R.N. Verma and Vishwadhar (1983c). Current status of soybean diseases in North east India. *Soybean Rust Newsletter*, 6(1): 14–19.

Maiti, S., A.K. Singh and H.S. Sahambi (1979). Occurrence of the perfect stage of *Rhynchosporium oryzae* in India. *Plant Dis. Repor.*, 63: 773–774.

Maiti, S., Vishwadhar and R.N. Verma (1983a). Soybean rust in North eastern hill region of India. Soybean rust in North eastern hill region of India. *Soybean Rust Newsletter*, 6(1): 22–24.

Maiti, S., Vishwadhar and R.N. Verma (1983b). Control of soybean rust in India. *Soybean Rust Newsletter*, 6(1): 8–13.

Raychaudhuri, S.P. and S.N. Chatterjee (1958). A preliminary note on the occurrence of a new virus disease of cardamom (*Amomum subulatum* Roxb) in Darjeeling district. *Mycological Research Workers Conference*, ICAR, Shimla, 20–23rd June 1958, pp. 174–176.

Raychaudhuri, S.P. and S.N. Chatterjee (1965). Transmission of chirki disease of large cardamom by aphid species. *Indian J. Ent.*, 27(3): 272–276.

Sangit Kumar and D.K. Jha (2002). Trichothecium roseum: a potential agent for the biological control of soybean rust. *Indian Phytopath.*, 55(2): 232–234.

Sharma, J.P., Sangit Kumar and R.N. Verma (1986). Relationship between rice blast and nitrogen fertilization in Nagaland. *Indian J. Hill Farming*, 1A(2): 75–76.

Singh, A.K. and M. Narayan Bhat (2002). Control of French bean rust through fungicides and a neem based fonuulation. *Indian Phytopath.*, 55(2): 241–243.

Srinivas Prasad, M., M.S. Lakshmi Prasad, A.K. Singh, and Y.P. Sharma (2001). Effect of Containers and fungicidal application on the storage ot Khasi mandarins. *Indian J. Hill Farming*, 14(1): 56–62.

Srinivas Prasad, M., Sangit Kumar and M.Santha Lakshmi Prasad (1998a). Weed hosts of rice blast fungus in Meghalaya. *Oryza*, 35(4): 384–385.

Srinivas Prasad, M., Sangit Kumar, M. Santha Lakshmi Prasad and Y.P. Sharma (2000). Rice blast disease: A farmer's foe in NEH region. *Indian Farming*, 50(4): 25–26.

Srinivas Prasad, M., M. Santha Lakshmi Prasad, A.K. Singh and Y.P. Sharma (2001). Reaction of citrus genu plasm to different fungal diseases in Meghalaya. *Indian J. Hill Farming*, 14(1): 113–116.

Srinivas Prasad, M., B.K. Sarma, Sangit Kumar and M.S. Lakshmi Prasad (1998b). Transmission tests and varietal screening for urd leaf crinkle virus in black gram (Vigna mungo). *Ann. Plant Prot. Sci.*, 6(2): 205–207.

Srivastava, L.S. and R.N. Verma (1989a). Leaf streak (C.O. Pestalotiopsis roynae): A new disease of large cardamom from Sikkim. *Curr. Sci.* 58: 682–683.

Srivastava, L.S. and R.N. Verma (1989b). Large cardamom: A new host for Pestalotiopsis versicolor (Speg) Steyaert. *Curr. Sci.*, 58: 971–972.

Verma, P.M. and S.P. Kapoor (1964). Foorkey disease of large cardamom. *Indian J. Agri. Sci.*, 34: 56–62.

Verma, R.N. and Sangit Kumar (1985). Efficacy of fungicides and their application methods for controlling blast. *IRRN*, 10: 2.

Verma, R.N. and A.K. Singh (1982). Sources of resistance to leaf scald disease. *IRRN*, 7(5): 8.

13

Eco-friendly Management of Wilt in Chickpea

S.N. Gurha, Vishwa Dhar *, Udit Narain** and Shubha Trivedi**

**Indian Institute of Pulses Research, Kanpur – 208 024*

***CSAUA&T, Kanpur – 208 002*

Chickpea (*Cicer arietinum* L.) is an important food legume in many parts of the world covering Asia, Africa, Australia, Europe, North America and South America. In India the crop has significance in meeting the protein demand of the people as well as in restoring the soil fertility. Inspite of best efforts of the researches and extension workers the area under chickpea is gradually declining (Table 13.1).

Table 13.1: Gradual Decline in Chickpea Area

Year	*Area M.hact.*	*Production M.ton*	*Productivity Kg/ha*
1998–99	8.47	6.80	803
1999–2000	6.31	5.08	806
2001–2002	6.10	5.27	865

The major reasons for this decline are biotic and abiotic stresses encounted by this crop. Diseases are most serious constraints to chickpea productivity causing upto 100 per cent losses. Chickpea crop is invaded by 67 fungi, 3 bacteria, 22 viruses and 80 nematodes causing more than 50 diseases (Nene and Reddy, 1987, Nene *et al.*, 1996). Environmental factors and intensity of abiotic stresses are known to compound the occurrence and severity of the diseases. For example, extent of yield losses due to wilt are more in the event of drought and high temperature in the country. Management of wilt through chemicals is not feasible to the farmers. Therefore, ecofriendly management of wilt is eagerly warranted.

Since the disease is both soil as well as seed borne, therefore for eco-friendly management, both these aspects need to be taken into consideration.

Eco-friendly management of wilt consists of the following components.

1. Resistant varieties
2. Cultural practices
3. Intercropping/crop rotation
4. Use of antagonistic fungi and bacteria (Biological control)
5. Judicious use of fungicides

Resistant Varieties

Growing wilt resistant varieties is no doubt a most economic and of farmers convenience method for the management of this disease. The existence of races (Haware and Nene, 1982 b) in the pathogen *F. oxysporum* f.sp. *ciceri* has posed a problem which has warranted for the development of varieties possessing broad based resistance such as multirace or multipathotype resistance.

Chickpea varieties having resistance against wilt have been recommended for different zones (Table 13.2).

Table 13.2: Wilt Resistant Chickpea Varieties Recommended for Different Zones

Zone	*Recommended Varieties*
NWPZ	Alok, Avrodhi, DCP 92-3, GNG 146, Haryana Chana 1, ICCC-32, Karnal Chana 1, Pusa 372, GNG 603, Pusa 212
CZ	Pusa 417, Gujarat Gram 1, Hirwa Chaffa, ICCC 32, JAKI 9218, JG 74, JG 315, JG 322, JGG 1, ICCC 37, PG 12, KAK 2, Vijay, Vishwal, Vaibhav and Pusa 372
SZ	JG 11, ICCC 37, Annegeri 1, ICCV 10, BDN 9-3
NEPZ	JG 74, KWR 108, Pusa 372, Avrodhi, GCP 105

Recently the need for identification of chickpea genotypes possessing race specific resistance has also attracted the attention of researchers. As an outcome of the efforts made in this direction chickpea genotypes having stable resistance or moderate resistance against race 2 (Kanpur isolate) of *F. oxysporum* f.sp.*ciceri* have been identified (Gurha and Chaturvedi, 2002, Gurha *et al.*, 2002, Gurha *et al.*, 2002).

Chickpea genotype with stable resistance/mod. Resistance against Race 2 of *F. oxysporum* f.sp. *ciceri.*

PG 95007, PBG 126, PG 703, ICC 10149, H 82-2, H 92-71, GNG 1000, ICC 11442, BG 372, GL 91061, BG 1084, WCG 95-50, FG 711, GL 96004, PBG 1, PBG 68, GL 90236, FG 702, IPC 99-1, IPC 99-10, IPC 99-34, FG 712, FG 694, GPF 133, ICC 2862, BG 209, PG 8602-10-12, ICCV 93217, GL 93011, H 92-17, BG 1043, IPC 94-93, IPC 94-98 and IPC 94-114

Attempts have also been made to locate chickpea lines having resistance against multiple isolates of *F. oxysporum* f.sp. *ciceri*. Some of the chickpea lines found either resistant or moderately resistant against Kanpur isolate (Reace 2) of the chickpea wilt pathogen were evaluated against 7 other isolates of *F. oxysporum* f.sp. *ciceri* from Badnapur, Dholi, Hisar, Delhi, Rahuri, Sehore and Junagadh. Some of these lines were found to have resistance against four or more isolates (Table 13.3). These lines may be used in breeding programme for incorporating broad based resistance into a good agronomic base.

Table 13.3: Multiple Isolate Resistance in Chickpea Against *F. oxysporum* f.sp. *ciceri*

Genotypes	*Number of Isolates Against which Found Resistant/Moderately Resistant*
GCP 9504	1,2,3,4,5,6 and 7
H 92-71	1,2,3,5,6 and 7
FG 711	-do-
FG 712	1,2,3,4,6 and 7
H 82-2	2,3,5,6 and 7
FG 703	1,2,3,5 and 7
IPC 97-1	1,2,3,5 and 6
WCG 95-50	2,3,4,5 and 6
PBG 34	1,2,5 and 6
GL 91061	-do-
H 95-30	2,3,5 and 7

Isolates: Badnapur (1), Dholi (2), Hisar (3), Junagadh (4), Delhi (5), Rahuri (6) and Sehore (7)

Cultural Practices

Cultural practices may play a vital role in the management of wilt disease in chickpea and also as an component of integrated plant disease management. However, this aspect has not gained much attention of the researchers.

Losses due to wilt may be minimised by avoiding cultivation of chickpea in a heavily infested soil. Since the pathogen survives for a longer period in the soil and in symptom less carriers (Haware and Nene, 1982 a) normal crop rotation will not be able to deliver the desired results. Deep ploughing in hot summer months and removal of host plant debris from the field helps in reducing the pathogen population in the soil. Soil solarization by covering the soil with transparent polythene sheet for 6–8 weeks during April-May effectively reduced pathogen population in the soil (Chauhan, *et al.*, 1988). However, this is quite expensive and cumbursome to the farmer, therefore, deep summer ploughing may be adopted. Delayed sowing and deep sowing (Dahiya *et al.*, 1988, Singh and Singh, 1984) and soil amendment with oil seed cakes are well known to reduce the wilt incidence in the field (Zakaria and Lockwood, 1980). Application of FYM (farm yard manure) @ 5 tonnes/ha considerably reduces the wilt incidence in the field. In the fields which are wilt sick, crops like wheat or barley should be grown alternate year. To avoid transmission of the pathogen through seed, the seeds for sowing should not be used from the plants which wilt after pod formation.

Intercropping/Crop Rotation

Intercropping or crop rotation with crops like wheat, barley, linseed and mustard which are not the host of *F. oxysporum* f.sp.*ciceri* may help to a great extent in the management of wilt. However, only limited information is available on this aspect. As the pathogen survives for a longer period in the soil, therefore, crop rotation with non host crops for a longer period may not be feasible to the farmers.

Root exudates of wheat and mustard at different concentrations effectively reduced the growth of six *F. oxysporum* f.sp. *ciceri* isolates *in vitro.*

Table 13.4: Per cent Reduction (Range) in Growth of F.O.C. Isolates at Different Dilutions of Wheat and Mustard Root Exudates

Crop	*Dilution of Root Exudates*		
	1:1	*1:2*	*1:3*
Mustard	4.3–47.2	36.4–77.1	37.9–91.8
Wheat	9.3–61.3	41.9–96.1	100.0

Fusarium oxysporum f.sp. *ciceri* isolates from Dholi (Bihar), Sehore (MP), Rahuri (Maharashtra), Ludhiana (Punjab), Kanpur (UP) and from Jammu (J&K). These observations convey a very clear message that use of these crops as intercrop may help in the management of wilt in chickpea.

Studies conducted at centres of AICRP on chickpea have shown that intercropping of chickpea with linseed has definite impact in reducing the wilt incidence in the field (Tables 12.5 and 12.6).

Table 13.5: Effect of Seed Treatment and Linseed Intercropping on the Incidence of Wilt at Hisar (Average of 2 years data)

Seed Treatments	*Percent Wilt Reduction Over Control*		
	Chickpea Sole	*Chickpea+Linseed (2:1)*	*Chickpea+Linseed (Mixture)*
T. viride+Vitavax (4+1 g/kg)	52.5	50.8	55.2
T. viride 4g/kg	35.0	31.7	3.5
Vitavax 1g/kg	32.7	33.5	37.4

Source: AICRP on chickpea Annual Report (2001–2003).

Table 13.6: Effect of Seed Treatment and Linseed Intercropping on the Incidence of Wilt at Dholi

Seed Treatments	*Percent Wilt Reduction Over Control*			
	Chickpea Sole	*Chickpea+Linseed (2:1)*	*Chickpea+Linseed (Mixture)*	
T. viride + Vitavax (4+1 g/kg)	50.4	49.4	48.4	47.2
T. viride 4g/kg	25.3	26.4	28.6	26.3
Vitavax 1g/kg	31.9	37.5	31.2	25.6

Source: AICRP on chickpea report (2002–2003).

The results of these studies have shown that seed treatment with 4 g *T.viride* + 1 g vitavax/kg seed sown with linseed as intercrop or in mixture may help to a greater extent in the management of wilt in chickpea.

Biological Control

Now a days the control of soil borne diseases in chickpea and many other crops with the use of fungi like *Trichoderma* sp., *Gliocladrum virens* and bacteria like *Bacillus subtilis* and *Pseudomonas fluorescens* is gaining importance (Mukhopadhyay, 1988, 1989). Due to hazardous effects of chemicals in soil and environment, the bioagents have now become an integral part of plant disease management. The studies at centres of AICRP on chickpea have shown the potential of these biocontrol agents in the management of wilt.

Trichoderma species are more commonly used now a days. But we have to consider the following issues in this direction.

1. Do *Trichoderma* alone can serve the purpose?
2. Whether one species of *Trichoderma* be equally effective against all the races/pathotypes/isolates of *F. oxysporum* f.sp. *ciceri?*
3. What should be the method of application and the amount of inoculum of *Trichoderma* to be used for expected results.

The studies made at Indian Institute of Pulses Research, Kanpur with 10 isolates of *F. oxysporum* f.sp. *ciceri* to evaluate the antagonistic efficiency of *T. viride* and *T. harzianum* clearly indicated that one species of *Trichoderma* may not be equally effective against all the isolates of the chickpea wilt pathogen. *T. viride* was found more efective against Junagadh, Kanpur and Jabalpur isolates, while, *T. harzianum* was found more effective against Sehore, Kopergaon, Rahuri, Ludhiana and Bharari isolates. Against Jammu and Dholi isolates both the species of *Trichoderma* were equally effective (Gurha, 2001 and Sekhar *et al.*, 2003) (Table 13.7).

Table 13.7: Effect of *Trichoderma* spp. on the Growth of *F. oxysporum* f.sp. *ciceri* Isolates

Isolates of FOC	*Per cent Inhibition in Growth*	
	T. viride	*T. harzianum*
Junagadh	90.13	75.80
Sehore	40.05	72.23
Kopergaon	27.99	69.13
Kanpur	75.96	67.9
Rahuri	50.07	87.9
Ludhiana	70.50	90.29
Jammu	70.50	70.58
Dholi	86.48	86.48
Bharari	73.02	84.58
Jabalpur	95.07	81.48

While, using *Trichoderma* sp. consideration should be made about the isolate of *F. oxysporum* f.sp. *ciceri* against which it has to be used. While, using *Trichoderma* the method of application is very important for achieving the desired results. Normaly 4 g *Trichoderma* + 1 g vitavax is recommended as seed treatment. It has been observed that soil application of these bioagents one week before sowing was more effective in reducing the wilt in chickpea than the seed treatment (Prasad *et al.*, 2002). These workers applied the bioagents to soil @ 5 g of powder formulation of bioagents mixed in one kg of farmyard manure in a 2 × 1.5 m plot, one week before sowing. These workers also used bioagents for seed treatment @ 10 g/kg seed. These observations clearly indicated that amount of *Trichoderma* powder formulation to be used also plays a vital role in effective management of wilt.

Judicious Use of Chemicals

Use of fungicides may not help much in the management of wilt under field conditions. However, seed treatment with fungicides may help in eradication or minimization of seed borne inoculum of the pathogen. Seed treatment with Benlate T (30 per cent Benomyl + 30 per cent Thiram) @ 1.5 g/kg seed was found to eradicate completely the seed borne inoculum of the chickpea wilt pathogen (Haware *et al.*, 1978). Seed treatment with carbendazim @ 2.0 g, Thiram + carbendazim @ (2+1) g or *T. viride + vitavax (4+1)g/kg* seed may also be done.(Gupta, 1995). Judicious use of fungicides as seed treatment before sowing may ensure better plant stand and may also protect the crop from infection of *F. oxysporum* f.sp. *ciceri* in the early stages of plant growth.

Feasibility of Integrated Management

Though it is difficult to introduce high level of resistance in a genotype with other desirable attributes. Therefore, it would be worth while to integrate available compatible control practices to develop viable and economic and eco-friendly management strategy particularly for the genotypes which have high yield potential with moderate resistance against wilt. Keeping this in view following module is suggested for effective management of wilt.

Soil = Deep summer ploughing

Seed = Treatment with 4 g *Trichoderma viride* + 1 g vitavax, or Bavistin + Thiram (1+2) g per kg of seed

Cropping system = Chickpea + Linseed mixed cropping or intercropping (1:1) row ratio

Resistant varieties = JG 315, Avrodhi, DCP 92-3, JG 74, BG 372 or others.

Conclusion

1. Molecular characterization of genetic diversity in *F. oxysporum* f.sp. *ciceri.*
2. Identification of chickpea genotypes having race specific or multirace resistance for insultating varieties against more number of races of the chickpea wilt pathogen.
3. Research on cropping system to find out more viable eco-friendly management.
4. Exploitation and integration of available wilt management components.
5. Evaluation of germplasm for identification of broad based stable resistance.

REFERENCES

Chauhan, Y.S., Y.L. Nen, C. Johanson, M.P. Haware, N.P. Saxena, S. Singh, S.B. Sharma, K.L. Sahrawat, J.R. Barford, O.P. Rupela, J.V. D.K. Kumar Rao and S. Sithanatham (1988). Effect of soil solarization on pigeonpea and chickpea. Research Bulletin No. 11. ICRISAT, Hyderabad, India, pp. 16.

Dahiya, S.S., S. Sharma and A.S. Paroda (1988). Effect of date and depth of sowing on the incidence of chickpea wilt in Haryana, India. *International Chickpea Newsletter* 18: 24–25.

Dey, R.K., R.G. Chaudhary and Naimuddin (1996). Compartive efficacy of biocontrol agents and fungicides for controlling chickpea wilt caused by *Fusarium oxysporum* f.sp. *ciceri. Indian J. Agric. Sci.*, 66: 370–373.

Dua, R.P., S.K. Chaturvedi and Shiv Sewak (2001). Reference varieties for IPR Regime. Indian Institute of Pulses Research, Kanpur, pp 6–24.

Gupta, S.K. (1995). Studies on the management of chickpea wilt complex. M.Sc. (Ag.) Pl. Pathology, thesis submitted to Rajendra Agricultural University, Pusa, Bihar, pp. 70.

Gurha, S.N. (2001). Effect of some *Trichoderma* spp. on the growth of different isolates of *F. oxysporum* f.sp. *ciceri in vitro. Ann. Pl. Protec. Sci.*, 14(1): 202–204.

Gurha, S.N. and S.K. Chaturvedi (2002). Isolation of stable resistance in chickpea (*Cicer arietinum* L.) breeding lines against Race 2 of *F. oxysporum* f.sp. *ciceri. Progressive Agri.*, 2(1): 90-91.

Gurha, S.N., R.A. Singh, Naimuddin and Atin Ghosh (2002). Stable and broad based resistance against wilt in chickpea. *Ann. Pl. Protec. Sci.*, 10(1): 88–90.

Gurha, S.N. and Rachana Srivastava (2002). Effect of root exudates of wheat (*Triticum aestivum*) and Indian mustard (*Brassica juncea*) on growth of *Fusarium oxysporum* f.sp. *ciceri* isolates. *Indian J. Agri. Sci.*, 72(12): 747–748.

Gurha, S.N., Naimuddin Vishwadhar and R.A. Singh (2002). Stable resistance in gram (*Cicer arietinum*) against Race 2 of *Fusarium oxysporum* f.sp. *cicer. Indian J. Agric. Sci.*, 72(2): 88–89.

Haware, M.P. and Y.L. Nene (1982). Races of *Fusarium oxysporum* f.sp. *ciceri. Plant Dis.*, 66: 809–810.

Haware, M.P. and Y.L. Nene (1982). Symptom less carriers of the chickpea wilt *Fusarium. Plant Dis.*, 66: 250–251.

Haware, M.P., Y.L. Nene and R. Rajeshwari (1978). Eradication of *Fusarium oxysporum* f.sp. *cicetri* transmitted in chickpea seed. *Phytopathology*, 68: 1364–1367.

Mukhopadhaya, A.N. (1989). Microbial control of soil borne plant pathogens by *Trichoderma* and *Gliocladium*. National Seminar and 17th workshop of AICRP on biological control, Lucknow, Oct. 23–25, 1989.

Mukhopadhyay, A.N. (1988). Biocontrol efficacy of *Trichoderma* spp. In controlling damping off in tomato and egg plants and wilt complex of lentil and chickpea. 5th *Int. Congress of Plant Pathology*. Ekato, Japan, 1988.

Nene, Y.L., M.V. Reddy (1987). Chickpea diseases and their control. In: *The Chickpea*, (Eds.) M.C. Saxena and K.B. Singh, CAB International Oxon, U.K., pp. 233–279.

Nene, Y.L., V.K. Sheila and S.B. Sharma (1996). *A World List of Chickpea and Pigeonpea Pathogens*. ICRISAT, Hyderabad, India.

Prasad, R.D., R. Rangeshwaran, C.P. Anuroop and H.J. Rashmi (2002). Biological control of wilt and root rot of chickpea under field conditions. *Ann. Pl. Protec. Sci.*, 10(1): 72–75.

Sekhar, Bandyopadhyay, N.D. Sharma and Subrota Dutta (2003). Screening of potential *Tridhoderma* strains against major not pathogens. *Ann. Pl. Protec. Sci.*, 11(1): 163.

Singh, U.P. and Singh, R.B. (1984). Effect of date of sowing on the sclerotinia stem rot and wilt of gram (*Cicer arietinum* L.). Paper presented in All India Rabi Pulses Workshop, Bhubaneshwar (Orissa).

Zakaria, M.A. and J.L. Lockwood (1980). Reduction in *Fusarium* population in soil by oil seed meal amendments. *Phytopathology*, 70: 204-243.

14

Predictive Management of Powdery Mildew Diseases

P. Sinha, D. K. Agarwal* and D. Prasad***

**Division of Plant Pathology, Indian Agricultural Research Institute, New Delhi – 110 012*

***Division of Nematology, Indian Agricultural Research Institute, New Delhi – 110 012*

Powdery mildews are well-defined family of obligate parasite, which grow abundantly on the foliage of angiosperms and cause damage to a wide variety of crops. They grow superficially on surface of host and draw nutrition through haustoria.

The damage caused by powdery mildews are generally associated with stunting and distortion of leaves, surface necrosis of infected tissue, a general decline in host growth, deformation of fruit, yellowing or chlororsis of leaves, and premature leaf fall. Cultivated plants, *viz.* apple, apricot, almond, barley, bean, clover, crucifers, cucurbits, citrus, cotton, grape, grasses, lettuce, mango, peach, peas, potato, rose, rubber, tobacco, tomato, *Ziziphus*, and wheat are attacked by powdery mildews. It is hypothesized that Erysiphaceae developed in either Europe or western Asia and is spreading eastwards. Distribution of *Erysiphe graminis* on cereals and grasses and of various *Erysiphe* and *Leveillula* spp. on Gramineae, Caryophyllaceae, Cruciferae, Umbelliferae, Convolvulaceae and Boraginaceae suggests the eastward migration of these powdery mildews.

Except for grape and mango, there is no appropriate forewarning information for powdery mildews on other crops. In absence of any forewarning information, growers suffer heavy yield loss. The most widespread and disastrous losses due to powdery mildews are recorded in grape (33 to 90 per cent), mango (up to 90 per cent), barley (up to 42 per cent), wheat, gooseberry (61 to 71 per cent), peach (up to 80 per cent), and in clover (75 per cent), apple and rose. The current paper discusses epidemiology emphasizing on forecasting, optimization and decision support in disease management in a broader sense.

Integrated Management of Powdery Mildews

Powdery mildews could be managed sustainably with timely spraying of effective fungicides and deploying resistant sources. Basic requirements or components of integrated disease management are surveillance and forecasting of disease outbreak, host resistance, use of appropriate bioagents, cultural practices for resisting spread and need based chemical application. Of course, for cultural practices there have not been many clues probably due to non-occurrence or uncertainty on existence of sexual stages in most of the powdery mildews affecting important crops. We emphasize here, mostly on epidemiology so that a rational management strategy is evolved for the disease.

Surveillance, Monitoring and Forecasting of Disease Outbreak

Surveillance implies information on the crop information, inoculum and other relevant data on decision making whether disease is likely to occur. Susceptible stages of the crops are known today and survival of powdery mildew inoculum also known for many cases. In ber, pathogen survives on leaves of plants grown in shade conditions as active mycelium (Parkash *et al.*, 1988) and in mango, perpetuates on leaves and malformed panicles and produces large number of conidia under favourable weather conditions (Munshi *et al.*, 1988; Gupta, 1988; Joubert, 1991). However, more often than not, development of powdery mildews is characterized by an initial slow growth when favourable conditions just set in during susceptible stage of host. But suddenly epidemic starts in a devastating appearance leaving no time for taking any appropriate control measures although effective fungicides are available. Sudden epidemic mostly goes unaware as most of the powdery mildews are very fast growing (4 to 5 days latent period) pathogens and a period of 7 to 10 days becomes vital for management decision. In absence of epidemiological information on epidemic rate and key weather parameters triggering massive infection within a short period no definite spray scheduling has been evolved. As a result, growers are either resort to uneconomic spray applications or suffer enormous yield losses. Obviously, monitoring and forecasting the exact time of outbreak of the disease becomes vital to management decisions.

Development of powdery mildew is highly influenced by weather factors, mainly temperature and high RH hours or dew wetness/leaf wetness duration. Sequential occurrence of favourable temperature and high relative (RH) above a threshold limit (mostly > 80 per cent) at a particular growth stage of plant are known to be conducive for growth and development of the pathogen as mildew is limited to those areas where environment is simultaneously favourable for a sufficient time period for pathogen and host. Applications of forecasting models include assessing impact of host growth and environmental factors on disease development, thus, facilitating optimization of disease management strategies (Jeger, 1984). Therefore, monitoring of temperature and RH during susceptible stages of the crops is essential to gaze the probable period of the disease occurrence at least with some measure of certainty. Forecasting or prediction by established relationships among disease incidence with weather parameters *viz.*, temperature, rainfall, relative humidity, sunshine hours etc. would aid to the growers for timely application of sprays.

Development of epidemic involves several overlapping and repetitive infection cycles within a short spell. An infection cycle may be summarized in phases: germination, penetration, colonization, sporulation and dispersal, which are influenced by weather factors. Therefore, most prediction/ forecasting models forecast the occurrence of disease symptom expression by monitoring weather factors conducive to infection cycles. Once favourable infection periods have occurred, subsequent expression is based on the assumption that certain weather factors favourable to colonization will proceed based on the weather trends. One basic assumption that sufficient inoculum to initiate an epidemic is available whenever the crop is present. Forecasting systems based on this assumption immediately incur a degree of inaccuracy. However, for powdery mildews, existence of inoculum in majority of cases is being certained.

Initiation of powdery mildews occurs either through conidia and/or ascospores under favourable weather conditions. Air-borne conidia are the major infective entities for powdery mildew epidemic. Development of epidemic involves several overlapping and repetitive infection cycles within a short spell. An infection cycle may be summarized in phases: germination, penetration, colonization, sporulation and dispersal, which are influenced by weather factors.

For powdery mildews, germination of conidia and subsequent growth pattern is an important element of infection cycle as conidia are found to be primary source of infection in majority of cases. Conidia germination and germ tube elongation pattern in terms of temperatures response for *Oidium mangiferae* (in mango) and *Uncinula necator* (in grapes) was explained by fitting polynomial functions (Sinha *et al.*, 2001; Chellemi and Marois, 1991). Gubler *et al.* (1999) incorporated temperature influence on conidia germination and infection efficiency in their forewarning scheme for grape powdery mildew. Effect of RH on germination is equally important epidemiological point of view. For *O. erysiphoides* in ber, optimum RH for germination and germ tube elongation found to be 80 per cent or above (Maheswari and Singh, 1997; Sinha *et al.*, 2001). Temperature and RH in nature is variable both in time and duration. For an idea or approximation about processes response surface is needed to facilitate estimation of development under temperature and RH not observed in the experiment. Therefore, information with exposure time in temperature and RH is required which is not taken into account in any of these studies, which require immediate attention.

In Pune, grape powdery mildew caused by *Uncinula necator*, disease incidence and multiplication was high in December and January when the climate was cool and humid (Mane *et al.*, 1996). Chellemi and Marois (1991) developed a simulation model studying effect of temperature and moisture factors on powdery mildew (*Uncinula necator*) on grapes. Sensitivity of population growth to changes in probability of conidia being deposited onto susceptible tissue is examined by varying probability over a range of temperatures. Only negative effect of water on conidia germination is considered. However, recent findings show that powdery mildew growth is highly influenced by high relative humidity, which is not accounted for Sall (1980) formulated a mathematical model, for grape powdery mildew including a host growth function, to examine the influence of seasonal weather and timing of initial infection on eventual colonization of leaves and fruits. Sinha (2003) studied the effect of temperature and hours of high relative humidity on powdery mildew of ber (*Ziziphus* sp) by developing a cubic model for temperature response and a sigmoid curve for relative humidity hours response on fruit infection (Figures 13.1 and 13.2). These equations are used to investigate effects of different temperature and RH hours on fruit infection. Singh *et al.* (1999) used multiple regression taking temperature and relative humidity to explain powdery mildew of ber. For grape powdery mildew (*Uncinula nectar*), Gubler *et al.* (1994) has derived a descriptive forewarning model or scheme based on temperature influences on conidia germination and infection efficiency. A portion of the model forecasts infection based on the occurrence of at least 12–15 hours of continuous leaf wetness (or RH> 90 per cent) with average temperatures between 10–15°C. Once infection has occurred, model switches to the risk assessment phase and is based entirely on effect of temperature on infection cycle for conidial infections. One drawback of this work is that decision rules do not have a sound quantitative basis.

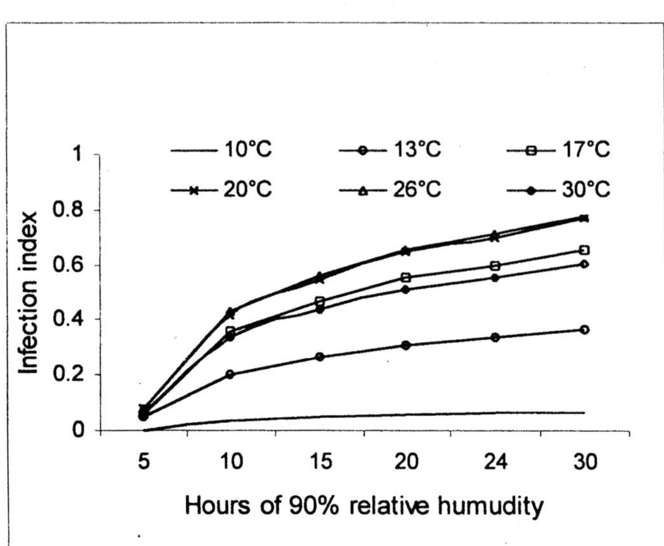

Figure 14.1: Effect of Temperature and Relative Humidity Hour on Powdery Mildew Infection

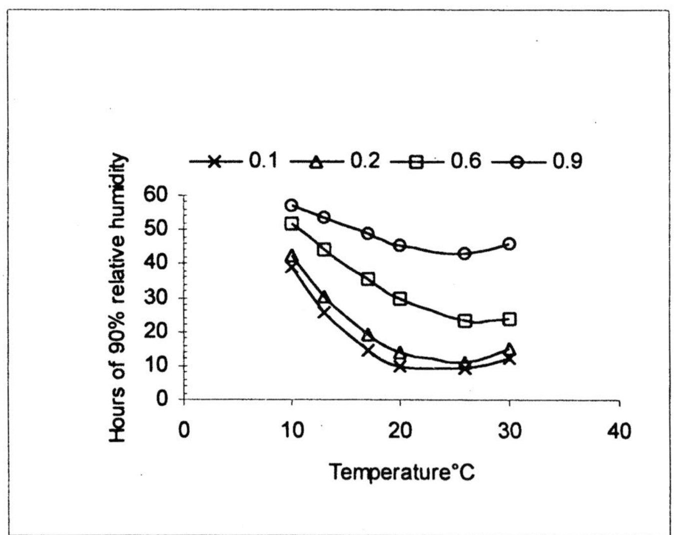

Figure 14.2: Powdery Mildew Severity in Relation to Relative Humidity Hour and Temperature

1. Conidial infections- risk assessment (risk index ranges from 0-100)
2. Requires 3 consecutive days with at least 6 hours between 21-30°C to trigger the index
3. Index increases 20 points each day with at least 6 hours between 21-30°C
4. Index decreases 10 points each day with less than 6 hours between 21-30°C

5. Index decreases 10 points on any day with a minimum temperature above 35°C
6. An index of 60-100 indicates the pathogen is reproducing every 5 days
7. An index of 0-30 indicates the pathogen is functioning minimally and reproductive rate is every 15 days or not at all

Kiraly (2000) established relationships among six epidemiological parameters (rates of sporulation SPOR, infection INF, colony growth CGR, incubation progress INC, latency progress LAT, infectiousness IFS) of *Erysiphe graminis* f. sp. *tritici* under fluctuating weather conditions. Out of these, SPOR and INF are most interesting epidemiological parameters for modelling powdery mildew outbreaks, whereas other parameters play a minor role. Further, the author developed a dynamic model simulating the effect of weather on powdery mildew (*Erysiphe graminis*) epidemics on winter wheat. In tomato, Whips and Budge (2000) studied effect of humidity on development of powdery mildew (*Oidium lycopersici*) in the glasshouse and concluded 80 per cent relative humidity was the major influencing factor.

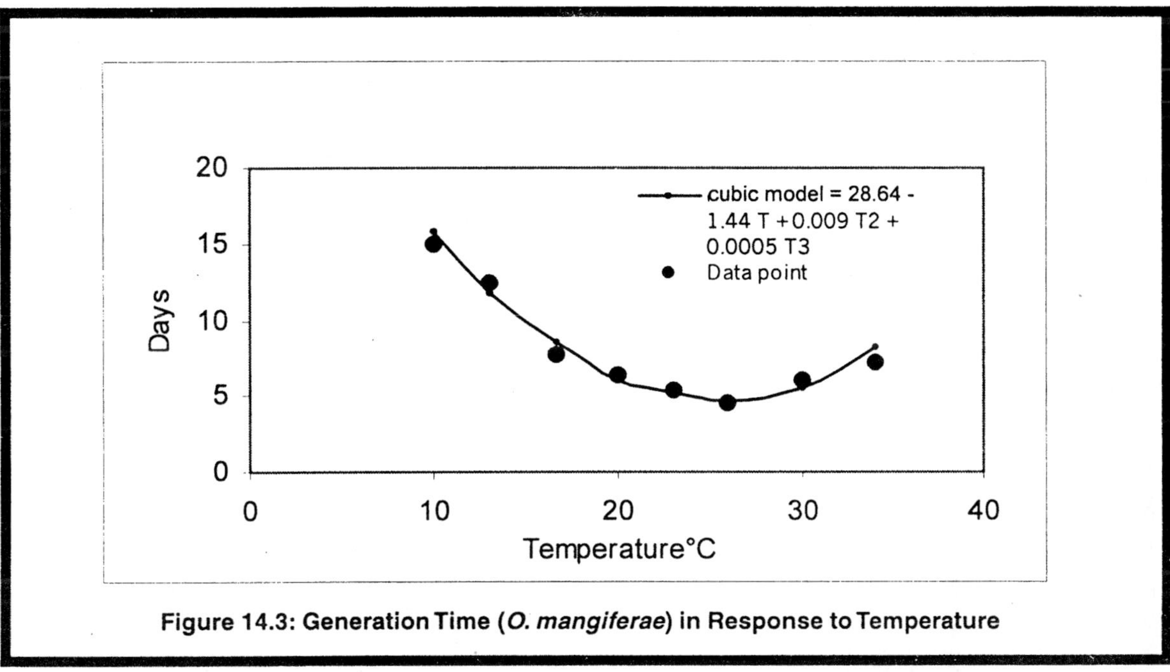

Figure 14.3: Generation Time (*O. mangiferae*) in Response to Temperature

Latent period is period from infection to sporulation by fungi. With each such period, pathogen multiplies and fresh inoculum becomes available. In spite of great interest that plant pathologists have in predicting this highly variable and crucial element of an epidemic, appropriate and specific research is still lacking. For a pathogen of high multiplication rate, latent period is major determinant of epidemic speed (Campbell and Madden, 1990). Generation time or latent period is influenced by temperature and characterized by decrease in duration as temperature rises from minimum to optimum and then increases duration with higher temperatures (Hau *et al.*, 1985). Third degree polynomial is generally fitted to express temperature response on latent periods. Sinha (unpublished) developed generation time or latent period as G_{50} (time to develop 50 per cent sporulating colonies, Figure 14.3) model. G_{50} ranged from 15.2 days at 10°C to 4.6 days at 26°C and best explained through cubic model,

$$G_{50} = 28.64 - 1.44\ T + 0.009\ T^2 + 0.0005\ T^3$$

within the range of temperature under controlled conditions. Pattern of generation time completion rate was best explained as a linear increase in rate up to temperature 28.6°C, then an exponential decline with temperature up to the 34°C the maximum temperature. Proportion of one generation period per half-hour was modeled as,

$$(0.000214\,T-0.000896)\{(1-\exp[0.2249\,(T-36)]\}$$

where, T is the average temperature during the half-hour period (Figure 14.4). Generation period predicted was compared with observed generation period in the orchard experiment. The generation period model could be used for giving agroadvisory to optimize fungicide use against the disease.

Kluge *et al.* (1989) developed ERYPROG for calculating regional differences in epidemic development of *Erysiphe graminis* in wheat and barley in GDR for determining the necessity for control under the environmental conditions prevailing in a specific field. Mildew development is calculated each week on the basis of weather forecasts and control thresholds are calculated on the basis of infection surveys, regional trends and field-specific factors (aspect, N application, crop density and tillering density). The model was found to be reliable both at regional and at local levels.

Progress of powdery mildews in time can be modelled as temporal graphs of epidemics. Biological growth functions derived from differential equations describe growth rates. Such models where parameters can be ascribed to have some epidemiological meaning are good for descriptive and predictive models. Jeger (1984) proposed a technique for relating disease progress (y) for powdery mildew (*Podosphaera leucotricha*) of apple with cumulative numbers of trapped spores (z). The relationship involves two epidemiological parameters, *viz.* disease asymptote (y_{max}) and the infection efficiency (k) of inoculum as follows:

$$y = y_{max} - c\exp[(-k/y_{max})\,z]$$

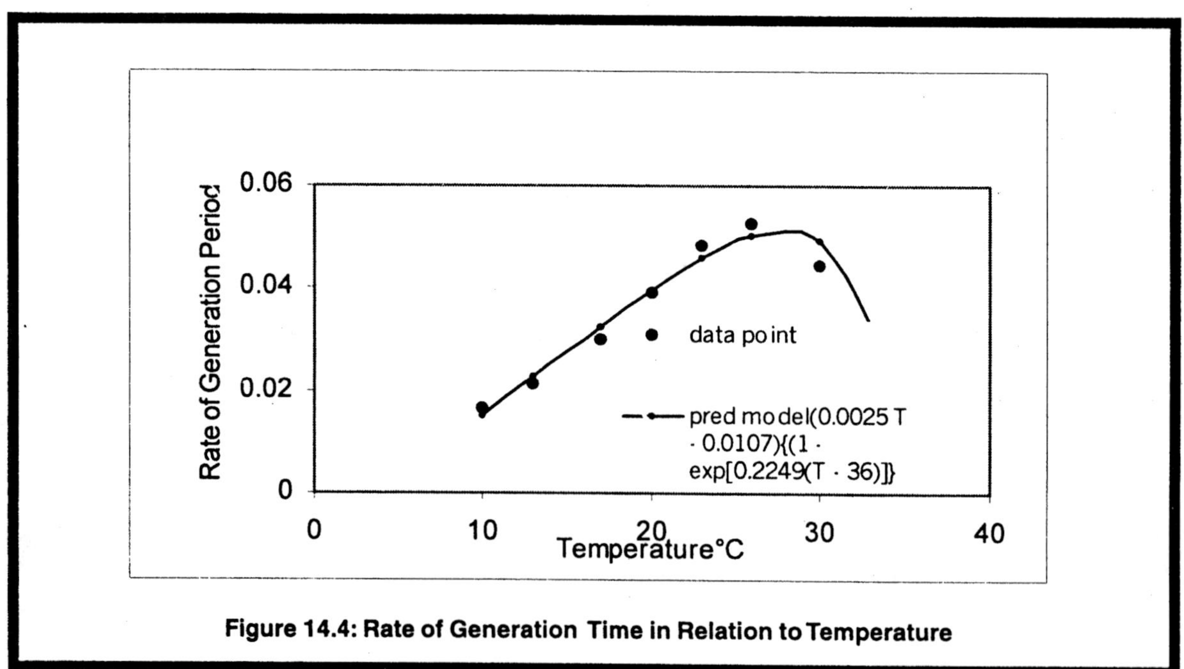

Figure 14.4: Rate of Generation Time in Relation to Temperature

However, this model is criticized in that the period of incubation is ignored. It would be more appropriate to relate amount of disease at a given time with number of spores trapped at some time previously. Unfortunately, incubation periods of powdery mildews are environmentally–dependent and use of arbitrarily fixed incubation periods is questionable. Sinha *et al.* (2002) fitted logistic and Gompertz models to describe the pattern of powdery mildew development in mango. Parameters concerning intrinsic infection rate and maximum mildew severity are estimated using nonlinear estimation procedures. Maximum rate of growth is found to be between 7 and 8 th weeks after initiation of disease.

However, growth models are purely phenological and do not provide clues about host-pathogen-environmental interaction. Therefore, they are inadequate for analysis of disease progress curves of widely different shapes. Lallancette *et al.* (1988) used linearized version of Richard's growth model to include the effect of temperature and duration of RH describing sporulation of *Plasmopara viticola* on grape leaves. When the asymptote, rate, and intercept parameters were represented as polynomial function of temperature, the model could explain 90 per cent of the variation in number of sporangia. For development of the model, first, an equation for predicting the upper asymptote of S, (k) was derived as

$$k = b_0 + b_1x + b_2x^2$$

By regressing the maximum S obtained for each temperature (T). Second, linearized version of Richard model was fitted using the data of temperature

$$\ln[1-(S/k)^{1-m}] = -rH + B$$

Where H= duration of high RH, k = estimated asymptote, m, r, and B are the shape, rate and intercept parameters respectively. Estimates of parameters, r and B were then regressed against temperature and polynomial equations derived for r and B. Finally, modelling S as a function of T and H, consisted of substituting the rate and intercept parameter functions for r and B respectively, in the Richard model:

$$\ln[1-(S/k)^{1-m}] = b_0 + b_1H + b_2HT + b_3HT^2 + b_4T + b_5T^2$$

Such modelling efforts are useful for the management decisions and for powdery mildews also work can be attempted.

Powdery mildews sporulate profusely under favourable conditions and produces clouds of air-borne inoculum. Strongly dependent on weather factors, sporulation distinctly fluctuates in time. On top of this, there is an overriding influence of time governed by increase in disease, rhythm of sporulation during lifetime of individual lesions, and age-stratification of population of lesions. Based on measurements, some mathematical models are developed which describe, analyze and/or predict sporulation as it changes with time. Kiraly (2000) reported sporulation rate as most important epidemiological parameter for modelling powdery mildew (*E. graminis*) outbreaks in wheat. Several sub-models, elaborated by regression analysis, are included in the model to calculate daily rates of sporulation using meteorological data. Patterns of spore release are examined for *Sphaerotheca pannosa* on rose, *Erysiphe pisi* on pea and *E. graminis* on *Bromus catharticus* (Adams *et al.*, 1986). No conidia of *S. pannosa* or *E. pisi* are released in dark or light (80 µE m-2/sec) in still air (ambient air velocity < 0.1 m/sec) at 100 per cent RH but when the RH is lowered rapidly, conidia of *S. pannosa* are liberated in light and darkness. *E. pisi* released conidia in still air when decreasing RH is accompanied by increasing temperature. Liberation of conidia of *E. graminis* from infected detached leaves was examined in a specially designed spore release apparatus using a linear air velocity of 0.5 m/sec. Conidia are liberated in darkness whenever RH was abruptly reduced or raised. Exposure to unfiltered infrared light also triggered discharge.

Management

Quite a large number of communications are available describing attempts to combat powdery mildew on cultivated species using a wide range of fungicides. Sulphur dusting has been found to be effective for a wide variety of crops like, cucurbits, grapes, cereals, mango, apple and ornamentals like rose, balsam, sunflower, zinnia, dahlia and other composits. Number of fungicides against powdery mildews has increased. Dinitrophenols as dinocap has been used against powdery mildew in cucurbits and some ornamentals. Fluotrimazole, halacrinate, nitrotalisopropyls used as sprays against powdery mildew in cerelas and apple. Piperalin is used for the control of powdery mildew of roses and ornamentals. Cholrothalonil also has been found to be effective agianst powdery mildew. Dimethirimol, ethirimol and bupirimates are common systemic fungicides used to treat as sprays on wheat, barley, roses, apples and cucurbits. In grape, excellent control of fruit infection, and adequate control of leaf infection, was achieved by two fungicide applications of myclobutanil targeted at the peak period of fruit susceptibility (Gadoury *et al.*, 2001). Schoeman *et al.* (1995b) have suggested that the fungicides can be applied when the first sign of disease appears usually at the 50 per cent flowering stage. But more precise knowledge of when to expect the first appearance of the disease and its subsequent progress, based on the weather conditions and susceptibility of different developmental stage, would aid growers in the timing of fungicide applications. Fungicidal trials conducted by many workers equivocally support the fact that several effective fungicides are available for the management of mango powdery mildew. But precise knowledge of when to expect first appearance of disease and its subsequent progress, based on weather conditions and susceptibility of different developmental stage, would aid growers in ascertaining optimum time of fungicidal applications (Schoeman *et al.*, 1996). In Israel, Reuveni and Reuveni (1995) reported foliar sprays of 0.025 M and 0.04 M solutions of K_2HPO_4 and KH_2PO_4 + KOH and systemic fungicides inhibited development of powdery mildew fungi on fruit clusters, flower clusters, fruits and leaves of field-grown grapevines, mango and nectarine. The effectiveness of phosphates in controlling *U. necator* on grapevines berries was similar to systemic fungicide pyrifenox. However, systemic fungicides diniconazole (Marit 12.5 per cent WP), myclobutanil (Sisthane 12E) and penconazole (Ophir), were more effective in controlling the disease on inflorescences of mango and fruits of nectarine, respectively, than either phosphate. However, alternating treatments of phosphate salt with each of these fungicides enhanced the inhibitory effect against the fungus in each crop. It is concluded that the inhibitory effectiveness of phosphate salts makes them useful 'biocompatible' fungicides and ideal foliar fertilizers for field application for disease control. Schoeman *et al.* (1995b) have suggested that the fungicides can be applied when the first sign of disease appears usually at the 50 per cent flowering stage. But more precise knowledge of when to expect the first appearance of the disease and its subsequent progress, based on the weather conditions and susceptibility of different developmental stage, would aid growers in the timing of fungicide applications. Fungicidal trials conducted by many workers equivocally support the fact that several effective fungicides are available for the management of mango powdery mildew. But precise knowledge of when to expect first appearance of disease and its subsequent progress, based on weather conditions and susceptibility of different developmental stage, would aid growers in ascertaining optimum time of fungicidal applications (Schoeman *et al.*, 1996).

Biological Control

Integrated management encourages use of potential biological agents. Fortunately, several such agents are reported against powdery mildews. Resistance of powdery mildews to the fungicides and demand for ecofriendly measures of control shifted the emphasis towards biocontrol. Several potential biological agents have been recognized for powdery mildews. *Ampelomyces quisqualis*, a pycnidial hyperparasitic fungus, occurs commonly on many species of powdery mildew, inhibits development of conidia and teleomorphs (Yarwood, 1939; Rogers, 1959; Sundheim and Krekling, 1982; Sztejnberg *et al.*, 1989; Jarvis and Slingsby, 1977; Odintsova, 1975). *Acremonium alternatum*, a pycnidium forming hyperparasite (Malathrakis, 1985) and *Stephanoascus* spp. an ascomycetous yeast that reduce disease development and conidial formation (Jarvis and Belanger, 1991). *Tilletiopsis* spp., a ballistospore-forming yeast that reduces powdery mildew development of both cucumber (Hijwegen, 1992; Hoch and Provvidenti, 1979) and barley (Klecan *et al.*, 1990). Both *Stephanoascus* and *Tilletiopsis* appear to be potential biological control agents for powdery mildew on rose and cucumber under commercial conditions in Canada (Jarvis and Belanger, 1991), the Netherlands (Hijwegen, 1992), and in the Denmark (Knudson and Skou, 1990).

Among these, *A. quisqualis* apparently is prominent agent that can be cultured on artificial media and is not so specialized. It has been used in the control of mildews on red clover (Yarwood, 1932), apple (Odintsova, 1975) and vegetable crops (Jarvis and Slingsby, 1977; Philipp and Cruger, 1979; Sundheim, 1982; Sundheim and Amundsen, 1982; Sztejnberg, 1979; Rayss, 1955). The fungus has been reported to tolerate a number of fungicides, acaricides and insecticides (Philipp and Cruger, 1979; Philipp *et al.*, 1984; Sundheim, 1982, 1984; Sundheim and Amundsen, 1982; Sztejnberg and Mazar, 1983). In an integrated management experiment, using combination of *A. quisqualis* and triforine, Sundheim (1982) and Sundheim and Amundsen (1982) reported that the growth of *S. fuliginea* was reduced.

Conclusion

Powdery mildew is the most common disease of vegetables, cereals, pulses, ornamentals, spices, fruit trees, sugar crops, and several other plants of economic importance are affected causing significant yield reduction. With the increasing trend in industrialization of agriculture, crop management practices are to be based on precise decisions. Therefore, predictive management of the disease is the only way to justify efficiency and economy of disease management practices. Majority of works on powdery mildews can be seen on resistance screening and management using fungicides. Large number of fungicides reported to be effective against the disease. However, issue, questioning when and how many sprays are to be given mostly being ignored. Epidemic pattern of powdery mildew in time and space influenced by a multiplicity of factors, like weather, host susceptibility, pathogen virulence, and human interference. Studies reveal that weather factors, mainly temperature and relative humidity above a threshold limit are key factors for occurrence of powdery mildew. Forewarning information is available for some cases where these factors are used as input. Information on inoculum, sporulation, latent period, infection, infection rates, weather factors, and disease progress and spatio-temporal variations are important epidemic parameters useful for quantitative estimation and forecasting of the disease, which in turn may prove vital for management decisions. In this paper, an attempt has been made to overview current status and future directions in management of powdery mildews with a special emphasis on epidemiology.

REFERENCES

Adams, G.C., T.R. Gottwald and C.M.Leach (1986). Environmental factors initiating liberation of conidia of powdery mildews. *Phytopathology*, 76, 1239–1245.

Campbell C.L., L.V. Madden (1990). *Introduction to Plant Disease Epidemiology*. John Wiley and Sons, New York.

Chellemi, D.O. and J.J.Marois (1991). Development of Demographic model for *Uncinula necator* by using a microcomputer spreadsheet program. *Phyotopathology*, 81, 250–254.

Gadoury, D.M., R.C. Seem, A. Ficke and W. F.Wilcox (2001). The epidemiology of powdery mildew on Concord grapes. *Phytopathology*, 91: 948-955.

Gubler, W.D., M.R. Rademacher, S.J. Vasquez, C.S. Thomas (1999). Control of powdery mildew using the UC Davis powdery mildew risk index. APS*net Plant Pathology on-line [http://www.apsnet.org/online/feature]/pmildew/Top.html*.

Gupta, J.H. (1985). Effect of temperature and RH on germination of conidia of powdery mildew of ber. *Indian J. Hort*. 42, 148–150.

Gupta J.H. (1988). Perpetuation and epidemiology of powdery mildew of mango. *Acta Horticulturae* 231, 528–533.

Hau, B. (1983). Epidemiologische Simulateron als heuristiche und operationale Instrumente der Systemanalyse mit besonderrer Berucksichtigung eines Modells des Gerstenmehltaus. Disserattion, University of Giessen, Giessen, Fed. Rep. of Germany.

Hau, B., S.P. Eisensmith, J. Kranz (1985). Construction of temporal models: II. Simulation of aerial epidemics. In: *Advances in Plant Pathology*, Vol. 3: *Mathematical modelling of crop disease*, (Ed.) C.A. Gilligan, Academic Press, Inc., pp. 44.

Hijwegen, T. (1992). Biological control of cucumber powdery mildew with *Tilleptiopsis minor* under greenhouse conditions. *Neth. J. Plant Pathol*. 98: 221–225.

Hoch, H.C. and R. Provvidenti (1979). Mycoparasitic relationships: Cytology of the *Sphaerotheca fuliginea-Tilletiopsis* sp. Infection. *Phytopathology*, 69: 359–362.

Jarvis, W.R. and R.R. Belanger (1991). Control of powdery mildews in cucumber and rose by *Stephanoascus* spp. *USDA Agric. Res. Serv. ARS*., 92. 324 pp.

Jarvis, W.R. and K. Slingby (1977). The control of the powdery mildew of greenhouse cucumber by water sprays and *Ampelomyces quisqualis*. *Plant Dis. Reptr*., 61: 728-730.

Jeger, M.J. (1980). Relating Disease Progress to cumulative numbers of trapped spores: apple powdery mildew and scab epidemics in sprayed and unsprayed orchard plots. *Plant Pathology*, 33: 517–523.

Kable, P.F., P.M. Fried and D.R. MacKenzie (1980). The spread of powdery mildew of peach. *Phytopathology*, 70: 601- 60.

Kiraly, Z.(2000). Modelling the effects of weather on wheat powdery mildew. *Acta-Phytopathologica-et-Entomologica-Hungarica*, 35: 323–332.

Kiraly, Z.(2000). Relationships between epidemiological parameters of *Erysiphe graminis* f.sp. *tritici* under fluctuating weather conditions. *Acta-Phytopathologica-et-Entomologica-Hungarica*, 35: 333–341.

Klecan, A.L., S. Hippe and S.C. Somerville (1990). Reduced growth of *Erysiphe graminis* f. sp. *Hordei* by *Tilletiopsis pallescans. Phytopathology*, 80: 325-331.

Kluge, E., V. Gutsche and G. Gunthur (1989). Computer-aided method for prediction and control of powdery mildew in winter wheat and winter barley (ERYPROG)–test results 1986 to 1988. *Nachrichtenblatt-fur-den-Pflanzenschutz-in-der-DDR*, 43: 161–165.

Knudson, I.M.B. and J.P. Skou (1990). Biological control of powdery mildew with *Tilletiopsis*. Report M-2891 from section of Plant Biology. Riso National Laboratory, Denmark. 80 pp.

Kranz, J. (1990). *Epidemics of Plant Diseases. Mathematical Analysis and Modelling*. Second edn. Springer-Verlag, Germany.

Lallancette, N., L.V. Madden and M.A. Ellis (1988). A quantitative model for describing the sporulation of *Plasmopara viticola* on grape leaves. *Phytopathology* 78, 1316–1321.

Linde, M. and T. Debener (2003). Isolation and identification of eight races of powdery mildew of roses (Podosphaera pannosa) (Wallr.: Fr.) de Bary and the genetic analysis of the resistance gene Rpp1. *Theoretical and Applied Genetics*, 107: 256–262.

Maheswari, S.K. and S.B. Singh (1997). Effect of temperature and relative humidity on development of powdery mildew of ber. *Ann. Pl. Protec. Sci.* 7, 105–107.

Malathrakis, N.E. (1985). The fungus *Acremonium alternatum* Linc. Fr., a hyperparasite of the cucurbits powdery mildew pathogen *Sphaerotheca fulginea. Z. Pflanzen. Schutz.* 92: 509–515.

Mane, B. S., S.V. Khandge, M.C. Varshneya and T.R.V. Naidu (1996). Role of micro-meteorological parameters on incidence and development of powdery mildew of grapes. *Journal of Maharashtra Agricultural Universities*. 21: 445-449.

Odintsova, O.V. (1975). Role of a hyperparasite *Cicinnobolus D. By. In suppressing powdery mildew on apple trees*. Mikol. Fitopat. 9: 337-339.

Parkash, V., G.P. Kaur and J.S. Jhooty (1988). Perpetuation of *Microsphaeria alphitoides f.sp. ziziphi* causing powdery mildew of ber. *Indian Phytopath.* 41: 141–143.

Pearson, R.C. and D.M. Gadoury (1992). Powdery mildew of Grape. *In: Plant Diseases of International Importance Vol III: Diseases of Fruit Crops*, (Eds.) J. Kumar, H.S. Chaube, U.S. Singh and A.N. Mukhopadhyay, Prentice Hall, Englewood Cliffs, New Jersey, pp. 129–146.

Pfender, W.F. (2001). A temperature-based model for latent period duration in stem rust of perennial ryegrass and tall fescue. *Phytopathology*, 91: 111–116.

Philipp, W.D. and G. Cruger (1979). Parasitismus Von *Ampelomyces quisqualis* auf Echten Mehltaus, *Sphaerotheca fuliginea* and *Erysiphe cichoracearum*. Nachrichtenbl. Dtsck. *Pflanzenshutzd. (Braunschweig)*, 28: 65–67.

Philipp, W. D., U. Grauger and F. Grossmann (1984). Erganzende unterschungen zur biologischen und integriet Glas durch *Ampelomyces quisqualis. Z. Pflanzen Schutz*. 91: 438–443.

Rayss, T. (1955). Nouvelle contribution a la connaissance des Deuteromycetes de Palestine. *Bull. Res. Conn.*, Israel, 5D: 37-46.

Reuveni, M. and R. Reuveni (1995). Efficacy of foliar sprays of phosphates in controlling powdery mildews in field-grown nectarine, mango trees and grapevines. *Crop-Protection* 14: 4, 311-314

Rogers, D.P. (1959). On *Cicinnobolus*. *Mycologia*. 51: 96–98.

Rubiales, D., J.K.M. Brown and A. Martin (1993). Hordeum chilense resistance to powdery mildew and its potential use in cereal breeding. *Euphytica*, 67: 215–220.

Sall, M. (1980). Epidemiology of grape powdery mildew: A model. *Phytopathology*, 70: 338–342.

Schoeman, M.H., B.Q. Manicom, M.J. Wingfield (1995). Epidemiology of powdery mildew on mango blossoms. *Plant Dis.*, 79: 54–528.

Seber, G.A.F., C.J. Wild (1989). *Nonlinear Regression.* John Wiley and Sons, New York, pp. 768.

Singh, R., R. Niwas and P.C. Gupta (1999). Prediction model for powdery mildew disease build up in ber based on weather parameters. *Ann. Agri. Bio. Res.*, 4: 11–15.

Sinha, P., U. Chakrabarti and A. Varma (2001). Critical weather factors for the development of mango powdery mildew. *Ann. Pl. Protec. Sci.*, 9: 264–267.

Sinha, P., Varma A. Prajneshu (2001). Growth models for powdery mildew development of mango. *Ann. Pl. Protec. Sci.*, 10, 84–87.

Sinha, P., Varma A. Prajneshu (2001). Studies on determining favourable conditions for the germination of conidia of *Oidium mangiferae. Indian Phytopath.*, 54, 197–200.

Sinha, P., N.V.K. Chakravarty, Prajneshu and A. Varma (1999). 'Prediction of mango powdery mildew'. In: *The National Workshop on Dynamic Crop Simulation Modeling for Agrometeorological Advisory Service*, National Centre for Medium Range Weather Forecasting, Department of Science and Technology, New Delhi, January 4-6,1999, pp.321–326.

Sundheim, L. (1982). Control of cucumber powdery mildew by the hyperparasite *Ampelomyces quisqualis* and fungicides. *Plant Pathol.*, 31: 209–214.

Sundheim, L. (1984). *Ampelomyces quasqualis*, a hyperparasitic fungus in biological control of powdery mildews on greenhouse cucumber. *Acta Hortc.*, 156: 229–236.

Sundheim, L. and T. Amundsen (1982). Fungicide tolerance in the hyperparasite *Ampelomyces quisqualis* and integrated control of cucumber powdery mildew. *Acta Agric. Scand.* 32: 349–355.

Sundheim, L. and T. Krekling (1982). Host parasite relationships of the hyperparasite *Ampelomyces quisqualis* and its powdery mildew host *Sphaerotheca fuliginea. Phytopathol. Z.*, 104: 202–210.

Sztenjberg, A. (1979). Biological control of powdery mildews by *Ampelomyces quisqualis. Phytopathology* 69: 1047.

Sztenjberg, A. and S. Mazar (1983). Studies on the hyperparasite *Ampelomyces quisqualis* and preliminary trials on biological control of powdery mildews. *Phytoparasitica*, 11: 219–220.

Sztenjberg, A., S. Galper, S. Mazar and N. Lister (1989). *Ampelomyces quisqualis for biological and integrated control of powdery mildews in Istail. Phytopathol. Z.*, 124: 285–295.

Whipps, J.M. and S.P. Budge (2000). Effect of humidity on development of tomato powdery mildew (Oidium lycopersici) in the glasshouse. *European J. Plant Pathol.*, 106: 395–397.

Yarwood, C.E. (1932). *Ampelomyces quisqualis* on clover mildew. *Phytopathology*, 22: 31.

Yarwood, C.E. (1939). Control of powdery mildew with a water spray. *Phytopathology*, 29: 283–290.

15

Viral Diseases of Fruits, Vegetables and Spices in North Eastern Himalayan Region of India and their Management

Kajal Kumar Biswas, D.R. Das** and Y. S. Ahlawat***

**Indian Agricultural Research Institute, Regional Station, Kalimpong, West Bengal – 734 301*

***Unit of Plant Virology, Division of Plant Pathology, Indian Agricultural Research Institute, New Delhi – 110 012*

Virus diseases affect most of the economic-crops and are responsible for far greater economic losses of crop productivity and quality not only in India but also worldwide. The extent of losses depends on many factors and may vary widely. Some viruses under appropriate conditions may infect a pant without producing any obvious signs of disease (latent), whereas, other may lead to rapid death of the whole plant (adverse affects). Between these extremes, a wide variety of diseases can be produced.

Various types of symptoms caused by virus infected plant. Symptoms expression obviously depends on the many factors like virus or virus strain, plants and plant genotypes, age of the plants, growth stages of the plant and several environmental and climatic factors. Viruses are transmitted by various means through mechanical sap, seed, pollen, graft, vegetatively propagative material and mostly insect vectors. North eastern Himalayan region is very much diversified in its climatic condition and produces many different kinds of fruits, vegetables and spices which are infected by virus diseases (Table 15.1). The economically important virus diseases in most economical crops and their possible control measure are described.

Table 15.1: Important Viral Diseases of Economically Important Fruits, Vegetables and Spices Crop in North Eastern Himalayan Region of India

Crop	*Species*	*Diseases*	*Mode of Transmission*
Fruits	Citrus	Citrus ring spot	Grafting
		Citrus tristeza	Vector, grafting
		Citrus mosaic	Grafting
	Papaya	Papaya ring spot	Sap, dodder, aphid
		Papaya leaf curl	Whitefly
	Banana	Banana bunchy top	Aphid
Vegetables	Potato	PVX	Seed, sap, beetle
		PVY	Seed, sap, aphid
		Potato leaf roll	Seed, aphid
	Tomato	Tomato mosaic	Seed, sap,
		Tomato leaf curl	Whitefly
	Radish	Radish mosaic	Sap, aphid
	Bhindi	Yellow vein mosaic virus	Whitefly
	Cucurbit	Cucumber mosaic	Sap, aphid
	Bean	Bean common mosaic	Seed, pollen. sap, aphid
	Chayote	Yellow mosaic	Whitefly
	Elephant foot yam	Mosaic disease	Seed
Spices	Chilli	Chilli mosaic	Aphid
		Chilli leaf curl	
	Large cardamom	Chirke	
		Foorke	Unknown

Fruits

Virus Diseases in Citrus

Citrus fruits are grown all over India as fresh fruit as well as its processed juice. In India position of citrus production is 3rd highest after banana and mango. Hills of West Bengal, Sikkim, Assam are well known states to produce commercial citrus fruits in Eastern Himalayan region of India. The major species of citrus grown in this region are mandarin orange (*Citrus reticulata* Blanco), and acid lime (*C. aurantifolia* Swingle). The productivity of the citrus fruit in this region is comparatively low due to many diseases of which virus disease play a very significant role. More than 30 virus diseases affecting citrus worldwide are reported. The three virus diseases which are mostly prevalent and

Figure 15.1: Variable Symptoms Caused by Plant Viruses in Different Kinds of Crops

(a) Vein clearing and flecking in younger leaves of Kagzi lime, (b) vein corking in sweet orange, (c) stem pitting in mandarin orange,(d) citrus decline caused by CTV; (e) ring spot on sweet orange by ICRSV;(f) yellow mosaic mottling in sweet orange by CYMBV; (g) mosaic/ringspot of papaya by PRSV; (h) leaf curl of papaya by PaLCV; (i) bunchytop of banana by BBTV.

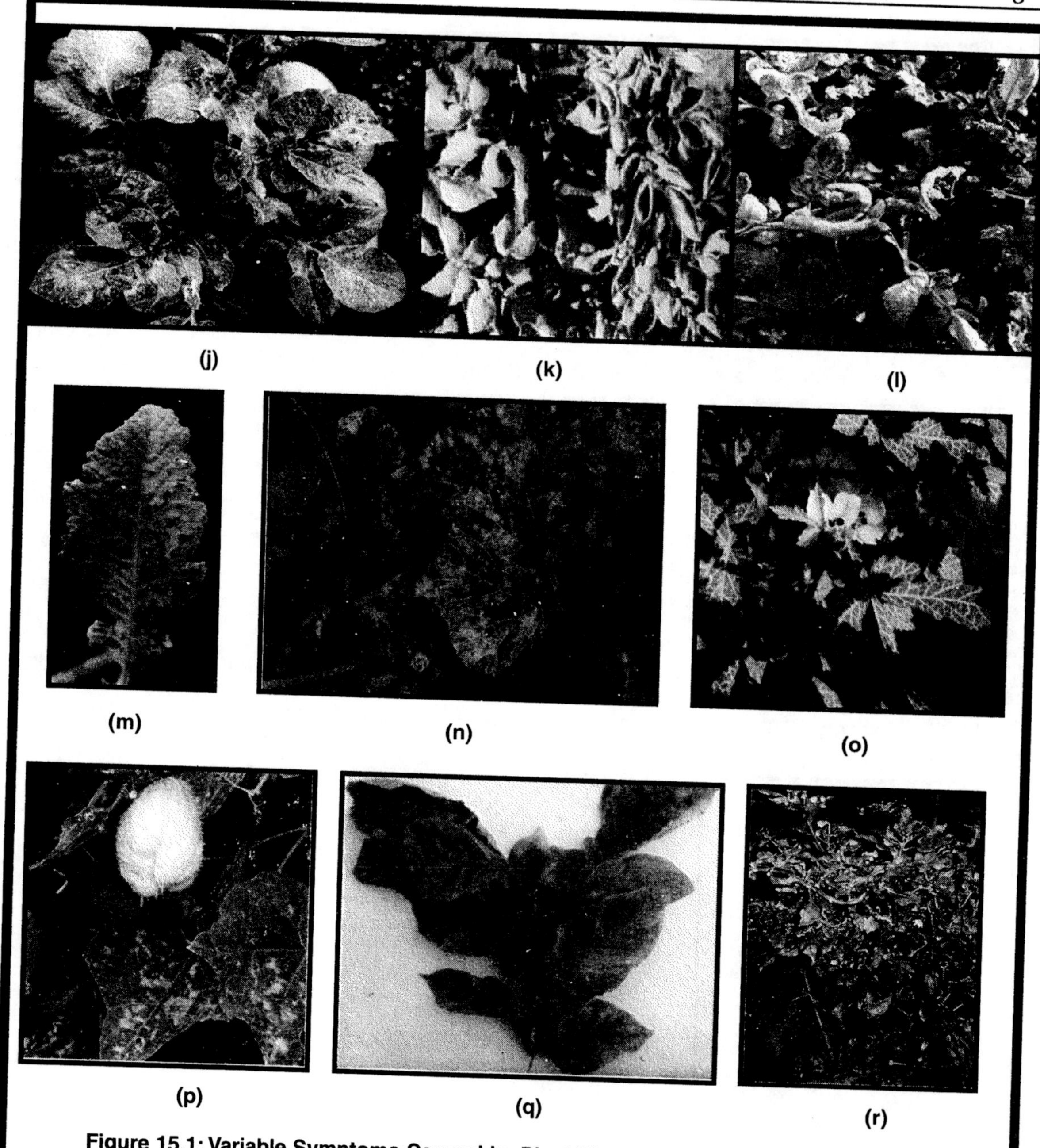

Figure 15.1: Variable Symptoms Caused by Plant Viruses in Different Kinds of Crops

(j) potato mosaic in Potato by PVY; (k) leaf roll of potato by PLRV; (l) Leaf curl in tomato by ToLCV; (m) mosaic of radish by TuMV; (n) Mosaic symptoms on cucurbits caused by CMV; (o) yellow vein mosaic on bhindi by Gemini virus; (p) yellow mosaic in chayote caused by gemini virus;(q) yellow mosaic in elephant foot yarn; (r) yellow mosaic in chili.

(s) (t) (u)

Figure 15.1: Variable Symptoms Caused by Plant Viruses in Different Kinds of Crops

(s) filli form leaf in chilli by chilli mosaic; (t) Chirke disease in large cardamom; (u) forke disease in large cardamom.

major limiting factor to yield losses region, are (CTV), Citrus yellow mosaic badna virus (CYMV) and Indian Citrus ring spot mandari virus (ICRSV).

Citrus Tristeza Virus

Citrus tristeza closterovirus (CTV) disease has been reported from almost all the citrus growing states of India. Tristeza means "sadness" in Portuguese and Spanish language to describe the rapid and wide spread decline and death of millions of citrus tree on sour orange root stocks in Argentina and Brazil during the years of 1930s. Later a virus like agent was established in 1946 which was aphid transmitted and named was citrus tristeza virus (CTV). Tristeza virus infects nearly all citrus species, citrus relatives and hybrids. It causes decline of trees grafted on sour orange (*Citrus aurantium* L) as originally described. It produces stem pitting on lime (*C. aurantifolium* Swingle and *C. latifolia* Tan), grape fruit (*C. Paradisi* Macf) and sweet orange (*C. sinensis* L.). Small acid lime like Kagzi lime (*C. aurantofolia*) are among the most susceptible hosts and show vein clearing, stem pitting causing CTV. Decline and stem pitting causing CTV isolates are genetically distinct.

Symptomatology

The characteristic symptoms in the ideal indicator host, Kagzi lime consist of clearing or flecking, corking of veins and veinlets. Infected leaf becomes corky and leathery in appearance. The stem pitting symptom *i.e.*, a longitudinal necrosis on the twigs or stem is occurred by infection of severe strain of CTV which is visible when the bark is removed. In mandarin orange the symptoms remain generally invisible but when the infected twigs grafted on Kagzi lime it produces vein flecking and later on vein corking etc. This disease characterized by sudden wilting and drying of the leaves, leaving behind the most of the fruits which get dried and fall later. The function of stem is drastically affected. After some time the diseased tree show partial recovery and such tree remain alive but die back of the twigs continues. The infected plants have heavy bearing but with small fruits. In mandarin orange the symptoms remain invisible but when the infected twigs grafted on Kagzi lime it produces vein flecking and later on vein corking. Decline symptom of CTV is due to necrosis of phloem in the bark of sour

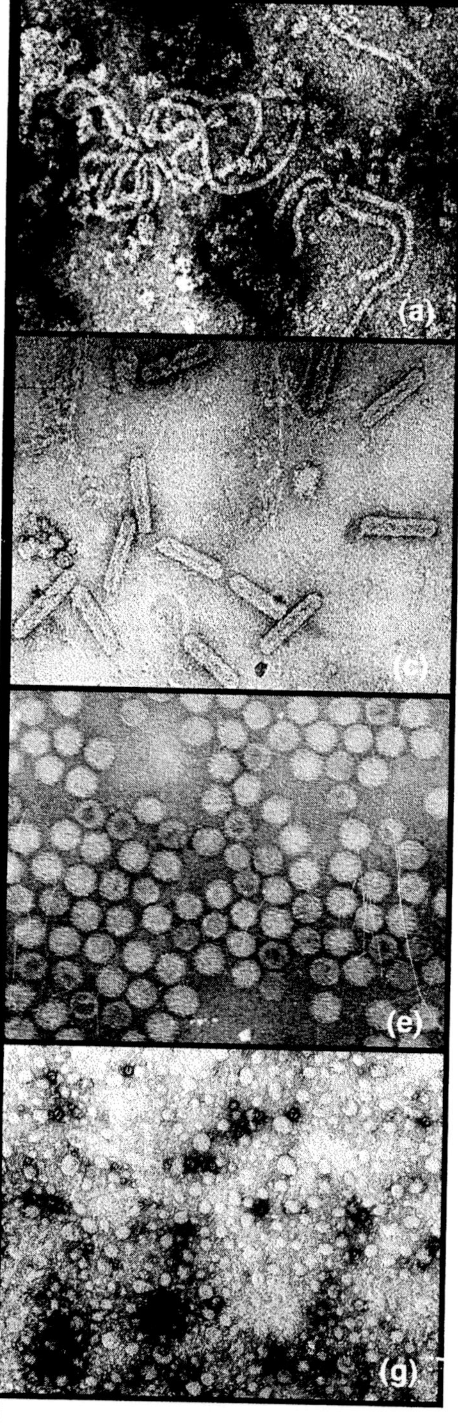

(a) CTV clostero virus causing tristeza disease in citrus, (b) ICRSV mandarin virus causing ringspot disease in citrus, (c) CYMBV badna virus causing yellow mosaic in citrus, (d). PVY Poty causing yellow mosaic in chilli, potato virus Y etc., (e) CMV cucumovirus causing mosaic in cucurbit, (f) gemini virus causing leaf curl in tomato and chilli, yellow vein mosaic in bhindi, (g) PLRV luteo-virus causing leaf roll in potato, (h) PVY and CMV causing mixed infection causing yellow mosaic in chilli.

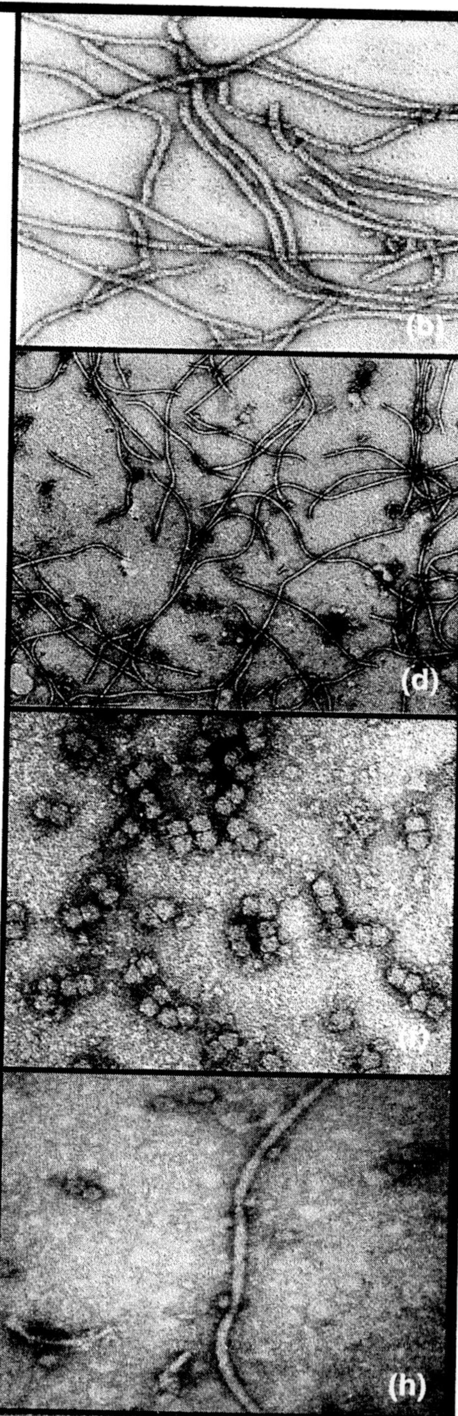

Figure 15.2 : Virus Particle Associated in Different Crop Plants

orange root stock just below the bud union. Movement of carbohydrates to the root system is prevented. Once the carbohydrate reserves in the root system exhausted, new fibrous roots can not be generated and root system degenerates resulting in tree decline. This decline is very fast and lethal and the symptoms appear within one to two years after infection and the disease is called quick decline.

Causal Factors and its Spread

CTV caused by thread like virions, 10–12 nm × 2000 nm. The virus is limited to phloem and single stranded RNA virus belongs to closterovirus group. This virus is transmitted by bud wood grafting. Brown plant aphid, *Toxoptera citiricidus* is the major vector to spread the disease in non-persistent manner but other aphids like *Apish gossypii, Myzus persicae, Apis craccivora* and *Dactynotus jaccae* are also reported to transmit the disease. In northern India the aphid, *Toxoptera* does not occur; hence tristeza disease is less common in this region. However, destructive strains are known to occur only in places wherever the vector Toxoptera is present and active. In north eastern Himalayan region vector population is very high. Recently occurrence of many CTV strains have been reported to cause different kinds of CTV symptoms different citrus species. (Biswas *et al.*, 2004). The dodder, *Cuscuta* spp., also reported to transmit CTV. No experimental host is still reported for this virus.

Indian Citrus Ring Spot Virus

Indian Citrus ring spot mandari virus (ICRSV) disease was first described in kinnow mandarin in north-west India but now this disease is known to occur in most of the citrus growing region of India.This disease infects several species of citruses in India. Maximum yield losses is observed in kinnow mandarin and yield loss in 7 years old tree is 20.5 to 98 per cent. The health of the infected trees deteriorated year after year and finally the infected trees are collapsed. Recent studies have shown that the citrus ring spot virus in India is distinct from other ring spot viruses and hence named as Indian citrus ring spot virus (ICRSV).

Symptomatology

The disease can be easily recognized in the field trees by it characteristics ring spot symptoms on leaves. Affected trees show conspicuous rings on mature leaves which may be one to several numbers per leaf. Most of the ICRSV affected trees show decline or die-back symptoms and the quantity and quality of fruits is greatly affected. The glasshouse inoculated sweet orange and mandarin orange plants show shock reaction after 2–3 days of inoculation. Young leaves of mosambi sweet orange plants also develop vein clearing symptoms on young leaves which persists till to the maturation of leaves.

Causal Factors and its Spread

A flexuous rod shaped virus known as a member of mandarin virus associated with most of the ICRSV. Except bud transmission, no other mode of natural spread of ICRSV is reported. The virus is mechanically sap transmissible to herbaceous hosts like *Chenopodium quinoa, Phaseolus vulgari* var. Saxa, Singhtamey, Gheusami and alpatri and some cultivars of *Glycine max*.

Citrus Yellow Mosaic

Citrus yellow mosaic badna virus (CYMV) disease has been reported in satgudi and chinni sweet orange (*C. sinensis* l) from south India and in pummelo (*C. decumana* L., syn. *C. paradisi* L) from north eastern India. Its incidence is 10–70 per cent, causing serious losses in citrus orchards, which are even abandoned. This disease can cause reduction of fruits up to 70 per cent. In view of the occurrence of this disease only in India and its wide spread and distribution and losses caused, it is studied in detail at Advance centre of plant virology, IARI, New Delhi.

Symptomatology

The characteristics symptoms of this disease in field infected sweet orange and pummelo trees consisted of bright yellow mottling of leaves and yellow flecking along the veins.

Causal Factors and its Spread

CYMV is identified as the DNA virus known as badna virus with bacilliform virions measuring 130 nm × 30 nm. It was transmitted to various citrus species by graft and to sweet orange and sugarcane by mechanical sap inoculation. This disease is transmitted by mealy bug, *Planococcus citri*.

Virus Diseases in Papaya

Papaya (*Carica papaya* L.) is one of the important fruit crops growing throughout India. Papaya is grown in area about 39.3 thousand hactres in India with 1.5 million tone ripe fruit production per year occupying a second highest position in the World after Brazil. Even the total area under papaya cultivation has been increased in the recent past, however, its cultivation has been affected badly due to numbers of diseases induced by viruses (Singh, 1996). In India Papaya ring spot (PRSV) and papaya leaf curl virus (PLCV) diseases are major concern causing considerable losses in yield and quality of papaya fruits.

Papaya Ring Spot Virus

Papaya ring spot virus (PRSV) disease is synonymous to papaya mosaic or watermelon mosaic virus-1 disease. PRSV is a destructive and a major limiting factor in production of papaya fruits in North Eastern hill regions of India causing near about 80 to 100 per cent incidence in susceptible cultivars. Occurrence of PRSV was first reported in India in the year 1998 and later on similar diseases were reported by several workers by the name of mosaic (Singh, 1996). Papaya mosaic reported from India was quite confusing and misleading since the name of the disease was based on symptoms in the host and its transmission.

Symptomatology

Leaves of affected field trees show a tendency to roll upward along the margin of the leaves. Malformation and reduction of leaf lamina with filliform appearance in infected plants are common symptom. Characteristic elongated dark green streaks develop on the petioles and upper portion of stem and sometimes on entire trunk. The main stem frequently develops dark green spots and streaks of an oily appearance. The symptoms are characterized by the development of a number of distinct round spots approximately 1/6 inch in diameter. In severe cases the spot coalesce to form a larger area which frequently appears elongated as streak. Infected plants become stunted, petioles are shortened, and fruits set become severely reduced as the disease progresses. Symptoms on mature green fruits provide the most striking and reliable yellow spots or yellow rings with solid green center. The number of spots or rings in a single fruit may vary from few to more. The spots never coalesce to form a large streak. When the infected mature and ripened fruits are peeled out characteristics broken round or elongated rings are seen on the surface.

Causal Factors and its Spread

The CRSV consists of flexuous filamentous particles with 12 × 760–800 nm in dimension. Infectious nucleic acids are double stranded RNA. Its particle morphology, aphid transmissibility, serological relationships and ability to induce pinwheel inclusion bodies in the host cells place this virus under the group of Poty virus. PRSV is readily transmitted by mechanical sap inoculation and by dodder, *Cuscuta reflexa*. Several aphid vectors also transmit PRSV in a non-persistent manner. The most important vectors are *Myzus persicae, Aphis gossypii, A. craccivora, Pentalonia nigronervosa, Lipaphis erysimi, Rhopalosiphum maidis* etc. Seed may transmit some strain of PRSV, yet to be confirmed.

The host range of PRSV is limited. Some plant species belonging family, Caricaceae, Chenopodiaceae, Cucurbitaceae have been reported, of which mostly squash watermelon, cucumber are most important natural hosts.

Papaya Leaf Curl Virus

Papaya leaf curl virus (PaLCV) disease is the second important viral disease after Papaya ring spot virus disease in papaya. PaLCV disease is first reported in Coimbatore and subsequently occurred in different states of India. In northern India PRSV causes about 60 per cent disease incidence in papaya. This disease is also referred as leaf crinkle disease which is widely spread in India.

Symptomatology

Papaya leaf curl disease in papaya causes severe curling, crinkling and distortion of leaves accompanied by vein clearing and reduction of leaf size. Leaf margins are rolled downward and inward in the form of inverted cup and the veins get thicken and dark green in colour. The petioles are severely twisted, become leathery and brittle. At time, enations are produced on the lower surface of the leaves. The affected plants fail to flower, in case of occasionally flowering they fail to fruit or bear few, small and deformed fruits. The fruits set before infection become rough and small in size. In advance stages, defoliation may take place and growth of the plant arrested. Leaf curl infected plant causes highly deformation of phloem and phloem parenchyma cells.

Causal Factors and its Spread

This disease is caused by whitefly transmitted gemini virus particle measuring about 18x30 nm in size. It consists of two single stranded DNA. PaLCV is transmitted by the cotton white fly, *Bemisia tabaci* in field conditions. Experimentally the virus is transmitted successfully by methods of grafting. The disease is not transmitted through sap or seeds. This virus infects all the cultivars of papaya irrespective of wild or cultivated species. The virus may perpetuate on a number of weed hosts, like tomato, tobacco, sunhemp, chilli, petunia, Zinnia, Datura etc. Although papaya is not a preferable host of whiteflies, it is assumed the PaLCV is occurred by whitefly from alternate hosts and then transmitted to papaya in a brief inoculation feeding.

Virus Diseases of Banana

Banana Bunchy Top Disease

Banana bunchy top virus (BBTV) disease is believed to have introduced in south India through infected suckers imported from Sri Lanka in the year of 1940. Since then, this disease has been the most important and serious disease affecting in banana all over the banana growing states in India. It causes lot of economic losses in banana industries in India. Incidence and spread of BBTV is faster in the neglected planting area especially when the banana is a subsidiary crop as compared with pure irrigated plantations. Planting of infected suckers causes about 100 per cent of crop losses. At present the disease is apparently restricted to particular area under those states of Kerala, Tamil Nadu, Assam, West Bengal, Bihar and Orissa through the legislation imposed by domestic quarantine.

Symptomatology

The plant may be infected at any stages of plant growth. The identification of the initial symptoms, however, is rather more difficult. The first evidence of BBTV disease is noticed on the leaves. Examination the leaves of diseased plant against light, shows irregular dark green streaks about 0.75 mm wide along with the secondary veins on the underside of the midrib. A few or several streaks may be present. The streaks may vary a series of dark green dots to a continuous dark green line, one inch or more in length with a ridged edge. The leaf may be smaller than the normal leaf, slightly chlorotic and the margin of the lamina becomes wavy and slightly rolled upward.

The presence of interrupted dark green streaks along with the secondary veins of the lamina or along the midribs or petioles, are considered to be the most definite and a reliable symptoms of BBTV disease. Leaves become progressively smaller in size, both in width and length. Petioles and leaves remain abnormally erect. When several abnormal leaves have emerged, an extreme congestion is appeared at crown and a rosette, a typical bunchy top with severely stunting growth develops. Bunchy top affected leaves give a distinct rigidity to the plant which are not readily shaken by winds and also show an apparently resistance to wilting.

Causal Factors and its Spread

BBTV disease is caused by isometric virus, the member of luteovirus particle measuring 18–28 nm in diameter. The virus particles are single stranded multi component RNA virus consisting of at least six single stranded RNA components, each of about 1000-1100 nucleotides in length. BBTV is not transmitted by mechanical means by seeds or sap. The banana aphid, *Pentalonia nigronervosa*, is the major vector to transmit the virus from infected plant to healthy plant. The most important factor to spread this virus is mainly infected suckers. The host range of BBTV is restricted in the genus *Musa* including many species under this genus. Many experimental hosts like *Cucumis sativus, Cucumis melo, Gynura aurantia* have been reported.

Vegetables

Virus Diseases in Potato

About 36 viruses and viroids affect potatoes worldwide. The most important viruses of potatoes which cause economic yield losses are potato virus Y (PVY), potato virus X (PVX) and potato leaf roll virus (PLRV).

Potato Virus X

Potato virus X (PVX) occurs worldwide and it is also known as latent or faint mottle or mild mosaic or interveinal mosaic. Although it causes light to moderate losses, but economically it is very important because of its universal present in potatoes. It causes severe crop losses when it is associated with other potato viruses.

Symptomatology

PVX causes only faint mottle or mild mosaic obviously in very young stage of the plant growth. In many cases the infected plants remain symptomless and occasionally hypersensitive reaction may occur. The reduction in yield ranges from 10–25 per cent depending upon the virus, cultivars and growth stages of the crop. PVX causes severe crop losses by causing crinkle mosaic in combination with potato virus A, and rugose mosaic in combination with potato virus Y.

Causal Factors and its Spread

PVX is RNA virus having filamentous particle of 515 × 13 nm in dimension. The virus is stable and occurs in high concentration in infected susceptible cultivars like Kufri Chandramukhi. It occurs worldwide with serotype PVX of common occurrence. PVX has a narrow host range infecting mainly potato, tomato and tobacco. PVX is mechanically transmitted by cutting of seed tubers or mechanically injury and by plant to plant or root to root contact. It is highly contagious and transmitted through field implements and garments of the field workers. PVX is not insect vector transmitted but it has been experimentally transmitted by the beetle *Ephilachna ocellata*.

Potato Virus Y

Potato virus Y (PVY) disease is also known as potato streak or potato severe mosaic or potato rugose mosaic disease. It occurs worldwide of great economic importance. It affects potato, pepper, tomato and tobacco, and causes severe losses on all these hosts. PVY may cause losses in the yield of tuber often well above 30 per cent and up to 85 per cent. Losses are naturally high when plants are co-infected especially with PVX.

Symptomatology

The primary symptoms are in the form of mild to severe veinal necrosis resulting from the necrotic lesions along the veins on the under side of the leaflets. Infected plats remain stunted. Veinal necrosis or necrotic streak may also be accompanied with mottling, yellowing of leaflets and premature death of leaves. The necrosis may extend to the petioles and reach to the stem. When PVY is associated with PVX it causes rugose mosaic in which the plants are dwarfed and tuber sizes are reduced. In this case lower leaves show veinal necrosis whereas the upper leaves are mottled with light green spots and the entire foliage is wrinkled.

Causal Factors and its Spread

PVY is a member Poty virus group and has flexuous long filamentous particles of about 740 nm long and 11 nm in diameter. Several distinct strains of PVY exist in nature as PVY-O, PVY-C, PVY-N etc. The strains PVY-O and PVY-C are prevalent in India. In nature rather than potato, PVY also infects many important crops like pepper, tobacco, tomato, some important ornamentals like Zinnia, Dahlia, Petunia and several weed hosts which serve as the source of infection. Mechanical field transmission of PVY is rare. It is transmitted through infected seed tubers. Many and about 25 species may transmit PVY naturally in a non-persistent manner. The aphid species, *Myzus persicae, Aphis gossypii* are the most important vectors.

Potato Leaf Roll Virus

Potato leaf roll virus (PLRV) disease is word wide in distribution and responsible for heavy losses of the crops. PLRV is the most important viral disease in potato because of its damaging nature in stem and tubers. PLRV may cause losses in tuber yield from 20 to 50 per cent by primary infection and 90-100 per cent by secondary infection.

Symptomatology

Symptoms of PLRV in potatoes are divided into two categories. First category of symptoms occurs after primary infection and second category of after secondary infection. In primary infections symptoms are appeared first in the upper leaves and then progress downward to the lower leaves. In secondary infection the symptoms develop first on the lower leaves and then progress towards upper leaves.

Primary infection causes chlorosis on leaves producing pale yellow or tinged purple or reddish coloration in leaves and the infected leaves remain erect. Secondary infection is more serious in view of the yield losses of tubers. In this case symptoms consist upward rolling of leaflets of the older leaves which turn chlorotic and the entire plant remains erect with stunted shoots. The infected leaves are also brittle and leathery. In some cultivars necrosis in phloem is occurred.

Causal Factors and its Spread

PLRV is a luteovirus having isometric virions measuring 24 nm in diameter and confined exclusively to phloem vessels. PLRV has only limited host range infecting mostly *Solanum* spp. It is transmitted through infected potato seed tubers. The virus is not mechanically sap transmitted. In the field this virus is transmitted very effectively by more than 10 species of aphids in the persistent manner, of which *Myzus persicae* is the principal vector.

Virus Disease in Tomato

In India more than 10 diseases have been reported caused by viruses. We will describe only two viral diseases, tomato mosaic and tomato leaf curl, which are most economically important in the north eastern this Himalayan region of India.

Mosaic Disease of Tomato

Symptomatology

The characteristics symptoms are mild to severe mosaic mottling with raised dark green areas. Severe filliform effect in the infected leaves are very common. The infected plants are stunted and bear a few deformed fruit. The deformity of fruits is very serious when plants get infected in very early stages of growth. Necrosis of stem and petioles is occasionally observed which kills the plant at any stages of growth.

Causal Factors and its Spread

Tomato mosaic virus (TMV) and cucumber mosaic virus (CMV) are two most common viruses causing mosaic disease of tomatoes while PVX, PVY are occasionally encountered. TMV in tomato is transmitted through contact, soil and to some extent by contaminated tomato seed. PVX is highly contagious by mechanical transmission but not transmitted by seed, soil and vectors. CMV and PVY are mechanically sap transmissible as well as aphid vector transmissible.

Tomato Leaf Curl

The most serious disease of tomatoes in present era throughout the world is tomato leaf curl disease (TLCV). TLCV disease has become the major limiting factor in tomato production causing yield losses of 28-92 per cent depending upon the age of the plants when it is infected and per cent of plant infection. In summer season and under irrigated condition TLCV is the major limiting factor to tomato cultivation.

Symptomatology

The main symptoms consist of vein clearing, stunting and marked reduction in leaf size. Affected plant are stunted, the shoots are erect and have short internodes. The leaves are small, curled leathery and chlorotic and most of the flowers drop prematurely. The leaflets which appear soon after infection are cupped downward. The severely infected plant show complete yellowing of interveinal area and puckering of leaves. The infected plants produce profuse lateral branches imparting the bushy appearance. Affected plants produce either no fruit or a few small fruits, depending on the time of infection.

Causal Factors and its Spread

It is caused by a gemini virus particle measuring 18–20 × 30 nm. The virus is transmitted by white fly, *Bemisia tabaci* in persistent manner. It is not transmitted by sap, seed and soil.

Virus Diseases in Radish

Radish (*Raphanus sativus*) is the most important vegetable in North Eastern hills of India having high economic values. The cultivation of this crops is hampered due to many viral and fungal diseases, of which mosaic disease is most important.

Mosaic Disease of Radish

Mosaic disease in radish is distributed not only in India but also in all the radish growing area in the world. This disease is widely prevalent in radish and oil yielding cruciferous crops in West Bengal, Sikkim, Assam causing losses 3–100 per cent in root weight and up to 90 per cent in seed weight depending upon the time of infection

Symptomatology

Symptoms appear as mosaic or mottling on young leaves with vein clearing, veinal chlorosis or vein netting. Circular inter veinal chlorotic areas are formed which gradually increases in number and sizes and finally coalesce and form irregular chlorotic big spot. The diseased plants are stunted and leaves are reduced in size. The stinted plants produced profuse uncommon weak branches. The flower buds proliferate in giving rise a rosette appearance. The flower buds turn yellow, do not open completely or fall off without settings any pod.

Causal Factors and its Spread

The causal virus has been identified as a member of Poty virus having particle size of 720 × 11 nm, which is a strain of Turnip yellow mosaic virus (TuMV) in India. (Ahlawat and Chenulu, 1984). Mixed infection of TuMV and CMV may also cause mosaic disease in radish. The TuMV is transmitted by seed sap and aphid vectors. Vector is the most important mode of transmission of this disease in India. About 50 species of aphids including *Lipaphis erysimi, Myzus persicae, Brevicoryne brassicae* etc have been reported to transmit the disease in non-persistent manner. Cruciferous crops are the major hosts of this virus. *Nicotiana glutinosa, Chenopodium amaranticolor,* ornamental cabbage (*Brassica pekinensis*) are the common experimental hosts for this disease.

Virus Diseases in Bhindi

Bhindi is widely cultivated vegetable crop throughout the tropical and sub-tropical regions in the world. However, a major limitation in the cultivation of bhindi is the whitefly transmitted yellow vein mosaic virus disease

Bhindi Yellow Vein Mosaic

Bhindi yellow vein mosaic (BYVMV) virus is perhaps the most severe disease of all the crops growing all over India and it causes about 80–90 per cent yield losses. If the plants become infected at early stages, the crop loss may be near to 100 per cent.

Symptomatology

Bhindi yellow vein mosaic virus (BYVMV) disease is characterized by vein clearing and vein chlorosis in leaves. Yellowing of the entire network of veins in the leaf blade is common in severely infected plants and fruits become dwarfed, malformed and yellowish green in colour. Younger leaves turn yellow and become reduced in leaf size and the plants become stunted much. The infection may start at any stages of the plant growth Due to the early infection, flowering of the plant may be restricted, the fruits, if formed may be smaller in size.

Causal Factors and its Spread

The virus causing BYVM is a single stranded DNA virus with typical twin particles measuring 18x30 nm in size. This virus is a member of gemini virus which is only whitefly transmitted by whitefly (*Bemisia tabaci*) in semi persistent manner.

Virus Disease of Cucurbit

Cucumber Mosaic Virus

Cucumber mosaic virus (CMV) disease is world wide in distribution, having a wide range of hosts in more than 40 plant families. Among the most important vegetables affected by this virus are cucumber, melon, squash, peppers, spinach, tomatoes, celery beets, bean crucifers etc. This virus affects the plants causing mosaic mottling or discoloration and distortion of leaves, flowers and fruits. Crop yields are reduced in quantity and also in quality. The plants are severely affected in the field as well as in the green house. In certain areas it may cause about 50 to 75 per cent crop losses in favorable conditions.

Symptomatology

The CMV may infect the young seedlings of about six weeks old by producing mottled, distorted and wrinkled leaves. Subsequently the growths of the plants are retarded and the plants appear dwarfed with results of shorter internodes, petioles and leaf lamina. Such infected plants produce few runners, flowers and fruits. The older leaves of infected plants develop alternate, irregular patches of green and yellow, which enlarge rapidly covering the entire leaf. The leaf becomes deformed, small, sometimes curled downward. The veins and veinlets in the entire leaf lamina turn yellow. In severely infected older plants the chlorotic patches turn to necrotic areas along the margins, and later spread over the entire leaf. The killed leaves hang down on the petiole or fall off, leaving plant or most of the older vine bare. In fruits, produced after infection, pale green or white areas intermingled with dark green, raised areas are developed, which cause distortion of the fruits.

Causal Factors and its Spread

Cucumber mosaic virus is a Cucumovirus having a polyhedral tripartite virus with 28 to 30 nm in diameter. All the three particles are essential to cause infection in the plants. CMV exists in numerous strains that differ in their hosts, symptomatology, transmission and other properties.

The virus is readily transmitted by sap. More then 60 aphid species can transmit this virus disease in the field in a semi persistent manner. The aphids *Myzus persicae, Aphis gossypii* are the most common vectors. Dodder species like *Cuscuta* can transmit the CMV experimentally and also in the field. CMV over winters in perennial weeds and other crop plants as it has wide host ranges. In growing season a few cucurbit plants become infected, later on insect vectors and humans during their practice of cultivation spread this virus to many other healthy plants.

Virus Disease of Bean

Beans are widely grown under diverse condition. The various beans which are cultivated in the hills of north eastern Himalayan region this region are French bean (*Phaseolus vulgaris*), board bean (*Vicia fabae*), cowpea (*Dolichos lablab*) etc. There are many viruses have been reported to cause diseases in beans. The most important virus disease in this region is bean common mosaic virus (BCMV).

Bean Common Mosaic Virus

Bean common mosaic virus (BCMV) is one of the earliest plant viruses described as pathogen of economically important plant species, *Phaseolus vulgaris*. Disease is high when bean is grown in dry and irrigated conditions.

Symptomatology

The general symptoms are mottling, blistering or puckering of leaves. Characteristics foliar symptom induced by BCMV consists of a mosaic with well defined dark and light green areas and often vein clearing and vein banding. BCMV can induce other symptoms in mosaic affected plants, such as leaf curling, dwarfing and chlorosis. A different syndrome of BCMV consists of top necrosis and death of the plants in severe cases.

Causal Factors and its Spread

The causal virus of BCMV is a member of the Poty virus group. It has filamentous particle about 750 nm long and 14 nm wide containing single stranded monopartite RNA genome. BCMV is transmitted by mechanical inoculation and several aphid species in a non persistent manner. This virus is also transmitted through seed and pollen. It may cause heavy losses in yield if the crop is infected at the early stages of growth. The disease also drastically affects the nodule formation.

Viral Disease of Chayote (Sqoosh)

Chayote (*Sechium edule*) belonging the family Cucurbitaceae, locally known as Sqoosh is a fruit vegetable commonly grown in the summer to late autumn season in the north eastern Himalayan region. This crop is considered as a poor man's potato for the hill people. Recently, a yellow mosaic disease occurring in this crop causes tremendous yield losses of this fruit.

Chayote Mosaic Disease

Chayote mosaic disease causes 20–100 per cent incidence in chayote plants throughout the hills. Production of fruits from this crop are markedly reduced and losses it economic importance due to this disease. In the higher altitude where the temperature is comparatively low, the presence of this disease is low.

Symptomatology

Symptoms initially appear as small yellow patches few in a number in the leaves and gradually increase the number and sizes of patches. Young leaves of severely infected plants show systemic yellow mosaic symptoms. In some susceptible plants leaf enation and upward curling of leaves also occurred. Infected plant produces small size and deformed fruits.

Causal Factors and its Spread

This disease is caused by single stranded DNA virus having twin gemini particle of 18–20 nm in diameter under Begomovirus group which is transmitted by whitefly, *Bemisia tabaci*. This disease is experimentally sap transmitted to bittergourd (*Luffa accutangula*) and tobacco (*Nicotiana benthamiana*). Some other viruses transmitted by other insect vector or seed may be involved in this disease yet to be confirmed.

Virus Disease of Elephant Foot Yam (*Amorphophallus* sp)

Amorphophallus consists of 90 species but widely cultivated species is *A. companulatus* popularly known as elephant foot yam. Though it is cultivated throughout the Indian subcontinent, the major cultivated area is in the north eastern Himalayan region of India.

Mosaic Disease

Mosaic disease has been reported from many other countries from its economic importance. Recently it has been reported from Assam state that this disease happens to cause about 80 per cent disease incidence.

Symptomatology

The symptoms include mottling, rolling and crinkling of leaves showing cup-like structure due to inward and upward rolling of leaf margins. Brownish chlorotic spots of 0.1 to 0.2 cm in diameter appear on the leaf lamina. The diseased leaves become leathery, brittle and severely reduced. The infected plants become stunted. The size and weight of corm reduced to 70 per cent.

Causal Factors and its Spread

This disease is caused by Carla- or Potyvirus or association of both the viruses. The particle size of Carlavirus is 725 nm long and Potyvirus is 610 nm long. Dasheen mosaic virus, a member of Potyvirus also have been reported earlier from mosaic infected elephant foot yam. The virus may spread through vegetatively propagated material inadvertently. Aphid vector may be the cause of its transmission but not yet confirmed.

Virus Diseases in Spices

Virus Diseases of Chilli

Cultivation of chilli (*Capsicum annuum* L) is hampered due to several biotic stresses induced by fungi, bacteria, Phytoplasma and viruses worldwide. More than 10 virus diseases are reported to cause diseases in chilli, which affects the yield of chilli. In this chapter the two most important diseases, yellow mosaic and leaf curl have been described.

Yellow Mosaic

Yellow mosaic disease (YMD) in chilli is most serious in the pot of crop losses in this region and it may cause 5-100 per cent incidence and 65-75 per cent yield losses. Incidence of the disease was noted up to 5 -100 per cent in the Eastern Himalayan region (Biswas *et al.*, 2004).

Symptomatology

Symptoms in chilli induced by chilli mosaic varied from area to area depending on their causal virus ranging from mild to severe mosaic mottling. Characteristic symptoms are dark green to light green mottling and systemic yellow mosaic of leaves with or without stunting. Infected leaves become leathery and brittle. In some cases vein clearing and vein banding are also noticed. Infected plants become stunted with bushy appearance. In extreme cases deformed leaves *i.e.*, shoe sting, filliform leaf symptoms are developed. In severe cases infected plant does not flower and bear fruits, if flowers it may be deformed without any fruits and if bear, fruits become deformed and unmarketable.

Causal Factors and its Spread

Causes of yellow mosaic of chilli are complex and about eight viruses are identified to cause this disease worldwide. The most common viruses are potato virus Y (PVY), pepper vein banding virus (PVBV), pepper veinal mottle virus (PVMV), chilli veinal mottle virus (CVMV), tobacco etch virus (TEV), tobacco mosaic virus (TMV), cucumber mosaic cucumovirus (CMV). Recently co infection of two viruses, CMV and PVY are reported in yellow mosaic infected chilli in north eastern Himalayan region (Biswas *et al.*, 2004). The virus was mechanically transmitted to the plant under family, solanaceae and cruciferae lime all type of chilli, tobacco, radish and rye (*Brassica juncea*). It is not seed or soil or pollen transmitted. Several aphid vectors are responsible to spread this disease depending on their causal viruses in chilli.

Chilli Leaf Curl

Chilli leaf curl virus (CLCV) disease is one of the most important virus diseases in summer and irrigated chilli crop not only in north eastern Himalayan region but also in many other countries. If infection of this disease is occurred in early stage of plant growth yield of this fruit is reduced to 100 per cent.

Symptomatology

The infected plants consist of curling of leaves and dwarfing of plants with intense deformation leaves and marked reduction of leaf size. The leaves become leathery, brittle and chlorotic. Most of the flowers drop prematurely. The fruits that set are usually smaller and deformed. In some cases complete yellowing of seedling is also observed. Puckering and leaf enation are also observed occasionally. The infected plants produce profuse lateral branches imparting the bushy appearance.

Causal Factors and its Spread

It is caused by a DNA virus, called geminivirus having particles measuring 18–20 × 30 nm. The virus is transmitted by whitefly, *Bemisia tabaci* in a persistent manner. It is not mechanically transmitted by sap, seed, pollen and soil.

Virus Diseases in Large Cardamom

Two species of large cardamom, *Ammomum sabulatum* known as Nepal cardamom and *A. aromaticum* referred as Bengal cardamom are grown as important cash crop in hills of West Bengal, Assam and Sikkim. Two viral diseases, Foorke/dwarf and Chirke/mosaic streaks are limiting factors to reduce the crop yields.

Foorke/Dwarf Disease

Foorke disease is reported first as an endemic disease in the hills of West Bengal and Sikkim in the year of early seventies. Recently the incidence and severity of this disease is increasing considerably. In some pockets of large cardamom plantation area this disease may cause about 20–90 per cent incidence of clumps.

Symptomatology

The most characteristic symptom of this disease is the severe dwarfing of the affected plants. The disease stimulates proliferation of large number of stunted shoots emerging from rhizome. The spikes that bear the fruits changes into a leafy structure (Phyllody), which suppress fruit production. The diseased clumps gradually turn to unproductive totally.

Causal Factors and its Spread

A virus having isometric particles of about 30–37 nm in diameter causes this disease, still this virus is unclassified. It is suspected that this virus may be a strain to banana bunchy top virus. This virus is not mechanically sap or seeds transmitted. It is transmitted by aphid species, *Micromyzus kalimpongensis* in a semi persistent manner. Although the transmission ability of this aphid is experimentally very less and it is only about 5 per cent, still this aphid is considered as major vector to transmit the disease in field. Infected propagative materials are the most important mode to spread the in a new plantation area.

Chirke/Mosaic Streaks

Chirke in large cardamom is first reported from Darjeeling District of West Bengal, which recently occurs in all the large cardamom growing area in the north eastern hill regions of India. The incidence of this disease varies from 5 per cent to 27 per cent.

Symptomatology

The initial symptoms appear on unfolding young leaves as pale spots which coalesce to form a long golden yellow streak, and latter develop into mosaic yellow streaks pattern. The affected leaves turn in to brown in colour and dried up. Flowering and fruiting are reduced considerably. Infected plants are stunted with less number of fruits.

Causal Factors and its Spread

A flexuous rod shaped virus particle causes the disease. Two types of virus isolates are present in this region on the basis of severity in mechanically sap transmitted host, one is mild and another is severe isolate.

This disease is transmitted mechanically by sap and by aphid species like *Rophalosiphun maidis*, *Rophalosiphun padi* in a non-persistent manner. Sorghum and ginger is suspected as alternate host of this disease.

Management of Virus Diseases

Fungicides and antibiotics when applied to plant infected by fungi and bacteria, protect them and reduce invasion by them and are widely practiced means to manage disease. Unfortunately there is no such direct method to cure or reduce diseases or latent infection caused by viruses. Most of the useful procedures are designated to prevent and reduce sources of the infection within or from the out side the crop with the interference of virus ecology so as to stop of delay the onset of virus incidence and to decrease the rate of progress of disease epidemic so that yield losses may be minimized. The essential of various approaches to crop protection against virus diseases are dealt in the Table 15.2. Usually the available procedures do not offer permanent solutions to a virus disease problem in particular area. Therefore, efforts should be made to develop permanent solution to control the viral diseases by giving emphasis on conventional breeding resistance to plant virus and to vector and mild strain cross protection (MSCP), and unconventional genetic engineering method like, biochemical, coat protein, replicase, antisense RNA, ribozyme, satellite RNA and antiviral mediated resistance.

Table 15.2: Some Useful Methods to Prevent or Reduce Detrimental Effect of Virus Infection in the Crops

Objectives	*Methods*	*Examples of Crop where employed*
Exclude virus from particular area or the states	Quarantine and inspections	Banana
Exclude virus from planting sites	Virus free propagated plant	Citrus
	Sowing virus free seeds	Beans, tomato, radish
Prevent infection	Grow cultivars resistant to virus or vector	Potato, chilli, tomato
	Cross protection	Citrus
	Transgenic cultivars	Vegetables
Reduce source of infection	Eliminate crop residues	Radish, potato
	Eliminate alternate hosts or weeds	Potato, papaya
	Rouge infected plants	Citrus, bean
	Eradication of plant from particular area	Banana, citrus
	Sanitation	Citrus
Reduce severity	Judicial NPK application	Chilli
	Nutrition application	Citrus
Avoid virus by cultural practice	Alter planting date	Radish, tomato
	Alter plant spacing	Vegetables
	Isolate field	Vegetables
Reduce spread of viruses	Control insect vector	Chilli, potato
	Reduce cultivation	Vegetables
	Yellow trapping and crop mulching	Tomato, bhindi
	Intercropping	vegetables
	Use physical barrier	Tomato, radish
	Careful intercultural operation and harvesting my man	Cucurbit, Chilli
Reduce impact of infection	Grow tolerant crop	Citrus, banana, papaya
	Selected root stock	Citrus

The table has been modified from the table developed by Waterworth and Hadidi (1998)

On the basis of the discussion on virus infection in different crop and their spread in nature in this chapter, it is clear that there is limitation of the individual control measure. It is obvious that a combination of measures will be more effective than any single measure. Single measures may not be effective in all seasons. Integrated control should be taken into account all the suitable measures and utilize them to reduce disease incidence and maintain it at the lowest possible level.

REFERENCES

Anonymous (1999–2003). *Annual Report*, Indian Agricultural Research Institute, New Delhi.

Ahlawat, Y S. (1997). Viruses, greening bacterium and viroids associated with citrus (citrus species) decline in India. *Indian J. Agri. Sci.*, 67: 51–57.

Ahlawat, Y.S. and V.V. Chenulu (1984). Radish mosaic: a new disease caused by turnip mosaic virus in India. *Tropical-Agriculture*, 61: 3, 188–192.

Ahlawat, Y.S., R.P. Pant and K.N. Bhagawati (2003). Association of Carla–and Poty–viruses with mosaic disease of elephant foot yam. *Indian Phytopath.*, 56: 300–301.

Ahlawat,Y.S and V.V. Chenulu (1982). Studies on the transmission of radish mosaic virus by the aphid, Lipaphis erysimi. *Indian Phytopath.*, 35: 4, 633–638.

Basu, N. and B.K. Giri (1993). *The Essentials of Viruses, Vectors and Plant Diseases*. Wiley Eastern Limited, 4835/24, Ansari Road, Daryaganj, New Delhi – 110 002.

Bidari, B.D. and H.R. Reddy (1990). Identification of naturally occurring virus on commercial cultivars of chilli. *Mrsore J. Agric Sci.*, 24: 42–51.

Biswas, K.K., R.P. Pant, D.R. Das, K.B. Pun, A. Chowdhuri and Y. S. Ahlawat (2004). Incidence, detection and management of chilli mosaic disese syndrome in capsicum annum cv Kalimpong local in the Hills of West Bengal. National Symposium on Molecular Diagnostics for the Management of Viral Diseases, pp. 89–90 (Abs.).

Biswas, K.K., K.B. Pun, R.P. Pant and Y.S. Ahlawat (2004). Mosaic Disease in Chilli *(Capsicum annuum)* cv Kalimpong Local in Darjeeling Hills of West Bengal and its management. *Indian Phytopath.*, (under publication).

Biswas, K.K., K.L. Manjunath, R.F. Lee and Y.S. Ahlawat (2004). Identification of diversified Citrus tristeza virus population isolates from Eastern Himalayan Region of India by heteroduplex mobility assay and nucleotide sequence analysis. National Symposium on Molecular Diagnostics for the Management of Viral Diseases, pp. 9 (Abs.).

Byadgi, A.S. and Y.S. Ahlawat (1995). A new viral ring spot disease of citrus (*Citrus* species) in India. *Indian J. Agric. Sci.*, 65: 673–70.

Clark, M. F. and M. Bar-Joseph (1984). Enzyme Immunosorbent assays in plant virology. *Methods in Virology*, 7: 51–85.

Dale, J.L. (1987). Banana bunchy top: An economically important tropical plant virus disease. *Adv. Virus Res.*, 33: 301–325.

Ghosh, D.K and Y.S. Ahlawat (1997). Filamentous virus associated with mosaic disease of garlic in India. *Indian Phytopath.*, 50: 266–276.

Green, S.K. and J.S. Kim (1991). Characteristics and control of viruses infecting peppers: a literature review. *AVRDC Technical Bull.*, 18: 60.

Handa, A. and M. Dutta Gupta (1993). Characterization of yellow mosaic virus of bhindi. *Int. J. Trop. Plant Dis.*, 11: 117–123.

Hull, R. (2002). *Matthews' Plant Virology*, 4th Ed., Academic Press. A Harcourt Science and Technology Company, Harcourt Place, 32 Jameston Road. London, NW1 7BY, UK.

Jeyaraman, S. and R. Balasubhramaniun (1988). Role of potassium treatment on yield and Incidence of pests and diseases. *J. Potassium Res.*, 4: 67–70.

Kaur, S. and J. Singh (1985). Resistance to fruit rot disease under field conditions in pepper (*Capsicum annuum* L). *Capsicum Newsletter*, 4: 67.

Kim, J. Soo., G.S. Choi, J.Y. Kim, M.K. Yoon, J.D. Cho, J S. Kim, G.S. Choi, J.Y. Kim, M.K. Yoon and J.D. Cho (1995). A new radish vein clearing potyvirus on radish in Korea. *RDA J. Agric. Sci. Crop Prot.*, 37: 2, 343–351.

Kiranmai, G., P. Sreenivasulu and M.V. Nayudu (1997). Characterization of cucumber mosaic cucumovirus isolates naturally infecting three Solanaceous vegetable crops in Andhra Pradesh. *Indian Phytopath.*, 50: 416–425.

Mandal, B., S. Mandal, S.S. Soharb, K.B. Pun, and A. Varma (2004). A new yellow mosaic disease of Chayote in India. *Plant disease* (Accepted).

Mukhopadhyay, S. (1985). The Dieback of mandarin orange in Darjeeling District: Pest and diseases, Bidhan Chandra Krishi Viswasvidyalaya, West Bengal, pp 1–12.

Mukhopadhyay, S. (1997). Medicinal values of citrus and integrated of its diseases. In: Indian Phytopathological Society-golden Jubilee International Conference, November, 10–15, 1997, New Delhi, pp. 147.

Mukhopadhyay, S., A. Gurung and T.K. Sarkar (1992). Management of Citrus greening disease in Darjeeling hills. In: *Management of Plant Disease Caused by Fastidious Prokaryotes*, (Eds.) S.P. Roychoudhuri and D.D. Teakle, Associated Publishing Company, New Delhi, pp. 35–48.

Mukhopadhyay, S., K. Mukhopadhyaya, N. Mukherjee, M.R. Ghosh, T.K. Bose, A.K. Choudhury, N.K. Chakrovorty, S. Dutta Roy, A. Konar and A. Gurung (1986). The state of mandarin orange cultivation in Darjeeling District. In: *Review of Tropical Plant Pathology*, (Eds.) S.P. Roychoudhuri and J.P. Verma, Vol 2.

Paul Khurana, S.M. and T.S. Thind (2002). Important biotic diseases of potato and their control. In: *Plant Pathology at Cross Road*, (Eds.) C. Sen, M.K. Dasgupta, S. Choudhuri, BCKV, Mohanpur.

Raychaodhuri, S.P. (1977). *A Manual of Virus Diseases of Tropical Plants*. The McMillan Co. of India Ltd, New Delhi.

Raychoudhury, S.P., T.K. Nariani and Y.S. Ahlawat (1977). Dieback of Citrus in India. *Proc. Int. Soc. Citriculture*, 3: 914–918.

Rishi, N. and P. Dhawan (1989). Characterization and identification of viruses causing mosaic diseases if *Capsicum* in nature. *Indian Phytopath.*, 42: 221–226.

Satya Prakash, S.J. Singh, R.K. Singh and P. P. Upadhyaya (2002). Distribution, incidence and detection of potyvirus on chilli from eastern Uttar Pradesh. *Indian Phytopath.*, 55: 284–298.

Saxena, S., V. Hallan, B.P. Singh and Sane (1997). Papaya leaf curl is a distinct whitefly transmitted geminivirus in India. *In: IPS Golden Jubilee Int. Conf. on Integrated Disease Management for Sustainable Agriculture*, (Abs.), IARI, New Delhi.

Singh, B.R., S.P. Gupta, R. Pandey and D. P. Tripathi (1990). Response of some chilli lines to mosaic and leaf curl diseases. *Indian J. Virol.*, 6: 113–115.

Singh, D. and S.J. Sing (2000). Chilli little leaf a new Phytoplasma disease in India. *Indian Phytopath.*, 53: 281–282.

Singh, S.J. (1993). Electron-microscopic study of cucumber phyllody. *Indian J. Plant Pathol.*, 16: 74–75.

Singh, S.J. (1996). Virus and mycoplasma diseases of papaya. In: *Advances in Diseases in Fruit Crops in India*, (Ed.) S.J. Singh, Kalyani publishers, Ludhiana, pp. 275–296.

Singh, S.J. (2003a). *Virus and Phytoplasma Diseases of Papaya, Passion Fruit and Pineapple*. Kalyani Publishers, Noida, pp. 230.

Singh, S.J. (2003b). *Viral Diseases of Banana*. Kalyani Publishers, Noida.

Singh, Amarjit, S.S. Cheema and J.S. Hundal (1990). Identification of virus(es) associated with chilli mosaic syndrome. *J. Res.*, PAU, 27: 75–83.

Sobita Devi, P.H. and H.R. Reddy (1995). Effect of insecticides on aphid transmission of pepper vein banding virus and cucumber mosaic virus on chilli (*Capsicum annuum* L.). *Mysore J. Agric Sci.*, 29: 141–148.

Tomlinson, J, A. (1987). Epidemiology and control of virus disease of vegetable. *Ann. Appl. Biol.*, 110: 661–681.

Varma, A. (2004). Changing pattern of viral diseases of plants: Diagnostics challenges. National Symposium on Molecular Diagnostics for the Management of Viral Diseases, pp. 30 (Abs.)

Verma, H.N. and V. Prasad (1985/86). Virus diseases in papaw (Papaya). *Rev. Tropical. Plant. Path.*, 2: 311–327.

16

Recent Advances in Tomato Leaf Curl Virus

Krishna Kumar, V.K. Razdan*, Efath Shahnaz*, Shahid Ahamad** and A.K. Tiwari**

**Division of Plant Pathology, Faculty of Agriculture, S.K. University of Agricultural Sciences and Technology, Udheywalla – 180002, Jammu*

*** RARS, Rajouri – 185 134, J&K*

Tomato (*Lycopesicon esculenlum* Miller) is one of the important and widely cultivated crops and is mainly consumed as fresh vegetable or used after processing (canned, soups, sauce and condiment). Its unsurpassable nutritional and health giving qualities are well known and recently it has also been extensively used in pharmaceutical and cosmetic industries. Being a photo-insensitive crop, it can be grown throughout the year except in regions experiencing extreme winters. However, its continuous cultivation over the years has led to build up of pest and disease vectors, and in the perpetuation of the inoculum of a large number of diseases. Among the major viral diseases of tomato, the mosaic and leaf curl are most predominant, followed by spotted wilt and shoestring (Sharma and Sharma, 2001). Tomato leaf curl virus (ToLCV) is widely distributed in many tropical and subtropical countries like South Africa (Mclean, 1940), Cairo (Nour-Eldin *et al.*, 1969), Philippines (Retuenna *et al.*, 1974), Sudan (Yassin, 1975), Thailand (Thanapase *et al.*, 1983), Somalia (Castellani *et al.*, 1984), Mexico (Brown *et al.*, 1993), Oman (Zouba, 1995) Australia (Behjatnia *et al.*, 1996), Panama (Engel *et al.*, 1998) and Nicaragua (Rojas *et al.*, 2000) and reports of its occurrence from other countries are pouring in. In India, the disease has been reported from Haryana, Punjab, Karnataka, Uttar Paradesh, New Delhi, Chattisgarh,

Tamil Nadu, J&K and West Bengal (Varma and Poonam, 1978; Butter and Rataul, 1981; Saikia and Muniyappa, 1989; Singh and Tripathi, 1991; Baranwal and Ahmad, 1997; Nainar and Pappiah, 1999; Singh *et al., 1999* ; Mahajan, 2001; Bhagat and Dey, 2001. Because of the wide spread occurrence of ToLCV epidemics and great economic significance of the disease, threatening tomato cultivation, research efforts have concentrated on the study of the factors involved in the host virus vector relationship in order to develop effective management strategies. Researches were also emphasized on molecular characterization of complete genome so that pathogen- derived resistance may be developed to target the pathogen as methods of investigation to date provide only partial management of the disease (Al-Musa, 1986; Ioannou, 1987).

Seasonal Incidence, Losses and Distribution

The disease causes extensive losses in both quality and quantity. In India, first report regarding variations in leaf curl incidence was noted by Vasudeva and Samraj (1948) who observed a low incidence in late planted tomatoes compared with high incidence (90 per cent) in the early planted crop. Sastry and Singh (1973) reported that the plants infected within 20 days of transplanting remain stunted in growth and produce fewer leaves and fruits than those infected 35 to 50 days following transplanting. Early infected plants resulted in 92.3 per cent less fruit as compared with 74.1 per cent to 28.9 per cent loss in plants infected 35 to 50 days after planting, respectively.

Incidence of disease observed in Punjab was greater in winter crop than in summer and the virus affected all the stages of crop causing 17.6–99.7 per cent loss in yield depending upon stage of the crop at the time of infection (Butter and Ratual, 1981). Saikia and Muniyappa (1986) reported that the disease incidence in Karnataka varied from 6.4 to 52.2 per cent in summer and 52 to 100 per cent in crop grown in the winter season. Similarly, Tripathi and Varma (2002) reported a higher degree of disease incidence (83 per cent) in winter crop planted in October as compared to summer crop planted in February where only 14 per cent incidence was recorded. Disease was also observed to cause 66.40, 67.00 and 88.00 per cent reduction in the yield of Pusa Ruby, Pusa Early dwarf and Maitri (hybrid) in Jammu, respectively (Shahnaz, 2003).

Symptomatology

Disease symptoms vary somewhat among tomato cultivars, but the major symptoms include curling and cupping of leaf lamina, stunting of infected plants and excessive branching. Affected tomato plants show considerable reduction in the size of leaf lamina which is more pronounced in the successive leaves and is accompanied by shortening of internodes with the result that the leaves are crowded together. Leaves developing soon after the infections in the early stage of crop growth are cupped downwards, whereas, leaves developing later are chloratic and deformed, with leaf margins curled upwards. If the infection occurs in the late stage of the crop, the emerging leaves are chlorotic and curled upwards. Puckering of the leaflets is a common symptom of the disease and is very severe in some plants. The terminal and axillary shoots remain erect in the infected plant and the plant has a greater tendency to produce stunted lateral branches, which impart a bushy appearance to the plant. The disease induces partial or complete sterility to the infected plant, depending on the stage at which infection has taken place. In case of late infection, the plant bears a few fruits but in earlier infections the symptoms being much more pronounced the few flowers that are formed bear reduced floral structures that never attain normal size. These flowers seldom open and soon drop off. The flowers that develop after infection are invariably sterile. The dropping of flowers and reduction in fruit yield in infected plants is common (Vasudeva and Samraj, 1948; Zouba, 1995; Srivastava *et al.*, 1995; Sharma and Sharma, 2001).

The Causal Virus

The tomato leaf curl virus has often been confused with tomato yellow leaf curl virus (TYLCV) due to similarity in symptom expression, transmission mode and host range. Some workers had considered both of them to be caused by the same virus (Yassin, 1989). However, recent studies have established them to be caused by different viral agents (Muniyappa *et al.*, 1991; Dry *et al.*, 1993; Chatchawankanphanich *et al.*, 1993). Coat protein analysis suggests that these viruses can be grouped into two branches that include the viruses from China, Malaysia, Taiwan, Thailand and Australia and those from Bangladesh, India and Sri Lanka (Zeiden *et al.*, 1998).

Molecular Characterization

ToLCV belongs to the begornovirus genus of the family Geminiviridae. These viruses have circular single stranded DNA molecules as their genome encapsidateci in icosahedral geminate particles. The virus particles are about 18–30 nm in diameter (Channarayappa *et al.*, 1992).

Generally, the begomoviruses are bipartite with respect to their genomic composition. About 80 per cent of the gerninivirus particle consists of protein. The viral capasid contains only a single type of protein of approximately 30 KD. The double icosahedra structure is composed of morphological subunits with a total of 22 capsomere particles. The only identified nucleic acid component of the geminivirus particle is DNA, encapsidated by the viral coat protein. The number of genomic components, one or two ssDNA molecules, varies according to the virus strain. Virus isolates from Israel Sardinia, Sicily, Egypt and Spain have only one circular ssDNA molecule of approximately 2.7 to 2.8 kb. ToLCV from South Indja is unique as only DNA–A component has been isolated and sequenced so far and thus, there is is no evidence of the presence of DNA–B component (Kirthi *et al.*, 2002). However, in North India, ToLCV like other whitefly-transmitted viruses infecting dicotyledonous plants has a bipartite genome. Both DNA–A and DNA–B are needed for infection and they have a common region of 166 bases with about 94 per cent nucleotide sequence homology, a characteristic of all gerniniviruses (Srivastava *et al.*, 1995; Padidam *et al.*, 1995).

Evolution of the Virus

The observations that geminivirus DNA replication functions are supported by the cellular machinery of *Agrobacterium tumefaciencs* provides evidences for the theory that these circular single stranded DNA viruses have evolved from prokaryotic episomal replicons (Rigden *et al.*, 1996). Genomic variation resuhing from mutation is amplified by acquisition of extra DNA component, pseudo-recombination, both inter and intra specific. Recent begmovirus epidemics reflect favourable conjections of plant, vector and viral factors resuhing in noble recombinant virus. In their pattern of variation and evolution, begmovirus differ greatly from plant virus with RNA genomes. Saunders *et al.* (2003) proved that single genomic component had been reported for eupatorium yellow-vein vjrus, the plant might be further infected with a DNA B satellite component that is similar to those described in ageratum (a wide spread weed) and cotton.

The Virus-Vector Relationship

ToLCV is transmitted by insect vector 'whitefly', *Bemisia tabaci* Genn. (Varma, 1963). The whiteflies are soft-bodied tiny sucking insects belonging to the family A1.eurodidae under the order Homoptera. Approximately 28 plant viruses are known to be transmitted by whiteflies. The most important whitefly species, which besides other diseases, transmits leaf curl disease of tomato, is *Bemisia tabaci*. Within the Aleyrodid vectors of plant viruses, this species parallels *Myzus persicae* in the aphid group (Costa,

1969). The whitefly requires 30 min. of acquisition and inoculation feeding periods to transmit the virus (Reddy and Yaraguntaiah, 1981a) though a minimum of 10 and 20 min., respectively, have also been reported (Muniyappa *et al.*, 1999). However, the transmission efficiency is greatly increased if these periods are increased to 24h. each. The acquisition feeding period is unaffected by temperature but inoculation access period is most effective at temperature of 33–39°C and is adversely affected at below 21 and above 44°C (Butter and Ratual, 1978).

Butter (1976) has reported that the virus does not require latent period in the vector but some other workers are of the view that a minimum latent period of 6h. is required by the virus (Reddy and Yaraguntaiah, 1981a). A starvation period of 1h. pre-acquisition and pre-inoculation increases the transmission efficiency but increase in the duration of starvation period decreases it. Females are able to retain the virus for a longer time than males and at the same time have better transmission rates (Butter and Ratual, 1977a). Various studies have shown that even a single whitefly can transmit the virus but five infective whiteflies are required to achieve cent per cent transmission of virus (Baranwal *et al.*, 1996; Muniyappa *et al.*, 1999).

The vector can acquire the virus more efficiently from the apical leaves than from cotyledonary leaves (Rataul and Brar, 1989) and from 2-month old virus source than from 11-month old virus source (Butter and Rataul, 1977a). Though the virus reduces the fecundity and fertility of whiteflies, but their longevity is not influenced by virus infection (Butter and Ratual, 1977b).

Under *in vivo* conditions, the frequency of transmission depends on distance of virus source and neighbouring plants that may harbour vector or virus or both (Shaheen, 1983), whereas, under *in vitro* conditions, the frequency of transmission from tomato to tomato depends on inoculation technique, whitefly culture and virus isolate (Pico *et al.*, 1998; McGrath and Harrison, 1995). It has been found that artificial cage inoculation, although more time consuming is the most efficient, adequate and reliable technique for transmission studies. Whitefly colonies raised on eggplant have been found to be more efficient than those from *Capsicum*, cotton, cowpea, tobacco or tomato (Butter and Rataul, 1977a). The virus is also transmissible by grafting (Reddy and Yaraguntaiah, 1979) and through dodder (Reddy and Yaraguntaiah, 1980). However, these methods are not of much economic importance.

Biochemical Changes in Diseased Plant

The ultra-structural modifications occurring in nuclei of tomatoes infected by ToLCV include hypertrophy of the nucleolus, segregation of nucleolar components into discrete granular and fibrillar regions, appearance of electron dense particles associated with ribbon-like structures and presence of virus particles as either loosely, compacted or hexagonally close packed symmetrical arrays. The thylakoid system gets fragmented and disorganized. In some chloroplasts starch grains are abnormally large. Excessive accumulation of osmiophillic bodies degenerating chloroplasts is prominent (Channarayappa *et al.*, 1992). The cambial activity is weaker, sclerenchyma rings are narrower and there are fewer narrow xylem vessels. The leaf blade is reduced in thickness, the cortical parenchyma is wider and formation of mechanical and conducting tissues is reduced (Gevovkyan *et al.*, 1976).

The infected leaves have less chlorophyll a and b, xanthophylls and beta carotene and a lower photosynthetic rate and catalase activity, whereas, respiration, peroxidase, catecholase and pectinesterare activities are higher than in leaves of healthy plants (Lodh *et al.*, 1973; Sasikumaran *et al.*, 1980). High phenol content has been associated with resistance (Banerjee and Kalloo, 1989) while the susceptible varieties have been found to have lower contents of reducing sugars and higher contents of non-reducing sugars.

Plants accumulating trunctated Rev-protein have been found to be susceptible to the virus (Brunetti *et al.*, 1997). Diseased tissue has less armmonical and non-protein nitrogen (Singh *et al.*, 1976). There is a reduction in the nitrate reductase activity (Mitra *et al.*, 1978). Affected fruits have more acidity, protein, pectin and ascorbic acid and less sugar and phenolics (Sasikumaran *et al.*, 1980).

Host Range

Whitefly

Bemisia tabaci is known to harbour on a wide range of cultivated plants as well as weeds, a good proportion of which are natural hosts of both vector and virus (Garneel, 1980). Whitefly population was observed to be less at the initial stage of the summer crop, whereas, during the winter crop, the whitefly population was very high at the similar stages of the crop (Tripathi and Varma, 2002). This variation in whitefly population was due to the differences in temperature and relative humidity.

The whitefly has a wide host range. Besides tomato and chilli a large number of plants such as bhindi, mungbean, urdbean, pigeon pea, soybean, cassava, cotton and tobacco are known to be the hosts in experimental studies. In nature, it has been observed on 142 plant species belonging to 23 different families (Saikia and Muniyappa, 1986).

Virus

ToLCV is known to infect *Nicotiana* sp., *Capsicum* sp., *Carica papaya, Datura* sp., *Zinnia elegana, Ageratum conyzoids, Vernonia cineria, Launia asplenifolia, Crotolariajuncea, Solanum nigrum, Sida rhobambifolia, Euphorbia* sp., *Scoparia dulcis, Shizanthus* sp., *Physalis* sp., *Acanthospermum hispidum, Galinosoga parviflora, Nicendra physaloides, Petunia hybrida, Centratherum anthelminticum* and *Solanum seaforthianum* (Maiappan and Narayansamy 1972; Saikia and Muniyappa, 1986; Reddy and Yaraguntaiah, 1982; Gupta *et al.*, 2001). Recently, ToLCV has also been observed infecting gaur, watermelon, cantaloupe and wax guard (Mansoor *et al.*, 2000; Samretwanich *et al.*, 2000; Khan *et al.*, 2002).

Management

Like other viral diseases, management of ToLCV is very difficult and reasonable, eco-friendly and economical control measures have eluded the scientists till date. Earlier workers concentrated more on the management of whiteflies by insecticides, oils, antibiotics, etc. but recently the emphasis has shifted on the development of resistant varieties, modifications in cultural practices and use of transgenic plants.

Chemical and Bio-chemical

Insecticides

Several insecticides have been used to control the vector *e.g.*, deltamethrin, pennethrin, fenvaterate, carbofuran, mephosfolan, aldicarb, disulfoton and monocrotophos (Rataul and Butter, 1975; Saklani and Mathai 1978; Reddy and Yaraguntaiah, 1981b; Mishra, 1984) with varying degrees of success. It has been found that the use of electro-dynamic technique for spraying is generally more effective than Knapsack sprayer (Kisha, 1986). However, the adverse effects of insecticides like pest resurgence, development of resistance, environmental pollution, health hazard and adverse effects on non-target insects has necessitated the need for the development of alternate strategies for the control of disease.

Antibiotics

Foliar sprays of a number of antibiotics have been reported to suppress disease symptoms and increase yields. Foliar application of 0.02 per cent Validomycin–A, 0.01 per cent Abomycin–A, 0.2 per cent guanidine hydrochloride, azaguanidine, 0.2 per cent azaadenine, DPP @75 ppm have given better results (Hayati and Varma, 1985; Thirumalachar *et al.*, 1973; Varma and Poonam, 1977). 8–azaguanine also suppresses the synthesis of a high molecular weight protein which is probably associated with virus multiplication.

Oils

The use of oil prevents acquisition and inoculation of the virus by whitefly. The main disadvantage of the use of oils is that if appropriate dilutions are not made, they may cause phytotoxicity and their efficiency is reduced at higher temperatures. Agricultural spray oil soon after transplanting or 20 days after transplanting and subsequent sprays at 10 days interval have been shown to reduce the population of vector and minimize the incidence of disease (Sastry *et al.*, 1978; Mote, 1979; Sastry, 1989).

Cultural Practices

Nursery Protection

Covering tomato nursery beds with nylon net or white muslin cloth reduces the disease incidence by preventing pest immigration and virus transmission. The white muslin cloth acts as a repellent due to its shining surface (Baranwal *et al.*, 1996). According to some workers, it completely replaces chemical management and thus allows bio-organic farming (Ali and Said, 1987; Berlinger *et al.*, 1988). However, most other workers are of the view that covering tomato nursery beds and one spray of dimethoate (Muniyappa and Saikia, 1983) or four sprays of nuvacron or fenom (Saikia and Muniyappa, 1989) delays the incidence of disease and increases yield.

Manipulation of Planting Dates

Adjustment in the planting time is helpful in avoiding excessive whitefly population and minimize disease incidence during the cropping period. In this way the plants may escape infection or become infected at a later stage of growth which does not adversely affect yield. Early transplanting in spring or late transplanting in winter is known to reduce the incidence and severity of infection (Ioannou, 1987; Chermiti *et al.*, 1993). The most effective time for planting of tomato has been recorded to be October to December followed by January to early March (Saklani and Mathai, 1978; Datar, 1984; Saikia and Muniyappa, 1989). The highest reduction in yield has been reported in the crop sown from March to August.

Irrigation

Crops frequently irrigated invite less oviposition by the whitefly. It has been found that plants grown under drip irrigation system are less succulent and attractive to *Bemisia tabaci* than plants grown under furrow or sprinkler irrigation (Sharaf *et al.*, 1984).

Border Cropping

The use of border or barrier crops like maize, jowar, bajra, sunhemp etc. minimize the disease spread by checking the viruliferous whiteflies from entering the tomato fields. Five or six rows of these crops all around the main tomato field should be sown 50–60 days before transplanting of tomato (Sastry *et al.*, 1977). The use of insecticides in association with barrier cropping further increases the yields, besides reducing the incidence of ToLCV.

Density of Plant Population

The lowest whitefly population, lowest disease incidence and highest yields can be obtained from plants planted at 90/15 m^2, rather than 30 or 60/15 m^2 (Sastry *et al.*, 1976).

Mulching

During the early crop stages, when the crop canopy is less and plants are succulent and susceptible to whitefly attack, mulching with straw, sawdust or preferably yellow polythene minimizes the ill-effects of whiteflies by acting as their death-trap. This is because whiteflies get attracted to such mulches and stay there long enough to be killed by reflected light. However, the protective effect of mulch lasts for only about 3 weeks and then declines (Cohen and Madjar, 1974). The use of polythene bags to cover tomato plants and polythene sheets as mulch was found most effective in reducing the disease incidence and promoting growth and yield of tomato plants (Tripathi and Varma, 2002).

Plant Bio-Regulators

Use of appropriate bio-regulators is known to increase the average yield in terms of weight and number of fruits. GA_3 @50ppm is reported to decrease the incidence of ToLCV (Mukherjee and Raychaudhuri, 1966). Similar results can be obtained by the application of chlormequat at 250ppm (Arora *et al.*, 1989).

Plant Extracts

Plant extracts are known to contain anti-viral proteins (AVPs). An AVP may be defined as a substance, when present in the inoculum or plant tissue is capable of either acting directly by inactivating the viral pathogen or acting indirectly by increasing the level of host resistance resulting in reduction of virus infection (Narayanasamy, 1990). Although plant extracts have been successfully used in the control of many viruses particularly sap-transmitted viruses, reports of their successful use in control of whitefly transmitted viruses particularly ToLCV, is scanty. Baranwal and Ahmad (1997) have shown that *Clerodendrum aculeatum* leaf powder, either as spray or soil application or both, enhance the growth of plants and give some protection against ToLCV infection by delaying the disease appearance.

Biological Control of Whiteflies

Whiteflies are attacked by a number of predators, parasitoids and parasites but they have not been fully exploited under field conditions. Nymphs of the insect are parasitized by *Encarsia* sp. and *Eretmocerus mundus* (Ramappa *et al.*, 1998) which can be used to develop bio -control strategy against the vector.

Resistant Varieties

Viral diseases are best managed by growing resistant varieties. Although, *Lycospersicon peruvianum* and *L. pimpinellifolium* have been shown to be resistant to ToLCV, breeding resistant commercial tomato lines has met with varying degrees of success. Many workers have screened a number of tomato varieties, lines and accessions against ToLCV (Table 16.1).

Guidelines for Future Research

1. The role of collateral and alternate hosts in the spread and perpetuation of virus inoculum and transmission efficiency of whiteflies from reservoir inoculum to main crop has to be investigated thoroughly to know the magnitude of disease development under natural recurrence.

Table 16.1: Sources of Resistance to ToLCV

Species/Accessions	*References*
Fiona, T.V. King, Top 21, Meghana	Razvi *et al.*, 2000
H–24, H–36	Singh *et al.*, 1999
BT–3, Arka Alok, Arka Abha	Mahanta *et al.*, 1998
Lycopersicon hirsutum (LA 1353), *Lycopersicon hirsutum f. glabratum* (LA 1223)	Ragupathy and Narayanswamy, 2001
H–24	Kalloo and Banerjee, 2001
Lycopersicon hirsutum (PI 390513, PI 390658, PI 390659, LA 386 and LA 1777)	Saikia and Muniyappa, 1989
Lycopersicon peruvianum (PI 127830 and PI 127831)	
Lycopersicon glandulosum, Lycopersicon peruvianum, Lycopersicon esculentum, Var. EC 104395	Varma *et al.*, 1980
BT 12–2 X KS2, BT 12–2 Z Dhansree, KS2 X Dhansree, BT 12–2 X Sel. –7, KS2 Z Pusa Sel.–4, Field King X Dhansree	Acharya and Ghosh, 1999
Brech, Kalohi, Piersol, Ronita, Sel.–152, VIN–8, Camphell–DI X Sel. 152, Patriot X Sel.–152, Piersot X Sel.–152, Ronita X Punjab Tropic, VFN–Bush X Punjab Tropic	Sharma *et al.*, 1984
Lycopersicon gladulosam (EC–66003)	Hayati and Varma, 1985
Lycopersicon hirsutum f. glabratum (B 6013), *Lycopersicon hirsutum f. trypicum* (A 1904)	Banerjee and Kalloo, 1987

2. Location specific intensive experimentation should be taken up to formulate an integrated disease management with the use of conventional and modern practices including cultural, biological and chemical approaches for sustainable crop production.
3. Stress should be given to characterize races and biotypes of viruses as well as whiteflies through the use of PCR and specific DNA probe target oriented management practices or even while developing transgenic plant resistant to ToLCV.
4. Long-term epidemiological studies on ToLCV in relation to vector and its population dynamics should be taken as a most challenging area of applied research.
5. Eco-friendly and safe alternative methods have to be one of the most important areas of research using potential anti-viral plants, which is a sustainable way of disease management. Anti-viral protein gene could be used for development of ToLCV resistant transgenic plant.

REFERENCES

Acharya, P. and I. Ghosh (1999). Assessment of tomato leaf curl virus in diallel cross analysis in tomato *Lycopersicon esculentum* Miller. *Env. & Ecol.*, 17: 53–56.

Ali, A.A. and N.M.H. Said (1987). Influence of seedling protection methods on infestation with tobacco whitefly and sequent infection with yellow leaf curl in the plastic houses. *Arab J. Plant Prot.*, 5: 24–30.

Al-Musa, A.M. (1986). Tomato yellow leaf curl virus in Jordan: Epidemiology and Control. *Dirasat*, 8: 199–208.

Arora, S.K., G. Kalloo and M.K. Banerjee (1989). Effect of cycocel and alar on vegetative growth, flowering, fruit set, yield and disease incidence in tomato (*Lycopersicon esculentum* Mill.). *Res. & Dev. Rep.*, 6: 65–71.

Banerjee, M.K. and G. Kalloo (1987). Sources and inheritance of resistance to leaf curl virus in *Lycopersicon*. *Theo. & Appl. Gen.*, 73:707–710.

Banerjee, M.K. and G. Kalloo (1989). Role of phenols in resistance to tomato leaf curl virus, Fusrium wilt and fruit bores in *Lycopersicon*. *Curr. Sci.*, 58: 575–576.

Baranwal, V.K. and N. Ahmad (1997). Effect of *Clerodendrum aculeatum* L. leaf extract on tomato leaf curl virus. *Indian Phyto.*, , 50: 297–299.

Baranwal, V.K., S. Kumar and S.N. Puri (1996). *Leaf Curl Disease of Tomato and its Management*. Technical Bulletin–1. National Centre for Integrated Pest Management, ICAR, New Delhi, 14 p.

Behjatnia, S.A.A., I.B. Dry, L.R. Krake, B.D. Conde, M.I. Connelly, J.W. Randles and M.A. Rezaian (1996). New potato spindle tuber viroid and tomato leaf curl geminiviroid from a wild *Solanum* sp. *Phytopathology*, 86: 880–886.

Berlinger, M.J., R. Dahan, and S. Mordechi. (1988). Integrated pest management of organically grown greenhouse tomatoes in Israel. *Appl. Agri. Res.*, 3: 233–238.

Bhagat, S. and B.K. De. (2001). Reaction of tomato varieties to mosaic and leaf curl viruses in the plains of West Bengal. *Env. & Ecol.*, 19: 865–867.

Brown, J.K., A.M. Idris and D.C. Fletcher (1993). Sinloa tomato leaf curl virus, a newly described geminivirus of tomato and pepper in West Coastal Mexico. *Pl. Dis.*, 77: 12.

Brunetti, A., M. Tarazza, E. Norais, R. Tavazza, G. Ancora, S. Crespi and P. Accotto (1997). High expression of truncated viral rep protein confers resistance to tomato yellow leaf curl virus in transgenic tomato plants. *Mol. Pl. Micro. Inter.*, 10: 571–519.

Butter, N.S. (1976). Studies on virus-vector relationship of tomato leaf curl virus and its vector, *Bemisia tabaci* Genn. Ph.D. Dissertation, Punjab Agricultural University, Ludhiana, India.

Butter, N.S. and H.S. Ratual (1977a). The virus vector relationship of the tomato leaf curl virus (TLCV) and its vector, *Bemisia tabaci* Genn. (Hemiptera : Aleyrodiadae). *Phytoparasitica*, 5: 173–186.

Butter, N.S. and H.S. Rataul (1977b). Effect of TLCV infection on *Bemisia tabaci*. *Entomol.*, 2: 163–164.

Butter, N.S. and H.S. Rataul (1978). Influence of temperature on the transmission efficiency and acquisition threshold of whitefly, *Bemisia tabaci* Genn. in the transmission of tomato leaf curl virus. *Sci. & Cul.* 44: 168–170.

Butter, N.S. and H.S. Rataul (1981). Nature and extent of losses in tomatoes due to tomato leaf curl virus (TLCV) transmitted by whitefly, *Bemisia tabaci*, Gen. (Alenyrodidae : Hemiptera). *Indian J. Ecol.*, 8: 299–300.

Castellani, E., A.M. Nur, and M.I. Mohamed, (1984). Tomato leaf curl in Somalia. *Annali de della Facotta di Scienze Agravi dell Univesite degli shidi di torino*, 11: 145–161.

Chatchawankanphanich, O., B.T. Choang, S.K. Green, S.J. Singh and D.P. Maxwell (1993). Nucleotide sequence of a geminivirus associated with tomato leaf curl from India. *Pl. Dis.*, 77: 1168.

Chermiti, B., J.C. Onillon, B.B. Aicha and Ben-Aicha (1993). Importance of tomato leaf curl virus (TYLCV), transmitted by the whitefly *Bemisia tabaci* (Genn.) (Homoptera : Aleurodidae) in late season cultivation in the irrigated area.

Cohen, S and V.M. Madjar (1974). Prevention of the spread of tomato yellow leaf curl vjrus transmitted by *Bemisia tabaci* (Genn.) in Israel. *Bull. Ent. Res.*, 64: 193–197.

Costa, A. S. (1969). White flies as virus vectors. In: *Viruses, Vectors, Vegetation*, Maramorosch, K. (ed). 95–119. Interscience New York, pp. 666.

Datar, V.V. (1984). Seasonal incidence of leaf curl in three varieties of tomato. *Indian Phytopath.*, 37: 540–543.

Dry, I.B., J.E. Rigden, L.R. Krake, P.M. Mullineaux, and M.A. Rezaian (1993). Nucleotide sequence and genome organization of tomato leaf curl geminivirus. *J. Gen. Virol.*, 74: 147–-I51.

Engel, M., O. Fernandez, H. Jeske and T. Frisch Muth (1998). Molecular characteristics of a new whitefly transmissible bipartite geminivirus infecting tomato in Panama. *J. Gen. Virol.*, 79: 2313–2317.

Gaeel, O.I. (1980). Annual report of the Gezira Research Station, Sudan.

Gevovkan, Z.G., K.G. Azaryan, R.S. Buniatryan and S.S. Papyan (1976). Morphological-anatomical changes in tomato plants infected with leaf curl. *Biologicheski Zhurnal Armenii*, 29: 99–103.

Gupta, D., S.C. Chowfla, P.D. Thakur, Pradeep Sharma, D. Gupta, and P. Sharma (2001). Host range of tomato leaf curl virus from Himachal Pradesh. *Ann. Biol.*, 17: 199–201.

Hayati, J. and J.P. Varma (1985). Effect of some chemicals on tomato leaf curl virus infection of tomato. *Indian J. Virol.*, 1: 152–156.

Ioannou, N. (1987). Cultural management of tomato yellow leaf curl disease in Cyprus. *Pl. Pathol.*, 36: 367–373.

Kalloo, G. and M.K. Banerjee (2001). H–24: Moderately leaf curl resistant variety of tomato *Lycopersicon esculentum* Mill. *Veg. Sci.*, 27: 117–120.

Khan, J.A., S.S. Sohrab, Aminuddin and R.K. Gupta (2002). Detection of a begomovirus affecting guar (*Cyamorsis tetragonoloba* (L.) Taub. in India. *Zeit Schrift-fur-Pflan-Zent rankheiten and Pflan Zenschutz.* 109: 68–73.

Kirthi, N., S. Maiya, M., R.N. Murthy and H.S. Savithri (2002). Evidence for recombination among the tomato leaf curl virus strains/species from Bangalore, India. *Arc. Virol.*, 147: 255–272.

Kisha, J.S.A. (1986). Comparison of electro dynamic spraying with use of a knapsack sprayer for control of whitefly on tomato in the Sudan Gezira. *Ann. Appl. Biol.*, 7: 36–37.

Lodh, S.B., Y. Selvaraj and N.G. Divakar (1973). Changes in pigments and oxidative and cellulolytic enzymes in tomato (*Lycopersicon esculentum* Mill.) infected with tomato leaf curl virus. *Indian J. Exp. Biol.*, 11: 210–212.

Mahajan, R. (2001). *Investigation on Leaf Curl Virus of Tomato* (*Lycopersicon esculentum*). M.Sc. Thesis in Plant Pathology, Sher-e-Kashmir University of Agricultural Sciences and Technology, Jammu, pp. 70.

Mahanta, I.C., A. Dhal and A.K. Mihanty (1998). Field performance of tomato cvs. in relation to wilt and leaf curl disease in North Central Plateau zone of Orissa. *Orissa J. Hort.*, 26: 34–39.

Mansoor, S., S.H. Khan, M. Hussain, N. Mushtaq, Y. Zaffar and K.A. Malik (2000). Evidence that watermelon leaf curl disease in Pakistan is associated with tomato leaf curl virus–India, a bipartite begornovirus. *Pl. Dis.*, 84: 102.

Mariappan, V. and P. Narayanasamy (1972). *Acanthospermum hispidum* DC a new host of tomato leaf curl virus. *Madras Agri. J.*, 59: 355–357.

McClean, A.P.D. (1940). Some leaf curl diseases in South Africa. *Science Bulletin of South Africa*, Department of Agriculture, 225: 1–70.

McGrath, P.F. and B.D. Harrison (1995). Transmission of tomato leaf curl geminivirus by *Bemisia tabaci*. Effects of virus isolate and vector biotype. *Ann. Appl. Biol.*, 126: 307–316.

Mishra, P.N. (1984). Studies on bio-efficacy of some insecticides against the pest complex of tomato, *Lycopersicon esculentum* Miller, var. Pusa Ruby. *Madras Agri. J.*, 71: 673–676.

Mitra, D.K., U.K. Sengupta and M. Majumdar (1978). Effect of some viral and mycoplasmal diseases on nitrate reductase activity in some solanaceous host plants. *Geobios*, 5: 74–75.

Mote, U.N. (1979). Effect of few insecticides alone and in combination with agricultural spray oil on the control of whitefly (*Bemisia tabaci* Genn.) population and incidence of leaf curl virus. *Indian J. Pl. Prot.*, 6: 19–22.

Mukherjee, A.K. and S.P. Raychaudhuri (1966). A note on the effect of some therapeutants on leaf curl disease of tomato. *Indian Phytopath.*, 19: 399–401.

Muniyappa, V., and A.K. Saikia (1983). Prevention of the spread of tomato leaf curl disease. *Indian Phytopath.*, 36: 183.

Muniyappa, V., H.M. Venkatesh, H.K. Ramappa, R.S. Kulkami, M. Zeidan, C.T. Tarba, M. Ghanim and H. Czosnek (1999). Tomato leaf curl virus from Bangalore (ToLCV–Ban 4): Sequence comparison with Indian ToLCV isolates, detection in plants and insects, and vector relationships. *Arch. Virol.*, 145: 1583–1598.

Muniyappa, V., M.M. Swanson, G.H. Duncan, and B.D. Harrison (1991). Particle purification, properties and epitope variability of Indian tomato leaf curl geminivirus. *Ann. Appl. Biol.*, 118: 595–604.

Nainar, P and C. M. Pappiah (1999). Assessment of yield loss in four tomato varieties due to tomato leaf curl virus (TLCV) disease. *South Indian Hort.*, 47: 300–301.

Narayanasamy, P. (1990). Antiviral principles for virus disease management. In: *Basic Research for Crop Disease Management*, Vidhyasekaran, P. (ed.), Daya Publishing House, New Delhi, India, pp. 139–150.

Nour-Eldin, F., H. Mazyad and M.S. Hassan (1969). Tomato leaf curl virus disease. *Agril. Res. Cairo*, 47: 49–54.

Padidam, M., R.N. Beachy and C.M. Fauquet (1995). Tomato leaf curl geminivirus from India has a bipartite genome and coat protein is not essential for infectivity. *J. Gen. Virol.*, 76: 25–35

Pico, B., M.J. Diez and I. Nuez (1998). Evaluation of whitefly mediated inoculation techniques to screen *Lycopersicon esculentum* and wild relatives for resistance to yellow leaf curl virus. *Euphytica*, 10: 259–271.

Ragupathy, N. and P. Narayanaswamy (2001). Screening of tomato germplasm to tomato leaf curl virus (TLCV) disease. *Madras Agril. J.*, 87: 715–717.

Ramappa, H.K., V. Muniyappa and J. Colvin (199 and). The contribution of tomato and alternative host plants to tomato leaf curl virus inoculums pressure in different areas of South India. *Ann. Biol.*, 133: 187–198.

Rataul, H.S. and J.S. Brar (1989). Status of tomato leaf curl research in India. *Trop. Sci.*, 29: 111–118.

Rataul, H.S. and N.S. Butter (1975). Effect of different systemic granular insecticides on the population of whitefly *Bemisia tabaci* Genn. (Aleyrodidae: Hemiptera), the vector of tomato leaf curl virus (TLCV). *J. Res.*, 12: 382–386.

Razvi, S.A., K.M. Azam and A. Zouba (2000). Screening of tomato cuhivators against whitefly, *Bemisia tabaci* (Gennadius) and tomato leaf curl virus. *Shaspa*, 7: 143–149.

Reddy, K.S. and R.C. Yaraguntaiah (1980). Transmission of tomato leaf curl virus through dodder. *Indian Phytopath.*, 32: 653.

Reddy, K.S. and R.C. Yaraguntaiah (1982). Host range of leaf curl virus of tomato. (*Lycopersicon esculentum* Mill). *South Indian Hort.*, 30: 25–26.

Reddy, K.S. and R.C. Yaraguntaiah (1979). Comparison of different methods of grafting for the transmission of leaf curl virus disease of tomato. *Curr. Res.*, 8: 196–197.

Reddy, K.S. and R.C. Yaraguntaiah (1981). Effect of granular insecticides on the tomato leaf curl virus disease. *Indian Phytopath.*, 34: 291–295.

Reddy, K.S. and R.C. Yaraguntaiah and K.S. Sastry (1981). Strains of leaf curl virus of tomato in India. *Zeitschroft fur Pflazen-krankheiten and Pflanzenschutz*, 88: 400–404.

Retuerma, M.L., G.O. Pableo and W.C. Price (1974). Preliminary study of the transmission of Philippines tomato leaf curl virus by *Bemisia tabaci* (Genn.) *Philippine J. Pl. Indus.*, 37: 45–49.

Rigden, J.E., I.B. Dry, L.R. Krake and M.A. Rezaian (1996). Plant virus DNA replication process in *Agrobacterium* insight into the origins of geminiviruses. *Proceedings of the National Academy of Sciences of United States of America*, 93: 10280–10284.

Rojas, A., A. Kvarnheden and J.P.T. Valkonen (2000). Geminiviruses infecting tomato crops in Nicaragua. *Pl. Dis.*, 84: 843–846.

Saikia, A.K. and V. Muniyappa. (1986). Epidemilogy and control of tomato leaf curl virus. *Abstract National Seminar on Whitefly Transmitted Plant Virus Diseases*, IARI, New Delhi, p. 30.

Saikia, A.K. and V. Muniyappa. (1989). Epidemiology and control of tomato leaf curl virus in Southern India. *Trop. Agril.*, 66: 350–354.

Saklani, U.D. and P.J. Mathai, (1978). Effect of insecticides on leaf curl incidence of tomato. *Pest.*, 12: 17–20.

Sarmetwanich, K., P. Chiemsombat, K. Kittipakom and M. Ikegami, (2000). Yellow leaf disease of cantaloupe and wax gourd from Thailand caused by tomato leaf curl virus. *Pl. Dis.*, 84: 200.

Sasikumaran S., T.K. Kandaswamy, P. Vidhyasekaran (1980). Physiology of tomato plants affected by leaf curl virus. *Indian Phytopath.*, 32: 352–359.

Sastry, K.S. (1989). Tomato leaf curl virus management by carbofuran plus oil treatment. *J. Turkish Phytopath.*, 18: 11–16.

Sastry, K.S. M., S.J. Singh and K.S. Sastry (1977). Effect of border cropping and the use of insecticide in relation to the incidence of tomato leaf curl virus (TLCV). *Indian J. Hort.*, 34: 319–322.

Sastry, K.S. M., S.J. Singh and K.S. Sastry (1978). Efficacy of Krishi oil in relation to the control of whitefly population and the spread of tomato leaf curl virus (TLCV). *Pest.*, 12: 28–29.

Sastry, K.S.M. and S.J. Singh (1973). Assessment of losses in tomato by tomato leaf curl virus. *Indian J. Myc. Pl. Pathol.*, 3: 50-54.

Sastry, K.S.M., S.J. Singh and K.S. Sastry (1976). Density of tomato plant population and application of insecticides in relation to incidence of tomato leaf curl virus. *Indian J. Myc. Pl. Pathol.*, 6: 8–13.

Saunders, K., D. Ian Badford, T. Yahara and J. Stanley (2003). The earliest recorded plant virus disease. *Nature*, 422: 831.

Shaheen, A.H. (1983). Some ecological aspects of the whitefly, *Bemisia tabaci* Genn. on tomato. *Bulletin de la Societe Entomologique d' Egypte*, 62: 83–87.

Shahnaz, E (2003). *Managenent of Leaf Curl Disease of Tomato* (*Lycopersicon esculentum*), M.Sc. Thesis, SKUAST, Jammu, pp. 92.

Sharaf, N., A. Al-Musa and I. Nazer (1984). The impact of three irrigation methods on the whitefly (*Bemisia tabaci* Genn.) population and the incidence of tomato yellow leaf curl disease in Jordan. *Dirasat*, 11: 109–117.

Sharma, O.P. and P.N. Sharma (2001). Virus and virus like diseases of vegetable crops. In: *Diseases of Vegetable Crops*, Gupta, V.K. and Y.S. Paul, (eds), pp. 181–199.

Sharma, O.P., R. Mahajan and S. Singh (1984). Field reaction of some root-knot resistance tomato varieties and hybrids to tobacco mosaic and tobacco leaf curl virus. *Indian J. Myc. Pl. Pathol.*, 14: 148-150.

Singh, A K., Shashank Singh and AK. Awasthi (1999). Studies on tomato leaf curl virus in Chattisgarh, Madhya Pradesh. *Indian J. Virol.*, 15: 115–117.

Singh, B. R and D. P. Tripathi (1991). Loss due to leafcurl and spotted wilt diseases of tomato. *Madras Agri. J.*, 78: 34–36.

Singh, R., R.B. Singh and R.P. Srivastava (1976). Deviation in nitrogen metabolism of tomato leaf curl infected tomato plant. *Acta Botanica Indica*, 4: 163–166.

Srivastava, K.M., V. Hailan, R.K. Raizada, G. Chandra, B.P. Singh and P.V. Save (1995). Molecular cloning of Indian tomato leaf curl virus genome following a simple method of concentrating the supercoiled replicative form of viral DNA. *J. Virol. Meth.*, 51: 297–304.

Thanapase, W., P. Poopol, T. Satabutra and S. Attathom (1983). Causal agent and some important characters of tomato leaf curl disease. *Kasetsart J. Nat. Sci.*, 17: 80–89.

Thirumalachar, M.J., H.C. Govindu, and R. C. Yaraguntaiah (1973). Chemotherapeutic control of tomato leaf curl virus disease. *Indian Phytopath.*, 26: 330–332.

Tripathi, S. and A. Varma (2002). Eco-friendly management of leaf curl disease to tomato. *Indian Phytopatho.*, 55: 473–478.

Varma, J.P. and Poonam (1977). Effect of 8–azaguanine on tomato leaf curl virus infection and protein constitution of leaves of two varieties of tomato *Lycopersicon esculentum* Mill. *Indian J. Exp. Biol.*, 15: 408–409.

Varma, J.P. and Poonam (1978). Proteins from *Lycopersicon* species resistant and susceptible to tomato leaf curl virus. *Indian J. Exp. Biol.*, 16: 527–529.

Varma, J.P., J. Hayati and Poonam (1980). Resistance in *Lycopersicon* species to tomato leaf curl disease in India. *Zeit Schhrift fur pflanzenkran Kheitan and Pflazenschutz*, 87: 137–144.

Varma, P.M. (1963). Transmission of plant viruses by whiteflies. *Bull. Nat. Instt. Sci.*, India Bulletin, 24: 11–33

Vasudeva, R.S. and I. Samraj (1948). A leaf curl disease of tomato. *Phytopath.*, 38: 364–369.

Yassin, A.M. (1975). Epidemics and chemical control of leaf curl virus disease of tomato in the Sudan. *Exp. Agri.*, 11: 161–165.

Yassin, A.M. (1989). Major disease problems of tomato production and their control in the Sudan. In: *Tomato and Pepper Production in the Tropics*. Griggs, T.O. and B.T. Mcleah (eds). In: Proceedings of the International Symposium on Integrated Management Practices, 21–26th March, Tainah, Taiwan, AVR DC Publication 89–317, pp. 561–572.

Zeiden, M., S.K. Green, D.P. Maxwell, M.K. Nakhla and K.H. Czosne (1998). Molecular analysis of whitefly -transmitted tomato geminiviruses from Southeast and East Asia Tropical. *Agri. Res. Ext.*, 1: 107–115.

Zouba, A (1995). A geminivirus causing tomato leaf curl disease in the Sultanate of and Oman. *Indian J. Myc. Pl. Patho.*, 25: 302–303.

17

Rice Blast: A Limiting Factor in Rice Production in J&K

*Shahid Ahamad**, *C.S. Kalha***, *K. Kumar***, *Mohd Ishaq** *and A.K. Tiwari***

**Regional Agricultural Research Station, SKUAST-J, Rajouri – 185 134*

***Division of Plant Pathology, Udheywalla – 180 002, Jammu, India*

Rice is one of the most important crop of J&K State. The J&K State grows rice over an area 2.70 lakh hectares and 60 per cent of this area is grown in Kashmir valley alone. Rice, in this state, is grown under sub-tropical conditions in a part of Jammu province, warm temperature conditions of Jammu hills, cool humid temperature conditions of lower elevations of Kashmir valley and sub-normal temperate conditions in high altitudes of the valley. Rice blast, the disease has also been called leaf blast, blast, Neck blast and rotten neck.

Casual Organism

Anamorph : *Pyricularia grisea* Cavara

Teleomarph

Margnaparthe grisea (Hebert) Barr.

Rice in the valley is grown at altitudes ranging from 1500 to 2250 meters above m.s.l. within 33° 25′ to 35° 25′ North latitude and 74°25′ East Longitude. The entire rice area in the valley is irrigated. The irrigation water for rice area comes from melting snow. The temperatures range from 8° to 36°C with an annual rainfall of 745 mm. The valley experiences long winter months from December to March, therefore the growing season in the valley is limited to 130 to 135 days at higher elevations (1650 to 2250 meters above m.s.l) and about 145 days at lower elevations (1500 above to 1650 meters. m.s.l) in a normal agro-climatic year. The adoption of high yielding varieties had great impact on the yield of rice crop. However as compared to Punjab and Haryana, the performance of J&K state with regards to production of rice per unit of land area has been encouraging. The main reasons responsible for this low productivity is the improper management of blast disease.

The rice blast disease has been known for centuries. Blast was known in Japan by the early 1700s, however, it was not reported in Europe or the Americas until the 19th century. A disease of rice called "brusone" was reported in Italy in 1828 by Astolfi, but it was not certain that it was blast (Ou, 1972). Cavara first formally described the fungus from diseased rice plants in Italy in 1891 Padwick. 1950). Fultan (1908) published the first reports of pathogenicity tests using pure cultures of the fungus. Metcalf (1907) was perhaps the one who first called the disease 'blast' in English language.

Padmanabhan (1965b) mentioned that the disease was first recorded in India in 1913 and a devastating epidemic occurred in 1919 in the Tanjore delta area of Tamil Nadu.

Rice blast occurs in almost all growing regions of the world and some 85 countries (throughout the world that have reported its presence (CMI distribution Maps of Plant disease 51, ed.6, 1981).

The Italian mycologist Cavara in 1891 placed the rice blast agent in the genus *Pyricularia* Sacc. and gava it the specific epithet *oryzae* (Padwick, 1950). There first has been considerable controversy on the taxonomy of the fungus casing rice blast. The first species of *Pyricularia* describing as causing disease similar to blast on monocotyledonous hosts (primary grasses was *Pyricularia grisea*). There was disagreement on whether the two species (*P. oryzae* and *P. grisea*) should be considered distinct species. However, with the discovery of the perfect stage of *P. oryzae* the rice blast pathogen is now accepted to be *Magnaporthe grisea.*

The perfect stage of *P. oryzae* was obtained by mating strains from rice with one another and/or with *Pyricularia* Spp. (Presumably *grisea*) from other graminaceous hosts. All crosses yielded the ascomycete *Magnaporthe grisea* (previously *Ceratosphaeria grisea*, Hebert, 1971, Kato *et al.*, 1976, Kato and Yamaguchi, 1962; Kato and Yamaguchi, 1982; Yagashi and Hebert, 1976 and Yacgashi and Udagawa, 1978). Thus, *Magnaporthe grisea* is now considered to be the casual agent of rice blast. Rossman *et al.* (1990) has recommended *P. grisea* as the correct name for the anamorphic stage of the fungus. The perfect stage has not been found in nature.

Importance

It is considered as the principal disease because of its wide distribution and its destructiveness under favourable conditions. Rice seedlings or plants at the tillering stage are often completely killed. Heavy infections on the panicles are often detrimental to rice yields. In Japan annual average yield losses over an eight year period (1953–1960) were estimated at 2.9 per cent (Goto, 1965) even through much of this area had been treated with chemicals to control *P. grisea*.

In India, Padmanabhan (1963 b) estimated that rice production was reduced by 98 per cent in J&K, 70 per cent loss in yield was reported in Tamil Nadu (Rangaswami and Subramanium, 1957) and a similar loss in yield due to blast infection was reported from Uttar Pradesh (Mathur *et al.*, 1974). Padmanabhan (1963b) reported that with 1 per cent neck infection, the reduction in yield varied, being 0.4 per cent, 1.4 per cent and 17.4 per cent in same cases.

Abeygunawardene (1966) reported the results of systematic survey of losses due to blast in the Anuradhpur district of Sri Lanka. An average of 28 per cent of the panicles were infected in paddy rice and 93 per cent of the panicle in upland rice. Although the use of regression analysis relating yield loss with incidence, is the most commonly method used (Goto 1965, Mathur *et al.* 1964, Padmanabhan, 1965), it seems that yield loss estimation is greatly affected by the method of evaluation of panicle infection. Further information on losses due to *P. grisea* are reported by Barr *et al.* (1975) and Zeigler *et al.* (1994).

Rice is the staple in Kashmir sheath blight and blast are the major diseases in rice grown in Kashmir. The blast disease appeared epidemic form during 1999–2000 and rice crop was almost wiped out (Anwar, 2000; Ahamad, 2001). The relevance of adverse weather conditions (Shah *et al.*, 2000) and inoculum potential with seeds quite often lead to the poor growth of seedlings or complete failure of the nurseries and thereby adversely affect overall productivity (Anwar, 2001).

The amount of loss by panicle blast is greatly influenced by the time of infection: the earlier the infection, the larger the loss. More losses also occur when the larger branches of the panicles are infected. Katsube of Koshimizu (1970) estimated both the yield and the quality losses due to neck blast in some detail. They reported that for every 10 per cent of neck blast, there was about a 6 per cent yield reduction and a 5 per cent increase in chalky kernels, which lowered the rice quality by one or two classes. The gross income loss was estimated at 7–9 per cent. For estimating losses from panicle blast. Kuribayashi and Ichikawa (1952) developed the fonnula $Y = 0.69\,X + 2.8$ where Y is the ratio of loss and X is the percentage of panicle affected by blast.

Padmanabhan (1965) estimated the yield losses from blast by comparing yields in (1) non-epidemic and epidemic areas; (2) resistant and susceptible cultivars which have similar potential yielding ability; and (3) Plots protected and unprotected by fungicides. Strict comparisons are not possible because many other factors are involved, but he reported that 1 per cent neck infection reduced the yield by 1–4 per cent in one case and by 17.4 per cent and 0.4 per cent in other cases.

Symptoms

All above ground parts of rice plant may be attacked by *P. grisea*, but symptoms on the leaves and neck are the most valuable for identification. The leaf spots are typically elliptical with more or less pointed ends. The centre of the spots is usually grey or whitish, and the margin is usually brown or reddish-brown. Environmental factors and varietals resistance may affect the size, color, and shape of the lesions. Under favorable conditions, the spots may become larger and coalesce, killing or blighting the leaves. Fully developed lesions reach 1–1.5 cm long 0.3–0.5 cm broad, and usually develop a brown margin. On highly resistant cultivars, only minute brown specks of pin-head size may be observed. While on susceptible cultivars growing under moist, shaded conditions show very little brown margin, but instead sometimes have a yellow halo around the spot. Probably the most severe damage occurs when the neck of the panicle is attacked. The neck is girdled by a grayish- brown lesion and the head or panicle falls over. If the attack occurs before the milk stage, no grain is formed, but in late attacks, poor quality grain may be formed. The nodes of the stem (culms) may be affected, becoming black and brittle. The fungus also causes brown spots on the spike, branches and grain.

The predominant symptoms of blast disease in any given area depend upon the climatic conditions. In the temperate regions, where long periods or drizzle or light rain occurs, leaf blast at the tillering stage is often severe and may be kill the plants completely. In the tropics seedlings in nurseries are often vulnerable, but after transplanting severe infections are seldom found. Moisture is primarily responsible for these differences in symptoms expression (Ou, 1985)

Host Plant

P. grisea infects a large number of grasses hosts including millets, wheat, barley, maize, and sugarcane. Some non-graminaceous hosts are also susceptible to *P. grisea,* including ginger, canna and banana. More than 38 grass hosts have been found susceptible. (Ou, 1985) summarized the host range of the fungus. Although the fungus can attack many grasses when artificially inoculated, the role of other grasses in the disease cycle of rice blast is not clear (Barromeo *et al.*, 1993 and Hamer *et al.*, 1989). Current evidence indicates that the pathogen is composed of a number of host-limited form, with isolates host from one generally unable to infect other hosts.

Life Cycle

In the rice growing regions of Kashmir and the Himalayas, however, the fungus survived the winter in the infected plant parts under snow. When the snow melted and with the onset of the first rains in April, the pathogen was seen to sporulate abundantly. It is very doubtful, whether the fungus could survive in the infected plant debris or in the soil through the hot and dry summer months in the principal rice growing regions in the plants of India.

The development of infection in nature is the result of interaction among the host, the pathogen and the environment in the back ground of time factor.

Misra and Misra (1965) reported that the conidia from cultures could remains viable upto 4 months and for 6 months, when harvested from the host.

The primary cycle is initiated after a period of seasonal inactivity when susceptible rice is exposed to inoculums under favourable environmental conditions. The primary cycle may be initiated by conidia or other fungal structures which over season. Over seasoning structures of *P. grisea* are conidia, mycelia, chlamydospores and perhaps the perfect stage. Mycelia has been known to survive one year in dry straw and two to four years on nodal lesions. The blast fungus can also over season with seed. Conidia can remain viable on both seed and straw, but not as long as mycelia.

Conidia are produced on lesion only when relative humidities are above 89 per cent with an optimum at 93 per cent R.H. Sporulation of the fungus in culture or on affected nodes of culms occurs at air temperature of 15–35°C, while the optimum for sporulation in culture is generally 28°C (Hashioka, 1965). Kato (1976) reports sporulation between 12°C and 34°C. Conidia are released into the air only when the relative humidity exceeds 90 per cent. Addition of water to conidiophores will result in the release of nearly all conidia within two minutes. In general, the more water retained on the leaves, the more spores released (Ou, 1972)

Spores release typically follows a diurnal pattern. Release begins around midnight, probably a result of a conidial maturation coinciding with dew deposition, and continue to daybreak, when it abruptly ceases. A second peak of spore release has been found in the tropics following afternoon showers conidial dispersal is primarily via wind and air current. Although most spores travel only a relatively short distance, they have been trapped in the air at elevations of 7,000 feet (Ou, 1972)

Conidia are formed on the tips of new branches arising just below the joints of the first conidium and conidiophores. One to 20 conidia may be found on a single conidiophore, conidiophores are produced in clusters from each stoma, rarely solitary with 2–4 septa. The conidia are variable in size and shape, pyriform to obclavate, base rounded, apex narrowed, 2-septate, rarely 1–3 septate, sometimes slightly constricted at the septa, almost hyaline to pale olive, usually 19–23 × 7–9 μm, with a small basal appendage 1.6–2.4μm (av. 2 μm). Conidia germinate from the apical or basal cell, and less frequently from the middle cell.

Conidial germination occurs only in free water and is most abundant at 25–30°C (Hashioka, 1965). Germinating conidia produce germ tubes from which apporessoria developed. Appressoria form penetration pegs which enter epidermal cells through the cuticle and cell wall. Within the epidermal cell the penetration peg develops a vesicle from which hyphae emerge. These hyphae then ramify through the host tissue both inter- and intra cellularly. The period required for invasion of host is 6 hours at 24°C, 8 hours at 25°C and 28°C and 10 hours at 32°C (Hashioka, 1965). Infection occurs at high relative humidities between 90–92 per cent (Ou, 1972) reported that lesions are produced in 4–5 days at 26–28°C, 5–6 days at 24–25°C, and 7–9 days at 17–18°C

Major factor that affect infection are soil moisture, soil fertility, air temperature and air moisture. Suzuki (1975) reviewed the literature on meteorological factors in the epidemiology of rice blast.

Pathogenic Races and Variability

The fungus is potentially very variable, and is capable of producing numerous races. The prevailing races in an area, however, depend upon the field cultivars growing in the area, as well as other factors.

The existence of strain of *P. oryzae* differing in pathogen city was first noticed by Sasaki (1922) who confirmed that rice cultivars resistant to his strain A were severely infected by strain B.

Pathogenic race studies were initiated in Taiwan in the middle of 1950s, sixteen cultivars have been used as differentials and 19 races were identified (Hung and Chin, 1961; Chiu *et al.*, 1965)

In view of the fact that different sets of cultivars have been used for the differentiation of races in various countries, the pathogenicity of races identified in one country can not be compared with that of races in other countries. In an attempt to simplify the situation, a co-operative effort was made in 1963 between Japan and the USA to develop an international set of differentials. During three years of co-operative studies, many hundreds of isolates collected in Japan and the USA were extensively tested on 39 differential cultivars which had been used in Japan, the USA and Taiwan. Eight cultivars were finally selected and 32 race groups characterized, based upon the isolates tested. The races were called international races and were designated as IA, 18 etc. to IH to indicate groups, followed by numerals to indicate race numbers (Atkins *et al.*, 1967). Yamada *et al.* (1976) proposed a new set of 9 differentials in Japan, each cultivars with a known resistance genes.

In India detailed studies on variability in *P. oryzae* were carried out at the Centre Rice Research Institute, Cuttack and elsewhere. Occurrence of 32 races of the pathogen has been found to be distributed throughout the country (Padmanabhan *et al.*, 1970) through IC 17 was found to be the most prevalent on in recent years. The fungus produces two toxins, *viz.* picolinic acid and piricularin.

Control Measures

Various Strategies including cultural practices, resistant varieties; chemical methods, and natural product may be applied for the management of rice blast. A degree of control may be achieved by (i) Reducing the amount of nitrogen fertilizer applied. However, this in its may reduce yields substantially. Other cultural method management are; (ii) early sowing and transplanting; (iii) A voidance of cold water irrigation; (iv) increasing the silica content of plants by the use of compost and calcium silicate; and (v) altering plant populations. Rice blast is generally less serve on flooded paddy rice than on upland rice.

Cultural Practices like Potash and phosphorus should be used (30 Kg/ ha each) along with nitrogenous fertilizers to reduce the epidemic.

Weeds like Panicum repens, Leersia, Echinochloa, Digitaria, etc. should be removed from the nearby areas of rice fields as these serve as alternate hosts for this pathogen.

Disease debris like rice straw should be burnt after the crop harvest.

Disease free seed should be used for sowing as seeds serve as a primary source of infection.

Effect of Nitrogen

Nitrogen fertilization favours the development of blast (Ou, 1985). The timing of N fertilizer application had more pronounced effects than those reported for drought prone upland conditions (Dossantos *et al.*, 1986). The timing of N application (early vs. late application) had much more important influence on leaf blast than the number of split applications of the same N dose. Leaf blast was consistently suppressed when N was not applied at planting (Kapoor and Sood, 2000).

The given rally high disease are intensity of leaf blast and neck blast might be attributed to increase susceptibility caused by water deficit (Gill and Ronman, 1998). Many workers had reported that an increase in the member of equal application of given quantity of N world reduce severity as had been reported for irrigated rice (Amin and Vankatarao, 1979). On the contrary, leaf blast tended to increase with three to four equal split application (Kapoor and Sood, 2000) Adjustment in the timing of splitting of application did nest significantly increase yield.

Orthogonal contrasts using means from inoculated plots showed that N application significantly increased leaf and neck blast compared with no-N treatment. Nitrogen applied twice, at planting and 30 DAS (early splits) gave significantly higher leaf blast than late split. Neck blast incidence did not differ significantly among treatments receiving N. From these studies it was found that the rate, frequency and timing of N application had a much greater influence on leaf blast than on neck blast. There was no consistent reduction in neck blast with any of the various N application treatments. However, the no-N treatment had less neck blast than the treatment where N was applied, and this result was consistent across years (Kapoor and Sood, 2000).

The potential to suppress blast disease through timing and splitting of N application was tested during Kharif (1996 and 1997 at two location in Palampur) India. The effect of 90 Kg N. per ha applied as two early two late and three and four equal split on disease progress and yield was studied. A no N check was included. In susceptible cv. Himalaya 741 leaf blast was made where N was applied in 3 and 4 equal splits. Every splits resulted more leaf blast as compared to late splits. No N resulted significantly very less leaf blast and neck blast but yield were very low. AUDPC values were maximum in three equal splits of N followed by two early and four equal splits. Similar trend was observed in neck blast also. The disease was high at three and four N splits and yield were reduced by more than 50 per cent of plots simultaneously protected with Beam @ 0.0 6 per cent.

Use of Herbicides

Like propanil, symetrin and nitrofen are reported to be effective if applied immediately *i.e.*, one day after disease appearance. Seed treatment may be done through fungorin @ 3 g/ kg and kept for 24 hours before sowing.

Use of Plant Product

Natural plant products especially from *A. marmelos* can effectively be utilized in controlling rice blast. The plant leaves can be utilized to control the disease both prophylactically and therapeutically.

Kamalakannal *et al.* (2001) reported that leaf extract of 20 plant species significantly reduced the mycelial growth of pathogen to varying degree. Among all *Prosophis juliflora* followed by *Zizyphus jujuba, A. indicun, Cynodan dactylon Cyperus rotundus* and *clerodendron inerme* recorded, maximum reduction in the mycetial growth to an extent 93.3, 82.0, 78.1, 77.8, 74.1 and 72.9 per cent over control respectively. Plant extract preparation from *P. juliflora* exhibited maximum inhibitions on the biomass production and spore germination of *P. grisea.* The present studies conclude that the leaf extracts of *P. juliflora* and *Z. jujube* which had higher levels of proteins, glycoprotein's, total phenol and tanin exhibited greater antifungal properties towards the rice blast pathogen whose potential can be effectively used for the management of disease under field conditions.

Use of Resistant Varieties

Breeding for resistance to blast has been described by many authors (Ahm, 1994; Khush, 1989 and Ou, 1985). Some authorities suggest the *P. grisea* is a highly variable organism (Ou, 1985) and that the existence of numerous physiologic races complicates breeding for resistance. Recently Valents and Chumley (1994) wrote "Although the extent of instability in race determining has been debated, it is clear that new races rapidly appear in the field when new blast resistance rice cultivars are introduced". They add that evidence is accumulating that the rice blast system is, in fact, a gine-for-gine system as defined by Flor (1971). Evidence for parasexual exchange of genetic information in laboratory studies and in the fields has been presented by Zeigler *et al.* (1997). This mechanism may explain how this apparently asexual organism maintains such a high degree of variation and may offer a means for it to reconstitute its genome. Rice cultivars resistants to *P. grisea* have generally had a short life. Previously several evaluation systems were used but in 1967 blast workers agreed on a uniform system of inoculation, differential cultivars and methods of taking notes (Atkins *et al.*, 1967).

The International Rice Research Institute cooperates with other entities in planting uniform blast nurseries in 50 testing stations in 22 different countries. Thus much valuable and standardized information on the occurrence and prevalence of races of *P. grisea* has been collected. More than 260 physiologic races have been reported (IRRI-1975). A recent important contribution to the understanding of pathotypic variation has been the development of near-isogenic lenes, each carrying one major resistance in a common back ground (Mackill and Bonman, 1992).

Resistance to *P. grisea* in rice is usually dominant and controlled by one or a few pairs of genes. At IRRI (1979) almost 100,000 lines and accessions were tested for resistance, and no single one was found to completely resistant to all races. Although sources of resistance effective against many races have been found.

1. In J&K following blast resistant entries *viz.*, IET–16455, IET–16868, IET–16465, IET–16456 may be used for rice cultivation and susceptible variety K–448 may be used in breeding programme as a check.
2. Use of resistant donors like Tetep, Tadukan, Zenith, Aichi, Ashahi, Raminad str. 3, Peta, Norin–22 and Trainan–3 etc. should be made in the breading programmes for blast resistant varieties.
3. Use of resistant varieties like Vandana, Pathara, Hira, IR–50 mutant, IR–36, Ratna, Sarasa help to get effective control of rice blast.

Singh *et al.*, 2002 evaluated 763 scented genotypes for resistance to blast. Out of 23 genotypes namely, Rasmati [illegible]la, Basmati nehtrah, Bar Chokura, Chandan, G–65 HKR 97–401, HKR–416, HKR 97–418, HKR [illegible]–42 Haryana Basmati No. 1, IR 841–85–1–1–2, Kalijira 8–1, MTU–4407, N–50,

NIRI–140–6, Pusa 743, RP–2144–19–6–2, RP 3120–15–12–9–4, RP 3135–17–12–8–8, RP–3238–33–15–7–1, RPU 41–62–1–1–3, Sonali and T.D were found consistently resistant.

Chemical Control

The copper fungicides were first used effectively. However because of the high phytotoxicity of copper a more alternative was sought. During 1950s organo-mercuric fungicides rapidly replaced the copper based compounds for blast control. These fungicides had very low phytotoxicity and gave excellent control of blast, however, they were highly toxic to mammals.

The first system antibiotic blasticidins was produced from Streptomyces griseachromogenes in 1961. Unfortunately, because of high phytotoxicity and mammalian toxicity, blasticidin S had to used with extreme care. Shortly after the development of blasticidine S a new antibiotic (Kasugamycin) produced by *Streptomyces Kayugaensis*, was discovered. Kasugamycin gave excellent control and had very low mammalian and rice toxicity. Beginning around 1970, in areas where the antibiotics had been used intensively and exclusively for blast control, populations of *P.grisea* began to show resistance to the antibiotic compound (Uesugi, 1978). However, after halting the use of the antibiotics in area with resistant populations, the proportion of resistant types in the populations dropped to near zero.

At about the same time the antibiotics were being developed. organo-phosphorus fungicides were introduced to control blast in Japan. In the late 1970s reports were emerging. It was found that resistant to one organo-phosphorus fungicides did not necessarily confer resistance to other specific fungicides, and it was suggested that rotating the use of fungicides of mixing them, rather than continuously relying on a single compound, could greatly reduce the risk of developing highly resistant populations (Uesugi, 1978) Webster and Gunnell (1992) noted that the development of systemic fungicides such as pyroquilon and tricyclazole opens new possibilities for chemical control of blast.

Chemical control practices in Japan involved chemical seed and seedling bed treatment to reduce the initial inoculum level followed by fungicide or antibiotics applications during the year.

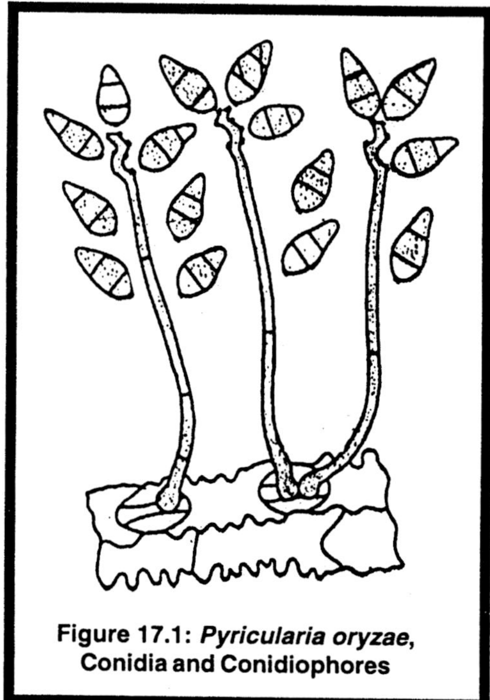

Figure 17.1: ***Pyricularia oryzae*, Conidia and Conidiophores**

Dubey (2000) used twelve different fungicidal spray schedules were evaluated against blast disease. The spray scheduled of beam (0.03 per cent)–indofil M–45 (0.2 per cent)–beam (0.03 per cent) proved most efficacious in which minimum neck and node infection and disease intensity resulting in highest grains yield were recorded during all the three years of experimentations followed by spray scheduled of Topsin M–Indofil M–45 Topsin-M in all respects. When the economics was taken into account (as per rupee returns), spray of Topsin M–Indotil M–45–Topsin M was highly economical with maximum economic return *i.e.*, Rs. 11.6 followed by Topsin M–Indofil M–45–Indofil M–45.

The mixture of beam and Indofil M 45 has been reported to be effective against blast, but it was not economical as Bavistin alone and mixture of Topsin M and Indofil M–45 (Dubey, 1997). Bavistein, Indofil M–45,

Kitazin and Topsin M have been found effective against blast of rice earlier (Rao and Murlidharan, 1983; Saikia, 1988; Sunder and Ahuja, 1990; Dubey and Mishra, 1992; Dubey, 1995).

Various fungicides have been found to provide effective control of rice blast when applied as spray or seed treatment some of these are mentioned here. Out of four fungitoxicants evaluated in a field trial during 1997 and 1998, carpropamid, tricyclazole and prochloroz + carbendazim significantly reduced neck blast incidence and enhanced the grain yield during both the years. Out of those, carpropamid (1 ml/l) proved most effective and provided 79.2 per cent disease control along with 29.7 per cent increased yield closely followed by tricylazole4 (Singh *et al.*, 2002).

In J&K a trial was conducted on rice blast management during Kharif 2000–01. Some new fungicides were evaluated at Khudwani, Pondicherry and Rewa (DRR, 2000). Results at all the three locations indicate that Bann (0.6g) was at par with the standard check fungicides (Beam 0.6g) both in containing the infection and in increasing grain yield, followed by Benlate, Swing, Saaf, Kasu Band Folicure (Table 17.1).

Table 17.1: Evaluation of Mew Fungicidal Formulations for Leaf Blast Control in Kharif, 2000, at "Khudwani, Pondicherry and Rewa"

Fungicides		*Leaf Blast Incidence (%)*						*Grain Yield (Kg/ha)*		
		Khudwani		*Pondicherry*		*Rewa*		*Khud-wani*	*Pondi-cherry*	*Rewa*
	Dosage/Litre of Water	*AT*	*O*	*AT*	*O*	*AT*	*O*			
Benlate	(1.0 g)	23.6	16.3	12.5	4.7	35.1	33.2	5083	4833	2807
Benlate	(1.5 g)	20.3	12.2	11.5	4.0	26.9	20.5	4917	5000	3020
Folicur	(1.5 ml)	25.8	19.0	16.1	7.7	33.4	30.4	4983	4633	2887
Folicur	(2.0 ml)	22.8	15.6	15.3	7.0	29.6	24.5	4733	4733	2933
Kasu–B	(2.0 ml)	27.6	21.5	18.7	10.3	34.8	32.6	5033	4133	2800
Saaf	(1.5 g)	27.8	21.9	17.4	9.0	32.6	29.1	5100	4200	2870
Swing	(2.0 ml)	23.7	16.2	20.0	11.7	24.7	17.5	4583	4100	3170
Baan	(0.6 g)	20.1	11.9	9.4	2.7	23.3	15.7	5600	5100	3170
Beam	(0.6 g)	18.0	9.6	8.7	2.3	18.5	10.1	5383	5233	3600
Check	(Untreated)	44.3	48.8	35.5	33.7	42.4	45.6	3867	3167	2293
CD (0.05%)		4.6		1.4		1.8		801	131	370
CV (%)		10.6		4.9		3.5		9.5	1.7	7.3
Test variety		K 448		IR 50		Basmati				
Sprays (No)		4		3		3				

AT: Aresin transformed means; O: Means of original values

Source: DRR Annual Progress Report 2000, Vol. 2: Pathology.

Some new fungicidal spray formulations was evaluated at Khudwani against blast control in transplanted and direct-seeded upland rice (DRR.2001). Leafblast was 36 per cent in the transplanted crop and 47 per cent in the upland crop. However, neck blast infection was very low (2 per cent) (Table 17.3).

Table 17.2: Evaluation of New Fungicidal Formulations for Leaf Blast Control in Transplanted Rice, Kharif, 2001, at "Khudwania."

Fungicides	*Dosage/Litre of Water*	*Leaf Blast Severity (%)*		*Neck Blast Severity (%)*		*Grain Yield (Kg/ha)*
		AT	*O*	*ST*	*O*	
Fongorene 50 WP (TIM)	0.4 g	26.7	20.1	0.9	0.9	5733
Fongorene 50 WP (TIM)	0.8 g	23.1	15.5	0.7	0.7	6500
Fongorene 50 WP (TIM)	1.2 g	24.1	16.7	0.9	0.9	6633
Fongorene 50 WP (TIT)	0.4 g	24.7	17.7	1.1	1.3	5933
Fongorene 50 WP (TIT)	0.8 g	24.3	16.9	1.1	1.2	5933
Fongorene 50 WP (TIT)	1.2 g	23.8	16.3	1.0	1.1	5883
Kasu-B SL	2.0 ml	25.3	18.3	0.6	0.6	6900
Saaf 75WP	1.5 g	24.7	17.4	0.9	0.9	7600
Swing 250 BC	2.0 ml	25.3	18.6	0.7	0.7	6300
Tricyazole (Sivic) 75 WP	0.6 g	19.2	10.8	0.9	0.9	9033
Check (Untreated)	–	37.1	36.4	1.4	1.9	5200
CD (0.05)		3.1		ns		1708
CV (%)		7.1		35.8		15.3
Test Variety			SKAU–27 (K–448)			
Sprays (No.)			4			

AT: Aresin transformed means; ST: SQRT transformed means; O: Means of original values

(*Source*: DRR Annual Progress Report 2001, Vol. 2: Pathology.)

Table 17.3: Evaluation of New Fungicides as Seed Treatment for Blast Control in Direct-Seeded Upland Rice, Kharif, 2001, at "Khudwania."

Fungicides	*Dosage/Litre of Water*	*Leaf Blast Severity (%)*		*Neck Blast Severity (%)*		*Grain Yield (Kg/ha)*
		AT	*O*	*ST*	*O*	
Fongorene 50 WP (TIM)	0.4 g	29.7	24.6	1.1	1.2	4933
Fongorene 50 WP (TIM)	0.8 g	27.8	21.7	1.0	1.1	5933
Fongorene 50 WP (TIM)	1.2 g	29.6	24.4	1.0	1.1	6367
Fongorene 50 WP (TIT)	0.4 g	31.2	26.8	1.1	1.2	5133
Fongorene 50 WP (TIT)	0.8 g	24.2	16.8	1.1	1.2	6167
Fongorene 50 WP (TIT)	1.2 g	28.1	22.2	1.1	1.2	6233
Tricyclazole (Sivic) 75WP	0.6 g	24.2	16.8	0.9	0.8	7233
Check (Untreated)	1.5 g	23.5	47.4	1.5	2.1	3933
CD (0.05)			1.2		0.2	557
CV (%)			2.3		9.6	5.6
Test Variety				SKAU–27 (K–448)		
Sprays (No.)				4		

AT: Aresin transformed means; ST: SQRT transfonned means; O: Means of original values

(*Source*: DRR Annual Progress Report 2001, Vol. 2: Pathology.)

The standard check fungicide. tricyclazole (Sivic 75wp) had sown its supremacy over all the other test fungicides. both in transplanted and upland conditions. in checking the blast infection and improving the grain yield. However, under the upland conditions, the imported technical (TIT) of pyriquilone (Fongolene 50 WP) @1.2 g/l was on par with the standard check.

The seed treatment with tricyclazole was significantly effective both in checking the blast infection and increasing the grain yield at Khudwani. Seed treatment with pyriquilone was however, significantly effective in checking the disease at all the test dosage, but the same was reflected in yield at certain dosage rates (Table 17.4).

Table 17.4: Evaluation of New Botanicals Against Location Specific Diseases in Kharif, 2001, at "Khudwania."

Fungicides Yield	*Dosage/Litre of Water*	*Leaf Blast Severity (%)*		*Sheath Blight Severity (%)*		*Panicles Affected with Blast (%)*		*Glume Discolouration (%)*		*Grain Yield (Kg/ha)*
		AT	*O*	*AT*	*O*	*ST*	*O*	*ST*	*O*	
Achook	5.0 ml	36.9	36.1	21.5	13.5	0.9	0.4	1.5	1.6	653
Neem Azal T/S	3.0 ml	39.5	40.5	24.7	17.4	1.3	1.3	1.9	3.1	647
Neem gold	20.0 ml	40.7	42.6	24.2	16.8	1.0	0.6	2.2	4.2	730
Spictaf	4.5 ml	30.4	25.6	23.1	15.4	1.1	0.7	1.6	2.1	650
Tricure	5.0 ml	21.6	13.6	18.5	10.1	0.9	0.3	1.2	0.9	697
Wanis	5.0 ml	38.3	38.4	27.0	20.6	1.1	0.8	1.9	3.0	577
Hinosan	1.0 ml	15.6	7.2	27.6	21.5	0.7	0.0	1.0	0.6	793
Tilt 25 EC	1.0 ml	38.8	39.3	25.1	18.0	0.9	0.4	1.9	2.9	730
Bavistin 50 WP	1.0 g	26.6	20.1	29.4	24.2	1.1	0.7	0.8	0.2	743
Check (Untreated)	–	45.7	51.2	38.2	38.2	1.5	1.9	2.6	6.3	443
CD (0.05)		2.0	2.0			0.4		0.3		109
CV (%)		3.5	4.6			21.9		9.9		9.5
Test Variety						SKAU–27 (K–448)				
Sprays (No)						4				

AT: Aresin transformed means; ST: SQRT transformed means; O: Means of original values

(*Source*: DRR Annual Progress Report 2001, Vol. 2: Pathology.)

Among the botanicals *viz.*, Achook, Neem Azal, Neem gold, Spictaf, Tricure, Wanis, Hinosan 50 EC. Tilt25EC and Bavistin 50WP, tested against blast, sheath blight. and increasing the grain yield (Table 17.5).

Ali *et al.* (2002) observed that blast lesions reduced in seed treatment-cum-spray with mancozeb (0.1 per cent) following Bavistin. Results revealed that all the fungicide tested reduced leaf blast incidence significantly over un-sprayed plots. Neck blast incidence did not occur in sprays of both the fungicides followed by seed treatment with bavistin and contaf.

Table 17.5: Management of Blast Diseases, Kharif, 2001, at "Khudwania."

Variety	*Leaf Blast Severity (%)*		*Neck Blast Incidence (%)*		*Grain Yield (Kg/ha)*	
	Disease Management	*No Disease Management*	*Disease Management*	*No Disease Management*	*Disease Management*	*No Disease Management*
Khutch (S)	67.3 (8.2)	86.8 (9.3)	9.5 (3.1)	23.5 (4.9)	875	400
K–39 (MR)	5.6 (2.4)	11.9 (3.4)	0.3 (0.8)	1.6 (1.4)	7475	6675
SKAU–105 (@)	1.6 (1.2)	2.6 (1.6)	0 (0.7)	0 (0.7)	7850	8025
K–143 (Hybrid R)	5.1 (2.3)	8.1 (2.9)	1.1 (1.3)	2.0 (1.6)	7400	6600
CD (0.05)						
Main	0.26		0.4		1087	
Sub-interaction	0.13		0.2		285	
M in S	0.32		0.48		NS	
S in M	0.26		0.39			
CV (5%)						
Main	6.1		20.0		17.7	
Sub	4.3		14.1		6.5	

Figures in parentheses are SQRT transformed values.

(*Source*: DRR Annual Progress Report 2001, Vol. 2: Pathology.)

REFERENCES

Abeygunawardena, D.V.W. (Ed.) (1966). *Proceedings of a Symposium on Research and Production of Rice in Ceylon*. Ceylon Assoc. Adv. Sci., Colombo, Ceylon.

Ahamad, S. (2001). Epidemiology and Integrated Diseases Management. In: *National Symposium on Plant Protection Strategies for Sustainable Agri-Horticulture*, SKUAST-Jammu, October 12–13, pp. 40–50.

Ali, Anwar (2000). Recent status of certain rice disease under field conditions of temperate zone on under Kashmir valley. *Nat Symp. Indian Phytopath Society*, Feb. 15–17, Karnal, pp. 16.

Ali, Anwar, G.N. Bhat and G.S Singhare (2002). Management of sheath Blight and blast is rice through seed treatment. *Ann Pl. Protec. Sci.*, 10: 285–287.

Ali, Anwar, G.N. Bhat, M.A. Zarger, N. Ahmed and G.A. Parry (2001). Effect of seed treatment through bioagents and blitox on incidence of leaf blast disease (*Magnaporthe grisea*) and seedling growth under temperate conditions. *Nat. Symp Agric. Environment*, SKUAST(K), Srinagar, pp. 55–57.

Amin, K.S. and G. Venkatrao (1979). Rice blast control by nitrogen management. *Phytopath. Z.*, 96: 140–145.

Atkins, J.G., A.I. Robert, C.R. Adair, K. Goto, T. Kozaka, R. Yanagida, M. Yamada and S. Matsumoto (1997). An international set of rice varieties for differentiating races of *Pyricularia oryzae.*

Barr, B.A., C.S. Koehler and R.F. Smith (1975). Crop losses: Rice; Field Losses to Insects. Diseases, Weeds and other Pests. Univ. Calif. Bereley/U.S. Agency for Int. Dev. Pest Management and Related Environ. Protec. Proj. Rep.

Barromeo, L.S., R.J. Nelson, J.M. Bonman, and H. Leung (1993). Genetic differentiation among isolates of *pyricularia* infecting rice and weed hosts. *Physiopathology*, 83: 393–399.

D.R.R. Annual Progress Report, 2000, Vol. 2. Pathology.

D.R.R. Annual Progress Report, 2001, Vol. 2. Pathology.

Dos santos, A.B, A.B Prabhu, A.R.L. DE Aquino and J.R.P. Decarvalin (1986). Effect of nitrogen rate time and method of application on blast and grain yield in upland rice. *Pesqui. Agropecui. Bras*, 21: 696–707.

Dubey, S.C. (1995). Evaluation of different fungicides against blast of rice. *Pl. Dis. Res.*, 10: 38–41.

Dubey, S.C. (1997). Efficacy of tricyclazole against rice blast. *J. Mycol. Pl. Pathol.*, 27: 335–337.

Dubey, S.C. (2000). Economical spray schedule of fungicide for blast management in rice. *Pl. Dis. Res.*, 5: 43–45.

Dubey, S.C. and B. Mishra (1992). Evaluation of fungicides against blast of rice. *Oryza*, 29: 399–401.

Flor, H.H. (1971). Current status of the gene for gene concept. *Ann. Rev. Phytopatho.*, 9: 275–296.

Fulton, H.R. (1908). Disease affecting in Louisiana. Louisiana *Agric. Exp Stn. Bull.*, pp. 105.

Gill, M.A. and J.M. Bonman (1980). Effect of water deficit on rice blast influences of water deficit on components of resistances. *J. Plant Proto. Trop.*, 5: 61–66.

Goto, K. and T. Kozaka (1965). A proposal for an international joint study for differentiation of the races of rice blast fungus, In: *The Rice Blast Disease*. IRRI, John Hopkins Press, Baltimore. pp. 447–450.

Hamer, J.E., L. Farral, M.J. Orbach, B. Valent and F. Chumley (1989). Host species-specific conservation of a family of repetitive sequences in the genome of a fungal plant pathogen. *Proc. Natl. Acd. Sci.*, USA, 86: 9981–9985.

Hashioka, Y. (1965). Effects of environmental factors on development of causal fungus, infection disease development, and epidemiology in rice blast disease.

Hung, C.S., C.C. Chien (1961). Investigation of physiological races of the blast fungus (*Pyriculria oryzae*). Special Publication of the *Taiwan Agricultural Research Institute*, 3: 31–37.

IRRI (1975). Annual Report IRRI, Los Banos, Philippines.

Kamalakannan, A., V. Shanmugan, M. Surrendran and R. Sunivasan (2001). Evaluation of plant extract against *Pyricularia grisea*, causing rice blast. *Pl. Protect Sci.*, 9: 68–72.

Kamalakannan, A., V. Shanmugan, M. Surrendran and R. Sunivasan (2001). Antifungal properties of plant extracts against *P. grisea* the rice blast pathogen. *Indian Phytopath.*, 54: 490–492.

Kapoor, A.S. and G.K. Sood (2000). Effect of time of application and splitting of nitrogen on rice blast. *Indian Phytopath.*, 53(3): 283–286.

Kato, H.T. Yamaguchi (1982). The perfect state of *Pyricularia* oryzae Cav. From rice plants in culture *Ann. Phytopathol. Soc.*, Japan, 48: 607–612.

Kato, H.T. Yamaguchi and N. Nishihara (1976). The perfect state of *Pyricularia* oryzae Cav. In culture *Ann. Phytopathol. Soc.*, Japan, 42: 507–510.

Katsube, T. and Y. Koshimim (1970). Influence of blast disease on harvests of rice plants. Effects of panicle infection on yield components and quality Bulletin of the *Tohaku Agricultural Experimental Station*, 39: 55–96.

Kuribayashi, K. and H. Ichikawa (1952). Studies on forecasting of the rice blast disease special report, *Nagano Agricultural Experimental Station*, 13: 1–229.

Mackill, D.J and J.M. Bonman. (1993). Inheritance of blast resistance in near isogenic lines of rice. *Phytopath.*, 82: 746–749.

Mathur, R.S., G.K. Bajpai, L.S. Chauhan and S.C Verma (1964). Assessment of losses caused by paddy blast. *Plant Dis. Report*, 48: 711–713.

Mathur, R.S., G.K. Bajpai, L.S. Chauhan and S.C. Verma (1974). *Plant Dis. Report*, 48: 711–718.

Metcalf, H. (1907). *The Pathology of the Rice Plant Science*, U.S.A., 25: 264–265.

Ou, S.H. (1972). *Rice Disease*, 2nd Edition. Commonw, Mycol. Inst., Kew Surrey.

Ou, S.H. (1985). *Rice Disease*, 2nd Edition. Commonw, Mycol. Inst., Kew Surrey.

Padmanabhan, S.Y. (1963a). *Oryza*, 1: 101–111.

Padmanabhan, S.Y. (1965). Estimating losses from the rice blast in India. In: *The Rice Blast Disease*, IRRI, John Hopkins Press Baltimore, pp. 203–221.

Padmanabhan, S.Y., N.K. Chakrabarti, S.C. Mathur and J. Veeraraghavan (1970). *Phytopathology*, 60: 1574–1577.

Padwick, G.W. (1950). *Manual of Rice Disease*. Commonwealth. Mycol. Inst., Kew Surrey.

Rangaswami, G. and T.V. Subramaniun (1957). *Sci. and Cult.*, 23: 192–193.

Rao, G.V. and K. Muralidharan (1983). Fungicides and control of leaf blast in dry paddy nursery Indian pry to path. 41: 274.

Rossman, A.Y., R.J. Howward and B. Valent (1990). *Pyricularia grisea*, the correct name for the race blast disease fungus. *Mycologica*, 82: 509–512.

Saikia, U.J. (1988). Combating rice blast through chemicals. *Indian Pytopath.*, 41: 274.

Sasaba, R. (1922) Existence of strains in rice blast fungus I. *J. Plant Protection*, Tokya 9: 631–644 (Ja).

Sharma, B.R. and A.S. Kapoor (2000). Management of rice blast through host resistance and chemicals. *Indian Phytopath.*, 53: 362.

Singh Ram, S. Sunder and D.S. Dodan (2002). Blast resistance in scanted rice genotype and its chemical control. *Indian Phytopath.*, 55(3): 394–395.

Sunder, S. and S.C. Ahuja (1990). Varieties modify activity of fungicides in neck blast control of paddy. *Pl. Pathl. Res. Prob. Progress IPS* Haryana April Univ. Hisar, Dec. 4–5, pp. 30–33.

Suzuki, H. (1975). Meteorological factors in the epidemiology of rice blast. *Annu. Rev. Phytopathol.*, 13: 239–256.

Uesugi, Y. (1978). Resistance of Phytopathogenic fungi to fungicides. *Japan Pesticide Information*, 35: 5–9.

Valent, B. and F.G Chumley (1994). Avirulences genes and mechanism of genetic instability in the rice blast fungus pp. 111–134. In: Zeigler, R.S., S.A. Leong and P.S. Teng, *Rice Blast Disease*, CAB Int. and IRRI, Wallingford, UK.

Yaegashi, H. and S. Udagawa (1978). The taxonomical identity of the perfect state of *Pyricularia grisea* and it allies. *Can. J. Bot.*, 56: 180–183.

Yaegashi, H. and T.T. Herbert (1976). Perithecial development and nuclear behaviour in *Pyricularia. Phytopath.*, 66: 122–126.

Yamada, M.S. Kiyosawa, T. Koboyashi, K. Kushibuchi, S. Watanabe (1976). Proposal of a new method for differential races of *P. Oryzae* Cavara in Japan. *Ibid*, 42: 216–219.

Zeigler, R.S., I. Tohme, R.I. Nelson, M. Levy and F.I. Correa-Victoria. (1994). Lineage exclusion a proposal to link blast population analysis to resistance breeding. In: *Rice Blast Disease*, Zeigler, R.S., S.A. Leong, and P.S. Teng. (eds). CABI Int. and IRRI, Wallingford, U.K., pp. 267–292.

Zeigler, R.S., R.P. Scott, H. Leung, A. Bordeos, and R.J. Nelson (1997). Evidence for sexual exchanges of DNA in the rice blast fungus (*Magnaporthe grisea*) challenges its exclusive clonality. *Phytopath.*, 87: 284–294.

18

Insect Pests and Diseases of Cold Arid Region, Ladakh

Ajay K. Pandey and Sanjai K. Dwivedi***

**Regional Agriculture Research Station, S.K. University of Agricultural Sciences and Technology Kashmir, Leh – 194 101*

***Field Research Laboratory, Defence Research Development Organization, C/o 56 APO*

The cold arid region of Ladakh geographically located between 32°–36° N latitude and 76°–79° E longitude. It covers approx 82,665 sq km area of which 37,555 sq km in under illegal occupation of Pakistan and China (SKUAST Leh, 1998). Hence area with India is 45,110 sq km representing 87.4 per cent of total cold arid region. However, Singh *et al.* (1998) has reported that Indian cold arid regions approximately 74,809 sq. km. of which major part is confined to Ladakh (68,321 sq. km), followed by Lahul Spiti (HP) (6,488 sq km) and only a very minimal part being found in Uttaranchal. The Ladakh region of J&K state characterized by extreme of seasonal temperature ranging from–40 to +40 °C, sharp temperature fluctuation, heavy influx of ultraviolet and infra red radiation, scanty rainfall, coarse soil texture with high permeability, short cropping season, and winter devoid any vegetation (Sharma, 2002). Temperature and moisture the two parameters mainly responsible for low vegetation coverage altitude also influenced the distribution of flora and fauna (Kachroo *et al.*, 1977).

The economy of Ladakh is dependent on agriculture including horticulture and forest plantations. Most of the agricultural practices in Ladakh are restricted to 8000–10000 feets. People of Ladakh grow a large number of field crops like cereals (wheat and barley), pulses (check pea, moong and lentil), oil seeds (mustard), vegetables (potato, onion, cabbage, pea, tomato, radish, turnip etc) and forest trees (poplar and willow) beside fruit corps like apricot, apple, walnut etc. The crop wise area under cultivation is given in Table 18.1.

Table 18.1: Cultivated Area in Ladakh Region and its Allocation to Different Field Crops

Agricultural Land (ha.)	*Leh*	*Kargil*	*Total*
Total area	11,813.0	12,392.0	24,205.0
Double cropping area	220.0	2,430.0	2,650.0
Net area sown	11,593.0	9,962.0	21,555.0
Crop-wise distribution			
Wheat	2,600.0	1,847.0	4,447.0
Barley	4,750.0	3,593.0	8,343.0
Pulses	200.0	634.0	834.0
Millets	450.0	88.2	538.0
Fodder	1,990.0	2,689.12	4,679.12
Vegetables	860.0	242.0	1,102.0
Fruit crops	863.0	608.12	1,471.12
Oilseed/Miscellaneous	100.0	2,689.62	2,789.62

Source: Statistical Handbook, Planing and Development, Deptt. Govt. of J&K, Leh, 1996.

As in entire low land ecosystem, in high altitude (Ladakh), the insect are also the dominant animals belonging to orders Coleoptera, Diptera, Lepidoptera, Orthoptera, Homoptera, Hemiptera and Thysanoptera. Most of the insect pest, which is found in this region, cause damage to cultivated as well as wild plants. Besides insect pests diseases are also of great importance which causes significant damage to the crops. Though, the problem of insect pests and diseases are very low due to ultra low temperature for over 5–7 months in a year, low annual precipitation and low atmospheric humidity. However, certain pests and diseases are making heavy loss to the crops in the region which been to be tackled scientifically. However, some important insect pests and diseases, which have been noticed in Ladakh, are discussed below.

Insect Pest of Fruits

The area under fruit production is given in Table 18.1. According to Dar *et al.* (1998) the fruit crops grown in Ladakh includes apricot (*Prunus armeniaca*), apple (*Malus domestica*), pear, (*Pyrus communis*), peach (*P. persica*), plum (*P. domestica*), cherry (*P. avium*) almond (*P. amygdalus*), strawberry (*Fragaria* sp.), walnut (*Juglans* sp.), mulberry (*Morus* spp.) and oleaster (*Elagenous hortensis*) etc. Most of these fruits are under cultivation since long back. These fruit plants are attacked by some insect which affected the production. Some of the major pests, which occur in Ladakh, are given below:

Codling moth (*Cydia pomonella* L.)

The codling moth is a major problem on fruits in the temperate area. It has also been reported from Europe, USA, Canada, South Africa, Australia, New Zealand, Afganistan, Pakistan and Russia (Sheldeshova, 1967; Anonymous, 1967). In India its distribution is restricted to Ladakh region of J&K state (Malik *et al.*, 1972; Wadhi and Sethi, 1975). The pest is believed to have introduced to Ladakh from North West Frontier province of Pakistan. In Pakistan, it is reported as a serious problem on deciduous fruits (Janjua, 1938; Pruthi, 1938; Janjua and Samual, 1941). This pest has remarkable ability to adopt a wide range of climate condition and being able to synchronize its life cycle with the fruiting period of its host.

In India, its distribution is restricted to Ladakh region of J&K state. Though in India, apple has been reported as main host of this pest but it has been found that the pest has now extending their host range to apricot, pear and walnut. Due to their occurrence in Ladakh region only and its great adoptability, the Government of J&K has imposed the domestic quarantine on the transportation of apple and apricot from Ladakh under J&K Plant Disease Act, 1973 to avoid the entry of this pest to other nearby apple growing areas of Srinager and Himachal Pradesh.

The adult female codling moth is approximately 10 mm long and grayish in colour. The male is lightly smaller than female. Main identification mark is dark gray colour near the base of wing. Initially larvae are cream in colour and later turn into light pink with brown head. Fully grown larva is 10-15 cm in length. The adult emergence takes place in the month of June. The female moth, after mating, start egg laying on fruits, leaves and twigs. After egg hatching larva come out and bore into the fruits, generally from the blossom on calyx end. The larva becomes mature in 21–30 days depending upon the climatic conditions. Mature larva come out from the infested fruits and from fallen fruits, it spins a silken cocoon and pupates inside. The pupal period lost 8–14 days. The duration of larval, pupal and time of adult emergence depends upon the temperature and humidity. Newly hatched larvae feed on the leaf for a while then burrows inside the fruit and feed on pulp. The infested apple becomes bright in colour and rips prematurely. These infested fruits cannot be marketed.

Control

1. Use of light trap is useful step to monitor the codling moth. Collection should made from the light trap and kill them.
2. Sex pheromone trap which act as mating disruptant, is effective in collection and destruction of female which reduced the further population built-up.
3. Dead barks of old trees provide shelter to hibernate the larvae during winter, so scrapping of dead bark and removal of dead branches is found effective to minimize pest population.
4. Bur lapping of apple trunk, before 10-15 days harvesting by using gunny bags, at the height of 2–3 feets with weekly collection and destruction of hibernating larvae found effective method.
5. Collection and destruction of infected fruits, it may fallen or still on tree, is useful.
6. Release of *Trichogramma embriophagum* has been found effective.
7. Two or three spray of insecticides like Chloropyrifos 20 EC (0.025 per cent) or Endosulfan 35 EC (0.05 per cent) at 10–15 days interval from peanut stage (fruits size 1.0 to 1.5 cm in diameter) is found effective.

The Gypsy moth (*Lymantria obfuscata* Walker)

The gypsy moth is the major distractive defoliating insect for fruit as well as for forest plantations in Ladakh. It attacks on apricot, willow, poplar, apple etc. in Ladakh region and also in Kashmir province (Masoodi, 1991).

The female moth has well developed wings but it cannot fly. She compensates this want by releasing a strong sex attractant from her abdomen that lures male moths to her location. After mating, the female deposits eggs in mass in months of June–July and dies shortly thereafter. An egg mass contain 75 to 350 eggs. After 3–10 days, eggs are hatched and larvae come out which are puff colour later turned into black. In favorable conditions, larvae leave egg mass and climb upon the tree in response to overhead light. They produce sulk from grand and lay down a trail of silk continually as they move. The larva pass through five instars and each larval instar losts 9 to 13 days and completes the development in 66–100 days depending upon the host and weather condition. Last instar larva pupates in the soil among the debris which almost 9 to 21 days. The longevity of male and female moths are 4 to 10 days and 11 to 31 days, respectively. One generation in a year has been recorded in Ladakh. The caterpillars are nocturnal in habit and feed gregariously on apple. They defoliate host tree completely which result in failure of fruits formation. Defoliation is higher in forest trees willow and poplar.

Control

1. Collection and destruction of egg mass at the time of pruning (from December to January) is effective method. This insect also lay their eggs in house roofs, debris and cracks so collection and destruction should made from these place also.
2. Foliar spray with 200 ml Endosulfan 35 EC or 500 ml Carbaryl 50 WP in 200 liters of water per acre is effective.

Blossem Thrips (*Thrips* spp.)

It is completely cosmopolitan insect that attacks a wide rang of fruits, vegetables and ornamental plants in Ladakh. It attacks on floral parts of apple, apricot including roses. It is very small, yellow, pale brown and black selendrical insect that has been noticed in the flowers and flower buds. The nymph are selender, yellow looks like adult but are wingless and slightly smaller than adults. It attacks on flower and buds. Infested flowers gives sickly and faded look. Adult and nymph lacerate floral parts resulting in brown spots lesion on *androecium* and *gymnasium* and deteriorates the fruits quality and fruits set. Damage on delicate leaves cause necrotic symptom and produce silvery appetence.

Spraying of Fenitrothion 50 EC (0.05 per cent) or Chlorpyriphos 20 EC (0.02 per cent) at early pink bud stage or 7–15 days before flowering is effective to control this pest. Do not spray when the flowers have opened.

Chaffere beetle (*Adoretus* spp.)

Beetle is polyphagous in nature and attacks a large number of fruits. It appears in the month of May-June and feed on foliage and developing fruits at dusk. It lays eggs in the soil near the plant. At night, only adult beetle cause damage by feeding on leaves, which get skeletonized. The maximum activity of beetle is recorded at slightly higher temperature. Larvae after hatching from eggs feed on soil humus, roots of grasses and other vegetations.

To control the beetle, collection and destruction of adult through using light trap is effective method. Spraying of Carbaryl 50 WP (0.01 per cent) in May-June, when beetle are observed in orchard at dusk, is also beneficial.

Two spotted Spider Mite (*Tetranychus* spp.)

It is a minor pest of apple and apricot in Ladakh. However, in severe infestation foliage become pale or bronzed in appetence. Damage on apple and apricot foliage by feeding on green matter and scraping cause loss of chlorophyll, bronzing, fallowed by leaf drops and weakening of buds. Feeding results in poor size and quality of fruits. Infestation of mite occurs some part of Ladakh region like Khaltsi block particularly on apricot plants.

To control the pest, spraying of Dicofol (0.05 per cent) or wetable sulfur (0.25 per cent) at bud stage and later when the mite population exceeds 15–20 per leaf, is advisable.

Apricot Chalcid (*Eurytoma samsonovi*)

Apricot chalcid is minor pest in Ladakh. It is a very small wasp. Female of this wasp lay their eggs inside the young fruits and grubs feeds on redeveloping seeds. Pupation occurs inside the fruits. To control this pest collection and destruction of fallen fruits is an effective methods.

Peach leaf curling aphid (*Brachycaudus helichrysi* Kalt.)

This pest is distributed in most of hilly area and is serious pest in Kullu valley, Simla hills and submounting regions of Uttaranchal. It attacks on a large number of fruit plants like peach, plum, almond, apricot and other temperate fruits. In Ladakh, it has been noticed on apricot which is the most important fruit. Nymphs are dark green in colour and adults are green that feed on leaves. It has also been reported that colour of this insect varied according the to host on which they feed. It appears in May-June. Nymph and adult suck the cell sap from leaves and affected leaves turn pale and curl up, blossom wither and fruit do not developed into normal size and drop prematurely.

To control the pest one pre bloom and a post bloom spray of 200 ml Monocrotophos 40 EC in 250 liters water (at 15–20 days interval) per acre has been effective.

Insect Pests of Vegetable Crops

There are more than 63 types of vegetables grown in the different regions of the cold arid region from May to October. Vegetable crops are among the most remunerative cash crop for the farmers of this region. However, the productivity of vegetable crops have influenced by a large number of factors including insect pests. Some of the important pest of vegetable crops are listed below:

Cabbage Butterfly (*Pieris brassicae* and *P. rapi*)

The larvae of butterfly are the serious pest for all the cole crops growing in Ladakh. It attacks on cabbage, cauliflower, knol-khol, radish, sarson, toria and other cruciferous plants. Female lays 5–250 yellowish-orange colour conical eggs in cluster on upper or lower surface of leaves. The eggs are hatched after 3–7 days. Early stage of larvae feed gregariously in groups later dispersed on same plant. The larvae mature in 14–21 days depending upon the temperature and humidity. The larvae pupate away from host. In Ladakh, pupal period goes under dormancy and adult emergence takes place in the month of April–May after winter. Damage caused by larvae that webs the leaves with silken strands and feed on the lower surface of the leaves.

For control this pest, hand picking and mechanical destruction of eggs and caterpillars during early stage of infestation may reduce the population. Spraying of Malathion 50 EC (0.03 per cent) has also been found effective to control the pest.

Diamond-back Moth (*Plutella xylostella* L.)

It is cosmopolitan pest that has also been noticed in Ladakh. It feeds on cabbage and cauliflower in Ladakh. The full grown larvae are pale yellow-green with fine black hairs all over its body. The adult moth is brown in colour with conspicuous white spots on the fore wings. The female lays minute yellowish white eggs singly under surface of leaves. The freshly hatched larvae crawl to the undersurface and penetrate the epidermis and mine in the tissues. Later three instar feed underside of leaf and make holes. Total larval period ranges between 10 to 20 days depending upon the temperature and humidity. Pupation take place inside the silken cocoon which is open at both the ends. Total life completes within 2–3 weeks. In Ladakh two to three generations in year has been recorded.

The larva causes damage to cabbage and cauliflower. However, it is not a serious pest in Ladakh. The freshly hatched larva feeds by scrapping the leaf tissue and as they grow older make wholes in them. Removal and destruction of stubbles and debris after harvesting of the crop is effective as preventive measure. In severe infestations, application of Malathion 50 EC (0.05 per cent) or Endosulfan (0.05 per cent) is effective to control the pest.

Cabbage Aphid (*Bravicoryne brassicae*)

It is a serious pest of cole crops grown for vegetable as well as for seed production. It causes 30 to 40 per cent loss to cabbage and cauliflower crops in Ladakh. Besides cabbage and cauliflower, it also attacks on beet leaf and spinach *which is winter leafy* vegetable. Nymph and adult aphids attack on all plant parts above the ground level. In severe infestation, aphid cover whole curd or head and make them inconsumable. This pest also invites secondary infection by honey-dew secretion. Destruction of highly infested plants is effective as preventive measures. Spraying of Malathion 50 EC (0.04 per cent) or Endosulfon 35 EC (0.05 per cent) has also been found effective to control the insect.

Gram Cut Worm (*Agrotis ipsilon* Hufn.)

It is a widely distributed pest and has been reported from Burma, Sri Lanka, Japan China, Hawai, Indonesia, New Zealand, Australia and many other countries including India. In Ladakh, it is a major problem for successful cultivation of potato, cabbage, peas, cauliflower, brinjal and onion crop. The wide range of hosts in Ladakh shows polyphegous nature of the pest.

This pest is active from June to September. The full grown larvae is dark brown in colour and greyish body. The adult has dull brownish fore wing with numerous wavy lines and spots. Female lays 150 to 250 creamy white dome shaped eggs in cluster under surface of leaves of food plants or in the soil. Larval period ranges from 35 to 45 days. Mature larval makes earthen chamber in soil and pupate inside. The insect completes one to two generation in Ladakh.

In Ladakh, it damages the plants at seedling stage. The worms feed on seedling at ground level and feed less and cause more damage. They are active in dry season and in light soil which is common in Ladakh. This pests hide under the soil during daytime and attack during night or morning hours. Application of Carbofuran 3 G at the rate of 30 Kg per hectare at the time of transplanting has been found effective to control this pest.

Tomato Fruit Borer (*Heliicoverpa armigera* Hubner)

This insect has worldwide distribution. In Ladakh, it is a minor pest which causes 5–10 per cent loss to tomato crops. The full grown larval are greenish with dark broken grey line on the side of body. The moth is large, brown in colour with dull black border on the hind wing. The larva is polyphagous in nature which attack on many crops like potato, cowpea, tomato and gram in Ladakh. They first feed on leaves and fruiting bodies and later on bore into fruits and pods (pea). They completely eat away the internal contents that made the produce unfit for marketing. For the control of this pest, spraying of Endosulfan 35 EC or Monocrotophos 40 EC at the rate of 0.03 per cent is effective.

Pea Pod Borer (*Etiella zinckenella*)

In Ladakh, this is important pest of pea. The damage by this pest varied in different regions of Ladakh. The infestation of pest is higher in warmer belt of Ladakh than cooler areas. It causes 2 to 10 per cent damage to green pea. The moth is nocturnal. Female lay eggs singly on various part of plant including developing pod. Larvae after emergence feed inside the pods and become fully grown in 10 to 25 days. Pupation takes place in soil. The actual damage is caused by the larvae which feeds on floral parts, pods and on developing seeds. Owing to feeding on developing seeds the reduction in yield goes upto 5 per cent in Ladakh. Spraying of systematic insecticides is effective to control this pest.

Onion Thrips (*Thrips tabaci* Lind.)

It is a cosmopolitan pest found throughout India. In Ladakh, it is a major threat for successful cultivation of onion. Besides onion, it also attacks on tomato, cabbage, apple etc. The nymph are selendrical in shape, brown in colour and looks like adult but they are wingless and slightly smaller than adult. The males are wingless however, the females have long narrow strap like wings. This insect becomes active from June to mid October depending upon the weather conditions which influence the length of the cropping season. They are generally congregated at the base of leaf or in the flower of onion. In cabbage and cauliflower, they feed on lower surface of leaves. The damage done by both nymphs and adults. The pest infests the crop throughout the cropping season. Attack of this pest results spotted appearance on leaves and later pale white patch gives slivery appearance due to drainage of sap. Due to drainage of the sap leaf tips becomes fad and gradually dries up. Infestation leads to production of low quality of bulbs and no seed setting in the flowers.

Spraying of Malathion 50 EC at the rate of 0.02 per cent with waiting period of at least 12 days before harvesting is effective to control this pest.

White Grub (*Holotrichia consanguinea* Blanchard)

It is the most destructive pest of cold arid region (Ladakh) of J&K. This pest attacks on cereal (wheat and barley), vegetables (brinjal, cabbage and chilly and capsicum etc). It has also been seen that some species of this pest attacks on forest plantation.

The grubs are "C' shaped and white colour having broad head and prominent legs. The adult beetles are dull brown or black in colour. Larval development completes in 50–70 days depending upon weather condition. Palpitation takes place in the soil. The grubs and beetles both damage the crops. The grubs eat away nodules, fine roots and may girdle the main root thereby leading to killing of plants. The damage becomes evident only when the entire plant dries up. Symptom can easily seen in crops like brinjal, cabbage, chilly and capsicum. It can be control by exposing larvae to predator by giving first deep ploughing. Collection and destruction of larvae is also effective preventive measures. Application of Carbofuran 3 G @ 10 kg per hectare in the soil is effective to control the insect.

Onion Maggot (*Hylemya antiqua*)

This is another pest of onion in Ladakh region, which is found in different onion growing area of Leh and Kargil district. The adult looks like housefly while maggot are small white in colour and devoid of legs. After hatching, maggots crawl upto the plant and enter inside the leaf sheath and reaches to the bulb. The attacked plant become flabby and yellowish. The maggots bore into the bulbs leaving outer sheath. It destroy 5–10 per cent of the crops. Infestation is not completed destruction to the bulb but its leads subsequent rotting in the storage. Affected bulbs rot in storage as infection leads to secondary infection by pathogenic organisms.

Adoption of crop rotation is useful to avoid the attack of this pest. In severe infestation, spraying of Malathion 50 EC at the rate of 0.05 per cent or application of Thimet 10 G in soil at the rate of 15 kg per hectare before sowing is effective.

Insect Pests of Cereal Crops

White grub, nematodes and thrips are major pests problem of cereal crops. Thirps and aphid suck the sap from leaf and ear while white grub attacks on fine roots of cereals, which result in drying of wheat and barley plants.

Insect Pests of Oil Seed Crops

In Ladakh, mustard in grown for oil purpose. Aphid, *Lipaphis erysimi*, and *B. brassicae* and painted bug are major pest problem. These insects suck the sap from plant and siliqua which reduce the yield. In Ladakh 5–10 percent yield loss has been recorded by these pests.

Insect Pests of Forest Plantation

Poplar blotch minor

Poplar is one of the most important tree found in cold arid region Ladakh. It is used for preparation of house roof and as timber. Among the large number of insect pests, poplar blotch minor is one of the important pest. This microlepidopterous insect cause damage to plant by mining the leaf and producing the blotch symptom. The female of this microlepidoptera lays eggs on lower surface of leaf which looks like drops of water. After hatching larva start feeding below the epidermis. It has been observed to be 3–30 larvae in one leaf. In severe infestation, whole leaf is infested with this pest and plant show light yellow appearance instead of green healthy look.

Willow scale insect (*Chionaspis salicis*)

Willow scale insect is major problem for willow in Ladakh. About 20–35 per cent willow plantation has been recorded to be infested with this scale insect. The infestation is varied in Ladakh according to the altitude and habitat. Some of the areas like Leh (low altitude), the infestation is higher while Neoma area (higher altitude) infestation is very low. The willow scale insect attacks on junction of branches which emerged from the main trunk. The average colony size range from 1.5 to 3 feets. This pest causes damage to willow tree by sucking the sap from the branches which results in reduction of wood and severe infestation leads to bark cracking and even death of the tree. This pest is active from June to September and rest of the time it remain in dormancy in egg stage.

To control this pest, at appropriate application time of insecticide is very important. Because after hatching, crawlers move for some time in search of suitable site. After that they settled down and start sucking the sap and secret wax from body which covers the insect and protect them which makes application of insecticide necessary just after hatching of eggs.

Willow leaf beetle (*Altica* sp.)

Willow leaf beetle is one of the most destructive pest of forest plantations in Ladakh. The adults are shiny blue in colour with a size of 4–8 mm in length while the grubs are black in colour. Adults and grub defoliate the willow trees leaving behind skeletoniged leaves. The grub feeds on green portion of leaves leaving veins. The infestation caused by this beetle varied in according to altitude. In warm and low altitude areas like in Khaltsi block, infestation is more as compared to cool and high altitude areas like around Leh.

Besides these forest pests, aphid (suck the sap from twigs), ermine moth (*Yponomeuta rorellus*) gypsy moth (by defoliation the plant), trunk borer (*Cossus cossus*) and gall forming insects are also important insects which cause the damage to the forest plantation.

Diseases of Different Crops in Ladakh

There are several diseases which attack on wheat in cold arid region. They cause substantial reduction in yield and deteriorate in quality of the grain (Sharma, 2002). Following major diseases noticed in Ladakh.

Yellow Rust of Wheat

It is one of the most important diseases of wheat in Ladakh. Rusts are caused by three species *Puccinia recondita* (brown rust) and *P. striiformis* (yellow rust). However, black rust has not been recorded in Ladakh. Brown rust occasionally make its appearance some time in sporadic but not always below economic level. Yellow rust is more regular in appearance some time in epidemic forms. For instance, in 1999 it was in epidemic form in Leh district and caused upto 80 per cent crop failure in the region. Due to epidemic attack, crop gives yellow look. Infection occurs on leave, but in severe conditions, it establish rapidly on leaf sheath, stalk and developing grains as well.

To control the disease growing of rust resistance wheat variety, application of potash which increase the resistance in crop and periodical spraying of Zineb, Maneb or Prazate (0.025 per cent) is effective measures.

Hill Bunt of Wheat

Recently this disease has been proved to be of economic significance. Its out brakes recorded in Nubra valley of Leh district and also in some part of Kargil district. There are two pathogen *viz. Tilletia caries* and *T. foetida* founed in Ladakh region which causes the disease in wheat. Symptom of this disease shows filling of powdery mass of black spore in spike which is easily detectable in field. Flour from infected produce gives fishy smell and is unfit for consumption. It is a soil born disease and thus is difficult to control. However, some measures like growing of resistance varieties, used of disease free seeds, treatment of seeds before sowing (with 2.5 g Agrosan GN or Vitavax 2 g per kg seed), avoidance of regular cropping of same crop are useful measures to minimize the disease.

Loose Smut of Wheat

Loose smut disease of wheat is caused by *Ustilago nuda tritici*, is prevalent in Ladakh. It gives sporadic appearance in all wheat growing field of Ladakh. The characteristic of this disease is black powder filled in place of developing grain. The black powder is infective (smut) part of fungus. Infective ear comes out earlier than those of normal ear.

Use of resistant variety, seed treatment and rouging of dried ear are some effective method to control the disease.

Ear Cockle *Anguina tritici*

Ear cockle disease of wheat has been reported from all wheat growing areas of Ladakh. It is caused by nematode, *Anguina tritici*. Severity of this disease depends upon the weather condition. In favorable conditions it causes economic loss. Plant infected with this nematode can be identified very easily with curved look and relatively smaller ear than normal. In place of grain, dark brown gall of nematodes occur. This disease spreads by seeds in which nematode is found in gall. To control this disease, rouging of infected plants, use of resistant variety (Kailash) and removal of infective body from seed are important measures.

Yellow Rust of Barley

Among the rusts *viz.* black, brown and yellow, only yellow is devastating while other two have not been reported so for in Ladakh (Sharma, 2002). The entire land races of barley are highly susceptible to yellow rust (Sharma *et al.*, 1999). This disease is caused by *Puccina glumarum*. The yellow coloured uredia appear on leaf blades and sheaths where as in severe infection it also attack on ear and even on developing grain. Infection results in production of shrivelled grains.

Use or resistant variety like Nurboo (SBL-4) and Sindhu which is found moderate resistant against yellow rust and spraying of Dithane Z-78, at the rate of two kg per hectare in 700–800 liter water at 15 days interval have been found effective.

Leaf Blight or Leaf Spot of Barley

This disease is caused by *Helminthosporium graminerum* which is widely distributed in Ladakh and cause significant economic loss in hulled and hull-less barley. Comparatively in hulled barley loss in more than hull-less. The loss caused by this disease has been workout under Ladakh condition but its severity is at second rank (Sharma, 2002). Low temperature with high humidity are conducive for severe infection. The symptom looks from tillering stage and onward. Severe infection of disease cause stunted growth, restricted spike emergence, shriveled poor quality of grain with less number.

Seed treatment through Agrosan GN @ 2g per kg seed and socking of seeds in tape water for 4 hours followed by sun drying for 7–8 hours is effective to kill the pathogen. In severe conditions two spraying of Dithane Z-78 at the rate of 2 kg in 1000 liter of water at 10–15 days interval is also effective.

Loose Smut of Barley

Loose smut of barley is caused by *Ustilago nuda*. This disease commonly observed in local variety of barley grown in Ladakh. In favorable condition, this disease caused 12 per cent incidence in local varieties. It is a seed born disease. The hyphe remain dormant in embryo which become active after sowing of infected seed. After sowing hyphe grow with growing point of the plant and cause infection in ear. Infected ear comes out earlier than healthy. Fungus produces infective spore in spikelets which covered by silvery membrane. After maturity, the membrane raptures before emergence of ear and release spore for further infection.

To control the disease, seed treatment with Vitavex or Bavisten @ 2–5 g per kg seed in effective. Use of resistant variety and rouging out the ear head before rapture of membrane is useful in reduction of disease in further cropping season. Besides these measures hot water treatment of seed before sowing is effective to avoid the presence of inoculum in seed.

Covered Smut of Barley

This disease is caused by *Ustilago hordei* and widely distributed in all cereals growing area of Ladakh. The loss caused by covered smut has not been studied so far. The disease is distinguishable from loose smut that head emerged out from the boot and black masses of spore in spikelts remain enclosed in rather breaking of membrane. The spores released during threshing operation and adhere to healthy seeds that act as further inoculum. To control this disease, seed treatment with 0.2 per cent Vitavex and adoption of resistant varieties are effective measures.

Wilt of Pea

The wilt of pea is very common diseased in Ladakh caused by *Fusarium oxysporium* f. *pisi* and spread through infected soil. The pathogen fallow water conducting tissue and result in blockage of water movement. Affected plant show yellowing and down ward curling of leaves. The affected seed, plant debris and soil carry the disease for next cropping season's infection. Adoption of crop rotation, growing resistant varieties and seed treatment are some effective measures to minimize the further attack.

Powdery Mildew of Pea

It is a seed born disease caused by *Erysiphe polygoni* in pea. This disease is noticed in all pea growing area of Ladakh. In warmer belt of Ladakh the disease is very prevalent. It is characterized by white powdery patches on both side of the leaves. In severe conditions it extent to tendril and stems. For spreading of this disease, hot dry weather is useful. Use of resistant variety, hot water treatment of seeds before sowing and field sanitation are effective practices to control the disease.

Early Blight of Potato

This is a major disease of potato in Ladakh which is caused by *Alternaria solani*. Symptom of disease first appears in small isolated light brown spots on leaves which later enlarge and turn into dark brown forming a rings of dead tissue. To control disease, burning of debris is effective way to avoid the further loss. In severe conditions, spraying of Dithan M-45 @ 2.5 kg per ha in 1000 liter water at 15–20 days interval is effective.

Damping off

This fungal disease is very common in Ladakh which found in crops like tomato, cabbage, cauliflower, chilly and capsicum etc. This disease is caused by fungus known as *Pythium* spp. It occurs in two phases *i.e.*, first pre-emergence damping off in which seedling are rotten before the emergence from soil. In post-emergence stage damping off the seedlings or freshly transplanted plant collapse in the field. Higher moisture and low temperature is conducive for damping off.

Late Blight of Potato

This disease has been observed in potato and tomato in Ladakh. It is caused by *Phytopthora infestans*. Leaves are attacked first subsequently infection spread to stem. Brown to purple black lesion enlarged in irregular shape caused the blight. High humidity and low temperature favour the disease spread.

REFERENCES

Dar,G. H., H. Wani, M.Y. Zargar and M.S. Mir (1998). Crop diversity in cold arid region of J&K. In: *Environment, Biology, Variety and Conservation*, (Eds.) M.A. Khan and S. Farook, University of Kashmir, pp. 45.

Janjua, M.A. (1938). Codling moth in Afghanistan. *Curr. Sci.*, 7: 125.

Janjua, M.A. and C.K. Samuel (1941). Fruit pests of Baluchistan, *ICAR Miscellaneous Bulletin*, 42: 1–41.

Kachroo, P., B.L. Sapru and U. Dhar (1977). *Flora of Ladakh: an Ecological and Taxonomical Appraisal.* International Book Distributors, Dehradun, pp. 100.

Malik, R.A., A.A. Punjabi and A.A. Bhat (1972). Survey and study of insect and non-insect pest in Kashmir. *Horticulturist (J&K)*, 3 (1-3): 29–44.

Masoodi, M.A.(1991). Biological studies on *Lymantria obfuscata* Walker (Lepidoptera: Lymentriidae) in Kashmir. *Indian Forester*, 117 (8)-: 644–651.

Pruthi, H.S. (1938). The distribution, status and biology of codling moth (*Cydia pomonella* L.) in Baluchistan with notes on some other insects infestion in apple. *Indian J. Agric. Sci.*, 9: 499–547.

Sharma, J.P. (2002). *Crop Production Technology of Cold Arid Region*. Kalyani Publication Ludhiana pp. 2.

Sharma, J.P., A.A. Mir and V. K. Yadav (1999). High yielding barley varieties for Ladakh region of J&K, *Indian Wheat Newsletter*, 5(2): 4–5.

Singh, B., S.K. Dwivedi, N. Singh and Eli Paljor (1998a). Sustainable horticulture practices for cold arid areas. Paper presented in international symposium on sustainable agriculture in hill areas. 29–31 Oct. 1998. HPKVV-Palampur. Abstract, pp. 35–36.

SKUAST, Leh (1998). A note on organization contributory towards agriculture development in Ladakh. SKUAST, RARS, Leh, pp. 19.

Wadi, S.R.A. and G.R. Sethi (1975). Eradication of codling moth- a suggestion. *J. Nucl. Agri. Biol.*, 4: 18–19.

19

Pest and Diseases Management in Bulbous Flowering Crops

A.K. Singh and Shiva Jauhari
Department of Horticulture, College of Agriculture,
G.B. Pant University of Agriculture and Technology, Pantnagar – 263 145, Uttaranchal, India

Internationally, bulbous flowering crops have got a pivotal place among the cut flowers. The importance of this group can be judged by the fact that their share in export from India was around Rs. 28.23 crores of the total export of Rs. 990.25 crores during 1998-99, which is about 2.85 per cent. The area of the cultivation of bulbous flowering crops had increased over 3,500 ha during 2000-2001 (Misra and Pathania, 2002). The wide range of climatic conditions in India provides a choice of various bulbous crops to grow them in different parts of country. Several bulbous crops like alstroemeria, dahlia, eucharis, gladiolus, halmanthus, hemarocallis, hippeastrum, lilium, tuberous begonia, tulip can easily be grown in the hills, while calla lily, tuberose are habitat to tropical region.

Although, this group provides a kind of flowers but the crops are infected with various pest and diseases at field level as well as during storage. Several control measures are available against the different pest and diseases. It is better to use an integrated approach to control them rather to apply a single control measure. This would not only directly be beneficial to increase the cost benefit ratio but indirectly it also leads to sustainability. The following content includes the description of major insect pest and diseases and their integrated management of some of the popular and commercially important bulbous flowering crops *viz.*, alstroemeria, amaryllis, hippeastrum, cyclamen, dahlia, gladiolus, lilium and tuberose.

Alstroemeria

Diseases

Root rot (*Pythium* sp. and *Rhizoctonia* sp.): *Pythium* sp. causes root rot by attacking juvenile or succulent tissue as the root tip, while *Rhizoctonia* sp. affects both, small and mature plants. After getting entrance to the root, the fungus may cause a rapid, black rot of the entire primary root and may even move up in to the stem tissues. Under continuous wet conditions more and more roots are killed and plant may wilt, stop growing or even collapse and die. Bulbs turn black, gradually desiccate and form a hard mummy.

By providing good drainage and water management and sterilization of growing media by steaming at 60°C for 30 min or chemically by Mefenoxam or formaldehyde at 2 per cent, root rot can be controlled easily. Drenching with copper fungicide can also be effective. Removal of diseased plants will reduce the further occurrence of disease. Use of properly composed pine bark at 20 per cent in potting mixture is reported to control *Pythium* sp. One mycoparasite, *Gliocladium virens*, is normally used as a *Pythium* biocontrol agent. The chemical Fosetyl-al (Aliette) applied at the rate of 3.0–4.0 g/l of water as a foliar spray.

Grey mold (*Botrytis cinerea*): This problem occurs when humidity is high. *Botrytis* does not affect the healthy green tissues unless an injured or dead area is present or it grows directly from fallen petals or leaf. The fungus sporulates and produces mass of grey or brownish spores.

The old flowers and leaves should be removed from field that may be the source of fungus. Humidity should be kept below 85 per cent by heating and ventilation. Overhead watering during blooming should be avoided. If it is the only method, irrigation should be done in the daytime. Wider plant spacing to increase the ventilation can also help to reduce disease incidence. Spraying of Fenhexamid at 1 g/litre of water could be effective after infection. Mancozeb 75 W at 1 g/litre of water should be sprayed just after infection.

Insect Pests

Aphids

They attack on young tissues and suck the sap from the growing shoots and buds. Thus they make the plant weak. Their exudates cause the secondary effects that result in fouling of stems and leaves with honeydew, this leads to the growth of sooty molds.

When the insects are seen, spray of Metasystox or Monocrotophos at 2 ml/litre water can be done to control the aphids. Some natural enemies of aphids like aphid midge (*Aphidoletes aphidimyza*), convergent lady beetle (*Hippodamia convergens*), lacewings (*Chrysoperla* spp.), microbial insecticide (*Beauveria bassiana*), minute pirate bugs (*Orius insidiosus* and *O. tristricolour*) and parasitic wasps (*Aphelinus mali*, *Aphidius* spp. and *Diaeretiella rapae*) are used to control their population.

White fly (*Trialeurodes vaporariorum*)

It damages the crop at adult and scale stages by sucking the sap. Adults, eggs and scales can be seen on the under side of leaves. The upper surface is usually covered with sticky honeydew that fouls and causes the sooty molds.

When the white flies are visible, spraying of Uttracid (0.1 per cent) or Actellic (0.075 per cent) twice can be done to control them. White flies can also be controlled by spray of Malathion, Rogor or Dimethoate. The use of yellow-orange plexiglass sticky traps was reported to control the insect (Nucifora and Vacante, 1981). This insect is biologically controlled by using natural enemies of white fly. They are lacewings (*Chrysoperla* spp.), microbial insecticide (*Beauveria bassiana*), parasitic wasps (*Encarsia, Eretmocerus* spp.) and predaceous lady beetle (*Delphastu pusillus*) and reported to control of white fly population.

Red Spider Mites (*Tetranychus* sp.)

The first symptom of their damage shows a very light spackling of pale yellow colour on the copper surface of leaves. In extreme conditions leaves get discoloured and show bronzing.

The mites can easily be controlled by spraying of Kelthane (0.2 per cent) or by Dimecron (0.1 per cent). Initially, increasing the humidity in greenhouse can also control population of red spider mite. The sprinkling of water washes away the insects from the lower surface of leaves.

Thrips

The insect causes the tomato spotted wilt disease, which caused by virus and damages the crop in the form of many yellow green patches on the leaves.

The control of mites is difficult in this crop. However, the spray of Nuvacron (0.04 per cent ai), Rogor (0.2 per cent) and Malathion (0.2 per cent) can be effective to control them. *Chrysoperla lucasina* has been found as an effective predator of *Thrips tabaci* in alstoemeria (Maisonneuve and Arrec, 1999).

Nematodes

Lesion nematode (*Pratylenchus penetrans* and *P. pratensis*)

This nematode feeds moving about inside the root. They form lesions as a result of cell death. This causes the lose of vigour and growth that ultimately leads to reduction in quality and yield of flowers. Small light brown colour lesions can be seen over the stems along the length.

Healthy plant material should be used. Carbofuran should be applied at 2 kg ai/ha to control this nematode.

Amaryllis and Hippeastrum

Diseases

Red leaf spot (*Stagonospora curtisii*)

This fungal disease starts in the form of spots on leaf that turn yellow and fall off (Leeuwen, 1974).

The bulbs should be subjected to warm water treatment at 43.5°C for 2 hours. Application of Zineb and Ferban at 2 g/litre of water is more effective than Captan and copper fungicides. In field, dipping of Aagliton (1 per cent) and Benlate (0.2 per cent) was found beneficial to control this disease. Dipping of hippeastrum bulbs of cultivars Diana and Masai, in Prochloroz (0.4 per cent) for 15 minutes gave better control of this disease than with a mixture of 0.2 per cent Benlate and 0.4 per cent Aagliton (Conijn, 1983).

Soft Rot (*Botrytis cinerea, Rhizopus* and *Sclerotium rolfsii*)

This disease appears in high humidity during storage of bulbs. Red brown sunken lesions are present on the bulbs.

The bulbs should not be placed under direct sun. Hot water treatment should be recommended along with dipping in Benlate (0.5 per cent) solution for 15–30 minutes. Some other fungal diseases are bulb decay (*Sclerotium rolfsii*), grey mold (*Botrytis cinerae*), root rot (*Pythium debaryanum*), black mildew (*Asterinella hippeastri*), leaf spot (*Cereospora amaryllidis*), brown leaf spot (*Aecidium amaryllidis*) and anthracnose (*Septoglocum amaryllidis*).

All these diseases can be controlled by the treatments used in controlling red leaf spot diease of amaryllis. These diseases is bulbs can be controlled up to some extent by harvesting of bulbs at full maturity, providing good ventilation during cleaning and curing, subjecting to treatment with copper fungicides or Dithane M-45 and storing them at optimum temperature and humidity.

Cucumber Mosaic Virus (Amaryllis Mosaic)

The symptom of this virus appears in the form of large yellowish green patches on the surface of leaf. The group of virus is cucumovirus and transmission takes place through seed and aphids. ELISA has successfully been used to detect this virus in amaryllis (Raizada *et al.*, 1993).

To control the vector Metasystox at 1 ml/litre water can be sprayed. Spraying of Malathion (0.2 per cent) in a regular interval of 15 days is also found effective.

Tomato Spotted Wilt

This viral disease is very similar to cucumber mosaic virus but is coarser. The virus is transmitted through thrips.

The vector can be controlled by spraying of Malathion or Rogor (0.2 per cent). Nuvacron (0.04 per cent ai.) is also gave the effective result. The biocontrol measures like greenhouse thrips parasite (*Thripobius semiluteus*), lacewings (*Chrysoperla* sp.) are used to control vector population.

Insect Pests

Caterpillar (*Polytela gloriosae*)

The extent of damage of caterpillar is so high that they eat all the leaves within 4–5 days. The insects attack the plants during June-August.

Malathion, endosulfan and parathion (0.2 per cent) are very effective insecticides that control the caterpillar within 48 hrs. Some egg parasites like *Trichogramma* spp. and entomopathogenic nematodes (*Steinermema carpocapsae* and *Heterorhabditis bacteriphera*) are used as biocontrol measures.

Thrips (*Taeniothrips simplex*)

Thrips suck sap from flowers, leaves and stems and cause damage in terms of small size of flowers.

Thrips can be controlled by soil application of Aldicarb (Temik) at 5.50–8.50 g ai/m^2 area, Thimet at 4–10 kg/ha or soil application with Furadan (Carbofuran) at 12–30 kg/ha can be effective to control thrips. However, some parasites of thrips like microbial insecticide (*Beauveria bassiana*), minute pirate bugs (*Orius insidiosus* and *O. tristicolour*) and predatory mites (*Amblyseius, Euseius, Iphiseius, Neoseiulus* spp. and *Hypoaspis mites*) are reported for controlling thrips.

Aphids, mealy bugs and scale insects: These insects infect the plant by puncturing and sucking the sap. Mealy bugs mainly affects the foliage and bulbs.

These insects can be controlled by the spray of nicotine sulphate (3 g/liter water) and soap chips at 5 g/liter of water or Malathion (0.2 per cent) at 10 days interval. Mealy bugs can be controlled by natural predators like citrus mealy bug parasite (*Leptomastix dactylopii*), lacewings (*Chrysoperla* spp.), mealy bug destroyer (*Cryptolaemus montrouzieri*) and microbial insecticides (*Beauveria bassiana*).

Mites

Various kinds of mites attack on the plant and they suck the sap from the lower surface of leaf or the petals. Such mites are spider mite (*Tetranychus bimacultus*), *Tagonemus laticeps*, *Rhizoglyphus echinopsus* and *R. solani*.

Dipping of bulbs in hot water at 45°C for 1 hour is useful to control *Rhizoglyphus* spp.

Nematode

Amaryllis is found highly resistant to root-knot nematode (Montasser, 1995).

Lesion Nematodes (*Pratylenchus penetrans and P. pratensis*)

These two nematodes can be identified by its infection, as small light brown colour spacing is present among the length of flower stem. This ultimately leads to the loss of vigour and growth. All stages from second juvenile stage to adult are parasitic.

These nematodes can be avoided by using healthy planting material. If symptoms appear, the nematicide, Carbofuran at 2 kg ai/ha would be effective in controlling their population.

Cyclamen

Diseases

Grey Mold (*Botrytis cinerea*)

This problem is generally faced in greenhouse cultivation under high humidity condition. The spots appear on the surface of petals. Under the leaf canopy, decaying of emerging flower stems and leaf petioles leads to the rotting of plant. Wooly grey fungal spores could be seen over the rotted tissues.

To reduce humidity, good air ventilation should be provided. Overhead watering can be replaced by drip irrigation. The old flowers and foliage should be removed. In chemical control, Mancozeb at 1 g/litre of water to be sprayed or Iprodione (Chipco 26019) can be drenched at seedling or transplanting stage to control this disease.

Spot Rot (*Erwinia chrysanthemi*)

Disease is favoured by high temperature (25°C and above). Bacteria are spreaded by splashing water and handling. Tubers become mushy and plants suddenly collapse.

This disease can be controlled by providing sanitary conditions. Removal of infected plants and avoiding excessive water splashing may reduce the occurrence of disease. A spray of plants with Streptomycin (0.2 per cent) after transplanting is recommended.

Root Rot (*Pythium* spp. and *Phytophthora* spp.)

Poor drainage and over watering enhance the growth of fungi. Plant shows stunted growth and roots become discoloured and rotten. Lower leaves may wilt and turn yellow.

Heat treatment of soil at 60°C for 30 minutes or fumigation with Formalin (2 per cent) kills the soil borne pathogens. Foliar spray of Fosetyl-al (Aliette) at 2–4 g/litre of water is more effective than soil drench.

Fusarium Wilt (*Fusarium oxysporum* f. sp. *cyclaminis*)

The progressive yellowing and wilting of leaves starts from the oldest leaf Browning of vascular tissues in tuber appears with the succession of disease. Disease is favoured by high temperature (25°C or above).

Pathogen survives in the planting material, therefore, the infected lot should not be used for planting. Good sanitation generally provides adequate control. Drenching with thiophanate methyl during early growth period is found beneficial. The soil pH should be maintained from 6.5 to 7.0. Seeds treatment with Bavistin (0.2 per cent) is also found beneficial to control at some extent.

Leaf Spots (*Phyllosticta cyclaminis, Cryptocline cyclaminis* and *Septoria cyclaminis*)

Several types of leaf spots are observed in cyclamen plants. Yellowish to brownish specks appear near the leaf margins is caused by *Phyllosticta cyclaminis*. Pale green distinctly zonate and circular spots present on the stems and leaves due to *Cryptocline cyclaminis*. This pale green colour changes in orange to pink coloured spores. *Septoria cyclaminis* produces red concentric spots appearance on the surface, which turns grey with real borders.

These diseases are also favoured by wet conditions. Reduced humidity and proper ventilation may be helpful in reducing the occurrence of fungi. Rests of the control measures are same as of grey mold.

Stunt (*Ramularia cyclamanicola*)

A conspicuous stunt can be seen due to the shortening of flower peduncles and flowers open below surrounding leaves. Reddish brown necrosis occurs in tubers. Frosty appearance on under side of lower yellowed leaves is the identical feature of this disease. The disease is favoured by moist and warm conditions.

Infected plants should be discarded. Cool and well air circulation reduce the growth of pathogen. The seedlings should be kept away from older plants. Protection of plants is done by spraying of thiophanate methyl or Bavistin (0.2 per cent).

Insects Pests

Thrips (*Frankliniella occidentalis, F. schultzi, F. fusca* and *Thrips tabaci*)

These thrips have been found as a vector of tomato wilt virus (Gofflot and Verhoyen, 1990). Release of high inoculation of predatory mites *Neoseiulus cucumeris* early in the flowering is able control thrips (Courcy, 2001). The chemical control of thrips in cyclamen is similar to other bulbous crops.

Vine weevil (*Otiorhynchus sulicatus*)

The larvae of weevil infest the root system of plant.

Vine weevil can be controlled chemically by treatment of growing medium. Fipronil at 1 g ai/m^2 should be applied along with compost to control this pest. Reduction of weevil population by 73 per cent was reported by a commercial population of entomopathogenic nematode *Steinernema carpocapsae* (Garnham and Mc Nab, 1990). Use of a microbial insecticide of fungus *Metarhizium anisopliae*, Bio-1020 is found effective to control weevil by 100 per cent when its granules applied as soil drench. This also has no toxicological risk or environmental hazard (Reinecke *et al.*, 1990 and Moorhouse *et al.*, 1993). A total control of black vine weevil was achieved with *Heterorhabditis* sp. kept at 20°C for 12 days, while optimum temperature for *Steinernema carpocapsae* was 20–25°C (Miduturi *et al.*, 1994). Granules containing 5 per cent Carbofuran at 5 g/m can effectively control larvae of weevil.

Sciarid Flies (*Bradysia* spp.)

A release of predatory mites *Hypoaspis mites* at the rate of 55 mites/pot is optimum for the control of these flies (Chambers *et al.*, 1993).

Mites (*Phytonemus pallidus*)

Four sprays of Dicofol at 6.7 ml/litre or Endosulfan at 1.9 ml/litre water provide a preventive control of mites (Goodwin, 1990).

Dahlia

Diseases

Powdery mildew (*Erysiphe polygoni* and *E. cichoracearum*)

This fungal disease infests the plant at the early stage and again at the end of the flowering season. This is responsible for rooting of the seed heads (Swami Vinyananda, 1985). The fungi grow on the upper surface of leaf in the form of greyish white powdery patches that cause distortion of leaves. The disease occurs in the warm humid condition.

Any sulphur containing fungicide in appropriate dose will be effective to control powdery mildew. The application of wettable sulphur (0.2 per cent), Morestan, Bavistin or Benlate (0.1 per cent) at fortnightly interval has found to control the growth of fungi. Some eradicants like stylet oil (6 g/litre water), neem oil can be applied to control powdery mildew.

Grey Mold (*Botrytis cinerea*)

The growth of fungus is favoured by cool and wet conditions. Brown, water soaked lesions appear on petals. Fungus may invade plant tissues that touches infected petals.

The disease can be avoided by the removal of withered or diseased flowers. During cool and wet weather, spraying of Fenhexamid (1 g/litre water) or drenching with Iprodione (0.4 g/litre water) should be applied to control the diseases.

Stem Rot (*Sclerotinia sclerotiorum* and *Sclerotium rolfsii*)

The growth of fungi is favoured by cool and moist condition. Water soaked stem cankers appear near the soil line. In severe infestation plants wilt and die. White cottony growth of mycelium on which black irregularly shaped sclerotia (2–10 mm) develop.

The field should be changed, where the disease has occurred previously. The soil treated with Terraclor at the rate of 1g ai/m, inhibits germination of sclerotia when incorporated into 5 cm of soil. Heavy soil can be made porous with addition of sand. The densely plantation should be avoided.

Smut (*Entyloma calandulae* f. sp. *dahliae*)

The disease is favoured by humid condition. The disease first appears on the older leaves in the month of June. Initially it forms circular to irregular spots of yellowish green colour, later they become brown and dry up. The infection can also be observed on petiole and stem. At last, the leaves become brown and die.

The infected soil should not be used for at least 5 years. The affected leaves should be burnt properly. The fortrightly spray of Mancozeb (0.2 per cent) and Zineb (0.2 per cent) to the foliage during June to October was suggested to get best result (Misra, 1978). Other fungicide like Benlate and Blitox (copper oxychloride) were also reported to be effective against fungus (Upadhyaya and Bhandari, 1985).

Charcoal Rot (*Rhizoctonia bataticola*)

The disease is favoured by low moisture in the soil and increasing temperature above 30°C. The disease symptom appears on main stem and branches near the plant base in the form of moist but firm rot. In the progression, drying of surface is observed and at the final stage, infected plant lodge and it branches and twigs drop as a result of breakdown of parenchymatous tissues.

The fungus can be controlled by treatment of tubers with PCNB (Quintozone) as a slurry dip for 10 minutes before planting (Lele and Ram, 1976).

Wilt (*Verticillium dahliae*)

Basal leaves wilt and turn yellow with the infection of this fungus. Dark discoloration of the vascular system occurs. Initially only one branch is affected and in severe cases entire plant dies. Symptoms are more visible during warm weather after a cool period.

The infected plants and roots must be destroyed. Soil solarization may be useful in sunny situation. Fumigation with Chloropicrin or methyl bromide (2 per cent) is necessary to kill the micro sclerotia.

Crown Gall (*Agrobacterium tumifaciens*)

Over growth or galls of tissues occur on the crown portion of plant (stem and root juncture). Infection appears as a stunted growth and spindly shoots.

The infested soil should not be used to grow this crop for 3 years. The infected roots must be destroyed. Rotation with non-woody crops such as cereals and legumes should be followed to avoid the infection. Roots should be dipped in streptomycin sulphate solution at time of planting to prevent the disease.

Bacterial Wilt (*Pseudomonos solanocearum*)

Infection starts with a rooting of stem at the base as it plugs the vascular bundles. When stems are cut, yellowish masses of bacteria releases from cut end.

Destruction of infected planting material must be followed. The fumigation of growing medium is also used to avoid the growth of pathogen. Other treatments of root are similar to control measures of crown gall disease.

Bacteriosis (*Erwinia carotovora* var. *carotovora*)

The disease causes the rotting of stem with foul smell. The layer of pith becomes black along with rot. Tubers are also affected by bacteriosis.

The infected plants and tubers must be destroyed. Tubers should be treated with bactericide, as in case of crown gall disease. Other management practices are also same.

Dahlia Mosaic (*Marmor dahliae*)

This viral disease develops as irregular green colour on the leaves. The bands adjacent to the veins remain pale green. Leaves become short. Flower colour is usually normal. This disease is transmitted by peach aphid (*Myzus persicae*).

Infected plants must be destroyed or not further used for planting. The vector may be controlled by spray of Metasystox (0.2 per cent). Other biocontrol measures against aphid can be practiced as discussed in alstroemeria.

Spotted Wilt (*Lethum australiense* or Tomato Spotted Wilt Virus)

This viral disease is also known as ring spot. A well defined mosaic mottle or irregular concentric rings or wavy lines in leaves appear. Symptom occurs on older leaves. Generally, distortion or stunting does not occur in ring spot. Thrips carry the virus.

The infected plant material should be avoided. Shoot tip culture is reported for its complete elimination. Control measures against thrips should be adopted as in alstroemeria or in amaryllis.

Insect Pests

Aphid (*Myzus persicae, Brachycaudus helichrysi* and *Aphid fabae*)

They suck the sap from the young tissues, which ultimately reduce the growth rate of plant.

Aphids can be controlled by the spraying of systemic insecticide like Metasystox (0.2 per cent), Metacid (0.1 per cent), Melathion (0.1 per cent) or Basudin (0.1 per cent). Biocontrol measures as mentioned in alstroemeria or amaryllis can be adopted for effective control.

European Corn Borer (*Pyrausta nubilalis*)

Tender tips, bud ends, flower parts and leaves are fed by its larvae. They generally appear in September. The affected tips die completely after some time of infestation, thus whole plant parts above the invasion on site become wilted.

For better control, the plants must be sprayed with Sevin (0.2 per cent) and Kelthane (0.1 per cent) at 7 days interval. *Trichogramma* is also used as biocontrol measure for reducing its population (Nguyen and Nguyen, 1982).

Leaf Hopper (*Empoasca fabae*)

The damage appears as discoloration of leaf from the margin that spreads to the mid vein. The infected portion become brittle and brown and the plant show stunted growth.

Spray of Malathion or Metosystox (0.2 per cent) at 10 days interval is effective control measure for leaf hoppers. The spray should be stopped before flowering.

Thrips (*Heliothrips haemorrhoidalis, Farnkeliniella tritici* and *Thrips tabaci*)

Thrips mainly attack on flowers. They rasp the surface and feed on exuding juice. This turns the lower surface whitish and withered.

The control measures are the same as for the leaf hopper.

Red Spider Mite (*Tetranychus urticae*)

This is a serious pest of dahlia. They feed the lower surface of leaf and small white speckling present on upper surface. In severe infestation, webs appear on leaves and buds over which very red small mites can be seen.

Increasing humidity in the microclimate of plant can control the population of red spider mite. By splashing of fresh water under side of the leaves will be effective to control the mites. Chemically it can be controlled by spray of wettable sulphur or Karathane (2 ml/litre).

Mulberry White Fly (*Tetraleurodes mon*)

An elliptical, jet-black larva of white fly adheres to the lower surface of the leaves.

Spray of Malathion (0.2 per cent) on the lower surface of leaves is effective to control the nymphs, whereas spray of Resmethrin can be done to control the adults and nymphs.

Root Knot Nematode (*Meloidogyne incognita*)

Second stage juveniles of nematodes penetrates the roots and damage the vascular tissue. This suppresses the growth of plant. Knots or small swellings appear on the roots.

Soil solarization may reduce the nematode population in the soil. This can be eliminated by use of Carbafuran or Phorate at the rate of 1 g ai/m. Aphids can easily be controlled by the spraying of systemic insecticide like Metasystox (0.2 per cent), Metacid (0.1 per cent), Melathion (0.1 per cent). Along with chemicals, some biocontrol measures are also effective to control aphids as described in alstroemeria or amaryllis.

Lesion Nematode (*Pratylenchus penetranus* and *P. pratensis*)

They cause a reduction in growth, which resulted clear loss of production and quality of flower. The nematodes cause small brown lesions present along the length of stem that leads to fade in colour and reduced plant vigour and ultimately plant dies.

Nematodes can be controlled by application of Carbofuran at 2 kg ai/ha. Healthy planting material should be selected.

Foliar Nematodes (*Aphelenchoides ritzemabosi*)

This nematodal infection causes the smaller shrunken leaves, yellow white spots, which gradually turned in brown colour on infected plants.

The infested leaves should be removed and burnt. Infected cuttings should be given hot water treatment at 50°C for 5 min. Basal application of Phorate 10G at 1.5 kg ai/ha, methyl parathion or spray of Chlorpyrifos (0.05 per cent) can be followed to reduce the population of this nematode.

Gladiolus

Disease

Fusarium wilt (*Fusarium oxysporum* f. sp. *gladioli*)

This is the most limiting factor in gladiolus production. The common symptoms of this fungal disease are rotting of corm, stunted, curved or bent leaves. In severe condition, plant suddenly wilts and dies prematurely. The disease in favoured by high temperature, high soil moisture and high level of nitrogen in the soil.

Use of healthy plant material and reduction of fungal growth are effective approaches. Soil solarization kills the soil borne pathogens. Instead of ammonical nitrogen, nitrates can increase the pH, which reduces the fungal growth. Calcium, potassium and sodium nitrates should be applied in cool and moist soil while ammonium nitrate in warm and dry soil. The infested soil should not be used for at least 3–4 years. Some cultivars like Albana, Apricot Glow, Sovenir, Hoppman's Glory, Sylvia, Dhiraj, White Friendship and White Prosperity are reported to be resistant to this disease. Non-pathogenic strains of *Fusarium* are reported as biological control of fusarium wilt (Alabouvette, 1993 and 1996).

Core or Spongy Rot (*Botrytis gladiolorum* and *B. cinerea*)

This rot is also known as soft corm rot, grey mold, neck rot, floral rot or botrytis blight. This disease is favoured by cool and humid conditions. Small rusty specks with red bordered are found over the leaves. *Botrytis* survives in corms, soil and plant debris. Translucent water-soaked lesions present on the petals and during transit, these infected flowers rot due to its spread. The infection on corms causes small brown, circular sunken spots and results in spongy rot during storage (Sunita, 1999).

. Curing and treatment should be given. For curing, the corms should be placed immediately after digging in shallow trays at 35°C and 80 per cent relative humidity and good air circulation is provided. The corms should be stored at 4.4°C. Before planting, the corms should be treated with Iprodione or Thiobendazole along with wetting agent. The water temperature should be 30°C. Spray of Vinclozolin, Iprodione and Benlate are suitable to control this disease.

Dry or Neck Rot (*Stromatinia gladioli* or *Sclerotinia gladioli*)

It is also known as root rot. The disease is favoured by cool moist weather and water logging conditions. The major symptoms are rotting of leaf sheath and stem at the soil level, which ultimately reduces the quality in term of short and small size of spikes and corms, respectively.

Corms should be cured first by dipping them in cold water for 24 hours and then in hot water at 54.4°C for 30 minutes. Corms may also be dipped in potassium permanganate (0.1 per cent) or Bavistin (0.2 per cent) for 20 minutes. Spraying of copper solution twice or thrice during the growing season at 15 days interval has been recommended.

Curvularia Blight (*Curvularia trifoli* f. sp. *gladioli* and *Curvularia brachyspora*)

The disease is favoured by warm (18.3–32.2°C) and humid weather. This fungus initially infects the young leaf, so the first appearance is severe yellowing and browning of leaf tips. Firstly the circular or oval spots become irregular. These spots later enlarge up to 12 cm or 20 cm with dark brown border on leaves. The quality of spikes reduces in the form of distorted flowers having brown spots on petals and young leaves.

Maneb is very effective in controlling this disease. Mancozeb and Miltox (copper oxychloride + Zineb) each at 0.2 per cent, sprayed at 15 days interval, may be effective in managing this disease.

Storage Rot (*funiculosum*, *P. gladioli* and *Rhizopus arrhizus*)

Penicillium funiculosum causes core rot, while *P. gladioli* is responsible for storage rot. A firm brown corm rot develops when corms are stored in a pile. If moist condition persists, greenish blue spore masses appear over rotted areas.

The corms should be treated for 20–30 minutes with Emisan-6 (0.2 per cent), Benlate (0.2 per cent), Bavistin (0.2 per cent), Captaf (0.2 per cent) and Calixin (0.1 per cent) as pre-planting and post-harvest treatments. Hot water treatment at 39°C along with fungicidal treatment with Bavistin, Benlate or Captan (0.2 per cent) at acidic pH (3–3.3) is more effective against this disease.

Leaf Spot (*Alternaria fasciculata, A. tenuis, Septoria gladioli* and *Stemphylium* sp.)

Alternaria fasciculata causes dirty dark brown to black irregular spots in long patches mostly on apical and marginal parts on the leaves. *Septoria gladioli* invades internodes of spikes and florets. The lesions are of circular purplish-brown coloured and containing black spores at the centre of older spot. *Stemphylium* sp. may kill leaves before flowering. Spray of Dithane M-45 or Miltox at 0.2 per cent, in fortnightly intervals was revealed effective against *Septoria gladioli* (Misra, 1977).

Rust and Smut (*Uromyces transversalis, Urocystis gladioli* and *Tuburcinia gladioli*)

Uromyces transversalis causes rust, where as *U. gladioli* and *Tuburcinia gladioli* cause smut disease on corm. Skin and leaves smut appears as blistering, shredding and necrosis of stem and leaf tissues, which leads to death of plants. Soil fumigation with chloro bromopropane, Formalin, Trapex (methyl iso thiocynate) or Mylone is satisfactory control measure.

Bacterial Scab Blight and Spots (*Pseudomonas marginata* and *Xanthomonas gummisudans*)

It causes neck rot and spots on leaves during rains. Leaves start drying from tip. Brown circular and sunken spots with black margins are found on corms.

Preplanting dipping of corms in 1:1000 corrosive sublimate solutions for 12 hours can control disease. Flowers should be harvested when foliage is dry. The cutting knife or blade must be sterilized.

White Break (Cucumber Mosaic Virus)

The symptoms appear as white discontinuous streaks on the foliage like that of severe attack by thrips. Yellow and white flowers show distortion. The disease also affects the plant height. Aphid (*Myzus persicae*) is supposed to be the carrier of this virus. A very effective method is the use of reflective surface (aluminium foils) around the plants to repel the insects.

Grassy Top (Aster Yellow Phytoplasma)

Symptoms of this disease appear as precocious sprouting, grassy top, weak shoots arrested root growth, stunting, spiral bending of spikes. Other symptoms are straw yellowing of younger leaves and spikes. Infected plants die from the tip down. This is carried by leaf hopper.

Healthy plant material should be taken. The infected plants should be rogued out and burnt. Control of leaf hopper may reduce its occurrence.

Mild Mosaic (Bean Yellow Mosaic Virus)

A faint leaf mottle and sometimes a pencil stripe colour break of blossoms. Blossoms may fail to open all the way.

Planting material should be used from disease free area.

Insect Pests

Aphids (*Aphis gossypii, A. craccivora, Myzus persicae, M. gossypii, Macrosiphum solanifolii* and *Anuraphis tulipae*)

The infected plant produces small flowers of lower quality spikes. Both adults and nymphs attack on plant by sucking the sap.

Spraying of Malathion 50 EC (0.2 per cent) or Metasystox 25 EC (0.3 per cent) is effective to control aphids.

Thrips (*Liothrips* sp., *Thrips hawaiiensis, Taeniothrips simplex* and *Franklinella occidentalis*)

Thrips damage the foliage, flowers and corms in storage by rasping and cause silvering and whitish streaks, which weaken the plants. Both nymph and adult stages are damaging.

Wet dipping of corms with Acephate (0.1 per cent) or Cypermethrin (0.05 per cent) and foliar spray of Methamidophos (0.1 per cent) was found to be effective against thrips (Labonowski and Mynett, 1990).

Caterpillar (*Agrostis segetum, Feltia subterranea, Prodenia dolichos* and *Heliothis zea*)

Larvae feed on leaves, flower buds and spikes and may completely destroy these parts.

Deep ploughing helps in exposes the pupae to natural enemy. Foliar spray of Quinolphos (0.05 per cent), Chlorpyriphos (0.05 per cent), Monocrotophos (0.05 per cent) or Dipel (0.05 per cent) applied at 15 days interval may control the insect.

Mites (*Rhizoglyphus echinopus, Tetranychus urticae, T. telarius* and *T. bimaculatus*)

White specks appear when mites feed on the lower surface of leaf under warm and shady conditions. In severe case leaves become dry and fall. *Rhizoglyphus* spp. feed on corms in the field and storage.

Integrated management using alternate cropping with rice, soil solarization, planting and ploughing in *Sesbenia* (green manuring), long term submergence of soil in water are effective to control the mite. Hot water treatment along with foliar spray of Ethion (0.05 per cent), Diazinon (0.04 per cent) will check the mite population.

Plant Bugs (*Poecilocapsus lineatus* and *Lygus lineolaris*)

Nymphs suck the sap from the young leaves, which causes sunken areas around the puncture. These areas finally become circular holes.

Spraying of neem oil (1 per cent) or Malathion (0.1 per cent) is effective to control plant bugs.

Mole Cricket (*Gryllotalpa gryllotalpa*)

The infestation may lead to a 30 per cent loss in production owing to damage to the bulbs and subsequent rotting observed. The female lays around 350–450 eggs in the nest, which is near to the stems of the plants.

Destroy these nests in May-June is more effective than to control individuals. Deep autumn ploughing in winter helps in larval mortality (Denisenko, 1986).

Snails

They are nocturnal in habit. They climb on the plant and feed the leaves. In cloudy weather, they feed in shady portions of the plant during daytime too.

Poison bait can be helpful in controlling these mollusks. Poison bait contains 250 g wheat barn + 50 g gaggery and 6–8 g Metaldehyde.

Nematodes

Root Knot Nematode (*Meloidogyne incognita*)

It causes gall on roots and establish near the vascular tissues. The infected plants show chlorotic, small and fewer leaves and plants are stunted (Khanna and Chandel, 1995).

Treatment of gladiolus plants with *Paecilomyces lilacinus, Trichoderma harzianum* + neem cake and *P. lilacinus* + *T. viride* + neem cake combination not only control nematode infection but also wilt disease by *Fusarium oxysporum* f. sp. *gladioli* and root knot complex (Nagesh *et al.*, 1998). Gladiolus cultivars Royal Jubilee, Orange Spire, Snow Princess, Sunny Boy, Pink Friendship, Hunting Song, American Beauty, Apolus, Darshan, White Friendship are resistant to *M. incognita*. Soil solarization is also effective in reducing their population.

Lilium

Disease

Grey Mold (*Botrytis elliptica*)

The growth of fungus is favoured by warm wet and humid conditions. Initially fungus causes circular lesions on leaves, which spread on flower buds. Infected leaves rotten and fall from the plant and flower buds may die.

Foliar spray of copper fungicide like Bordeaux mixture at 2 per cent has been found useful. Other integrated management practices like reducing humidity inside the polyhouse by opening the vents and providing good aeration can be adopted. Avoiding overhead watering and following strict phytosanitation along with chemical spray will be effective to control fungus.

Soft Bulb Rot (*Rhizopus stolonifer*)

Symptoms first appear on the bulb. The bulbs become rotten, soft and covered with a thick mat of mycelial growth.

The bulbs should not be damaged or wounded during harvesting. Soil sterilization with Formalin (2 per cent) may be effective.

Bulb Rot (*Fusarium oxysporum* f. *lilii*)

The infected plants show stunted growth and lower leaves become yellow or purple and dry prematurely. The disease is confined to the base of the scales, which are detached from the basal part of the bulb.

Soil sterilization with formaldehyde (2 per cent) has found effective against this fungus. The infected plants should be removed. Bulbs should be stored at low temperature.

Foot Rot (*Phytophthora cactorum*)

It attacks stems just below the soil surface. Bulbs are not damaged and infected parts become shrunken. Plants wilt and topple as the infection advances.

The infection can be avoided by proper drainage and destroying the infected plants. Application of Bordeaux mixture is helpful in reducing the infection. Soil fumigation with Formalin or application of Metalaxyl or Fosetyl-al gives effective control of the disease.

Bacterial Soft Rot (*Pectobacterium carotovorum* and *Bacillus lilii*)

The disease mainly occurs during storage and transit. The bulbs become soft and watery decay takes place. The infection spreads out when the bulbs are wounded. The infected and wounded bulbs must be discarded. Care should be taken to avoid injury during the lifting of corms.

Mosaic (*Marmor mite, M. cucumeris* and *Adelonosus lilii*)

The infestation of viruses is transmitted by melon aphids. The infected plants show stunted growth with mottled leaves. Plants become dwarf and crippled.

The infected plants should not be used for further planting purpose and the whole plant should be burnt. The vector can be controlled by spraying of Metasystox (0.2 per cent) or Monocrotophos (0.2 per cent).

Rosette

The infected leaves show pale green or yellowish appearance without mottling or streaking and plant shows stunted growth. Bulbs become flat, small and tend to split. The virus is transmitted by *Aphis gossypii* and *Macrosiphum lilii*.

The management and control of this disease is similar to that of mosaic disease.

Insect Pests

Aphids (*Aphis gossypii, Aulacorthum circumflexum* and *Rhopalosiphum nymphaease*)

Aphids pierce and suck the sap from the tender plant parts. Very severe infestation leads to wilting and dieing of plants. If the plant shows withering of tender shoots, premature falling of buds, lose of elegance and presence of sooty mold on the infested twig, the aphid infestation must be there.

Some natural predators of aphids like aphid midge, ladybird beetle, lacewings, microbial insecticide, minute pirate bugs and parasitic wasps were found effective to control aphid population. Foliar spray of Malathion (0.1 per cent) is effective.

Lily Beetle (*Lilioceris lilii*)

Both larvae and adults feed on the leaves, stems and seedpods of lily. This causes severe damage and decreases flower quality.

Foliar spray of Malathion (0.2 per cent) is reported as an effective control measure. The parasitoids of lily beetle, ichneumoids (*Lemophagus pulcher, L. errabundus* and *Diaparis jucmda*) are used biologically in reducing the insect population (Haye and Kenis, 2004).

Thrips (*Liothrips vaneeckei* and *Taeniothrips simplex*)

Lily bulbs are mainly attacked by these thrips. The symptoms appear as rust coloured, sunken areas at the base of the outer scales, which become soft and break off easily.

The infested bulbs should be treated with Tedion and store below 10°C. Hot water treatment at 44°C for one hour can be followed to control this pest.

Bulb Mite (*Rhizoglyphus echinopus*)

The mites feed around the basal plate of the bulb and destroy the roots. The leaves and stems suddenly turn yellow after infestation.

Proper care should be taken in the drainage of the field. Crop rotation is also helpful in controlling the mite population. The affected bulb may be treated in wettable Tedion 25 per cent (1g/l water) for 24 hours and this will control mite.

Nematode

Foliar Nematode (*Aphelenchoides fragariae*)

The symptoms appear as strips or rectangular panels on the leaf. Off colour areas on leaves can be easily seen and these areas progress to leaf blight. This leads to the reduction of plant growth. Infected buds do not open. Plants may eventually die, resulting in huge losses.

Dipping of bulbs either in neem seed kernel extract or in garlic clove extract along with neem cake, as an organic amendment is beneficial to reduce the infestation to a good extent. Hot water treatment of bulbs at 50°C for 5 minutes also controls the nematodal population.

Root Lesion Nematodes

The symptoms appear as a premature yellowing of easter lily foliage. Affected plants show stunted growth. Several dead spots or lesions are present on the roots.

Application of Aldicarb (30 kg/ha) reduces population of nematodes as well as increases the bulb yield (Boontjes *et al.*, 1978). Carbofuran at 2 kg ai/ha effectively reduces damage of nematode. Nemagon is recommended as soil application (0.8–1 ton/ha in granular form or 100–150 litre/ha as a liquid) before 2–5 weeks of planting for their control.

Tuberose

Diseases

Stem Rot (*Sclerotium rolfsii*)

This fungal disease is also known as basal rot, foot and tuber rot, wilt and stem rot. The initial symptom shows flaccidity and drooping of leaves. The fungus reaches to the plants as the lower leaves touch with the wet soil. The fungus causes rotting of leaf base and stem and it grows as a fan shaped white cottony mycelium over the rotting tissues. The disease is favoured by high soil moisture.

By proper drainage and planting of bulbs at wider spacing can reduce the occurrence of this disease. The field should be cleaned by removing of senescent and dried leaves. The disease can be chemically controlled by copper fungicide or Benzimidazoles or by soil application with dusting of Brassicol (20 per cent) at 30 kg/ha.

Blossom Blight (*Fusarium equiseti*)

The disease appears as light brown lesions develop on petals, which soon darken and tissues become dry. Infection is also noticed as collapsed flower stalk. This blight is more common in double varieties of tuberose. The blight starts from the margin of the petal and later spreads over the entire inflorescence. The infected spike becomes weak and cannot support the flowers, which ultimately droop down (Roy, 1984).

Foliar spray of Benomyl (0.2 per cent) or Thiophanate methyl (0.2 per cent) is recommended to reduce the occurrence of disease.

Leaf Spot (*Alternaria polyanthi*)

With the infection of this fungal disease, brown concentric spots develop on the mid rib, which spread to the entire leaf in severe condition. The leaves and peduncles become dry and necrotic with the severity of infection.

Application of Mancozeb (0.2 per cent) or Iprodione (0.2 per cent) at 10 days interval as a foliar spray reduces the occurrence of this disease.

Flower Bud Rot (*Erwinia* sp.)

The disease mainly appears on young flower bud and cause dry rotting with brown scorched. This leads to the shrivelling and drying of buds.

The diseased plants should be selected and destroyed completely. Streptocyclin (0.01 per cent) should be sprayed to control this disease effectively.

Insect Pests

Aphids

These attack on the young and tender flower buds and suck the sap, which causes the wilting and drooping of flower buds. In severe infestation buds are unable to open.

The insect population can be effectively controlled by the spraying of Monocrotophos (0.05 per cent), Malathion (0.1 per cent) or Dimethoate (0.06 per cent). Natural enemies can be used as biocontrol measures of aphids. Detail of natural enemies has been given in alstroemeria crop.

Thrips

It is the most severe pest of tuberose. Both adults and nymphs damage the leaves, flowers and flower stalks by sucking sap. Sometimes it is associated with bunchy top that produces malformed inflorescence.

Application of Endosulphan (0.2 per cent) or Malathion (0.15 per cent) as foliar spray can be effectively controlled this pest.

Caterpillar (*Helicoverpa armigera*)

After hatching, the larvae enter into the young tender buds by puncturing or making tiny holes and feed inside the flower bud. The affected flower bud gets completely destroyed.

Foliar spray of Monocrotophos (0.05 per cent), Endosulphon (0.07 per cent) or Deltamethrin (0.05 per cent) controls the pest.

Weevil (*Myllocerus* sp.)

It causes damage by feeding on leaves and shoot. The cutting habit of this insect is specific as it controls notched effect by feeding at the edge of the leaf. The larvae feed on the roots and make tunnel in side the bulb.

Foliar spray of Malathion (0.1 per cent) or Carbaryl (0.2 per cent) in the evening can be effective against weevil. Generally entopathogenic nematodes (*Steinernema carpocapsae* and *Heterorhabditis bacteriophora*) are used to control weevil biologically.

Red Spider Mite (*Tetranychus urticae*)

Yellow strips and streaks appear on the leaves as mites suck the sap that leads to silvery or bronze and distorted leaves.

Foliar application of Kelthane (0.2 per cent) has been found effective against mite.

Nematodes

Root Knot Nematode (*Meloidogyne incognita, M. javanica* and *M. arenaria*)

Meloidogyne incognita causes 13.25, 9.87, 14.3, 13.78 and 28.58 per cent reduction in plant weight, number of florets, spike length, spike weight and number of bulblets, respectively (Khan and Reddy, 1994). These nematodes cause the formation of knots or galls on the roots. After penetration they establish near by vascular tissue. The high infestation results in suppression of spike emergence and rotting of bulbs that causes the total loss of flower yield.

No nematode resistant varieties have been found. Application of *Paecilomyces lilacinus* at 2x10 spores/ml with 5 per cent neem leaf or caster leaf extract results in reduced gall numbers and improve fresh plant height and flower yield (Nagesh *et al.*, 1997). Soil application of Carbofuran at 2.5 kg ai/ha or neem cake at 1 ton/ha is effective to its control. Integration of Vesicular Arbuscular Mycorrhiza (VAM) fungi like *Glomus mossae* and *G. fasciculatum* with neem cake provides protection against this nematode (Khan and Reddy, 1994). Soil solarization during May and June is also effective.

Reniform Nematode (*Rotylenchulus reniformis*)

The infested plants show reduced root system leaf chlorosis, overall stunting of host plants and reduced yield with plant longevity.

Soil solarization has been found effective against this nematode.

Foliar Nematodes (*Aphelenchoides bessyi*)

The symptoms of damage appear as rough, crinkled and stunted flowers. The flower buds fail to bloom. Brown streaks develop on leaves, bracts and base of petals. In severe condition, the flowering stalks rot and become brittle and may become blind (Khan and Pal, 2001).

A combination of bulb dipping either in neem seed kernel extract or in garlic clove extract along with neem cake as an organic amendment controls the disease to a good extent (Chakraborty and Dutta, 1997). The infested leaves must be cleaned and burnt. Hot water treatment at 50°C for 5 minutes or at 44.4°C for 30 minutes of infected cuttings or stools helps in reducing the incidence of its infestation.

REFERENCES

Albouvette, C., P. Lemanceau and C. Steinberg (1993). Recent advances in the biological control of Fusarium wilts. *Pest Sci.*, 365–373.

Albouvette, C., P. Lemanceau and C. Steinberg (1996). Biological control of fusarial wilts: Opportunities for developing a commercial product. In: *Principles and Practices of Managing Soil-borne Plant Pathogens*, (Ed.) Robert Hall, The American Phytopathological Society. St. Paul. Minnesota, USA, 330 p.

Boontjes, J., P. Mantel and H. Brinkman (1978). The effect of aldicarb (Temik 10 G) against root lesion nematodes in lily planting stock. *Bloembollencultuur*, 88(41): 994–995.

Chakraborty, H.S. and P. Dutta (1997). Phytochemical control of floral malady of *Polianthes tuberosa* L. *J. Mycopath. Res.*, 35: 111–113.

Chambers, R.J., E.M. Wright and R.J. Lind (1993). Biological control of glasshouse sciarid flies (*Bradysia* spp.) with the predatory mite, *Hypoaspis mites* on cyclamen and poinsettia. *Biocontrol Science and Technology*, 3(3): 285–293.

Conijn, C.G.M. (1983). New material for controlling *Stagonospora* in hippeastrum (amaryllis). *Bloembollencultuur*, 93 (40): 1039.

Courcy, W.M.E. (2001). Biological control of thrips on ornamental crops: Interaction between the predatory mite *Neoseiulus cucumeris* (Acari: Phytoseiidae) and western flower thrips, *Frankliniella occidentalis* (Thysanoptera: Thripidae), on cyclamen. *Biocontrol Sci. Tech.*, 11(1): 41–55.

Denisenko, M.K. (1986). An effective method of control of mole cricket. *Zaschita Rastenii*, 10:51.

Garnham M. and H. Mc Nab (1990). The use of entomogenous nematodes for the control of black vine weevil in cyclamen. In: *Forty Third New Zealand Weed and Pest Control Conference, Proc.* pp. 70–72.

Gill, J.S. and N.K. Sharma (1979). Additional hosts of the foliar nematode *Aphelenchoides ritzemabosi* from India. *Indian J. Nemat.*, 6: 169–171.

Gofflot, A. and M. Verhogen (1990). Speedy development of tomato spotted wilt virus infection transmitted by *Frankliniella occidentalis* in ornamental greenhouse cultivars in Belgium. *Parasitica*, 46(2-3): 85–88.

Goodwin, S. (1990). Controlling cyclamen mite. *Australian Hort.*, 88(1): 56.

Haye, T. and M. Kenis (2004). Biology of *Lilioceris* spp. (Coleoptera: Chrysomelidae) and their parasitoids in Europe. *Biological Cont.*, 29(3): 399–408.

Khan. M.R. and A.P. Pal (2001). Plant parasitic nematodes associated with tuberose (*Polianthes tuberosa* L.) in West Bengal: *Ann. Plant Prot. Sci.*, 9: 357–359.

Khan. R.M. and P.P. Reddy (1994). Nematode problems in ornamental crops and their management. In: *Floriculture Technology, Trades and Trends,* (Eds.) J. Prakash and K.R. Bhandari, Mohan Primalini for Oxford and IBH Publishing Co. Pvt. Ltd. 66. Janpath, New Delhi, pp. 468–472.

Khanna, A.S. and S.S. Chandel (1995). Effect of various population level of *Meloidogyne incognita* on growth patterns of Gladiolus L. variety Sylvia. *Curr. Nemat.*, 6: 13–15.

Labanowski, G. and M. Mynett (1990). Control of the gladiolus thrips, *Taeniothrips simplex* on gladiolus crop. *Acta Hort.*, 266:547–551.

Leeuwen, A.J. Van. (1974). Bulb disinfection for hippeastrum (amaryllis). Good alternatives to mercury compounds. *Vakblad voor de Bloemisterij*, 29(4): 15.

Lele, V.C. and A. Ram (1976). Charcol rot of dahlia and its control. *Pesticides*, 10: 39–40.

Maisonneuve, J.C. and C. Arrec (1999). The potential of *Chrysoperla lucasina* for IPM programmes in greenhouses. *In*: Integrated control in glasshouse, Brest, France, 26–29 May, 1998. Proceedings edited by J.C. Van Lanteren, p. 294.

Miduturi, J.S., R. de Clercq, H. Casteels and A. de. Grisse (1994). Effect of temperatures on the infectivity of entomopathogenic nematodes against black vine weevil (*Otiorrhynchus sulcatus* F.). *Parasitica*, 50(3–4): 103–108.

Misra, R.L. (1978). Combating dahlia Smut. *Prog. Hort.*, 10: 43–46.

Misra, R.L. and N.S. Pathania (2002). Production of commercial bulbous flower. In: *National Symposium on Indian Floriculture in the New Millennium, Bangalore*, 25–27 February, 2002. Souvenir. 26 p.

Misra, R.L. (1977). Control of leaf spot disease of gladioli. *Prog. Hort.*, 9: 55–59.

Montessar, S.A. (1995). Reaction of certain flower bulb plants to root knot nematode, *Meloidogyne incognita*. *Pakistan J. Nemat.*, 13(2): 99–102.

Moorhouse, E.R., A.T. Gillespie and A.K. Charnley (1993). The development of *Otiorhynchus sulcatus* (Fabricius) (Coleoptra: Curculionidae) larvae on a range of ornamental pot–plant species and the potential for control using *Metarhizium anisopliae*. *J. Hort. Sci.*, 68(4): 627–635.

Nagesh, M., P.P. Reddy and M.S. Rao (1997). Integrated management on tuberose (*Polianthes tuberosa* L.) using *Paecilmyces lilacinus* in combination with plant extracts. *Nematologia Mediterranea*, 25:3–7.

Nagesh, M., P.P. Reddy, N. Ramachandran and U.K. Mehta (1998). Integrated management of *Meloidogyne incognita* and *Fusarium oxysporum* f. sp. *gladioli* in gladiolus using antagonistic fungi and neem cake. In: *International Symposium on Nematology Challenges and Opportunities in 21st Century*, Coimbatore, 16–19 April, (1998. Proc. pp.263–266.

Nguyen, I.T. and S.T. Nguyen (1982). The use of *Trichogramma* in Vietnam. *Zashchita Rastenii*, 1: 52.

Nucifora, A. and V. Vacante (1981). Integrated control of *Trialeurodes vaporariorum* (Westw.) pest of gerbera. *Colture Protette*, 10: 33–36.

Raizada, R.K., K.M. Srivastava and B.P. Singh (1993). Detection of amaryllis cryptic virus by ELISA. *Indian J. Exp. Biol.*, 21: 229–300.

Reinecke, P., W. Andersch, K. Stenzel and J. Hartwig (1990). Bio 1020, a new microbial insecticide for use in horticultural crops. *Brighton Crop Protection Conference, Pests and Diseases*, 1: 49–84.

Roy, A.K. (1984). Two new Fusarium diseases in Karbi Anglong. Assam, India. *Current Sci.*, 53(24): 1304–1304.

Smith, F.F. (1967). Controlling Insects of Flowers, Agriculture Information Bulletin No. 237, U.S. Dept. of Agriculture. pp.30.

Sunita, S.C. (1999). Fungal and bacterial diseases of bulbous ornamentals. In: *Diseases of Horticultural Crops: Vegetables, Ornamentals and Mushrooms*, (Eds.) L.R. Verma and R.C. Sharma, Indus Publishing Co., New Delhi, pp. 501–531.

Swami, Vinyananda (1985). *Dahlia Growing*. Associated Publication Co., New Delhi, 92p.

Upadhyaya, J. and T.P.S. Bhandari (1985). Fungicidal control of leaf smut of dahlia caused by *Entyloma dahiae*. *Indian Phytopath.*, 38:338–339.

20

Status and Management of Pests and Diseases in Nut Fruits

S.S.Singh, D. Prasad** and H.Lal**

**Horticultural Research and Extension Centre, Chaubattia – 263651, Almora (Uttaranchal)*

***Division of Nematology, Indian Agricultural Research Institute, Pusa, New Delhi – 110 012*

Horticulture has, in the recent years, emerged as an important and viable diversification option in agriculture and has transformed the subsistence level farming system into a high value commercial enterprise. Horticultural crops provide food and nutritional security, generate rural employment, maintain ecological balance and provide raw materials for agro-processing industries, thereby improving the economy of the people. As a consequence, a phenomenal progress has been achieved in the production of fruits.

Fruits, as backbone of horticulture, are grown over an area of 3.89 million hectares in the country with a production of 45.37 million tonnes per year. India accounts for 10 per cent of the total world fruit production and ranks first in the production of mango, banana, sapota and acid lime. In India, temperate fruits are cultivated in more than ten states, but the quality production is in the hill states of the Himalayan region only. This region comprises more than one-eighth of the total land area of the country extending from J&K in the west to Arunachal Pradesh in the east. Though these areas are sparsely populated, yet are the main centres of temperate fruit production, in which apple is the leading fruit crop occupying nearly 85 per cent of total area under temperate fruits.

The importance of horticulture, particularly the fruits, in the economy of Uttaranchal needs no emphasis. Due to inherent problems like difficult terrain, undulating topography, lack of assured irrigation and limited availability of solar energy for crop production, the scope for intensification of traditional agriculture is limited. However, the varied climatic and topographical conditions of the state have provided ideal conditions for the development of a wide range of fruits from temperate to subtropical ones.

Uttaranchal came into existence as 27th state of India on 9th November, 2000, is located at 28°69'-31°28' north latitude and 77°49'-80°60' east longitude and covers a total geographical area of 53,483 square km with a population of 84,79,562. This region is gifted with different agro-climatic conditions. The temperature ranges from 29.4 to 40°C and goes below freezing point during winter. Normally, the rainy season commences from the 2nd week of June and lasts till the end of September. The average rainfall of this area is 1525 mm. The terrain of this region is abrupt and has steep to gentle slopes. There are many flat valleys where the climate is warmer as in the plains. This region comprises of 13 districts, *viz.*, Almora, Bageshwar, Nainital, Pithoragarh, Champawat, Udham Singh Nagar in Kumaon Division and Pauri, Tehri, Dehradun, Chamoli, Uttarkashi, Rudra Prayag, Haridwar in Garhwal Division.

The critical study of the meteorological data and the elevation which extends from 1000 meters to 3500 meters in area clearly shows that this region has different climatic conditions right from arctic to sub-tropical, thereby suggesting the suitability for growing almost all kinds of fruit. Temperate fruits like apple, pear, peach, cherry, walnut, almond, pecan nut, chest nut, hazelnut, apricot, etc., are grown at the elevation of 2000 to 3000 meters. Such areas are also abundantly found in the parts where soft fruits like strawberry, gooseberry and other berries can be grown successfully. Both tropical and sub-tropical types of fruits are grown successfully in the valley.

Agro Climatic Zones

The Tarai and Bhabar Sub-region 100–300 meters Above Mean Sea Level (msl)

The Tarai area extends over a narrow belt in Nainital and Pauri Districts. Its soil is deep and alluvial, rainfall is copious, around 150 cms and the mean temperature is about 30°C. The Bhabar area adjoins the Tarai towards the hills. It is a waterless tract along the foothills, characterized by gravel and silt brought by rivers from the hills. Beneath the gravelly bed is excellent soil. The water table is low and the area is rich in forests.

The Low Hills

These range from 300–1,000 meters above msl. They have a tropical to sub-tropical climate. They contain deeper soils and most of the irrigated horticulture in the hilly areas is practiced in low hills.

The Mid Hills

The altitudes range from 1000–1500 meters above msl. The climate is tropical to temperate. In this sub-region most of the horticulture is rainfed though there are some irrigated areas too.

High Hills

The altitude varies from 1500–2000 meters above msl. The climate is tropical to temperate. The agriculture is mainly rainfed though much of the land is suitable for horticulture.

Very High Hills

The altitude varies from 2000–3000 meters above msl. The climate is temperate to sub-arctic. Though agriculture is pursued in some isolated pockets, yet the area is suitable for temperate fruits and livestock production.

The walnut (*Juglans regia)* is the most important temperate nut fruit of India which is not only consumed on large scale within the country but also earns foreign exchange. Walnut is grown extensively in almost all the temperate regions of Uttaranchal. The area under walnut cultivation in the state is 16723 ha yielding 9099 MT nuts. The walnut has a tremendous export potential provided international standards are maintained in production and marketing technology. The good quality thin-shelled cultivars are to be propagated vegetatively for planting instead of seedlings, which come into bearing very late and do not produce nuts of uniformly good quality. In most parts of Uttaranchal, seedling plants are planted; hence thick shelled and poor quality nuts are found. Keeping the unlimited export potential of walnut and favourable climatic conditions in view, it is highly imperative to give proper emphasis on plantation of export oriented thin shelled cultivars, which are vegetatively propagated.

Almond (*Prunus dulcis*) is a commercial nut in India and commanding widespread popularity throughout the world because almond kernels are rich source of oil, protein, minerals and vitamins. Almonds are highly acclaimed for their dietic, cosmetic and pharmaceutical quality. Its cultivation is confined to only 2120 ha in Uttaranchal with a production of 768 MT. Although, climatic conditions of the state is quite favourable for almond cultivation, but systematic efforts have not been made so far to exploit its potentiality in the state. Temperature is the most important climatic factor and almond requires a mean winter cold for a regular bud sprouting in spring; has a moderate tolerance to low temperature during winter and possesses a low tolerance to early spring frost because of its early blossom. The commercial cultivation of this nut in Uttaranchal could not spread much on account of intrinsic problems like damage by rains, wind, frost and hails during the blossoming time. Another very important limiting factor in the cultivation of almonds is the lack of proper pollinizer as almost all the commercial cultivars are self-incompatible. Problem of gummosis is also one of the constraints for its cultivation which affect the fruits during its developmental stage, but it can be controlled when proper management practices are adopted in integrated manner.

The pecan nut (*Carya illinoensis*) was introduced from U.S.A. early in 1950. The pecan plants raised from the seed introduced in Almora in 1954 by Mrs. Gertrude Emeison Sen, American wife of Dr. Boshi Sen, founder Director of Vivekanand Laboratory, Almora became source of seedling plantations in Uttaranchal hills. It is closely related to walnut belongs to Juglanadaceae family. It can be grown successfully in areas having an elevation of about 914–1829 m above mean sea level, free from spring frost and characterize a mild temperate climate. Its large trees are quite suitable for planting in road side or as windbreak, the trees grows from 70 to 100 feet tall with deeply ridged, brown-gray bark. The leaves are yellowish-green with finely toothed leaflets. The deep brown fruit is encapsulated in a thin and brittle shell and husk. The nut kernel is sweet to the taste, rich source of fat, protein, carbohydrate, minerals and vitamins. It is being cultivated in Pithoragarh, Uttarkashi, Chamoli, Almora, Pauri Garhwal districts in limited areas. The major problem in pecan cultivation is the plantation of seedling plants. Seedlings have problem of late bearing, low density, low productivity and great variation in cropping and produce inferior quality fruits. The vegetatively propagated plants reduce the juvenile phase and variation in cropping as well as produce very good yield of quality fruits.

The hazel or European filbert or Bhotia badam (*Corylus avellana*) is essentially an introduced nut fruit in this country even though a few thousand plants of native hazel (C. *colurna*) occur in remote extremely cold forest areas of J&K and Himachal Pradesh and may be in similar areas of Uttaranchal hills as well. The production from these is considerably low probably because of poor fruit set and inaccessibility to the trees. In hills of Uttaranchal, the seedling trees of hazelnut (*Corylus colurna*) locally known as 'Kapasi', are found in bearing with uneconomic importance because of its

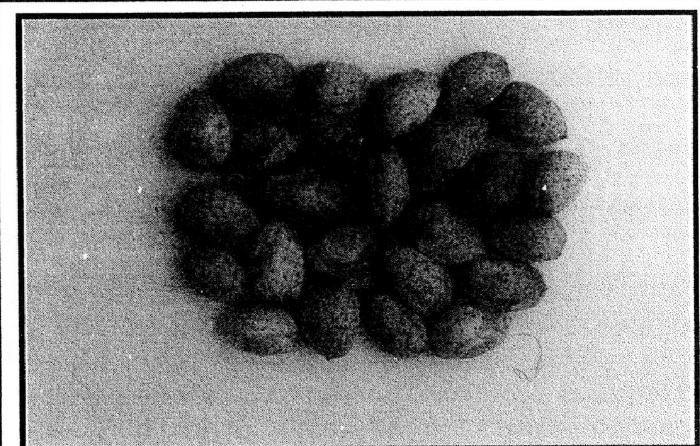

Plate 20.1: View of Mature Almond Variety IXL

Plate 20.4: Walnut Leaves Infested with Hairy Caterpillar

Plate 20.2: A Fruit Developmental Stage of Walnut Variety Chaubattia Kagzi

Plate 20.5: A View of Mature Walnut Variety Chaubattia Kagzi

Plate 20.3: A View of Green Almond Variety IXL

unacceptable nut quality. In recent years, some improved cultivars have been introduced from abroad and northern hills of other states for evaluation under Uttaranchal hills conditions on the basis of their economic importance. Its kernels rich values in energy and contain fat, protein, minerals, vitamins and starch. It is eaten fresh or, more commonly, dried. They are used in confectionery for making particular kinds of chocolate and nougat candies. When subjected to a pressing process, the filberts yield a fair amount of oil, which although not one of the best from a food point of view, is often used in the preparation of cosmetics and perfumes. Very thin crepes filled with a cream made of lightly roasted, finely ground hazelnuts.

Pistachio nut (*Pistacia vera*) is one of the most valuable nut with high demand for its delicacy. It is a member of the family Anacardiaceous. It is a small spreading type tree. Leaves are leathery with pinnate leaflets. The bearing habit is lateral on previous season's wood. It is a dioecious plant and produces huge flowers of both pistil late and staminate type. Wind pollination is the rule. It is mainly used as an ingredient of sweetmeat, confectionary, ice-creams and also used as dessert. Fruit husks are utilized as marmalade. The kernels are often processed for their oil, before the nuts are sold for commercial use; the shell is either bleached white or dyed red. The composition of nut (per 100 g) is as follows: 16.2 g carbohydrate, 19.8 g protein, 53.5 g fat, 2.8 g minerals, 2.1 g fiber, 140 mg calcium, 430 mg phosphorus, 7.7 mg iron, 0.67 mg thiamine, 0.28 mg riboflavin, 2.3 mg niacin, 144/µg carotene, 5.6 g water. Besides, potassium, magnesium and ascorbic acids are also present. The kernel yields about 56 per cent oil. The fatty acids of the oil include myristic, palmtic, stearic, oleic and linoleic.

The sweet or Spanish chestnut (*Castanea sativa*) is one of the hardiest nut trees, native to southern Europe. Large wild chestnut forests can still be found in Corsica, France and Italy. The trees eventually grow to a huge size-up to 115 feet height. Its contain somewhat less than 60 per cent water and as much as 37 per cent sugar and starch. The rest is provided by nitrogen compounds, cellulose, ash, lipids and mineral salts. So there is a large amount of carbohydrates, while the protein is present only in traces. Chestnuts were considered a staple food and inferior flour can be made of them. In Corsica they are made into a kind of polenta, a thick, dry paste which is poured onto a floured cloth, then cut into slices with a thread (as are certain cheeses), and either eaten as they are, or fried or grilled Mont Blanc, by comparison, is a very sophisticated dessert, and can be easily prepared by using a tin of sweetened chestnut puree, whipping it up with brandy or any liqueur, and serving it chilled, in individual dishes with stiffly beaten cream piled on top. It is a tiring and painful process, involving burnt fingers, to make the puree oneself, although the empty can always be hidden in the trash can and the impression given that the hostess is a cordon-bleu. Unsweetened chestnut puree, also available commercially, is a good accompaniment to Turkey and Brussels sprouts. When roasting chestnuts, a small cut should be made on the domed face of the shell, so as to avoid explosive popping. Another use for chestnut flour is as castagnacci, thick fritters with a chestnut flour base. Chestnuts can also be made into preserves particularly that true delicacy, the candied chestnuts known as morons glaces. Chestnuts have undoubtedly a high food value.

Pest Monitoring and Forecasting Systems

Necessity, economy and efficacy of control measures can be determined if there is a system of prognosis or forecasting the extent and time of occurrence of plants diseases and pests. Forecasting involves all the activities in ascertaining and notifying the farmers in an area/region that, (*i*) conditions are sufficiently favourable for certain diseases and pests causing economic losses; that, (*ii*) application of control measures most acceptable under integrated diseases/pests management will result in economic gain and (*iii*) the amount of diseases/pests expected is unlikely to be enough to justify the expenditure of time, energy and money to control. This concept of forecasting requires complete knowledge of biology and epidemiology on insect-pest and pathogen *i.e.*, development of disease/ pest in plant population under the influence of the factors associated with the host, the pathogen/pest and weather (Thakur, 1999).

The accumulated losses through these diseases are more than 40 per cent annually in all the nut fruits growing States of India. Likewise, insect-pests account for nearly 35 per cent of the total damage at pre and post harvest stages. Though the magnitude of pests incidence varies from region to region and orchard to orchard but in all, 24 insect-pests have been rated economically important of which about 12 account for the major share of the damage throughout the states. Among the insect-pests usually walnut weevil, codling moth, naval orange moth, walnut husk fly are the most dominant and principal insect-pests, affecting 82 per cent orchards each and followed by hairy caterpillar and hickory shuck worm. These insect-pests are being managed through high potency pesticides every year to maintain the economic threshold level (Verma *et al.*, 1986 and Singh and Lal, 2002).

Now, the question is that how long this practice will go on and what would be the consequences of these chemicals on human and plant health, soil fertility and crop productivity and quality? Yes, they are. They affect the soil microbial population, which transfer and make the nutrients available to the plant system. However, answer to this problem is to adopt need-based use of pesticides aided by an effective forecasting system through monitoring weather parameters for predicting diseases and pests outbreaks as well as production. This system has taken over the chemical management practices to judicious use of chemicals on temperate fruit crops world over.

Agro Ecosystem Analysis (AESA)

Agro Ecosystem Analysis is an approach, which can be gainfully employed by extension functionaries and farmers to analyse field situation with regard to pests, defenders, soil conditions, plant health, the influence of climatic factors and their interrelationship for growing healthy crop. Such a critical analysis of the field situations will help in taking appropriate decision on management practices. The basic components of AESA are:

1. Plant health at different stages.
2. Built-in-compensation abilities of the plants.
3. Pest and defender population dynamics.
4. Soil conditions.
5. Climatic factors.
6. Farmers past experience.

Methodology of Agro Ecosystem Analysis

Field Observations

1. Enter the orchard atleast 15–20 feet away from road/bund. Select fruit tree of medium size randomly.
2. Record the visual observations on the following parameters:
   (a) Flying insects (both pests and defenders).
   (b) Close observation on pest and defenders which remain on the plant.
   (c) Observe the appearance of major pests/diseases and their intensities and defenders like Lady Bird beetle, Green Lace wing, Anthocorid, Bugs, Spiders, Predatory Ants, Phystoseiid mites and syrphids.

3 Frequently visit the orchard during peak period of pests/diseases incidence and also record their level of incidence to adopt timely management strategies.
4. Record soil conditions *viz.* dry, wet or water logged.
5. Observe Rodent live burrows.
6. Repeat the steps on atleast 5 trees in a diagonal row or cris cross fashion across the orchard to have overall average assessment of the orchard.
7. Record the climatic factors like sunny, partially cloudy, cloudy, rainy etc. for the proceeding week.

Drawing

First draw the plant with actual number of main limbs in the center of a chart. Then draw pests diseases on left side and defender on the right side. Indicate the soil condition, rodent damage, etc. Give natural colours to all the drawings, for instance draw healthy plant with green colour, diseased plant/leaves with yellow colour. While drawing the pests and the defenders on the chart care should be taken to draw them at appropriate part of the plant, where they are seen at the time of observation. The common name of the pests, diseases and defenders and their population count should also be given along with diagram. The weather factor should be reflected in the chart by drawing the diagram of sun just above the plant, if it is sunny. If cloudy, the clouds may be drawn in place of sun. In case of partially sunny, the diagram of sun may be half masked with clouds.

Group Discussion and Decision Making

The observations recorded in the previous and current charts should be discussed among the farmers and apprentice trainees by raising questions relating to change in pest and defender population in relation to crop stages, soil conditions, weather factors, such as rainy, cloudy or sunny etc. The group may evolve a strategy based upon week AESA, ETL and corresponding change in pests, diseases and defenders ratio and take judicious decision for specific pest and disease management practices.

AESA by Extension Functionaries

The extension functionaries during the regular visit to the village should mobilize the farmers, conduct AESA and critically analyse the various factors such as the pest population viz-a-viz defender population and their role in natural suppression of the pest, the influence of prevailing weather condition/soil conditions on the likely build-up of defender/pest population. They may also take the decision based on the AESA, with IPM components like release of defenders, application of neem based formulations/safe pesticides are to be used for specific pest situation. Such an exercise may be repeated by the extension functionaries during every visit to the village and motivate the farmers to adopt AESA in the fields.

AESA by Farmers

After a brief exposure during IPM demonstrations/field trainings, farmers can practice AESA in their own fields. Wherever trained farmers are available, their experiences could be utilized in training their fellow farmers in their own villages. Thus a large group of farmers could be made efficiently competent in undertaking weekly AESA, thereby empowering themselves in decision making on any specific pest/disease situation. Farmer-to-farmer training approach will go a long way in practicing IPM on a large area on sustainable basis.

Pests and Diseases Management of Nut Fruits in Nursery

Nursery is one of the important components of nut fruits production. There is lot of demand for quality, genuine and pest and disease free planting material. Nursery is a place where young crop plants are raised under intensive care for later transplanting to the field. The use of quality nursery plants is the pre-requisite for the success of any orcharding enterprises. Therefore, the production of healthy and pests/disease free planting material of high quality deserves special attention. The number of genuine and quality walnut plants varieties produced by Agricultural Universities, State Departments of Horticulture, Research Institutions and Private Sectors are very meager to meet the demand of large number of growers through out the country. For this reason, most of the nut fruits growers, orchardists are depending on private nurseries and nurserymen for their requirement of fruit plants. Hence, there is a greater scope to prosper for the nurserymen, who can supply genuine and high quality pests and disease free planting material of different varieties of walnut plants produced on scientific lines. Such high quality planting material could be produced by careful planning, establishment and management of nurseries (Singh and Mishra, 2003). The success of any nursery depends largely upon several factors *viz.*, proper selection of site, careful planning, selection of varieties which are in good demand and procuring the progeny from the reputed institutions/nurseries, elite gardens, establishment of mother plants block/scion banks, acquiring scientific information of each variety and skillful management in production of necessary scions as well as marketing the same.

Following pest and disease management practices should be adopted at nursery stage or during establishment of nut fruits nursery.

1. Raise nursery only in solarised soil. In hilly regions the area which has been identified for nursery raising should be moistened and covered with polythene sheet during May up to sowing period which is done in November. It is also important to cover all 4 sides of polythene with soil or any other material in order to make it air tight and also to increase the temperature of covered area. This helps in controlling all kinds of soil borne pathogens and pests.
2. Before four weeks of sowing of seeds in nursery during November, apply *Trichoderma viridae* @ 1 g per kg of FYM and *Beauveria bassiana* @ 1 g per kg of FYM separately in same day. The powder of both the antagonists should be thoroughly mixed in FYM separately. This technique is very effective in controlling soil borne diseases and pests in nursery. Besides, FYM also useful for providing nutrition to the plants for ideal vegetative growth (Bhalla and Gupta, 1993).
3. For the management of seedling blight in nursery, apply *Trichoderma viridae* or *T. harizianum* @ 0.4 per cent in the soil or drench the soil with thiram 0.3 per cent or aureofungin 0.04 per cent.
4. Crown gall disease can be managed by molecular characterization of agrocin and non-agrocin producing strains of nonpathogenic *A. tamefaciens* strain K-84 biocontrol agent.
5. Do not grow the nursery at the same site every year.
6. Destroy the affected seedlings.

Production of Virus-free Nursery Plants in Nut Fruits

Among viruses black line disease is known to cause hugs losses ranging from 30-50 per cent in walnut crop in several countries like USA, China, France, Italy, Turkey, Poland, Yugoslavia, Rumania, Japan, Ukraine and Austria, reducing productive age of the plant in addition to yield loss. The importance of this disease is increased many fold when the virus goes untested due to their latent nature. In India this virus has not reported but we should take proper care in production of nursery plant and planting materials of these countries should not be imported.

The presence of viruses in the mother stock can be tested by highly sophisticated techniques like Electron microscopy, Enzyme Linked Immunosorbant Assay (ELISA) or by Biological Indexing using indicator hosts. The procedure for maintenance of virus free nursery stock is given below (Atwal and Dhaliwal, 1997).

1. Use only virus indexing propagating material.
2. Establish and maintain virus tested mother stock.
3. Determine the presence of virus vector in or near nursery area and apply control measures, if necessary.
4 Avoid re-grafting of rootstock where first bud-take has failed.
5. Avoid top working the mother plants.
6. Maintain recommended isolation and adjust free spacing to minimize the possibility of natural root grafting.
7. Remove wild hosts growing in or near virus indexed plantings.
8. Use only sterilized propagation tools.

Pest and Disease Management Practices During Establishment of Orchard in Nut Fruits

Integrated Pest Management requires through preplanning even before establishing the orchard which includes selection of suitable site, soil type, regular pest and weather monitoring in the orchard, balanced use of fertilizers, water management, proper training and pruning, improved cultural practices, augmentation and conservation of parasitoids and predators application of bio-pesticides, neem based formulation and need based application of selective pesticides (Singh, 1964 and Singh *et al.*, 2004).

1. For plantation of fruit plants in the orchard, digging of pit should be done of 1x1x1m size in the month of September-October and plantation should be made in January. Before one or two days planting, the pit must be thoroughly treated with carbendazim @ 10g per pit. The soil which is to be filled in the pit should also be treated by mixing carbendazim 20g in the soil which is needed for one pit.
2. Before plantation, the nursery plants must be treated with chlorpyriphos solution at 0.05 per cent for 20 minute to check the spread of soil borne pests.
3. Plants which are infected with Armillaria root, crown gall, crown rot or any other soil borne diseases should not be planted in the orchard.
4. Before plantation in the orchard, all the nursery plants must be dipped in copper sulphate solution of 1 per cent for one hour. This treatment takes care from soil borne diseases like crown gall, Armillaria root and crown rot.

Pest and Disease Management Practices in Established Orchard

1. Orchard sanitation should be given top priority. Sanitation is a simple, but important preventive measure of pest and disease control. Even simple removal of alternate hosts, weed hosts and crop residues greatly reduce the infesting potential.

2. During pruning of walnut trees in December-January, also prune and destroy the diseased and infested twigs and highly affected thin branches. Use Chaubattia Paste or copper oxychloride 0.3 per cent after pruning on the cutting portion. Correct pruning allows adequate air circulation round the tree canopy, thereby modifying the microclimate and reducing pest and disease development.
3. Digging the basins of the tree with spades during winter (October-November) to expose the diapausing pupae and spores to adverse climatic conditions.
4. Over wintering caterpillars should be collected from the loose bark, other debris, bands used for trapping and destroyed.
5. Apply lime sulphur @ 1:20 (one litre lime sulphur in 20 litre of water) every year in December or January after pruning. It helps in managing fungal diseases.
6. Mulching with dry grass or hay (10–15 cm in thickness) in the tree basins during March should be done to conserve the moisture during fruit setting and fruit development period.
7. Avoid excessive use of nitrogen and application of nitrogen should be made in split doses only.
8. Apply white wash impregnated with glue regularly on the tree trunks to avoid sun burn and winter injury.
9. If plants/trees exhibit the deficiency of micro nutrients go for the application of the same on the basis of soil and leaf analysis.
10. Two to three Bee colonies per ha should be provided in each orchard at the time of flowering to increase pollination.
11. In case of naval orange worm infestation, prompt harvest, hulling and fumigation of nuts can reduce losses substantially. In case of severe infestation, apply imidaclopride @ 0.01 per cent on need basis.
12. The grubs of walnut weevil can be managed by collecting and destroying the fallen fruits in May-June. It also helps in minimize the damage of walnut husk fly. If infestation is serious, spraying of endosulfan 0.07 per cent can be done. The spraying should be commenced from the time of 1st attack.
13. Application of *Trichogramma embryophagum* (parasitoid) @ 2000 adults per infested tree at the interval of one week is found effective in checking the population of codling moth.
14. Growing of flowering plants, especially marigold and maize on the peripheries and legumes as inter-cropping help in conservation of both predators and parasitoids, especially Chrysoperlla and Anthocorids and also help in nematode management.
15. Walnut blight can be controlled by spray of copper containing fungicide like copper oxychloride 0.3 per cent just before flowering and before rainy season, depending upon the level of incidence.
16. The collection and burning of infected leaves are found effective against anthracnose disease, while the infestation is severe; apply captan 0.1 per cent at fortnightly interval. The second spray should be done only on need basis.
17. Proper drainage, avoidance of contamination during cultural operations, keeping of graft union sufficiently high while planting and use of Paradox rootstock are some effective measures to control the phytophthora root and crown rot.

18. Fallen walnut fruits should be collected throughout the season and either buried deep in the soil or destroyed otherwise.
19. All loose bark of walnut trees should be scraped off to remove over wintering sites for the caterpillars. Vicinity of trees should also be kept clean of packing cases and all other debris which are likely to shelter the over wintering caterpillars of codling moth during August to mid October.
20. In areas where natural population of predatory coccinellids is high during summer (April-June), application of toxic chemicals should as far as possible be avoided.
21. For management of Armillaria root and crown rot diseases, the basin of the infected tree is brought to a moisture level of 30–40 per cent and 15–20 cm deep holes at a distance of 30 cm are dug with a crow bar all over the basin. These holes are closed after filling with 0.1 per cent suspension of carbendazim. The treatment must be repeated 3 to 4 times every year during the rainy season.
22. Application of *Trichoderma viridae* @ 40 g per tree has shown good result against Armillaria root and crown rot at initial stage.

Pest and Disease Management Practices in Almond Orchard

1. A single spray of copper oxychloride 0.3 per cent or Carbendazim 0.1 per cent or Captan 0.3 per cent at pink bud stage (10 days before flowering) provides nearly perfect control of leaf curled disease.
2. Effective control of stigmina blight is possible by spraying captan 0.3 per cent, once at leaf fall in autumn and second time at bud-swell stage in spring.
3. Almond blight can be controlled by spraying copper oxychloride at 0.3 per cent at pink, petal fall and preharvest stages.
4. For the control of blossom wilt in almond, spray carbendazim 0.1 per cent just fefore flowering and repeat it after 7 days, if weather is warm and wet.
5. Gummosis is a major problem in almond cultivation, which can be managed through spraying of lime sulphur 1:20 at the time of pruning of the plants in January. After this a combined spray of copper oxychloride 0.3 per cent and borax 0.3 per cent, just after the fruit setting and second, 20 days after the first spray.
6. Collect and destroy grubs of root borer during December-January coinciding with the preparation of tree basins.
7. Adult beetles of root borer can be collected from the light source (light trap) at night and killed in kerosenized water.
8. Treat the tree basin around the trunk with chlorpyriphos 0.05 per cent or monocrotophos 0.04 per cent in infested tree basin of root borer in December–January.

Pest and Disease Management Practices in Pecan Nut Orchard

1. For control of hickory shuckworm, all the shucks after harvest should be burnt after collection. It is difficult to control the pest by insecticidal sprays because of overlapping generations and reinfestation from adjoining unsprayed trees. However, first application of thiomethoxam 0.05 per cent should be done at the beginning of shell hardening followed by another one application after 15 days, according to need.

2. Insecticidal spray of endosulfan 0.07 per cent is found quite effective in controlling pecan nut case bearer. The application should be done when about 10 per cent moths have emerged after pupation or tips of small nuts have become brown.
3. Early maturing cultivars are more attacked by pecan weevil than late maturing ones. Therefore, farmers should try to promote the cultivation of early maturing varieties. However incase of infestation of pecan weevil, first spray should be applied as soon as the nuts begin to harden.
4. If incidence of black pecan aphid is found during monitoring of the crop, spray the crop with methyl demeton 0.04 per cent or thiomethoxam 0.05 per cent.
5. Collect and destroy grubs of root borer during December-January coinciding with the preparation of tree basins.
6. Adult beetles of root borer can be collected from the light source (light trap) at night and killed in kerosenized water.
7. Treat the tree basin around the trunk with chlorpyriphos 0.05 per cent or monocrotophos 0.04 per cent in infested tree basin of root borer in December-January.

General Pest and Disease Management Practices in Established Nut Fruits Orchard

Cultural Practices

1. Orchard sanitation should be given top priority. Sanitation is a simple, but important preventive measure of pest control. Even simple removal of alternate hosts, weed hosts and crop residues greatly reduce the infesting potential.
2. Cutting and burning of severely infested and dried branches.
3. Digging the basins of the tree with spades during winter (December–January) to expose the diapausing pupae and spores to adverse climatic conditions.
4. Fallen fruits should be collected throughout the season and either buried deep in the soil or destroyed otherwise.
5. Over wintering caterpillars should be collected from the loose bark, other debris, bands used for trapping and destroyed.
6. Collection and destruction of fallen leaves from the orchard floor during winter are found effective.
7. Correct pruning during winter allows adequate air circulation round the tree canopy, thereby modifying the microclimate and reducing disease development.
8. Plant early maturing varieties of peach and almond in case of peach fruit fly and gummosis endemic areas.
9. Make use of balanced dose of chemical fertilizers.
10. Mulching with dry grass or hay (10–15 cm in thickness) in the tree basins during March should be dome to conserve the moisture during fruit setting and fruit development period.
11. Avoid excessive use of nitrogen and application of nitrogen should be made in split doses only.
12. Apply white wash impregnated with glue regularly on the tree trunks to avoid sun burn and winter injury.

13. If plants/trees exhibit the deficiency of micro nutrients go for the application of the same on the basis of soil and leaf analysis.
14. Growing of flowering plants especially marigold and maize on the peripheries will help in conservation of both predators and parasites.
15. Two to three Bee colonies per ha should be provided in each orchard at the time of flowering to increase pollination.
16. Proper irrigation schedule should be adopted.
17. Trees should neither be forced to drought nor water logging conditions.
18. Rogue out infected plants.
19. Maintain vigour of the tree to keep away shot hole/pin whole borers, bark beetles and other pest infestation.
20. Grow marigold, berseem and mustard in between lines of Plantation to check the nematode population.
21. Do not delay harvesting of fruits.
22. Keep the trees as free as possible from mechanical wounds, winter injury, crotch separation and cankers.
23. Pruning cuts should be made close to the branches leaving no stubs.
24. Cut wounds should be covered with superior white lead paint or any other paint.
25. Remove and destroy dead and drying fruit trees to ward off borer infestation.
26. Clean the stem borer holes with flexible wire and apply the recommended chemical.
27. Cover the exposed part of the stem to sun with dry grass or gunny bags soaked with methyl parathion 0.1 per cent once a month in March till October.
28. Install a light trap near the orchard to collect and kill the beetles in kerosenized water.
29. Shake non bearing trees over a cloth sheet at dusk, collect and destroy the collected beetles (defoliating and fruit eating beetles).
30. Clip off terminal shoots with unshed cluster of dry leaves in winter for the destruction of shoot borers.
31. Remove the dead bark and frass and apply water proof paint on hard wood to avoid borer attack.
32. Add well rotten manure in case of termite infestation.
33. Destroy termitaria in the vicinity of the orchards.
34. Apply Chaubattia paste after cleaning the weeping wounds at the time of dormancy break for the control of bacterial gummosis.
35. Cover fresh wounds with Chaubattia paste or copper oxychloride paint.
36. Remove the infected roots and apply Chaubattia paste on the cut ends.
37. Collect and destroy the grubs of root borers while preparing basins.
38. Ensure complete collection and destruction of foliage and pruned wood in the orchards itself after leaf fall.
39. Clip off mildewed twigs and destroy them.

Table 20.1: Major Pests of Nut Fruits in Indian Himalayas

Common Name	*Scientific Name*	*Period of Damage*	*Distribution*	*Host Range*	*Nature of Damage*
Walnut weevil	*Sphenoptera lafertei* Thanpson	April-September	Himachal Pradesh, Uttaranchal, J&K	Walnut	Grubs bore into the fruit and reduce the kernels into a black rotten mass. Attacked fruits drop prematurely with grubs inside
Codling moth	*Cydia pomonella* Linnaeus	April-June	J&K	Walnut	The newly hatched larvae bore into the blossom end of the immature nuts causing premature abscission and emerge to pupate in late spring and early summer
Naval orange worm	*Paramyelois transitella*	June-August	J&K, Himachal Pradesh	Walnut	Moths of naval orange worm emerge from mummy nuts on the tree or ground in late spring and lay their eggs on nuts damaged by codling moth. A second generation emerges during hull split and attacks on sound nuts causing wormy unmarketable nuts
Walnut husk fly	*Rhagoletis cingulata*	August-September	J&K, Himachal Pradesh	Walnut	It causes damage to only mid and late season cultivars feeding on the fleshy tissues of the husk. The nut shell gets stained rendering the nut in shell unfit for sale
Hairy caterpillar	*Lymantria obfuscata* walker	April-June	J&K, Uttaranchal	Walnut, Almond	Damage on leaves, green fruit nuts and if serious infestation causes total defoliation of the trees. Leaves are eaten away from the margin by the caterpillar
Hickory shuckworm	*Laspeyresia caryana*	May-June	Himachal Pradesh, J&K	Pecan nut	This moth is a serious pest of pecan and other hickories which lays eggs on foliage and green nuts. Upon hatching, the cream coloured small caterpillars attack the immature nuts which prematurely fall to the ground
Pecan nut case bearer	*Acrobasis caryae*	May-June	Himachal Pradesh, J&K	Pecan nut, Chestnut	The pest causes great damage to pecan nuts as a single larva can destroy the entire cluster. The moths overwinter as larvae in small cocoons located near the buds on the stem. As the buds begin growth in spring, the larvae activate and feed on them and later bore into new shoots and pupate there

Contd...

Table 20.1–Contd...

Common Name	*Scientific Name*	*Period of Damage*	*Distribution*	*Host Range*	*Nature of Damage*
Pecan weevil	*Curculio caryae*	August-September	Uttaranchal, Himachal Pradesh, J&K	Pecan nut, Almond	The weevil damage the nuts both in the orchards and storage. The type of damage includes premature nut drop, black spots on the kernel, shriveled kernel and destruction of kernel by larvae
Black pecan aphids	*Myzocallis caryaefolia*	April-May	J&K	Pecan nut, Almond	It feeds on both sides of leaves and appearance of large yellow angular blotches is the first sign of its injury. The blotches later turn brown and further damage causes premature leaf fall adversely affecting the current as well as subsequent crops
Peach leaf curl aphid	*Brachycaudis helichrysi*	February-March	Uttaranchal, Himachal Pradesh, J&K	Almond	It damages floral and vegetative buds by sucking sap due to which unfolding leaves curl up, remain smaller, get much distorted and later on turn pale. Flowers appear sickly and dull; and inside numerous aphids and nymphs sucking sap can be seen
Root borer	*Dorysthenes hugelii redt.*	Round the year	Uttaranchal, Himachal Pradesh, J&K	Walnut, Almond, Pecan nut Chestnut	In badly infested orchards, grubs of all stages can be seen feeding on root zone and from a tree basin as many as 100 or more grubs can be collected. The grubs either bore into or girdle around roots. Consequently roots get severed and branches wither and die
Stem and shoot borers	*Apriona cinerea chevrolet*	Round the year	Uttaranchal, Himachal Pradesh, J&K	Almond, Chestnut, Walnut	Larvae bore into branches, stem and tree trunk just under the bark and sometimes feed on the heart wood. The larva cuts circular holes on the bark to eject frass of the tunnel through it and such holes are at a distance of 10–15 cm on the branches.

Table 20.2: Major Diseases of Nut Fruits in Indian Himalayas

Common Name	*Scientific Name*	*Period of Damage*	*Distribution*	*Host Range*	*Nature of Damage*
Walnut blight	*Xanthomonas campestris* pv. jugalandis	July-September	Uttaranchal, Himachal Pradesh, J&K	Walnut, Pecan nut	The earliest visible symptoms in the spring appear on the leaf lamina as water soaked spots which develop into brown necrotic areas. Early infection causes nut drop while late infection results in nut shriveling and discolouring
* Black line disease	Viral	Not found	Not found	Walnut	It is caused by a walnut strain of the cherry leaf roll virus and results in a lethal girdle at the graft union of Persian walnut grafted on hyper sensitive rootstock
Phytophthora root and crown rot	*Phytophthora* spp.	June-September	Uttaranchal, Himachal Pradesh, J&K	Walnut, Pecan nut	The rot is characterized by small chlorotic leaves, premature defoliation, decreased yield, die back of terminal shoots and subsequent collapse and death of the tree
Deep bark canker	*Erwinia rubrifaciens*	March-October	Uttaranchal, J&K	Walnut, Almond	This disease is characterized by large longitudinal cracks in the bark which ooze a reddish brown mixture of bacteria, dead cells and plant sap
Die back	*Glomerella cingulata*	May-July	Himachal Pradesh, J&K	Walnut	Irregular dark brown spots appear on leaves which involve the entire leaf causing defoliation. Blighting of twigs is also common and young plants may die
Crown gall	*Agrobacterium tumefaciens*	Throughout the year	Uttaranchal, Himachal Pradesh, J&K	Walnut, Pecan nut, Almond	This soil borne bacterium causes gall that can eventually girdle a walnut tree
Anthracnose	*Gnomonia leptostyla*	May-July	Uttaranchal, Himachal Pradesh, J&K	Walnut Almond	It attacks leaves, young shoots and fruits. Initially small, circular light brown spots develop into large blotches involving a major part of leaf lamina
Armillaria root and crown rot	*Armillaria millea*	Throughout the year	J&K	Walnut Pecan nut	The symptoms are similar to *Phytophthora* rot, posing a problem of identification. Signs of *Armillaria* infection are rhizomorphus, mycelial fans under the bark and distinct mushroom like odour can be used to distinguish the two diseases

Contd...

Table 20.2–Contd...

Common Name	*Scientific Name*	*Period of Damage*	*Distribution*	*Host Range*	*Nature of Damage*
Peach leaf curl	*Taphrina deformans*	February-March	Uttaranchal, J&K, Himachal Pradesh	Almond	Symptoms of leaf curl are confined to current year's growth of leaves, twigs, blossoms and fruits. The leaf blade along the midrib becomes thickened and puckered. Infected twigs become swollen and are killed. Infected flowers and fruits drop rapidly
Pecan scab	*Cladosporium caryigenum*	March-June	Himachal Pradesh, J&K	Pecan nut	The characteristic symptoms of scab are irregular, olive-green to black spots on leaves or young shoots and small dark brown to black circular lesions on the nuts
Leaf blotch	*Mycospharella dendroides*	April-June	J&K, Himachal Pradesh	Pecan nut	The disease is only serious on neglected, declining trees and nursery plants and the latter are usually the source of infection. The symptoms include the appearance of olive-green, velvety tufts on the undersurface in early summer and yellow spots on the upper surface of leaflets in late season
Stigmina bligh	*Stigmina carpophylla*	March-June	J&K, Himachal Pradesh, Uttaranchal	Almond	The disease attacks twigs, blooms, leaves, fruits and unopened buds, Light yellow to reddish coloured spots appear on leaves, which subsequently fall down to form shot holes. Fruits develop dull brown measles on the outer surface
Almond blight	*Crytosporiopsus sp.*	March-June	Uttaranchal, J&K, Himachal Pradesh	Almond	The attack is on twigs, leaves, petioles, blossoms and even developing hulls, causing their premature drying or death and other effects are such as curling of leaves, defoliation and blossom blight
Blossom wilt	*Monilinia laxa*	March-May	Uttaranchal, J&K, Himahcal Pradesh	Almond	Flowers wither to give a typical blighted and clustered appearance. These remain attached to tree, with considerable dieback of the spurs. Wet weather with moderate temperature favours disease development. The blighted blossoms bear smoky or dull grey fungal tufts
Bacterial canker or Gummosis	*Pseudomonas syringae*	May-August	Uttaranchal, J&K, Himachal Pradesh	Almond	In almond the formation of gum on fruit and bark of the trees are suppored to be complex phenomena. In the rainy season a wax like substance is oozing through the affected site. The bark of the affected site starts cracking and ultimately the plants become dry

* *Note*: Since black line disease is a devastating problem in other walnut growing countries like in USA, China, France, Italy, Turkey, Poland, Yugoslavia, Rumania, Japan Ukraine and Austria, therefore planting material from these countries should not be imported in India.

40. Remove all the mummified fruits, dead fruits and pruning from the orchards.
41. Pruning of suckers and water sprouts be made regularly.

Biological Practices

Conservation

1. Conserve the predators like Lace wings, Lady bird beetles, Carabids, Syrphids, Anthocorid bugs, Mirid bugs, Nobid bugs, Capsid bugs, Spiders, Predatory ants, Phytoşeiid mites, Parasitoids like *Encarsia, Aphytis, Triohogramma, Telenomus* etc. in orchard.
2. Growing of flowering plants, especially Marigold and maize on the peripheries and legumes as inter-cropping help in conservation of both predators and parasitoids, especially *Chrysoperlla and Anthocorids* and also help in nematode management.
3. Collection of egg masses of hairy caterpillars and putting them in a fine meshed cage for emergence of egg parasitoids, protect the parasitoids from orchard spray and predation. The larvae hatched may be destroyed.
4. Collect the different larval instars of hairy caterpillars (suspected parasitized) from orchards and place them in wire meshed cages for emergence of larval parasitoids like *Exorista* sp., *Diriona* sp., *Sarcophage* sp., *Pollenia* sp., *Helina* sp., *Anilastus* sp., *Anthomyia* sp., *Euplectrus* sp., to protect them from scheduled spray effects and bird predation.
5. Collect the pupae of hairy caterpillars from the orchards and place them in wire mesh cages for emergence of pupal parasitoids like *Brachymeria* sp., *Monodontomerus* sp., *Hyposoter* sp., *Exorista* sp. and *Pimpla* sp. to protect the parasitoids from scheduled spray effects and predation.

Augmentation

Monitor the incidence of pests like San Jose scale, woolly aphid, thrips, peach leaf curl aphid, mites, codling moth etc. and make releases of the following.

Predators

Lady bird beetles–*Chilochorus bijugus, Pharoscymmus horni, Coccinella septumpunctata, Chilochorus tristis, Adalia bipunctala, Synharmonia* sp., *Exochomus quadripustulantus, Hippodamia convergens, stethorus* sp. @ 30–50 adults/infested tree (Singh, 2002).

Green Lace Wings

Chrysoperla sp. and *syrphus* sp. 10-20 1 instar larvae/tree.

Parasitoids

Encarsia perniciosi and *Aphytis* spp. @ 2000 per infested tree at least 15 days of insecticidal sprays and 10 days after fungicidal sprays against scale insect. *Trichogramma embryophagum* (egg parasitoid) @ 2000 per infested tree at weekly interval. Aphelinus mali @ 1000 adults per infested tree. The dosage may vary depending upon the intensity and extent of damage.

Chemical Control Measures

1. Need based, judicious and safe application of pesticides are the most vital segments of chemical control measures under the IPM programme. It involves developing IPM skills to play safe with environment by proper crop health monitoring. Observing ETL and conserving

natural biocontrol potential before deciding in favour of use of chemical pesticides as last resort.

2. Following suggestions have important bearing for success of control measures in the context of IPM strategy:
   (a) The number of insecticidal/fungicidal sprays recommended can be minimized as per need after proper surveillance and pest intensity considering both biotic and abiotic factors.
   (b) Avoid mixing of two or more pesticides/tank mixing as it may alter effectiveness of pesticides.
   (c) Repeated application of same pesticides should be avoided as it may develop resistance in the target pest.
   (d) Avoid use of synthetic pyrethroids which may cause resurgence of sucking pests.
   (e) Use safer pesticides during early fruiting phase of crop growth.
   (f) Encourage the use of neem based formulations and biopesticides which are comparatively environmental safer.
   (g) Proper spray equipments should be used to achieve good result.

Strategies of IPM Implementation

An integrated strategy for the management of major pests and diseases has been possible by, (*i*) breeding new varieties with built-in resistance, (*ii*) evolving efficient methods of pest control through pest surveys and monitoring and (*iii*) biological control of pests with the help of conservation and augmentation of natural enemies like parasites, predators and insect pathogens. Economically viable integrated pest management strategies have been developed for the control of major pests in rice, cotton, pulses, vegetable, many fruit crops, sugarcane etc. Control of Pyrilla and top borer of sugarcane, mealy bug of coffee, lepidopterous pests affecting cotton, tobacco, coconut, sugarcane, etc. are a few examples where success has been achieved through the release of bio-control agents. A major achievement has been the development of mass rearing technology for bio-control agents such as *Trichogramma* spp., *Chrysoperla* spp. and nuclear polyhedrosis viruses (NPV) of *Helicoverpa* and *Spodoptera*. Additionally, crop diversity, intercropping and the fact that most agricultural holdings are relatively small and intensively managed favour the implementation of IPM at the farm level. Indian scientists and extension workers are all aware of the problems that can result from over-use of pesticides and the concept of economic threshold for pesticides use is well understood. The Department of Biotechnology is also providing financial assistance to various State Agricultural Universities and Research Centres for developing and producing biopesticides, biocontrol agents *viz.* NPV, GV, *Trichogramma*, *Chrysoperla* and *Trichoderma*. Under this programme biopesticide production units and plant protection clinical centres at regional research stations have been established and strengthened. The use of biopesticides and bio-control agents in India is on the increase but not up to the desired level of growth. Bio-pesticides as a component of IPM are cheaper than pesticides by 50 per cent. Once applied they help and aid healthy growth throughout and are easy to apply. Besides being eco-friendly, they have higher cost-benefit ratio (19 to 360) and do not pose risk of resistance development. It appears that the concept of using biopesticides and bio-control agents among the farmers is still in infancy. Only 1 per cent of 143 million hectares crop area and only about 2500 villages out of over 6 lakh villages in the country have been covered so far under IPM. Thus, there is a vast potential and need to synthesize, validate and promote appropriate location specific IPM modules by all concerned.

Although, IPM has been accepted as the most attractive option for the protection of crops from the ravages of pests (insect-pests, diseases, nematodes, weeds etc.), yet implementation at the farmers level has been rather limited. Pesticides continue to dominate and their injudicious use represents the greatest threat to IPM.

Table 20.3: Efficacy of Different Treatments Against Mealy Bug Infesting Fruit Crops

Treatment	*% Decline in Mealy Bug Population Days After Applications*		
	7	*15*	*21*
Arkur Test Spray Oil 3%	37.7	3234	34.8
	(37.9)	(34.7)	(36.0)
Arkur Test Spray Oil 5%	58.5	56.5	65.0
	(49.3)	(48.7)	(53.7)
Fish Oil Resin Soap 1%	66.3	58.0	63.5
	(54.7)	(49.6)	(52.9)
Fish Oil Resin Soap 2%	76.5	73.9	77.2
	(61.0)	(59.5)	(61.2)
Malathion 0.15%	76.9	75.3	82.3
	(61.3)	(60.2)	(65.1)
Neem Seed Kernel Extract 5%	72.5	68.4	69.7
	(58.6)	(55.8)	(56.7)
Parathion Dusting 2% @ 300 g/tree	49.2	38.8	45.9
	(44.5)	(38.5)	(42.6)
Release of *Cryptoleamus* 4 Beetles an Grubs/Tip	62.6	65.2	72.8
	(52.3)	(53.9)	(58.6)
Dichlorvos 0.1%	71.9	68.2	79.0
	(58.0)	(55.8)	(62.8)
S.E. ±	0.9	2.3	2.4
C.D. at 5% (P = 0.05)	2.8	7.0	7.3

Figures in parentheses are arcsin transformed values.

For an effective implementation strategy, it is necessary to identify the obstacles to its dissemination.

Major Obstacles Dissemination of IPM

1. Low awareness and innovativeness of extension personnel and target groups.
2. Inadequate interaction between research and extension agencies.
3. Problem of timely and adequate supply of quality inputs, including biocontrol agents and biopesticides.
4. Complexity of IPM vs Simplicity of chemical pesticides.
5. The dominant influence of pesticide industry.
6. Non-availability of location specific IPM modules for many crops.

Essentials for Implementation and Dissemination of IPM Approach

1. Availability of location specific IPM modules, which are ecologically sound, economically viable and socially acceptable.
2. High level of target group participation.
3. An elaborate dissemination strategy.
4. Removal of obstacles to dissemination of IPM.
5. Measuring, evaluating and publicizing the impacts of IPM.

Conservation of natural enemies of pests and their augmentation to avoid the use of insecticides, occurrence of residual toxicity and resurgence of pests is of prime importance. The intrinsic property of renew ability, reversibility and resilience of botanicals and biopesticides make them most dependable tools for sustainable IPM. Hence, to conserve ecological balance and to manage the pests as well, the use of bio-agents and biopesticides/botanicals must receive priority attention.

The IPM packages tested at several centres vis-à-vis the farmer's practice indicate the superiority of the farmer for obtaining higher net returns as compared to the latter. IPM practices enabled reduction in number of chemical sprays and consequent pollution. The adoption of IPM technology resulted in 73.7 per cent and 12.4 per cent reduction in number of insecticide sprays for the control of sucking pests and bollworms, respectively, on cotton in Punjab. IPM system also resulted in increase of natural enemies by three fold, reduced the insecticide and environmental pollution by 50.3 per cent and 53.4 per cent, respectively.

The case studies in Andhra Pradesh, Karnataka, Maharashtra (Ashta) and Uttar Pradesh (Shikopur) indicate that it is possible to reduce consumption of pesticides by 50 per cent by adopting IPM strategies involving field sanitation, cultural practices, use of resistant varieties, crop pest monitoring and need based use of pesticides along with bioagents and biopesticides.

Conclusion

From the last few years, considerable progress has been made in the development and promotion of IPM in nut fruits at national level under the various projects and policies. There is quite a lot of information available about IPM in Indian Himalayas where these fruits are grown and wherever IPM practices adopted in true letter and spirit, farmers harvested good quality fruits with minimum use of pesticides. It is also true that IPM packages developed are mainly based on combination of improved cultural practiced and need based application of safer and selective pesticides. The studies conducted under the World Bank funded project of Diversified Agriculture Support Project show that most of the IPM practices including pesticides in walnut are done during dormant stage (October-March) of the tree and if these practices are adopted by the farmers, the incidence of pests and diseases during flowering, fruit setting and fruit developmental period of walnut (April-September) is very less for which 2–3 applications of appropriate pesticides on need basis are quite enough to manage pests and diseases during these crucial stages. Thus, implementation of given IPM strategies does not cause any adverse impact on non target organisms and undesirable residue in the fruits.

REFERENCES

Atwal, A.S. and G.S. Dhaliwal (1997). Pest of Temperate Fruits. In: *Agricultural Pest of South Asia and their Management*, Kalyani Publishers, New Delhi, pp. 274–286.

Bhalla, O.P. and P.R. Gupta (1993). Insect-Pests of temperate fruits. In: *Advances in Horticulture: Fruit Crops, Vol.3,* (Eds.) K.L. Chadha and O.P. Pareek, Malhotra Publishing House, New Delhi, pp. 1557–1589.

Singh, C. (1964). *Temperate Fruit Pests in Entomology*. In: E.S.I, New Delhi, pp. 266.

Singh, S.P. (2002). Role of Biocontrol Agents in Pest Management in the Millennium. In: *Crop Pest and Disease Management: Challenges for the Millennium,* (Eds.) D. Prasad and S.N. Puri, Jyoti Publishers, New Delhi, p. 1–17.

Singh, S.S. and S.D. Lal (2002). Integrated pest management in temperate fruits. In: *Crop Pest and Disease Management: Challenges for the Millennium,* (Eds.) D.Prasad and S.N. Puri, Jyoti Publishers, New Delhi, p. 39–50.

Singh, S.S. and P.N. Mishra (2003). Sitoshna Phalo me Ekikrit kit Prabandhan. *Phal Phool,* 26 (2): 17-19.

Singh, S.S., U.S. Singh and N.C. Nainwal (2004). Present Status and Management of Pests and Diseases of Temperate Fruits in Indian Himalayas. In: *Advances in Plant Protection Sciences,* (Eds.) D.Prasad and Amerika Singh, Akansha Publishing House, New Delhi, p. 305–329.

Thakur, V.S. (1999). Temperate Disease and Pest Management Through Weather Monitoring and Forecasting systems Horti–Pestvision 2020. In: *Proc. National Seminar on Himalayan Horticulture,* (Eds.) Jagmohan Singh, R.P. Sharma and V.K. Sharma, *Hortivision,* p. 346–356.

Verma, K.D., V.K. Gupta and H.S. Verma (1986). Role of fungicides in the control of fruit spots and storage rots of apple. In: *Advances in Research on Temperate Fruits,* (Eds.) T.R. Chadha, V.P. Bhutani and J.L. Kaulm, Dr. Y.S. Parmar University of Horticulture and Forestry, Solan, p. 411–414.

21

Application of Molecular Markers in Identification of Disease Resistance Genes in Plants

A.K. Mukherjee, S.K. Lenka and L.K. Acharya

DNA Fingerprinting Laboratory, Division of Plant Biotechnology, Regional Plant Resource Centre, Nayapalli, Bhubaneswar – 751 015, Orissa

After domestication of plants for their needs human beings often encountered with crop loss due to different stresses like biotic (attacks by different pathogens) and abiotic (physiological and environmental). Out of these, the pathogens were reported to be responsible for a loss of about 12 per cent of crop production globally (James *et al.*, 1990) in spite of continuous release of resistant varieties and pesticides. Furthermore the evolution of new races of the pathogens made the task of the plant breeders and pathologists more difficult to combat with the challenge. So, they are always in search of new techniques, which would be helpful for them to develop new resistant varieties in a short time. Recently developed molecular markers like restriction fragment length polymorphism (RFLP), random amplified polymorphic DNA (RAPD), DNA amplification fingerprinting (DAF), simple sequence repeats (SSR), inter simple sequence repeats (ISSR), expressed sequence tag sites (ESTS), amplified fragment length polymorphism (AFLP) etc. have opened new avenues to the breeders for efficient screening of resistant varieties in a short span of time. These molecular markers are also helpful for identification and isolation of resistance genes.

Recent developments in the field of DNA technology provide exciting approaches for identification of disease resistance genes. DNA contains an individual's genetic blue print. The order of nucleotides in DNA of an individual is unique and thus determines its identity. It is stable and not influenced by developmental or environmental factors. These features impart greater precision, more accuracy and high degree of reproducibility to the assessment of polymorphism between individuals. The current emphasis, therefore, is on the use of DNA profiling systems, which reveal variation in the nucleotide sequence of DNA, thereby generating large number of molecular markers for identification of disease resistance genes.

Tremendous progress has been made in mapping and tagging of many agronomically important genes with molecular markers, which form the foundation of Marker Assisted Selection (MAS) in crop plants. These molecular markers are advantageous over the traditional phenotypic markers. These markers are not influenced by the environmental changes and detectable in all the stages of plant development.

Molecular Markers

The molecular markers detected at genetic level are less affected by the environment and hence can able to distinguish the varietal difference in crops. The various molecular markers discovered so far ultimately depend upon the genomic DNA of an individual. Though the other molecular markers have their special identity and relevance, we now a days generally use DNA marker due to its simplicity and reproducibility.

DNA Markers

DNA markers are alleles of loci at which there is sequence–or polymorphism–that is neutral in terms of phenotype (Jones *et al.*, 1997). Some advantages of DNA markers are that they are not influenced by the environment, are expressed in all tissue and can be scored at all stages of plant growth. A large number of markers have been developed. A survey of the literature reveals description of many DNA marker systems. All these are the modifications of two basic techniques. The non-PCR based (RFLP Southern blot hybridization) and the PCR (polymerase chain reaction) based (RAPD, DAF, AFLP, SSR, ISSR, STS, SNP and SCAR etc.).

Restriction Fragment Length Polymorphism (RFLP)

This is the first molecular marker system to be conceived and developed (Botstein et *at.*, 1980) which is based on Southern hybridisation technique. It employs restriction endonucleases, which recognise four to six nucleotide long sequences called target sites in the DNA molecules to be cleaved. It also utilises single/low copy number DNA sequences as probes. The size (length) of the DNA fragments homologous to the probe DNA depends upon the location of target sites. Suppose that there are two Eco RI sites A and B located 1000 bases (*i.e.*, 1 kb} apart in variety X. In variety Y, however, a point mutation event has created an internal target site C which is 800 bases away from A and 200 bases away from B. Restriction digestion of DNA with enzyme Eco RI will therefore generate a 1 kb AB fragment in variety X and two (0.8 kb AC and 0.2 kb BC) fragments in variety Y. If the probe DNA is homologous to a segment of DNA between A and C target sites, on hybridisation it will bind to the AB fragment in variety X and AC fragment in variety Y, producing 1 kb and 0.8 kb bands respectively. This difference in size of the restricted DNA fragments from homologous sites is called RFLP, which has been used extensively to characterize plant pathogens and host genes. RFLP profiles are highly reproducible and it is possible to determine allelic diversity and frequency in a population.

Highly repetitive DNA sequences called microsatellites < 6 nucleotide long sequence (repeated up to 100 times at a locusand minisatellites < 100 nucleotide long sequence repeated up to 1000 times at locus have also been used as probes in Southern hybridisation to generate highly polymorphic DNA profiles. Use of minisatellites for identification of human individuals has given birth to the concept of multi-loci DNA profiling, appropriately named as DNA fingerprinting (Jeffreys *et at*. 1985 a, b}. It is easy to identify disease resistance genes employing multi-locus DNA probes.

Random Amplified Polymorphic DNA (RAPD)

This marker system is based on polymerase chain reaction. It employs a single decamer primer of arbitrary sequences, which is annealed to the template DNA typically at 37°C (Williams *et al.*, 1990). The RAPD profile is obtained by resolving the amplified products on an agarose gel. The variation in RAPD profile in the form of presence or absence of band results from variation in primer binding sites. A large number of RAPD markers can be generated by the use of a variety of decamer primers. A major limitation of this marker system is non-reproducibility due to low annealing temperature. However, the utility of a desired RAPD marker can be enhanced by sequencing the ends of the polymorphic fragment, designing longer (*e.g.*, 24 nt.) primers for specific amplification through PCR. Paran and Michelmore (1993) have named such a modification that imparts the required reproducibility as sequence characterised amplified region (SCAR).

Sequence Tagged Sites (STS)

This marker class was first conceived and used in human genome mapping (Olson *et al.*, 1989) and recently got included in genome mapping in plants. STS markers are generated by an unmodified PCR. The primers are designed based on the sequence of cDNA, random genomic DNA and ends of large genomic DNA fragment cloned in cosmid, lambda or artificial chromosomes. Besides, sequences flanking the microsatellites can be used to generate PCR primers for assaying the variation in the number of tandem repeats in a given repeat motif at homologous sites. This marker system has all the advantages of PCR technique. A limitation of STS is that considerable efforts and time are needed to generate DNA sequence data for designing suitable primers. Besides, the fragments amplified by the primers may not show length variation between individuals. This problem however, is circumvented by restriction digestion of the PCR products, which detects polymorphism in the target site of a restriction enzyme internal to the two primer binding sites. This modification has led to a new marker system called cleaved amplified polymorphic sequence (CAPS) (Konieczyn and Ausubel, 1993).

DNA Amplification Fingerprinting (DAF)

DAF is another PCR based marker, which employs shorter (< 10 nt.) primers of arbitrary sequence (Caetano-Anolles *et al.*, 1991). The primer annealing temperature is also lower (typically 30°C) as compared to RAPD. The amplified products are separated on polyacrylamide gels and detected by silver staining. This gives a better resolution that is essential for proper visualisation and scoring of large number of bands usually obtained in DAF. The number of polymorphisms scored per run is high. Since the length of primer is short and primer-annealing temperature is low, DAF suffers from the problem of non- reproducibility of some of the amplified fragments hence generally it is not used in molecular marker assisted selection.

Amplified Fragment Length Polymorphism (AFLP)

The AFLP markers are generated by selective amplification of DNA fragments obtained by restriction enzyme digestion (Vos *et al.*, 1995). High molecular weight DNA is digested by two restriction enzymes, one hexacutter (*e.g., Eco RI*) and one tetra cutter (*e.g., Mse I*). Adapter molecules are ligated to the ends of DNA fragments. Two primers possessing sequence complementarity to the adapters as well as few (typically 3) extra random nucleotides at their 3'ends are used for selective amplification of fragments employing PCR. The amplified products are separated on sequencing gels or even on ordinary PAGE and visualised by silver staining. Alternatively, the primers are labeled either by radioisotope or fluorescent dye so that the AFLP profile can be obtained by autoradiography or by using image analysis. The highest number of amplified products (50–100) is produced in AFLP among the entire DNA profiling systems. This increases the probability of detecting polymorphism many folds. The technique is, at present, lengthier and costlier than other PCR based techniques. It requires good quality of DNA for ensuring complete digestion by enzymes. Partial digestion of DNA may result in non-reproducible variation in DNA profiles.

Simple Sequence Repeat (SSR)

SSR are also known as microsatellite marker or variable number of tandem repeat markers. These are short 2-8 nucleotide repeats such as CA or GATA etc., which are repeated in tandem up to hundreds of times at many independent loci, and are ubiquitous in eukaryotic genomes (Lagercrantz *et al.*, 1993). They are generally very highly polymorphic, mainly based on the number of tandem repeat units. SSR markers can be assayed in a similar way like RFLPs using a short oligonucleotide a short probe such as $(CA)_{20}$ to hybridize to blots. However to speed up and simplify the process, SSR markers may be sequence tagged (Morgante and Oliveiri, 1993). This requires sequencing the flanking regions of a specific SSR locus and designing primers, which will amplify the SSR. Polymorphism will be based on either the flanking sequence or, more usually, the number of repeats. SSRs are very powerful markers in that they are single locus, co-dominant and multi-allelic. They do not require radioactivity for detection although this is some times used on polyacrylamide gels to detect accurately alleles, which differ, by one repeat unit (as little as 2 bp).

Inter Simple Sequence Repeat (ISSR)

ISSR fingerprinting was developed such that no sequence knowledge was required. Primers based on a repeat sequence, such as $(CA)_n$ can be made with a degenerate 3' $(CA)_8$RG or $(AGC)_6$TY. The resultant PCR amplifies the sequence between two SSRs yielding a multilocus marker system useful for fingerprinting. PCR products may be radiolabelled with ^{32}P or ^{33}P via end labelling PCR-incorporation, and separated on a polyacrylamide sequence gel prior to auto radiographic visualization. A typical reaction yields 20–100 bands per lane depending upon the species and primer (Godwin *et al.*, 1997a). These markers can also be separated on agarose gel with ethidium bromide visualisation or polyacrylamide gels with silver staining techniques.

Single Nucleotide Polymorphisms (SNPs)

Single nucleotide polymorphisms (SNPs) are DNA sequence variations that occur when a single nucleotide (A, T, C, or G) in the genome sequence is changed. Most SNPs, actually about two of every three SNPs, involve the replacement of cytosine (C) with thiamine (T). SNPs occur every 100 to 300 bases along the human genome. SNPs are stable from an evolutionarily standpoint, not changing much from generation to generation, making them easier to follow in population studies.

SNP scoring is done using SNP-IT (Single-nucleotide Polymorphism Identification Technology), primer-extension SNP scoring technology. In SNP Identification Technology primer extension is a method of isolating the precise location of the site of a suspected single-nucleotide polymorphism and utilizing the inherent accuracy of DNA polymerase to determine the SNP's presence or absence. In SNP- IT primer extension, a specially synthesized DNA primer is bound to the sample DNA to expose the DNA site of interest where a SNP may be present. DNA polymerase, a naturally occurring molecule whose design is specifically tailored to accurately and reliably insert the appropriate complementary base to a chain of DNA, is then added to extend the DNA chain by one base at the suspected SNP location. This single base extension is then detected by one of several conventional methods, including fluorescence, optical density, and electrophoresis and mass spectroscopy. The result is a direct read-out method of detecting SNPs that creates a simple binary "bit" of genetic information representing the presence of a SNP in a DNA sample.

Identification of Disease Resistance Genes

Two types of disease resistance have been reported in host plants so far. They are monogenic (vertical) resistance and polygenic (horizontal) resistance. The vertical resistance is also known as a gene-for-gene system and shows simple Mendelian pattern of monogenic inheritance and a single locus or gene is responsible for resistance. The horizontal resistance, on the other hand, is polygenically inherited where a group of resistance genes or loci are present. Both categories of host resistance have been characterised by molecular markers in many different crop species.

Mapping of genes for monogenic resistance in crop plants has involved three distinct approaches:

1. Use of a complete linkage map,
2. Use of near isogenic lines and
3. Bulked segregant analysis of F_2 population.

The first approach essentially requires availability of a complete genetic linkage map. Such maps need to be constructed based on segregation and linkage analysis of a large number of molecular markers using a mapping population (F_2 backcross, doubled haploid or recombinant inbred lines), which is generated from the F_1 hybrid of two divergent parent varieties/lines. For the purpose of mapping a gene for disease resistance, markers are selected from all the chromosomes such that the maximum distance between any two adjacent markers does not exceed 20 centimorgan (cM). Co-segregation of any of these markers with trait phenotype in the mapping population reveals linkage. This facilitates immediate localisation of the gene on the linkage map, since marker positions are already known. Although laborious, it is the most directed method of gene mapping.

The other two approaches do not require a linkage map. For instance, in case of the use of near-isogenic lines (which are developed by repeated backcrossing of the F with the recipient parent and therefore, differ for a single gene), DNA markers distinguishing the two lines (resistant vs. susceptible) are first identified. Co-segregation of these markers with the trait is analysed in a mapping population, often derived from the cross of the two isogenic lines (Martin *et al.*, 1991). Development of near-isogenic lines, however, takes many years. This method of gene mapping therefore is best suited to crops, in which intensive backcross breeding has been already practiced to generate isogenic lines.

Use of F_2 population in conjunction with a technique called bulked-segregant analysis, is the most rapid way of mapping genes (Michelmore *et al.*, 1991). In this approach, DNA from ten homozygous resistant F_2 plants is bulked to constitute a resistant bulk. Similarly, a susceptible bulk is made out of DNA from ten homozygous susceptible F_2 plants. The DNA markers, which distinguish these two bulks, are considered to have linkage with the target gene. Even though, molecular linkage map is not a pre-requisite for its success, use of markers with known map position imparts greater efficiency to gene mapping in this approach.

The charaterisation of polygenic resistance is totally different from that of monogenic resistance because of the difficulty in categorising the individual plants of a mapping population into distinct phenotypic classes due to interaction of many genes and a significant role of environment in trait expression. Therefore, different statistical methods have been devised to deal with such traits (Tanksley, 1993), which include:

1. Single marker analysis,
2. Interval mapping and
3. Composite interval mapping.

Of these, composite interval mapping is considered most precise. Unlike other two methods, this method eliminates the variance caused by other loci for resistance while identifying a quantitative locus for disease resistance. However, composite as well as simple interval mapping requires information of a complete linkage map, whereas, single marker analysis can be employed prior to extensive genome mapping. All the three methods require data from replicated trials for an accurate estimation of environment effects. This necessitates, generation of permanent mapping population such as recombinant inbred lines or doubled haploids, which can be replicated over seasons and locations.

Marker Assisted Selection for Disease Resistance Genes

The molecular markers have brought revolution in the plant breeding. The molecular marker aided selection (MAS) is employed not only in identification of desirable traits in plants but also in the isolation of desirable genes on the basis of map based cloning. A number of authors have reported the utility of MAS in plant breeding (Madan Mohan *et al.*, 1999; Mohapatra *et al.*, 2002; Young, 1996; Mukherjee and Acharya, 2003). Traditionally, the desirable segregant is selected based on expression of the phenotype in an appropriate selection environment. Creation of artificial epiphytotic conditions for large segregating populations routinely over generations is often difficult. In addition, identification of segregant in early generations requires greater efforts and time, when the resistance is recessive to susceptibility. Moreover, it is impossible to pyramid genes for resistance for many races of a pathogen based on resistance scoring if presence of only one gene offers some degree of resistance to all the races. Under these circumstances, markers can play very significant role. Tightly linked markers located on either side of the target gene allow most efficient selection. Although, this method is yet to be integrated with conventional crop improvement schemes, its potential use has been demonstrated in rice by Dr. G. S. Khush and his colleagues working at IRRI, Philippines. Individual F_2 segregants carrying four genes, *Xa4, Xa5, Xa13* and *Xa21*, which offer resistance to different races of leaf blight pathogen *Xanthomonas oryzae*, could be efficiently identified using RFLP and STS markers (Huang *et al.*, 1997). With the development of saturated linkage maps and identification of PCR based flanking markers, marker assisted selection is going to be routinely practised in breeding for disease resistance (Voort *et al.*, 1999; Struss and Plieske, 2000). Numbers of disease resistance genes have been reported from different countries in several crops using various molecular markers. Some of the examples are sited in the Table 21.1.

Table 21.1: Identification of Disease Resistance Genes in Crops Using Marker Assisted Selection

Name of the Pathogen	*Name of the Crop*	*Molecular Marker System*	*Objective of the Investigation*	*Reference*
Alternaria solani	Tomato	RFLP	To map QTL	Zhang *et al.*, 2003
Heterodera avenae	Wheat	RFLP	To map QTL	Williams *et al.*, 2003
Fusarium gramminearum	Wheat	Microsatellite	To map QTL	Somers *et al.*, 2003
Uromyces viciae-faba	*Vicia faba* L.	RAPD	Uvf-I gene	Avila *et al.*, 2003
Orobanche race E	Sunflower	RFLP, SSR	*Or(5)* gene	Tang *et al.*, 2003
Orsiolea oryzae	Rice	AFLP, SCAR	*C1M7* gene	Sardesai *et al.*,2002
Potato *virus X*	Potato	AFLP	*Nb* gene	Marano *et al.*, 2002
Coletotrichum gloeosporioides	*Dioscoria alata*	RAPD	Anthracnose resistance gene	Mignouna *et al.*, 2002a
Yam mosaic virus	*Dioscoria rotundata*	RAPD	Virus resistance gene	Mignouna *et al.*, 2002b
Alternaria solani	Tomato	RFLP	To map QTL	Foolad *et al.* 2002
Puccnia triticina	Wheat	Microsatellite marker	Molecular mapping of *Lr39* gene	Raupp *et al.*, 2001
Phialophora gregota	Soybean	RFLP	To develop molecular marker	Klos *et al.*, 2000
Phytophthora rojae	Soybean	RFLP	To detect resistance marker	Hegstad *et al.*, 1998
Pyricularinia grisea	Rice	RFLP	To map QTL	Zenbayashi *et al.*, 2002
Meloidogyne javanica	Carrot	RAPD	Linkage map for resistance locus	Boiteux *et al.*, 2000
Puccinia graminis	Barley	RAPD/RFLP	To develop stem rust resistance marker for *rpg4* gene	Borovkova *et al.*,1995
Leptosphaeria maculans	*Brassica*/Mustard	RFLP	Resistance loci	Fereira *et al.*, 1995a
Albugo candida	*Brassica juncea*/ Mustard	RAPD	Resistance locus	Mukherjee *et al.*, 2001
Venturia inequalis	Apple	AFLP, RAPD, RFLP	Linkage mapping of resistance loci	King *et al.*, 1998
Dysaphis devecta	Apple	RAPD,RFLP	Marker assisted selection	Roche *et al.*, 1997
Albugo candida	*Brassica*/Mustard	RFLP	Resistance loci mapping	Ferreira *et al.*, 1995b
Pericularia grisea	Rice	RFLP	Molecular mapping	Yu *et al.*, 1996
Heterodera glycines	Soybean	RAPD	QTL associated with resistance	Schuster *et al.*, 2001
Powdery mildew	Wheat	AFLP	Identification of resistance gene	Hartl *et al.*,1999
Pyricularia grisea	Rice	RFLP, AFLP STS-PCR	Molecular mapping of resistance gene	Chen *et al.*,1999
Cucumber mosaic virus	Tomato	RFLP	Genetic mapping	Stamova *et al.*,1999
Barley yellow dwarf	Barley	QTL	QTL mapping	Toojinda *et al.* 2000

Contd...

Table 21.1–Contd...

Name of the Pathogen	Name of the Crop	Molecular Marker System	Objective of the Investigation	Reference
Erysiphe graminis	Tomato	AFLP	Mapping of powdery mildew resistance gene *Ol-1*	Huang *et al.*, 2000a
Blumeria graminis	Wheat	RFLP	Specific resistance selection	Pillard *et al.*, 2000
Erysiphe graminis	Barley	RFLP	High resolution genetic map	Kurth *et al.*, 2000
Plasmodiophora brassicae	Brassica	RFLP, AFLP	Gene mapping	Voorrips *et al.*, 2000
Phytopthora	Potato	RFLP, RAPD	Chromosome mapping	Naess *et al.* 2000
Puccinia hordei	Barley	RFLP	Molecular mapping	Brunner *et al.*, 2000
Cercospora beticola	Sugar beat	AFLP, RFLP	QTL analysis of leaf spot resistance	Nilsson *et al.*, 1999
Nilaparvata lugens	Rice	RAPD, STS	Map based cloning	Renganayaki, *et al.*, 2002
Pyricularia grisea	Rice	PCR amplification	Dominant resistance gene marker	Jia *et al.*, 2002
Premia letucae	Lattuce	RFLP,isozyme	Disease resistance and morphological marker	Landry *et al.*, 1987
Tilletia indica	Wheat	RFLP	Chromosomal location of gene	Nelson *et al.*, 1998
Albugo candida	*Brassica*/Mustard	RFLP	Mapping resistance locus	Kole *et al.*, 1996
Sarocladium oryzae	Rice	RFLP	QTL analysis of sheath rot resistance	Srinivasachary *et al.*, 2002
Pyricularia oryzae	Rice	RFLP	RFLP mapping	Wang *et al.*, 1994
Albugo candida	*Brassica*/Mustard	RAPD	Marker for linked white rust resistance	Prabhu *et al.*, 1998
Erysiphe graminis	Wheat	AFLP	Molecular mapping of *pm24* gene	Huang *et al.*, 2000b

Map Based Cloning of Genes for Disease Resistance

Map-based cloning of genes is otherwise known as positional gene cloning. It involves:

1. Identification of markers genetically linked to the target gene,
2. Determination of relationship between genetic and physical distance in the target region,
3. Chromosome walking or jumping starting from a tightly linked marker to the gene and
4. Identification of genomic and/or cDNA clones carrying the target gene through complementation of the recessive phenotype.

Tightly linked markers are identified as described previously under the section on gene mapping. The physical distance between the markers, which flank the target gene, is determined to know the actual distance in base pairs and thus, to select the appropriate genomic DNA library for long distance walks or jumps along the chromosome. The smallest possible genomic DNA fragment hybridising with the flanking markers is assumed to be carrying the target gene. This DNA fragment is either directly introduced into the plant through genetic transformation using *Agrobacterium* to see if resistance reaction is expressed by the transgenic plant. Alternately, the DNA fragment is used as a probe to identity cDNA clones, which are subsequently tested, for complementation.

Many genes for disease resistance have been isolated using map-based gene cloning strategy, which include *pto,* in tomato (Martin *et al.,* 1993) and *Xa21* in rice (Song *et al.,* 1995), *Vf* resistance gene in apple (Barbieri *et al.,* 2003), leaf rust resistance gene Lr21 in bread wheat (Huang *et al.,* 2003). The *pto* gene that confers resistance to *Pseudomonas syringae,* is the first such gene to be isolated. Intensive efforts are being made to isolate genes involved in complete plant resistance mainly in *Arabidopsis,* rice and tomato, due to the small size of genomes in these species. Larger genomes pose problems in chromosome walk due to presence of highly repetitive sequences. Alternative strategies based on the syntenic relationship with small-related genomes are being designed to deal with large genomes.

Map-based gene cloning is a powerful gene isolation strategy, particularly for genes, which are known only by their phenotypic effects but not by the products they encode. It is most directed; each step yields useful results; progress can be monitored and potential problems can be identified. Availability of high-density molecular linkage maps in most crop plants has propelled the enthusiasm and scientific curiosity to new height in this area. Clustering of genes for both monogenic and polygenic resistance for different pathogens in specific genomic regions provides a unique opportunity to isolate a number of genes at a go. Cloning of loci particularly those involved in quantitative resistance based on saturated linkage maps will facilitate elucidation of the molecular basis of its expression and lead to genetic engineering of durable disease resistance in crop plants.

Conclusion

The genes that control disease resistance in plants have long been too difficult to identify or characterize precisely. But the DNA marker systems developed so far are making the task easier to some extent. Still the need for better forms of disease resistance gene identifying marker system is experienced. So there is a continuous requirement for refinement of DNA marker system, to make them more users friendly, effective and economical, still keeping high reproducibility across the laboratories. Identification of disease resistance gene will greatly benefit the farmers and the huge loss to crop due to pathogens may be avoided, as the genetic resistance is cost effective and sound approach to disease control. It will also be easier to engineer resistant crop varieties using these advanced techniques. Hence these above techniques have lots of potential to bridge the gap between the demand and production of twenty first century. Traditional breeders should combine molecular markers with traditional breeding approach to speed up the selection procedure.

REFERENCES

Avila, C.M., J.C. Sillero, D. Rubiales, M.T. Moreno and A.M. Torres. (2003). Identification of RAPD markers linked to the Uvf–1 gene conferring hypersensitive resistance against rust (*Uromyces viciae-fabae*) in Vicia faba L.*Theor. Appl. Genet.,* 107: 353–358.

Barbieri, M., E Belfanti, S Tartarini, BA Vinatzer, S Sansavini, E SilfverbergDilworth, L Gianfranceschi, D Hermann, A Patocchi, and C Gessler. (2003). Progress of map–based cloning of the Vf–resistance gene and functional verification: Preliminary results from expression studies in transformed apple. *Hort. Sci.,* 38: 329–331.

Boiteux, L.S., J.G. Belter, P.A.Roberts, and P.W. Simon. (2000). RAPD linkage map of the genomic region encompassing the root-knot nematode (*Meloidogyne javanica*) resistance locus in carrrot. *Theor. Appl. Genet.,* 100: 439–446.

Borovkova, G., B.J. Steffenson, Y. Jin, J.B. Rasmussen, A. Kilian, A. Kleinhofs, B.G. Rossnagel and K.N. Kao (1995). Identification of molecular markers linked to the stem rust resistance gene *rpg4* in barley. *Genetics,* 85: 181–185.

Botstein, D., R.L. White, M.H. Skolnic and R.W. Davis. (1980). Construction of a genetic map in man using restriction fragment length polymorphism. *Am. J. Hum. Genet.*, 32: 314–331.

Brunner, S., B. Keller and C.Feuillet. (2000). Molecular mapping of the *Rph7.g* leaf rust resistance gene in barley *(Hordeum vulgare L.) Theor. Appl. Genet.*, 101: 783–788.

Caetano-Anolles, G., B.J.Bassam and P.M. Gresshoff (1991). DNA amplification fingerprinting using very short arbitrary oliginucleotide primers. *Biotech.*, 9: 553–557.

Chen, D.H., M. delaVina, T. Lunkai, D.J. Mackill, P.C. Ronald, and R.J. Nelson (1999). Molecular mapping of the blast resistance gene, *Pi44(T)*, in a derived from a durably resistant rice cultivar. *Theor. Appl. Genet.*, 98: 1046–1053.

Chowdappa, P., D. Brayford, J. Smith and J. Flood. (2003). Molecular discrimination of *Phytophthora* isolates on cocoa and there relationship with coconut, black pepper and bell pepper isolates based on rDNA repeat and AFLP fingerprints. *Current Sci.*, 84:1235–1238.

Craenen, K. and R. Ortiz (1996). Effect of the black sigatoka resistance locus *bs1* and ploidy level on fruit and bunch traits of plantain–banana hybrids. *Euphytica*, 87:97–101.

Ferreira, M.E., S.R. Rimmer, P.H. William and T.C. Osborn (1995a). Mapping loci controlling *Brassica napus* resistance to *Leptosphaeria maculans* different screening conditions. *Genetics*, 85: 213–217.

Ferreira, M.E., P.H. Williams and T.C. Osborn (1995b). Mapping of locus controlling resistance to *Albigo candida* in *Brassica napus* using molecular markers. *Genetics*, 85:218–220.

Foolad, M.R., L.P. Zhang, A.A. Khan, D.N. Liu and G.Y. Lin (2002). Identification of QTLs for early blight (*Alternaria solani*) resistance in tomato using backcross populations of a *Lycopersicon esculentum x L–hirsutum cross. Theor. Appl. Genet.*, 104: 945–958.

Grube, R.C., J.R.Blauth, M.S. ArnedoAndres, C. Caranta, M.K. Jahn (2000). Identification and comparative mapping of a dominant potyvirus resistance gene cluster in *Capsicum. Theor. Appl. Genet.*, 101: 852–859.

Hartl, L., V. Mohler, F.J. Zeller, S.L.K. Hsam and G. Schweizer (1999). Identification of AFLP markers closely linked to the powdery mildew resistance genes *Pm1c* and *Pm*4a in common wheat (*Triticum aestivum* L.).*Genome*, 42: 322–329.

Hayes, A.J., Y.G.Yue, and M.A. Saghai-Maroof (1999). Expression of two soybean resistance gene candidates shows divergence of paralogous single-copy genes. *Theor. Appl. Genet.*, 101: 789–795.

Hegstad, J.M., C.D. Nickell and L.O. Vodkin (1998). Idenntifying resistance to *Phytophthora sojae* in selected soybean acceessions using RFLP techniques. *Crop Sci.*, 38: 50–55.

Helguera, M., I.A.Khan, and J.Dubcovsky (2000). Delopment of PCR markers for wheat leaf rust resistsance gene *Lr47. Theor. Appl. Genet.*, 101:625–631,

Huang, N., A. Parco T. Mew, G. Magpantay, S.McCouch, E. Guiderdoni, J.Xu, P.subudhi, E.R. Angeles and G.S.Khush (1997). RFLP mapping of isozymes, RAPD and QTLs for grain shape, brown planthopper resistance in a doubled haploid rice population. *Molecular Breeding* 3:105–113,

Huang, C.C., Y.Y. Cul, C.R.Weng, P.Zabel, and P.Lindhout (2000a). Development of dignostic PCR markers closely linked to the tomato powdery mildew resistance gene *Ol–1* on chromosome 6 of tomato. *Theor. Appl. Genet.*.101: 918–924,

Huang, L., S.A. Brooks, W.L. Li, J.P. Fellers, H.N. Trick, and B.S. Gill (2003). Map-based cloning of leaf rust resistance gene *Lr21* from the large and polyploid genome of bread wheat. *Genetics*, 164: 655–664.

Huang, N., E.R. Angeles, J. Domingo, G. Magpantay, S. Singh, G. Zhang, N. Kumaradivel, J. Bennett, G.S. Khush (1997). Pyramiding of bacterial blight resistance genes in rice: marker-assisted selection using RFLP and PCR. *Theor. Appl. Genet.*, 95: 313–320.

Huang, X.Q., S.L.K. Hsam, F.J. Zeller, G. Wenzel, V. Mohler (2000b). Molecular mapping of the wheat powdery mildew resistance gene *Pm24* and marker validation for molecular breeding. *Theor. Appl. Genet.*, 101: 407–414.

James, W.C., P.S.Teng, and F.W. Nutter (1990). Estimated loss of crops from plant pathogens. In: *CRC Handbook of Pest Management, Vol. 1*, (Ed.) D. Pimentel, Boca Ratoon, FL.CRC Press, pp. 15–50.

Jeffreys, A.J.,V. Wilson and S.L. Thein (1985 a). Hypervariable minisatellites regions in human DNA. *Nature*, 314: 67–73.

Jeffreys, A.J.,V. Wilson and S.L. Thein. (1985 b). Individual specific "fingerprints" of human DNA. *Nature*, 314: 76–79.

Jeon, Y.H., S.N. Ahn, H.C. Choi, T.R. Hahn and H.P Moon (1999). Identification of a RAPD marker linked to a brown planthopper resistance gene in rice. *Euphytica*, 107: 23–28.

Jeong, S.C., S. Kristipati, A.J. Hayes, P.J. Maughan, S.L. Noffsinger, I. Gunduz, G.R. Buss and M.A. Saghai Maroof (2002). Genetic and sequence analysis of markers tightly linked to the Soybean mosaic virus resistance gene, *Rsv3*. *Crop Sci*.42:265–270.

Jia, Y., Z. Wang and P. Singh. (2002). Development of dominant rice blast *Pi–ta* resistance gene markers. *Crop Sci.*, 42: 2145–2149.

Jorge, V., M.A. Fregene, M.C. Duque, M.W. Bonierbale, J.Tohme and V. Verdier (2000). Genetic mapping of resistance to bacterial blight disease in cassava (*Manihot esculenta* Crantz) *Theor. Appl. Genet.*, 101: 865–872.

King, G.J., F.H. Alston, L.M. Brown, E. Chevreau, K.M. Evans, F. Dunemann, J. Janse, F. laurens, J. Janse, F. Laurens, J.R. Lynn, C. Mailiepaard, A.G. Manganaris, P. Roche, H. Schmidt, S. Tartarini, J. Vehaegh and R. Vrielink (1998). Multiple field and glasshouse assessments increase the reliability of linkage mapping of the *VF* source of scab resistance in apple. *Theor. Appl. Genet.*, 96: 699– 708.

Klos, K.L.E., M.M. Paz, L. Fredrick Marek, P. B. Cregan and R.C. Shoemaker (2000). Molecular markers useful for detecting resistance to brown stem rot in soybean. *Crop Sci.*, 40: 1445–1452.

Kole, C.R., Teutonico, A. Mengistu, P.H. Williams and T.C. Osborn (1996). Molecular mapping of a locus controlling resistance to *Albugo candida* in *Brassica rapa. Genetics*, 86: 367–369.

Konieczyn, A. and F.M. Ausubel (1993). A procedure for mapping Arabidopsis mutations using co–dominant ecotypes specific PCR-based markers. *Plant J.*, 4: 403–410.

Kurth, J. R.Kolsch,V. Simons, and P. Schulze-Leefert (2000). A high-resolution genetic map and a diagnostic RFLP marker for the *Mlg* resistance locus to powdery mildew in barley. *Theor. Appl. Genet.*, 102: 53–60.

Lagercrantz, U., H. Ellegren and L. Andersson (1993). The abundance of polymorphic microsatellites motifs differs between plants and vertibrates. *Nucleic Acids Research*, 21: 1111–1115.

Landry, B.R., H. Kesseli, Lueng and R. Michelmore (1987). A genetic map of lettuce (*Lactuca sativa L.)* with restriction fragment length polymorphism, isozyme, disease resistance and morphological markers. *Genetics*, 116: 331–337.

Madan Mohan, Suresh Nair, A. Bhagbat, T.G. Krishna, Masahiro Yano, C.R. Bhatia and Takuji Sasaki (1997). Genome mapping, Molecular markers and marker–assisted selection in crop plants. *Mol. Breed.*, 3: 87–103.

Manzanares-Dauleux, M.J., R. Deelourme, F. Baron, and G. Thomas (2000). Mapping of one gene and QTLs involved in resistance to club root in *Brassica napus. Theor. Appl. Genet.*, 101:885–891.

Marano, MR.I., Malcuit, W. DeJong and D.C. Baulcombe (2002). High-resolution genetic map of Nb, a gene that confers hypersensitive resistance to *Potato Virus* X in *Solanum tuberosum. Theor. Appl. Genet.*, 105: 192–200.

Martin, G.B., J.G.K. Williams and S.D. Tanksley *et al.* (1991). Rapid identification of marker linked to a Pseudomonas rersistance gene in tomato by using random primers and near isogenic lines. *Proc. Natl. Acad. Sci. (USA)*, 88: 2336–2340.

Martin, G.B., S.H. Brommunschenkel, J. Chunwongse, A. Frary, M.W. Ganal, R. Spivey, T. Wu, E.D. Earle and S. D. Tanskley. (1993). Map based cloning of a *protein kinase* gene conferring disease resistance in tomato. *Science*, 262: 1432–1436.

Micheletto, S., R. Boland and M. Huarte (2000). Argentinian wild diploid solanum species as sources of quantitative late blight resistance. *Theor. Appl. Genet.*, 101: 902–906.

Michelmore, R.W., I. Paran and R.V. Kesseli (1991). Identification of markers linked to disease resistance genes by bulked segregant analysis: A rapid method to detect markers in specific genomic regions by using segregating populations *Proc. Natl. Acad. Sci. USA*, 88: 9828–9832.

Mignouna, H.D., M.M. Abang, A. Onasanya and R. Asiedu. (2002a). Identification and application of RAPD markers for anthracnose resistance in water yam (*Dioscorea alata*) *Ann. Appl. Biol.* 141: 61–66.

Mignouna, H.D., M.M. Abang, A. Onasanya, B. Agindotan and R. Asiedu (2002b). Identification and potential use of RAPD markers linked to Yam mosaic virus resistance in white yam (*Dioscorea rotundata*). *Annl. Appl. Biol.*, 140: 163–169.

Mohapatra, T., A.K. Mukherjee, K. Srinivasan and R.P. Sharma (2002). Molecular markers for characterisation of pathogen population and host genes. In: *Molecular Approaches in Plant Disease Management*, (Eds.) G.S. Shekhawat, B.P. Singh and R.A. Singh, Malhotra Publishing House, New Delhi, India, pp. 38–53.

Mohler, V., A. Klahr, G. Wenzel and W. Schwarz (2002). A resistance gene analog useful for targeting disease resistance genes against different pathogens on group 1S chromosomes of barley, wheat and rye. *Theor. Appl. Genet.*, 105: 364–368.

Morgan, R.N., J.P.Wilson, W.W. Hanna, and P. Ozias-Akins (1998). Molecular markers for rust and pyricularia leaf spot disease resistance in pearl millet. *Theor. Appl. Genet.*, 96: 413–442.

Morgante, M. and A.M. Oliveiri (1993). PCR–amplified microsatellites as marker in plant genetics. *Plant J.*, 3: 175–182.

Mukherjee, A.K. and L.K. Acharya (2003). Application of Random Amplified Polymorphic DNA in Plant Breeding. In: *Proc. Ind. Sci. Cong. Assoc. Bhubaneswar Chapter*; November 9, 2003. Held at the PG Deptt. Of Botany, Utkal University, Bhubaneswar, Orissa, India, pp. 65–70.

Mukherjee, A.K., T. Mohapatra, A. Varshney, R. Sharma and R. P. Sharma (2001).Molecular mapping of a locus controlling resistance to *Albugo candida* in Indian mustard. *Plant Breeding*, 120: 483–487.

Mullis, K.B., F. Faloona, S. Scharf, R. Saiki, G. Horn and H. Elrich (1986). Specific enzymatic application of DNA in vitro: the polymerase chain reaction. *Cold Spring Harbor Symp. Quanti. Biol.*, 51: 263–273.

Naess, J.M., S.K. Bradeen, S.M. Wielgus, G.T. Haberlach, J.M. McGrath and J.P. Helgeson (2000). Resistance to late blight in Solanum bulbocastanum is mapped to chromosome 8. *Theor. Appl. Genet.*, 101: 697–704.

Narayanan, N.N., N. Baisakh, C.M. Vera Cruz, S.S. Gnanamanickam, K. Dutta and S.K. Dutta (2002). Molecular breeding for development of blast and bacterial blight resistance in rice cv. IR50. *Crop Sci.*, 42:2072–2079.

Neeraja, C.N., N. Vijayabhanu, V.V. Shenoy, C.S. Reddy and N.P. Sharma (2002). RAPD Analysis of Indian isolates of rice sheath blight fungus *Rhizoctonia solani*. *J. Plant Biochem. Biotech.*, 11:43–48.

Nelson, J.C., J.E. Autrique, G. Fuentes-Davila and M.E. Sorrells (1998). Chromosomal location of genes for resistance to karnal bunt in wheat. *Crop Sci.*, 38: 231–236.

Nilsson, N.O., M. Hansen, A.H. Panagopoulos, S. Tuvesson, M. Ehlde, M. Christiansson, I.M. Rading, M. Rissler and T. Kraft (1999). QTL analysis of Cercospora leaf spot resistance in sugar beat. *Plant Breeding*, 118: 327–334.

Olson, M., L. Hood, C. Cantor and D. Botstein (1989). A common language for physical mapping of the human genome. *Science*, 245: 1434–1435.

Paran, I. and R.W. Michelmore (1993). Development of reliable PCR-based markers linked to downy mildew resistance gene in lettuce. *Theor. Appl. Genet.*, 85: 985–993.

Peng, J.H., T. Fahima, M.S. Roder, Y.C. Li, A. Dahan, A. Grama,Y.I. Ronin, A.B.Korol, and E. Nevo (1999). Microsatellite tagging of the strip-rust resistance gene *yrH52* derived from wild emmer wheat, Trificum dicoccoides, and suggestive negative crossover interference on chromosome 1B. *Theor. Appl. Genet.*, 98: 862–872.

Pillard, S., I.Goldringer, J.Enjalbert, G.Doussinault, C. de vallavieillepope, and P. Brabant. (2000). Evaluation of resistance against powdery mildew in winter wheat populations conducted under dynamic management. I–specific seedling resistance selected. *Theor. Appl. Genet.*, 101: 449–456.

Prabhu, K.V., D.J. Somers, G. Rakow and R.K. Gugel (1998). Molecular marker linked to white rust resistance in mustard *Brassica juncea*. *Theor. Appl. Genet.*, 97: 865–870.

Rajendraprasad, M.N., M.V.C. Gowda and G.K. Naidu (2000). Groundnut mutants resistant to tobacco cutworm (*Spodoptera litura* F.).*Current Science*, 79: 158–160.

Raupp, W.J., S. Singh, G.L. Brown-Guedira and B.S. Gill (2001). Cytogenetic and molecular mapping of the leaf rust resistance gene *Lr39* in wheat. *Theor. Appl. Genet.*, 102:347–352.

Renganayaki, K., A.K. Fritz, S. Sadasivam, S. Pammi, S.E. Harrington, S.R. McCouch, S.M. Kumar and A.S. Reddy (2002). Mapping and progress towards map based cloning of brown planthopper biotype-4 resistance gene introgressed from *Oryza officinalis* into cultivated rice, *O. sativa*. *Crop Sci.*, 42: 2112–2117.

Roche, P., F. H. Alston, C. Maliepaard, K.M. Evans, R. Vrielink, F. Dunemann, T. Markussen, S. Tartarini, L.M. Brown, C. Ryder and G.J. King (1997). RFLP and RAPD markers linked to the rosy leaf curling aphid resistance gene (*Sd1*) in apple. *Theor. Appl. Genet.*, 94:528–533.

Rommens, M. Caius and G.M. Kishore. (2000) Exploiting the full potential of disease resistance gene for agricultural use. *Curr. Opinion in Biotech.*, 11:120–125.

Sanchez, A.C., L.L. Ilag, D. Yang, D. S. Brar, Frederick M. Ausubel, G.S. Khush, M. Yano, T. Sasaki, Z. Li, and, N. Huang (1999). Genetic and physical mapping of *Xa 13*, a recessive bacterial blight resistance gene in rice.*Theor. Appl. Genet.*, 98:1022–1028.

Sanchez, G., S. Restrepo, M.C. Duque, M. Fregene, M. Bonierbale, and V. Verdier (1999). AFLP assessment of genetic variability in cassava accessions (*Manihot esculenta*) resistant and susceptible to the cassava bacterial blight (CBB). *Genome*, 42: 163–172.

Sardesai, V., A. Kumar, K.R. Rajyashri, S. Nair and M. Mohan (2002). Identification and mapping of an AFLP marker linked to GM7, a gall midge resistance gene and its conversion to a SCAR marker for its utility in marker aided selection in rice. *Theor. Appl. Genet.*, 105: 691–698.

Schuster, I., R. Abdelnoor, S. Marin, V. Carvalho, R. Kiihl, J. Silva, C. Sediyama, E. Barros, M. Moreira. (2001). Identification of a new major QTL associated with resistance to soybean cyst nematode (*Heterodera glycines*). *Theor. Appl. Genet.*, 102: 91–96.

Simon, P.W., W.C. Matthews, P.A. Roberts (2000). Evidence for simply inherited dominant resistance to *Moloidogyne javanica* in carrot. *Theor. Appl. Genet.*, 100: 735–742.

Somers, D.J., G Fedak and M. Savard (2003). Molecular mapping of novel genes controlling Fusarium head blight resistance and deoxynivalenol accumulation in spring wheat.*Genome*, 46: 555–564.

Song, W.Y., G.L. Wang, L.L. Chen, H.S. Kim, L.P. Pi, T. Holsten, J. Gardner, B. Wang, W.X. Zhai, L.H. Zhu, C. Fauquet and P. Ronald (1995). A receptor kinase-like protein encoded by the rice disease resistance gene *Xa21*. *Science*, 270: 1804–1806.

Southern, E.M. (1975). Detection of specific sequences among DNA fragments separated byn gel electrophoresis. *J. Mol. Biol.*, 98: 503–517.

Srinivasachary, S. Hittalmani, K.G.Kumar, H.E. Shashidhar, M.G. Vaishshali (2002). Identification of quantitative trait loci associated with sheath rot resistance (*Sarocladium oryzae*) and paicle exsertion in rice (*Oryza sativa* L.). *Curr. Sci.*, 82:133–135.

Stamova, B.S., R.T. Chetelat (2000). Inheritance and genetic mapping of cucumber mosaic virus resistance introgressed from *Lycopersicon chilense* into tomato. *Theor. Appl. Genet.*, 101: 527–537.

Struss, D. and J. Plieske (2000). Development of STS markers linked to blackleg resistance genes of b genone for marker assisted selection in oilseed rape. In: *Plant and Animal Genome VIII Conference.* Jan 9–12 San Diego, California, USA.

Tang, S.X., A. Heesacker, V.K. Kishore, A. Fernandez, E.S. Sadik, G. Cole and S.J. Knapp (2003). Genetic mapping of the Or(5) gene for resistance to *Orobanche* Race E in sunflower. *Crop Sci.*, 43: 1021–1028.

Tanksley, S.D. (1993). Mapping polygenes. *Annu. Rev. Genet.*, 27: 205–233.

Toojinda, T., L.H. Broers, X.M. Chen, P.M. Hayes, A. Kleinhofs, J. Korte, D. Kudrna, H. Leung, R.F. Line, W. Powell, L.Ramsay, H. Vivar and R. Waugh. (2000). Mapping quantitative and qualitative disease resistance genes in a doubled haploid population of barley (*Hordeum*)

Van der Voort, J.N.A.M.R., G.J.W. Janssen, H. Overmas, P.M. Van Zanvoort, A.van Norel, O.E. Scholten, R. Janssen and J. Bakker (1999). Development of a PCR–based selection assay for root-knot nematode resistance (*Rmcl*) by a comparative analysis the *Solanum bulbocastanum* and *Solanum tubersom* genome. *Euphytica*, 106:187–195.

Voorrips, R.E., M.C. Jongèrius, H.J. Kanne (1997). Mapping of two genes for resistance to clubroot (*Plasmodiophora brassicae*) in a population of doubled haploid lines of *Brassica oleracea* by means of RFLP and AFLP markers. *Theor. Appl. Genet.*, 94: 75–82.

Vos, P., R. Hogers, M. Bleeker, M. Reijans, T. Van de Lee, M. Hornes, A. Fritjers, J. Pot, J. Peleman, M. Kuiper and M. Zabeau (1995). AFLP: a new technique for DNA fingerprinting. *Nucleic Acids Res.*, 23: 4407–4414.

Wang, G.L., D.J. Mackill, J.M.Bonman, S.R. McCouch, M.C. Champoux and R.J. Nelson (1994). RFLP mapping of genes conferrring complete and partial resistance to blast in durably resistant rice cultivar. *Genetics*, 136: 1421–1434.

Williams, J.G.K., A.R. Kubelik, K.J. Livak, J.A. Rafalsky and S.V. Tingey (1990). DNA polymorphism amplified by arbitrary primers are useful as genetic markers. *Nucleic Acids Res.*, 18: 6531–6535.

Williams, K.J., J.G. Lewis, P. Bogacki, M.A. Pallotta, K.L. Willsmore, H. Kuchel and H. Wallwork (2003). Mapping of a QTL contributing to cereal cyst nematode tolerance and resistance in wheat. *Austral. J. Agril. Res.*, 54: 731–737.

Winter, P., A.M. Benko Iseppon, B. Huttel, M. Ratnaparkhe, A. Tullu, G. Sonnante, T. Pfaff, M. Tekeoglu, D. Santra, V.J. Sant, P.N. Rajesh, G. Kahl and F.J. Muehlbauer (2000). A linkage map of the chichpea (*Cicer arietinum* L.) genome based on recombinant inbredlines from a *C.Arietinum´C.reticulatum* cross: Localization of resistance genes for *Fusarium* wilt races 4 and 5. *Theor. Appl. Genet.*, 101: 1155–1163.

Xu, M.L. and S.S. Korban (2000). Saturation mapping of the apple scab resistance gene *Vf* using AFLP markers. *Theor. Appl. Genet.*, 101: 844–851.

Young, N.D. (1996). QTL Mapping and quantitative disease resistance in plants. *Annu. Rev. Phytopathol.*, 34: 479–501.

Yu, Z.H., D.J. Mackill, J.M. Bonman, S.R. McCouch, Guiderdoni, J.L. Notteghem and S.D. Tanksley (1996). Molecular mapping of genes for resistance to rice blast (*Pyricularia grisea* Sacc.) *Theor. Appl. Genet.*, 93: 859–863.

Zenbayashi, K., T. Ashizawa, T. Tani and S. Koizumi. (2002). Mapping of QTL (quantitative trait locus) conferring partial resistance to leaf blast in rice cultivar *chubu* 32. *Theor. Appl. Genet.*, 104: 547–552.

Zhang, L.P., G.Y. Lin, D. NinoLiu and M.R. Foolad (2003). Mapping QTLs conferring early blight (*Alternaria solani*) resistance in a *Lycopersicon esculentum x L. hirsutum* cross by selective genotyping. *Molecular Breeding*, 12: 3–19.

22

Integrated Disease Management in Linseed: Present Status and Future Thrust

Jyoti Singh
Project Coordinating Unit (Linseed),
C.S. Azad University of Agriculture and Technology, Kanpur – 208 002 , U.P.

Linseed or flax (*Linum usitatissimum* L.) is an important oilseed crop for its oil and fibre cultivated in almost all continents of the world. In India linseed is cultivated primarily for its oil and occupies fifth place *i.e.*, after groundnut, rapeseed and mustard, soybean and sesame. Major linseed growing areas in the country are Madhya Pradesh, Uttar Pradesh, Maharashtra, Bihar, Rajasthan, Orissa and Karnataka. Also grown in Assam, Himachal Pradesh, J&K and Andhra Pradesh in a lesser area (Srivastava *et al.*, 1997).

India ranks first in area (754 thousand ha) but the productivity is very low on global level. Average national productivity of this crop is about 324 kg/ha, which is much lesser to other linseed growing countries like Germany (1699 kg/ha), U.S.A. (1244 kg/ha) Canada (1078 kg/ha) China (933 Kg/ha), Argentina (780 Kg/ha) or 778 kg/ha at global level (Damodaran and Hegde, 2002). Among several constraints responsible for its low productivity, diseases are a major one causing significant yield losses. The crop suffers from several diseases (Table 22.1) of which rust (*Melampsora lini* Lev.) wilt, (*Fusarium* oxysporum f. *lini*), Alternaria blight (*Alternaria lini* Dey) and powdery mildew (*Oidium*

Table 22.1: Major Disease of Linseed in India

Sl.No.	*Disease*	*Causal Organism*	*Yield Losses*	*Prevalence/Reported from*	*Reference*
1.	Rust	*Melampsora lini* (Eenrenb.) Lev.	70–100% (Hora *et al.*, 1961)	Hilly and cooler regions of the country, prevalent in North	Butler (1914, 1918); Howard (1921); Pearl (1923); Plymen (1933); Chowdhury (1951); Mishra and Prasad (1966); Saharan (1978)
2.	Wilt	*Fusarium oxysporum* f. sp. *lini* (Bolley) Snyder and Hansen	80% (Sattar and Hafiz,1952) 87% (Sharma and Mathur,1971) 100% (Vasudeva,1961)	Throughout the country but serious in Central, semi arid region	Butler (1918); Pearl (1923); Mc Rae (1928); Alam (1931); Anon (1949); Verma (1945); Chowdhury (1951); Merh and Kulkarni (1961); Merh (1967); Asthana (1957); Bedi *et al.* (1961)
3.	Alternaria blight	*Alternaria lini* Dey	26–60% (Dey, 1933) upto 90% in serious bud infection (Chauhan and Srivastava, 1975)	Serious in Northern part of India in high humidity areas	Dey (1933); Arya and Prasad (1953); Kalia *et al.* (1965); Singh *et al.* (1974); Ji (1975), Garg (1982)
4.	Powdery mildew	*Oidium lini* Skoric	Heavy losses (Pavgi and Singh 1965) 2–10% (Kushwaha and Chand, 1971)	Thorough out the country Serious in Central and Peninsular region	Padwick (1941); Patel, *et al.* (1949); Mundkur and Ahmad (1946); Porwal (1964), Shukla and Pathak (1984); Sandhu and Chandwahi (1965); Pavgi and Singh (1965); Kushwaha and Chand (1971); Bhargava and Shukla (1980)
5.	Seed rot	*Alternaria* sp. *Alternaria alternata* *A. brassicae*	NR	Sporadic	Mc Rae (1928); Kadian and Surranarayanan (1971); Khanna and Chandra (1981); Kumar *et al.* (1985).
		Rhizoctonia sp.	—	—	Asthana (1957)

Contd...

Table 22.1–Contd...

Sl.No.	Disease	Causal Organism	Yield Losses	Prevalence/Reported from	Reference
6.	Seedling rot/ seedling blight and root rot	*Rhizoctonia bataticola/ Macrophomina* rot *Sclerotium rolfsii*	Up to 80% 10–25% at Hoshangab, 2–15% at Raipur sometime, 32-61%	Kangra Hoshangabad, Sagar and Kanpur Sclerotium rot is an endemic problem at Chattisgarh (1995–2002) Aslo reported from Kangra	Bedi, *et al.* (1961); Misra and Sinha (1982) Ramkrishnan(1931), Kadian and Surya-narayana (1971); Khatri *et al.* (1980); Siddaramaiah and Desai (1980) AICRP Report (1994–2003)
7.	Anthracnose	*Colleto trichum lini*	0–35%	Varanasi,	Rathaiah and Pavgi (1973)
	Die back			Kanpur	AICRP Report (1999)
8.	Cercospora leaf spot	*Cercospora linicola* Pavgi and Rathaiah		Varanasi	Rathaiah and Pavgi (1970, 71, 73)
9.	Sclerotinia rot	*Sclerotinia selerotiorum*	5–20%	Kangra (1995–97) Raipur (1999)	AICRP Reports (1995–97, 1999)
10.	Botrytis rot	*Botrytis cinerea*	32–61%	Raipur	AICRP Reports (1994, 1995)
11.	Tobacco leaf curl	Tobacco leaf curl virus	—	—	Vasudeva (1958)
12.	Phyllody		8–12%	Berhampore (1997–98) Kanpur (1998–99) Faizabad	AICRP Linseed Reports (1997–99)
13.	Doddar or *Cuscuta*	*Cuscuta epithelium, C. hylina*	10–18%	Regular at Chattisgarh and adjoining Part of M.P., Orissa and Maharashtra. Also recorded at Mauranipur (1997–98) and Kangra (1999–2000)	Rai (1982); AICRP Reports (1991–2003)

lini Skoric) are most important under Indian conditions causing considerable damage to the crop. The losses under epidemics of these diseases are estimated in the range of 70–100 per cent due to rust in susceptible cultivars (Hora *et al.*, 1962); losses upto 80 per cent due to wilt was reported by Sattar and Hafiz (1952) under favourable condition. Sharma and Mathur (1971) recorded 87 per cent wilt incidence in Rajasthan and a total loss in severe cases by Vasdeva (1962). Losses in the range of 28–60 per cent was reported by Dey (1933) at Kanpur due to Alternaria leaf and bud blight, yield losses upto 90 per cent in case of severe bud infection was reported by Chauhan and Srivastava (1975). A serious damage to linseed crop due to powdery mildew has been reported by Pavgi and Singh (1965). The overall yield can be increased by 23–26 per cent if diseases are controlled.

Earlier the principal method of management of crop was the use of pesticides. The growing awareness of harmful effect of pesticides world over has diverted the interest of scientists towards Integrated Pest Management, which is gaining great importance in recent years. It is appropriate to develop disease management strategy as an integrated approach, combining different methods of control which are economically sound and socially acceptable. Based on researches in past including AICRP (Linseed) trials following IDM strategy for linseed diseases may be followed.

Legislative Approach

Regulatory measures are the foremost and essential component of the management of diseases and it should be strictly observed. There are a number of diseases not reported from India as yet but occurring in other countries. The seeds imported from those countries should be strictly checked before transportation, to avoid the entry of the disease in the country. The pathogens likes *Septoria linicola* (*Mycosphaerella linorum*) and *Cuscuta* (a flowering plant parasite) are of quarantine significance and their entry into an uninvaded geographical area should be kept under check through plant quarantine rules (Anonymous, 1978). Cuscuta (Doddar) was a conventional problem at Chattisgarh, Orissa and adjoining Maharashtra but has entered the uninfected area *i.e.*, Mauranipur (Anon 1997-98) and Kangra (Anon 1999–2000). Probably it has entered alongwith seeds supplied for testing under AICRP trials.

Host Resistance

The most practical and least input approach for disease management is host resistance. Resistance breeding is of top priority in National Coordinated Research Programme. A lot of work has been done on resistance breeding and a number of resistant varieties were released time to time (Kerkhi *et al.*, 1999). The diseases are location specific, therefore, there is need to select and cultivate varieties which have tolerance to the predominant pathogens/diseases in a particular locality.

A large number of genotypes have been identified as resistant to rust disease in India and abroad which is difficult to mention here, therefore, only Indian work have been cited. Vasudeva (1962) found some of the NP varieties *e.g.*, NP (RR)-5, RR-9, RR-10, RR-37, RR-38, RR-40, RR-45, RR-197, RR-236, RR-262, RR-267 and RR-272 to be immune or resistant to all prevalent races of rust in India. Kotasthane *et al.* (1973) found about 29 linseed varieties to be resistant to *M. lini*. In some of the lines as EC 77959 rust resistance genes are known to be linked tightly with resistance to powdery mildew caused by *Oidium lini* (Misra and Shukla, 1981; Misra and Pandey, 1981).

A comprehensive screening of a large number of genotypes under artificial inoculation conditions against all the prevalent races of *M. lini* in India indicated LC-216, LC-255, LC-256, Ottawa, 770 B (L), Dacota (M), Victory A (M^4), Tower (L8), Marshall (N^2) and Pale blue crimped (L^3) as an excellent sources of resistance (Saharan and Singh, 1978; 1980). Most of them carried single or a few dominant genes for resistance. Shukla and Misra (1983) and Singh (1984) reported a large number of genotypes free from rust.

Intra specific crossing have also been attempted to incorporate rust resistant genes in *Linum usitatissium* from related species such as *L.africanum* and *Linum anguistifolium* (Tammes, 1928, Gill and Yermanos, 1967, Wicks and Hammond, 1978).

A great deal of work has been done in selecting and breeding flax for resistance to wilt. Cultivar Sabour was reported to be resistant to wilt by McRae (1928) and Alam (1931); ST-6 by Asthana (1957) and K_1 and K_2 by Goel and Swarup (1964). Genotypes RR-8, NP-12, NP-21, NP-124, 58, RR-80, RR-82, 439, 440, 5, 10, 37, 38, 62, 197, 199, 200, RR-202 (1), 272, 33B, B-5128, NP (RR) 5 and NP (RR)-12 were identified as wilt resistant (Kulkarni *et al.*, 1966; Merh and Kulkarni, 1958, 1961). NP-124, 439, 440, NPRR-9 (Shrama *et al.*, 1972); BR-9, BR-29, Indore-1, Malvi-1, NP-12, Bison-69, Bison-70 and Canadian Western No.1 were found resistant to wilt by Chandwani and Srivastava (1968).

Identification as sources of resistance against Alternaria blight is most challenging field. Most of the recommended cultivars in linseed grown in different parts of the county are susceptible to Alternaria blight. The breeding programme is therefore, focused mainly on this disease. However cultivars like LC-54, Himalini, Nagarkot, Surabhi, Jeevan, Gaurav, Sheela, Rashmi are moderately resistant to the disease. Dey (1933) suggested cultivation of late varieties for wet tracts of India. Two cultures *viz.*, K_2 and El-5665 have been found to be resistant to *A. lini* (Kalia *et al.*, 1965). Singh *et al.* (1974) and Ji (1975) found four genotypes *viz.*, FRW-12, EC-4162, My-1 and *Linum strictum* to be resistant to this disease. In AICRP (Linseed) trials the genotypes *viz.*, Accession no. 2921, Ayogi, BAU-610A and ES-44 were found to have resistance to this disease during several years testing under Uniform Disease Nursery Trials (Anon 1991-2003).

Several workers have screened a number of flax cultures to breed cultivars for resistance to powdery mildew disease. EC-22587 and EC-22684 genotypes were reported to be free from powdery mildew under M.P. conditions (Agarwal, 1975). LC-216, LC-255, LC-256 and LC-269 were found highly resistant both in seedling and adult plant stages (Singh and Saharan, 1979). Some of the most promising powdery mildew resistant cultures reported from India are EC-9832 (Sandhu and Chandwani 1965; Singh and Kaurav, 1973; Kaushal and Srivas, 1974 and Gupta and Jain, 1980), K-5835 (Sandhu and Chandwani, 1965 and Dutt *et al.*, 1975) and EC-5663 (Dutt *et al.*, 1975). There are several varieties which have resistance to this disease (Table 22.2).

Cultural Approach

The cultural practices like clean cultivation, summer ploughing, planting time, crop density, cropping sequence, fertilizer doses and organic amendments have immense impact on pathogen population and disease incidence in linseed crop.

Use of Clean Healthy Seeds

The healthy clean seeds support good crop stand, therefore, larger pathogen propagules may be separated manually or mechanically before treatment with specific fungicides. The transmission of rust infection by means of seed and stem fragments could be avoided by thorough cleaning, and burning of screenings (Straib, 1939; Cass Smith, 1942; Ware, 1943 and Granhall, 1944).

Seeds of flax varieties (Giza and Baldi), when treated at 45°C temperature for 5 minutes, have shown improvement in seed germination and better seedling growth and reduced the wilt caused by

Fusarium oxysporum f. *lini* (Abd-El-Rehim *et al.*, 1969). Hot water treatment of seeds (48–50°C) was found effective against Pasmo (Newhook, 1942; Luchiana and Bedevich, 1979) and stembreak (Baylis, 1941). Seed injury during threshing should be avoided to reduce the infection of gray mold (De Tempe, 1962).

Table 22.2: Sources of Resistance Against Major Diseases of Linseed

Sl.No.	*Diseases*	*Resistant/Tolerant Varieties**	*Sources of Resistance*
1.	Rust [*Melampsora lini* (Ehrenb.) Lev.]	K2 LC-54, LC-185, Himalini, Jawahar -7, Jawahar-17,R-552, JLS (J)-1, Nagarkot. Surabhi Jeevan, J-23, Garima, Shubhra, Sweta, Gaurav, Sheela, Shekhar, T-397	LC-216, LC-255, LC-256, Ottawa 7 70 B (L), Dacota (M), Victory A (M), Tower (L8) Marshal (N2) and Pale blue crimped (L3) (Sharan and Singh, 1976,80). RR Numbers 5, 9, 10, 37, 38, 40, 45, 197, 204, 236, 262, 267 and 272 (Vasudeva, 1962); T-1192-2 (Dey, 1948); K-2, LC-185 (Gill, 1958)
2.	Wilt (*Fusarium oxysporum f. sp. lini* Bolley)	K2,LC-185, LC-54 Kiran, Jawahar-23, R-552, Sheetal, Surabhi Himalini, Jeevan LC-54, LC-185, T-397	Sabour (Mc Rae, 1928. Alam, 1931) ST6 (Asthana, 1957); RR-439, 440 (Merh and Kulkarni, 1961); K1,K2 (Goel and Swarup, 1964); NP-124, 439, 440, NP (RR)-9 (Sharma *et al.*, 1972); Indore -1, Malvi-1, NP-12, Bison-70, Canadian western (Chandraswami and Srivas, 1968).
3.	Alternaria blight *Alternaria lini* Dey	LC-54, Himalini, Jeevan, Gaurav, Nagarkot, Surabhi, Sheela, Shekhar, Rashmi, Meera, Parvati, Sweta	K2, EI 5665 (Kalia *et al.*, 1965); FRW-12, EC-4162, MY-1 and *Linum strictum* (Singh *et al.*, 1974; Ji, 1975).
4.	Powdery mildew (*Oidium lini* Skoric)	LC-54, Himalini, J-23, R-552, Kiran, Jeevan, Nagarkot Rashmi Surabhi, Sheela, Sheetal, Shekhar, Meera, Shubhra, Sweta	EC-22587, EC-22684 (Agarwal, 1975); LC-216, LC-255, LC-256 and LC-269 (Singh and Saharan, 1979)
5.	Rhizoctonia wilt (*Rhizoctonia solani* Kuhn)	Not known	NP-5 (Mishra and Sinha, 1982)

* *Source*: Anon (1986), Saharan (1988) and Kerkhi *et al.* (1999).

Field Sanitation

Proper field selection and preparation also play important role in disease development. Several pathogens perpetuate in the soil and plant debris, which may be destroyed by deep summer plaughing, burning plant residues and destroying the wild alternate hosts. To reduce the rust infection all infected straw/stubble should be removed before sowing and the infected straw should not be dragged into adjacent fields (Frederiksen, 1948).

Crop Rotation and Cropping Sequences

A crop rotation with non host crop is recommended for control of wilt and other soil borne diseases.

Three to four years crop rotation with corn or leguminous crop is recommended for rust (Hart, 1923, 1926 and Menzel, 1952). However crop rotation was not found feasible in case of flax wilt on account of long survival of the pathogen in the soil. It was observed that growing of rapeseed crop in sequence with flax reduced the population of *Fusarium oxysporum* f *lini*, eliminating the pathogen from the infested soil by following 6-years rotation (Zarzycka, 1977). Wilt incidence was reduced to only half when flax was sown after oat. The root extract of oats inhibited conidial germination of *F.lini* (Luchiana, 1979). The incidence of wilt was reduced under maize-linseed and soybeen–linseed cropping sequence at Palampur AICRP (Linseed) Coordinated trials (1999–2002).

Intercropping

Intercropping linseed with crops like chickpea, lentil, mustard, wheat and safflower has resulted in reduction of diseases with higher monetary returns in All India Coordinated trials on linseed. Lower percentage of wilt was recorded when mustard was intercropped with linseed in 5:1 ratio (Anon 1994), whereas minimum incidence of Alternaria blight and powdery mildew was recorded when wheat was sown as inter crop with linseed (4:2). The incidence of Alternaria blight was however, higher when linseed was intercropped with mustard in 4:2 ratio (AICRP, Linseed Reports 1996–1999). The lower incidence of wilt in linseed + mustard intercropping may be due to the toxic effect of root exudates of mustard, while higher incidence of AB in same intercropping may be due to rise in relative humidity on account of smothering effect of mustard.

Sowing Time

Proper adjustment of sowing time helps in escaping from several diseases. Early sowing is recommended for escape from rust (Ware, 1943 and Bogulawski, 1961). Early sowing favours escape from rust and powdery mildew while lower percentage of wilt and Alternaria blight is reported in late sowing. Minimum incidence of Alternaria blight was recorded when linseed was sown during second week of November (Singh *et al.*, 2000).

Sowing time also affects the development of Anthracnose (Vinogradov, 1962). More incidence of crinkle virus is reported in late planting (Frederiksen, 1962, 1964), while incidence of Curly Top virus was more severe in early sowing (Barnett and Houston 1944).

Seed Rate

Higher seed rate and dense population result into higher disease incidence in general. Lower intensity of Alternaria blight and powdery mildew was recorded at lower seed rate (AICRP reports 1997-99). For grey mold also the sowing should not be too dense (Vander Spek, 1956, 1958).

Fertilizers and Irrigation

Excessive use of nitrogen fertilizers is not recommended as it favours several diseases like *Rhizoctonia solani* (Frederiksen, 1965), rust (Menzel, 1952) Alternaria blight and powdery mildew (AICRP Reports 90–92). The intensity of Alternaria blight and powdery mildew were low at 50 per cent RDF and minimum at zero level (AICRP report 1990–91, 1999–2002). However application of fertilizers especially NPK to light soil increased resistance to *Fusarium* (Bolley 1934; Sodowskii, 1968).

For resistance to rust the crop should be supplied with potash and phosphorus (Menzel, 1952). Dunnin and Mladinov (1968) reported that higher concentration of zinc in the soil increased resistance to wilt. Dastur and Bhatt (1974) observed that the wilt resistant plants contained more potassium as compared to susceptible ones. Application of potassium fertilizers alone or in combination with zinc or molebdanum were found to be effective in reducing the Anthracnose infection (Karpunina, 1976). The application of Farm Yard Manure (FYM) was reported to reduce wilt incidence in linseed (Goel and Swaroop, 1964).

Not much work has been done on effect of irrigation on disease incidence in linseed, However, mortality due to wilt was reported to be higher in unirrigated fields at Raipur (AICRP Report 1994).

Biological Approach

Biological approach in disease management in linseed is less explored area. However, there are indications of success in management of seed and soil borne diseases.

The suppressiveness of Fusarium wilt in sandy loam soils was found to be due to antagonistic bacteria like *Pseudomonas putida* (Anon, 1967; Chiou and Yang, 1980; Scher and Baker 1980, 1982; Tu and Chang, 1982; Dupler and Baker, 1984) and *Clostridium* sp. (Scher and Baker 1980, 1982; Dupler and Baker, 1984). *P. putida* also shows stimulatory effect on plant growth and is termed as rhizobacteria or plant growth promoting rhizobacteria (PGPR) (Kleopper *et al.*, 1980; Teintze and Leong, 1981; Zeintze *et al.*, 1981). Some fungi like *Trichoderma lignorum* Syn. *T. viride* (Fedorinchik and Vander flaas, 1958; Fedorinchik, 1960, 1963 and 1970, Vodyanaya, 1962), *Asperillus* flavus and *Arachniotus* sp. (Akhter *et al.*, 1982) show antagonistic effect against *F. oxysporum f. lini.*

The saprophytic species of *Fusarium* have been found to increase suppressiveness in the activity of *F. oxysporum* f. sp. *lini* in soil (Tamiett and Alobouvette, 1986).

Introduction of *Trichoderma lignorum* into soil has been found to reduce the infection of flax caused by *Colletotrichum lini* (Lachançe, 1951; Fedorinchik and Vander flaas, 1958, Fedorinchik 1960, 1963 and 1970). *F. oxysporum f. lini* is reported to be antagonisted to Collectotrrichum *lini* (Laciocowa, 1979). The seedling infection caused by *C. lini* can be reduced considerably by bacterization of flax seeds with *P. fluorescence* and *B. mesentericus* (Novogrundskii *et al.*, 1937).

In Taiwan mycophagos nematode *Aphelenchus avenae* has been found to be antagonistic to *Rhizoctonia solani* causing seed and seedling rot in linseed (Cheng and Tu, 1974).

The seed treatment of flax seeds with *Pseudomonas fluorescence* (Novotna, 1986, 1990), *Agrobacterium* (Shaprio, 1992) and non parasitic *Fusarium* sp. (Nagao et al., 1990 and Lemanceau and Albouvette, 1991) reduced the fusarial wilt. Application of non pathogenic *F. oxysporum* is highly effective in reducing Fusarial wilt incidence (Edel *et al.*, 1997; Steinberg *et al.*, 1997). Microbial antagonism at root level is involved in the suppression of *Fusarium* wilt by the combinations of non pathogenic *Fusarium oxysporum* and *Pseudomonas putida* (Duijff *et al.*, 1999).

Seed treatment or spray of *Trichoderma* spp. or *Epicocum nigrum* has been found to be effective against Atternaria blight and Phoma rot (Mercer and Papadopolous, 1990; Mercer et al., 1993).

The *Zygorhyncus molleri* present in compost has antagonistic effect on soil borne pathogens like *Pythium* spp. *Rhizoctonia solani* and *Fusarium oxysporum* f. sp. *lini* causing seedling blight, stem rot and witt diseases (Brown, 1987). The anaerobically fermented pig slurry is reported to reduce fusarial witt (Reinsinger, *et al.*, 1992). Use of FYM or compost therefore, have a great scope in reducing wilt incidence in linseed crop.

The adult larvae of coccinalids, *Thea cincta* (*Psylobra cincta*) predate on the powdery mildew pathogen, cleaning the infected part from fungus (Prasad and Rai 1990).

The seed treatment with *Trichoderma harzianum* reduced wilt (*F. oxysporum lini*) incidence to the tune of 67.4 per cent with highest grain yield at Kanpur. It was superior to *T. viride* and Thiram However, better disease control was observed with *T. viride* at Raipur. The *Trichoderma* cultures used were same but the pathogen strains were different (AICRP Report 2000–2002).

Chemical Approach

In IPM the use of chemical is not totally prohibited but its use is restricted to only when absolutely necessary to keep the pests below economic threshold level with minimum environmental damage. Their use should be need based, pest specific and at right time, for which there should be proper disease forecasting. The minimum quantity of selective pesticides can be utilized more effectively in seed dressing. Selective pesticides and botanicals with least environmental pollution and toxicity to biotic agents need to be opted for the use in IPM programmes. A number of fungicides have been tested as seed dressing and spray for control of linseed diseases (Table 22.3). They can be utilized judiciously when utmost necessary.

Table 22.3: Chemical Control of Major Diseases of Linseed

Diseases	*Fungicides/Mode of Treatment*	*References*
	(I) Soil treatment	
Rhizoctonia seedling blight	Quintazene	Frederiksen (1965); Hulea and Mujdaba (1971)
Aster yellow	Phorate	Frederiksen (1962); Westdal and Richardson (1972)
	(II) Seed treatment	
Wilt	Thiram	Zvyagintseva (1959); Anspok (1961); Soloveva (1962); Mladinov (1934)
	Organomercurials	Burnett and Reddy (1931); Carerra (1942); Anspok (1961); Pietkiewicz and Czyzenwska (1961)
	Bentate, Vitavax, Bavistin	Haware (1972)
	Benlate and Thiobendazole	Ondrez (1975)
	Thiobendazole,Thiobendazole + Flutriabol (vincit); Metalaxyl + mancozeb (Ridomil); Iprodione	Andruszenska *et al.* (1995)
Alternaria blight	Bavistin, Agrosan G.N., Topsin M	AICRP Reports (1984-85)
	Iprodione, Mancozeb	Mercer (1984)
	Prochloraz	Fitt *el al.* (1991)
Seed rot, seedling blight	Organomercurials, Rhizoctol combi, Spergon	Henry (1943); Cruick shank and Jacks (1953); Martin (1953) Michael *et al.* (1973)
Anthracnose	Thiram	Muskette and Colhouns (1941); La Chance (1945-47); Krzystalowska (1959); Goryacheva (1966); Karpunina (1976)
Stem break	Thiram, Cupric chloride	Ignatovich (1958)
	Organomercurials	Eglitis (1932); Anon (1944); Muskette and Colhouns (1944); Todorova (1958); Levitin and Kvitko (1971)
Foot rot	Organomercuriais	Vander Spek (1957) Anselme (1958, 1958)
Pasmo	Organomercurials	Newhook (1942); Nattrass (1943); Borgas (1946); Butler (1949); Anselme (1952) Luchina and Bedevich (1978, 1979)
Grey mold	Organomercurials Captan, Thiram, Quinone	Schoevers (1923); Anon (1924); Vander Spek (1957); Anselme (1967)
	Thiram	De Tempe (1958, 1968); Rataj (1961)

Contd...

Table 22.3–Contd...

Diseases	Fungicides/Mode of Treatment	References
	(III) Spray	
Rust	Fermasol (74 per cent) Feridimethyl dithio carbamate)	Frederiksen (1956)
	Parazate liquid, Dithane M-22	Vir and Agarwal (1964)
	Dithane S-41, RH-539 (Nickel sulphate and Dithane M22)	Upadhyay and Verma (1972)
	Mancozeb, Cuman, Miltox	Agarwal and Kotasthane (1973)
	Plantavax, Vitavax	Froiland and Little field (1972); Goel *et al.* (1973)
	Benomyl, Calixin	Krzysztalowska (1974, 1977)
	Benodanil	Khatua *et al.* (1977)
Alternaria blight	Bordeaux mixture	Vasudeva (1962)
	Dithane M-45, Mixture of Calixin (0.5%) and DM-45 (0.2%)	AICRP repots (1985–86)
	Prochloraz, Iprodione	Fitt *el al.* (1991)
	Prochloraz or Prochloraz + fenpropimorph	Davis and Fitt (1992)
	Iprodione	Mercer *et al.* (1992); Mercer and Hardwick (1993)
	Mancozeb, Iprodione (Rovral)	AICRP reports (1996); Singh (2000)
Powdery mildew	Elosal, Afugan Morestan, sulphur dust wettable sulphur and Karathane	Sharma and Khosla (1979)
	Morestan, Carbendazeme	Vyas *et al.* (1982)
	Tridemorph, sulphur Karatharie	AICRP Report (1986)

Integrated Pest Management

Attempts were made to manage a few diseases of linseed by integrating 2–3 factors under AICRP (All India Coordinated Research Projects on Linseed). A cent percent control of powdery mildew was achieved by using resistant variety protected by three sprays of wettable sulphur saving 23.81 per cent yield loss at Nagpur, (AICRP Report 1999–2000).

Integrated Management of Alternaria blight in linseed was tried at three locations *i.e.*, Faizabad, Kanpur and Kanke with three IPM Modules under AICRP (Linseed) programme. Module including tolerant variety, sown during second week of November with optimum fertilizer dose, seeds treated with Topsin-M (0.2 per cent) and one spray of mancozeb (0.25 per cent) scored minimum disease intensity with highest grain yield and additional return of 472 to 8,373 rs/ha over traditional farmers practice (Singh *et al.*, 2003).

Future Thrust

1. The foremost prerequisite for developing long term and sustainable strategy is to develop mapping of major diseases with respect to their occurrence and severity in different agroclimatic regions. This is to be updated regularly to enable to develop need based strategy

for resistance breeding, exploiting the available infrastructure facilities and resources in the country.

2. Identification of key factors which influence the outbreak of diseases or proper disease forecasting system.
3. Study of the survival and perpetuation of the pathogens.
4. To study the races/biotypes of the major pathogens.
5. To exploit the sources of disease resistance especially the multiple disease resistant sources. Biotechnology and genetic engineering may also be helpful in developing the resistant genotypes against diseases like Alternaria blight for which there is no source of resistance.
6. Practical application of IPM modules and improvement of IPM modules from time to time keeping in view the changed cultural practices.
7. A joint venture and team work of Breeders, Pathologists, Agronomists and Biotechnologists is required.

REFERENCES

Abd-El-Rehim, M.A., S.H. Michail, H. Elarosi, F.A. Kairy (1969). Hot water treatment of seed as a method for decreasing the incidence of certain cotton and flax seedling diseases *Ann. Appl. Biol.*, 63(1): 95.

Agarwal, S.C. (1975). Observations on *Oidium lini* Skoric on linseed. *Oilseeds J.*, 5: 19–20.

Agarwal, S.C. and S.R. Kotasthane (1973). Control of linseed rust by fungicides under artificially inoculated condition. *Sci. & Cult.*, 39: 508–509.

Akhter, C.M., C.M. Amin and C.M. Salim (1982). Investigation of the biological control of *Rhizoctonia solani* and *Fusarium oxysporum* f. *lini* causing root rot and wilt disease in cotton and linseed crop respectively. *Ayub. Agric. Res. Inst., Faisalabad, Pak.*, 2: 63.

Alam, M. (1931), Rep. Agric. Dep. Bihar and Orissa for the period from 1st April 1930 to 31st March 1931, 42.

Andruszenska, A., M. Oryjanska and I. Wysakowska (1995). The influence of seed dressing on health and yield of flax. *Materially -Sesjs Instytutu Ochrony Roslin*, 35: 2.

Anonymous (1924). Reports of the activities of Phytopathological service in the year 1923. Versl. en, Meded. Plantenzie Ktenkunigen Dienst te Wageningen, 34: 66

Anonymous (1924). Survey of the prevalence of plant diseases in Dominion of Canada 1923. *Fourth Ann. Rep. Canada Dep. Agric. Exper. Farms Branch*, 125 (mimeographed).

Anonymous (1944). Symposium of seed borne diseases *Proc. Can. Phytopathol. Soc.*, 12: 18.

Anonymous (1949). Administrative Report of the Department of Agriculture, United Provinces for the year 1947–48, pp. 89.

Anonymous (1967). Nutritional interactions between plants and fungi, General report. *Abstr. Rev. Appl. Mycol.*, 46: 1843e.

Anonymous (1978). *A Handbook of pests, diseases and weeds of quarantine significance*, (2nd ed.), Trans. from Russian, Amerind Publ. Co. Pvt. Ltd., New Delhi.

Anonymous (1986). Linseed. Linseed production technologies for better crop, 28, ICAR, New Delhi.

Anonymous (1985–86, 1990–2003). Annual Progress Reports on Linseed, All India coordinated research project on linseed, Directorate of Oilseeds Research, ICAR, Rajendra Nagar, Hyderabad.

Anselme, C. (1952). La maladie sur le pasmo des lins a huil. *Ann. Inst. Nat. Rech. Agron.*, 11 (2): 201,

Anselme, C. (1958). A comparative study of various products used for the disinfection of flax seeds, *Phytiatrie Phytopharm.*, 7 (1): 3

Anselme, C. (1958). The sanitary condition of the seed and the cultivation of flax, *Bull. Tech. Inq. Serv. Agric.*, 129: 203.

Anselme, C. (1967). Disinfection of flax seeds, Abstr., *Rev. Appl. Mycol.*, 46: 638.

Anspok, P. (1961). The use of microelements with insectofungicides and herbicides in the conditions of Latvian S. S. R., *Latv. Zinst. Akad. Vest.*, 6 (167): 117.

Arya, H.C. and R. Prasada (1952). Alternaria blight of linseed *Indian Phytopath.*, 5 (1): 33.

Asthana, R.P. (1957). Incidence of wilt disease of *Linum usitatissimum Nagpur Agric. Coll. Mag.*, 31(1–4): 16.

Barnette, H.H. and B.R. Houston (1944). Culy top of flax and other flax diseases in Central California in 1944, *Plant Dis. Reptr.*, 28 (15): 507

Baylis, G.T.S. (1941). Stem break and browning (*Polyspora lini*) of flax in New Zealand *N.Z.J.Sci. Technol.* A., 23 (1): 1

Bedi, K.S., C.S. Paracer and S.P.S. Bedi (1961). Studies on wilt disease of linseed in Punjab, Part-I, *Indian Oilseeds J.*, 5(3): 208

Bhargava, S.N and D.N. Shukla (1980). Losses of oil content due to fungal invasion *Nat. Acad. Sci.*, 3 (5): 141.

Bogulawskii, W. (1961). Influence of sowing period on the rust *Melampsora lini (Pers.) Lev.* infection in flax. Abstr. *Rev. Appl. Mycol.*, 40: 170.

Bolley, H.L. and Manns, T.F. (1932). Fungi of flax seed and flax sick soil. *N.D. Agric. Exp. Stn. Bull.* 259: 57.

Borgas, M.L.V. (1946), Uma doenca do Linho nova para Portugal *Broteria*, 15 (3): 129.

Brown, A.E. (1987). Activity of *Zygorhynchus molleri* in relation to antagonism against some soil borne plant pathagenic fungi. *J. Phytopathology*, 120 (4): 298–309.

Burnette, L.C. and C.S. Reddy (1931). Seed treatment and date of sowing experiments with six varieties of flax. *Phytopathology*, 21(10): 985.

Butler, E.J. (1914), Notes on some rusts in India, *Ann. Mycol.*, 12: 76.

Butler, E.J. (1918). *Fungi and deseases in plants*. Thacker Spink and Co., Calcutta.

Butler, F.C. (1949). Pasmo disease of linseed. *Agric. Gaz. N.S.W.*, 60 (6): 291

Carrera, C.J.M. (1942). La 'fusariosis' o marchitamiento del Lino en la Republica Argentina debida al *'Fusarium lini'* Bolley Publ, Misc. Minist. Agric. B. Aires, 113: 4.

Cass-Smith, W.P. (1942). Flax rust, *J. Dep. Agric. W. Aust.* Ser. 2 (1): 56.

Chandwani, G.H. and Y.C. Srivastava (1968). Varietal resistance of linseed to wilt (*Fusarium oxysporum lini*) *Indian J. Agric. Sci.*, 38(6): 1021.

Chauhan, L.S. and K.N. Srivastava (1975). Estimation of loss of yield caused by blight disease of linseed *Indian J. Farm Sci.*, 3: 107–109.

Cheng. Y.H. and C.C.Tu (1974). Studies on the control of Rhizoctonia damping off of flax and jute by mycophagus nematode, *Aphelenchus avenae J. Taiwan, Agric. Res.*, 23(1): 75.

Chiou, Y.Y. and S.S. Yang (1980). Water activity in relation to chlamydospore germination of *Fusarium oxysporum* sp. *lini., Proc. Natl. Sci. Counc. Repub, China*, 4(1): 87.

Chowdhary, S. (1951). Linseed rust in Assam, *Plant Prot. Bull.*, New Delhi, 3(4): 119.

Cruickshank, T.A.M. and H. Jacks (1953). Seed disinfection. VII. Effect of seed treatments on emergence of linseed and linen flax *N.Z.J. Sci. Technol. Sect A.*, 35(1): 33.

Damodaran, T. and D.M. Hegde (2002). *Oilseed Situation–A Statastical Compendium* (ICAR), Directorate of Oilseeds Research, Rajendra Nagar, Hyderabad, India.

Dastur, R.H. and J.G. Bhatt (1964). Relation of Potassium to Fusarium wilt of flax. *Nature*, London, 201: 1243–1244.

Davis, H. and B.D.L. Fitt (1992). Effect of fungicide timing on fungal diseases, growth and yield on linseed. *Test of Agrochemicals and Cultivars*, 13: 68–69.

De Tempe, J. (1958). Botrytis infection of flax seed: Evaluation and control. *Tijdschr. Piziekt.*, 64 (4): 281.

De Tempe, J. (1962). Hot water treatment as a means for determining seed weakness. Comparison of methods for seed health testing. *Proc. Int. Seed Test Assoc.*, 27(3): 773, 815.

De Tempe, J. (1968). The quantitative evaluation of seed borne pathogenic infection. *Proc. Int. Seed Test Assoc.*, 33 (4), 573.

Dey, P.K. (1933). An Alternaria blight of linseed plant. *Indian J. Agric. Sci.*, 3: 881–96.

Dey, P.K. (1948) *Plant Pathology*, Administration Report, Department of Agriculture, pp. 43–46.

Duijff, B.J., G. Recorbet, P.A.H.M. Bakker, J.C. Loper and P. Lemanceou (1999). Microbial antagonism at root level is involved in the suppression of Fusarium wilt by the combination of non pathogenic *Fusarium oxysporum* FO47 and *Pseudomonas putida* wes 358. *Phytopathology*, 89(11): 1073–1072.

Dupler, M. and R. Baker (1984). Survival of *Pseudomonas putida*, a biological control agent in soil. *Phytopathology*, 74(2): 195.

Dutt, S., A.S. Chahal, K.S. Lobana and M. Singh (1975). Evaluation of genetic stock of linseed (*Linum usitatissimum* L.) against powdery mildew in the Punjab, *J. Res.*, Punjab Agric Univ., Ludhiana, 12(1): 6.

Dunin, M.S. and M.R. Mladenov (1968). Effect of zinc on resistance of flax to Fusarium disease. *Vest. Selkhoz. Nauki Mosk.*, 13: 29.

Edel-V, Steinberg, C., N. Gautheron and C. Albouvette (1997). Population of non pathogenic *Fusarium oxysporum* associated with roots of four plant species compared to soil borne populations. *Phytopathology*, 87(7): 693–697.

Eglitis, H. (1932). Flax diseases and flax seed disinfection experiments during the period 1929-31 *Acta Inst. Defens. Plantarum letviensis. Riga*, 2: 5.

Fedorinchik, N.S. (1960). The effectiveness of he bioperpetuation trichodermin. *Abstr. Rev. Appl. Mycol.*, 39: 281.

Fedorinchik, N.S. (1965). The use of *Trichoderma* for the control of plant diseases. *Abstr. Rev. Appl. Mycol.*, 44: 1393.

Fedorinchik, N.S. (1970). Methods of enriching soil with the saprophytic fungus *Trichoderma viride*, an antagonist to pathogenic microorganisms. Abstr. *Rev. Plant Pathol.* 49: 681.

Fedorinchik, N.S. and L.K. Vanderflaas (1958). Effect of the antagonistic activity of the soil fungus *Trichoderma lignorum* Harz. on increase in yield of Agricultural crops. Abstr. *Rev. Appl. Mycol.*, 37: 24.

Fitt. B.D.L., A.W. Ferguson, U., Dhua and S. Berhenne (1991). Epidemiology of Alternaria species of linseed. *Aspects of Appl. Biol.*, 28: 95-100.

Fitt, B.D.L., H. Coskum and D. Schmechel (1991). Biology of Alternaria species on linseed, A comparison. *Aspects of Appl. Biol.* 28: 101-106.

Frederiksen, P.S. (1948). Horrustens Udbredelse skadvirkning og Bekaempelse En Fare for vol. spindherdyrkning. Vyrkning. Ugeskar Landm., 18: 8.

Frederiksen, R.A. (1962). Reduction of Aster yellow in flax via systemic insecticides *Plant Dis. Reptr.*, 46 (1): 813.

Frederiksen, R.A. (1964). Aster yellows of flax *Phytopathology*, 54 (1): 44.

Frederiksen, R.A. (1965). Rhizoctonia seedling blight of flax, Abstr. *Phytopathology*, 55 (1): 498.

Froiland, G.E. and I.J. Little Field (1972) Systemic protectants and eradicate, chemical control of flax rust *Pl. Dis Reptr.*, 56: 737-39.

Garg, S.K. (1982). Studies on leaf spot and black point disease of linseed caused by *Alternaria* spp. Ph.D. Thesis, C.S.A.U. Agri. and Tech., Kanpur, 165 pp.

Gill, K.S. (1958). Powdery mildew, a new disease of linseed in Punjab, *Indian Oilseeds J.*, 2: 16–17.

Gill, K.S. and D.M. Yermanos (1967). Cytogenic studies on the genus *Linus*, I. Hybrids among taxa with 15 as the haploid chromosome number. *Crop Sci.*, 7: 623.

Goel, L.B., P. Bahadur, V.C. Sinha and S. Singh (1973). Systemic fungicides in the control of different rusts. *Indian Phytopath.*, 26 (2): 349.

Goel, L.B. and G. Swarup (1964). Some observations on linseed wilt. *Indian Phytopath.*, 17: 133–137.

Goryacheva, L.V. (1966). Influence of microelements in mixture with fungicides on the increase in resistance in fibre flax to disease. In brief summaries of research work on Plant Protection in the Baltic zone of U.S.S.R., Abstr. *Rev. Appl. Mycol.*, 45: 9405.

Granhall, I. (1944). Linrosten-en sponadslinets flende, Lantamannen, Uppsala, 28 (32): 733–735.

Hart, E. (1925). Factors affecting the development of *Melampsora lini* (Pers.) Lev. *Phytopathology*, 15 (1): 53.

Hart, E. (1926). Factors affecting the development of flax rust. *Phytopathology*, 16 (3): 185–205.

Haware, M.P. (1972). Effect of systemic fungicides on *Fusarium lini in vitro* and their uptake by linseed seedlings. *Indian J. Mycol. Pl. Pathol.*, 2 (2): 136

Hora, T.S., V.V. Chenuleu and R.L. Munjal (1962). Studies on assessment of losses due to *Melampsora lini*. *Indian Oilseeds J.*, 61: 196.

Howard, A. (1921). Influence of soil factors on disease resistance. *Ann App. Biol.*, 7: 373.

Hulea, A. and M. Mujdaba (1971). Seedling rot and stem break of Rhizoctonia disease of flax in Rumania. *Rev. Pl. Path.*, 51: 1548.

Ignatovich, G. (1956). A chemical method of controlling Polysporosis of flax *Zemledelle Moscow*, 4(1): 93 (*Rev Appl. Mycol.* 35: 609).

Ji, T. (1975). Reaction of linseed varieties to Alternaria leaf spot. *Indian J. Mycol Pl. Pathol.*, 5: 103.

Kadian, O.P. and D. Suryanarayana (1971). Studies on seed mycoflora of linseed (*Linum usitatissimum*). *Indian Phytopath.*, 24 (3): 487.

Kalia, H.R., J.N. Chand and B.S. Ghai (1965). Inheritance of resistance to Alternaria blight of linseed. *J. Res.* Ludhiana, 2: 104–105.

Karpunina, Yu T. (1976). Vredonosnost antraknozlina *Zashchita Rastenil*, 6: 23.

Kaushal, P.K. and S.R. Shrivas (1974). Inheritance of resistance to powdery mildew (*Oidium lini*) in linseed, (*Linum usitatissimum*). *Curr. Sci.*, 43: 353-54.

Kerkhi, S.A., R.L. Srivastava, Jyoti Singh, Y.P. Malik, K. Hussain, S.D. Dubey and J. Rai (1999). *Linseed Technology for Increasing Production* (Ed.) Hegde *et al.*, Directorate of Oilseed Research, Rejendra Nagar, Hyderabad, p. 9–16.

Khanna, K.K. and S. Chandra (1981). Evaluation of fungicides and antibiotics in the control of some leaf blights caused by *Alternaria alternata. Indian J. Mycol. Pl. Pathol.*, 11 (1): 110

Khatri, R.K. and D.R. Saxena (1986). Studies on factors influencing the mechanism of resistance in linseed to rust (*Melampsora lini). Indian Phytopath.*, 39: 151.

Khatua, D.C., S. Maiti, S. Poi, S. Bandopadhyay and S. Sen (1977). Field assessment of fungicides for flax rust control *Plant Dis. Reptr.*, 61 (3): 185.

Kloepper, J.W., J. Leong, M. Teintze and M.N. Schroth (1987). Enhanced plant growth by siderophores produced by plant-growth promoting rhizobacteria *Nature* (London), 286: 885.

Kotasthane, S.R., J.L. Merk, S.C. Agarwal and K.G. Singh (1973). Varietal resistance in linseed to rust (*Melampsora lini*). *Oilseeds J.*, 3: 1–2.

Krzysztalowska, H. (1959). The effect of seed treatment on flax yield *Abstr. Rev. Appl. Mycol.*, 38: 600.

Krzysztalowska, H. (1974). Badanie skutecznosci chemicznego zwalez aniardzy Inu *Melampsora liniperda* koernicke (Palm) Ckyl II (1969–71), *Pr. Inst. Krajow. Wlok. Nature*, 21: 87.

Krzysztalowska, H. (1977). Causes of the appearance of flax rust, *Melampsora lini*perda Koernicke (Palm) and the possibility of its control. *Referativnyl Zhurnal*, 8: 55.

Kulkarni, N.B., B.B. More and P.L. Patil (1969). Occurrence of a new race of *Fusarium lini* (Bolley) Snyder and Hansen inciting linseed wilt in India *Mycopath. Mycol. Appl.*, 30 (3): 243.

Kumar, K., S.K. Garg and H.K. Saksena (1985). Seed borne nature of *Alternaria alternata* in linseed and its detection *Indian Phytopath.*, 38 (1): 162.

Kushwaha, U.S. and J.N. Chand (1971). Efficacy of fungicides for control of powdery mildew of linseed (*Linum usitatissimum* L.) caused by *Oidium lini Indian Phytopath.*, 24 (1): 200.

Lacicowa, B. (1979). Badania nad podatnosoia roznych odmian Inu na fusarioze (*F. oxysporum f.S.P. lini*), *Akademia Rolnicza Lublin.* Poland, 241–250 (Abtr. *Rev. Plant Pathol.*, 58: 4210).

Lachance, R.O. (1947). Action de divers desinfectants sur la viabilities de *Colletotrichum lini* dans la graine de Lin, Rep. Quebec. *Soc. Prot. Pl.*, 1945–1947, pp. 76.

Lachance. R.O. (1951). Antagonisme des microoganismes du sol envers, le *Colletotrichum linicola* agent devl, anthracnose du lin. *Can. J. Bot.*, 29 (4): 433.

Lemanceau, P. and C. Alabouvette (1999). Biological control of Fusarium diseases by fluorescent *Pseudomonas* and non pathogenic *Fusarium. Crop Protection*, 10(4): 279–286.

Levitin, M.M. and K.V. Kvitko (1971). Izuchenie vol. 3 buditelya polisporoza, V. Asepticheskol Kulture Prorostkov L' na, Mikol. i. *Fitopatol.*, 5 (4): 398.

Luchina, N.N. (1979). Effect of field crops on the suppression of soil infestation by *Fusarium oxysporum* f sp. *lini. Rev. Plant Pathol.* 58: 2259.

Luchina, N.N. and G.V. Bedevich (1978). Disinfection of flax seed against Pasmo. *Zashchita Rastenii*, 7: 41.

Luchina, N.N. and G.V. Bedevich (1979). Biological features of the pathogen of pasmo in Belorussion SSR. Abstr. *Rev. Plant Pathol.* 58: 719.

Martin, H. (1953). Chemical methods of crop protection. *Agricultual Institute of Canada, Review*, 8 (2): 5, 9.

Mc Rae, W. (1920). Report of the Imperical Mycologist, *Sci. Rep. Agric. Res. Inst.*, Pusa, 1926–27, p. 45.

Menzel, K.C. (1952). Zurr frage des Rostbefalles Von Flachs, Faserforsch. *Textil Tech.*, 3(3): 105.

Mercer, P.C. (1984). Linseed diseases. In Annual Report on Research and Technical Work of the Department of Agriculture for Northern Ireland, Befast, Northoesin, Ireland.

Mercer, P. C. and N. N. Hardwick (1993). Chemical control of seed brone diseases of linseed, Aspects of Appl. Biol. in the U.K. *Bulletin Oil SROP*, 10(9): 73–83.

Mercer, P.C. and S. Papadopolous (1990). Biological control of seed borne diseases of linseed, *Proc. Biological Control of Pests and Diseases*, (Eds.) A.R. McCracken and P.C. Mercer, Department of Agriculture for Northern Ireland, pp. 19–26.

Mercer, P.C., A. Ruddok and H.C. Mc Gimpsey (1992). Evaluation of Iprodione and *Trichoderma viride* against *Alternaria linicola. Test of Agrochemicals and Cultivars*, 31: 20–21.

Mercer, P.C., A. Ruddock, E. Mee, S. Papadopolous (1993). Biological control of Alternaria diseases of linseed and oilseed rape. *Bulletin 0ILB/SROP*. 16 (9): 89–99.

Merh, J.L. (1967). Short note on varietal resistance of Indian linseed to wilt (*Fusarium oxysporum* f. *lini*). *Indian J. Agric. Sci.*, 37 (1): 45.

Merh, J.L. and S.N. Kulkarni (1961). Varietal resistance of linseed to wilt and rust II. *Indian Oilseeds J.* 5: 25–26.

Michail, S.H., H. Elarosi, M.A. And-El Rehim and E.A. Khairy (1973). Hot water treatment and fungicidal dressing of seed for the control of damping off of cotton and flax. *Phytopathologia Medeterranea*, 13: 111.

Mishra, B. and S.K. Sinha (1982). Studies of wilt of linseed caused by *Rhizoctonia bataticola. Indian Phytopath.*, 35: 555–557.

Mishra, D.P. and R.N. Pandey (1981). Linkage of regenes for powdery mildew and rust resistance in linseed. *Indian J. Agric. Sci.*, 51: 559–562.

Misra, D.P. and R.R. Prasad (1966). Status of linseed rust races in India and sources of resistance. *Indian Phytopath.*, 19: 184-188.

Misra, D.P. and A.K. Shukla (1981). Genetic identity for rust resistance in some linseeds. *Indian J. Agric. Sci.*, 51: 444–446.

Mladinov, M. (1974). Flax seeds as a source off infection in the transmission of Fusarium wilt. *Rastitelna Zashchita*, 22 (2): 29.

Mundkur, B.B. and S. Ahmed (1946). Revision of the additions of Indian fungi. *II Mycol. Rap. Imp. Mycol. Inst.*, 18: 11.

Muskett, A.E. and J. Colhouns (1944). The prevention of seed borne diseases of flax by seed disinfection. II. Comparison of the dusting, short wet and fixation method of treatment. *Ann. Appl. Biol.*, 31 (4): 295.

Muskett, A.E. and J. Colhouns (1947). The diseases of flax plant (*Linum usitatissimum* L.) Belfast, The Bursar, Queen's Univ.

Nattrass, R.M. (1943). The pasmo disease of flax in Kenya, *Sphaerella linorum* Wollenwcher. *E. Afr. Agric. J.*, 8 (4): 223.

Newhook, F.J. (1942). Pasmo (*Sphaerella linorum*) on flax in New Zealand. *N. Z. J. Sci. Technol. A* 24 (2): 102.

Novogrundiskii, D., E. Berezova, M. Nakhimovskaya and M. Perviskova (1937). The influence of Bacterization of flaxseed on the susceptibility of seedling to infection with parastic fungi. *G.R. Acad Sci. URRSNS*, 14: 385.

Novotna, J. (1990). Antagonistic activity of *Pseudomonas fluorescence* against some pathogenic and saprophytic fungi on rape flax, Sborndk, UVTIZ. Ochrana Rostlin, 26 (2): 113–122.

Ondrez, M. (1975). Comparison of the promising fungicidal preparations against pathogenic fungi of flax (*Alternaria, Botrytis Colletotrichum, Fusarium* and Kabatiella Abstr. *Rev. Plant Pathol.* 54: 165.

Padwick, G.W. (1941). Report of the Imperical Mycologist. Rep. Agric: Res. Inst. 1939–40, p. 94

Patel, M.K., M.N. Kamat and V.P. Bhinde (1949). Fungi of Bombay, Suppement I. *Indian Phylopath.*, 2: 142.

Pavgi, M.S. and U.P. Singh (1965). Parasitic fungi from North India. IV, V. *Mycopath. Appl.*, 27 (1–2): 8.

Pearl, R.T. (1923). Report of the mycologists to the Government of Central Provinces of Berar. Rep. Dep. Agric. Central Provinces and Berar for the year ending 30th June, 1922.

Pietkiewicz T. and S. Czyzesiskii (1961). Effect of seed treatment on the infection of flax by fungi of the genus *Fusarium*. Abstr. *Rev. Appl. Mycol.*, 40: 683.

Plymen, F.J. (1933). Reports of the working of the department of Agriculture of the Central Provinces for the year ending 31st March 1932 and 31st March, 1933.

Porwal, S. (1964). *Oidium* spp. from Rajasthan. *Univ. Udaipur Res. Stud.*, 2: 913.

Prasad, R. and M. Rai (1988). Record of beetle, *Thea cineta* Fabr., feedingon powdery mildew of linseed. *Farm Sci. J.*, 3 (1): 89-90

Prasad, R., M. Rai and S.A. Kerkhi (1988). Resistance of linseed (*Linum usitatissimum*) germplasm to rust (*Melampsora lini*) and powdery mildew (*Oidium lini*), *Indian J. Agric. Sci.*, 58 (7): 548.

Rai, M. (1982). Increasing linseed productivity. *Indian Fmg.*, 32(8): 67.

Ramakrishnan, T.S. (1931). A wilt of Zinnia caused by *Sclerotium rolfsii. Madras Agric. J.*, October 1930 (Abtr.) *Rev. Appl. Mycol.*, 10: 246.

Rataj, K. (1961). Physiological biotypes of the fungus *Colletotrichum linicola*. Abstr. *Rev. Appl. Mycol.*, 40: 418.

Rathaiah, Y. and M.S. Pavgi (1970). Evaluation of fungicides against some leaf spot fungi parasitizing on oilseed crops. *Hindustan Antibiotic Bull.*, 13(1–2): 20.

Rathaiah, Y. and M.S. Pavgi (1971). A Cerscospora leaf spot of linseed. *Riv. Pathol. Veg. Pavia Ser.* 47 (4): 249.

Rathaiah, Y. and M.S. Pavgi (1973). Perpetuation of species of *Cercospora* and *Ramularia* parasitic on oilseed crops. *Ann. Phytopathol. Soc. J.*, 39(2):103

Sharan, G.S. (1978). Studies on linseed rust. *Ph.D.* Thesis, Himachal Pradesh University Palampur, 75pp.

Saharan, G.S. (1988). Plant disease management in linseed. *Rev. Trop. Pl. Pat.*, 5: 119-140.

Saharan, G.S. and B.M. Singh (1978). Inheritance of resistance in linseed to *Melampsora lini. Indian Phytopath.*, 31: 542–544.

Saharan, G.S. and B.M. Singh(1980). Inheritance of resistance in linseed to rust caused by *Melampsora lini* under field conditions. *Indian J. Mycol. Pl. Pathol.*, 10: 1'36-141.

Sandhu, R.H. and G.H. Chandwani (1965). Occurrence of powdery mildew of linseed at Hoshangabad (M.P.) *Indian Oilseeds J.*, 9: 102–103.

Sattar, A. and A. Hafiz (1952). Research on plant diseases of Pubjab. *Mongr. Pak. Ass. Sci.*, 1: 1–55.

Scher, F.M. and R. Baker (1980). Mechanism of biological control in *Fusarium* suppressive soil. *Phytopathology*, 70 (5): 412.

Scher, F.M. and R. Baker (1982). Effect of *Pseudomonas putida* and a synthetic iron chelator on induction of soil suppressiveness to Fusarium wilt pathogens. *Phytopathology*, 72 (12): 1567.

Schoevers, T.A.C. (1923). Flax seed disinfection, *Rep. Int. Conf. Phytopathol. Econ. Entoral. Holland*, 116.

Shaprio, Y.A.S. (1992). Bacterization of flax seed and Fusarial wilt. *Zashchita Rastenii* (*Morkva*), 8: 28.

Sharma, H.C. and H.K. Khosla (1979). Fungicidal control of powdery mildew of linseed (*Linum usitalissimum*) in relation to losses. *JNKVV Res. J.*, 10: 161–162.

Sharma, H.C. and R.L. Mathur (1971). Variability in first single spore isolates of *Fusarium oxysporum f. lini* in Rajasthan, *Indian Phytopath.*, 24: 698–704.

Sharma, L.C., R.L. Mathur and S.C. Gupta (1972a). Linseed wilt incidence in Rajasthan and loss in yield. *Rajasthan J. Agric. Sci.*, 2(1): 48.

Sharma, L.C., R.L. Mathur and S.C. Gupta (1972b). Performance of linseed varieties against Fusarium wilt. *Indian phytopath.*, 25: 303.

Shukla, A.K. and D.P. Misra (1983). Sources of rust resistance in linseed. *Indian J. Agric. Sci.*, 53: 454-456.

Shukla, D.M. and N.N. Pathak (1967). Occurrence of *Oidium lini* Skoric on different varieties of *Linum usitatissimum* Linn. *Labdev. J. Sci. Technol.*, 5: 338.

Siddaramaiah, A.L. and S.A. Desai (1980). Fungicidal control of foot rot of linseed. *Curr. Res.*, 9 (10): 175.

Singh, B.M. and G.S. Saharan (1979). Inheritance of resistance to *Oidium lini* Skoric in linseed (*Linum usitatissimum* L.). *Euphytica*, 28: 531–532.

Singh, B.P. and V.L.P. Kaurav (1973). Inheritance of resistance of powdery mildew (*Oidium lini* Skoric) in linseed (*Linum usitatissimum). Curr. Sci.*, 43: 353–54.

Singh, B.P., B.N. Shukla and S.P. Singh (1974). Susceptibility of linseed varieties of Alternaria bud blight at Jabalpur. *Proc. Nat. Acad. Sci.*, 44: 11–12.

Singh, Jyoti (2000). Effect of some spray fungicides against foliar diseases of linseed. *Ann. Pl. Protec. Sci.* 10(1): 169–170.

Singh, N. (1984). Reaction of exotic cultivars of linseed (*Linum usitatissimum*) to Indian physiologic races of *Melampsora lini* (Pers.) Lev. *Madras Agric. J.*, 40: 319–320.

Singh, R.B., A. Muheet, U. Chaudhary, Jyoti Singh and S.A. Kerkhi (2003). Integrated Management of Alternaria blight. Proc. ISOR National Seminar-Stress Management in Oilseeds Jan. 28–30, 2003, pp. 138–139.

Singh, R.B., R.K. Srivastava and Y.S. Chauhan (2000). Effect of date of sowing on severity of Alternaria blight and seed yield of linseed. In *Proc. Int. Conf. on IDM, Indian Phytopath. Soc.*, Vol. 2: 923–924.

Sodowskii, S. (1968). Observations on the influence of mineral fertilization on infection of flax by fungi of the genus *Phoma Acta agrobot*, 21: 143–147.

Solovena, G.P. (1962). Effect of timely seed treatment on the yield of fibre flax. *Mem. Inst. Agron. Leningr*, 84: 116.

Srivastava, R.L., Jyoti Singh, Karam Hussain, Y.P. Malik, S.D. Dubey, J. Rai and M. Vajpai (1997). Linseed. Efficient Management of Dryland Crops in India: Oilseeds. *Indian Society of Oilseeds Research*, Hyderabad, pp. 228-272.

Steinberg, C., C. Albouvette, N. Gantheron and A. Farine (1992). Relation between soil structure, Fusarium location and soil respectively to the Fusarium wilt of flax. Preliminary results. *Bulletin Oil B/SROP*, 15 (1): 156–158.

Steinberg, C., V. Edel and C. Albouvette (1997). The effect of formulation on survival and antagonistic activity of agents of biological control used against Fusariosis of cultivated plants. *Cryptogamic Mycologie*, 18: 139–143.

Straib, W. (1939). Zum epidemischen Aufracten des Leinrostes Leinrostes in Ostpreussen *NachrBl. dtsch. Pfl. Sch Dienst.*, 19: 49.

Tamiett, G. and C. Albouvette (1986). Disease resistance of soils XII. Role of non pathogenic *Fusarium oxysporum* in mechanism of resistance of Noirmoutier soil to vascular Fusarium diseases. *Agronomie*, 6 (6): 541-548.

Tammes, T. (1928). The genetics of genus *Linum Biologr. Genet.*, 4(1).

Tedorova, V. (1958). Polysorosis of flax. Abstr. *Rev. Appl. Mycol.*, 37: 410.

Teintze, M. and J. Leong (1981). Structure of pseudobactin A, a second siderphore from plant growth promoting *Pseudomonas 10. Biochem.*, 20: 6457.

Tu, C. and Y.H. Chang (1982). Soil microbial activity in relation to Fusarium wilt–suppressive soil and conducive soil in Taiwan, DAIS Scientific Meeting Report 1981, Taiwan, Taiwan Agricultural Improvement Station, 1, 1982.

Upadhyay, A.R. and M.L. Verma (1972). Control of linseed rust (*Melampsora lini*) with application of fungicides. *Mysore J. Agric. Sci.*, 6(2): 196.

Vander Speck, J. (1956a). Seedling infection by *Botrytis cinerea* in flax. Abstr. *Tijdschr. Plziekt*, 62(1): 28.

Vander Speck, J. (1956b). The influence of environment on the effect of linseed disinfection. Abstr. *Rev. Appl. Mycol.*, 37: 722.

Vander Speck, J. (1960). Control of Botrytis infection of flax seed. *Tijdschr. Plziekt*, 66 (2): 91.

Vasudeva, R.S. (1958). Report of the Division of Mycology and Plant Pathology, Sci. Rep. Agric. Res. Inst. New Delhi, 1955-56: 85.

Vasudeva, R.S. (1962). Diseases of linseed. In *Linseed*, Richaria, R.H. (Ed). Indian Central Oilseeds Committee, Hyderabad, Chap. 8.

Vinogradov, A.K. (1962). The effect of seed sowing times on the resistance of flax seed to diseases and pests. Abstr. *Rev. Appl. Mycol.*, 41: 155.

Vir, D. and J.S. Grewal (1964). Efficacy of different fungicides. VII Field trials on linseed rust. *Indian Oilseeds J.*, 8(2): 139.

Vodyanaya, L.A. (1962). The effect of *Trichoderma viride* on the phytotoxic mycoflora of flax sick soils. Abstr. *Rev. Appl. Mycol.* 41: 442.

Vyas, S.C., K.V.V. Prasad, I. Kotwal and A.C. Jain (1982). Fungicidal control of powdery mildew of linseed. *Pesticides*, 16: (1) 24.

Ware, W.M. and H.H. Glasscock (1943). Flax rust. *J. Minist Agric.*, 1(1): 16.

Westdal, P.H. and H.P. Richardson (1972). Control of the Aster yellow leaf hopper in relation to incidence of Aster yellows and effects on seed yield of barley. *Can. J. Plant Sci.*, 52(2): 177.

Wicks, Z.W. and J.J. Hammond (1978). Screening of flax species for new sources of genes resistant to *Melampsora lini* (Ehrenh). *Lev. Crop Sci.*, 18(1): 7.

Zarzycka, H. (1977). Seed dressing with fungicides as a protection against soil-borne *Fusarium* infection of flax. *Acta Agrobotanica*, 30: 317.

Zvyagintseva, E.I. (1959). New compound for flax seed treatment–50 per cent thiram. Abstr. *Rev. Appl. Mycol.*, 38: 85.

23

Induced Resistance: Practical Approach for Disease Management

S.K. Biswas and Udit Narain

Department of Plant Pathology, C.S.A. University of Agriculture and Technology, Kanpur–208 002

"Desperate diseases require desperate remedies" the quotation was attributed to Hippicrates which applies both to plant diseases as well as human diseases. From that time, several control measures have been adopted to minimize the disease. But most of the conventional chemical, biological and plant protection procedures as well as use of disease resistance cultivars are tend toward direct control of pests and diseases by their elimination. However, in case of monocotyledonous crops like rice, wheat, maize, barley etc., these practices raise problems in development of pesticide resistance strain in pathogen by continuous use of chemicals which are hazardous to environment also. The current global concern about the environment, indicates need to limit application of chemicals for plant disease control. Obviously, there is a, demand for procedures which can supplement the use of resistance breeding and chemical control in plant protection with new strategies *i.e.*, induced resistance (Chester, 1933).

Induced resistance is the phenomenon in which a plant once appropriately stimulated exhibits an enhanced resistance upon challenge inoculation with a pathogen. To a certain degree the susceptible plants can be altered in their resistance against the pathogens without genetics manipulation. This enhancement of resistance in response to an extrinsic stimulus without a known alteration of the genome is called *"induced resistance."*

Types of Induced Resistance

Induced resistance can take several forms, which are as follows:

Cross Protection

It can be defined as "when a mild strain of virus is injected into a plant, such plant becomes resistance against relative viruses called cross protection" as described by Mckinney (1929). The multiplication of second virus is prevented, but not always abolished. When strongly reduced, symptoms are masked. Cross protection can be a very effective form of induced resistance, but it is specific for the infecting virus. Moreover, it is dependant on the presence of the virus in the cells. If the second virus is able to spread in the plant beyond the zone infected with first virus, it regains activity and does incite disease. Cross protection therefore, is a cellular phenomenon, depending on the interaction between the first and second virus with the cells of the host.

Localized Acquired Resistance

A different form of acquired resistance was found by Ross (1961) when searching for the mechanism of hypersensitive resistance in infected tobacco plants by TMV. At temperature below 28°C, tobacco cultivars containing the N-gene react to infection with TMV with the formation of necrotic lesion, which become visible about 2 days after infection and enlarge but in another 2–5 days there is no enlargement of lesions Ross (1961) argued that the limitation of lesion spread must be caused by a mechanism altering the cells that the tissue surrounding an infection site would become highly resistant or immune. He found a high degree of resistance developed in a 1–2 mm zone around each TMV induced lesion. Following a second inoculation with TMV, no lesion or at most a few tiny ones, were seen to develop in this ring of tissue. The resistance zones increased in size and degree of protection for about 6 days coinciding with the original lesion was enlarging (Van Loon, 2000). He also reported that development of resistant zone was not affected by relative humidity or light before or after the second inoculation. However, resistance was not detectable at temperature near 30°C, where hypersensitivity breaks down. Thus, this type of induced resistance was clearly dependent on the expression of an active virus localizing mechanism. Ross (1961) concluded acquired resistance in localized. infection may be due to two mechanisms: one that is specific for the virus inducing it and another that is not. This type of induced resistance was called localized acquired resistance (LAR).

Systemic Acquired Resistance (SAR)

Ross (1961) coined the term SAR to refer this induced resistance. When Ross used TMV on a local lesion host (Xanthi-nc-tobacco) are found that subsequent infections were reduced in severity of diseases. It can be defined, as a distinct signal transduction pathway that plays an important role to confer long-lasting, broad spectrum resistance systematically in tissues remotely located from the point of initial treatment. SAR activation results in the development of a broad spectrum systemic resistance (Hunt and Ryals, 1996; Neuenschwander *et al.*, 1996).

Extending work on SAR to fungi, Cruickshank and Mandryk (1960) showed high level resistance in field grown tobacco against blue mold (*Peronospora tabacina*) when plant were injected in the stem with a spore suspension of the same fungus. Cohen and Kuc (1981) subsequently showed that resistance takes approximately three weeks to develop. Heat killed conidia and other pathogens of tobacco failed to induce SAR when inoculated in the same fashion.

Localized Induced Resistance (LIS)

Induction of living or heat killed pathogenic or saprophytic bacteria into tobacco leaves has been reported to induce resistance against subsequent infection by compatible pathogenic bacteria, incompatible bacteria and TMV (Sequeira, 1983).

Localized necrosis ensued in the area of the infiltration with living, *Pseudomonas solanacearum* which is a virulent on tobacco but when heat killed cells or only lipopolysaccharIde of bactenum were injected into the leaves, no necrosis occurred. The LIS has been little studied and apart from microbial determinant triggering LIR, nothing is known about the mechanism involved (Van Loon, 2000).

Systemic Induced Resistance

Kuc (1982) coined the term "Induced immunity" to refer systemic protection against pathogenic fungi, bacteria and viruses. But in plant systems, protection has none of the characteristics of animal immunity and plants do not become immune to infection but the effects of the infectious agent on plants are reduced. The term induced immunity is replaced by general term induced resistance (Van Loon, 1997).

The inoculation of hypocotyls of bean plant with avirulent pathogen of *Colletotricum lagenarium* provides protection of bean plant against anthracnose caused by *C. lagenarium*. The treated plant developed minute lesions typical of a hypersensitive response and the reaction was sufficient to reduce subsequent infection with pathogenic race (Kuc, 1982). The SIR exhibits a higher levels of resistance, toward different types of pathogenic organisms including fungi, bacteria, viruses and viroids and in few cases insects also. Resistance was expressed as a reduction in disease severity but level was dependent on the specific interaction between pathogen and its host.

Induced Systemic Resistance

Biotically induced resistance has been associated almost invariably with limited necrotic infection. However, certain strains of non-pathogenic and plant growth promoting Rhizobacteria can induce resistance without inciting any visible symptoms in the host. As for example, colonization of roots of carnation by *Pseudomonas fluorescens* strain WCS417 reduced the numbers of plants developing wilting symptom caused by *Fusarium oxysporium* f. sp. *dianthi* (Van Peer *et al.*, 1991), but no *Pseudomonas* bacterium was detected in the stem, indicating the effect must be latent mediate. Since the effect of Rhizobacteria extended systematically, this phenomenon has been termed induced systemic resistance. Similarly, ISR has been found in radish, tobacco, cucumber and Arabidopsis against both soil borne and foliar fungi, bacteria and viruses (Van Loon *et al.*, 1997, 1998). Rhizobacteria mediated-ISR was invariably expressed as a reduction in symptom of the challenging pathogen and appeared phonotypical similar to SAR.

However, distinction among SAR, ISR and SIR can be made. In the SAR induced by pathogen, causing hypersensitive necrosis, resistance mechanism to challenge inoculation becomes operative earlier or operates at a high level than in non-induced plant. The expression of SAR is dependent on the accumulation of salicylic acid (SA) and associated with the induction of PRs or with anti-pathogen activities. In contrast, Rhizobacteria mediated ISR dose not require SA and can occur without production of PRs and is dependent on ethylene and jasmonic acid for signaling. Both SAR and ISR can be defined as a state in which the plant is to some extent protected against the effect of challenging pathogen through previous biotic stimulation.

In systemic induced resistance require necrosis for induction of resistance, involving production of protinase inhibitor against herbivores also require JA. Synthesis of JA is initiated or perception of polypeptide systemin, which is released on wounding of tomato and potato and act as a mobile signals.

Criteria for Induced Resistance

Induced resistance is distinguished from conventional chemical as well as biological procedure in plant protection as follows:

1. Lack of toxicity of the inducing agents toward the pathogens.
2. Protection of plant is based on the activation of plant defense mechanism or enhancement of their activities rather than elimination of the pathogens.
3. Non-specificity of protection.
4. Local or systemic protection.
5. Absence of typical dose response correlation known for toxic compound.
6. Induced resistance is the genes for resistance or defense reaction exists in all plants. These genes are not expressed until after resistance inducing treatments activate or enhance their expression.
7. Necessity of time interval between application of induce and the onset of protection in plants.
8. Suppression of inducing resistance by previous application of specific inhibitors such as actinomycin D, which affects gene expression of plants.

So, induced resistance is considered to be a biological plant protection procedure in which the plant is the target of the procedure, not the pathogens.

Inducing Agents

Pathogenic Organisms as Inducers

To achieve successful protection, plant must recover from initial infection. Resistance against blue mold of tobacco can be achieved by stem injection with live sporangiospores of the causal organism of blue mold, *Peronospora tabacina* (Cruickshank and Mandryk, 1960; Tuzun and Kuc, 1985). The systemic protection of tobacco against blue mold is somehow unique since it was first reported in Australia. These tests indicate that immunized plant had reduced lesion number and size compared to the non-treated control. Jetiyonan *et al.* (1993) reported that cabbage plants were successfully protected against black rot pathogen, *Xanthomonas campestris* pv. *campestris* (XCC) under field conditions by prior application of attenuated strains of XCC.

Non-pathogenic Organisms as Inducers

There are several research groups which have been used non-pathogenic organisms as inducers to induce resistance in plants. (Schmidt 1988). Pre-inoculation with avirulent or virulent pathogens induce local resistance against powdery mildew in barley and wheat (Ouchi *et al.*, 1974; Kunoh *et al.*,1985). Similarly, Skipp and Deverall (1973) applying a droplet of an avirulent strain of *Colletotrichum lindemuthianum* prevented the development of avirulent. strain on bean after 3 days of inoculation. Biswas (2001) found that pre-inoculation with avirulent strain of *Drechslera sorokiniana* reduced the number of lesion caused by virulent strain of *D. sorokiniana* in wheat. Some examples are shown in Table 23.1.

Table 23.1: Resistance Induced by Non-pathogenic Strain Against Virulent Pathogen (Singh *et al.*, 2002)

Crop Plants	*Non-pathogenic Inducing Agent*	*Effective Against Pathogen(s)*
Spot blotch of wheat	*Drechslera sorokiniana*	*D. sorokiniana*
Blast of rice	*Magnaporthe grisea*	*M. grisea*
Powdery mildew of barley	*Erysiphe graminis tritici*	*Erysiphe graminis tritici*
Alternaria blight of mustard	*Alternaria brassicae*	*Alternaria brassicae* *Peronospora parasitica*
White rust of mustard	*Albugo candida*	*A. candida, P. parasitica*
Anthracnose of bean	*Colletotrichum lindemuthianum*	*B. lindemuthianum, C. lagenarium*
Late blight of potato	*Phytophthora infestans*	*P. infestans, P. cryptogea*
Wilt of tomato	*Fusarium. oxysporium* f. sp. *cucumerinum*	*F. o.* f. sp. *Lycopersici*
Wilt of water melon	*Fusarium, oxysporium* f. sp. nivum	*Fusarium. oxysporium* f. sp. nivum
Bacterial wilt of tobacco	*Ralstonia solanacearum*	*Ralstonia solanacearum*

Biocontrol Agents as Inducer

Biocontrol agents play an important role in plant disease control. Abundantly present in rhizosphere and phyllosphere of plants are effective against several plant pathogen. It has been found that the biocontrol agents triggered the induced resistance in plant against pathogens directly and also stimulate plant growth which is indirect means of plant disease control. Some examples are shown in Table 23.2.

Table 23.2: Biocontrol Agent Mediated Induced Systemic Resistance in Plants

Diseases	*Pathogens*	*Biocontrol Agents*	*References*
Blossom blight of apple	*Erwinia amylovora*	*Pseudomonas fluorescens*	Van Wees *et al.* (1997)
Grey mold of bean	*Botrytis cineria*	*Trichoderma harzianum* T39 *Pseudomonas fluorescens*	De Mayer *et al.* (1998) Van Loon *et al.* (1998)
Grey mold of cotton	*Botrytis cineria*	*Trichoderma harzianum* T39	De Mayer *et al.* (1998)
Anthracnose of cucumber	*Colletotrichum orbiculare*	*Trichoderma harzianum* T203 *P. auriofaciens* 25–33 *P. putida* 89B–27	Yedidia *et al.* (1999) Van Wees *et al.* (1997)
Cucumber mosaic	Cucumber mosaic virus	*P. putida* 89B–27	Van Loon *et al.* (1998)
Bacterial wilt of cucumber	*Erwinia tracheiphila*	*P. putida* 89B–27	Van Loon *et al.* (1998)
Vascular wilt of cucumber	*Fusarium. oxysporium* f. sp. *cucumerinum*	*P. putida* 89B–27	Van Loon *et al* (1998)
Wilt of Muskmelon	*F. o.* f. sp. *melonis*	*Aspergillus niger* AN27	Sen (2000)
Vascular wilt of tomato	*F.* o. f. sp. *lycopersici*	*P. fluorescens* WCS417	Van Loon *et al.* (1998)
Mottling of tomato	Tomato mottle virus	*Bacillus amyloquefaciens* 937	Murphy *el al.* (2000)
Tobacco mosaic	TMV	*Bacillus uniflagellatus*	Van Loon *et al.* (1998)
Tobacco necrosis	Tobacco necrosis virus	*P. fluorescens* CHA0	Van Loon *et al.* (1998)
Sheath blight of rice	*Rhizoctonia solani*	*P. fluorescens* 7-14i	Krishnamoorthy and Gnanamanickam (1997)
Vascular wilt of radish	*F. o.* f. sp. raphani	*P. fluorescens* WCS374	Leeman *et al.* (1995)
Spot blotch of wheat	*Drechslera sorokiniana*	*Chaetomium globosum*	Biswas (2001)

Plant Growth Promoting Rhizobacteria (PGPR) as Inducers

The first systemic resistance by plant growth promoting Rhizobacteria was published for the three plant pathogen systems in 1991.

1. In carnation, application of *Pseudomanas* sp. strain WCS417 to rockwool cubes resulted in protection from wilt caused by *Fusarium oxysporium* f. sp. *dianthi* (Van Per *et al.*, 1991).
2. In bean, seed treatment with *P. flurescens* PGPR strain led to reduction in the number of foliar lesions caused by subsequent inoculation with *P.* s. pv. *phaseolicola* (Alstrom, 1991)
3. In cucumber, 94 known PGPR strains were examined for ability to control *C. orbiculare* (Wei *et al.*, 1991).

Several studies have indicated that specific PGPR provided systemic resistance in plants against pathogens due to stimulate production of biochemical compounds associated with host defense.

Chemical Activators of Induced Resistance

For the first time in induced resistance was described as a response to infection by pathogen. Subsequently, it has been found that the treatment of plant with low molecular weight molecules, can also induce induced resistance. Activation of induced resistance by chemical should provide novel alternative approach for disease management in future.

But the following criteria should have to qualify chemical as an inducer:

1. The compound or its significant metabolites should not exhibit direct antimicrobial activity in plant.
2. The agent modifies the plant-pathogen interaction, so that it resembles phenotypically an incompatible interaction, which includes defense-related mechanisms induced prior to or after challenge.
3. The agent should protect plant against same spectrum of pathogens as in biologically activate induced resistance. It should induce the expression of the same marker genes as evident in pathogen activated SAR (Kessmann *et al.*, 1994) or induced resistance.

Inorganic Compounds as an Inducer

Several inorganic compounds like phosphate, salts (dipotassium/sodium or tripotassium/sodium phosphate) silicon, SiO_2 etc. have been reported as potential activators of resistance (Gottistein and. Kuc, 1987; Mucharro-Mohand and Kuc, 1989; Waiters and Murray, 1992), but have failed to fulfill the above criteria (Kessmann *et al.*, 1994). Reuveni *et al.*, (1994) reported phosphate salts induce SAR in cucumber, bean and maize provides defense response against *Puccinia sorghii.* In cucumber and tobacco, powdered SiO_2 preparation induced SAR accompanied by enhanced activities of chitinase, glucanase, peroxidase, and polyphenolxidase (Schneide and Ullrich, 1994).

Plant and Fungal Extracts as Inducer

There are several reports of plant and fungal extracts having resistance inducing activity in plant (Doke and Chai, 1985; Doubrara *et al.*, 1988, Giri and Sinha, 1979; Salt *et al.*, 1986, Yamada *et al.*, 1990; and Biswas *et al.*, 2002). Doubrava *et al.* (1988) found that treatment of first leaves with spinach extracts provide induced resistance against *C. orbiculare* in cucumber. Aqueous extract of healthy barley leaves induced papillae formation which in turn resistance against of powdery mildew in barley seedlings (Yokoyama *et al.*, 1991). Biswas *et al.* (2002) found that crude extract of *Chaetomhlm globosum* provided induce resistance in wheat against spot blotch pathogen. The chitosan oligomer, a fungal origin molecules was found to provide induced resistance in tomato against *Fusarium* sp. (Benhamou *et al.*, 1994). Hrp Zp_{ss}, the product of Hrp gene *Pseudomonas syringae*-61 can both induce PR gene and SAR against several pathogens in cucumber. In this case Hrp Zp_{ss} seems to trigger an initial HR that *infestans* turns on SAR (Strobel *et al.*, 1996), Some other examples are shown in Table 23.3.

Table 23.3: Resistance Induced by Microbial Metabolites (Hammerschmidt and Kuc, 1995)

Inducing Compound	*Effective Against Disease(s)*	*Causal Organism*
Lipo-glycoprotein (LPG) complex	Late blight of tomato Early blight of tomato	*Phytophthora infestans* *Alternaria solani*
	Root knot of tomato	*Meloidigyne incognita*
Lipo-polysaccharides from *Ralstonia solanacearum*	Wilt of potato	*Ralstonia solanacearum*
Extract from various bacteria	Rust of bean	*Uromyces phaseoli*
Culture filtrate of *Verticillium albo-atrum*	Verticillium wilt of cucumber	*Verticillium albo-atrom*

Synthetic Compounds

The chemicals 2,6-dichloro isonicotinic acid and it methyl ester were the first synthetic compounds shown to activate SAR, which provide good protection against several fungal and bacterial pathogens on cucumber, rice and other crops (Metraux *et al.*, 1991; Vernooij *et al.*, 1995), The BABA, (DL-3-aminobutyric acid) a synthetic compound has no anti fungal activity *in vitro* but local treatment with BABA protect tomato, potato tobacco systemically against *Phytophthora infestans, Peronospora tabacina*, respectively (Cohen, 1994; Cohen and Gisi, 1994), Several other chemicals such as fosetyl-A1, metalaxyl, and triazole are resistance inducing compound, that provided various induced defense response in rice plant (Agrios, 1997).

Signaling

The Role of Ethylene

The role of ethylene appears complex because it has shown to promote leaves yellowing and senescence (Aharoni and Lieberman, 1979) but when a single leaf of a healthy tobacco plant was enclosed in an atmosphere containing ethylene, other leaves of same plant showed more resistance response. The peak of ethylene production occurs by the time lesions become visible. The burst of ethylene occurring at early HR was responsible for initiating a signaling pathway leading to SAR and its associated biochemical alterations. Though, ethylene is a stress hormone that is produced at high level during slow progressive of necrosis and many of conditions found to induce PRs in absence of apparent necrosis do increase ethylene production. Thus, the physiology of ethylene production appears consistent with a possible role of its singling (Vanloon, 1983).

There are also evidences that ethylene is not primarily responsible for the induction of SAR. Application of inhibitor of ethylene synthesis such as amino-ethoxyvinyly-glycine or inhibitor of ethylene action such as norbomadiene to TMV infected tobacco leaves reduced the level of resistance attained in upper leaves (Vanloon and Pennings, 1994). These results strongly suggest that ethylene contributes to SAR possibly by facilitating the release, synthesis or transport of the mobile signal.

Role of Salicylic Acid

In 1979, White accidentally observed that aspirin (acetyl salicylic acid) can decrease the disease symptom caused by TMV in tobacco and can lead to accumulation of pathogenesis related proteins (PRs). Of all hydroxylated derivatives of benzoic acid only 2-hydroxy benzoic acid (salicylic acid) and 2,6-dihydroxybenzoic acid acted as inducer (Vanloon, 1983). The effect of SA on inducing PR and resistance is not systemic when SA is sprayed in or injected into individual leaves. In contrast, the effect was systemic when SA was applied as a soil drench (Vanloon and Antoniew, 1982). In the latter conditions, SA appeared to be taken up by the root and transported into the shoot. When SA injected into a leaf as quickly conjugated to glucose and stored in the vacuole (Metraux *et al.*, 1990, Ryals *et al.*, 1996). Under this conditions, SA might be at sufficient concentration to act locally, but too little amount might be transported elsewhere to have an effect systemically. Vanloon and Antoniew (1982) speculated that ethylene, by stimulating phenyl propanoid metabolism, would lead to the synthesis of an aromatic compound resembling to SA that acted as inducer of SAR.

SA was identified as the signal compound as well as a hormonal role in plants (Raskin *et al.*, 1987). Indeed SA was found to be present in tobacco plants at nanogram concentration and is to be increased to be about 1 μg/gm of fresh weight upon TMV inoculation associated with induction of PR–1 gene expression (Malamy *et al.*, 1990). Metraux *et al.* (1990) independently identified SA appearing on the phloem sap of cucumber plants inoculated with either C. *lagenarium* or TNV- suggesting that SA itself could function as the endogenous signal in the transmission of SAR but no increase of SA detected when leaves were mechanically or chemically injured.

Several observations cast doubt on SA being the 'signal.' Pathogen infection results in significant amounts of SA in the phloem sap of both cucumber and tobacco (Metraux *et al.*, 1990; Yalpani *et al.*, 1991). Additionally *in vivo* studies of SA provide evidence that SA produced in the leaves of TMV infected tobacco or TNV infected cucumber is transported throughout the plants and accumulates in uninfected tissue (Shulaev *et al.*, 1995; Molders *et al.*, 1996). As much as increased levels of SA in the uninfected tissue of pathogen inoculated plants result from translocation of SA from infected leaves to uninfected leaves. Several reports suggested that SA may be the signal that translocated from an infection sites to activate SAR elsewhere in the plant. However, two lines of evidence suggested that SA is not the long distance signal. (John *et al.*, 1996; Vemooij *et al.*, 1994).

To clarify the question, reciprocal grafts were made between non-transformed Xanthi-nc tobacco (wt) and transformants containing the NahG gene (NahG). The latter gene codes for SA-hydroxylase that rapidly converted inducing SA in the plant to inactive catechol. Root stock leaves of grafted plants were inoculated with water are with TMV and the scion leaves were challenge inoculated with TMV 7 days later. TMV lesion on NahG root stock were larger than lesion on wt root stock indicating that lack of SA not only abolishes SAR but also reduces N-gene mediated resistance to TMV. In the four root stock/scion combination *i.e.*, (wt/wt, wt/NahG, NahG/NahG and NahG/wt) lesion diameter was reduced in wt/wt combination but on the other cases no reduction was seen. wt/wt root stock are expected to produce the mobile signal but NahG scion did not express SAR indicated that the signal requires the presence of SA in tissue distant from the infection site. NahG root stock expected not to be able to provide signal to wt scion, because NahG root stock was unable to accumulate SA. Hence, the expression of SAR, the scions should have an increase in endogenous SA for induction of resistance. On the other hand, SA is not translocated signal for induction of SAR in tobacco.

H_2O_2

It has been proposed that H_2O_2 acts as a second messenger of SA in SAR signaling. SA binds and inhibits catalase activities which could lead to an increase in the endogenous level of H_2O_2 (Chen, *et al.*, 1993, 1995). A soluble high affinity SA-binding protein ($SABP_2$) was identified which reversibly binds biologically active but not inactive in a analogs of SA *in viva* (Dumer and Klessig, 1995) and it role is not yet clear.

Another report indicated that H_2O_2 is not a second messenger of SA in SAR, signaling (Bi *et al.*, 1995; Leon *et al.*, 1995; Summermatter *et al.*, 1995). It functions as a signaling agent of SA. According to them, if it is second messenger for the induction of defense response, H_2O_2 levels should increase in uninfected leaves of tobacco plants during SAR activation, but it has been found that in the uninfected leaves of inoculated plants, SAR gene expression and the establishment of SAR did not correlate with an increase in H_2O_2 levels. On the other hand, accumulation of PR and mRNA as a result of infiltration of 1M H_2O_2, a concentration that also caused severe tissue damage. However, in NahG plants, 1M H_2O_2 did not induce PR_1 significantly, indicating a requirement for SA in H_2O_2 mediated PR_1 expression (Neuenschwander *et al.*, 1995). In fact, high concentration of H_2O_2 was, found to induce SA synthesis in both tobacco and arabidopsis (Neuenschwander *et al.*, 1995). Summermatter *et al.* (1995) suggesting that H_2O_2 may induce PR_1 accumulation through induction of SA. These results indicate that H_2O_2 is not playing a second messenger of SA in the signal establishment of SAR.

Jasmonates

Jasmonates (JA) are produced in all higher plants from linolenic acid by a lip oxygenase (Lox)-mediated oxygenation process (Hamberg and Gardner, 1992). In plants JA and its methyl ester move easily in both the liquid and vapour phases (Farmer and Ryan 1990) and affect a variety of physiological processes including root growth, tuber formation, tendril coiling, senescence of leaves and stomatal opening. (Staswick, *et al.*, 1992; Koda, 1992; Weiler *et al.* 1993; Sembdner and Parthier, 1993). The role of JA in SAR is not yet clear, however low concentration of JA has been found to induce proteinase inhibitors, thionin, osmotin (PRs) proline rich cell wall protein and different enzymes involved in plant defense reaction such as chalcone synthase, PAL, lipoxygenase (Farmer and Ryan, 1990; Epple *et al.*, 1995; Xu *et al.*, 1994; Creelman *et al.* 1992; Gundiach *et al.*, 1992; Bell and Mullet, 1991). Ryan (1992) proposed that these molecules act as secondary messengers in SAR system. Mitchel and Walters (1995) reported induced systemic protection against *Erysiphe graminis* f. sp. *hardie* in barley by treating the seedling leaves with methyl jasmonate, They also found an increase activities of PAL, peroxidase and lipoxygenase in the treated plant. Induction of resistance by JA and MeJA have also been shown in potato and tomato against *Phytophthora infestans* (Cohen *et al.*, 1993).

It has been shown that inoculation of *A. thaliana* with an avirulent race of *Alternaria brassisicola* leads to an increase of endogenous JA in treated as well as in the upper untreated leaves (Penninckx *et al.*, 1997). On the other hand, treatment of plant with methyl jasmonate (MeJA) induces accumulation of an anti-fungal defensin protein, but not of PR-1 unlike application of SA. It suggests, the presence of two different pathways, both inducible by pathogens, but one is a SA to PR-protein expression and other is jasmonates to the accumulation of defensin (Penninckx *et al.*, 1997).

Systemin

Systemin is a signal molecule of 18 aa peptide which induces *de novo* synthesis of proteinase inhibitors after wounding (Pearce *et al.*, 1991). Upon wounding, there is an increase level of Abscisic acid (ABA) and JA (Pena-Cortes *et al.*, 1991, 1993). Systemin, ABA and JA induce proteinase II inhibitor (pin2) protein and gene expression in the treated leaves and in systemic leaves (Farmer and Ryan 1992; Pena-Cortes *et al.*, 1995). Proteinase inhibitors inhibit the activity of digestive proteases leading to the malnutrition, reduced growth and sometimes death of the feeding insects (Ryan, 1990).

Translocation of systemin probably occurs through the phloem (Navaez-Vasquez *et al.* 1995). Systemin is synthesized from a precursor protein of 200 amino acid, prosystemin which stored in cytoplasm (Schaller and Ryan, 1996). Prosystemin mRNA is present throughout tomato plant except root and accumulates in leaves upon wounding (McGurl *et al.*, 1992). Over expression of prosystemin gene in transgenic tomato plants results in constitutive expression of proteinase inhibitors in the absence of wounding (Mc Gurl *et al.*, 1994). Overexpression of prosystemin gene in tomato leaves also shown enhanced levels of polyphenol oxidase (Constabel *et al.*, 1995).

Abscisic Acid (ABA)

Spraying on the foliage of ABA-deficient potato plants with ABA, the level of ABA increases in distal non-sprayed tissue and pin2mRNA accumulates (Pena-Cortes *et al.*, 1995). ABA treatment of tomato plant induces little proteinase inhibitor accumulation as compared to systemin treatment. It seems that ABA is required for the wounds response but would not behave as the primary wound signal (Pena-Cortes *et al.*, 1995).

In another set of experiment, it has been found that ABA–deficient tomato and potato plants do not accumulate Pin2 mRNA upon wounding which suggest that ABA is involved at some stage during induction of Pin2 by wounding. Treatment of potato leaves with JA induces Pin2 mRNA accumulation in a wild type and ABA deficient mutant plant (Pena-Cortes *et al.*, 1995). These results suggest that JA is involved at a step downstream of ABA in the wound signal transduction pathway leading to pin2 gene activation. Systemin induces pin2 gene expression in wild type but not in ABA-deficient potato and tomato plants (Pena-Cortes *et al.*, 1995). Since in both cases, systemin induces pin2 gene expression and lead to an increase in endogenous ABA and JA levels, systemin might control the endogenous levels of ABA and JA.

Based on the observations that ABA treatments lead to an increase in JA but JA treatment does not lead to an increase of ABA, it has suggested that systemin and ABA act upstream of JA (Pena-Cortes *et al.*,1995).

Electrical Signal

Mechanical wounding of the cotyledons of young tomato plants leads to slow (1-4 mm/sec.) transmission of an action potential out of the cotyledons and into the first leaf (Wildon *et al.*, 1992). They also reported that the systemic electrical activity correlates with the induction of pin2 mRNA.

Hydraulic Signals

Hydraulic signals are triggered by wounding. At wound sites leaf cells are broken and the sap is released in the apoplasm. The sap and its solutes are drawn into nearby intact xylem vessel (Malone *et al.*, 1994). The hydraulic signals propagate changes in water pressure. In wheat the hydraulic signals travels at the rate of 10 cm/sec. and there is concomitant development of electrical changes (Malone *et al.*, 1994). Hydraulic signal itself can not induce proteinase inhibitors but carry substances rapidly through the plant. However, it is not known if the hydraulic pressure wave is required for the translocation of systemin.

Mechanisms of Resistance

The mechanisms of induced resistance are the indirect method of plant disease. management. It does not affect spore germination, appressorium formation and penetration is drastically reduce infection due to systemically induce resistance to tissue probably, as a result of formation of papillae like materials, callose and lignin deposition (Vance *et al.*, 1980; Kauss, 1987), synthesis of phytoalexin (Dixon, 1986) novel proteins (Bot *et al.*, 1980), cell death and oxidation brust (Low and Merida, 1996). But exact fundamental process involved in induced resistance is not yet known well understood. However, large number of biochemical, physio-chemical and molecular changes have been noted in relation to resistance response.

Histopathological Changes

Lignin

Lignin is the most abundant and widely distributed phenolic polymers found in plants (Brown, 1964). Lignin is mostly synthesized by shikimic acid pathway. Lignifications have been observed in many plant species following attempted infection by pathogenic organisms such as fungi, bacteria, viruses and nematodes.

Lignification may contribute to resistance in many different ways, although surprisingly little is known about its role in induced resistance. Firstly, incorporation of lignin into plant cell wall makes it strengthen in both mechanically and resistant to degradation by enzymes secreted by an invading pathogen. Lignified cell wall could also constitute a barrier preventing free nutrients movement and therefore help to starve a pathogen. Lignin precursors (PAL and cinnamic acid) themselves might have toxic effect on pathogen or by binding to fungal cell walls make them more rigid and impermeable, thus hindering further growth or uptake of water and nutrients; for example: coniferyl alcohol is highly toxic for fungi at low concentration *in vitro* (Hammerschmidt and Kuc, 1982). Polymerization of lignin precursors by free radical in the intercellular space might also lead to lignification of pathogen structures (Friend, 1985). The mycelium of *Colletotrichum lagenarium* and *Cladasporium cucumerinum* become lignified *in vitro* in presence of coniferyl alcohol, H_2O_2 and crude peroxidase preparation from immunized plants.

The formation of papillae at the sites of attempted penetration is well known in both monocot and dicot as a means of restricting fungal penetration to epidermal cell. In cucumber and potato, induction of systemic resistance lead to higher rate of papillae formation (Stroemberg and Brishammar, 1993; Kovats *et al.*, 1991). An important role of callose in the expression of resistance is supported by the resistance *Arabidopsis* mutants (Dietrich *et al.*, 1994). After infection, strengthing of the cell wall can also occur by peroxidase catalyzed cross linking of hydroxyproline-rich structural cell wall glycoproteins (Bradley *et al.*, 1992). Glycine rich proteins which are also found in the cell wall of many plants, accumulated systemically in tobacco plants infected with TMV (Van Kan *et al.*, 1988).

Biochemical Changes

Peroxidases

Early investigation revealed a systemic increase in peroxidase activity in tobacco plants with localized TMV infection (Simons and Ross, 1971; Van Loon, 1976). Similarly, in cucumber plants, the induction of systemic induced resistance by localized C. *lagenarium* infection is also associated with enhanced peroxidase activity (Hammerschmidt *et al.*, 1982). The enhanced peroxidases are acidic, extracelluar and most probably localized in the cell wall (Smith and Hammerschmidt, 1988). Peroxidases have several function in inducing resistance to plants, which are as follow:

1. Oxidative polymerization of hydroxycinnamyl alcohols to form lignin, a process that is commonly associated with defense mechanism in plants (Vance *et al.*, 1980).
2. Oxidative cross-linking of pre-existing hydroxyproline-rich structural proteins in the cell wall, making the cell wall more resistant to degredation by microbial enzymes (Bradley *et al.*, 1992).
3. Peroxidase generated hydrogen peroxide may function directly as an antimicrobial (Peng and Kuc, 1992).

Phytoalexin

Phytoalexins are synthesized and accumulate locally around infection sites, notably in incompatible interactions, but have never been found in systematically protected tissues before challenge inoculation. Kuc (1982) has found that the bean hypocotyls induced by avirulent pathogen of *C. lagenarium*, accumulated phaseolin at sites of infection of challenge pathogen, *C. lagenarium*. Carnation induced by the rhizobacterial strain GCS417, increased accumulation of dianthramide phytoalexins around the inoculation sites (Van Peer *et al.*, 1991). Such enhanced production of phytoalexins may contribute to the increase resistance against fungi and bacteria. However, viruses are not affected by phytoalexins and additional mechanism must be responsible for broad spectrum protection observed. The rapid accumulatioin of the phytoalexin phaseollin to antifungal levels in bean expressing resistance to a virulent strain ofC. *lindemuthianum*.

PR-Proteins

Antoniew *et al.* (1980) identified more proteins and proposed the name pathogenesis-related (PR) proteins. These proteins accumulate in the uninfected as well as the infected tissues of inoculated plants. They are initially defined as acid soluble, protease resistance, localized in the extra cellular space. Later, basic form proteins were also identified. In tobacco 11 families of proteins were identified and were grouped into five classes–PR1 (PR 1a, 1b and 1c), PR 2(β 1,3 glucanase), PR 3 chitinase, PR 4 (PR 4a and 4b) and PR 5 (5a and 5b) thaumatin like protein. Chitinase and β-1,3-glucanases involved in the defense response of plants against fungi, bacteria and also by their action on the cell walls of invading pathogens directly (Boiler, 1985).

Chitinase hydrolyzes the β-1,4-N-acetylglucoseamine linkages of chitin, a major component of the cell wall of most classes of fungi, except Oomycetes. The peptidoglycan polymer of bacterial cell walls is also cleaved by most of the chitinase. The β-1,3-glucanases hydrolyze β-1,3-glucan polymers, another major component of fungal cell walls. The defense-related function of these two hydrolases may be to release elicitor active molecules from the surface of pathogens and thereby increase the speed and magnitude of a defense response (Hadwiger and Beckman, 1980; Keen and Yoshikawa, 1983). Biswas (2001) found that some new proteins of different molecular weights are synthesized in wheat seedling due to pre application of crude extract of *C. globosum*, imparting resistance in wheat against spot blotch pathogen. These proteins may be PR proteins.

As described, PR-proteins expressions are correlated with increased resistance in several plants, but the exact functions of the PRs in the defense reaction against viruses, bacteria and fungi remain to be defined more exactly.

Phenols

Role of phenols are more important in induced resistance for signal transduction than for directly inhibiting pathogens. The best studied is salicylic acid, a compound derived from cinnamic acid.

Systemic changes in the level of phenolic compounds have been examined by Simons and Ross (1971). They measure the level s of orthodihydroxyphenols, chlorogenic acid and total phenolic compounds in the upper leaves of tobacco and increased after inoculation of the lower leaves with a local lesion strain of TMV. Increase amount of phenol contents were also found in leaves of wheat seedling treated with crude extract of C. *globosum* (Biswas, 2001). Phenols provide defense response in plant by conversion of phenolic compound to quinon or phenol to lignin which is deposited on the cell surface.

Generation of Active Oxygen

It is now well documented that in early event of defense response, active oxygen (super oxides, hydrogen peroxide and hydroxyl ions) is generated which inhibit the pathogenic attack in plant. It has speculated that the active oxygen involved in the rapid cross-linking of cell wall polymers which provided protection against cell wall degrading enzyme, secreated by pathogens. Active oxygen itself has an inhibitory effect on pathogen.

Lipids

Little is known about the possible systemic changes in the concentration of lipids in plant expressing induced resistance to disease. But many changes in membrane lipids are known to occur in plants at the site of infection. Losel (1991) reported that infection of cucumber leaves with *Sphaerotheca fuliginea* increased systhesis of sterols at about three fold over initial. In tobacco, an increase in the saturation of fatty acids contained in the tissue while linolenic acid content decreased by 9 per cent at the time of hypersensitive response to TMV (Ruzicska *et al.*, 1983). The change in fatty acid composition to several fold increase in lipoxygenase activity of the infected tobacco tissues. This enzyme has a key role in converting linolenic acid to jasmonic acid, producing a signal that activates the systhesis of proteinase inhibitors in tomato, tobacco and alfalfa (Farmer and Ryan, 1992).

Applicability to Induced Resistance

Induced resistance offers a significant potential for control of plants diseases. Currently its plays other roles alongwith disease control, which are as follows:

Effect on Disease Severity

Induced resistance caused by prior infections with challenge inoculum have a practical importance in agriculture. Polycultures of cereals and mixed cultivation of crops have a important role in induced resistance (Ibenthal *et al.*, 1985). Villich-Meller and Weltzien (1989) reported that 50 per cent reduction in disease incidence of powdery mildew and *Drechslera avenae* occurred in the field with mixture of spring barley and oat. Application of microbial metabolites of *Bacillus subtilis* as a inducer provided induced resistance in monocots. Repeat treatment of winter wheat with microbial inducers, reduced disease severity, comparable to the treatment with highly effective systemic fungicides (Schonbeck *et al.*, 1982; Dehne *et al.*, 1984). Foliar spray with crude extract of C. *globosum* provided protection against *D. sorokiniana* in wheat which was characterized by decrease in the disease severity (Biswas *et al.*, 2001).

Effect on Yield Response

Induced resistance can only be applicable in plant protection procedure, if yield response of plant is not influenced negatively due to additional energy consumption of the resistance reactions.

Very little information has been generated on the effect of induced resistance on yield capacity of induced plants. Disease-yield loss relationships were determined for induction of resistance by metabolites of *B. subtilis* in winter wheat and in winter and spring barley (Dehne *et al.*, 1984; Steiner *et al.*, 1988; Oerke *et al.*, 1989). The treatment with the inducer altered host susceptibility as well as influenced the yield. The induction of resistance by microbial metabolites seem to stimulate the ability of plants to compensate the damaging effects of pathogen on plant metabolism. This lead to prolonged maintenance of assimilation rates. The effect on disease severity and influence on yield may result from reduced damage caused by the pathogens. Similar type results were recorded by Newton *et al.*, (1993) using extract derived from yeast cell wall against powdery mildew.

Conclusion

International scientific communities are acuately aware of both public perception of chemical hazard and customers desire to protect their livelihood. This has been stimulated wide range techniques to control the pests and diseases, induced resistance is one of them. It increased concern for health and the environment increase restrictions on the use of chemicals pesticides, the withdrawal of many effective pesticides from use and the increased cost of developing, testing and marketing the new products. The use of chemicals as inducers for systemic resistance offers an exciting possibility for disease control since their use would be compatible with existing foliar spray and soil application equipment. Chemical used as inducers must be safe for animal health and the environment, otherwise we have merely substituted one chemical problem for another.

Induced resistance readily integrates into a programme utilizing disease resistant plants, biocontrol, genetic engineering and appropriate pestcides. Induced resistance could not only decrease the quantity of pestcide needed for disease control but could, also increase the effectiveness and useful life of those on the market. It does deseve to be extensively and vogorously explored for its potential in providing an effective, inexpensive, natural, consumer-friendly and environmentally-friendly technology for plant disease control.

REFERENCES

Agrios, G.N. (1997). *Plant Pathology*, 4th Edition. Academic Press, pp. 635.

Aharon and Linerman (1979). Ethylene as a regulator of senescence in tobacco leaf discs. *Plant Physiol.*, 64: 801–804.

Alstrom, S. (1991). Induction of disease resistance in common bean susceptible to halo blight bacterial pathogen after seed bacterization with rhizosphere Pseudomonads. *J. Gen . Appl. Microb.*, 37: 495–501.

Antoniew, J.F., C.F. Ritter, W.S. Pierpoint and L.C. Van Loon (1980). Comparison of three pathogenesis-related proteins from plants of two cultivars of tobacco infected with TMV. *J. Gen. Virol.*, 47: 79–87.

Bell, E. and J.F. Mullet (1991). Lipoxygenase gene expression is modulated in plants by water deficit, wounding and methyjasmonate. *Mol. Gen. Genet.*, 230: 456–462.

Bemhamou, N., P.J. Lafontaine and M. Nicole (1994). Induction of systemic resistance of Fusarium crow and root rot in tomato plants by seed treatment with chitosan. *Phytopath.*, 84: 1432–1444.

Bi, Y.M., P. Kenton, L. Mur, R. Darby and J. Draper (1995). Hydrogen peroxide does not function downstream of saliocylic acid in the induction of PR-protein expression. *Plant. J.*, 8: 235–245.

Biswas, S.K. (2001). *Biochemical Changes in Wheat Induced by Chaetomium globosum Against Spot Blotch Pathogen*. Ph.D. Thesis, IARI, New Delhi, pp. 1–87.

Biswas, S.K., K.D. Srivastava, R. Aggarwal and D.V. Singh (2002). Evolution of *Chaetmium globosum* as an inducer of resistance against spot blotch pathogen of wheat. *Indian Phytopath.*, 54(4): 570–572.

Boiler, T. (1985). Induction of hydrolases as a defense reaction against pathogens. In: *Cellular and Molecular Biology of Plant Stress*, (Eds.) J.L. Key and T. Kosuge, UCLA Symposia on Molecular and Cellular Biology, New Series, Volume 22, Alan R. Liss, Inc., New York, pp. 247–262.

Bot, J.F., H.J.M. Linthorst and B.J.C. Comelisssen (1980). Plant pathogenesis-related proteins induced by virus infection. *Annu. Rev. Phytopathol.*, 28: 113–138.

Bradley, D.J., P. Kjellbom and C.J. Lamb (1992). Elicitor-induced and wound-induced oxidative cross-linking of a praline-rich plant cell wall protein: A novel, rapid defense response. *Cell*, 70: 21–30.

Brown, S.A. (1964) Lignin and tannin biosynthesis. In: *Biochemistry of Phenol Compounds*, (Ed.) Harbone, Academic Press, London and New York, pp. 361–395.

Chen, Z.X., J. Malamy, J. Henning, U. Conrath and P. Sanchez-Casas (1995). Induction, modification, and transduction of the salicylic acid signal in plant defense responses. *Proc. Natl. Acad Sci., USA*, 92: 4134–4137.

Chen, Z.X., H. Silva and D.F. Klessig (1993). Active oxygen species in the induction of plant systemic acquired resistance by salicylic acid. *Science*, 262: 1883–1885.

Chester, K. (1933). The problem of acquired physiological immunity in plants. *Quarterly Rev. Biol.*, 8: 129–154; 275–324.

Cohen, Y. (1994). 3-Amionbiutyric acid induces systemic resistance against *Peronospora tobacina. Physiol. Mol. Plant Pathol.*, 44: 273–288.

Cohen, Y. and U. Gisi (1994). Systemic translocation of ^{14}C–D–L–3–aminobutyric acid in tomato plants in relation to reduced resistance against *Phytophthora infestans. Physiol. Mol. Plant Pathol.*, 45: 441–456.

Cohen, Y. and J. Kuc (1981). Evaluation of systemic resistance to blue mold induced in tobacco leaves by prior stem inoculation with peronospora tabacina. *Phytopathology*, 71: 783–787.

Cohen, Y., U. Gisi and T. Niederman (1993). Local and systemic protection against *Phytophthora infestans* induced in potato and tomato plants by jasmonic acid and jasmonic methyl ester. *Phytopathology*, 83: 1054–1062.

Constabel, C.P., D.R. Bergey and C.A. Ryan (1995). Systemin activates synthesis of wound inducible tomato leaf polyphenol oxidase via octadecanoid defense signaling pathway, *Proc. Natl. Acad. Sci., USA*, 92: 407–411.

Creelman, R.A., M.L. Tierney and J.E. Mullet (1992). Jasmonic acid/methyl jasmonate accumulate in wounded soybean hypocotyls and modulate wound gene expression. *Proc. Natl. Sci.,USA*, 89: 4938–4941.

Cruickshank, I.A.M. and M. Mandryk (1960). The effect of stem infection of tobacco with *Peronospora tabacina* on foliage reaction to blue mold, *J. Australian Instt. Agric. Res.*, 26: 369–372.

De Mayer, G., J. Bigirimana, Y. Elad and M. Hofte (1998). Induced systemic resistance in *Trichoderma harzianum* T 39 biocontrol of *Botrytis cineria. Eur. Plant Pathol.*, 104: 279–286.

Dehne, H.W., K. Stenzel and F. Schonbeek (1984). The efficiency of induced resistance under practical culture condition III. Reproduction of powdery mildew on induced resistant plants. *Plant Dis. Protec.*, 91: 258–265.

Dietrich, R.A, T.P. Delany, S.J. Uknes, E.R. Ward, J.A. Ryals and J L. Dangl (1994). Arabidopsis mutants stimulating disease resistance response. *Cell*, 77: 565–577.

Dixon, R.A.(1986). The phytoalexin response: Elicitation, signaling and control of host gene expression. *Biol. Rev. Camb. Philos. Sec.*, 61: 239–292.

Doke, N. and H.B. Chai (1985). Activation of superoxide generation and enhancement of resistance against compatible races of *Phytophthora infestans* in potato plants treated with digitonin. *Physio. Pl. Pathol.*, 27: 323–334.

Doubrava, N.S., R.A. Dean and J. Kuc (1988). Induction of systemic induced resistance to anthracnose caused by *Colletotrichum lagenarium* in cucumber by oxalate and extracts from spinach and rhubarb leaves, *Physio. and Mol. Pl. Pathol.*, 33: 69–79.

Durner, J. and D.F. Klessig (1995). Inhibition of ascorbate peroxidase by salicylic acid and 2,6-dichloroisonicotinic acid, two inducers of plant defense responses. *Proc. Natl. Acad Sci., USA*, 92: 1312–1316.

Epple, P., K. Apel and H. Bohlmann (1995). *An Arabidopsis thaliana* thionin gene is inducible via a signal transduction pathway different from that for pathogenesis related proteins. *Plant Physiol.*, 109: 813–820.

Farmer, E.E. and C.A. Ryan (1990). Interplant communication: Airborne methyl jasmonate induces synthesis of proteinase inhibitors in plant leaves. *Proc. Natl. Acad Sci., USA*, 87: 7713–7716.

Farmer, E.E. and C.A. Ryan (1992). Octadecanoid precursors of jasmonic acid. activate the synthesis of wound–inducible proteinase inhibitors. *The Plant Cell*, 4: 129–134.

Friend, J. (1985). Phenolic substances and plant disease. In: *The Biochemistry of Plant Phenolics* (ed). C.F. van Sumere, P.J. Lea, Oxford: Clarendon, pp. 367–392.

Giri, D.N. and A. K. Sinha (1979). Effect of nontoxic chemical on brown spot disease in rice seedlings. *International Rice Research Newsletter*, 4: 10–11.

Gottstein, H.D. and J.A. Kuc (1987). Induction of systemic resistance to anthracnose in cucumber by phosphates. *Phytopathology*, 79: 176–179.

Gundiach, H., M.J. Muller, T.M. Kutchan and M.H. Zenk (1992). Jasmonic acid is a signal transducer in elicitor–induced plant cell cultures. *Proc. Natl. Acad Sci., USA*, 89: 2389–2393.

Hadwiger, L.A. and J. M. Beckman (1980). Chitosan as a component of pea-Fusarium solani interactions. *Plant Physiology*, 66 : 205–211.

Hamberg, M. and H.W. Gardner (1992). Oxylipin pathway to jasmonates-biochemistry and biological significance. *Biochim. Biophys Acta*, 1165: 1–18.

Hammerschmidt, R. and J. Kuc (1982). Lignification as a mechanism for induced systemic resistance in cucumber. *Physiol. Pl. Pathol.*, 20: 61–71.

Hammerschmidt, R., E.M. Nuckles and J. Kuc (1982). Association of enhanced peroxidase activity with induced systemic resistance of cucumber to *Colletotricum lagenarium. Physiol. Pl. Pathol.*, 20: 72–82.

Hunt, M. and J. Ryals (1996). Systemic acquired resistance signal transduction. *Crit. Rev. Plant Sci.*, 15: 583–606.

Ibenthal, W.D., Zu Meier, H. Beerentrup and F. Nabizadehr (1985). Ertragsniveau und krankheitsbefall von sommergerste in sortenmischungen, *Zeitschrift fur Pjlkrankheiten und pflanzenschut.*, 92: 37.

Jetiyonan, K., P.A. Backman and S. Tuzun (1993). Use ofpolysilicone surfactant for delivery of resistance inducing strains of *Xanthomonas campestris* for control of black rot of cabbage, *Proc. 6th International Congress of Pl. Pathol.*, pp. 65.

John, A., Ryals, Urs H. Nenens Chwander Michael G. Willits, Antrnio Molina, Henry-york Steiner and D. Hant Michelle (1996). Systemitric acquired resistance. *The Plant Cell.*, 8: 1809–1819.

Kauss, H. (1987). Some aspects of calcium-dependent regulation in plant metabolism. *Annu. Rev. Plant Physiol.*, 38: 47–72.

Keen, N.T. and M. Yoshikawa (1983). β 1,3-endoglucanase from soyabean releases elicitor active carbohydrates from fungus cell walls. *Plant Physiology*, 71: 460–465.

Kessmann, H., T. Staub, C. Hofmann, T. Maetzke and J. Herzog (1994). Induction. of systemic acquired disease resistance in plants by chemicals. *Ann. Rev. Phytopathol.*, 32: 439–459.

Koda, Y. (1992). The role of jamonic acid and related compounds in the regulation of plant development. *Int. Rev. Cytol.*, 135: 155–199.

Kovats, K., A. Binder and H.R. Hohl (1991). Cytology of systemic induced resistance of tomato to *Phytophthora infestans. Planta*, 183: 491–496.

Krisnamoorthy, K. and S.S. Gnanamanickam (1997). Biological control of sheath blight of rice: Induction of systemic resistance in rice by plant associated *Pseudomonas* sp. *Curr. Sci.*, 72: 331–334.

Kuc (1982). Induced immunity to plant diseases. *Bioscience*, 32: 854–860.

Kunoch, H., K. Kuroda, A. Hayashimoto and H. Ishizaki (1985). Induced susceptibility and enhanced resistance at the cellular level in barley coleoptiles. II. Timing and localization of induced susceptibility in a single coleoptite cell and its transfer to an adjacent cell. *Canad. J. Bot.*, 64: 889.

Lawton, K.A., S.L. Potter, S. Uknes and J. Ryals (1994). Acquired resistance signal transduction in Arabidopsis is ethylene dependent. *Plant Cell*, 6: 581–588.

Leeman, M., J.A. Van Pelt, Den, F.M. Ouden, M. Heinsbroek, Pahm Bakker (1995). Induction of systemic resistance against Fusarium wilt of radish by lipopolysaccharides of *Pseudomonas fluorescens. Phytopathology*, 85: 1301–1305.

Leon, J., M.A. Lawton and I. Raskin (1995). Hydrogen peroxide stimulates salicylic acid biosynthesis in tobacco. *Plant Physiol.*, 108: 1673–1678.

Losel, D. M. (1991). Synthesis and functioning of membrane lipids in fungi and infected plants. *Pesticide Science*, 32: 353–362.

Low, P.S., J.R. Merida (1996). The oxidative burst in plant defense: function and signal transudation. *Physiol. Plant Pathol.*, 96: 533–542.

Malamy, J., J.P. Curr, D.F. Klessig and I. Raskin (1990). Salicylic acid a likely endogenous signal in the resistance response of tobacco to viral infection, *Science*, 250: 1002–1004.

Malone, M., J.J. Alarcon and L. Palumbo (1994). An hydraulic interpretation of rapid, long distance wound signaling in the tomato. *Planta*, 193: 181–185.

McGurl, B., M. Otozco-Cardenas, G. Pearce and C.A. Ryan (1994). Over expression of prosystemin gene in transgenic tomato plants generates a systemic signal that constitutively induces proteinase inhibitors synthesis, *Proc. Natl. Acad. Sci.*, USA, 91: 799–802.

McKinney, H.H. (1929). Mosaic diseases in the Canary islands, West Africa and. Gibraltar. *J. Agric. Res.*, 39: 557–578.

Metraux, J.P., Aho-Goy, P., Staub, T., Speich, J. and Steikmann, A. (1991). Induced systemic resistance in cucumber in response to 2,6-dichloro-isonicotinic acid and pathogens. In: *Advances in Molecular Genetics of Plant-Microbe Interaction* (ed.) Hplunnecke, D.P.S. Verma, Amsterdam: Kluwer, pp. 432–439.

Metraux, J.P., H. Singer, Ryals, E. Ward and M. Wyss-Benz (1990). Increase in salicylic acid at the onset of systemic acquired resistance in cucumber. *Science*, 250: 1004–1006.

Mitchell, A. and D. Waiters (1995). Systemic protection in barley against powdery mildew infection using mathyljasmonate. *Aspects Appl. Soil.*, 42: 323–326.

Molders, W., A. Buchala and J.P. Metraux (1996). Transport of salicylic acid in tobacco necrosis virus–infected cucumber plants. *Plant Physiol.*, 112: 787–792.

Mucharro-Mohan and M. Kuc (1989). Oxalic acid, K_2HPO_4 as inducers of systemic resistance against disease caused by fungi, bacteria and viruses in cucumber; *Phytopathology*, 79: 1144.

Murphy, J.E., G.W. Zehnder, D.J. Schuster, E.J. Sikora, J.E., Polston and J.W. Kloepper, (2000). Plant growth promoting rhizobacterial mediated protection in tomato against tomato mottle virus. *Plant Dis.*, 84: 779–784.

Navaez-vasquez, J., G. Pearce, M.L. Orozco-Cardenas, V.R. Franceschi and C.A. Ryan (1995). Autoradiographic and biochemical evidence for the systemic translocation of systemin in tomato plants. *Planta*, 195: 593–600.

Neuenschwander, U., K. Lawton and Ryals, J. (1996). Systemic acquired resistance in plant-microbe interactions Vol. 1. G. Stacey and N.T. Keen (eds.), New York: Chapman and Hall, pp. 81–106.

Neuenschwander, U., B. Vernooij, Friendrich, L., Uknes, S. and Kessmann, H. (1995). Is hydrogen peroxide a second messenger of salicylic acid in systemic acquired resistance? *Plants*, 8: 227–233.

Oerke, E.C., U. Steiner and F. Schonbeck (1989). The efficiency of induced resistance under practical culture conditions V. Powdery mildew and grain yield of winter and spring barley in relation to nitrogen fertilization. *Zeitschrift fur Pflanzenkrankheiten and pflanzenschutz*, 96: 140–153.

Ouchi, S., H. Oku, Hibino and C., I. Akiyama (1974). Induction of accessibility and resistance in leaves of barley by some races of *Erysiphe graminis. Phytopath. Z.*, 79: 24.

Pearce, G., D. Strydom, S. Johnson and C.A. Ryan (1991). A polypeptide from. tomato leaves induces wound inducible proteinase inhibitor proteins. *Science*, 253: 895–898.

Pena-cortes, H., T. Albrecht, S. Prat, E.W. Weiler and L. Willmitzer (1993). Aspirin prevents wound induced gene expression in tomato leaves by blocking jasmonic acid biosynthesis. *Planta*, 191: 123–128.

Pena-cortes, H., J. Fisahn and L. Willmitzer (1995). Signals involved in wound induced proteinase inhibitor II gene expression in tomato and potato plants, *Proc. Natl. Acad Sci.* USA, 92: 4106–4113.

Pena-Cortes, H., L. Willmitzer and J.J. Sanhez-Serrano (1991). Abscisic acid mediates wound induction but not development specific expression of the proteinase inhibitor-II gene family. *Plant Cell*, 3: 963–972.

Peng, M. and J. Kuc (1992). Peroxidase–generated hydrogen peroxide as a source of antifungal activity in vitro and on tobacco leaf disks, *Phytopathology,* 82: 696–699.

Penninickx, I.A.M.A., K. Eggermont, F.R.G. Terras, B.P.H.J. Thomma and G.W. Samblanx (1997) Pathogen-induced systemic activation of a plant defensin gene in Arabidopsis follows a salicylic acid–independent pathway involving components of the ethylene and jasmonic acid responses. *Plant Cell,* 8: 2309.

Raskin, L., A. Ehmann, W.R. Melander and B.J.D. Meeuse (1987) Salicylic acid: A natural inducer of heat production in *Arurn lilies. Science,* 237: 1601–1602.

Reuveni, R., V. Agapov and M. Reuveni (1994). Foliar spray of phosphates induces growth increase and systemic resistance to *Puccinia sorghii* in maize. *Plant Pathol.,* 43: 245–250.

Ross, A.F. (1961). Systemic acquired resistance induced by localized virus infections in plants. *Virology,* 14: 340–358.

Ruzicska, P., Z. Gombos and G.L. Farkas (1983). Modification of fatty acid composition of phospholipids during the hypersensitive reaction in tobacco. *Virology,* 128: 60–64.

Ryals, J., U. Neuenschwander, M. Willits, A. Molina and Hy Steiner (1996). Systemic acquired resistance. *Plant Cell,* 8: 1899–1919.

Ryan, C.A. (1990). Proteinase inhibitors in plants genes for improving defense against insects and pathogens. *Annu. Rev. Phytopathol.,* 28: 425–449.

Ryan, C.A. (1992). The search for the proteinase inhibitor-inhibitor-inducing factor, PI IF. *Plant Mol. Biol.,* 19: 123–33.

Salt, S.D., S. Tuzun and J. Kuc (1986). Effect of beta–income and abscisic acid on the growth of the tobacco and resistance to blue mold. Mimicry of effects of stem infection by *Peronospora tabacina. Physio. and Mol. Pl. Pathol.,* 28: 287–297.

Schaller, A. and C.A. Ryan (1996). Systemin: A polypeptide defense signal in plants. *BioEssay.,* 18: 27–33.

Schmodt, D. (1988). Prevention of bacterial wilt of grasses by phylloplane bacteria. *J. Phytopathol.,* 122: 253–260.

Schneider, S. and W.R. Ullrich (1994). Differential induction of resistance of enhanced enzyme activities in cucumber and tobacco caused by treatment with various abiotic and biotic inducers. *Physiol. Mol. Plant Pathol.,* 45: 291–304.

Schonbeck, F., H.W. Dehne and H. Balder (1982). Zur Wirksarnkeit induzierter Reistenz under Praktischern Anbaubedingungon. In: *Echter Mehttau an Rebon, Gurken und Weren. Zeitschaift fur Pflanzenkrankheiten und Pflanzenschutz,* 87: 177.

Sembdner, G. and B. Parthier (1993). The biochemistry and physiological and molecular actions of jasmonates. *Plant Mol. Biol.* 44: 569–89.

Sen, B. (2000). Biological control: A success story. *Indian Phytopathol.,* 53: 243–249.

Sequeria, L. (1983). Mechanisms of induced resistance in plants, *Ann. Rev. Microbiol.,* 37: 51–79.

Shulaev, V., J. Leon and I. Raskin (1995). Is salicylic acid a translocated signal of systemic acquired resistance in tobacco? *Plant Cell,* 7: 1691–701.

Simons, T.J. and A.F. Ross (1971). Metabolic changes associated with systemic induced resistance to tobacco mosaic virus in Samsun NN tobacco. *Phytopathology*, 61: 293–300.

Singh, U.S., D.S. Mishra, Rashmi, Rohilla, Singh, Amita, Vishwanath and B.P. Pal (2002). Induced resistance: Present status and future prospects as disease management strategy In: *Biopesticides and Pest Management*, Vol. 1 (Eds.), Opender Koul, G.S. Dhaliwal, S.S. Marwaha and J.K. Arora, pp. 262–302.

Skipp, R.A. and B.J. Deverall (1973). Studies on cross protection in the anthracnose disease of bean. *Physiol. Plant Pathol.*, 3: 279–314.

Smith, J.A. and R. Hammerschmidt (1988.). Comparative study of acidic peroxidases associated with induced resistance in cucumber, muskmelon and watermelon, *Physiol. and Mol. Pl. Pathol.*, 33: 255–261.

Staswick, P.E., W.P. Su and S.H. Howell (1992). Methyl jasmonate inhibition of root growth an induction of a leaf protein are decreased in an *Arabidopsis thaliana* mutant. *Proc. Natl. Acad Sci.* USA, 89: 6837–6840.

Steiner, U., E.C. Oerke and F. Schonbeck (1988). The efficiency of induced resistance under practical culture conditions. IV. Powdery mildew grain yield of winter barley cultivars with induced resistance and fungicide treatment. *J. Plant Diseases and Protect.*, 95: 506–517.

Stolen, O., J.E. Hermansen and J. Lohde (1981). Varietal mixtures of barley and their ability to reduce powdery mildew and yellow rust diseases. *Kgl. Vet.-og landbohojsk. Arsskr.*, 109.

Strobel, N.E., C. Ji, S. Gopalan, J.A. Kuc and S.Y. He (1996). Induction of systemic acquired in cucumber by *Pseudomonas cyringae* pv *syrinage* 61HrpZ(Pss) protein. *Plants J.*, 9: 431–439.

Stroemberg, A. and S. Brishammar (1993). A histological evaluation of induced resistance to *Phytophthorta infestans* (Mont.) de Bary in potato leaves. *J. Phytopathol.*, 137: 15–25.

Summermatter, K.. L. Sticher and J.P. Metraux (1995). Systemic responses in *Arabidopsis thaliana* infected and challenged with *Pseudomonas syringae* pv. *syringae*. *Plant Physiol.*, 108: 1379–1385.

Tuzun, S. and J. Kuc (1985). Movement of a factor in tobacco infected with *Peronospora tabacina* Adam which systematically protects against blue mold. *J. Physiol. Plant Pathol.*, 26: 321–330.

Van Kan, J.A.L., B.J.C. Comelissen and J.F. Bol (1988). A virus-inducile toacco gene encoding a glycie–rich protein shares putative regulatory elements with theriulose bisphosphate carboxylase small su-uit gee. *Mol. Plant-Microbe Interact.*, 1: 107–112.

Van Loon, L.C. and J.F. Antoniw (1982). Comparison of the effects of salicylic acid and etherphon with virus-induced hypersensitivity and acquired resistance in tobacco. *Neth. J. Plant Pathol.*, 88: 237–56.

Van Loon, L.C. (1976). Systemic acquired resistance, peroxidase activity and lesion size in tobacco reacting hypersensitivity to tobacco mosaic virus. *Physiol. Plant Pathol.*, 8: 231–242.

Van Loon, L.C. (1983). The induction of pathogenesis related proteins by pathogen and specific chemicals. *Neth. J. Plant Pathol.*, 89: 265–273.

Van Loon, L.C. (1997). Induced resistance in plants and the role of pathogenesis–related proteins. *Eur. J. Plant Pathol.*, 103: 753–765.

Van Loon, L.C. (2000). Systemic induced resistance In: *Mechanism of Resistance to Plant Diseases*, (eds.) Slusarenko, A., Fraser, R.S.S. and Van Loon, I.C., Kluwer Academic Publishers, Netherlands, pp. 521–574.

Van Loon, L.C. and Pennings (1994). Involvement of ethylene the inducti,on of system is acquired resistance in tobacco. In: *Mechanisms of Plant Defense Responses,* (eds.) B. Fritig, M.M. Legrand, Kluwer Academic Publishers, Dordrecht, Bostan London, p. 156.

Van Loon, L.C., P.A.H.M. Bakker and C.M.J. Pieterse (1998). Systemic resistance induced by rhizosphere bacteria. *Ann. Rev. Phytopathol.,* 36: 453–483.

Van loon, L.C., Bakker, P.A.H.M. and C.M.J. Pieterse (1997). Mechanisms of PGPR-induced ressitance against pathogens In: *Plant Growth Promoting Rhizobacteria, Present Status and Future Prospects,* (eds.) Ogosht, A. Kobayashi, K. Homma, Y. Kodama, F. Kondo, N. and Akino, S., Faculty of Agriculture, Hokkaido University, Sapporo, pp. 50–57.

Van Peer, R., G.J. Niemann and B. Schippers (1991). Induced resistance and phytoalexin accumulation in biological control of Fusarium wilt of carnation by *Pseudomonas* sp. Strain WCS 417. *Phytopathology,* 81: 728–734.

Van Wees, S.C.M., C.M. J. Pieterse, A. Trijseenaar, Y. Van't Westende, F. Hartog (1997). Differential induction of systemic resistance in Arabidopsis by biocontrol bacteria. *Mol. Plant-Microbe Interact.,* 10: 716–724.

Vance, C.P., T.K. Kirk and R.T.Sherwood (1980). Lignification as a mechanism of disease resistance. *Annu. Rev. Phytopathol.,* 18: 259–288.

Vernooij, B., L. Friedrich, A. Morse, R. Reist and R. Kolditz Jawahar (1994). Salicylic acid is not the translocated signal responsible for including systemic acquired resistance but is required in signal transudation. *Plant Cell,* 6: 959–965.

Vernooij, Friendrich, L., P. Ahi-Goy, T. Staub, H. Kessmann, and J. Ryals (1995) 2,6-Dichloroisonicotinic acid-induced resistance to pathogens does not require the accumulation of salicylic acid. *Mol. Plant-Microbe Interact.,* 8: 228–234.

Villich-Muller, V. and H.C. Weltzien (1990). Resistenzinduktion in Gerste und Hafer dunc Vorinokulation mit Apathogenenein befalsmindernder Mechanismus in gemischten Getreidebestenden. *Zeitschrift fur pflanzenkrankheiten und pflanzenschutz,* 97: 532.

Waiters, D.R. and D.C. Murray (1992). Induction of systemic resistance to rust in *Viciafaba* by phosphate and EDTA: effects of calcium. *Plant Pathol.,* 41: 444–448.

Wei, G., J.W. Kloepper and S. Tuzun (1991). Induction of systemic resistance of, cucumber to *Colletotrichum orbiculare* by select strains of plant growth promoting i rhizobacteria. *Phytopathology,* 81: 1508–1512.

Weiler, E.W., T. Albrecht, B. Groth, Z.Q. Xia and M. Luxem (1993). Evidence for the involvement of jasmonates and their octadecanoid precursors in the tendril coiling response of *Bryoniadioca. Phytochemistry,* 32: 591–600.

White, R.F. (1979). Acetyl salicylic acid (aspirin) induces resistance to tobacco mosaic virus in tobacco. *Virology,* 99: 410–412.

Wildon, D.C., J.F. Thain, P.E.H. Minchin, I.R. Gubb, A.J. Reilly, Y.D. Skipper, H.M. Doherty, P.J. O'Donn and D.J. Bowles (1992). Electrical signaling and systemic proteinase inhibitor induction in the wounded plant. *Nature,* 360: 62–65.

Xu, Y., P.F.L. Chang, D. Liu, M.L. Narasimhan and K.G. Raghothama (1994). Plant defense genes are synergistically induced by ethylene and methyl jasmonate. *Plant Cell.,* 6: 1077–1085.

Yalpani, N., P. Silverman, T.M.A. Wilson, D.A. Kleier, and I. Raskin (1991). Salicylic acid is a systemic signal and an inducer of pathogenesis-related proteins in virus infected tobacco. *Plant Cell,* 3: 809–818.

Yamada, T., T. Hiramoto, R. Tobimatsu, T. Shiraishi and H. Oku (1990). Elicitor like substances present in barley and wheat seeds. *J. Phytopathology,* 128: 89–98.

Yedidia, I., N. Benthamau and I. Chet (1999). Induction of defense responses in cucumber plant (*Cucumis sativus* L.) by the biocontrol agent *Trichoderma harzianum. Appl. Environ. Microbiol.,* 65: 1061–1070.

Yokoyama, K., J.R. Aist and C.J. Bayles (1991). A paplilla-regulating extract that induced resistance to barley powdery mildew. *Physiol. and Mol. Pl. Pathol.,* 35: 166–175.

24

Advances in Rice Blast Epidemiology and Management

M. Surulirajan and P. Srinivas

Division of Plant Pathology, Indian Agricultural Research Institute, New Delhi – 110 012

The green revolution is one of the biggest success stories of India cited globally, which enabled the country to convert its nightmarish "begging bowl" status to that of "self-sufficiency."

India is the seventh largest country in the world by area, with 329 million hectares (M ha). It is also the second-most populous country (1 billion people). Demographers indicate that by 2012 India's population will reach 1.2 billion. Rice being the staple food of our country contributes consistently around 45 per cent of India's cereal production. Though India, the world's second largest rice producer registered a record 91.6 mt in 2001–02, the next year the production declined by due to bad spell of monsoon (Annonymous, 2002). By 2006–7 India needs 100 million tonnes to feed its growing population and the demand for rice in India is projected at 128 million tonnes (Mt) for the year 2012 which renders requisite production level to reach 3,000 kg/ ha significantly greater than the present average yield of 1,930 kg/ha (Tiwari, 2002). But this demand in increase in rice production faces serious impediments from many diseases and pests. One among the major diseases which threats rice production world over is the rice blast. This disease stands first in its destructive nature where the annual loss due to this disease is nearly 4–5 million tonnes, Understanding the epidemiology and various management practices will be crucial for designing novel environmentally sound control strategies to protect world food supplies from this devasta .ing disease.

History

Blast has been worked out almost in all the countries in the world especially in Japan, India, Taiwan, China and USA. The disease is caused by *Pyricularia oryzae* Cav (Class: Denteromycetes; Order: Moniliales; Family: Moniliaceae). The teleomorph stage is *Magnaporthe grisea* (Hebert) Barr. The disease was first reported in China by Soong Yini-Shin in the year 1637. Later in 1896 Shirai described the fungus in Japan. Rosman *et al.* (1990) renamed the fungus as *Pyricularia grisea*. In the United States Metcalf recorded the disease in South carolina in 1876 and called it as Blast. In India it was first recorded in Thanjavur (Tanjore) delta of South India in 1913 by MacRae. An epidemic occurred in 1919 in Tanjore area. Severe outbreaks of *Pyricularia oryzae* during January–April 1990 in the Gangetic plains were reported by (Chakraborti and Pramanick, 1990).

Geographical Distribution

The disease is present in almost all rice-growing tracts in the world except California, in the US. The disease is adaptable to adverse environmental conditions of widely fluctuating temperatures and relative humidity. It appears in irrigated low land, rain-fed upland rice as well as in rice grown submerged or in deep waters. In desert conditions, especially in the Middle East, where high temperatures prevail with very low humidity, the blast fungus infects the nodes just along the field water level.

Yield Losses

Blast is considered as the principal disease of rice owing to its wide, distribution and destructiveness and remains the world's most economically important malaise. Early seedling infection often results in death of the plants while neck infection leads to half- filled and totally unfilled panicles, which reduce the yield.

Kuribayashi and Ichikawa estimated losses caused by blast by applying the regression equation or non-linear curve as:

Y = 0.69–2.8

Y = Yield loss ratio

X = per cent panicle blast incidence

Later in India Padmanabhan (1965) estimated the yield losses from blast by comparing yields in

1. Non-epidemic and epidemic areas.
2. Resistant and susceptible varieties having similar potential yielding ability.
3. Plots protected and unprotected by fungicides.

Torres and Teng (1991) have proposed a modified single hill technique. At max. tillering significant relationships of leaf blast and panicle blast incidence with yield loss were estimated by regression analysis collectively and separately. The equation obtained suggested that in the absence of panicle blast, 1 point on the leaf blast scale at max. tillering stage causes c. 0.4 per cent yield loss and in the absence of leaf blast, 1 per cent panicle blast at harvest results in c. 0.3 per cent yield loss. This technique allows a preliminary estimate of yield loss in situations where it may be difficult to experiment with many blast severities.

An eco-physiological simulation model for crop growth reported by Bastiaans (1991) could determine the effect of leaf photosynthesis on a rice crop and on subsequent yield. Simulation indicated that a reduction in crop growth before flowering might affect final yield. Model computations showed the reduction in leaf photosynthesis resulting from lower N content accumulation, which could affect final yield.

Another model was proposed by Ding-KeJian *et al.* (1999) for estimating yield loss caused by leaf last, neck blast and mixed infections based on systematic monitoring of disease development and the yield per single plant as well as per plot experiments conducted in a farm in Anhui Province, China. Difference in damage caused by rice neck blast among cultivars was also observed which could be attributed to the time of infection (Ding-KeJian *et al.* 1999).

Symptomatology

Rice blast is characterized by the appearance spots or lesions on leaves, nodes and different parts of the panicle and the grains, and rarely on the leaf sheath in nature. The leaf spots are typically elliptical of' spindle-shaped with more or less pointed ends. The centre of the spot is usually grey or whitish, and the margin is usually brown or reddish brown. The spots usually begin as small, water soaked, whitish, greyish or bluish dots. Under conducive conditions, lesions on the leaves of susceptible lines expand rapidly and tend to coalesce, leading to complete drying of infected leaves. Fully developed lesion reach 1–1.5 cm long, 0.3 to 0.5 cm broad and usually develop a brown margin, but instead sometimes have a yellow halo around the spot. On highly resistant varieties, only minute brown specks of pinhead size may be observed indicative of a hypersensitive reaction. These small brown lesions have been called as chronic lesions while those of greyish colour and those which enlarge considerably are called as acute lesions.

Yoshii (1937) differentiated three Zones in lesions:

Tenenate Zone

This is the outer portion of the lesion, formed by the infiltration of toxic substances secreted by the fungus, and consists of a yellowish Zone which blends in to healthy tissues.

Necrotic Zone

This is usually a brown, narrow streak along the inner side of the venerate zone. The cell inclusions and cell, walls are degenerated and discoloured in the necrotic zone.

Disintegrated Zone

In this zone the cell inclusions and cell walls are completely broken down and destroyed.

When the node is infected, sheath pulvillus rots and turns black and on drying often breaks apart, remaining connected by the nodal septum only. All parts above the infected node die. Panicles are infected producing brown lesions. Areas near the panicle base are often attacked, causing the 'rotten neck' or 'neck rot' symptoms, and glumes are also attacked.

Morphology of the Pathogen

Shirai has described the fungus on rice in Japan in, 1896:

Mycelium

Mycelium in cultures aerial or submerged, hyaline or olivaceous, 1.5–6.0 µm in width, septate and branched.

Conidiophores

One to many, fasiculate, simple or rarely branched, 2–4 septate, at first monoporic, then pleuroigenous on sympodoium, olivaceous to fuligenous, base swollen, dark coloured, becoming lighter colour towards apex.

Conidia

Variable in size and shape, terminal, pyrifonn to obclavate, base rounded, apex narrowed, 2-septate rarely 1–3 septate. Conidia germinate from apical or basal cell and less frequently from middle cell. Germinating hyphae hyaline, mayor may not be constricted at septa, branched, 3–5 µm in width.

Disease Cycle

Primary Source of Inoculum

The blast fungus is reported to persist in the infected seed, diseased straw, infected seedlings or even in the soil.

Secondary Source of Inoculum

Diseased tissue.

Infected seedlings mostly serve as a source of primary inoculum and, when. transplanted, two phases of disease development, namely, the multiplying phase on a single tiller and the dispersing phase among neighbouring ones are commonly observed.

Predisposing Factors Favouring Infection

Hahioka (1965) reported that several factors affect the development of the blast. The major ones are:

1. Humidity
2. Temperature
3. Excessive N-fertilizer
4. Host susceptibility

Continuous wetness on the leaves is necessary to ensure infection. Conidia fail to germinate when the relative humidity drops below 92 per cent. The invasion of the host by the blast fungus is possible only at 92–96 per cent RH. Temperatures at 21–28°C are most favourable for the initial stages of infection. Nitrogen fertilizer have a remarkable effect on the susceptibility of rice to blast infection. A linear increase in infection in susceptible types with an increase in the level of N-fertilizer has been reported by De Faria *et al.* (1982).

Rice plants especially the upper (younger) leaves, are considered more susceptible than the middle or lower leaves. Plants are more susceptible in the seedling stage than in the tillering and booting stage. There are three distinct stages of growth when rice plants are highly susceptible to infection. They are seedling, rapid tillering (foliar infection) and ear emergence stages (nodal and neck infection).

Kim (1982) correlated the relationship between panicle blast incidence and infected leaves at the booting stage as follows:

$Y = 2.84x + 19.3\ (r = 0.8)$

Y = % of panicle blast incidence

X = Number of blast infected leaves at the booting stage

Mode of Infection and Spread

Penetration of the host by germ tube may be through the stomata or directly through the epidermis, while appressoria are formed from the germ rubes from which slender penetrating hyphae are extruded. These pierce the outer layer of the epidermis and enter the cell cavity.

Host Penetration

Successful hyphal penetration of the epidermal cells in plants growth under low night temperatures have been observed more frequently, while penetration is rare in plants grown under high night temperature. In the latter case, although the fungus succeeds in entering the host, its subsequent spread was restricted. The average time for penetration of appressorium was around 14 hours at 18–24°C. Water is necessary for spore germination, appressorium formation and penetration of the peg into the epidermal cells. Dew has a key role where host penetration occurs with in 3 hours after formation of the peg. Lesion growth depends on temperature along with leaf wetness and water status of the plant. The longer the period of wetness will increase the number of lesions. Wind speed also influences initial infection indirectly by its effect on the time and the amount of dew formation,. but there is no statistical correlation between wind speed at inoculation and number of lesion formed subsequently.

Penetration pegs produced *in vitro* by appressoria of *M. grisea* examined by electron microscopy revealed that peg cytoplasm consisted primarily of a zone of exclusion, excluding even ribosomes, and was continuous with a similar region in the appressorium. Apical vesicles were, however, observed in short, presumably elongating pegs (Bourett and Howard, 1992).

Bourett and Howard (1992) also used Immunofluorescence microscopy to demonstrate binding of a monoclonal anti-actin to penetration peg cytoplasm, which showed that the monoclonal probe bound predominantly to a single band with a molecular weight similar to that of rabbit muscle actin. The actin may have a role to play in the formation of the zone of exclusion, during the peg formation and penetration.

Cytological Aspects of Host Response

Lipid bodies in rice leaves get discoloured during pathogenesis. Kurosaki observed low water content in the brown tissue surrounding blast lesion and interpreted that the inhibited supply of water and water soluble nutrients to the fungus enclosed inside the lesion perhaps deprive the fungus further growth similar to that of a hypersensitive reaction that reflects the inability of the plant and the pathogen to co-exist. Epidermal cells in leaves of wild rice species namely, *Oryza latifolia* and *O. minuta*, reacted sharply against penetration by the blast fungus by secreting a resin-like material around the invaded hyphae, inhibiting its further growth.

Yaegashi *et al.* (1990) made observations on the specific interaction of cell reaction between rice and blast fungus. The leaf cell reaction at host-parasite encounter sites were classified into six types according to the fungal development and host response.

Lesion Enlargement

Lesion expansion over time is also a significant component of disease, progress. Although infection occurs most easily in darkness, light has been reported to, promote lesion enlargement. Lesion expands slowly at 16–20°C than at 25–32°C. High temperature affects the host by rapid senescence of leaves and indirectly restricts lesion expansion.

Epidemiology

Studies concerned with the complex effects of the environmental conditions on disease development are of prime importance for predicting any epidemic. The intensity of the epidemic is determined mainly by the abundance of the conidia, the influence of environmental conditions on infection processes, and the resistance of the host plants. These three components are influenced hot only by meteorological factors, but also by the host cultivars and cultural practices used. The meteorological factors influence the host as well as the pathogen.

The blast epidemic will be mainly governed by the disease proneness of the host, environmental condition such as temperature, duration of leaf surface wetness relative humidity (RH) and duration of dew.

Perennation and Survival of the Pathogen

The carryover of the blast inoculum from one season to the next is an important problem. However, the fungus survives the winter in the infected plants parts under snow in Himalayan region. When the snow melts and with the onset of the first rains in April, the pathogen sporulates abundantly and causes infection.

A few collateral hosts have also been reported by the various workers. Singh and Singh (1988) reported that weed hosts *viz., Leeisia hexandra, Cypreus rotundus, C. compressus, C. irria* harboured *Pyricularia oryzae* blast fungus, which on artificial inoculation to rice plant produced typical blast symptoms. Srinivas Prasad (1998) reported that *Digitaria ciliaris* and *D. marginalia* act as collateral host for the blast fungus.

Factors Influencing the Development of Infection

The development of infection in nature is the result of interaction among the host the pathogen and the environment in the background of the time factor. The different predisposing factors of the host are:

Nutritional Factors

Even under normal conditions of growth, the increased application of nitrogen has a remarkable effect on the susceptibility of rice to blast infection. Long *et al.* (1997) reported that application of nitrogen above the recommended amount for a cultivar significantly increased severity and incidence of leaf blast on susceptible cultivars. In studies can:ied out at CRRI on the effect of the, application of fertilizers on the incidence of blast under very high levels of fertility, it was found that the higher levels of N, P and K (200–240 kg/ha) either alone or in combination brought about an increase in infection.

Age

The susceptibility of the rice plant to leaf infection due to blast varies with its age. Rice is susceptible to leaf infection a few days after transplanting, and at the rapid tillering stage and is less susceptible, as it approaches heading, the plant is most susceptible to nodal infection, especially at the node below the rachis.

Bio-meteorological Factors

High and low temperatures with long periods of rain (drizzling), lack of sunshine, slow winds, high humidity and drought not only increase the disease proneness of the host but also influence the life cycle of the pathogen.

Lesions developed on the plants raised at night temperature of 20°C were larger than those developed on plants raised at 25°C and 30°C. The occurrence of low, night temperatures of 26°C and below, relative humidity of 90 per cent and above, dew at least for 6 hours during October–January results in an epidemic. Lesion growth was reported to be optimum at 28°C day temperature, 24°C night temperature and 97–98 per cent relative humidity. Further it was noted that growth decreased with day temperatures above 28°C but increased with decreasing night temperature (Tripathi, *et al.* 1998).

Mao, *et al.* 1997 reported that continuous low temperature decreased disease resistance of the plants reduced PAL (phenylalanine ammonia lyase) activity in rice plants by 14.9–16.4 per cent. However, a longer period of low temperature (5 days or more) inhibited disease development.

The influence of weather parameters, including rainfall, temperature and relative humidity, on the incidence of rice blast were determined by Tripathi, *et al.*, 1997. They recorded maximum percentage disease index (85.5 per cent) in the second fortnight of October followed by the first fortnight of November when the maximum and minimum temperature varied between 31.5°C and 18.1–16.6°C and RH 90–40 per cent.

Sharma *et al.*, 1993 observed that crops sown early, in April or May in Nagaland, India, escaped or showed very low levels of infection and produced the highest grain yields. Disease incidence was observed to be highest in July and August when minimum temperature were 20.75–22.29°C and rainy days were most frequent (19.3 d). Of the 7 meteorological factors studied, the frequency of rainy days were reported to have the greatest influence on disease development.

Spore Production

Spore production and/or dissemination are important in the blast disease cycle. Spore formation occurs at 15–30°C with the optimum production at 25–28°C. The maximum spore formation occurs at 20°C. Good sporulation has been observed at relative humidity 93–100 per cent, while there is no sporulation at lower relative humidity than 93 per cent.

Spore production improves with increasing light intensity 6–8 hours of darkness followed by light favours sporulation. Cultures exposed to a period of constant darkness followed.by continuous light produced large number of spores. Continuous radiation with near UV light (366 nm) shortened the time required to reach maximum sporulation but the rate slowed down remarkably under darkness. Maximum sporulation is observed between 7 to 12 days after inoculation although sporulation continued for 60 days. Substantial variation is observed in the liberation of spores from lesions of the same size, but differing in colour. Maximum spore production is seen in grey centred lesions with dark purple to brown margin.

Krishnan *et al.* (1992) had studied the spore releasing capacity of various types of blast lesions using Spore Releasing Monitor (SRM) against weather parameters. The SRM was found to be ideal for collecting spore print from lesions on the coated micro slide under field conditions at rice canopy level. Purple grey lesions produced highest number of spores while spore density was least with brown coloured lesions.

Spore Discharge

The release of conidia from the conidiophore is called spore discharge. This process begins after mid-night (2 am) and increases till daybreak (6 am) while it ceases at sunrise. The presence of water droplet is quite essential for spore discharge and takes place within 2 minutes and especially to a short distance by the bursting of the minute stalk-cell attached to the conidophore. A typical lesion produces as many as 2000–6000 conidia per day and condial production continues for 10–14 days. It is estimated that in extreme cases, 15 million conidia may be deposited in a square meter of rice field a day.

Kim *et al.* (1992) studied the conidia release phase critically on three different aged lesions (3 days, 4 days and 7 days). All the lesions exhibited 2–3 peaks on the 1st and 2nd day respectively, but the 7 days old lesions revealed only one peak on the first day. The studies revealed that the conidial release pattern seems to have a regular rhythm and most peaks appear between I am and 7 am.

Spore Dispersal

Strong winds leads to greater spore flight to longer distances. The released conidia float under the canopy of rice plants and then are scattered into the air above the canopy. The conidia are then carried upward by convection currents to the top of the plant layer, and from there further disseminated by the winds. Abundant conidial release occurs at higher wind velocities (3–5 m/sec) under 80–100 per cent relative humidity.

Spore Deposition

The position of the leaf in relation to the leaf angle influences the number of spores deposited on it. In general, the spores fall perpendicularly in the atmosphere with a weak wind and are deposited more on horizontal than on non-horizontal plant surfaces.

The number of conidia deposited (N) can be estimated by the following equation:

N = Number of conidia sampled × revised spore ratio × leaf area index

Height using Suzuki's formula × coefficient of deposition × Area of the sampler

Epidemics of rice neck blast could be predicted from the airborne spore density and it fluctuation, and a parallel was observed between the field disease intensity of rice blast and the airborne spore load by Xu *et al.* (1990). The intensity of rice neck blast and leaf blast were positively related to the amount of airborne spores of the pathogen.

Forecasting

Forecasting is done using either mathematical/statistical methods (simple, single equation type) or simulation modelling approaches (system type). Japan has developed an elaborate and effective forecasting system for determining the optimum time for application of fungicides to derive maximum control of the blast disease. This enabled the farmers to use minimum fungicidal sprays before the spread of the disease.

Manibhushanrao and Krishnan (1991) developed a computer based forecasting system, to stimulate the rice leaf blast based on multiple criteria. Using stepwise regression analysis they evaluated the best fitting model to estimate the number of *P. oryzae* spores over the paddy field as

$Y = 123.068 - 3.08X_1 + 0321.X_2 - 0.3X_3$

Y = estimated number of spores

X_1 = daily maximum temperature

X_2 = daily maximum relative humidity

X_3 = daily minimum relative humidity

Studies on the effect of different weather parameters on the rice blast disease severity in Kerala by Premanathan, *et al.* (1999) revealed that bright sunshine hours was positively correlated, while rainfall and number of rainy days (RD) were negatively correlated with disease severity. Relative humidity (RH) exerted highest positive direct effect. The prediction equations derived through multiple regression analysis was also formulated for effective prediction of blast severity.

Simulation Models

The computer based simulation modelling is a recent concept in modern sciences. Various techniques of simulation are being used currently to have a better understanding of host-pathogen complex. The resultant models enable us to explore the factors that govern disease epidemics and to design the control system to minimise yield loss.

Leaf Blast Models

First leaf blast model BLASTL was developed in Japan by Hasimoto *et al.* (1984) which is a system analytic model that stimulates the epidemic. This model has been verified under field experiments for disease progress and used to predict efficient time for the fungicide application. Ishiguro, (2001), validated this model with the incorporation of the components of fungicide and N fertilizer application. N application at the early phase of epidemics was predicted to particularly enhance disease progression. The model was also found to estimate lesion density with relative errors within a reasonable range. A polycyclic model PYRICULARIA as developed by Gunther (1986) and this was later modified by Tastra *et al.* (1987) for upland rice farming in Indonesia and was subsequently called PYRNEW. The BLASTAM system was developed by Sakai *et al.* (1990). This uses the weather data of the automated meteorological data acquisition system. (AMcDAS) to predict weather condition favourable for infection.

Later Manibhusanrao and Krishnan (1991) developed EPIBLA to stimulate the leaf blast incidence in tropics. CERES-Rice (a rice growth simulation model) and BLASTSIM (a rice leaf blast epidemic simulation model) were combined by Luo *et al.*, 1997 by linking the effects of leaf blast on rice leaf photosynthesis and biomass production. And the results demonstrated that simulated and observed distribution of vapour pressure which was needed to estimate daily humidity were consistent. Simulation outputs included daily disease severity, the area under disease progress curve (AUDPC), and yield loss. Temperature was found to be a sensitive variable in the combined model influencing rice growth and leaf blast epidemics.

Panicle Blast Models

The first developed model in Japan for panicle blast was by Takasahi (1982). He treated spore deposition and penetration as a stochastic process and subdivided each panicle into small infection site units. This model does not account for secondary infection. Ishiguro in 1986 developed the most comprehensive stochastic model for panicle blast (P BLAST).

Management

Attempts to control rice blast have been directed at the use of resistant cultivars, chemical protection using economic spray schedules, biocontrol and cultural practices. Ensuring adequate water management, observing field sanitation, judicious use of N-fertilizers on high yielding cultivars are some of the prophylactic measures that may contribute in checking the disease to some extent.

Breeding for Resistance

Blast resistance has consistently been one of the important objectives of rice breeders in most of the rice growing countries. Systemic rice breeding work started as early as 1927 in Japan, and since then a number of blast resistant cultivars have been released. The important ones are the Norin series, Norin 6, Norin 8, Norin 22, Norin 23, Koganenami and Tacho.

In India, considerable work on breeding for resistance to blast began in Tamil Nadu in early 1920s and two promising cultivars Co4 and TMK 9 were developed and replaced the then popular cultivar ADT 10. Multineal lines and horizontal or field resistance developed based upon polygenic resistance has been used to overcome the problem of variability of the fungus.

The promising blast resistant varieties in India are:

Vikas	H PPU741	Rasi	VL Dhan 81
IR64I	VL 8	Tulsi	Neeraja
Nidhi	Pant Dhan		

Ashizawa *et al.* (1999) studied the effect partial resistance on rice blast with cultivar mixtures. They found that the mixtures of the susceptible and resistance cultivars reduced leaf and panicle blast. The effectiveness in the blast control was greater for mixtures containing cultivars with low level of partial resistance.

Lai *et al.* (1999) evaluated several rice cultivars for resistance to rice blast in upland nurseries. They found that the cultivars with intermediate susceptibility under upland culture appeared to benefit to most from flooded culture. Manandhar *et al.* (1998) reported that the resistance to rice blast can be induced by the application of ferric chloride, di-potassium hydrogen phosphate and salicylic acid. The greatest reduction in blast was obtained with ferric chloride.

In greenhouse tests, humic acid powder extracted from lignite and dissolved in KOH (pH 7.2) added to the soil, 50 d after sowing, containing plants inoculated with *P. oryzae*, reduced rice blast severity and was most effective when applied at the rate of 20 kg/ha (Thangvelu and Ramabadran, 1992).

Cultural Practices

Cultural control includes the manipulation of various farming practices that favour the crop at the expense of the pathogen. Soil fertility, land improvement, plant spacing and split application of nitrogen fertilizer, flooding reduce the blast disease in specific situations.

A voidance is considered as shifting of sowing or date of planting which enables to prevent the susceptible stage of the plant and the favourable conditions of the fungus from coinciding. Exclusion is to prevent the new races of the fungus from reaching an uninfected area and avoiding contact between a new race and the crop. As in California, USA there is no record of blast although rice is grown commercially.

Eradication involves elimination of the pathogen from infected areas by rouging sanitation, crop rotation removal of collatral hosts etc. Hooda and Srivastava (1998) studied the impact of neem coated urea (NCU) and potash on the incidence of rice blast. They found that the NCU (@ 60 Kg /ha or 90 Kg/ha) effectively reduced the leaf and neck blast and also increased the yields. The potash could not influence the incidence of rice blast as well as yield.

Hooda and Srivastava (1996) studied the role of silicon in the management of rice blast. They conducted an experiment using burnt rice husk, sodium silicate, sodium silicate (AR) and commercial sodium silicate. All the compounds reduced rice blast but the burnt rice husk and commercial sodium silicate were found to be most effective and economically viable.

Chemical Control

The majority of the fungicides used in blast control are protectants and eradicants. The most important groups and chemicals used for blast control are:

Copper Compounds

Metcalf (1906) claimed that spraying with Borqeaux mixture (5–5–50) just before the emergence of ear heads almost entirely controlled leaf blast. Due to the phytotoxic nature of these compounds these were later banned from use. However, under environmental and physiological conditions highly favourable for blast, copper fungicides alone do not control blast.

Mercuric Compounds

Ogawa *et al.* (1953) discovered that phenyl mercuric acetate (PMA) is less phytotoxic to rice plants. This was used as dust, wettable powder or emulsifiable concentrate for the effective control of blast. The PMA compounds are therapeutic in action and have a strong preventive action against blast fungus in inciting infection and sporulation on lesions. Although this compound have slight or no toxic action on *Japonica* rice cvs, they are toxic to *Indica* rice cvs.

Organophosphorus Compounds

Quite a good number of fungicides which are promising ones are present in this group. Some of the promising fungicides are Kitazin P (IBP), Hinosan (edifenphos), Oryzemate (probenazole), Fujione (isoprothiolane). Rabcide (pthalide), Topsin M (thiophanate methyl Nf–44). EL–291 (tricyclazole), Inezin (phenyl phosphonothioate) and Benlate (benomyl).

Sood and Kapoor (1997) evaluated 7 fungicides for the control leaf and neck blast which included Cyproconazole (Atemi 50 SL), Bitertanol (Baycor 25 WP), Tricyclazole (Beam 75 WP), Chlorothalonil (Chlorothalonil 40 SC), Hexaconazole (Contaf 5 EC) and Propiconazole (Tilt 25 EC). It was foun.d that tricyclazole (Beam 75 WP) was the most effective and reduced the leaf and neck blast by 89.2 per cent and 94.5 per cent respectively. Chlorothalonil and hezacanazol were also comparatively effective in controlling the disease.

Muhammad Saifulla *et al.* (1998) confirmed the above results *i.e.*, tricyclazole is the most effective chemical to control leaf and neck blast. Hdeverding *et al.* (1998) reported that the fungicide AC 382042 [N-(1-cyano-, 2-di metthypropyl)-2-(2,4- dichlorophenoxy) propionamide] acts through the inhibition of melanin biosynthesis in *P. oryzae* and there by preventing the appressorium formation. Tsuda *et al.* (1998) reported that the packed formulation containing pyroquilon was effective against rice blast.

Thiron *et al.* (1999) reported the effectiveness of carpropamid, a root application systemic fungicide, for the control of rice blast. Application of carpropannid, a novel melanin biosynthesis inhibitor brings about a strong inhibition of penetration by the pathogen. Melior *et al.* (1999) studied the fungicidal activity of some heteroaryl thiophene carlamates against *P. oryzae*. They observed that heteroaryl group comprising either 2-pyridyl or 2-pyraximyl groups had the highest levels of fungicidal activity when compared to ester and amide derivations.

Biocontrol

Though chemical control of the disease has been followed by majority of farmers world wide, intensified use of agro chemicals has resulted in the accumulation of toxic compounds hazardous to humans and environment besides causing built up of resistance in the pathogens. This necessitates the need for innovative disease control strategies such as the use of biocontrol agents as a viable alternative in modern disease management programmes.

Gnanamanickam (1992) reported that *P. fluorescens* strain 7–14 was most effective against rice blast. He found that the crude extract prepared from this strain provided 70–100 per cent inhibition of conidial germination at 1.0 ppm. Krishnamurthy and Gnanamanickam (1998) studied the persistence and migration of Pfl-14 on the rice plant with the aid of *laz* genes inserted into the bacterium. The bacterium survived through out the crop season but migration was limited, indicating the need for multiple foliar application of the bacterium to sustain the bacterial population for effective suppression of rice blast.

Rosales *et al.* (1995) reported that two strain of *P. cepacia*, In -b- 6854 and in -b- 6858 produce large inhibition zones on pigment production medium (PPM), King's medium B (KMB) and Yeast Malt (YM). Later it was found that these strains contain pyrrolnitrin which resulted in the inhibition of the blast fungus.

Beckennan (1997) reported that the factor pheromone of *Saccharromyces cerevisiae* blocked appressorium formation by the pathogen arid protected the plants from infection. Vidhyasekaran *et al.* (1997) developed a powder formulation from strain of Pfl of *P, jluorescens* for control of rice blast. This inhibited the growth of the blast pathogen *in vitro.* A talc-based formulation was developed which proved effective for both seed treatment and foliar spray.

Manandhar *et al.*, (1998) reported that the avirulent isolates of *P. oryzae* and isolates of *Bipolaris sorokiniana* could suppress the rice blast. They suggested that induced resistance might be involved in the suppression of disease. Kim-Byungsup *et al.* (1998) reported that a myxobacterial isolate KRO 25 showed good antifungal activity against *P. oryzae.* The bacterium was identified as *Myzococus fulvus* and the anti fungal substance as myxothiazole. This compound controlled more than 97.0 per cent of rice blast in the field.

Antibiotics

Yoshii (1949) was the first to report that a culture filtrate of Cephalothecium sp. could completely inhibit conidial germination of the blast pathogen and called it "cephalothecin." Several antibiotics have been produced since then, such as

Antiblastin
Blasticidin
Blasticidin–A
Blasticidin–S
Blastimycin

Except blasticidin–S none of these antibiotics can be put to use because of their instability and high toxicity to fishes. Thirumalachar (1968) reported two antibiotics 'hamycin' and 'thiolutin' which are effective against rice blast. Kim-Hyunju *et al.* (1999) reported that the antibiotic podoblastin extracted from the TP–4 isolate *Helminthosporium* sp. exhibited potent control of rice blast.

Botanical Fungicides

Plant extracts are less toxic to plants, systemic in nature and easily biodegradable. Ohno *et al.* (1983) tested new antifungal natural products, podoblastin and found it effective against the rice blast disease. Tewari and Dath (1984) reported that leaf extracts of henna (*Lawsonia inermis*) contain 2-hydroxy 1, 4-napthoquinone (Lawsone) is toxic to blast. Iwata *et al.* (1988) reported that phosphatidylethanolamine (PE) present in lecithin in soybean inhibits the penetration by hyphal cells. Manasi Mishra and Tewari (1990) reported that the ethanolic extracts of the leaves of *Calorropis procera, Azadirachta indica* and *Datura stramonium* effectively inhibited the *P. oryzae* conidia and mycelial growth.

A new antifungal naphthopyran derivative (ED50 value of 0.4 p.p.m.) having potential of 82.3 per cent inhibition of rice blast disease at 100 ppm, was isolated from the extract of stems and leaves of *R. nasutus* in Thailand. Its structure was elucidated by spectroscopic methods as 3,4-dihydro-3,3-dimethyl-2H-naphtho[2,3-b]pyran-5,10-dione (Kodama *et al.*, 1993).

Tewari and Mandakini Nayak (1991) reported that leaf extracts of *Piper betle, Nyclantlzes arbor-traistis* and citrus lemons were effective in reducing the radial *in vitro* growth of *P. oryzae.* Tewari (1995) reported the leaf extracts (aqueous. ethanolic and an essential oil) from *Ocimum sanctum* as potent fungicides against rice blast. He found that ethanolic extract was most effective among the three.

Future Challenges

1. The role of seed-borne inoculum should be investigated thoroughly especially in northern India.
2. There is an urgent need to develop Near Isogenic Lines (NIL) in India which can differentiate Indian isolates of the fungus.
3. Role of silica on the incidence of the disease in different soils and ecosystems should be studied to manage the disease.
4. Development of a simple summary model is necessary for disease forecasting which may be used by field and extension workers.
5. Blast of rice cannot be managed by the scientists in a state or region working in isolation. A team of Plant Pathologists should be identified with a lead centre and the other scientists located at representative centres with clear cut mandate and objectives of the work.

REFERENCES

Annonymous (2002). Business Line, International edition Financial Daily from The Hindu Group Monday Oct 14, 2002.

Ashizawa, T., K. Zenbayashi and S. Koizumi (1999). Effect of partial resistance on rice blast control with cultivar mixtures. *Phytopathology,* 89(6) Supplement: S3.

Bastiaans, L. (1991). Quantifying the effect of rice leaf blast on leaf photosynthesis, crop growth, and final yield using an ecophysiological model of crop growth. *Rice Blast Modeling and Forecasting.* Selected papers from the International Rice Research Conferences, 27-31st August 1990, Seoul, Korea Republic, pp. 89–99.

Beckerman, J.L., Fred Naider and Daniel J. Ebbole (1997). Inhibition of pathogenicity of the rice blast fungus by *Saccharomyces cerevisiae*–factor. *Science,* 276: 1116–1119.

Bourett, T.M. and R.J. Howard (1992). Actin in penetration pegs of the fungal rice blast pathogen, *Magnaporthe grisea. Protoplasma,* 168: 1–2, 20–26.

De Faria, J.C., A.S. Prabhu and Y.J.P. Zimmerman (1982). Effect of nitrogen fertilization and fungicidal sprays on blast and yield of upland rice (*O. Sativa* cv. *IAC–47*). *Pesqn. Agropec. Brasl.,* 17: 847–852.

Ding, KeJian, G. Tan, Wu-Jian, K.J. Ding, G.J. Tan and Wu-J (1999). A study on yield loss caused by rice blast. *Acta-Phytophylacica-Sinica*, 26: 1, 60–64.

Gnanamanickam, S. and Twangwahmew (1992). Biological control of blast disease of rice (*Oryza sativa* L.) with antagonistic bacteria and its mediation by a *Pseudomonas antibiotic. Annals of Phytopathological Society of Japan*, 58: 380–385.

Hashioka, Y. (1965). Effects of environmental factors on development of causal fungus disease. In: *The Rice Blast Disease*, Johns Hopkins Press, Baltimore and Maryland, pp. 153–161.

Hooda, K.S. and M.P. Srivasatava (1999). Impact of neem coated urea and potash on the incidence of rice blast. *Plant Disease Res.*, 13: 28–34.

Hooda, K.S. and M.P. Srivastava (1996). Role of silocon in the management of rice blast. *Indian Phytopath.*, 49: 2631.

Ishiguro, K. (2001). Improving management strategies for rice blast disease using a simulation model of rice leaf blast epidemics. *Bulletin of the Tohoku National Agricultural Experiment Station*, 99: 1–110.

Kim, C.H. and C.K. Kim (1992). Rice blast development under different water management, practices in fields in Korea. *Korean. J. Pl. Pathol.*, 6: 51–57.

Kim, C.K. (1982). Improved methods for rice blast forecasting. *Korean J. of Plant Prot.*, 21: 19–22.

Kim-Hyunju, Kim-Byungsup, Kim-Hongci, Cho-Kwang Yun, H.J. Kim, J.C. Kim, B.S. Kim, H.C. Kim and K.Y. Cho (1999). Antibiotic and phytotoxic activities of ophiobolins from *Helminthosporium* species. *Plant Pathology J.*, 15: 14–20.

Kodama, O., H. Ichikawa, T. Santisopasn Akatsuka, V. Kato, Y.A. Hayashi (1993) Isolation and identification of an antifungal naphthopyran derivative from *Rhinacanthus nastus. J. Nat. Prod.*, 56: 2, 292–294.

Kuribayashi, K. and H. Ichikawa (1952). Studies of the forecasting of the rice blast disease. *Bull. Nagano Prof Agric. Exp. Stn.*, 13: 1–229.

Lai, X.H., M.A. Marchetti and H.D. Peterson (1999). Comparative slow-blasting in rice grown under upland and flooded blast nursery culture. *Plant Dis.*, 83: 681–684.

Long, D.H., D.O. Te Beest and F.N. Lee (1997). The effect of nitrogen fertilization on the epidemiology of rice blast disease in small plots in Arkansas. *Research series of Arkansas Agricultural Experiment Station*, 456: 94–108.

Luo, Y., P.S. Teng, N.G. Fabellar and D.O. TeBeest (1997). A rice-leaf blast combined model for simulation of epidemics and yield loss. *Agricultural-System*, 53: 1, 27–39.

Mahadev, P., N.K. Chakraborti and M. Pramanick (1990). Severe incidence of rice blast in West Bengal during 1990. *Indian J. Myc. & Plant Path.*, 20: 3, 256–226.

Makoto Kimura and Isamu Yamagochi (1996). Recent development in the use of Blasticidin S: A microbial fungicide, as a useful reagent in molecular biology. *Pesticide Biochemistry and Physiology*, 56: 243–248.

Manandhar, H.K., H.J.L. Jorgemsen, S.B. Mathur and V. Smedegaard Peterson (1998). Resistance to rice blast induced by ferric chloride, dipotassium hydrogen phosphate and salicilic acid. *Crop Prot.*, 17: 323–329.

Manandhar, H.K., H.J. Lyngs Jorgennsen, S.B. Mathur and V. Smedegaard Peterson (1998). Suppression of rice blast by preinoculation of avirulent *Pyricuaria oryzae* and the non rice pathogen *Bipolaris sorokiniania. Phytopathology,* 88: 735–739.

Manasi Mishra and S.N. Tewari (1990). Ethanolic extract toxicity of three botanicals against five fungal pathogens of rice. *Nat. Acad. Sci. Letter,* 13: 409–412.

Manibhushan Rao, K. (1994). *Rice Blast Disease,* pp. 180.

Manibhushan Rao, K. and P. Krishnan (1991). Epidemiology of blast (EPIBLA): A simulation model and forecasting system for tropical rice in India. In: *Rice Blast Modelling and Forecasting.* Int. Rice Res. Inst. Manila, Philippines, pp. 33–38.

Mao-J., He-Ming, He-Zhong Quan, Wu-Hong, Liu-Yong, Mao-JH, He-M, Hem-ZQ, Wu-H, and Liu-Y (1997). A study of the effects of continuous low temperature on rice blast (*Pyricularia oryzae* Cav.). *J. Southwest Agril. Univ.,* 19: 3, 228–230.

Mellor, M., D.R. Mitchell, S.B. Obsoum and P.D. Riordan (1990). Synthesis and fungicidal activity against *Pyricularia oryzae* of some heteroryl thiophene carbomates. *Pesticide Sci.,* 55: 326–330.

Muhammad Saifulla, H.K. Ramappa, N. Shiva Kumar and N.M. Poonacha (1998). A promising fungicide for the control of rice blast disease. *Oryza,* 35: 188–189.

Ou, S.H. (1972). *Rice Diseases,* pp. 368.

Padmanabhan, S.Y. (1965). Estimating losses from rice blast in India. In: *The Rice Blast Disease.* Proc. Symp. at IRRI, July, 1963, pp. 203–221.

Premanathan, T., G.S.L.H.V.P. Rao, N.K. Mahana (1999). Forecasting of rice blast disease epidemic in northern region of Kerala. *Oryza,* 36: 4, 364–368.

Rice Research in India (1990). Published by Publication and Information Division, ICAR, New Delhi, 726 pp.

Rosales, A.M., L. Thomashow, R.J. Cook and T.W. Mew (1995). Isolation and identification of antifungal metabolites produced by rice-associated antagonistic *Pseudomonas* spp. *Phytopathology,* 85: 1028–1032.

Sharma, J.P., S. Kumar and R.N. Verma (1992). Relationship between rice blast and meteorological factors in Nagaland. *Indian Phytopath.,* 46: 1, 78–80.

Sieverding, R., T. Hirooka, T. Nishiguchi, Y. Yarnamoto, V.J. Spadatom and H. Hasui (1998). *AC 382042: A New Rice Blast Fungicide.* Proceedings of an International Conference, Brighton, UK, pp. 359-366.

Singh, N.I. and K.H.U. Singh (1988). Unrecorded weed hosts for *Pyricularia oryzae* Cav. In India. *International Rice Research Newsletter,* 13(4): 31–32.

Singh, R.A. (1997). Current status of rice blast in India and challenges ahead. *Indian Phytopath.,* 50: 186–191.

Sood, G.K. and A.S. Kapoor (1997). Efficacy of new fungicides in the management of rice blast. *Plant Dis. Res.,* 12; 140–142.

Srinivas Prasad, M., Sangit Kumar and M.S. Lakshmi Prasad (1998). Weed hosts of rice blast fungus in Meghalaya, India. *Oryza,* 35: 384–385.

Tewari, S.N. (1995). *Ocimum sanctum* L.: A botanical fungicide for rice blast control. *Trop. Sci.*, 35: 263–273.

Tewari, S.N. and Mandakini Nayak (1999). Activity of four plant leaf extracts against three fungal pathogens of rice. *Trop. Agril.*, 68: 373–375.

Thangavelu, R. and R. Ramabadran (1992). Effect of humic acid (HA) on severity of rice blast (BI). *International Rice Research Newsletter*, 17: 3, 18.

Thiron, M. (1999). Development of a new rice blast controlling agent, carpropamid, *J. Pest. Sci.*, 24: 214–216.

Tiwari, K.N. (2002). Rice Production and Nutrient Management in India, *Better Crops International* Vol. 16, Special Supplement, May 2002, pp. 18–22.

Torres, C.Q. and P.S. Teng (1991). Using single hills to determine an equation for estimating yield loss caused by rice blast. *International Rice Research Newsletter*, 16: 5, 19–20.

Tripathi, P., R.S. Mishra and S.K. Tomar (1998). Effect of temperature and relative humidity on the infection of rice blast (*Pyricularia oryzae*). *Ann. Plant Prot. Sci.*, 6: 1, 22–24.

Tripathi, S.K., R.P. Patel and K.L. Tiwari (1997). Influence of weather factors on rice blast (*Pyricularia oryzae* Cav.) and its chemical control. *Adv. Plant Sci.*, 10: 2, 89–93.

Tsuda, M.H., Y. Takahi, K. Tanizawa, A. Kunitomo, Y. Tsnjino, H. Tanaka, T. Miura and P. Karto (1998). Rice blast control with a throw in packed formulation containing pyroquifon. *J. Pesticide Sci.*, 23: 230–234.

Vidhyasekharan, P., R. Rabindran, M. Muthamilan, K. Nayar, K. Rajappan, N. Subramanian and K. Vasumathi (1997). Development of a powder formulation of *Pseudomonas fluorescens* for control of rice blast. *Plant Pathology*, 46: 291–297.

Xu, H.G., C.X. Zhu, X.M. Tong, Y. Luo and F.K. Xu (1990). The relationship between the airborne spore load of *Pyricularia oryzae* and *P. grisea* to the field intensity of rice blast disease. *Acta Agriculturae Universitatis Zhejiangensis*, 16(2): 107–112.

Yaegashi, H., T. Higashi and H. Koga (1990). Varietal resistance to blast: A Japanese Perspective. In: *Int. Rice Res. Conf. Seoul*, S. Korea, p. 13.

25

Chitinases in Defense Against Phytopathogenic Fungi

V. Shanmugam

Hill Area Tea Science Division, Institute of Himalayan Bioresource Technology, Palampur – 176 061, Himachal Pradesh, India

Plant pathogen interactions are highly complex involving series of reactions in disease development. Plants despite lacking a definite immune system of their own for protection against pathogen attack are endowed with other constitutive defense mechanisms in the form of defense genes. Plants accumulate chitinases (EC 3.2.1.14), a pathogenesis related (PR) protein, as part of self-defense mechanisms against biotic, and abiotic stresses to degrade the cell wall components of fungal pathogens (Lamb *et al.*, 1989, Bowles, 1990). These chitinases inhibit fungal growth both, directly and indirectly by acting on chitin, a homopolymer of β-1,4-N-acetyl-glucosamine (Deshpande, 1986), and an important component of fungal cell walls (excluding oomycetes), and arthropods or nematode exoskeletons (Boiler, 1988) but not found in higher plants (Gooday, 1990a). Chitinases catalyse the hydrolysis of β-1-4 linkages of the polymer into N-acetyl-chitooligosaccharides into oligomers, which are then rapidly hydrolysed into the dimer, chitobiose with retention of configuration (Roberts and Selitrennikoff, 1988; Schlumbaum *et al.*, 1986). During the process, these chitinases release elicitors from the fungal cell walls and these elicitors induce various defense responses in plants (Ren and West, 1992). It has been found that some plant chitinases may also display varying levels of lysozyme (EC.3.2.1.17) activity and confer resistance to bacterial pathogens (Jolles and Jolles, 1984).

Chitinases are present in all plants analyzed to date and many have been shown to inhibit fungal growth either *in vitro* or when expressed in transgenic plants (Hamel *et al.*, 1997). Naturally occurring or transgenic cultivars with increased pathogen resistance showed accelerated chitinase induction (Rasmussen *et al.*, 1992a) or increased expression of chitinases (Broglie *et al.*, 1991). Genes encoding these enzymes have been cloned and characterised from a variety of plant sources. Currently, there has been a considerable amount of recent research in delineating the molecular events from pathogen recognition to the expression of these genes and subsequent exploitation in disease management. The nature, and the role of chitinases in host defense mechanisms against phytopathogenic fungi are discussed.

Classification of Chitinases

Chitinases have been assigned to various classes, depending upon the primary structures (Collinge *et al.*, 1993) and each class of chitinases has its own substrate specificity (Brunner *et al.*, 1998). The acidic forms accumulate in apoplast while the basic forms accumulate in host cell vacuoles. However, a few acidic chitinases accumulate in vacuoles as well. Among the chitinases, basic forms alone have been shown to have fungitioxic action (Woloshuk *et al.*, 1991). Lawrence *et al.* (1996) observed that a basic chitinase purified from *Alternaria salani* inoculated tomato plants effectively inhibited the pathogen *in vitro* and virtually no inhibitory effect was exhibited by an acidic chitinase.

The basic class I chitinases contain N-terminal cysteine-rich domain with putative chitin-binding properties linked, by a variable proline rich hinge region, to a highly conserved catalytic domain. Most of these display a C-terminal extension required for vacuolar localization of the protein (Di Sansebastiano *et al.*, 1998). Basic forms of chitinases from tobacco, potato, bean, *Arabidopsis*, and rice (Vidhyasekaran, 1997) belong to class I and are primarily located in the central vacuole. Class II chitinases are mainly acidic isoforms located in the apoplast and resemble class I chitinase, but lack the N-terminal cysteine-rich domain. This class includes chitinases from tobacco, petunia, barley, rice, and potato (Vidhyasekaran *et al.*, 1997; Kim *et al.*, 1998, and Buchter *et al.*, 1997). Class IV chitinases display four deletions in the catalytic domain also sharing high homology with the class I chitinase, and are extra cellular. Chitinases from *Arabidopsis thaliana* (Gerhardt *et al.*, 1997), grape berries (Robinson *et al.*, 1997), and bean (Lange *et al.*, 1996) belong to class IV. Class V chitinases are characterised by the duplication of the N-terminal chitin-binding domain. In contrast, class III chitinases located apoplastically is structurally unrelated to the other types of plant chitinase possessing a domain with weak similarity to prokaryotic chitinases (Watanabe *et al.*, 1993), and display lysozyme activity (Bruener *et al.*, 1998). Acidic chitinases from cucumber, tobacco, chickpea, *A. thaliana*, and soybean lacking the cysteine-rich domain are class III chitinases (Vidhyasekaran, 1997; Horita *et al.*, 1996). Basic class III chitinase structurally unrelated to acidic class III chitinase was reported in tobacco (Ward *et al.*, 1991) and *Trichosanthes kirilowii* Maxim (Shih *et al.*, 1997).Finally, class VI chitinases are similar to bacterial chitinases for *e.g.*, *Bacillus circulans*, *Serratia marcescens* or *Streptomyces plicatus* (Melchers *et al.*, 1994).

While class I, II and IV belong to PR–3 protein, class III belong to PR–8, and class V belong to PR–11 group of protein. The recognized families of chitinases along with their properties are highlighted in Table 25.1. Later, to correspond to the nomenclature of PR–proteins and to distinguish the classes from families, Neuhaus *et al.* (1996) revised the nomenclature for chitinase genes into four families, two of which are further, divided in classes. Those from the PR–3 family are designated as Chi a, PR–8 as Chi b, PR–11 as Chi c and PR–4 as Chi d. Chitinases of both acidic and basic isoforms have been reported, in many, if not in all of them. Hamel *et al.* (1997) studied the evolutionary relationships between chitinase classes by comparing the nucleotide sequences of the catalytic domains of chitinases from various angiosperms. Class I and IV form discrete and well-supported monophyletic groups derived from a common ancestral sequence that predates the divergence of dicots and monocots. In contrast, other sequences included in class I and II, lacking one or both types of auxiliary domains, were nested within class I sequences, indicating that they have a polyphyletic origin, Based on phylogenetic analyses and the calculation of evolutionary rates, it is concluded that these chitinases probably arose from different class I lineages by relatively recent, deletion events.

Table 25.1: Recognized Families of Plant Chitinases

Class	*Groups*	*Properties*
I	PR–3	Basic, antifungal
II	PR–3	Mainly acidic, antifungal
III	PR–8	Both acidic and basic, Lysozyme activity
IV	PR–3	Antifungal
V	PR–11	Antifungal

Chitinases as Effectors of Defense

Constitutive Chitinases

Grassman *et al.* (1934) first reported chitinases in extracts of sweet almond (*Prunus amygdalus*). Since then chitinases have been reported in several plants tissues (reviewed by Vidhyasekaran, 1997). Of late, chitinases have also been purified and characterised in gladiolus bulbs (Yamagami *et al.*, 1997), papaya latex (Azarkan *et al.*, 1997), bean seeds (Ramos *et al.*, 1998) and all parts of sweet potato (Hou *et al.*, 1998). Martin (1991) reported constitutive chitinases during specific steps of plant development. In *Brassica napus*, the synthesis of a class I chitinase transcript was regulated during development and was also induced in roots by wounding and ethephon (Hamel and Bellemare, 1995). Addition of a purified bean chitinase to actively growing cultures of *Rhizoctonia solani*, the rice sheath blight pathogen, developed inhibition zone at the point of application (Broglie *et al.*, 1991). In an *in vitro* assay, a chitinase (20 kDa) and a β-1,3 glucanase (26 kDa) isolated from cowpea seeds were shown to deter the development of *Colletotrichum lindemuthianum* and *Colletotrichum musae* causing anthracnose diseases (Gomes *et al.*, 1996). Chitinases purified from sorghum seeds exhibited high antifungal activity against *Fusarium moniliforme*, *Curvularia lunata* and *Aspergillus flavus* in combination with other antifungal proteins by *in vitro* assays (Seetharaman *et al.*, 1997). Taira *et al.* (2002) studied the antifungal activities of rye seed chitinases, RSC–a (class I) and RSC-c (class II) against *Trichoderma* sp., with the cell walls prepared from its mycelia. The chitin-binding domain in RSC-a assists in inhibiting the. fungal growth more effectively by binding the lateral walls and septa, consisting of mature cell walls, and degrading mature chitin fibre, while RSC-c binds only to the hyphal tip followed by degradation of only nascent chitin.

Induction and Expression of Chitinases

Pathogen Infection

Chitinase induction in plants is one of a coordinated, often complex, and multifaceted defense mechanism. Chitinases purified from bean, induced upon challenge inoculation with *Uromyces phaseoli*, readily inhibited the growth of *Trichoderma viride* even at 2 μg/ml (Schlumbaum *et al.*, 1986). Tomato plants inoculated with Tobacco Necrosis Virus (TNV) exhibited systemic acquired resistance (SAR) against *Phytophthora infestans* inoculated subsequently (Anfoka and Buchenauer, 1997). Potato leaves infected by *P. infestans* or treatment with fungal. elicitor leads to a strong increase in chitinase activity.

The induction occurred locally at the site of infection, and systemically in upper uninfected leaves (Buchter *et al.*, 1997). A combination of chitinases and β-1,3 glucanases isolated from cucumber leaves infected with *Colletotrichum lagenarium* caused lysis of hyphal tips and abnormal growth of the pathogen (Ji and Kuc, 1996).

Differential expression of a basic class I (*VCHIT 1b*), and class III (*VCH3*) chitinase was recorded in grapevine tissues in response to fungal challenge, and commercial systemic acquired resistance (SAR) activators. While both commercial activators, and *Plasmopara viticola* inoculation exclusively enhanced *VCH3*, leaf inoculation with spore suspension of the pathogen alone systemically induced the protein in susceptible, and locally in resistant plants (Busam *et al.*, 1997). Similarly, Robert *et al.* (2002) observed the induction of an extra cellular class III chitinase (*Vvchit 1b*) alone in unripe berries infected with *Plasmopara viticola*.

Enhanced chitinase activities upon pathogen infection have also been reported in *B. napus* (L.) sp. *oleifera* infected by *Phoma lingam* (Rasmussen *et al.*, 1992b); cotton seeds infected with *Verlicillium dahliae* (Liu *et al.*, 1995), bean leaves infected with *C. lindemuthianum*, carrot infected with *Pythium violae* (Zamski and Peretz, 1996); chestnut stems infected by *Cryphonectria parasitica* (Murr.) Barr (Schafleitner and Wilhelm, 1997); populus infected with *Dothiorella gregaria* (Hu *et al.*, 1997), and wheat infected with scab (Li *et al.*, 2001).

Elicitors

Hosts cells sense or recognize the presence of microorganisms, irrespective of whether they are compatible or incompatible pathogens, and produce defense chemicals to suppress them. Most of the fungal pathogens, whether compatible or not, are known to have elicitors, particularly in their cell walls to elicit defense mechanisms in plants. Contact of the fungal cell wall with plant cell enables release of elicitors to activate defense mechanisms.

Transcripts of a rice chitinase (*RCH10*) accumulate to a high level in roots of suspension-cultured cells, when treated with *Phytophthora megasperma* pv. *gycinea* elicitors (Zhu and Lamb, 1991). Xu *et al.* (1996) observed that transcripts of a basic chitinase gene, *RC24* rapidly accumulated within 1 h after fungal elicitor treatment of suspension-cultured cells, and the levels continued to increase for at least 9 h. *RC24* transcript accumulation was also observed in intact leaf tissues upon wounding. Similarly, dramatic increase of chitinase transcript level occurred by treatment with a combination of glycol chitin and fungal elicitor derived from *Magnoporthe grisea* (Kim *et al.*, 1998).

Treatment of tobacco cell culture with heat-released soluble cell-wall elicitors from *Phytophthora nicotiana* cell walls induced a multicomponent dynamic system including chitinases and glucanases with different protective mechanisms having complementary roles in the overall expression of the defense response (Oelofse and Dubery, 1996). Similar type of induction was reported in cotton leaf discs elicited by *Verlicillium dahliae* derived signal molecules (Dubery and Slater, 1997). Suspension-cultured white lupin (*Lupinus albus*) cells treated with fungal elicitor preparations obtained from purified cell walls of yeast and *C. lindemuthianum* showed enhanced chitinase, polygalacturonase inhibiting protein, and germin/oxalate oxidase activities (Wojtaszek *et al.*, 1997). Similar type of chitinase induction in cell suspension cultures upon treatment with elicitors has also been reported in sunflower (Siefert and Grossmann, 1997), and *R. solani* (Velazhahan *et al.*, 2000) respectively.

In some cases, cells directly challenged or stimulated by pathogen derived elicitors release diffusible secondary signal molecules that orchestrate the induction of complementary defense responses in neighbouring cells. Chappell *et al.* (1997) observed that treatment of tobacco cell- suspension cultures with an elicitin protein from *Phytophthora crytogea* resulted in the release of a diffusible factor capable of inducing rapid accumulation of transcripts encoding chitinase, and other defense proteins against direct treatment of the cultures with pure elicitin protein.

The action of endogenous suppressor in host plants on the activation of PR-proteins and on the production of phytoalexins is species-specific. The concomitant presence of the endogenous suppressor in host plants to the fungal elicitor may result in the suppression of enzyme activation. Nasu *et al.* (1995) reported an endogenous suppressor in healthy pea plants quite similar to the exogenous suppressor from *Mycosphaerella pinodes* with respect to its biological activities, and suppress the accumulation of elicitor-induced pisatin and chitinase. In contrast, the endogenous suppressor alone activated these enzymes in *P. vulgaris* and soybeans to which M. *pinodes* is non-pathogenic.

Elicitation by Abiotic Agents

Chitinase induction in plants is not considered solely as an antifungal resistance mechanism. Plant chitinases can be induced by various abiotic factors as well and there is some circumstantial evidence to suggest a morphogenetic role despite the apparent absence of the substrate in plant cells (reviewed by Sahai and Manocha, 1993). Chitinase activity increased to several folds in plants upon treatment with ethylene (Broglie *et al.*, 1986; Boller, 1988; Botha *et al.*, 1998), salicylic acid (Derckel *et al.*, 1996), ethephon (Jiang and Jiang, 1997), and mercuric chloride (Kim *et al.*, 1996). Similarly, wounding of host tissues also accumulated chitinases (Van Pelt-Heerschap and Smit-Bakker, 1999). However, evidences for the role of such induced chitinases in plant defenses have been demonstrated only in selected cases. Spraying β-amino-n-butyric acid strongly protected pepper plants against *Phytophthora capsici* by inducing synthesis, and accumulation of β-1,3 glucanases and chitinases in the stem tissues (Hwang *et al.*, 1997). Cucumber hypocotyls treated with 2,6-dichloro isonicotinic acid acquired resistance against C. *lagenarium* infection due to enhanced chitinase production. The protein was expressed strongly only in resistant hypocotyls soon after germination of the fungal spores (Kastner *et al.*, 1998). Extracts of *Lychnis viscaria* L. seeds containing brassinosteroid, enhanced the levels of different PR-proteins including chitinases in cucumber, upon challenge inoculation with the powdery mildew pathogen (Roth, 2000). Ben-Shalom *et al.* (2002) observed that spraying cucumber and bean plants with chitin and chitosan oligomers or with chitosan polymers having 5 or 30 per cent degree acetylation in general elicited the induction of chitinase as well. Increasing the degrees of acetylation of chitosan from 5 to 30 per cent appreciably increased the elicitation of the PR-proteins.

Elicitation by Plant Growth Promoting Rhizobacteria

Plant growth promoting rhizobacteria (PGPR) are known to induce systemic resistance (ISR) in different crops against plant pathogens (Van Loon *et al.*, 1998). Viswanathan *et al.* (2000) attributed the induction of chitinases as one of the mechanisms of ISR against the pathogens. Gooday (1990a) reported *Pseudomonas* spp., *Bacillus* spp., and *Streptomyces* spp. as chitinolytic. Seed inoculation of chickpea with either *Pseudomonas fluorescens* or *Bacillus* spp. or application of cell free extracts on seeds significantly inhibited the growth of *Fusarium oxysporum* f. sp. *ciceris*, and *Rhizoctonia bataticola* causing wilt. The disease suppressive activity was mainly due to the production of chitinase, glucanase, and siderophores (Khot *et al.*, 1996). The level of antagonism exhibited by P. *fluorescens* strains against *R. solani* is positively correlated with chitinase production (Velazhahan *et al.*, 1999). Application of *P. fluorescens* strains exhibited chitinase induction in sheath blight susceptible rice cultivar (IR 50) upon challenge inoculation with *R. solani* and the induced chitinase was found to be associated with Induced Systemic Resistance (ISR) (Nandakumar *et al.*, 2001). Nandakumar *et al.* (2002) demonstrated the antifungal activity of a 35-kDa chitinase induced by P. *fluorescens* in rice plants against mycelium and sclerotia of *R. solani*. The production of antibiotics, antifungal compounds, bacterial lytic enzymes and the induction of plant chitinase were correlated with antifungal activity against *R. solani* sclerotia. Similar type of induction due to *P. fluorescens* was reported in ragi upon inoculation with M. *grisea* and the treated plants exhibited reduced blast incidence under glasshouse conditions (Ramanathan *et al.*, 2002) (Table 25.2). Inoculation of cotton and bean seedlings with *Bacillus cereus* or *Bacillus pumilus* or *B. subtilis* reduced the disease indices by more than 50 per cent (Pleban *et al.*, 1995). β-1,3 glucanase and chitinase activities were induced locally and systemically after spraying coffee leaves with an aqueous suspension of *Bacillus thuringiensis.* The treatment induced local and systematic resistance to *Hemileia vastatrix* (Guzzo and Martins, 1996).

Table 25.2: Effect of Foliar Application of *P. fluorescens* on Induction of Chitinase

Treatments	*μ mol Glu Na $h^{-1}g^{-1}$ Fresh Weight (Days after treatment)*			
	0	*2*	*4*	*6*
PF 1	0.586g	2.321f	3.826g	3.321bc
PF 2	0.564f	2.290e	3.541f	3.520cd
FP 7	0.533e	2.295e	3.48ge	3.418d
PFKO 2	0.502d	2.228d	3.365d	3.242bc
FP 1	0.479b	2.208c	2.333b	2.995b
Carbendazim	0.490c	2.121b	2.987c	2.928b
Control	0.253a	0.260a	0.226a	0.222a

Data followed by the same letter in a column are not significantly different from each other according to Duncan's multiple range test (P=0.05).

[*Source*: Ramanathan *et al.* (2002)].

The tissue specific expression during development and consistent localization in the apoplast as well as in the vacuolar compartment and their differential induction by endogenous and exogenous signaling compounds suggests that chitinases may have important functions extending beyond their apparently limited role in plant defense.

Chitinases in Disease Resistance

Chitinase induction seems to be tissue-specific and developmentally regulated (Gao and Gao, 1996) and the magnitude of induction depends upon the host tissues and its susceptibility. In order to determine the relationship between resistance genes and hydrolytic enzyme induction, Sayler *et al.* (1995) analysed the interaction between 2 bean cultivars (Sierra and Olathe) and 2 *Uromyces appendiculatus* races (47 and 55). Resistance of Sierra to race 55 was expressed as hypersensitive flecking, whereas the resistance of Olathe to race 47 was not associated with macroscopic symptoms and was classed as immunity. The Sierra/race 47 and Olathe/race 55 interactions were compatible in which the hydrolytic enzyme activities increased only slightly over time. In contrast, the β-1,3-glucanase and chitinase activity was significantly greater for the incompatible interactions. Thus, induction of hydrolytic enzyme activity may be associated with hypersensitive, but not immune resistance in the bean/bean rust interaction.

Tomato leaves infected with *A. solani* exhibited higher chitinase activity compared to uninfected ones, and the increase being greater in resistant than in susceptible plants (Fernandez *et al.*, 1995). High level of constitutive chitinase and β-1,3 glucanase activities were detected in pepper (*Capsicum annuum* L.) leaves resistant to P. *capsici* (Mozzetti *et al.*, 1995).

Sugar beet cultivars tolerant to *Cercospora beticola* accumulated a basic class IV chitinase faster than a susceptible one and the protein was spatially distributed over a comparatively larger area. The symptoms appeared a few days later, and the number of necrosis was reduced, and, smaller than the sensitive cultivar (Nielsen *et al.*, 1996). Chitinase activity pattern was similar in compatible and incompatible interactions of *Puccinia striiformis* f.sp. *tritici* on wheat, up to 72 h after inoculation. However, the specific activity increased rapidly in the incompatible reaction thereafter. In susceptible reaction, chitinase activity gradually declined after 72 h post-inoculation, reaching a level similar to that in the control plants two weeks after inoculation (Mojtaba *et al.*, 2002).

Kang and Buchenauer (2002) studied the involvement of β-1,3-glucanase and chitinase accumulation in the *Fusarium culmorum-infected* wheat spike for resistance to pathogen spread in the host tissue. Sub cellular localization of β-1,3-glucanase and acidic chitinase by immunogold labelling technique in uninoculated healthy and infected spike demonstrated similar distribution of the enzymes. The enzymes were localized mainly in the cell walls of different tissues, while the cytoplasm and organelles of cells in these tissues showed almost no labelling. However, the accumulation of β-1,3-glucanase and chitinase in the infected wheat spikes differed distinctly between resistant and susceptible wheat cultivars. The labelling densities for the two enzymes increased only slightly as compared to the corresponding uninoculated healthy tissues, whereas higher labelling densities were found in the infected tissues of wheat spikes from the resistant cultivar over uninoculated healthy tissues. Furthermore, the labelling of β-1,3-glucanase and chitinase also occurred over the cell walls of the hyphae in the infected wheat spike, but not over the hyphal cytoplasm. In addition, labelling for the two enzymes was often detected over the cell wall appositions and the electron-dense material located between the host cell and the hyphal cell in the infected tissues of the resistant wheat cultivar.

In a study on the expression of various chitinase homologues in response to pathogen- associated signals in members of the genus *Pinus*, Davis *et al.* (2002) separated a putative extra cellular class II chitinase (*psch4*) secreted into liquid medium by pine cells and by transgenic tobacco cells that ectopically expressed *pschi4* was separated by isoelectric focusing. The *pschi4* transcript and protein accumulated in pine seedlings challenged with the necrotrophic pathogen *Fusarium subglutinans* f.sp. *pini*, with the protein reaching detectable levels in susceptible seedlings concomitant with the onset of visible disease symptoms. Additional chitinase transcripts assigned to classes I and IV based on primary sequence analysis were also induced by pathogen challenge. Jasmonic acid induced class I and class IV but not class II chitinase, whereas salicylic acid induced all three classes of chitinase. These results show that multiple chitinase homologues are induced after challenge by a necrotrophlc pathogen and by potential signalling molecules identified in angiosperms.

The possibility of binding actin filaments by chitinases to cause cytoplasmic aggregation was observed in resistant reactions of potato against incompatible race of *P. infestans*, or after treatment with hyphal wall components prepared from *P. infestans* (T akemoto *et al.*, 1997).Spatial localization of chitinases and N-acetyl-D-glucosamine oligomers in host plants were related to the compatibility of pathogens. Upon infection with *Piptocephalis virginiana*, chitinases occurred in cytoplasmic vacuoles of the uninfected susceptible host, *Choanephora cucurbitarum* and in cell walls and vacuoles of the resistant host, *Phascolomyces articulosus*. In both the hosts, infection by the pathogen destroyed vacuolar integrity and dispersed chitin into the cytoplasm (Manocha and Zhonghua, 1997).

An early induction of gene transcription encoding apoplastic chitinase and β-1,3-glucanase, and their accumulation in tomato leaves inoculated with *Cladosporium fulvum* was correlated with resistance. Injection of purified race specific elicitors, *A VR4* and *A VR9* induced differential expression of acidic chitinase and β-1,3 glucanase in the tomato genotypes *Cf4* and *Cf9* respectively. The induction, observed most frequently in resistant genotypes, corresponds with differences in gene expression previously observed for *C. fulvum-tomato* interactions (Wubben *et al.*, 1996).

Positive correlation between chitinase induction, and disease resistance has also been reported in sugarcane against *Puccinia melanocephala* (Diaz *et al.*, 1996), in chickpea against *Ascochyta rabei* (Nehra *et al.*, 1997), and in tomato against *Fusarium* wilt (Fuchs *et al.*, 1997). However, the reduction of induced resistance at times when peroxidase and chitinase activities are high also suggests that the activities of these enzymes at the time of challenge are not major factors for the restriction of *C. lagenarium* in induced cucumber plants (Dalisay and Kuc, 1995).

Differential Response of Host Plants Against Pathogenic and Symbiotic Interaction

Crop plants show differential responses during pathogenic and symbiotic interactions. Dassi *et al.* (1996) studied chitinases from endomycorrhiza-resistant mutant, and wild type pea (*Pisum sativum* cv. Frisson) to characterise plant hydrolases specific to pathogenic *Aphanomyces euteiches* and *Thielaviopsis basicola* or mycorrhizal (*Glomus mosseae*) root interactions. Stimulation of constitutive, and induction of new chitinase activities was detected for acidic proteins in both pea genotypes inoculated with pathogenic fungi. However, a different additional chitinase, isoform was induced in *G. mosseae*-colonized roots. The isoform was also not elicited in chemically stressed roots, confirming its mycorrhizal specificity. On the contrary, Salzer *et al.* (1997) observed during ectomycorrhizal formation of the fungi, *Hebeloma crustuliniforme, Amanita muscaria and Suillus variegatus* on *Picea abies* cells, only some of the chitin-derived elicitors reach their receptors at the plant plasma membrane, initiating hypersensitive response in the host cells. The remaining fungal elicitors were degraded to varying extents by wall-localized chitinases of the host root, reducing the defense reactions of the plant and allowing symbiotic interactions of both organisms. Mohr *et al.* (1998) investigated the responses of bean roots to infection by the arbuscular mycorrhizal fungus *G. mossae* and the pathogen *Fusarium solani* f. sp. *phaseoli*. While the pathogen changed chitinase activities even at the level of transcription, little change was detected in the symbiotic interaction.

The Structural and Functional Basis of Chitinase-chitin Interaction

Chitinases inhibit fungal growth by hydrolyzing the chitin component of cell walls to cause lysis (Roberts and Selitrennikoff, 1998; Schlumbaum *et al.*, 1986) rather than binding the chitin or release elicitors from the fungal cell walls by their enzymatic action to induce defense responses in plants (Ren and West, 1992). The hydrolysis of chitin is accomplished randomly (endochitinase) giving diacetylchitobiose as the major product (Robbins *et al.*, 1988) with preferential action on the dimmers (N-acetyl glucosamini.dases) (Gooday, 1990b), and progressive release of diacetylchitobiose units from the non-reducing ends of chitin chains (exochitinase). Roberts and Selitrennikoff (1988) observed that endochitinases had a stronger effect on hyphal extension than exochitinases. Wurms *et al.* (1997) assessed the activity of chitinase in kiwifruit infected with *Botrytis cinerea*, and concluded that whilst endochitinase activity originates from plant, exochitinase is associated with the pathogen.

Chitin oligomers generated upon hydrolyzing fungal cell walls by endochitinase can induce defense responses or related cellular responses in many monocots and some dicots (Ishihara *et al.*, 1996; Minami *et al.*, 1996). Doernenburg and Knorr (1994) induced chitinases and lysozyme activities in *Morinda citrifolia* cell cultures by supplementing a water-soluble chitin derivative. Hence, it is probable that breakdown products of fungal chitin are capable of inducing chitinases *in planta* during pathogenesis. In other words, pathogens themselves could be inducing resistance *in planta* against themselves. In rye seed chitinase, the N-terminal cysteine-rich chitin-binding domain and a catalytic domain are connected by a flexible linker and it was suggested that the chitin binding domain increases the hydrolytic action of catalytic domain towards inducible chitin derivatives by binding to them (Yamagamai and Funatsu, 1997).

Yamada *et al.* (1993) showed induction of phytoalexin biosynthesis by purified chitin fragments in suspension cultured rice cells even at very low (nM) concentrations. Inui *et al.* (1997) reported induction of class III chitinase in rice suspension cultures on treatment with chitin oligosaccharides. The structural requirement of the chitin fragment for the induction of various cellular responses was found to be the same suggesting that all of these responses may be mediated through a single receptor. The presence of such receptor was reported in rice microsomal (Shibuya *et al.*, 1993), as well as plasma membrane preparation (Shibuya *et al.*, 1996) of suspension cultured rice cells. Affinity labeling of plasma membrane preparation and. autoradiography study was used to identify the receptor ligand. The presence of a 75-kDa protein corresponds well with those of the high affinity-binding site detected in the plasma membrane. The 75-kDa binding protein, a putative receptor molecule for chitin oligosaccharides elicitor has been purified. Cloning of the corresponding gene and the generation of transgenic plants containing such a gene will clarify whether the 76-kDa protein really functions as a receptor for the elicitor. The rapid and transient natures of some of these cellular responses suggest their involvement in the signal transduction cascade (Viswanathan *et al.*, 2000).

Surface modification of fungal cell walls by chitin de-N-acetylation is discussed as a fungal strategy to protect cell walls of pathogenic hyphae from enzymatic hydrolysis by host chitinases, and to avoid generation of an autocatalytic defense response system in the invaded host tissue. In a study on the conversion of surface-exposed chitin to chitosan in cell walls of *in vitra*, and *in viva* differentiated infection structures of two rust fungi, the wheat stem rust fungus *Puccinia graminis* f.sp. *tritici*, and the broad bean rust fungus *Uromyces viciae-fabae*, and the causal agent of maize anthracnose, *Cottetotrichum graminicota*, fiuorescence-labelled lectin wheat germ agglutinin (WGA) revealed that surfaces of infection structures formed on the plant cuticle expose chitin, whereas surfaces of structures formed after invading the host do not. Antibodies recognizing de-N-acetylated chitosan labeled only those infection structures that differentiate inside the plant, indicating that chitosan is exposed on cell wall surfaces post penetration (EI-Gueddari *et al.*, 2002).

Cloning and Characterisation of Chitinase Genes

As an approach to identifying means for possible enhancement of fungal resistance in crop plants through the use of genetic engineering technology, cDNAs corresponding to chitinase genes have been isolated from plants. Chitinase cDNA clones were isolated from bean by antibody screening of a λggt11 cDNA library containing sequences complementary to poly (A) + RNA from elicited cells. The identity of these clones was confirmed by analyzing the nucleotide sequence and by comparing the deduced amino acid sequence with the amino terminal sequence of bean chitinase (Susan *et al.*, 1988). Genes encoding chitinases were identified and characterised in rice (Zhu and Lamb, 1991; Nishizawa and Hibi, 1991, Zhu *et al.*, 1994). Li *et al.* (1997) analysed a basic chitinase genomic clone, *RCHB* isolated from the rice cv. IR36 genomic library using a bean chitinase gene fragment as a probe. The deduced amino acid sequence shares identity with other three basic chitinase genes reported in rice. However, a chitinase cDNA clone encoding a class I, chitinase with a 35 kDa molecular weight was isolated from *R. solani* infected IR58 rice plant (Anuratha *et al.*, 1996). The deduced amino acid sequence N as significantly different from that of other rice chitinases described so far.

Hudsperth *et al.* (1996) isolated cDNA clones representing mRNAs encoding chitinase and β-1,3 glucanase in cotton leaves. The chitinase clones encode a 28-kDa protein with 71 per cent amino acid sequence similarity to the *SK2* chitinase from potato. The chitinase is encoded by a small gene family of which two members *chi2;1* and *chi2;2* share 97 per cent sequence identity in their transcribed regions. The 5′ flanking regions of *chi2;1* and *chi2;2* exhibit a large degree of similarity, and may contain sequence for gene response to chemical agents and fungal attack.

A class III chitinase gene of strawberry encodes a polypeptide of 31.6 kDa and translation product of this gene has highest similarity to chitinases from rice, *Hevea brasitiensis* and *A, thatiana* (Khan *et al.*, 1999). A class III chitinase gene, *Chi3-P1*, cloned from pepper (*C. annuum*) encodes a chitinase of 295 amino acid residues with a molecular mass of 32 kDa and shows sequence similarities to the tobacco clone P29060 (*Nicotiana tabacum*). A comparison of *Chi3-P1* DNA sequences among five pepper-inbred lines revealed single nucleotide polymorphisms in the protein-coding region (Cheng *et al.*, 2002).

Ohnuma *et al.* (2002) purified a recombinant rye seed chitinase *RSC-c* (rRSC-c) after expression in *Escherichia coli* Origami (DE3). *rRSC-c* had almost the same chitinase activity toward glycolchitin and antifungal activity against *Trichoderma* sp. as the authentic RSC-c did. RSC-c mutants were subsequently constructed and characterised with respect to their chitinase and antifungal activities. Mutation of Glu67 to Gin completely abolished the chitinase activity and diminished the antifungal activity. Considerable decreases in both activities were observed in the mutations of Trp72, and Ser120 to Ala, and Glu89 to Gin.

Chitinase genes were also cloned from cucumber (Metraux *et al.*, 1989), potato (Gaynor, 1988), and tobacco (Payne *et al.*, 1990). In angiosperms, chitinases seem to be encoded by a relatively small number of genes since only a limited number of distinct sequences have been recovered, even from plant genomes subjected to extensive molecular survey (Hamel *et al.*, 1997). (Levorson and Chlan, 1997) analysed 86 chitinase sequences from 29 crop species and one populus hybrid, obtained from the on-line GenBank nucleotide database. The sequences were grouped into five families and no conserved nucleotide sequence existed between them to generate primers and probes to target chitinase genes of all families.

Transgenics Expressing Chitinases in Disease Resistance

The expression of PR-proteins in plants is correlated with resistance in a number of plant-pathogen interactions (Van Loon, 1997). It has been postulated that constitutive, high-level expression of chitinases in transgenic plants can enhance resistance to a variety of pathogens (reviewed by Datta *et al.*, 1999).

Lin *et al.* (1995) developed transgenic indica rice constitutively expressing a 1.1 kb rice-derived chimerical chitinase gene, *chi11*, with the 35S Cauliflower Mosaic Virus (CaMV) constitutive promoter. The transgenic chitinase is expressed in leaves, sheath and roots of *R. solani* inoculated and uninfected plants. The degrees of resistance displayed by the transgenic plants against sheath blight, upon challenge inoculation with *R. solani* is correlated to the level of chitinase expression. Subsequently, Xu *et al.* (1996) developed transgenic rice plants expressing a new basic chitinase gene (*RC24*) for tissue-specific expression. Transgenic japonica rice expressing a cDNA clone, *cht-2* and a genomic clone, *cht-3* of the rice chitinase gene was developed through *Agrobacterium* mediated gene transfer method under the control of 35S CaMV virus promoter. The transgenic plants constitutively overexpressed the chitinase against control, to exhibit higher resistance against the rice blast pathogen, *M. grisea,* in a non-race specific manner (Nishizawa *et al.*, 1999). Subsequently, Datta *et al.* (2001) introduced a PR 3 rice chitinase gene (*RC7*) isolated from *R. solani* infected rice plants, into indica rice cultivars by biolistic and PEG-mediated transformation system. The transformants synthesized different levels of chitinase proteins constitutively and progeny from the plants containing the chitinase gene showed different levels of enhanced resistance when challenged with *R. solani.*

Chai *et al.* (2002) developed transgenic creeping bentgrass (*Agrostis stolinifera* var. *palustris*) expressing a plasmid KYLX71-pHS2 with the cDNA clone (*hs2*) encoding a chitinase, from a triploid Dutch elm disease resistant American elm (*Ulmus americana* NPS-3-487). The transgenics exhibited resistant to *Rhizoctonia solani*, causing brown patch disease.

Kellmann *et al.* (1996) cloned and sequenced two class II chitinase genes, *A.h.Chi2;1* and *A.h.Chi2;2*, from peanut. In peanut cell suspension cultures, mRNA levels of *A.h.Chi2;2* increased after ethylene or salicylate treatment and in the presence of conidia from *B. cinerea*. In contrast, *A.h.Chi2;1* was only expressed after treatment with the fungal spores. Transgenic tobacco plants containing the complete peanut *A.h.Chi2;1* gene exhibited essentially the same expression pattern in leaves as observed in peanut cell cultures. Transcriptional fusion of a 1.2 kb 5'-upstream region of *A.h.Chi2;1* to the GUS reporter gene conferred expression of GUS in leaves of transgenic tobacco plants after treatment with *P. megasperma* elicitor or *B. cinerea* spores.

Grison *et al.* (1996) introduced a hybrid endochitinase gene under a constitutive promoter by *Agrobacterium-mediated* transformation into a winter-type oilseed rape inbred line. Progeny from transformed plants was challenged with 3 different fungal pathogens *viz.*, *Pyrenopeziza brassicae*, *Phoma lingam* and *Sclerotinia sclerotiorum* in field trials at 2 different geographical locations. These plants exhibited an increased tolerance to disease as compared with the non-transgenic parental plants.

Transgenic cucumber plants expressing a rice chitinase cDNA (*RCC2*) under the control of the CaMV 35S promoter exhibited higher resistance against *B. cinerea* than controls and completely inhibited disease spread. The gene inherited in a Mendelian manner (Tabei *et al.*, 1997). Later, Kishimoto *et al.* (2002) selected three transgenic lines representing different resistance levels, namely the most resistant line (CR32), an intermediate-resistance line (CR3), and a susceptible line (CR20) and studied for the correlation between *RCC2* expression and disease resistance. ELISA analysis showed that the rice chitinase levels of *CR32* and *CR3* were -higher than that of *CR20* with *CR32* expressing homogeneous distribution. Rice chitinase was localized within cells but was not detected between cells. Optical microscopy of *B. cinerea* infection behaviour revealed that none of the transgenic lines nor any of the non-transgenic lines inhibited the penetration of *Botrytis cinerea*, However, fungal growth within leaf tissue was suppressed in the resistant lines but not in the non-transformed or *CR20* leaf tissue. Thus, high expression and intracellular localization of rice chitinase may be involved in enhancing the resistance of transgenic plants to gray mould.

Isolation of a *Brassica juncea* cDNA encoding *BjCHI1*, a novel chitinase with two chitin-binding domains, has been shown that its mRNA is induced by wounding and methyl jasmonate treatment, By the presence of two chitin-binding domains, *BjCHI1* resembles the precursor of UDA (*Urlica dioica* agglutinin) but, unlike UDA, *BjCHI1* retains its chitinase catalytic domain after post-translational processing. Transgenic tobacco plants were developed with a construct expressing the *BjCHI1* cDNA from the CaMV 358 promoter as well as *BjCHI2*, a deletion derivative of *BjCHI1* lacking one chitin-binding domain. Colorimetric chitinase assays showed no significant differences between the activities of *BjCHI1* and *BjCHI2*, suggesting that chitinase activity, due to the catalytic domain, is not enhanced by the presence of a second chitin-binding domain. Both *BjCHI1* and. *BjCHI2* showed *in vitro* antifungal activity toward *T. viride*, causing reductions in hyphal diameter, hyphal branching, and conidia size (Fung *et al.*, 2002).

Transgenic sugarbeets expressing a pumpkin chitinase gene developed by *Agrobacterium* mediated gene transformation were tested for enhanced resistance to *R. solani* causing seedling blight and root rot. Chitinase assays of protein extracts indicated an increased enzyme activity in the genetic transformants. An infection assay on leaves infected with *R. solani* showed that the disease symptoms were suppressed in some genetic transformants compared with the control plants (Hashimoto and Shimamoto, 2002). Transgenic silver birch (*Betula pendula*) expressing high levels of chitinase showed higher levels of resistance to *Pyrenopeziza betulicola* causing leaf spot, than transgenics with intermediate or low mRNA levels or a non-transgenic control plant (Pappinen *et al.*, 2002),

Problems Encountered in Chitinase Gene Expression

Fungal Inhibitor Protein

In certain cases, though purified chitinases inhibit the growth of saprophytic fungi, apparently no effect is observed on the growth of several phytopathogenic fungi, even though they have chitin in their mycelial walls. Purified chitinases from thorn-apple, tobacco, wheat and *Arabidopsis* inhibit the growth of several saprophytic fungi, but have no effect on the growth of several phytopathogenic fungi (Kyung *et al.*, 1997). Purified chitinases and β-1,3-glucanases from tomato do not inhibit the growth of the tomato fungal pathogen *Cladospoium fulvum*, although the enzymes do inhibit the growth of the fungus *T. viride*, which is not a pathogen (Joosten *et al.*, 1995). Constitutive chitinases purified from the intercellular fluids of uninoculated carnation tissues effectively inhibited the growth of *T. viride* but not *Fusarium oxysporum* f. sp. *dianthi* causing wilt. (Van Pelt-Heerschap and Smit-Bakker (1999). Furthermore, some fungi those are sensitive to β-1,3-glucanases and/or chitinase become resistant to these enzymes after several hours of exposure to them (Ludwig and Boiler, 1990), indicating an ability of the fungi to adapt to the proteins. Kyung *et al.* (1997) postulated the secretion of certain inhibitors by fungi against some or all plant chitinases and β-1,3-glucanases, just as plants produce inhibitor proteins against fungal enzymes that degrade plant cell walls.

Gene Silencing

It has been shown that constitutive over expression of chitinases did not lead to increased resistance of transgenic plants (Neuhaus *et al.*, 1991). Basic chitinases and β-1,3 glucanases from tomato showed high antifungal activity, but *C. fulvum*, the tomato pathogen, was insensitive to these PR proteins (Joosten *et al.*, 1995). Transgenic *Nicotiana sylvestris* plants expressing a tobacco basic vacuolar chitinase gene *chn48* under the control of the CaMV 35S promoter though accumulated high levels of chitinases in their leaves, showed only slight increase in resistance to *Cercospora nicotianae*, the frog eyespot pathogen (Neuhaus *et al.*, 1991). This may be possible due to adaptation of the fungus to the host defense mechanisms. Besides, the RNA expression signals may inactivate its own expression as well as expression of homologous host genes at the mRNA level. A comparison of the silencing of chitinase and β-1,3 glucanase transgenes with same expression signals, and breeding experiments with transgenes inserted in different chromosomes, showed that chitinase gene silencing depends on transcribed sequence homology and transgene dose. The *chn48* transgene silenced host class I but not class II chitinases, suggesting that greater than about 60 per cent identity of sequences encoding the catalytic domain is required for silencing. Although steady-state mRNA contents were greatly reduced in silent plants, the levels of nascent chitinase transcripts detected in nuclear run-on experiments were the same in nuclei from high-expressing and silent plants. This strongly suggests that chitinase gene silencing is a post-transcriptional phenomenon. Silencing and resetting to high-level expression of chitinase genes is developmentally regulated. Detailed analyses of the chitinase expression patterns indicate that the gene silencing is not a clonal event. It is likely that communication between cells rather than the lineage of the cells is important in stabilizing the silent state (Kunz *et al.*, 1996).

Stress Proteins

Chitinases on the other hand are also reported to be stress proteins induced upon infection, and not involved in host defense mechanisms. Several isoforms of chitinases accumulated in both compatible and incompatible interactions of pepper stem tissues inoculated with *P. capsici* (Kim and Huang, 1994) and soybean hypocotyls and leaves with *P. megasperma* f. sp. *glycinea* (Vi *et al.*, 1996).

Strategies Employed to Overcome Problems Associated with Chitinase Expressing Transgenics

Co-expression of PR-proteins

Several reports envisaged the inability of chitinase and β-1,3-glucanases from pea do not inhibit the growth of fungi, especially when used individually. Purified chitinases or β-1,3-glucanases from pea do not inhibit the growth in culture of most of the fungi tested when used individually, whereas a combination of these enzymes is effective in inhibiting the growth of most of those fungi (Mauch *et al.* 1988). Simultaneous expression of a tobacco class I chitinase and a class I β-1,3-glucanase gene in tomato resulted in increased fungal resistance, whereas transgenic tomato plants expressing either one of these genes were not protected against fungal infection (Jongedijk *et al.*, 1995). Offsprings of transgenic tobacco cultivars expressing chitinase and β-1,3-glucanase genes exhibited good resistance to *Colletotrichum* and *Phytophthora* (Li *et al.*, 2002).

Zareie *et al.* (2002) isolated and identified five major barley intercellular washing fluid (IWF) *viz.*, 32-kDa β-1,3-glucanase (Pr32), a 25-kDa chitinase (Pr25), and three 22-kDa thaumatin-like (TL) proteins (Pr22-1, Pr22-2, and Pr22-3) proteins based on N-terminal amino acid sequencing, peptide mapping, and determination of mass and isoelectric point. Various combinations of the five proteins had a greater effect on *Rhynchosporium secalis* spores than did the individual proteins. Anand *et al.* (20C3) evolved four fertile transgenic wheat lines, which stably and constitutively express one or both of the PR-protein genes, a chitinase and a glucanase. Of the various combinations of genes encoding PR-proteins tested, only one line expressing a rice thaumatin like protein (TLP) and a specific combination of a wheat chitinase and a wheat, glucanase had partial resistance to scab initiated by *Fusarium graminearum*, under greenhouse conditions. However, field trails of the same lines failed to show any improved resistance under heavy inoculum loads.

Expression of Non-host Chitinases

Transgenic plants harboring heterologous chitinase genes could not be overcome by the invading fungus (Lamb *et al.*, 1992). Broglie *et al.* (1991) evolved transgenic tobacco and *arassica napus* plants that constitutively expressed a bean vacuolar chitinase gene under the control of CaMV constitutive promoter. The transgenic tobacco plants exhibited resistance to *R. solani*, which was evident by means of reduced symptom development. A drought induced chitinase gene (*pcht28*) obtained from *Lycopersicon chilense* was transferred to cultivated tomato through *Agrobacterium* mediated gene transfer techniques. Both self-pollinated (R0) and (R1) seeds of the transgenic plants demonstrated high level of chitinase activity and exhibited significant resistance against race 1 and 2 of *V. dahliae* causing wilt disease even under field conditions (Zohreh *et al., 1997*).

Transgenic tobacco plants expressing *chiA* gene from *S. marcescens* showed enhanced tolerance to *R. solani* (Jach *et al.*, 1992). However, constitutive co-expression of chitinase and glucanase genes in transgenic tobacco showed enhanced protection against fungal attack (Zhu *et al.*, 1994). A chitinase gene *chi1* involved in cell autolysis of *R. oligosporus* was introduced into tobacco through *Agrobacterium* leaf disc system. Chitinase activity in young leaf extracts of the individual homozygous progeny, *Tch1-1* and *Tch2-1* expressing *chi* gene product was 3 to 4 fold higher than that from the control plants. Besides, disease symptoms incited by *S. sclerotiorum* and *B. cinerea* were remarkably suppressed (Terekawa *et al.*, 1997). Carstens *et al.* (2003) constitutively overexpressed a chitinase gene, *CTS1-2* from *Saccharomyces cerevisiae* in tobacco to assess for endogenous resistance against *B. cinerea*. Leaf extracts from the transgenic lines inhibited *B. cinerea* spore germination and hyphal growth by up to 70 per cent in a quantitative *in vitro* assay, leading to severe physical damage on the hyphae. Challenge inoculation of the F1 progeny lines with the pathogen, in a detached leaf infection assay, showed a decrease in susceptibility ranging from 50 to 70 per cent with higher chitinase activities.

Conclusion

Plant defense mechanisms due to chitinases though well established, its induction in plants can't be considered solely as an antifungal resistance mechanism. Since, many chitinases have been identified as enzymes, disappearance of the substrates from the infection court would suppress the action of these enzymes. Besides, phytopathogenic fungi may either adopt themselves to the defense response of host plants or suppress them by secreting inhibitor proteins. An early exclusion of chitin from the hyphal tip is a typical example of the method by which pathogens escape the fungitoxic action of host chitinases. Hence, it is necessary to understand the mechanisms by which fungal pathogens overcome chitinase induced defense mechanisms. Though genetic engineering offers an excellent opportunity to complement conventional breeding by enhancing and redirecting expression of chitinase genes through the use of appropriate promoters for further crop improvement, the efficacy can be influenced by the plant species used, the type of chitinase expressed, and the characterization of fungal pathogens. The level of protection observed in transgenic plants harboring heterologous chitinase to overcome gene silencing might vary, depending on the particular combination of enzyme, plant and pathogen tested. Hence, a proper understanding of the various classes of chitinases, their function, and mode of action is very important for managing modern biotechnology based crop improvement. Breeding for resistance in transgenic plants with a multigene combination of PR-proteins implying different modes of action, and with other antimicrobial genes effective against the target organisms may be a good strategy for achieving better and durable resistance to a range of diseases.

One of the most widespread mechanisms used by plants to resist phytopathogens is the expression of the so-called pathogenesis-related (PR) proteins. Among the PR–proteins, chitinases are of particular interest, because many pathogenic fungi contain chitin as major structural cell wall component. Chitinases enzymatically degrade chitin and elicit host defense responses through its degraded products. The antifungal activities of chitinases render these proteins ideal candidates, for the enhanced resistance of plants, through constitutive over expression of these normally inducible defense genes. While doing so, a proper understanding on the nature of chitinases, and adaptive response of fungal pathogens to the proteins is essential to encounter problems associated with chitinas expression. Co-expression of PR–proteins implying different modes of action against the target organisms, and expression of non-host chitinases might be a good strategy for achieving better and durable resistance to a range of diseases. The nature, and possible role of chitinases in plant defense mechanisms in bringing about an eventual disease resistance against phytopathogenic fungi are discussed.

Acknowledgements

The author is thankful to the Director, IHBT for extending necessary support and encouragement in preparing the article.

REFERENCES

Anand, A., T. Zhou, H.N. Trick, B.S. Gill, W.W. Bockus and S. Muthukrishnan (2003). Greenhouse and field-testing of transgenic wheat plants stably expressing genes for thaumatin-like protein, chitinase and glucanase against *Fusarium graminearum*. *J. Exp. Bot.*, 54: 1101-1111.

Anfoka, G. and H. Buchenauer. (1997). Systemic acquired resistance in tomato against *Phytophthora infestans* by pre-inoculation with tobacco necrosis virus. *Physiol. Mol. Plant Pathol.* 50: 85–101.

Anuratha, C.S., K.C. Zen and K.C. Cole. (1996). Induction of chitinases and β-1,3-glucanases in *Rhizoctonia solani* infected rice plants: isolation of an infection related chitinase cDNA clone. *Physiol. Plant*, 97: 39–46.

Ben-Shalom, N.C. Aki, R. Ardi and R. Pinto. (2002). Elicitation effects of chitin oligomers and chitosan sprayed on the leaves of cucumber (*Cucumis sativus*) and bean (*Phaseolus vulgaris*) plants. *Israel J. Plant Sci.*, 50: 199–206.

Boiler, T. (1988). Ethylene and the regulation of antifungal hydrolases in plants. *Plant Mol. Cell Biol.*, 5: 145–171.

Botha, A.M., Nagel, M.A.C., Westhuizen-AJ-van-der, F.C. Botha and A.J. van-der-Westhuizen (1998). Chitinase isoenlymes in near isogenic wheat lines challenged with Russian wheat aphid, exogenous ethylene, and mechanical wounding. *Bot. Bull. Academia Sinica*, 39: 99–106.

Bowles, D.J. (1990). Defense-related proteins in higher plants. *Ann. Rev. Biochem.*, 59: 873–907.

Broglie, K.E., J.J. Gaynor and R.M. Broglie (1986). Ethylene regulated gene expression: molecular cloning of the genes encoding an endochitinase from *Phaseolus vulgaris*. *Proc. Natl. Acad. Sci.*, USA, 83: 6820–6824.

Broglie, R., K. Broglie, I. Chet and R. Benhamou (1991). Transgenic plants with enhanced resistance to *Rhizoctonia solani.* Third International Congress of the International Society of Plant Molecular Biology, Tucson, Arizona, USA, October 5–11, pp. 1364.

Brunner, F., Stintzi, B. Fritig and M. Legrand (1998). Substrate specificities of tobacco chitinases. *Plant J.*, 14: 225–234.

Buchter, R., A. Stromberg, E. Schmeller and E. Kombrink (1997). Primary structure and expression of acidic (class II) chitinase in potato. *Plant Mol. Biol.*, 35: 749–761.

Busam, G., K.T. Junghanns, R.E. Kneusel, H.H. Kassemeyer and U. Matern (1997). Characterization and expression of caffeoyl-coenzyme a 3-O-methyltransferase proposed for the induced resistance response of *Vitis vinifera* L. *Plant Physiol.*, 115: 1039–1048.

Carstens, M., M.A. Viver and I.S. Pretorious (2003). The *Saccharomyces cerevisiae* chitinase, encoded by the CTS-1 gene, confers antifungal activity against *Botrytis cinerea* to transgenic tobacco. *Transgenic res.* PC1150: 1–12.

Chai, B., S.B. Maqbool, R.K. Hajela, D. Green, J.M.Jr. Vargas, D. Warkentin, R. Sabzikar and M.B. Sticklen (2002). Cloning of a chitinase-like cDNA (hs2), its transfer to creeping bentgrass (*Agrostis palustris* Huds.) and development of brown patch (*Rhizoctonia solam*) disease resistant transgenic lines. *Plant Sci.*, 163: 183–193.

Chappell, J.A., Levine, R. Tenhaken, M. Lusso and C. Lamb. (1997). Characterization of a diffusible signal capable of inducing defense gene expression in tobacco. *Plant Physiol.*, 113: 621–629.

Cheng, C.M., A. Palloix, V. Lefebvre and C.M. Cheng (2002). Isolation, mapping and characterization of allelic polymorphism of Chi3-P1, a class III chitinase of *Capsicum annuum* L. *Plant Sci.*, 163: 481–489.

Collinge, D.B., K.M. Kragh, J.D. Mikkelsen, K.K. Nielsen, U. Rasmussen and K. Vad (1993). Plant chitinases. *Plant J.*, 3: 31–40.

Dalisay, R.F. and J.A. Kuc (1995). Persistence of reduced penetration by *Colletotrichum lagenarium* into cucumber leaves with induced systemic resistance and its relation to enhanced peroxidase and chitinase activities. *Physiol. Mol. Plant Pathol.*, 47: 329–338.

Dassi, B., E. Dumas-Gaudot, A. Asselin, C. Richard and S. Gianinazzi (1996). Chitinase and β-1,3-glucanase isoforms expressed in pea roots inoculated with arbuscular mycorrhizal or pathogenic fungi. *European J. Plant Pathol.*, 102: 105–108.

Datta K., S. Muthukrishnan and S.K. Datta (1999). Expression and function of PR-proteins genes in transgenic plants. In: *Pathogenesis: Related Proteins in Plants,* (Eds.) Datta S.K. and S. Muthukrishnan, Boca Raton, FL: CRC Press, pp. 261–277.

Datta, K., J. Tu, N. Oliva, I. Ona, R. Velazhahan, T.W. Mew, S. Muthukrishnan and S.K. Datta (2001). Enhanced resistance to sheath blight by constitutive expression of infection–related rice chitinase in transgenic elite indica rice cultivars. *Plant Sci.*, 160: 405–414.

Davis, J.M., H.G. Wu, J.E.K. Cooke, J.M. Reed, K.S. Luce and C.H. Michler (2002). Pathogen challenge, salicylic acid, and jasmonic acid regulate expression of chitinase gene homologs in pine. *Mol. Plant Microbe Interactions* 15: 380–387.

Derckel, J.P., L. Legendre, J.C. Audran, B. Haye and B. Lambert (1996). Chitinases of the grapevine (*Vitis vinifera* L.): five isoforms induced in leaves by salicylic acid are constitutively expressed in other tissues. *Plant Sci. Limerick*, 119: 31–37.

Deshpande, M.V. (1986). Enzymatic degradation of chitin and its biological applications. *J. Sci. Indus. Res.* 45: 273–287

Di Sansebastiano, G.P., N. Paris, S. Marc-Martin and J.M. Neuhans (1998). Specific accumulation of GFP in a non-acidic vacuolar compartment via a C-terminal properptide-mediated sorting pathway. *Plant J.* 45: 449–457.

Diaz, M., S. Hernandez and F. Fric (1996). Enzymatic activities of chitinases and proteases in rust sugarcane interaction. *Revista de Proteccion Vegetal*, 11: 25–27.

Doemenburg, H. and D. Knorr (1994). Elicitation of chitinases and anthraquinones in *Morinda citrifolia* cell cultures. *Food Biotechnol.*, 8: 57–65.

Dubery, I.A. and V. Slater. (1997). Induced defense responses in cotton leaf disks by elicitors from. *Verticillium dahliae. Phytochemistry*, 44: 1429–1434.

El-Gueddari, N.E., U. Rauchhaus, B.M. Moerschbacher and H.B. Deising (2002). Developmentally regulated conversion of surface-exposed chitin to chitosan in cell walls of plant pathogenic fungi. *New Phytologist*, 156: 103–112.

Fernandez, A., E. Solorzano, B. Peteira and E. Fernandez (1995). Chitinase induction in tomato leaves with different degrees of susceptibility to *Alternaria solani. Revista-de-Proteccion-Vegetal*, 10: 259–264.

Fuchs, J.G., Y. Moenne-Loccoz and G. Defago. (1997). Nonpathogenic *Fusarium oxysporum* strain Fo47 induces resistance to *Fusarium* wilt in tomato. *Plant Dis.*, 81: 492–496.

Fung, K.L., K.J. Zhao, Z.M. He, M.L. Chye, K.L. Fung, K.J. Zhao, Z.M. He and M.L. Chye (2002). Tobacco-expressed *Brassica juncea* chitinase BjCHI1 shows antifungal activity *in vitro*. *Plant Mol. Biol.*, 50: 283–294.

Gao, B. and Gao, B.D. (1996). Molecular biology of plant chitinases. *J. Hunan Agric. Univ.*, 22: 587–602.

Gaynor, J.J. (1988). Primary structure of an endochitinase mRNA from *Solanum tuberosum*. *Nucl. Acids Res.* 16: 5210.

Gerhardt, L.B. de A., G. Sachetto-Martins, M.G. Contarini, M. Sandroni, R. de. P. Ferreira, V.M. de Lima, M.C. Cordeiro, D.E. de Oliveira, M. Margis-Pinheiro, V.M. De-Lima and D.E. De-Oliveira (1997). *Arabidopsis thaliana* class IV chitinase is early induced during the interaction with *Xanthomonas campestris*. *FEBS Letters*, 419: 69–75.

Gomes, V.M, A.E.A. Oliveira and J.A. Xavier-Filho (1996). Chitinase and a β-1,3-glucanase isolated from the seeds of cowpea (*Vigna unguiculata* L Walp) inhibits the growth of fungi and insect pests of the seed. *J. Sci. Food Agric.*, 72: 86–90.

Gooday, G.W. (1990a). The ecology of chitin degradation. In: *Advances in Microbial Ecology* II (Ed.) Marshall, K. C., Plenum Press, New York, pp. 387–430.

Gooday, G.W. (1990b). Physiology of microbial degradation of chitin and chitosan. *Biodegradation*, 1: 177–190.

Grassmann, W., L. Zechmeister, R. Bender and G. Toth (1934). Uber die chitin spaltung dorch Emulsin–preparate. III *Mitteil uber enzymatosche spaltung fur polysacchariden Berichte*, 67: 1–15.

Grison, R., B.G. Besset, M. Schncider, N. Lucante, L. Olen, J.J. Leguay and A. Toppan (1996). Field tolerance to fungal pathogens of *Brassica napus* constitutively expressing a chimeric chitinase gene. *Nature Biotechnol.*, 14: 643–646.

Guzzo, S.D. and E.M.F. Martins (1996). Local and systemic induction of β-1,3-glucanase and chitinase in coffee leaves protected against *Hemileia vastatrix* by *Bacillus thuringiensis*. *J. Phytopathol*, 144: 449–454.

Hamel, F. and G. Bellemare (1995). Characterization of a class I chitinase gene and of wound-inducible, root and flower-specific chitinase expression in *Brassica napus*. *Biochimica et Biophysica Acta, Gene Structure and Expression*, 1263: 212–220.

Hamel, F., R. Boivin, C. Tremblay and G. Bellemare (1997). Structural and evolutionary relationships among chitinases of flowering plants. *J. Mol. Evol.*, 44: 614–624.

Hashimoto, R. and Y. Shimamoto (2002). Transgenic sugarbeet plants harboring a pumpkin chitinase gene demonstrating improved resistance to *Rhizoctonia solani*. *Proc. Japanese Soc. Sugar Beet Technol.* 2001, No. 43, 24–28.

Horita, K.T. Araki, M. Kuramoto and T. Torikata. (1996). Amino acid sequence of acidic class III chitinase from soyabean (*Glycine max* L.) seeds. *Proc. of Faculty of Agric.*, Kyushu Tokai University, pp. 71–79.

Hou, W.C., Y.C. Chen, Y.H. Lin, W.C. Hou, Y.C. Chen and Y.H. Lin. (1998). Chitinase activity of sweet potato (*Ipomoea batatas* [L.] Lam var. Tainong 57. *Botanical Bulletin* of *Academia Sinica* 39: 93–97.

Hu, J.J., W. Zhu, J.J. Hu and W. Zhu (1997). Chitinase and β-N-acetyl-D-glucosaminidase induced by *Oothiorella gregaria* in poplars. *Acta Phytopathologica Sinica*, 27: 181–185.

Hudspeth, R.L., S.L. Hobbs, D.M. Anderson and J.W. Grula (1996). Characterization and expression of chitinase and β-1,3-glucanase genes in cotton. *Plant Mol. Biol.*, 31: 911–916.

Hwang, B.K., J.Y. Sunwoo, J.Y. Kim, B.S. Kim, B.K. Hwang, J.Y. Sunwoo, Y.J. Kim and B.S. Kim (1997). Accumulation of β-1,3-glucanase and chitinase isoforms, and salicylic acid in the DL-β-amino-n-butyric acid-induced resistance response of pepper stems to *Phytophthora capsici*. *Physiol. Mol. Plant Pathol.*, 51: 305–322.

Inui, H., Y. Yamaguchi and S. Hirano (1997). Elicitor actions of N-acetylchitooligosaccharides and laminarioligosaccharides for chitinase and L-phenylalanine ammonia-lyase induction in rice suspension culture. *Biosci. Biotechnol. Biochem.*, 61: 975–978.

Ishihara, A., H. Miyagawa, Y. Kuwahara, T. Ueno and S. Mayama (1996). Involvement of Ca^{2+} ion in phytoalexin induction in oats. *Plant Sci.*, 115: 9–16.

Jach, G., S. Logemann, G. Wolf, A.B. Oppenheim, I. Chet, J. Schell and J. Logemann. (1992). Expression of abacterial chitinase leads to improved resistance to transgenic tobacco plants against fungal infection. *Biopractice*, 1: 33–40.

Ji, C. and J. Kuc (1996). Antifungal activity of cucumber β-1, 3-glucanase and chitinase. *Physiol. Mol. Plant Pathol.*, 49: 257–265.

Jiang, Y.M. and Jiang, Y.M. (1997). Incidence of anthracnose in relation to chitinase, β-1,3-glucanase and dopamine of banana fruits after harvest. *Acta Phytophysiologica Sinica.*, 23: 158–162.

Jolles, P. and Jolles. J. (1984). What is new in lysozyme research? *Biochemistry*, 63: 165–189.

Jongedijk, E., H. Tigelaar, S.C. vanRoekel, S.A. Bres-Vloemans, I. Dekker, P.J.M. van den Elzen, J.C. Cornelissen, L.S. Melchers, S.C. Van-Roekel, P.J.M. Van den Elzen, and A.C. Cassells (1995). Synergistic activity of chitinases and β-1,3-glucanases enhances fungal resistance in transgenic tomato plants. *Euphytica*, 85: 173–180.

Joosten, M.H.A., H.M. Verbakel, M.E. Nettekoven, Van Leeuwen, R.T.M. Van der Vossen and P.J.G.M.De Wit. (1995). The phytopathogenic fungus *Cladosporium fulvum* is not sensitive to the chitinase and β-1,3-glucanase defense proteins of its host, tomato. *Physiol. Mol. Plant Pathol.*, 46: 45–49.

Kang, Z. and H. Buchenauer (2002). Immunocytochemical localization of β-1,3-glucanase and chitinase in *Fusarium culmorum* infected wheat spikes. *Physiol. Mol. Plant Pathol.*, 60: 141–153.

Kastner, B., R. Tenhaken and H. Kauss (1988). Chitinase in cucumber hypocotyls is induced by germinating fungal spores and by fungal elicitor in synergism with inducers of acquired resistance. *Plant J.* 13: 447–454.

Kellmann, J.W., T. Kleinow, K. Engelhardt, C. Philipp, D. Wegener, J. Schell and P.H. Schreier. (1996). Characterization of two class II chitinase genes from peanut and expression studies in transgenic tobacco plants. *Plant Mol. Biol.*, 30: 351–358.

Khan, A.A., J. Wu and D.S. Shih (1999). Cloning and sequence analysis of a class III chitinase gene (Accession No. AF 134347) from *Frngaria ananassa* Dutch. (PGR99–069). *Plant Physiol.*, 120: 340.

Khat, G.G., P. Tauro and K.R. Dadarwal. (1996). Rhizobacteria from chickpea (*Cicer arietinum* L.) rhizosphere effective in wilt control and promote nodulation. *Indian J. Microbial.*, 36: 217–222.

Kim, C.Y., S.W. Gal, M.S. Choe, S.Y. Jeong, S. Cheong, Y.H. Lee, S.H. Lee, Y.J. Choi, C.D. Han, K.Y. Kang and M.J. Cho (1998). A new class II rice chitinase, Rcht2, whose induction by fungal elicitor is abolished by protein phosphatase 1 and 2A inhibitor. *Plant Mol. Biol.*, 37: 523–534.

Kim, Y.J., and B.K. Huang (1994). Differential accumulation of β-1,3-glucanse and chitinase isoforms in pepper stems infected by compatible and incompatible isolates of *Phytophthora capsici. Physiol. Mol. Plant Pathol.*, 45: 195–209.

Kim, Y.J., B.K. Hwang, Y.J. Kim and B.K. Hwang (1996). Purification, N-terminal amino acid sequencing and antifungal activity of chitinases from pepper stems treated with mercuric chloride. *Physiol. Mol. Plant Pathol.*, 48: 417–432.

Kishimoto, K., Y. Nishizawa, Y. Tabei, T. Hibi, M. Nakajima and K. Akutsu (2002). Detailed analysis of rice chitinase gene expression in transgenic cucumber plants showing different levels of disease resistance to gray mold (*Botrytis cinerea*). *Plant Sci.*, 162: 655–662.

Kunz, C., H. Schob, M. Stam, J.M. Kooter and F. Jr. Meins (1996). Developmentally regulated silencing and reactivation of tobacco chitinase transgene expression. *Plant J.*, 10: 437–450.

Kyung, S.H., C.W. Sheng, A.G. Darvill and A. Peter (1997). Fungal pathogens secrete an inhibitor protein that distinguishes isoforms of plant pathogenesis-related endo β-1,3-glucanases. *The Plant J.*, 11: 169–179.

Lamb, C.J., J.A. Ryals, E.R. Ward and R.A. Dixon. (1992). Emerging strategies enhancing resistance to microbial pathogens. *Biotechnology*, 10: 1436–1445.

Lamb, C.J., M.A. Lawton, M. Dron and R.A. Dixon (1989). Signal transduction mechanisms for activation of plant defenses against microbial attack. *Cell*, 56: 215–224.

Lange, J., U. Mohr, A. Wiemken, T. Boiler and R. Vogeli-Lange. (1996). Proteolytic processing of class IV chitinase in the compatible interaction of bean roots with *Fusarium solani. Plant Physiol.*, 111: 1135–1144.

Lawrence, C.B., M.H.A. Joosten and S. Tuzun (1996). Differential induction of pathogenesis related proteins in tomato by *Alternaria solani* and the association of a basic chitinase isozyme with resistance. *Physiol. Mol. Plant Pathol.*, 48: 361–377.

Levorson, J. and C.A. Chlan. (1997). Plant chitinase consensus sequences. *Plant Mol. Biol. Reporter*, 15: 122–133.

Li, D.J., K.J. Zhao, Q.M. Zhou, K.D. Luo, J.L., K.J. Zhao, Q.M. Zhou and K. Luo (2002). Transgenic tobacco containing chitinase and β-1,3-glucanase. *J. Human Agric. Univ.*, 28: 211–213.

Li, H., Q. Zhu, V.V. Bai, H. Li, Q. Zhu and V.V. Bai (1997). DNA structure analyses of chitinase gene clone RCH8 from rice. *Acta Phytophysiologica Sinica*, 23: 391–398.

Li, W.L., J.D. Fairs, S. Muthukrishnan, D.J. Liu, P.D. Chen and B.S. Gill (2001). Isolation and characterisation of novel cDNA clones of acidic chitinases and β-1,3–glucanases from wheat spikes infected with *Fusarium graminearium. Theoretical Appl. Genet.*, 102: 353–362.

Lin, W., C.S. Anuratha, K. Datta, I. Potrykus, S. Muthukrishnan and S.K. Datta (1995). Genetic engineering of rice for resistance of sheath blight. *Biotechnology*, 13: 686–691.

Liu, R.J., H.F. Li, C.V. Shen and W.F. Chiu. (1995). Detection of pathogenesis-related proteins in cotton plants. *Physiol. Mol. Plant Pathol.*, 47: 357–363.

Ludwig, A. and T. Boiler (1990). A method for the study of fungal growth inhibition by plant proteins. *FEMS Microbiology Lett.*, 69: 61–66.

Manocha, M.S. and Z. Zhonghua. (1997). Immunocytochemical and cytochemical localization of chitinase and chitin in the infected hosts of abiotrophic mycoparasite, *Piptocephalis virginiana. Mycologia*, 89: 185–194.

Martin, M. (1991). The latex of *Hevea brasiliensis* contains high levels of both chitinases and chitinases/ lysozymes. *Plant Physiol.*, 95: 469–476.

Mauch, F., L.A. Hadwiger and T. Boiler (1988). Antifungal hydrolases in pea tissue. I. Purification and characterization of two chitinases and two β-1,3-glucanases differentially regulated during development and in response to fungal infection. *Pl. Physiol.*, 87: 325–333.

Melchers, L.S., M. Apotheker-de Groot, J.A. van der Knaap, A.S. Ponstein, M.B. Sela-Buurlage, J.F. Bol, B.J.C. Cornelissen, P.J.M. van den Elzen and H.J.M. Linthorst (1994). A new class of tobacco chitinases homologous to bacterial exochitinases displays antifungal activity. *Plant J.*, 5: 469–480.

Metraux, J.P., M. Burkhart, M. Moyer, S. Dincher, W. Middlesteadt, S. Williams, G. Payne, M. Cames and J. Ryals (1989). Isolation of a cDNA encoding a chitinase with structural homology in a bifunctionallysozyme/chitinase. *Proc. Natl. Acad. Sci.*, USA 86: 896–900.

Metraux, J.P., W. Burkhaart, M. Moyer, S. Dincher, W. Middlesteat, S. Williams, G. Payne, M. Carnes and J. Ryals (1989). Isolation of a complementary DNA encoding a chitinase with structural homology to a bifunctional Iysozyme/chitinase. *Proc. Natl. Acad. Sci.*, USA, 86: 896–900.

Minami, E., K. Kuchitsu, D.Y. He, H. Kouchi, N. Midoh, Y. Ohtsuki and N. Shibuya (1996). Two novel genes rapidly and transiently activated in suspension cultured rice cells by treatment with N-acetylchitoheptaose, a biotic elicitor for phytoalexin production. *PI. Cell Physiol.*, 37: 563–567.

Mohr, U., J. Lange, T. Boiler, A. Wiemken and R. Vogeli-Lange. (1998). Plant defense genes are induced in the pathogenic interaction between bean roots and *Fusarium solani*, but not in the symbiotic interaction with the arbuscular mycorrhizal fungus *Glomus mosseae. New Phytologist*, 138: 589–598.

Mojtaba, M., R. Ramin and T. Mohammad (2002). Induced chitinase activity in resistant wheat leaves inoculated with an incompatible race of *Puccinia striiformis* f. sp. *tritici*, the causal agent of yellow rust disease. *Mycopathologia*, 154: 119–126.

Mozetti, C., L. Ferraris, G. Tamietti and A. Matta (1995). Variation in enzyme activities in leaves and cell suspensions as markers of incompatibility in different *Phytophthora pepper* interactions. *Physiol. Mol. Plant Pathol.*, 46: 95–107.

Nandakumar, R., S. Babu, R. Viswanathan, T. Raguchander and R, Samiyappan (2001). Induction of systemic resistance in rice against sheath blight disease by P. *fluorescens Soil Biol. Biochem.* 33: 603–612.

Nandakumar, R.S. Babu, R. Radjacommare, T. Raguchander and R. Samiyappan. (2002). *Pseudomonas fluorescens* mediated antifungal activity against *Rhizoctonia solani* causing sheath blight in rice. *Phytopathologia Mediterranea*, 41: 109–119.

Nasu, K., H. Yoshioka, Y. Ichinose, T., Yamada, H. Oku and T. Shiraishi (1995). Suppression of the activation of chitinase and β-1,3-glucanase in pea epicotyls by endogenous suppressor from *Pisum sativum. Ann. Phytopathol. Soc. Japan* 61: 13–17.

Nehra, K.S., S. Anil, S. Rajendra, B. Praveen, S. Dhillon, A. Sindhu, R. Singh and P. Batra (1997). Induction and characterization of exochitinase from chickpea (*Cicer arietinum* L.) infected with *Ascochyta rabiei*. *Ann. Biol.*, 13: 305–308.

Neuhaus J.M., P. Ahl-Goy, U. Hinz, S. Flores and F.J. Meins (1991). High-level expression of a tobacco chitinase gene in *Nicotiana sylvestris* susceptibility of transgenic plants to *Cercospora nicotianae* infection. *Plant Mol. Biol.*, 16: 141–151.

Neuhaus, J.M., B. Fritig, H., J.M. Linthorst, F. Jr. Meins, J.D. Mikkelsen and J.A. Ryals. (1996). A revised nomenclature for chitinase genes. *Plant Mol. Biol. Reporter*, 14: 102–104.

Nielsen, J.E., K.K. Nielsen and J.D. Mikkelsen. (1996). Immunohistological localization of a basic class IV chitinase in *Beta vulgaris* leaves after infection with *Cercospora beticola*. *Plant Sci. Limerick.*, 19: 191–202.

Nishizawa, Y. and T. Hibi (1991). Rice chitinase gene: cDNA cloning and stress induced expression. *Plant Sci.*, 76: 211–218.

Nishizawa, Y., Z. Nishio, K. Nakazono, M, Soma, E. Nakajima, U. Ugaki and T. Hibi (1999). Enhanced persistence to blast (*Magnaporthe grisea*) in transgenic japonica rice by, constitutive expression of rice chitinase. *Theor. Appl. Genet.* 99: 383–390.

Oelofse, D and I.A. Dubery (1996). Induction of defense responses in cultured tobacco cells by elicitors from *Phytophthora nicotianae*. *International J. Biochem. Cell Biol.*, 28: 295–301.

Ohnuma, T., M. Yagi, T. Yamagami, T. Taira, Y. Aso and M. Ishiguro. (2002). Molecular cloning, functional expression, and mutagenesis of cDNA encoding rye (*Secale cereale*) seed chitinase-c. *Biosci. Biotechnol. Biochem.*, 66: 277–284.

Pappinen, A., Y. Degefu, L. Syrjala, K. Keinonen; K. von. Weissenberg, K and von Weissenberg (2002). Transgenic silver birch (*Betula pendula*) expressing sugarbeet chitinase 4 shows enhanced resistance to *Pyrenopeziza betulicola*. *Plant Cell Reports*, 20: 1046–1051.

Payne, G., P. Ahl, M. Moyer, A. Harper, J. Beck, F. Meins and J. Ryals (1990). Isolation of cDNA clones encoding pathogenesis-related proteins P and Q two acidic chitinases from tobacco. *Proc. Natl. Acad. Sci. USA*, 87: 98–102.

Pleban, S., F. Ingel and I. Chet. (1995). Control of *Rhizoctonia solani* and *Sclerotium rolfsii* in the.greenhouse using endophytic *Bacillus* spp. *European J. Pl. Pathol.*, 101 : 665–672.

Ramanathan, A., Shanmugam, V., Raguchander, T. and Samiyappan, R. (2002). Induction of systemic resistance in ragi against blast disease by *Pseudomonas fluorescens*. *Ann. Pl. Prot. Sci.*, 10: 313–318.

Ramos, A.L., G. Seijo, C. Isshiki and T. Iwamoto. (1998). Distribution of defense–related enzymatic activities in the quiescent organs in *Phaseolus vulgaris* L. seeds. *Biosci. Biotechnol. Biochem.*, 62: 54–59.

Rasmussen, U., H. Gieze and J. D. Mikkelsen (1992a). Induction and purification of chitinase in *Brassica napus* sp. *oleifera* infected with *Phoma lingam*. *Planta* 187: 328–334.

Rasmussen, U., K. Bojsen, D. B. Collinge (1992b). Cloning and characterisation of pathogenesis–induced chitinase in *Brassica napus*. *Plant Mol. Biol.*, 20: 277–287.

Ren, Y. and C.A. West (1992). Elicitation of diterpene biosynthesis in rice (*Oryza sativa* L.) by chitin. *Plant Physiol.*, 99: 1169–1178.

Robert N., K. Roche, Y. Lebeau, C. Breda, M. Boulay, R. Esnault and D. Buffard (2002). Expression of grapevine chitinase genes in berries and leaves infected by fungal or bacterial pathogens. *Plant Sci.*, 162: 389–400.

Roberts, W.K. and C.P. Scliterennikoff (1988). Plant and bacterial chitinases differ in antifungal activity. *J. Gen. Microbiol.*, 134: 169–176.

Robinson, S.P., A.K. Jacobs and I.B. Dry (1997). A class IV chitinase is highly expressed in grape berries during ripening. *Plant Physiol.*, 114: 771–778.

Roth, U., A. Friebe, and H. Schnabl (2000). Resistance induction in plants by a brassinosteroid containing extract of *Lychnis viscaria* L. *Z. Naturforsch.*, 55c: 552–559.

Sahai, A.S. and M.S. Manocha (1993). Chitinases of fungi and plants: their involvement in morphogenesis and host–parasite interaction. FEMS *Microbiol. Rev.*, 11: 317–338.

Salzer, P., G. Hebe and A. Hager. (1997). Cleavage of chitinous elicitors from the ectomycorrhizal fungus *Hebeloma crustuliniforme* by host chitinases prevents induction of K^+ and Cl^- release, extracellular alkalinization and H_2O_2 synthesis of *Picea abies* cells. *Planta*, 203: 470–479.

Sayler, R.J., J.D. Ewing and P.E. McClean (1995). Monogenic and epistatic resistance to bean rust infection in common bean. *Physiol. Mol. Plant Pathol.*, 47: 173–184.

Schafleitner, R. and E. Wilhelm (1997). Effect of virulent and hypovirulent *Cryphonectria parasitica.* (Murr.) Barr on the intercellular pathogen related proteins and on total protein pattern of chestnut (*Castanea sativa* Mill.). *Physiol. Mol. Plant Pathol.*, 51: 323–332.

Schlumbaum, A., F. Mauch, U. Vogoeli and T. Boiler (1986). Plant chitinases are potent inhibitors of fungal growth. *Nature*, 324: 365–367.

Seetharaman, K., E. Whitehead, N.P. Keller, R.D. Waniska and L.W. Rooney (1997). *In vitro* activity of sorghum seed antifungal proteins against grain mold pathogens. *J. Agric. Food Chemistry*, 45: 3666–3671.

Shibuya, N., H. Kaku, K. Kuchistsu and M.J. Maliark (1993). Identification of a novel high affinity-binding site for N-acetylchitooligosaccharides elicitor in the microsomal membrane fraction from suspension cultured rice cells. *FEBS Lett.*, 329: 75–78.

Shibuya, N., N. Ebusu, Y. Kamada, H. Kaku, J. Cohn and Y. Lto (1996). Localization and binding characteristics of a high affinity binding site for N-acetyl chitooligosaccharide elicitor in plasma membrane from suspension cultured rice cells suggest a role as a receptor for the elicitor signal at the cell surface. *Plant Cell Physiol.*, 37: 894–898.

Shih, N.J.R., K.A. McDonald and N.J. Remi-Shih. (1997). Purification and characterization of chitinases from transformed callus suspension cultures of *Trichosanthes kirilowii* Maxim. *J. Fermentation and Bioengineering*, 84: 28–34.

Siefert, F. and K. Grossmann (1997). Induction of chitinase and β-1,3-glucanase activity in b, sunflower suspension cells in response to an elicitor from *Phytophthora megasperma* f.sp. *glycinea*. Evidence for regulation by ethylene and 1-aminocyclopropane-1-carboxylic acid. *J. Exp. Bot.*, 48: 2023–2029.

Susan, A.H., N.B. John, T. Boller and J.C. Lamb (1988). Chitinase cDNA cloning and mRNA induction by fungal elicitor, wounding and infection. *Plant Physiolm.* 86: 182–186.

Tabei, Y., S. Kitade, Y. Nishizawa, N. Kikuchi, T. Kayano, T. Hibi and K. Akutsu (1997). Transgenic cucumber plants harboring a rice chitinase gene exhibit enhanced resistance to gray mold (*Botrytis cinerea*). *Plant Cell Rep.*, 17: 159–164

Taira, T., T. Ohnuma, T. Yamagami, Y. Aso, M. Ishiguro and M. Ishihara (2002). Antifungal activity of rye (*Secale cereale*) seed chitinases: the different binding manners of class I and class II chitinases to the fungal cell walls. *Biosci. Biotechnol. Biochem.*, 66: 970–977.

Takemoto, D., K. Furuse, N. Doke and K. Kawakita. (1997). Identification of chitinase and osmotin-like protein as actin-binding proteins in suspension-cultured potato cells. *Plant Cell Physiol.*, 38: 441–448.

Terakawa, T., N. Takaya, H. Horiuchi, M. Koike and M.A. Takagi. (1997). A fungal chitinase gene from *Rhizopus oligosporus* confers antifungal activity to transgenic tobacco. *Plant Cell Reports*, 16: 439–443.

Van Loon, L.C. (1997). Induced resistance in plants and the role of pathogenesis related proteins. *European J. Plant Pathol.*, 103: 753–765.

Van Loon, L.C., P.A.H.M. Bakker and C.M.J. Peiterse (1998). Systemic resistance induced by rhizosphere bacteria. *Ann. Rev. Phytopathol.*, 36: 453–483.

Van Pelt-Heerschap, H. and O. Smit-Bakker (1999). Analysis of defense–related proteins in stem tissue of carnation inoculated with a virulent ands avirulent race of *Fusarium oxysporum* f. sp. *dianthi. European J. Plant Pathol.*, 105: 681–691.

Velazhahan, R., R. Samiyappan and P. Vidhyasekaran (1999). Relationship between antagonistic activity of *Pseudomonas fluorescens* isolates against *Rhizoctonia solani* and their production of lytic enzymes. *J. Pl. Dis. Prot.* 106: 244–250.

Velazhahan, R., R. Samiyappan and P. Vidhyasekaran (2000). Purification of an elicitor-inducible antifungal chitinase from suspension cultured rice cells. *Phytoparasitica* 28.

Vidhysekaran, P. (1997). *Fungal Pathogenesis in Plants and Crops*. Marcel Dekker, Inc. New York, pp 552.

Viswanathan, R., Nandakumar, R., Babu, S., Raguchander, T. and Samiyappan, R. (2000). Role of pathogenesis related (PR) proteins in rice (*Oryza sativa*) disease management. In: *Advances in Agricultural Research in India*, (Eds.) R.D. Sharma, P. Gahlot and M. Gahlot, International Book Distributors, Dehradun.

Ward, E.R., G.P. Payne, M.B. Moyer, S.C. Williams, S.S. Dincher, K.C. Sharkey, J.J. Beck, H.T. Taylor, P. Ahl-Goy, F.Jr. Meins and J.A. Ryals (1991). Differential regulation of β-1,3-glucanse messenger RNAs in response to pathogen infection. *Plant Physiol.*, 96: 390–397.

Watanabe, T., K. Kobori, K. Miyashita, T. Fujii, H. Sakai, M. Uchida and H. Tanaka (1993). Identification of glutamic acid 204 and aspartic acid 200 in chitinase A1 of *Bacillus circulans* WL-12 as essential residues for chitinase activity. *J. Biol. Chem.*, 268: 1856–1857.

Wojtaszek, P., M. Stobiecki, G.P. Bolwell (1997). Changes in the composition of exocellular proteins of suspension–cultured *Lupinus albus* cells in response to fungal elicitors or $CuCl_2$. *J. Exp. Bot.*, 48: 2015–2021.

Woloshuk, C.P., J.S. Meulenhoff, M. Sela-Buurlage, Van Den P.J.M. Elzen and B.J.C. Cornelissen (1991). Pathogen-induced proteins with inhibitory activity towards *Phytophthora infestans. Plant Cell.*, 3: 619–628.

Wubben, J.P., C.B. Lawrence and P.J.G.M. De-Wit (1996). Differential induction of chitinase and 1,3-β-glucanase gene expression in tomato by *Cladosporium fulvum* and its race–specific elicitors. *Physiol. Mol. Plant Pathol.*, 48: 105–116.

Wurms, K., P. Long, D. Greenwood, K. Sharrock and S. Ganesh. (1997). Endo- and exochitinase activity in kiwifruit infected with *Botrytis cinerea*. *J. Phytopathol.*, 145: 145–151.

Xu Y., Zhu, Q., W. Panbangred, K. Shirasu and C. Lamb (1996). Regulation, expression and function of a new basic chitinase gene in rice (*Oryza sativa* L.). *Plant Mol. Biol.*, 30: 387–401.

Yamada, A., N. Shibuya, O. Kodama and T. Akatsuka (1993). Induction of phytoalexin formation in suspension cultured rice cells by N–acetylchitooligosaccharides. *Biosci. Biotechol. Biochem.*, 57: 405–409.

Yamagami, T. and G. Funatsu (1997). Involvements of Trp23 in the chitin–binding and of Trp131 in the chitinase activity of rye seed chitinase-a. *Biosci. Biotechnol. Biochem.*, 61: 1819–1825.

Yamagami, T., Y. Mine, Y. Aso and M. Ishiguro (1997). Purification and characterization of two chitinase isoforms from the bulbs of gladiolus (*Gladiolus gandavensis*). *Biosci. Biotechnol. Biochem.*, 61: 2140–2142.

Yi, S.Y., B.K. Hwang, S.Y. Yi and B.K. Hwang (1996). Differential induction and accumulation of β-1,3-glucanase and chitinase isoforms in soybean hypocotyls and leaves after compatible and incompatible infection with *Phytophthora megasperma* f.sp. *glycinea*. *Physiol. Mol. Pl. Pathol.*, 48: 179–192.

Zamski, E. and I. Peretz (1996). Cavity spot of carrots: II. Cell-wall-degrading enzymes secreted by *Pythium* and pathogen-related proteins produced by the root cells. *Ann. Appl. Biol.*, 128: 195–207.

Zareie, R., D.L. Melanson and P.J. Murphy (2002). Isolation of fungal cell wall degrading proteins from barley (*Hordeum vulgare* L.) leaves infected with *Rhynchosporium secalis*. *Mol. Plant Microbe Interactions*, 15: 1031–1039.

Zhu, Q. and C.J. Lamb (1991). Isolation and characterization of a rice gene encoding a basic chitinase. *Mol. Genet.*, 226: 289–296.

Zhu, Q., E.A. Maher, S. Masoud, R.A. Dixon and C.J. Lamb (1994). Enhanced protection against fungal attack by constitutive co-expression of chitinase and glucanases genes in transgenic tobacco. *Biotechnology*, 12: 807–812.

Zlarkan, M., A. Amrani, M. Nijs, A. Vandermeers, S. Zerhouni, N. Smolders and Y. Loole. (1997). *Carica papaya* latex is a rich source of a class II chitinase. *Phytochemistry*, 46: 1319–1325.

Zohreh, T., A. Zahra and P. Vaino. (1997). Transgenic tomato plants expressing *Lycopersicon chilense* gene demonstrates resistance to *Verticillium dahliae*. Abstract No. 1564 of papers presented at American Society of Plant Biologists on Transgenics and Biotechnology.

26

Characterization of Pathotype Diversity in Nine Isolates of *Ascochyta rabiei* Based on DNA Polymorphism

Atul Kumar, Neena Mitter and Rashmi Agarwal

Division of Plant Pathology, Indian Agricultural Research Institute, New Delhi – 110 002

Pulses form an integral part of the vegetarian diet in the Indian sub continent. *Ascochyta rabiei* is notorious for its variability in morphology, pathogenicity and phytotoxin production. Due to breakdown of existance by shift in pathogen virulence, it is imperative to have markers, which may give useful information regarding strainal variability. A set of highly descriminating and reliable criteria for genotypic identification is required for evaluating genetic diversity. Characterization of isolates by their pathogenecity on differential host genotypes have been done in India, but this pathotyping may not give repetitive results as it requires controlled environmental conditions and uniform set of differential host genotypes.

The need of the hour is to compliment traditional markers with DNA polymorphism based studies in the analyses of genetic identity, variability and relatedness in this fungi. The scenario is ripe with the work progressing on the molecular characterization of *A. rabiei*. Two methods have received most attention

1. DNA fingerprinting with multilocus probes that detect numerous restriction fragment length polymorphisms (Weising *et al.*, 1991; Kaemmer *et al.*, 1992.) and
2. PCR based techniques (Fischer *et al.*, 1995; Geistlinger *et al.*, 1997; Udupa *et al.*, 1997).

The study envisages to utilize RAPD of *A. rabiei* isolates collected from different geographical regions of India in order to obtain a clear and lucid variability spectrum of this important pathogen.

Nine single spore culture of a pathotypes of *A. rabiei* collected from different geographical regions of India were used in the study (Table 26.1). Single spore culture of nine isolates of *A. rabiei* were grown on modified Richards medium (sucrose 30 gm, KH_2PO_4 5 gm, $KMnO_3$ 2 gm, $MgSO_4$ 2.5 gm, $FeCl_3$ 20 ppm and distilled water to make 100 ml). Mycelial mats were harvested after 15 days incubation at 20°C by filtering through Whatman paper no.1 filter paper. To remove excess of salts adhering to mycelial mats they were repeatedly washed with sterile distilled water.

Table 26.1: Isolates of *Ascochyta rabiei* Used in the Present Investigation

Sl.No.	*Isolate*	*Source*	*Place of Origin*
1.	ArH_1	Infected chickpea samples	Hissar
2.	ArH_2	—do—	—do—
3.	ArH_3	—do—	—do—
4.	ArD_1	—do—	New Delhi
5.	ArD_2	I.T.C.C. No. 4638, Plant Pathology, IARI	—do—
6.	ArK_1	I.T.C.C. No. 4636	Kashmir
7.	ArK_2	I.T.C.C. No. 4785	—do—
8.	ArG_1	I.T.C.C. No. 4637.	Gurdaspur
9.	ArL_1	I.T.C.C. No. 4634	Ludhiana

Table 26.2: Chickpea Differential Genotypes Used for Virulence Analysis

Sl.No.	*Chickpea Genotype*	*Source*
1.	Pusa 256	Pulse Laboratory, Division of Genetics, IARI
2.	Pusa 362	—do—
3.	C 235	—do—
4.	ILC 249	ICRISAT, Hyderabad
5.	ILC 482	—do—
6.	ILC1929	—do—
7.	ILC3279	—do—
8.	P 1343-1	NBPGR

Table 26.3: Disease Scoring of 9 isolates of *A. rabiei* on 8 Differential Chickpea Genotypes

Chickpea Genotype	*Isolates of Ascochyta rabiei*								
	ArH_1	ArH_2	ArH_3	ArD_1	ArD_2	ArK_1	ArK_2	ArG_1	ArL_1
Pusa 256	9	9	9	9	9	7	3	9	9
Pusa 362	9	9	9	9	9	5	1	9	1
C 235	9	9	7	9	9	3	1	9	9
ILC 249	9	7	1	7	7	1	1	9	7
ILC 482	3	3	1	7	7	3	1	9	7
ILC1929	5	9	5	7	9	1	1	9	7
ILC3279	3	3	1	3	3	1	1	7	3
P 1343-1	5	7	7	7	7	3	1	9	3

The disease scoring was done 10 days after inoculation based on 1-9 scale of Grewal, 1984.

Table 26.4: Disease Reaction of 9 Isolates of *A. rabiei* on 8 Differential Chickpea Genotypes

Chickpea Genotype	*Isolates of Ascochyta rabiei*								
	ArH_1	ArH_2	ArH_3	ArD_1	ArD_2	ArK_1	ArK_2	ArG_1	ArL_1
Pusa 256	S	S	S	S	S	S	R	S	S
Pusa 362	S	S	S	S	S	S	R	S	S
C 235	S	S	S	S	S	R	R	S	S
ILC 249	S	S	R	S	S	R	R	S	S
ILC 482	R	R	R	S	S	R	R	S	S
ILC1929	S	S	S	S	S	R	R	S	S
ILC3279	R	R	R	R	R	R	R	S	R
P 1343-1	S	S	S	S	S	R	R	S	R

Disease rating of 1–3 is scored as resistance (R) and 5–9 as susceptible(S).

DNA extraction was done as per the methods of Zolan and Pukilla (1985) with slight modifications. Mycelium was ground in liquid N_2 with pre-chilled pestle and mortar and the powder was transferred into centrifuge tubes carrying 25ml of preheated (65°C) 2 per cent CTAB extraction buffer (0.7 M NaCl, 50mM Tris. Hydrochloride and 10mM EDTA, pH 8). The tubes were incubated at 65°C for an hour followed by addition of 15ml chloroform: isoamyl alcohol (24:1).

Samples were centrifuged at 10,000 rpm for 10 minutes at room temperature. Upper aqueous amplification assay contained: template DNA 25mg; *Taq* DNA polymerase (genes)–0.5 unit; $MgCl_2$–2.5 mM; dNTP (genes)–100 μm each of dATP, dGTP, dTTP, dCTP; Primer (Operon technology)–1.0 μm Assay Buffer (genes–1X in reaction volume of 25 μl. The polymerase chain reaction was performed using Bio-Rad Gene Cycler TM with the following temperature profile: 94°C for 2 min. followed by 45 cycles of 92°C for 1 min and 37°C for 1 min and 72°C for 2 min, with final elongation at 72°C for 10 minutes. Primer survey was carried out and 40 primers from OPM, OPN, OPP and OPS series (Operon Technologies, Inc., USA) were screened using DNA from all the nine isolates. The primers that gave reproducible and scorable amplifications were used in the analysis of all the isolates.

Amplified products obtained after the PCR amplification were loaded into 1.2 per cent agarose gel in TAE (0.04 M Tris acetate, 0.001 M EDTA, pH 8.0) buffer. Electrophoresis was carried out at 60 volts for 3 hours. Gels were stained with Ethidium Bromide (1μg/ml) and observed in a trans-illuminator under UV light. Lambda DNA double digested with EcoRI and HindIII, was used as a marker.

Data entry was done in matrix form in which all observed bands or characters were listed. The RAPD pattern on each isolate was evaluated assigning character state to all bands that could be observed in gel with certain primers. The similarity co-efficients were subjected to unweighted pair-group method on arithmetic averages (UPGMA) and cluster analysis was done to group the isolates based on their overall similarities. NTSYSPC was used for cluster analysis and dendrogram preparation.

Nine isolates were divided into 2 major groups, one comprising ArH_1, ArH_2, ArH_3, ArD_1, ArD_3, ArG_1 and ArL_1, and the other comprising of ArK_1 and ArK_1. The present study elucidates the presence of pathogenic variability in *A. rabiei* and provides us with molecular markers which can be used by various researchers working on *A. rabiei*.

Out of 40 primers screened for amplification of DNA of isolates of *A. rabiei* 23 were found to give reproducible and scorable bands with high percentage of polymorphisms. Hence the numerical taxonomic analysis included the results from these 23 primers.

PCR amplification with each primer was done twice before scoring for presence and absence of bands. Number of amplification products obtained were specific to each primer and ranged from 5 to 17 with fragment size varying from 5 kb–0.5 kb. The selective primers gave a total of 230 amplification products. But of these 194 were polymorphic in nature.

A high degree of polymorphism, in general, was obtained with most of the primers especially OPM6, OPM9, OPM12, OPN18, OPP2, OPP3, OPP11, OPP12, OPP14, OPS1 and OPS 5 indicating a wide range of diversity at DNA level among the isolates. The selection of primers has great significance in pathotype identification and characterization and molecular mapping of genome. These primers can also serve as a reference for integral comparison in future studies. The usefulness of this techniques was first established in studying pathogenic variability of 30 Iranian isolates of *A. rabiei* by Fischer *et al.* (1995) using three 10 mer random primers. In our study with each of the primers several amplification products were obtained which were common to all the isolates as has also been observed by Fischer *et al.* (1995) and Udupa and Weigand (1997). Banding profiles obtained with 23 random primers for nine isolates of *A. rabiei* was analyzed on the basis of presence or absence of the band. Jaccard's similarity co-efficients between nine isolates were calculated. The values of this index ranged from 0.41 to 0.78, indicating the presence of wide range of genetic diversity at molecular level among nine isolates. The present study has also resulted in identification of useful primer markers for separation of isolates of *A. rabici*. Isolates can be separated from all others using different primers. The most virulent isolate

ArG_1 can be separated from others by using primer OPP3 while least virulent isolate ArK_2 gives a distinct banding pattern with primer OPP2 (Figures 25.1 and 25.2). Similarity coefficients were subjected to unweighted pair group method on arithmetic averages (UPGMA). The dendrogram obtained after cluster analysis divided nine isolates into two major clusters separating the least virulent isolates ArK_1 and ArK_2 from all others. Isolates ArH_1 and ArH_2 were grouped together. The other isolates like ArD_1, ArD_2, ArL_1, ArG_1 and ArH_3 were further separated from ArH_1 and ArH_2 as well as each other (Figure 26.3). The work has yielded significant information in terms of pathotype diversity and markers, which can be used to separate isolates varying in their disease reaction.

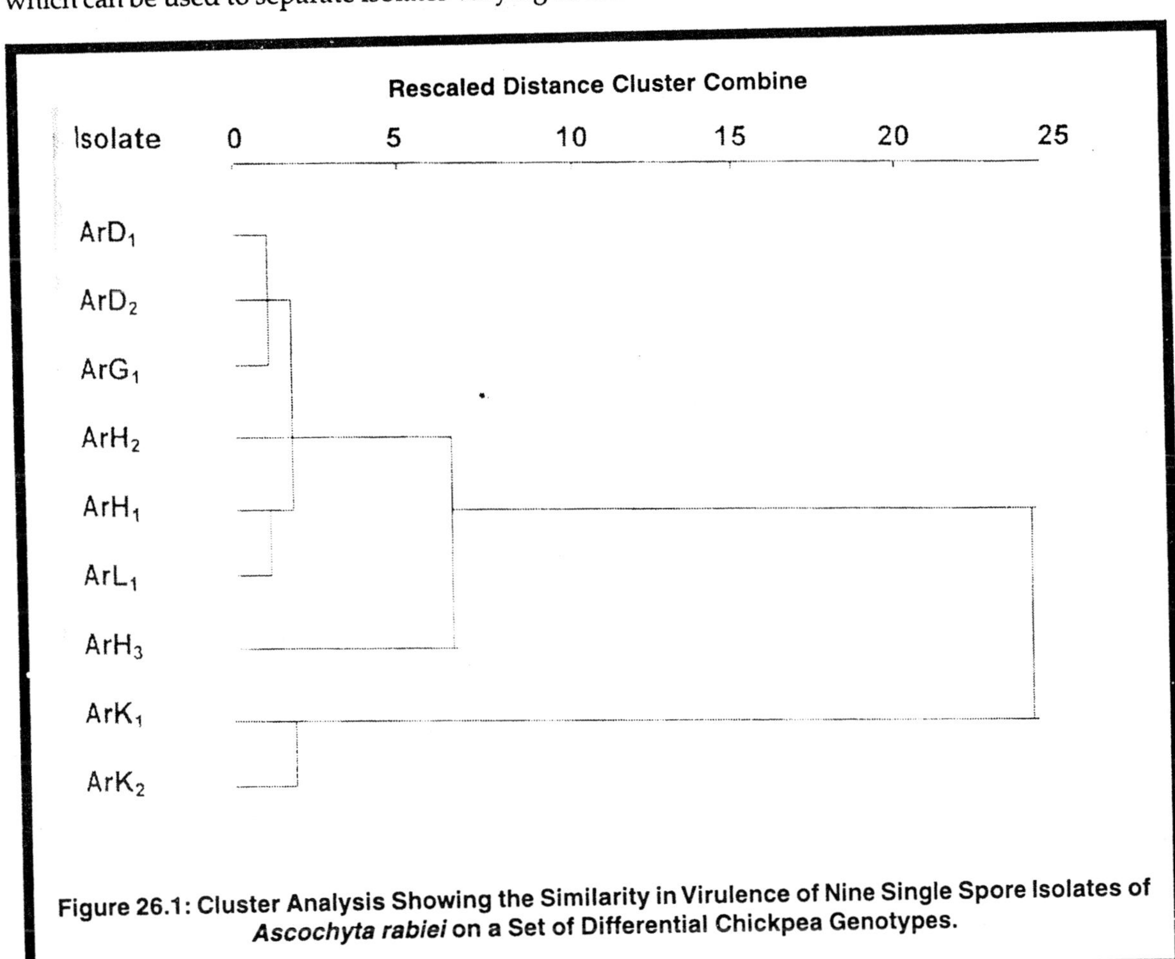

Figure 26.1: Cluster Analysis Showing the Similarity in Virulence of Nine Single Spore Isolates of *Ascochyta rabiei* on a Set of Differential Chickpea Genotypes.

Conclusion

Chickpea (*cicer arietinum* L.) is the major crop accounting for around 45 per cent of total production of pulses in India. The most destructive among the biotic stresses is the blight caused by *Ascochyta rabiei*. Variability in isolates of *Ascochyta rabiei* in terms of molecular characterization was studied. Rapid amplification of polymorphic DNA was carried out based on UPGMA clustering analysis, the isolates were divided into two major groups. The least virulent isolate ArK_1 and ArK_2 were separated

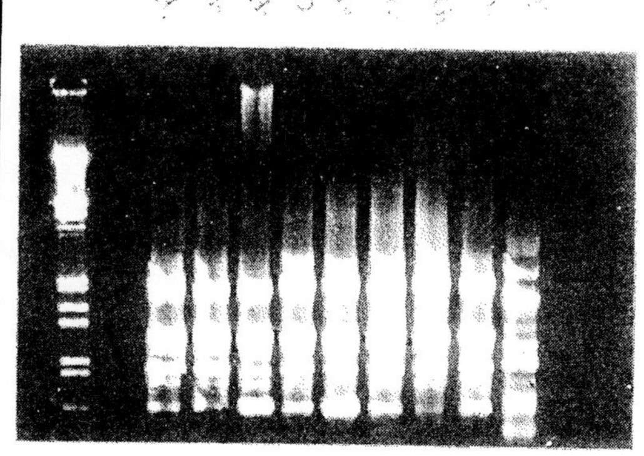

Figure 26.2(a): RAPD Amplification of Isolates of *A. rabiei* with Primer OPP2 in 1.2% Agarose gel. Lane M is marker DNA digested with *Hin*D III and *Eco*R I.

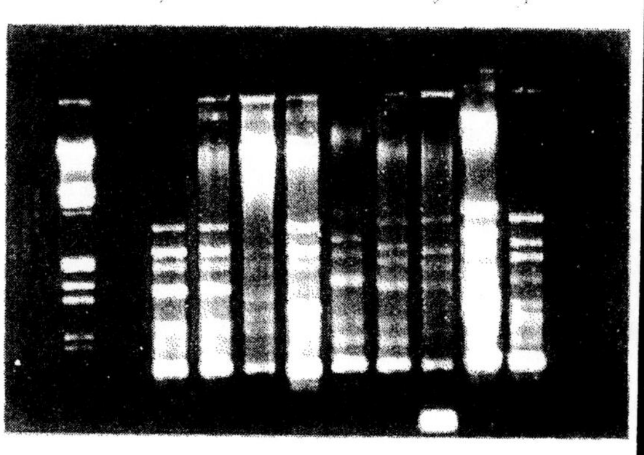

Figure 26.2(b): RAPD Amplification of Isolates of *A. rabiei* with Primer OPP3 in 1.2% Agarose gel. Lane, M is marker DNA digested with *Hin*D III and *Eco*R I.

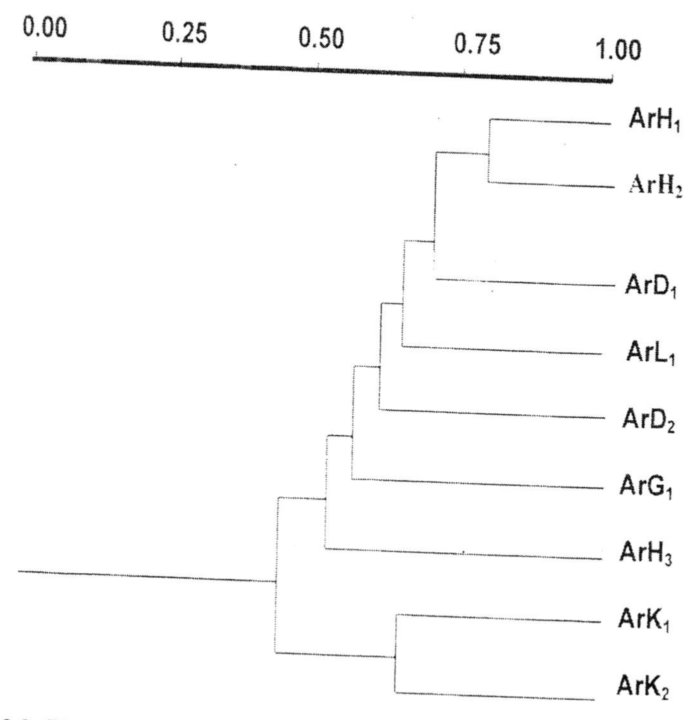

Figure 26.3: Phenogram Derived by UPGMA Cluster Analysis of Nine Single Spore Isolates of *Ascochyta rabiei* Based on Random Amplification of Polymorphic DNA with 23 Decamer Primers

from rest of the isolates which shows that correlation exists between virulence analysis and RAPD pattern. The investigation also resulted in identification of useful molecular markers for separation of different isolates.

Acknowledgements

The authors are thankful to Head Division of Plant Pathology for providing necessary facilities related with research work. The financial help given by ICAR in the form of JRF is duly acknowledged.

REFERENCES

Fischer, C., A. Porta-Puglia and W. Barz (1995). RAPD analysis of pathogenic variability in *Ascochyta rabiei*. *J. Phytopathology*, 143: 601–607.

Geistlinger, J., K. Weising, W.J. Kaiser and G. Kahl (1997). Allelic variation at a hyprevariable compound microsatellite locus in the ascomycete *Ascochyta rabiei*. *Mol. Gen. Genet.*, 256: 298–305.

Kaemmer, D., J. Ramser, M. Schon, F. Weigand, M.C. Saxena,A.J. Driessel, G. Kahl and K. Weising (1992). DNA fingerprinting of fungal genomes: A case study with *Ascochyta rabiei*. *Adv. in Mol. Gent.*, 5: 255–270.

Udupa, S.M. and F. Weigand (1997). Pathotyping of *Ascochyta rabiei* isolates from Syria. Pages 39-48 In: DNA Markers and Breeding for Resistance to Ascochyta Blight. In Chickpea. Proceedings of the Symposium on Application of DNA Fingerprinting for Crop Improvement: Marker–assisted Selection of Chickpea for Sustainable Agriculture in the Dry Areas. (S Udupa and F. Weigand, eds). 11-12 April 1994, Alleppo Syria. ICARDA, Alleppo, Syria.

Weising, K., D. Kaemmer, J.T. Eppen, F. Weigand, M.C. Saxena and G. Kahl (1991). DNA fingerprinting of *Ascochyta rabiei* with synthetic oligonucleotides. *Curr. Genet.*, 19: 483–489.

Zolan, M.E., P.J. Pukilla (1986): Inheritance of DNA methylation in Coprinus cinerius. *Mol. Cell. Biol.*, 6: 195–200.

Part III
NEMATOLOGY

27

Interaction Phenomena Between Air and Soil Pollutants on Plant Parasites

Shefta T. Chandel, Saqia F. Chandel**, Babu Mansuri*** and D. Prasad*****

**Department of Crop Protection, Faculty of Agricultural Sciences, AMU, Aligarh – 202 002*

***Department of Crop Protection, IIPR, Kanpur – 208 024*

****Women's College, AMU, Aligarh – 202 002*

***** Division of Nematology, Indian Agricultural Research Institute, Pusa, New Delhi – 110 012*

A rapid place of industrialization, coupled with uncontrolled exploitation of nature, has caused continuous dumping of industrial by-products, hazardous chemicals and nuclear wastes, deforestation, in vast areas of the earth and pollution of the river basins, lakes and seas.

Environment is regarded as "the sum or total of all conditions and influences that affect the development and life of organisms". Each living organism from the lowest to the highest has its own environment and this is affected by changes in natural cycles. Environmental pollution is one of the major problems, threatening the growing the population of the world during the past two decades. Rapid industrialization and urbanization has resulted in the accumulation of sewage sludge, industrial wastes and fertilizer impurities containing various heavy metals and organic pollutants. The term 'heavy metals' is often used where there are connotations of toxicity. The translocation of some heavy metals such as Hg, Cd, Cu, Co, Ni and Zn is affected by organic pollutants, *i.e.*, methanol, ethanol, propanol, acetone, formaldehyde, benzaldehyde, ethyl methyl ketone and cyclohexone. These organic compounds are also added to the soil as a result of decomposition of organic matter in the presence of microorganisms. The important organic compounds released are carboxylic acids, formic acid, oxalic acid, phenols, etc. These organic compounds form complexes with soil organic matter, humic acid, fulvic acid, etc. and affect the mobility and toxicity of heavy metals. The presence of organic and inorganic pollutants in sewage wastes restricts their use in agriculture due to toxic metals accumulation in soil.

Heavy metals find their way into the environment through various natural and man made sources, and lead to several undesirable effects on water, air and soil components of the environment. Volatilization of metals from metallurgical industries can give rise to localised pollution by metals initially in the atmosphere and then on plants, soil and water by wet or dry deposition of metallic compounds. Toxic metals to a large extent are dispersed in the biosphere. They can be carried to places far away from the source by wind depending on whether they are in gaseous or particulate form. Metallic pollutants are ultimately washed out by rain on to land or the surface of water ways. Metals contained in industrial effluents constitute a major source of metallic pollution of hydrosphere. This discharge into river, water leads to concentration of toxic elements within food chains in marine eco-system.

Metals entering the soil constitute a more lasting form of pollution. The potential pathways of the metals present in the soil are uptake by plants, movement with water to ground or surface water sources, and immobilization in the soil matrix. Heavy metals applied to soil may pass through the soil unchanged, react with organic or inorganic compounds to form soluble or insoluble compounds, be adsorbed on soil colloids, volatize from the soil or be taken-up by plants (Epstein and Chaney, 1978). Uptake of heavy metals by plant can continue long after the pollution ceased. Electroplating, metal processing and refining, storage battery manufacturing and tanning industries, etc. are mainly responsible for heavy metal pollution.

"Pollution" is the contamination of air, water and soil with harmful chemicals and radiations. The chemicals and radiations whose presence causes pollution are called "pollutants". Wood (1968) classified air pollutants into two categories based on their origin, *viz.* primary air pollutants and secondary air pollutants. Primary air pollutants are those that originate directly from the sources. Such air pollutants may be in gaseous or particulate forms. Gaseous air pollutants are oxides of nitrogen, NH_3, SO_2, fluoride, CO_2, carbonmonooxide etc. The particulate air pollutants are cement dust, soil dust, suspended particulate matter, flyash, coal dust, etc. The secondary air pollutants are produced by the reaction of primary gaseous air pollutants, *e.g.*, O_3 and peroxyacetyl nitrate (PAN). Some gaseous air pollutants like SO_2 and NOx in high humid conditions are converted into acid (H_2SO_4 and HNO_3) which fall on the ground in the form of 'acid rain' during the atmospheric precipitation (Oden, 1968).

The role of air pollutants on crop plants are being realised now in different parts of the world (Heck *et al.*, 1986). Growth, biomass, yield of the crop plant and trees are adversely affected by air pollutants (Mudd and Kozlowski, 1975; Heggested, 1988; Heck *et al.*, 1986). Gaseous air pollutants enter the leaves through stomata which cause injury directly in the leaf tissue or interfere in the bio-chemical reactions (Pell, 1979 and Prasad, 1991).

In nature, interaction between parasites (biotic pathogens) and air pollutants (abiotic pathogens) may also develop under specific conditions (Pell, 1979). Air pollutants produce visible symptoms like necrosis, stunting, early senescence and chlorosis; etc. and also affects the physiology and biochemistry of the plant (Heagle, 1973). Another way in which air pollutants may influence plant population and community composition is by modifying the plant -parasitic biota and virus disease activity. Biotic components of the environment, the microflora and microfauna may be far more sensitive to pollution than the higher plants. Fungi, bacteria, viruses, nematodes and insects are all important species to bring into equilibrium with its environment and determining the ultimate plant population. Nickle toxicity is a recognized problem in agricultural areas with high soil nickle levels (Frank *et al.*, 1976; Hunter and Vargnano, 1952; Lagerwerff, 1967).

Several reports of effects of pollutants on plant pathogens have been published (Heagle, 1973), but relatively few have concerned nematodes. Plant-parasitic nematodes are minute, vermiform, multicellular, and non-segmented invertebrate animals with simple body organization. They are cosmopolitan in distribution and are one of the important limiting factors of plant growth and productivity. They cause great destruction to plants either singly or in collaboration with other pathogens and parasites. Nematodes incurred losses not only in the form of reduced plant growth and yield but also in the marketable quality of the products. It is difficult to assess economic losses they cause to crop plants.

Areas of forest severely damaged by sulfur dioxide (SO_2) and alkaline particulates were found to have higher nematode populations than less damaged areas (Bassus, 1968). Ozone and SO_2 inhibited reproduction and development of certain nematodes species on soybean (Weber *et al.*, 1979). Studies with cysts nematodes (*Heterodera* spp.) indicated that several inorganic ions, including Ni^{++} stimulated hatching of eggs (Clarke and Shcpherd, 1966).

Air Pollutant

Sulphurodioxide

Plant Fungi

Sulphurdioxide (SO_2) is reported to influence a number of fungal plant pathogens (Table 27.1) and quality of air (Prasad and Chaudhary, 1992)

It has been observed that fungal hyphae show resistance to SO_2, especially when protected by the host tissue. SO_2 is used to inhibit post-harvest decay of grapes by *Botrytis cinerea*, *Cladosporuim herbarum* and species of *Alternaria* and *Stemphylium* but concentration of SO_2 required for this effect are more than 200 times those usually found in polluted ambient air (Heagle, 1973). The spores of the most fungi tested have demonstrated great resistance to direct exposure to SO_2, even at high doses (Couey and Uota, 1961; Hibben, 1966). Couey and Uota (1961) reported that only 20 per cent conidia of *B. cinerea* germinated when exposed to 36 ppm SO_2 for 1/2 h. However, Hibben (1966) observed no effect on the germination of spores of 10 saprophytic and parasitic fungi exposed on agar to 10ppm SO_2 for 1–6h. Moist spores are comparatively more sensitive to SO_2 than dry spores. The rust *fungus*, *Uromyces phaseoli* parasitising bean leaves was inhibited by SO_2 (Wenistein *et al.*, 1975). Sulphurdioxide SO_2 also inhibited *Puccinia graminis* on wheat when exposed to SO_2 (Laurence *et al.*, 1979). Kock (1935) reported the absence of oak posery mildew caused by *Microsphaera alni* near a paper mill. Laurence *et al.* (1979) obtained 38 per cent decrease in number of lesions caused by *Helminthosporium maydis* on maize exposed to 0.15ppm SO_2 for 14 h daily for 8 days before inoculation. Weinstein *et al.* (1975) observed that exposures that decrease parasitism of bean by *U. phaseoli* did not affect parasitism of tomato leaves by *Alternaria solani*, Scheffer and Hedgcock (1955) recorded decreased parasitism by the species of *Cronartium*, *Coleosporium*, *Melamspora*, *Peridermium* and *Puccinia* where trees were injured by SO_2. With increasing distance from SO_2 sources, decrease in plant injury and increase in incidence of fungi were observed. Bevan and Greenhalgh (1976) recorded negative correlations between the number of *Rhytisma acerinum* caused spot lesions/unit leaf area and mean annual concentrations of SO_2.

Table 27.1: Interaction of SO_2 and Plant Pathogenic Fungi

Dose	*Fungi*	*Host*	*Effect*	*Reference*
50 ppm for 2 min at moist condition of 100 ppm for 24 min., at 98 per cent	*Alternaria*	–	60 per cent decrease in spore germination	Covey, 1965
Ambient	*Armilloria mellea*	Trees	Incidence increased	Donaubawer, 1968; Jancar, 1961; Kundela and Novakava, 1962
Ambient	*Coleosporium* sp.	Trees	Disease incidence decreased	Linzon, 1958; Scheffer and Hedgcock, 1955
Ambient	*Cronatrium*	Trees	Disease incidence decreased	Linzon, 1958; Scheffer and Hedgcock, 1955
0.01 and 0.04 ppm for 2 days after inoculation	*Deplocorpon rosae*	Rose	Diseased leaflets area slightly increased at 0.01 ppm more than the 0.04 m	Saunders, 1966
Ambient	*Glocophyllum*	Trees	Increased incidence	Jancarik, 1961
0.15 ppm for 14 h/d. 8 days before inoculation	*Helminosporium maydis*	Maize	Decreased in lison number	Laurence *et al.*, 1979
Ambient	*Hypodermella* sp.	–	Decreased in disease incidence	Scheffer and Hedgcock, 1955
Ambient	*Hysterium pulicare*	Alder birch	Decreased in disease incidence	Skye, 1968
Ambient	*Lophodermium piceae*	–	Increase in disease incidence	Kundela and Novakova, 1962
Ambient	*Melampsora* sp.	–	Increase in disease incidence	Scheffer and Hedgcock, 1955
0.30–0.40 ppm for 24–72 h before inoculation of conidia	*Miecrosphaera alni*	Lilac	Decreased in conidial germination and penetration	Hibbon and Taylor, 1975
90 ppm SO_2 in nutrient Sol.	*Penicillium* sp.	–	Slight Stimulation in growth	Saunders, 1966
Ambient	*Puccinia*	Trees	Decrease in disease incidence	Scheffer and Hedgcock, 1955
Ambient	*P. graminis*	Wheat	Less disease incidence	Johansson, 1954
0.10 ppm for 100 h from 2 days for inoculation	*P. graminis*	Wheat	Decrease in lesion no.	Laurence *et al.*, 1979
Ambient	*P. struformis*	–	Decrease in germination of uredospores	Sharp, 1967
Ambient	*Rhizosphaera kalkhoffii*	Redpine	Increase in disease incidence	Chiba and Tanaka, 1968
Ambient	*Rhytisma acerinum*	Sycamarea	Decrease in disease incidence	Bevan and Geenhelgh, 1976
0.20 ppm for 6 h. 5 days after inoculation	*Scirrhia acicola*	Scots pine	Increase in lesion number	Weidensaul and Darling, 1979
1.0 ppm for 4 h.	*S. acicola*	Agar media	Normal growth	Ham, 1971
Ambient	*Trametes serialis* and *T. heteromopha*	Trees	Increase in disease incidence	Jancarik, 1961
0.13 ppm for 24 h/d (on 8 days before or 7 days after inoculation)	*Uromyces phaseoli*	Bean	Decrease in pustule no., spore size and germination	Weidensoul and Darling, 1979

Plant Bacteria

Bacterial plant pathogens are also influenced by SO_2. While studying the effect on the parasitism of bacteria Laurence and Aluiso (1981) found that SO_2 reduced the rate of lesion development and lesion size, when SO_2 exposures were sufficient to increase sulphur content of leaves, *Corynebacterium nebraskense* was inhibited when maize plant were exposed continuously to 0.20 ppm SO_2 for 24h daily, 5 days before inoculation, 2 days after inoculation and both pre and post-inoculation. However, maximum inhibition occurred at 2 days post-inoculation exposure. In further studies, Laurence and Aluiso (1981) observed SO_2 effect on *Xanthomonas phaseoli*. Inhibition in *X. phaseoli* occurred when soyabean plants were exposed to 0–10 ppm SO_2 for 24h 5 days before incoulation, 5 days after inoculation and both 5 days before and after inoculation. Inhibition of bacteria was maximum at pre- and post inoculation exposure (Table 27.2).

Table 27.2: Interaction of SO_2 and Plant Pathogenic Bacteria and Viruses

Treatment	*Bacteria/Viruses*	*Host*	*Effect*	*Reference*
0.20 ppm for 24 h daily 5 days before inoculation or both	*Corynebacterium nebraskense*	Maize	Inhibitory	Laurence and Aliusio, 1981
0.10 ppm for 24 h daily for 5 days before inoculation 5 days after inoculation or both	*Xanthomonas phuseoli*	Soyabean	Inhibitory effect	Laurence and Aliusio, 1981
0.10 ppm for 24 h for 7 days	*Southern bean mosaic virus*	Bean	Increase in virus conc.	Laurence *et al.*, 1981
0.10 ppm for 24 h for 7 days	*Maize dwarf mosaic virus*	Maize	Increase in virus conc., infection and severity of symptoms	Laurence *et al.*, 1981

Plant Virus

Some plant viruses are known to be affected by SO_2. SO_2 has been found to increase viruses in beans and maize (Laurence *et al.*, 1981). An increase in Southern bean mosaic virus in bean leaves was noted when plants were exposed to 0.10 ppm SO_2 for 7 days continuously with pre and post-inoculations. Increase of maize dwarf mosaic virus (MDWV) leading to greater infection and more severe symptoms has also been observed when plant were exposed continuously to SO_2 pre- or post-inoculations of plants (Laurence *et al.*, 1981) (Table 27.2).

Fluoride

Plant Fungi

Hydrogen fluoride accumulates in plant tissues and can affect plant and microbial pathogen interaction. C.D. McCune, L.H. Weinstein, J. Mancini and P. Van Leuken (Pers. Comm.) indicate that fluoride in leaves at levels above those found in the field decrease infection and invasion of bean and tomato plants by *Uromyces phaseoli, Erysiphe polygoni* and *Alternaria solani*. Industrial emissions may contain small amounts of fluoride, but the relative impact of fluoride form most industrial sources on vegetation is probably minor compared with that from SO_2. Nevertheless, fluoride may contribute to effects on parasitism that have been attributed primarily to SO_2. Numerous points sources of fluoride exist but there are no field reports that specifically link fluoride to effects on parasitism. Colony growth of fungi was inhibited on agar containing sodium fluoride (NaF). More NaF was required to inhibit *Verticillium alboatrum* and *Helminthosporium sativum* than to inhibit *Phythium debaryanum*. Growth of *B. cinerea* and two *Colletotrichum* spp. was stimulated slightly at low concentration (Treshow, 1965). It is possible that smaller amounts of fluoride in leaf tissue could cause stimulation of disease similar to the stimulation of fungus growth on agar containing NaF (Treshow, 1965). Significantly higher levels of flourides in air inhibit most of the fungi (McCune *et al.*, 1973). Effects of hydrogen fluoride (HF) on fungal plant pathogens have been studied in controlled conditions (McCune *et al.*, 1973; Weinstein *et al.*, 1975). McCune *et al.* (1973) exposed bean plants to hydrogen fluoride before and after inoculation with *U. phaseoli*. Both number and growth of uredia in leaves were inhibited. In second experiment, pustule numbers were reduced by HF exposure before inoculation while in third experiment both pre- and post- inoculations caused inhibition of the rust. Khan *et al.* (1988b) observed very low incidence and intensity of powdery mildew caused by *Sphaerotheca fluliginea* on *cucurbits* like *Cucumis sativus, Cucurbita moschata, Cucumis melo* and *C. meto var. utilissimus* grown around the Ceramic and Pottery Industrial Units, Khurja. This effect was more apparent and common within 2 km from the industries where HF was present in higher concentration. Incidence and intensity of powdery mildew was higher beyond 3 kms where ambient concentration of HF was 0.073 ppm (Table 27.3).

Table 27.3: Interaction of Fluorides and Plant Pathogenic Fungi and Viruses

Treatment	*Fungus/Viruses*	*Host*	*Effect*	*Reference*
HF, before inoculation	*Alternaria solani*	Tomato	Development of fungus inhibited when HF level in leaves were 200–500 ppm	McCune *et al.*, 1973
NaF in high dose	*Helminthosporium sativum*	Media	Inhibition of colony growth	Heagle, 1973
HF in high dose pre and post-inoculations	*Uromyces phaseoli*	Bean plants	Inhibition in number and growth of Uredia	McCune *et al.*, 1973
HF post-inoculation	*Phytophthora infestans*	Tomato	HF inhibited fungal growth	McCune *et al.*, 1973
100–300 ppm fluoride	TMV	Bean leaves	Number of TMV lesions increased	Treshow *et al.*, 1967
HF	TMV	Bean leaves	Reduction in TMV particles and infection when leaves contain more than 500 ppm	Dean and Treshow, 1965
Ambient conc. HF 0.073 ppm	*Sphaerotheca flulginea*	Cucurbits	Intensity and incidence of powdery mildew was higher	Khan *et al.* (1988b)

Plant Bacteria

According to Heagle, 1973 fluoride may decrease the disease caused by *Pseudomonas phaseolicola*. However, Heagle (1982) observed that disease caused by *P. phaseolicola* in bean leaves was not affected by HF exposure at pre-or post- inoculation even at the concentration where 700 ppm fluoride accumulated in leaves. Infection by *P. phaseolicola* also decreased in bean leaves high fluoride levels.

Plant Viruses

Fluoride in bean leaves can increase a decrease tobacco mosaic virus (TMV) lesions depending on fluoride concentration (Dean and Treshow, 1965; Tershow *et al.*, 1967). If leaf tissue contained fluoride more than 500 ppm, reduction in TMV particles and infection was observed (Dean and Treshow, 1965), The mechanism of action of fluoride and ozone on disease caused by TMV are unknown but both pollutants apparently can increased disease under certain, condition (Table 27.3).

Ozone

Plant Fungi

Most of the parasitic fungi are inhibited by ozone. Faculative parasites are affected more than facultative saprophytes (Heagle, 1973). Ozone is more effective than SO_2 in decreasing spore germination. Percent germination of large multicelled conidia of *Macrosporium* spp. was found greater than that of smaller, single-celled spores of *Sclerotiniea fructicola* and *Penicillium expansum* at 0.60 ppm O_3 for 1–3 h freshly cut apples (Smock and Watson 1941). Dry spores are more sensitive than wet spores to ozone (Heagle, 1973). Heagle (1970) exposed 10 varieties of oat infected with *Puccinia coronata* at 10 ppm O_3 for 6 h in light after 10 days of inoculation. Slight injuries in all plants due to O_3 and significant reduction in growth of ureaia was observed. Heagle and Strucklan (1972) reported 84 per cent inhibition in sporulation and 8 per cent and 75 per cent in penetration of *Erysiphe graminis* in barley leaves when exposed to 0.05, 0.10 and 0.15 ppm O_3 respectively. Similar effects were also observed when barley leaves having mature conidia were exposed to 0.10, 0.20 and 0.30 ppm O_3 for 24 h during incubation. Heagle and Key (1973) studied the effects of low concentrations of O_3 on various phases of the uredial cycle of the wheat stem rust fungus, *Puccinia graminis.* Daily 6h O_3 exposures of infected plants significantly decreased the growth of hyphae and the number of uredospores produced. Kochhar *et al.* (1982) noticed greater reduction in biomass of clover by *Rhizoctionia solani* in the presence of a single acute O_3 exposure. However, in the presence of multiple O_3 exposures the synergistic effect on clover biomass disappeared and the stress from *R. solani* was lost. Ozone suppressed growth in both cultivars only in the pre inoculation fumigation treatment, indicating protective effect of powdery mildew infection, by itself did not affect dry weight. Bounty was more resistant to infection by *E. p. pisi* than Alaska and was also less sensitive to foliar ozone injury colonization of leaves by the fungus resulted in suppressed foliar ozone in both the cultivars (Rusch and Laurence, 1993) (Table 27.4).

Plant Bacteria

There is some evidence of pollutant effects on bacteria. Bacteria are resistant to direct attack by pollutants in concentration likely occur in ambient air. Laurence and Wood (1978) observed decreased number of lesions of *Pseudomonas glycinea* on soyabean leaves at 0.08 or 0.25 ppm O_3 for 4h from 16 days to 1 h before inoculation or 1 h to 1 day after inoculation, but not after 2 days of inoculation. Similarly, O_3 at 0.20 ppm for 4 h decreased the number of lesions caused by *Xanthomonas fragariae* on wild strawberry.

Protective effect is offered to some lost plant by bacteria from O_3 injury (Heagle, 1982). Temple and Bisessar (1979) observed that *X. phaseoli* did not provide such protection to navey bean leaves from ambient oxidants. However, in controlled conditions, *Xphaseoli* did not provide such protection to navey bean from ppm O_3 over 8h for 1, 2 or 4 days after inoculation (Olson and Saettler, 1979). Pell *et al.*, (1977) reported that acute foliar injury caused by *Pseudomonas* spp. would protect soyabean leaves from acute injury caused by O_3 and *vice-versa*. Howell and Graham (1977) also observed antigonistic interaction between O_3 and *X. alfalfa* on alfalfa plants. They suggested that disease on alfalfa caused by *X. alfalfa* was decreased due to O_3 exposure and that bacterial infection reduced O_3 injury (Table 27.5).

Table 27.4: Interaction of Ozone (O_3) and Plant Pathogenic Fungi

Treatment	*Fungi*	*Host*	*Effect*	*Reference*
Ambient	*Botrytis cinerea*	Potato	Severe infection on O_3 injured leaves of potato	Manning *et al.*, 1969
Very Low Doses	*B. gladiolorum*	Gladiolus Flower	Inhibit development of disease	Magdycz and Manning,1973; Magie, 1963
–	*Botrytis* spp.	Chrysanthemum	Decreased disease petals	Magie, 1963
0.07–0.10 ppm 10h/day for 15 days before inoculation	*B. cinerea*	Potato	90–100 per cent leaf showed symptoms of fungal disease	Manning *et al.*, 1970
Large doses	*B. cinerea*	Straw berries	Invasion not decreased, growth of aerial mycelium suppressed	Spalding, 1966
–	*Fusarium oxysporum*	Pinto bean plant	20 per cent more fungal colonies on root of plant with leaves injured by O_3	Manning, *et al.*, 1971b
0.10 ppm 8 h/day for 10 weeks	*Fusarium oxysporum f.* sp. *Conglutinans*	Cabbage seedlings	No significant effect	Manning *et al.*, 1971a
0.06 and 0.12 ppm 6 h/day for 6 days before inoculation	*Helminthosporium maydis*	Maize	Sporulation increased	Heagle, 1973
Ambient	*Heterobasidion annosum*	Pine trees	O_3 injured needles had more fungal infection	James *et al.*, 1980
0.5 ppm continuously	*Monilinia fructicola*	Peaches	Decreased fungal disease	Spalding, 1966
0.10 ppm 6 h, 10 days for inoculation	*Puccinia coronata*	Oat	Uredia became significantly smaller	Heagle, 1970
0.06, 0.12, 0.18 ppm 8 h/day	*P. graminis*	–	Decreased sporulation and growth of Hyphae	Heagle and Key, 1973
1 ppm for 15 days	*P. expansum*	Oranges	Less sporulation, invasion and decay	Harding, 1968
1 ppm for 15 days	*P. digitatum*	Lemons	Less invasion, sporulation and decay	Harding, 1968
0.25 ppm–6h	*Rhizopuns stolonifer*	–	Spore germination was inhibited	Hibben and Stotzky, 1969
0.10 ppm–14 h/day one day after inoculation	*Uromyces shaseoli*	Beans	Uredia became significantly smaller in size and plants became resistant to fungus	Heagle, 1973

Large doses of O_3 (0.12–0.15ppm for 15 days) over a 3 week period decreased the number of *Rhiobium* nodules on soybean roots (Reinert *et al.*, 1971). Smaller doses of O_3 (0.06ppm–8 hr/days-20–60days) decreased the number, size, and weight of *Rhizobium* nodules on pinto bean roots (Manning *et al.*, 1972).

Table 27.5: Interaction of Ozone (O_3) and Plant Pathogenic Bacteria

Treatment	Bacteria	Host	Effect	Reference
0.08 or 0.25 ppm 4 h from 1 h–16 h days before inoculation	*Pseudomonas glycinea*	Soybean leaves	Decreased number of lesions	Laurence and Wood, 1978
1–3 days after inoculation	*P. glycinea*	Soybean	Severe bacterial infection protected soybean from acute O_3 injury antagonistic effect	Pratt and Krupa, 1979
High dose	*P. glycinea*	Soybean	Severe bacterial infection protected soybean from acute O_3 injury antagonistic effect	Pell *et al.*, 1977
0.20 ppm–4 h	*Xanthomonas fragariae*	Strawberry	Decreased number of lesions	Laurence and Wood, 1978
Ambient	*X. phaseoli*	Bean leaves	Antagonistic effect	Temple and Bisessar, 1979
0.24 ppm–8 h, 1, 2, 3, days after inoculation	*X. phaseoli*	Navy beans	No effect	Olson and Saettler, 1979
O_3	*X. alfalfa*	Alfalfa	Disease decreased and bacterial infection reduced O_3 injury	Howell and Graham, 1977
O_3 (0.12–0.15 ppm for 15 days) over 3 wk pd	–	Soybean roots	Decreased the nos. of Rhizobuim nodules	Reinert *et al.*, 1971
O_3 (0.06 ppm–8 h/ days-20–60 days	–	Pinto bean leaves	Decreased nos. size 2wk of Rhizobuim nodules	Manning *et al.*, 1972

Plant Viruses

Synergistic or antagonistic interactions between ozone and plant viruses have been reported. Pollutant effects on disease caused by viruses have concerned tobacco mosaic virus (TMV) (Table 27.6). Yarwood and Middleton (1954) observed that rust infected (*Uromyces phaseoli*) portion of bean leaves were not injured by natural smog in California when non-rusted leaves were seriously injured. Brennan and Leone (1970) observed more local tesions on pinto bean leaves inoculated with tobacco mosaic viruses (TMV) and exposed to O_3 (0.10ppm–3h) than on the controls. The response occurred when exposure began 3 or 24 hr after inoculation but was absent when exposure occurred 0 or 48 hr after inoculation. More TMV lesions developed on pinto bean leaves when Tobacco plants were exposed to O_3 (0.30ppm, 6 h) before using them to obtain inoculum than when the inoculum came from nonexposed Tobacco plants (Brennan and Leone, 1970). Several other viruses (tobacco ring spot, tomato ring spot, alflalfa mosaic and tobacco mosaic) also provided some protection from O_3 injury to primary leaves of pinto beans when plants were inoculated 5 days before exposure (0.25 ppm for 4 h) (Davis and Smith, 1974a, 1974 b). Bissesar and Temple (1977) observed 60 per cent less O_3 injury due to O_3 in field grown tobacco infected by TMV than non-infected virus field. Tobacco etch virus protected tobacco leaves from O_3 (0.25 ppm for 4 h) when plants were exposed 9 days after inoculation (Moyar and Smith, 1975). Vargo *et al.*, (1978) exposed soybean plants to 0.40 ppm 4 h after 6, 8, 12 days of inoculation with tobacco ring spot virus. They observed less O_3 injury in inoculated plants. However, Reinert and Gooding (1978) reported that systemic infection by Tobacco Streak Virus caused Tobacco plants to suffer significantly more injury of O_3 than non-inoculated one when exposed to 0.30 ppm 03 for 3 h on for 2 days at 3 weeks after inoculation (Table 27.6).

Table 27.6: Interaction of Ozone (O_3) and Plant Viruses

Treatment	*Virus*	*Host*	*Effect*	*Reference*
0.25 ppm–4 h, 4, 5, 5 days after inoculation	Bean, common mosaic virus	Pinto bean	Less infection	Davis and Smith, 1974 a, b
0.10 ppm–3 h, 3 or 24 h after inoculation	TMV	Pinto bean leaves	More local lesions	Bernnen and Leone, 1970
Ambinet	TMV	Tobacco	TMV reduced O_3 injury by 60 per cent	Bisessar ànd Temple, 1977
0.25 ppm–4 h, 9 days after inoculation	Tobacco etch virus	Tobacco	Tobacco etch virus protected tobacco leaves from O_3	Moyer and Smith, 1975
0.30 ppm–3h for 1 or 2 days after 3 weeks of inoculation	Tobacco streak virus	Tobacco	Significantly more O_3 injury	Reinert and Gooding, 1978
0.40 ppm–4h, 6,8,10, days after inoculation	Tobacco ring	Soybean	TRSV reduced O_3 injury of Soybean	Vargo *et al.*, 1978

Acid Rain

Plant Fungi

Stimulated acid rain caused inhibition of *Cronartium fusiforme* on leaves of yellow oat inoculated with aeciospoers. The number of infections and telia were decreased when sulphuric acid at pH 3.2 was applied on each of 14 days before and after inoculation (0.63 cm rainfall applied over 10 min per day) (Heagle, 1982).

The number of *H. maydis* (Race T) lesions on the maize were increased when conidiea were incubated in water at pH 3.5 and leaves were subsequently treated with *Rain* at pH 3.5 (0.63 cm over 10 min/day for 21 days). Shriner (1978) treated kidney bean plants infected by *Uromyces phaseoli* and of *M. hapla*, three times weekly with stimulated rain at pH 6.0 or 3.2 *Rain* at pH 3.2 caused temporary inhibition of rust severity. Shafer *et al.* (1985) observed decreased number of sporangia of *phytophthora cinnamomi* on radicles of blue, lupine seedlings with increasing solution acidity (pH 5.6,4.0,3.2 and 2.4). However, release of 200 spores from sporangia incubated in soil extracts was unaffected by the acidity of solutions. They further observed that seedlings grown in soil, inoculated with 200 spores and exposed to simulated rainfall (2.4 cm, 1h) at pH 2.4 had 44 per cent fewer reduction sites on roots than did the seedlings exposed to rain at pH 5.6 (Table 27.7).

Plant Bacteria

Acid rain at levels that can occur in the field might significantly affect bacterial parasitism (Heagle, 1982). Shriner (1978) observed increased disease symptoms in kidney bean by *P. phaseolicola* when plants were stressed by stimulated acid rain at pH 3.2 before inoculation (0.63 cm over 10 min/day for 10 days).

The effect of simulated acid rainfall (H_2SO_4, pH 3.5) on infectivity and invasion of kidney bean by *Pseudomonas phaseolicola* suggest that the bacterium is resistant but that the host may be effected in ways that predispose it towards greater susceptibility (D.S. Shriner, personal communication) (Table 27.7).

Table 27.7: Interaction of Acid Rain with Plant Pathogens (Fungi and Bacteria)

Treatment	Pathogen	Host	Effect	Reference
H_2SO_4 of pH 3.2 14 days pre- and post-inoculation	*Cornartium fusiforme*	Yellow Oat	Number of Telia and infection decreased	Heagle, 1982
Simulated acid rain at pH 3.2 (0.63 cm over 2 min/day for 21 days)	*Uromyces phaseoli* + *Meloidogyne hapla*	Kidney beans	Fungal and nematode infection diseases decreased	Shriner, 1978
Solution acidity pH 5.6, 4.0, 3.2 and 2.4	*Phytophthora cinnamomi*	Radicles of blue lupine seedlings	Decreased nos. of sporangia	Shafer *et al.*, 1985
Simulated acid rain at pH 3.2 (0.63 cm over 10 min/day for 10 days)	*Pseudomonas phaseolicola*	Kidney beans	Increased bacterial disease	Shriner, 1978

Pollutants on Insects

Sulphurdioxide and Ozone

Insects play an important role in the transmission and development of many plant disease. Thus, any alteration of population or activities of insects by air pollution may also affect plant diseases or the amount of injury caused by the insects themselves.

Most reports have resulted from field observations of coniferous trees severely affected by SO_2. A common finding is that trees injured and weakened by pollutants are more likely to be attacked by insects that normally require weakened trees for successful reproduction. Thus, the incidence of some species of bark beetles was increased in trees injured by SO_2 (Bosener, 1969; Donaubauer, 1968; Kudela and Novakova, 1962; Scheffer and Hedgcock 1955) or O_3 Stark *et al.* (1968). Templin 1962 reported the decreased incidence of two bark-infesting insects, *Tomicus* (=*Myelophilus*) *piniperda* and *Melanophila* (=*Phaenops*) *cyanea* in trees injured by SO_2 pollution. Cobb *et al.* (1968) suggested that even though *Ponderosa* pines that are severely injured by O_3 may be attacked by large population of bark beetles, the trees may be too decadent for establishment of successful broods. Less injury from the white pine weevil, *Pissodes strobi*, occurred in trees injured SO_2 (Linzon, 1958). But a high incidence of *Pissodes* sp. weevils has been reported on pine and spruce injured SO_2 (Kudela and Novakova 1962).

The larvae of *Acantholyda nemoralis*, are exposed for long periods and while feeding on contaminated foliage near an SO_2 source the incidence of this pest was decreased (Sierpinski, 1967). On the other hand, larvae of *Exoteleia dodecella* remain concealed within the buds for a long period and its incidence was relatively unaffected by SO_2 (Sierpinski, 1966, 1967) (Table 27.8).

Fluoride

In fir damaged by fluoride, bark beetle and weevil populations were decreased, possibly as a result of toxic concentrations of fluoride In plant tissue, while several other species were not harmed (Pfeffer, 1963). On the other hand, accumulation of a pollutant or a substance resulting from pollutant exposure, might directly or indirectly stimulate certain insects. The incidence of a spruce gall louse,

Adelges (=*Sacchiphantes*) *abietis* was greater in an area where spruce trees were damaged fluoride (Wentzel, 1965) (Table 27.9).

Table 27.8: Interaction of Sulphur dioxide (SO_2) and Ozone (O_3) and insects

Treatment	*Insects*	*Host*	*Effect*	*Reference*
Injured by SO_2	Sp. bark beetles	Trees	Increased	Bosener, 1969; Donaubauer, 1968; Kudela and Novakova, 1962; Scheffer and Hedgcock, 1955
Injured by O_3	Bark beetles	Trees	Increased	Stark *et al.*, 1968
Injured by SO_2	*Tomicus piniperdo* and *Melanophila cyanea*	Trees	Decreased incidence	Templin, 1962
Injured O_3	Bark beetles	*Ponderosa pines*	Severely injured	Cobb *et al.*, 1968
Injured SO_2	(White pine Weevil) *Pissodes strobi*	Trees	Less injury	Linzon, 1958
Injured SO_2	*Pissodes* sp.	Pine, Spruce	High incidence	Kudela and Novakova, 1962
SO_2	Larvae of *Acatholyda nemoralis*	Contaminated foliage	Pest decreased	Sierpinski, 1967
SO_2	Larvae of *Exoleleia dodecella*	Concealed within the buds	Unaffected	Sierpinski, 1966, 1967

Table 27.9: Interaction of Fluoride and Insects

Treatment	*Insects*	*Host*	*Effect*	*Reference*
F	Bark beetle and weevil population	Fir	Decreased	Pfeffer, 1963
F	*Adelges* (*Sacchiphantes abietis*)	Spruce trees	Greater	Wentzel, 1965

Pollutants on Nematodes

Sulphurdioxide and Ozone (SO_2 and O_3)

Bassus (1968) reported that saprophagus and predacious nematodes were severely damaged by SO_2 and alkaline particulate materials in the forest area. Weber *et al.* (1979) studied the effect of O_3 and SO_2 on the reproduction of five species of plant-parasitic nematodes with different feeding habits. Exposure of infected soybean plants to O_3 and O_3–SO_2 mixture inhibited and reproduction and development *Heterodera glycines* and *Paratrichodorus minor,* while *Belonolamimus longicaudatus* remained unaffected. Exposure of soybean host plants to SO_2 enhanced the reproduction of *Pratylenchus penetrans* compared with that exposed to the charcoal filtered air (control) or to O_3. Foliar injury of begonia by O_3 or O_3–SO_2 mixture inhibited the increase of *Aphelenchoides fragariae*. The suppressive effects of *A. fragariae* were greater in leaves pre-exposed to O_3 or O_3–SO_2 mixture before rather than after leaves were inoculated with nematodes. Growth of nematodes in infested soybean plants and leaves of begonia was inhibited by O_3 and O_3–SO_2 mixture compared with that of similar control plants grown in presence of nematodes and charcoal filter air. Ozone and O_3–SO_2 mixture suppressed the nodulation of soybean plants inoculated with *Belonolaimus longicaudatus* and *P. minor.* The inhibition of nodulation of soybean by *H. glycines* was extensive. Ozone and O_3–SO_2 mixture inhibited the growth of soybean both in the presence and absence of nematode. These pollutants also caused clorosis and necrosis of leaves and increase in the abscission of trifoliate leaves. The effects of pollutant-nematode combinations on nodulation of soybean by *H. glycines.* The number of nodules from roots parasitized this nematode did not differ among the pollutant treatments the harmful effect of industrial dust on the nematode *Panagrolaimus rigidus* was observed by Kozlowska (1981). Shew *et al.* (1982) studied the response of tomatoes to ozone, sulphur dioxide and infection by *Pratylenchus penetrans.* They observed synergistic interaction at 0.2 µl O_3/1, 0.2 µl SO_2/l singly and in combination. However, a mixture of 0.2 µl O_3 and 0.8 µl SO_2/l of air showed antagonistic reaction that caused less change in leaf and shoot dry weight that could be predicted by the main effects of O_3 or SO_2. The presence of *P. penetrans* attacking the roots enhanced the negative effects of O_3 plus SO_2 on the leaf growth (dry weight), but suppressed the inhibitory effects of O_3 plus SO_2 on axillary shoot dry weight. Tomato fruit weight was reduced by 0.8 µl SO_2/l of air but the amount of reduction was antagonised by the presence of O_3. Khan (1989) experimented SO_2–O_3 mixture at different concentrations and *M. incognita* race on tomato and observed that SO_2 and O_3 at 0.2 ppm and *Meloidogyne incognita* alone significantly reduced plant growth, yield parameters and leaf pigment contents of tomato, Sulphur dioxide (0.2 ppm) and 03 (0.1, 0.2 ppm) mixture acted synergistically and caused greater reduction than the sum of their individual effects (Table 27.10).

Sulphur dioxide and O_3 acted synergistically with the nematode and caused greater reduction in growth and yield parameters than the reductions recorded in nematode inoculated unexposed plants or uninoculated exposed plants. Synergistic reduction in growth of *Meloidogyne incognita* race 1 inoculated plant exposed to SO_2 plus O_3 at 0.2 plus 0.2 ppm and 0.2 plus 0.1 ppm were also noticed. Fruit count was synergistically reduced by SO_2 plus O_3 at 0.2 ppm each plus nematode.

Ozone (O_3)

The effect of ozone was worked out on tobacco cv. 'Virginia-115' inoculated with *Meloidogyne* hapla (Bisessar and Palmer, 1984). They observed that ambient O_3 inhibited growth and yield of tobacco regardless of inoculation or non-inoculation with *M. hapla.* Tobacco inoculated with the nematode suffered more O_3 injury than uninoculated ones. Inoculated plant sprayed with EDU (aethylenediurea) produced 20 per cent less galling than non sprayed ones (Table 27.10).

Flyash

Singh (1989) reported that flyash suppressed the juveniles of root-knot nematode in the ambient and artificial polluted soil. *Meloidogyne incognita* suppressed more than *M. javanica* by the application of flyash. The hatching of *M. incognita* and *M. javanica* juveniles decreased with the increasing level of flyash mixed in amended soil and these juveniles were inhibited totally at 80 and 90 per cent flyash level (Table 27.10).

Singh *et al.* (1991) found that the population and development of the root-knot nematode decreased with an increase in the concentration of flyash up to 750 t/ha. The availability of N, K and DTP A extractable Cu, Mn, Zn and Fe was more in the nematode inoculated and the flyash amended soils.

Table 27.10: Interaction of Air Pollutants and Plant Parasitic Nematodes

Treatment	Nematodes	Host	Effect	Reference
Industrial Dust	*Panagrolaimus rigidus*	–	Harmful effect on nematodes	Kozlowska, 1981
O_3 and $O_3 + SO_2$	*Heterodera glycines*	Soybean	Inhibition in reproduction and development of nematode	Weber *et al.*, 1979
O_3 and $O_3 + SO_2$	*Belonolaimus longicaudatus*	Soybean	Unaffected	Weber *et al.*, 1979
SO_3	*Paratylenchus penetrans*	Soybean	Enhanced reproduction of nematode	Weber *et al.*, 1979
O_3 and $O_3 + SO_2$	*Paratrichodorus minor*	Soybean	Reproduction and development of nematode inhibited	Weber *et al.*, 1979
O_3 and $O_3 + SO_2$	*A. fragariae*	Begonia	Decreased foliar injury by nematode	Weber *et al.*, 1979
O_3 and $O_3 + SO_2$	*H. glycines* + *Rhizobium*	Soybean	Inhibition in number of nodules	Weber *et al.*, 1979
–	*B. Langicaudatus* + *Rhizobium*	Soyabean	Inhibition in number of nodules	Weber *et al.*, 1979
–	*P. minor* + *Rhizobium*	Soyabean	–	–
0.24 g O_3/lit. 0.2 μg SO_2/lit. singly or in combination	*P. penetrans*	Tomato cv. "Walter"	Synergistic interaction	Shew *et al.*, 1982
0.2 μg O_3/lit + 0.8 μg SO_2/lit	*P. penetrans*	Tomato cv. "Walter"	Antagonistic interaction	Shew *et al.*, 1982
80 ppm O_3 (Ambient)	*M. hapla*	Tobacco cv. "Virginia–115"	Plants with nematodes suffered marc O_3 injury	Bisessar and Phamer, 1984
Simulated acid rain of pH 6.0 or 3.2 3 times weekly	*M. hapla*	Kidney beans	Decreased nematodes infection at pH 3.2	Shriner, 1978
Flyash	*Root-knot nematode*	Ambient and artificial polluted soil	Suppressed J_2	Singh, 1989
Volcanic ash different concentrations of ash	*Root-knot nematode*	Tomato	No effect to nematode infection and reproduction	O' Bannon and Santo, 1981

Volcanic Ash

O'Bannon and Santo (1981) conducted an experiment to determine the influence of volcanic ash on the survival and infectivity of root-knot nematode. Ash incorporation had no deliterious effect on root-knot nematode. Different concentrations of 0, 0.5, 1.0, 2.0, 4.0, 8.0, 25, 50 and 100 per cent ash showed no effect to the nematode infection and reproduction on tomato (Table 27.10).

Ammonia

Eno *et al.* (1955) showed great reduction of plant-parasitic nematode by all levels of anhydrous NH_3 from 136 to 741 ppm. Mojtahedi and Lownsbery (1976) reported that the fertilizer which generated NH_3 was detrimental to *Criconemoides xenoplax*. The fertilizer was lethal to nematodes *in vitro* only when it was accompanied by urease positive bacteria or partially purified urease. The detrimental action of fertilizer- urease mixture was more rapid at pH 8 than pH (Table 27.10).

Khan (1989) reported that NH_3 at 0.2 ppm caused greater reduction in pre-inoculation exposure. Synergistic interactions between nematode and air pollutants were most consistent in the treatments. In other treatments antagonistic interactions were also noticed in some parameters of tomato. Disease intensity (no. of galls/roots) enhanced in exposed plants. Reproduction of the nematode (no. of eggmass/root) was also inhibited.

Acid Rain

Shriner (1987) studied the effects of simulated rain acidified with sulphuric acid on host parasite interaction in plant diseases. Reproduction of root-knot nematodes (*Meloidogyne hapla*) on field-grown kidney beans (*Phaseolus vulgaris*) exposed to simulated rain of pH 3.2 was inhibited by 66 per cent (Table 27.10) Bolla and Huber (1988) treated pine seedlings with simulating acid rain and inoculated with the White pine specific pathotype of *Bursaphelenchus xylophilus* VPSt-1. The oleoresin concentration slightly increased while carbohydrate concentration decreased in all seedlings when treated with simulated acid rain (SAR). The changes significantly increased after inoculation of SAR-treated White and Scots pine seedlings with VPSt-1. Wilting was delayed and nematode reproduction decreased in SAR-treated white 50–60 treated. White pine seedlings inoculated with VPSt-1. Simulated acid rain (SAR) treated Austrian pine seedlings were resistant to VPSt-1 but SAR-treated Scots pine seedlings lost tolerance to VPSt-1 and wilted 50–60 days after inoculation.

The influence of simulated acidic rain on interactions of root- knot (*Meloidogyne hapla, M. incognita*) or cyst (*Heterodera glycines*) nematodes with soybean plants (*Glycine max*) was investigated (Shafer *et al.*, 1992). Three days later, plants and soil were exposed to simulated rain to pH 5.3, 4.3, 3.3 or 2.3. After three rains per week for 8 weeks shoot dry weight and production of cyst nematode eggs were suppressed by 80 per cent and 90 per cent respectively at pH 2.3. Characteristics of polynomial dose-response relationships indicated that the effects of simulated acid rain on plants and nematodes were nematode species dependent. Dose-response relationships for many dependent variables versus rain pH differed between cyst nematodes and root-knot nematodes, but most dose response characteristics for the two *Meloidogyne*. spp. were similar. Acid deposition can influence nematode plant interactions, but the acidity of simulated rain caused major changes.

Soil Pollutants on Nematodes

Fresh water resources, though limited in the world, get polluted due to increasing urbanization, industrialization and modern/agricultural, practices. Waste water contain objectionable and hazardous constituents which exert direct toxicity or bring about physico-chemical and biological changes in the aquatic environment. Inorganic pollutants like heavy metals, acids, bases and inert substances are hazardous when discharged into water. Free chlorine, chloramines, ammonia, sulphides, even in small concentrations may cause death to fish. Heavy metal salts such as Cd, Cu, Pb, Hg, Ni, Se, Th, V, Zn and As, etc. are highiy toxic to fish, aquatic organisms, animals, microorganisms and man. These metal concentrations range up to 50,000 mg/l. Some of the waste water deteriorate fresh water quality by altering its taste and colour and some others may be detrimental to plants growth and plant yield at their different levels of concentration (Prasad, 2001).

Industrial effluents wastes, present in the soil and water, contain toxic materials which cause harmful effects on plant life (Schraufnagel, 1962; Ajmal and Khan, 1984; Ajmal *et al.*, 1984). Hutchinson and Whltby (1974) and Lagerwerff and Brower (1974) reported soil pollution due to heavy metals near the industries. Gotoh *et al.* (1979) reported that mercury is also found in sufficient amount to cause pollution in the soil which is derived from the igneous rocks. Katsuhiro and Shigenori (1969) found that mercury retained in the soil by its colloidal particles. The roots absorb the heavy metals which are translocated to different parts of the plant (Haghiri, 1973; Lagerwerff, 1971). The uptake of these soil pollutants is influenced by soil factor like cation exchange capacity, pH, availability of phosphorus, etc. (Miller *et al.*, 1976). Heavy metals also affect the plants (Tyler, 1974; Vaituzis *et al.*, 1975). Many workers have reported harmful effects on the growth and yield of crop plants due to industrial effluents, waste materials and heavy metals (Froster, 1954; Hassett *et al.*, 1976; Beckett and Davis, 1978; Hale *et al.*, 1985; Ajmal *et al.*, 1984).

Industrial dairy processing effluent reduced the growth of kidney bean and pearl millet (Ajmal *et al.*, 1984). Ajmal and Khan (1984) irrigated wheat and pea with the effluent from Mohan Meakin Breweries Ltd., Ghaziabad, U.P. and found reduction in the germination and growth when irrigated with 100 per cent effluent. However, irrigation with 50 per cent was beneficial to plant growth. Goodbold and Hutterman (1985) observed that mercury was much toxic in inhibiting root elongation in comparison of zinc and cadmium. Similar results were also observed by Graft and Schwantes (1983).

Free-living nematodes are used in toxicological research on environmental pollutants and pesticides by (Samoiloff, 1980; Samoiloff *et al.*, 1.980; Haight *et al.*, 1982; Mudry *et al.*, 1982; Doelman *et al.*, 1984; Samoiloff and Bogaert, 1984; Coomans and Vanderhaeghen, 1985; Vranken and Help, 1986; Vranken *et al.*, 1985; Frey, 1971; 1976; Neuschulz and Kampfe, 1980; Simpkin and Coles, 1981; Ohba and Ishibashi, 1984). Nematodes have already been used in a bioassay to determine the relative toxic effects of sediments (Howell, 1982; Tietjen and Lee, 1984) or fractions of sediments (Samoiloff *et al.* 1983).

Bisessar (1982) found higher concentrations of lead, arsenic, cadmium, copper and low population of bacteria, fungi, nematodes and actinomycetes near lead smelter plant. There are several reviews concerning the interaction between pollutants and plant diseases caused by bacteria, fungi and viruses (Manning, 1975; Heagle, 1973; 1982; Huttunen, 1984), however, this type of information is meagre in case of nematode and pollutant interactions (Hodda and Nicholas, 1986; Van Kessel *et al.*, 1989). Few reports are available with respect, to the effect of heavy metals and plant parasitic nematodes.

Van Gundy and Thomson (1962) pointed out that the concentration of 0.5 ppm copper killed *Trichodorus christiei* within 24h while *Hemicycliophora arenaria* tolerated concentrations of 4 ppm copper for 48h without apparent injury. Clarke and Shepherd (1965) showed the zinc ions stimulated hatching of seven species of cyst nematodes (*Heterodera* spp.). Zinc chloride stimulated more hatching than root diffusate with six species. It was most active at a concentration of 3mM and was still significantly active at 0.2 mM.

Clarke and Shepherd (1966) showed that some inorganic ions inhibited and some promoted hatching to the *Heterodera* spp. The spontaneous hatch of larvae from eggs in cysts was influenced by the inorganic ions including Ni^{++} in the soil. Toxicity of copper to *Trichodorus pachydermus* was demonstrated by adding copper salts in known quantities to water. The quick death of the nematodes in copper dishes was striking, while they died more slowly in nickle-plated dishes. Under experimental conditions, viable nematodes were obtained by using nickle-plated supporting sieves sprayed with plastic (Hafkenscheid, 1971). Pitcher and McNamara (1972) investigated the toxicity of low concentrations to silver and cupperic ions to three species of plant-parasitic nematodes. *In vitro* study revealed that *Pratylenchus penetrans* was most susceptible to silver and *Xiphinema diversicaudatum* to copper while *Aphelencoides ritzemabosi* was least affected by either metal. In a greenhouse bioassay root-knot nematode (*Meloidogyne hapla*) retarted growth of lettuce leaves by 50 per cent in metal contaminated soil and 30 per cent in control soils (Temple and Bisessar, 1981) (Table 27.11).

Table 27.11: Interaction of Oil Pollutants and Plant Parasitic Nematodes

Treatment	*Nematodes*	*Host*	*Effect*	*Reference*
Heavy metals	*M. hapla*	Celery	Increased the incidence and severity of root-knot disease	Bisessar *et al.*, 1983
Zinc Salt	*H. glycines*	–	Stimulant of egg hatching	Paul and Zeon, 1984
Sewage sludge and heavy metals	*Nematodes*	Arabal soil	Total nematode numbers were highest in treated site and in control site found in small quantity	Weiss and Zarink, 1991
Copper sulphate and leadacetate	*B. xylophilus*	Pines	Increase in oleoresin production stressed energy homeostasis and induces rapid wetting of the seedings	Bolla and Fitzsimmons, 1989
1 μm $CdCl_2$	*Caenorhabditis elegans*	Organisms	Growth and reproduction of organisms was significantly reduced	Van Kessel *et al.*, 1989
Level of 160 and 320 μm	*Caenorhabditis elegans*	Organisms	Growth was retarded at early juvenile stage and the organisms did not reach the adult stage and could not reproduce	Van Kessel *et al.*, 1989
Cu, Chromium and Cd	*Panagrellus*	–	Blocked growth at all developmental stages	Haight *et al.*, 1982
Ni and lead (highest soluble concs)	*Panagrellus*	–	No affect on growth of the worms	Haight *et al.*, 1982
Zn (7500 mg/l)	*Panagrellus*	–	Partially blocked growth	Haight *et al.*, 1982
Hg	*Panagrellus*	–	Lethal (20 mg/l) or ineffective in blocking growth (10 ml/l)	Haight *et al.*, 1982
Heavy metals (increased)	*P. silusiae*	Kinetics of pharyngeal pumping	Proportion of feeding worms decreased	Haight *et al.*, 1982
Copper	*T. pachydermus*	–	Quick death	Hafken scheid, 1971
Nickle	*T. pachydermus*	–	Died more slowly	Hafken scheid, 1971
Toxicity of low concs. Silver and cupperic ions	*P. penetrans*	–	Most susceptible to silver	Pitcher and McNamara, 1972
–	*X. diversicaudatum*	–	Most susceptible to copper	Pitcher and McNamara, 1972
–	*A. ritzemabosi*	–	Least affected by either metal	Pitcher and McNamara, 1972
Metal contaminated soil	*M. hapla*	Lettuce leaves	Started growth by 5 per cent and 30 per cent in control soil	Temple and Besessiar, 1981

Haight *et al.* (1982) obtained Lc50 values for each heavy metal and found *Panagrellus* to be highly resistant. The chronic effect of seven heavy metals revealed that copper, chromium and cadmium effectively blocked growth at all developmental stages. Nickle and lead at highest soluble concentrations did not affect growth of the worms. The high concentrations of zinc (7500 mg/l) partially blocked growth and mercury was either lethal (20 mg /1) or ineffective in blocking growth (10 ml/1). In another experiment, they studied the effect of heavy metals on the kinetics of pharyngeal pumping in *Panagrellus silusiae*. They found that as the duration of heavy metal increased, the proportion of feeding worms decreased (Table 27.11).

Howell (1982) explained that one probable reason of the survival of nematodes at lower concentrations of the heavy metal may be the secretion of mucus which caused the binding of heavy metal that influenced the intake. Recent work. (Howell, 1982a, 1982b, 1983, 1984) has indicated that some species of nematodes are very suitable for studies concerning the physiology of accumulation of heavy metals. Howell and Smith (1983) studied the binding of heavy metals by the marine nematode *Enoplus brevis* and revealed that the molecular weight of the proteins were estimated to be 29,000 and 63,000 Daltons under denaturing conditions, and under non-denaturing conditions they were estimated to be 28,000 and 450,000 Daltons.

Bisessar *et al.* (1983) explained the effects of, heavy metals and *Meloidogyne hapla* on celery growth on organic soil near nickle refinery. He observed that soil contaminated with heavy metal increased in the incidence and severity of root-knot disease on celery (Table 27.11).

Zinc salts were reported as a stimulant of egg hatching in the soybean cyst nematode, *Heterodera glycines* (Paul and Leon, 1984) (Table 27.11). Concentrations of calcium chloride, magnesium chloride and manganese chloride had no effect on hatching of eggs, but reduced hatching at higher concentration by osmotic influences. Hatching of eggs increased as the time of exposure to zinc chloride increased and was maximum at 28°C and pH of 5.3. Picrolonic acid increased *Helerodera glycinenes* hatching while sodium metavanadate had no metal through the marine nematode *Enoplus brevis*. Cadmium as $CdCl_2$ was first seen in the cuticle. Polyacrylamide gel electrophoresis demonstrated that the metal was bound in the cuticle by a protein with molecular weight (MW) of 450,000 Daltons. Cadmium (^{109}Cd) was detected in the hypodermis and muscle layer where it was associated with protein of MW 28,000 Daltons and finally it accumulated in the gut, associated with a third protein of MW, 200,000 Daltons. The meiofaunal nematode populations in mangrove mud-flats adjacent to steel works and chemical factories in the Hunter River Estuary in Australia have been analysed and compared with populations in unpolluted mud-flats nearby. They found that the polluted areas showed more diverse taxonomically, though seasonal variations in population density and other environmental factors complicate the comparison (Hodda and Nicholas 1986). Sturhan (1986, 1988) found that the soil nematode fauna can be affected by heavy metals and other elements; certain nematode taxa may be susceptible and different nematode taxa may react to particular elements in different ways. Nematodes may be suitable as bioindicators for certain environment chemicals. Zullini and Peretti (1986) reported that the nematode community was sensitive to lead pollution. The total biomass of nematodes, the number of species, and the Shannon diversity index significantly decrease for an increase of the lead content in the moss, while the density of individuals does not seem to be affected by pollution. The nematode suborder Dorylaimina is found to be most sensitive to lead.

Alphey and Brown (1987) studied the effect of pollutants on plant-parasitic nematodes and found that the *Xiphinema, diversicaudatum* and *Longidorus elongatus* population was significantly decreased by watering with $CuSO_4$ solution and tap water containing 0.5 μg ml^{-1} Cu^{++}. Numbers of trichodorid nematodes were not significantly affected by either of the treatments containing copper. They concluded that small quantities of pollutants adsorbed by the soil particles decreased the toxic effect of these materials up on the soil inhabiting nematodes.

Bolla and Huber (1988) investigated the effect of heavy metals on the susceptibility of pines to pine wood heavy metals on the susceptibility of pines to pine wood nematode. They found that heavy metal stressed Scots and Austrian pine type VPSt-1. Both the pines infected with heavy-metal-stress and VPSt-1 decreased the concentration of carbohydrate and produced phytotoxic oleoresins. This suggests that the introduction of pinewood nematode, where pines often grow under environmental-stress conditions, may lead to establishment of the nematode and seriously alter forest productivity (Prasad, 1992 & 2001).

Bolla and Fitzsimmons (1989) studied the effect of copper sulfate and lead acetate on infection of pines with *Bursaphelenchus xylophilus* and found increase in oleoresin production, stressed energy homeostasis and induced rapid wilting of the seedlings. Scots pine lost tolerance and pine lost resistance tò VPSt-1 after the seedlings were treated with either copper sulfate or lead acetate (Table 27.11). These results suggests that environmental pollution may significantly affect susceptibility of pines to *B. xylophilus* and may have a role in establishment of this nematode in uninfested areas. The studies on saprophytic nematode *Caenorhabditis elegans*, various concentrations of $CdCl_2$ revealed that the growth and reproduction of the organisms was significantly reduced from a level of 1 μM $CdCl_2$. At levels of 160 and 320 μM, growth was retarded at the early juvenile stages and the organisms did not reach the adult stage and could therefore not reproduce (Van Kessel *et al.*, 1989) (Table 27.11) Weiss and Larink (1991) studied the influence of sewage sludge and heavy metals on nematodes in an arable soil and reported that total nematode numbers were highest in the site treated with sewage sludge and heavy metals but in control site it was found in small quantity. The plant-parasitic nematode genera showed different patterns of abundance depending on the sludge treatment and heavy metals content. For the mycophagic and bacteriophagic nematodes number increased with the amount of sludge especially in the sites with a higher heavy metal content. The family Rhabditae was the most numerous group in the sludge plus heavy metal treatment. In contrast to these findings the omnivorous nematodes were very rare in the sludge treated plots and were completely absent in plots treated with sludge plus heavy metals, whereas predatory nematodes were numerous only after the application of sludge alone.

Conclusion

India with a face-lift in industrial development in the present times is facing problems in pollution control. Advances in pollution control technology and the enforcement of new and stricter regulations governing industrial and automotive emissions have significantly contributed to the fight for cleaner air. Air pollution includes a number of pollutants which are now becoming permanent component of our biosphere. A number of new substances are formed due to accumulated pollutants in the stratosphere by simple chemical reactions.

Air pollutants affect growth, reproduction and virulence of certain fungus parasites, but the effects on host predisposition and, mechanism of the host-parasite interaction. The effect pollutants on fungus parasitism of plants is limited. Data relating ozone to parasitism by fungi are just beginning to emerge and little in known of the effect sulphur dioxide or fluoride. Only reports are available on the effects of particulate pollution on diseases caused by fungi infection by *Cercospora beticola* was greater on beet leaves covered with limestone dust than on non-dusted leaves (Schöenbeck, 1960). Data on pollutant effects on diseases caused by bacteria, nematodes, viruses and insects are almost nonexistent.

Plants and plant parasites display a wide range of sensitivity to pollutants. The degree and type of pollutant effect appears to depend upon the pollutants, pollutants dose, host species and parasites. Some pollutant mixtures are known to interact synergistically to increase plant injury. Pollutants might also act synergistically to effect parasitism.

REFERENCES

Ajmal, M. and A.U. Khan (1984). Effects of brewery effluent on agricultural soil and crop plants. *Environ. Pollut.*, 33(A): 341–351.

Ajmal, M., M.A. Khan and A.A. Nomani (1984). Effect of industrial dairy processing on soil and crop plants. *Environ. Pollut.*, 33: 97–106.

Alam, M.M. (1985). A simple method for *in vitro* screening of chemicals for nema-toxicity. *Int. Nematol. Network Newsl.*, U.S.A., 2(1): 6.

Alphey, T.J.W. and D.J.F. Brown (1987). Effect of pollutants on plant-parasitic nematodes. *Test Agrochem. Cult. Ann. Appl. Biol.*, 110(Supplement): 38–39.

Anon. (1971). Estimated crop losses due to plant-parasitic nematodes in the United States. *Sppl. J. Nematol. Special Publ.*, 1: 7 p.

Antonio, H. (1988). Avalicao das, perdas causadas for *Meloidogyne incognita* raea a no cultivar BR 4 desoja. *Nematologica Brasileira*, 12: 29–34.

Backett, P.H.T. and K.I. Davis (1978). The activity of the toxic effects of copper, nickle and zinc in young barley. *New Phytol.*, 81: 155–173.

Bassus, W. (1968). On the effects of industrial emissions on the population of nematodes in the soil of pine forests. *Pedobiologia*, 8: 289–295.

Bevan, R.J. and G.H. Greenhalgh (1976). *Rhytisma acerinum* as a biological indicator of pollution. *Environ. Pollut.*, 10: 271–285.

Bisessar, S. (1982). Effect of heavy metals on microorganisms in soil near a secondary lead smelter. *Water, Air and Soil Pollut.*, 17: 305–308.

Bisessar, S. and P.V. Temple (1977). Reduced ozone injury on virus infected tobacco in the field. *Plant Dise. Rep.*, 61: 961–963.

Bisessar, S. and K.T. Palmer (1984). Ozone antioxidant spray and *Meloidogyne hapla* effects on tobacco (*Nicotiana tabacum*) cv. Virginia–115. *Atmos. Environ.*, 18: 1025–1028.

Bisessar, S., R.J. Rinne and J.W. Potter (1983). Effect of heavy metals and *Meloidogyne hapla* on celery grown on organic soil near a nickle refinery. *Plant Dis.*, 67(1): 11–14.

Bolla, R.I. and K. Fitzsimmons (1986). Effect of simulated acid rain on *Bursaphelenchus xylophilus*: infection of pine seedlings. *J. Nematology*, 20(4): 590–598.

Bolla, R.I. and M.C. Huber (1988). Heavy metal effects on pine susceptibility to pinewood nematode. *J. Nematology*, 20(4): 629–630.

Bosener, Rolf (1969). Zum Vorkommen rivdenbritender schadinsecten in rauchgeschadigten Kiefernund Fichtenbestanden. *Arch. Forstwes. Ed.*, 18: 1021–1026.

Brennan, E. and I.A. Leone (1970). Interaction of tobacco mosaic virus and ozone in *Nicotiana sylvestris*. *J. Air Pollut. Control Assoc.*, 20: 470.

Chiba, O. and K. Taneka (1968). The effect of SO_2 on the development of pine needle blight caused by *Rhizosphera kalkhofful* Bubok (I). *J. Jap. Forest Soc.*, 50: 135–139.

Clarke, A.J. and A.M. Shepherd (1965). Zinc and other metallic ions as hatching, agents or the beet cyst nematode *Heterodera schachtii* Schm. *Nature Land*, 208: 502.

Clarke, A.J. and A.M. Shepherd (1966). Inorganic ions and the hatching of *Heterodera* spp. *Ann. Appl. Biol.*, 58: 497–508.

Cobb, F.W. Jr., D.L., Wood, R.W. Stark, J.R. Parmeter Jr. (1968). Photochemical oxidant injury and bark beetle (Coleoptera : Scolytidae) infestation of Ponderoser pine IV. Theory on the relationships between oxidant injury and bark beetle intestation. *Hilgardia*, 39: 141–152.

Coomans, A. and R. Vanderhaeghen (1985). *In vitro* testing toxicities. Exercise 29. In: *Plant Nematology Laboratory Manual*, (Ed.) B.M. Zuckemann, W.E. Mai and M.B. Harrison, University of Massachusetts, Agricultural Experiment Station, Amherst, M.A., pp. 172–176.

Couey, H.M. (1965). Inhibition of germination of *Alternaria* sports by SO_2 under various moisture conditions. *Phytopathology*, 51: 815–819.

Couey, H.M. and M. Uota (1961). Effect of concentration, exposure time, temperature and relative humidity on the toxicity of sulphur dioxide to the spores of *Botrytis cinerea*. *Phytopathology*, 51: 815–819.

Davis, D.D. and S.H. Smith (1974b). Reduction of ozone sensitivity of pinto bean by bean common mosaic virus. *Phytopathology*, 74: 383–385.

Davis, D.D. and S.H. Smith (1974a). Reduction of ozone sensitivity of pinto bean by.virus induced local lesions. *Plant Dise. Rep.*, 60: 31–34.

Dean, G. and M. Treshow (1965). Effects of fluoride on the virulence of tobacco mosaic virus *in vitro*. *Proc. Utah. Acad. Sci. Arts Lett.*, 42: 236–239.

Doelman, P., G. Nieboer, J. Schrooten, and M. Visser (1984). Antagonistic and synergistic toxic effects of lead and cadmium in a simple food chain: Nematode feeding on bacteria and fungi. *Bull. Environ. Contam. Toxicol.*, 32: 717–723.

Donaubauer, E. (1968). Sekundar-schaden in Osterreichischen Rauchschadensgebieten. Schwierigkeiten Der. Diagnose. Und Bewertung. Nied-zynarodowej Knof. Wplyw Zanie czyszczen Powietrza na Lasy, 6th Katowice, Poland.

Doncaster, C.C. (1962). A counting dish for nematodes. *Nematologica*, 7: 334–336.

Dropkin, V.H (1980). *Introduction to Plant Nematology*. John Wiley and Sons, New York, 298 pp.

Eno, C.F., W.G. Blue and J.M. Good (1955). The effect of anhydrous ammonia on nematode, fungi, bacteria and nitrification in some Florida soil. *Proceedings Soil Science Society of America*, 19: 55–58.

Epstein, E. and R.L. Chaney (1978). Land disposal of toxic substances and water related problems. *J. Wat. Pollut. Cont. Fed.*, 6: 518 –522.

Feldmesser, J., D.I. Edwards, J. M. Epps, C.M. Held, W.R. Jenkins, H.J. Jonsen, B. Lear, C.W. McBeth, E.L. Nighi, and V.G. Perry (1971). Estimated crop losses due to plant-parasitic nematodes in the United States. *Supp. J. Nematol.* Special publication No. 1. Report of Soc. Nematologists Committee on Crop losses, 1970.

Frank, R.I., K. Ishida and P. Suda (1976). Metals in agricultural soils of Ontario. *Can. J. Soil Sci.*, 56: 181–196.

Frey, F. (1971). Uber die Eignung van *Acrobeloides buetschlii* (Cephalobidae) fur nematologische Laboruntersuchungen. *Nematologica*, 17: 474–477.

Frey, F. (1976). Untersuchungen uber die wirkung von im obstbau verwendeten Herbiziden auf den Testnematoden *Acrobeapides buetschlii* (de Man 1884) Steiner and Buhrer 1933. *Z.Pflanzenkrankh Phlansenschutz*, 83: 434– 441.

Froster, W.A. (1954). Toxic effects of heavy metals on crop plants grown in soil culture. *Ann. Appl. Biol.*, 41: 637–651.

Goodbold, D.L. and A. Hutterman (1985). Effect of zinc, cadmium and mercury on root elongation of *Picea abies* (Karst.), seedling and the significance of these metals to forest, die-back. *Environ. Pollut.*, 38: 375–381.

Gotoh, S., S. Tokudome and H. Koga (1979). Mercury in soil derived from igneous rock in northern Kyushu, Japan. *Soil Sci. & Plant Nutr.*, 24: 391–406.

Graft, H.J. and H.O. Schwantes (1983). Der rinflu Bvon Cadmium, Zinc, Beluind Quecksilbe aufdie Atmung unct Garung von. *Saccharomyces cerevisiae. Angew. Bot.*, 57: 31–43.

Hafkenscheid, H.H.M. (1971). Influence of Cu^{++} ions on *Trichodorus pachydermus* and an extraction method to obtain active specimens. *Nematologica*, 17: 535–541.

Haghiri, F. (1973). Cadmium uptake by plants. *J. Environ. Qual.*, 2: 93–96.

Haight, M., T. Mudry and J. Pasternak (1982). Toxicity of seven heavy metals on *Panagrellus silusiae:* The efficacy of the free-living nematode as an *in vivo* toxicological bioassay. *Nematologica*, 28: 1–11.

Hale, J.C., P.J. Ormrod, P.J. Laffey and O. Allen (1985). Effect of nickle and copper mixtures on tomato in sand culture. *Environ. Pollut.*, 39: 53–59.

Ham, D.L. (1971). The biological interactions of SO_2 and *Scirrhia acicola* on loblolly pine. *Ph.D. Thesis*, Duke University Durban, N.C. 74 pp.

Harding, R.P. (1968). Effect of ozone on *Penicillium* mold decay and sporulation. *Plant Dis. Reptr.*, 52: 245–247.

Hassett, J.J., J.E. Miller and D.E. Koeppe (1976). Interaction of lead and cadmium by roots. *Environ. Pollut.*, 11: 297–302.

Heagle, A.S. (1970). Effect of low level ozone fumigations on crown rust of Oat. *Phytopathology*, 60: 252–254.

Heagle, A.S. (1973). Interaction between air pollutants and plant parasites. *Ann. Rev. Phytopathology*, 11: 365–388.

Heagle, A.S. (1982). Interaction between air pollutants and parasitic plant disease. In: *Effects of Gaseous Air Pollutant in Agriculture*, (Eds.) M.H. Unsworth and D.P. Ormrod, Butterworth Scientific, London, pp. 333–348.

Heagle, A.S. and A. Strickland (1972). Relation of *Erysiphe graminis* f. sp. *hordei* to low levels of ozone. *Phytopathology*, 62: 1144–1148.

Heagle, A.S. and L.W. Key (1973). Effect of *Puccinia graminis* f. sp. *tritici* on ozone injury in wheat. *Phytopathology* 63: 609– 613.

Heck, W.W., A.S. Heagle and D.S. Shriner (1986). Effects on vegetation: Native, Crops, forests. In: *Air Pollution*, (Ed.) A.S. Stern, Academic Press, New York, 6: 247–350.

Heggested, H.E. (1988). Reduction of soybean seed yields by ozone air pollution. *J. Air Pollut. Control. Assoc.*, 38: 1040–1041.

Hibben, C.R. and G. Stotzky (1969). Effects of ozone on the germination of fungus spores. *Can. J. Microbiol.*, 15(1): 1187–1196.

Hibben, C.R. and M.P. Taylor (1975). Ozone and sulphur dioxide effects on the lilac powdery mildew fungus. *Environ. Pollut.*, 9(A): 107–114.

Hibben, C.R. (1966). Sensitivity of fungul spores to sulphur dioxide and ozone. *Phytopathology*, 56: 880 (Abstract).

Hodda, M. and W.L. Nicholas (1986). Nematodes diversity and industrial pollution in the hunter river estuary, NSW, Australia. *Marine Pollut. Bull.*, 17(6): 254–255.

Howell, R. (1982). The secretion of mucus by marine nematode (*Enoplus* spp.): A possible mechanism influencing the uptake and loss of heavy metals pollutants. *Nematologica*, 28: 110–114.

Howell, R. (1982a). The incidence of heavy metal pollutants in *Enoplus brevis* (Bastian) and *Enoplus communis (Bastian)* with some observations on toxicity accumulation, depuration and metal binding by proteins. *Nematologica*, 28: 110–114.

Howell, R. (1982b). Levels of heavy metal pollutants in two species of marine nematodes. *Mar. Pollut. Bull.*, 13: 396–398.

Howell, R. (1983). Heavy metals in marine nematodes: Uptake, tissue distribution and loss of copper and zinc. *Mar. Pollut. Bull.*, 14: 263–268.

Howell, R. (1984). Acute toxicity of heavy metals to two species of marine nematodes. *Mar. Environ. Res.*, 11: 153–161.

Howell, R. and L. Smith (1983). Binding of heavy metals by the marine nematode *Enoplus brevis* Bastian, 1865. *Nematologica*, 29: 39–48.

Howell, R. and L. Smith (1985). Flux of cadmium through the marine nematode *Enoplus brevis* Bastian, 1865. *Nematologica*, 8(1): 45–51.

Howell, R.H. and J.H. Grahm (1977). Interactions between air pollutants and parasitic plant diseases by A.S. Heagle. In: *Effects of Gaseous Air Pollutants in Agriculture* (Eds.) M.H. Unsworth and D.P. Ormrod, 1982, Butterworth Scientific, London, pp. 333–348.

Hunter, J.G. and O. Vergnano (1952). Nickle toxicity in plants. *Ann. Appl. Biol.*, 39: 279–284.

Hutchinson, M.T., J.P. Reed, H.T. Streu, A.A. Edward and P.H. Schroeder (1961). Plant-parasitic nematodes of New Jersey. *New Jersey Agric. Stat. Rutsers Bull.*, 796.

Hutchinson, T.C. and L.M. Whitby (1974). Heavy metal pollution in the sudbury mining and smelting region of Canada I: Soil and vegetation contamination by nickle, copper and other metals. *Environ. Conserv.*, 1: 123–132.

Huttunen, S. (1984). Interactions of disease and other stress factors with atmospheric pollution. In: *Air Pollution and Plant Life*, (Ed). M. Treshow, New York, pp. 346–390.

James, R.L., F.W. Cobb Jr., P.R. Miller and J.R. Parmeter Jr. (1980). Effects of oxidant air pollution on susceptibility of pine roots to *Fames annosus. Phytapathalay.*, 80: 560–563.

Jancark, V. (1961). Vyskyt drevokaznych hub Kourem Poskozovant oblasti Krusnych hor. *Lesnietvi*, 7: 677–692.

Johansson, O. (1954). Report over ett studuim av luft och nederbord omkring svenska klifferolje Aktiebologets anlaggninger vid Kavarntarp ay syayel och dess inoerka pa Vaxtarna. *Lic. Dvh. Vid. K. Laintbruk Shogskolan*, Uppsala, John Wiley and Sons.

Katsuhiro, I. and A. Shigenori (1969). Retention of mercury by soil colloids. *Soil, Sci. & Plant Nutr.*, 15: 86–91.

Katznelsoll, H. and V.E. Henderson (1963). Ammonium as an attractant for a soil nematode. *Nature Land*, 198: 907.

Khan, M.R. (1989). Studies on root-knot nematodes in relation to environmental pollution on vegetable crops. *Ph.D. Thesis*, A.M.U., Aligarh, pp 269.

Khan, M.R., A.A. Khan and M.J. Pasha (1988b). Studies on powdery mildews and root-knot nematodes in relation to air pollution. (Abstract) *Proc. 40th Annual Meeting of Indian Phytopathological Society*, March 1988, Jaipur, India (In Press).

Kochchar, M., R.A. Reinert and U. Blum (1982). Effect of fescue *Festuca arundinacea* and or clover *Trifolium repens* debries and fescue leaf leachate on clover as modified by ozone and *Rhizoctonia solani*. *Environ. Pollut.*, 28: 255–264.

Kozlowska, J. (1981). Effect of industrial dusts on the nematode *Panagrolaimus rigidus* in laboratory nematode *Panagrolaimus rigidus* in laboratory conditions. *Pollut. Ecol. Stud.*, 7: 121–126.

Kudela, M., E. Novakova (1962). Lesni skudei a skody zveri-v-Iesich poskozvanych Kourem. *Lesnictvi*, 6: 493–502.

Lagerwerff, J.V. (1967). Heavy metal contamination of soils. In: *Agriculture and Quality of Our Environment*, (Ed.) N.C. Brady, Am. Assoc. Adv. Sci. Washington DC, 85: 353–360.

Lagerwerff, J.V. (1971). Uptake of cadmium, lead, zinc, by radish from soil and air. *Soil Sci.*, 111: 129–133.

Lagerwerff, J.V. and D.L. Brower (1974). Effect of smelter on the agricultural conditions in the surrounding environment. *Proc. Symp. Trace Substances in Environmental Health*, 8th Univ. of Missouri, Columbia, pp. 203–212.

Laurence, J.A. and A.L. Aluiso (1981). Effect of sulphur dioxide on expansion of lesions caused by *Corynebacterium nebraskernse* soybean. *Phytopathology*, 71: 445–448.

Laurence, J.A. and F.A. Wood (1978). Interactions between air pollutants and parasitic plant diseases by A.S. Beagle, 1982. In: *Effects of Gaseous Air pollutants in Agriculture*, (Eds.) M.H. Unsworth and D.P. Ormrod, 1982, Butterworth Scientific, London, pp. 333–348.

Laurence, J.A., A.L. Aluiso, L.H. Weinstein and D.C. McCune (1981). Effects of sulphur dioxide on Southern bean mosaic and maize dwarf mosaic. *Environ. Pollut.*, 24: 185–191.

Laurence, J.A., L.J. Weinstein, D.C. McCune and A.L. Aluiso (1979). Effects of sulphur dioxide on Southern corn leaf blight of maize and stem rust of wheat. *Plant Dise. Rep.*, 63: 975–978.

Linzon, S.N. (1958). The influence of smelter fumes on the growth of white pine in the sudbury region. *Can. Dep. Agr. Publ. Ontario Dep. Lands Forests*, 45 pp.

Magdycz, W.P. and W.J. Manning (1973). *Botrytis cinerea* protects broad beans against visible ozone injury. *Phytopathology*, 63: 204 (Abstract).

Magie, R.O. (1963). *Botrytis* disease control on gladiolus carnations and chrysanthemums. *Proc. Florida State Hort. Soc.*, 76: 458–461.

Manning W.J., W.A. Feder and P.M. Papia (1972). Influence of long term low level of ozone and benomyl on growth and nodulation of pinto bean plants. *Phytopathology*, 62: 497 (Abstract).

Manning, W.J. (1975). Interactions between air pollutants and fungal, bacterial, and viral plant pathogens. *Environ. Pollut.*, 9: 87–90.

Manning, W.J., W.A. Feder and I. Perkins (1970). Ozone injury increases infection of geranium leaves by *Botrytis cinerea. Phytopathology*, 60: 669–670.

Manning, W.J., W.A. Feder, I. Perkins and M. Glickman (1969). Ozone injury and infection of potato leaves by *Botrytis cinerea. Plant Dis. Rep.*, 53: 691–693.

Manning, W.J., W.A. Feder, P.M. Papia and I. Perkins (1971a). Effect of low levels of ozone on growth and susceptibility of cabbage plants to *Fusarium oxysporum* f. sp. *Conglutinans. Plant Disease Rep.*, 55: 47–49.

Manning, W.J., W.A. Feder, P.M. Papia and I. Perkins (1971b). Influence of foliar ozone injury on root development and root surface fungi on pinto bean plants. *Environ. Pollut.*, 1: 305–312.

Mathess, G. (1972). Bleigehalte in Gestein, Boden and Grundwrasser. In: *Blei und Umwelt. Berlin Dahlem (Ver Wasser, Boden–Lufthygiene)*, (Ed.) K. Aurand, pp. 21–27.

Mathess, G. (1974). Heavy metals as trace constituents in natural and polluted Groundwaters. *Geol. Mijnbouw* 53: 149–155.

Miller, J.E., J.J. Hassett and D.E. Koeppe (1976). Uptake of cadmium by soybean as influenced by soil cation exchange capacity pH and available phosphorus. *J. Environ. Qual.*, 5: 157–160.

McCune, D.C., L.H. Weinstein, J.F. Mancine and P. VanLeuken (1973). Effect of hydrogen fluoride on plant-pathogen interactions. *Proc. Int. Clean Air Congr.*, 3rd, A146–A149.

Mojtahedi, H. and B.E. Lownsbery (1976). The effects of ammonia generating fertilizer on *Criconemoides xenoplax* on pot cultures. *J. Nematology*, 8: 306–309.

Moyer, J.W. and S.H. Smith (1975). Oxidant injury reduction on tobacco induced by tobacco etch virus infection. *Environ. Pollut.*, 9: 103–106.

Mudd, J.B. and T.E. Kozlowski (1975). *Responses of Plants of Air Pollution*. Academic Press, New York, pp. 383.

Mudry, T., M. Haight and J. Pasternak (1982). The effects of some heavy metals on the kinetics of pharyngeal pumping in *Panagrellus silusiae. Nematologica*, 28: 12–20.

Neuschulz, N. and L. Kampfe (1980). Das Reporduktions potential von *Caenorhabditis briggsae* unter unterschiedliche nematizide rinftusse. 5. *Vortr. tag. Zu Aktuel'en Problenlen der Phytonematology Rostock, GDR* 1980, S. 46–55.

O'Bannon, J.H. and G.S. Santo (1981), Influence of volcanic ash on infectivity and reproduction of two species of *Meloidogyne. J. Nematology*, 13(3): 409–412.

Oden, S. (1968). The acidification of air and precipitation and its consequences on the natural environment. Nat. Sci. Res. Coun., Bull No. 1, Ecology Committee, Swedish National Science Research Council, Stockholm.

Ohba, K. and N. Ishibashi (1984), A nematode *Caenorhabditis, elegans* as test organism for nematocide evaluation. *Nippon Noyaku Gakkaishi* 9(1): 91–96.

Olson, B. and A.W. Saettler (1979). In: *Proceedings of the 9*[th] *International Congress of Plant Protection, Washington, D.C.* Aug. 5–11, 1979; p. 192 (Abstract).

Ouden DEN, H. (1958). A new method for culturing plants enabling the observation of nematodes on growing roots. *Tydschr. Plziekt.*, 64: 269–272.

Pansey, V.G. and P.V. Sukhatme (1978). Statistical methods for agricultural workers (Revised by P.V. Sukhatma and V.V. Amble), ACAR, New Delhi.

Parvatha Reddy, P. and D.B. Singh (1981). Assessment of yield loss in okra, brinjal, french bean and cowpea due to root-knot nematode. *Proc. 3*[rd] *Intn. Symp. Pl. Path.*, New Delhi, pp. 93–94.

Paul, M.T. and W.B. Leon (1984). Zinc-mediated hatching of eggs of soybean cyst nematodes, *Heterodera glycines. J. Chemical Ecol.*, 10: 361–372.

Pell, E.J. (1979). How air pollutants induce disease. In: *Planl Disease: An Advanced Treatise,* (Eds.) J.G. Horsefall and E.B. Cowling), Academic Press, New York San Francisco, London, 4: 273–292.

Pell, E.J., F.L. Lukezic, R.G. Levine and W.C. Weissberger (1977). Effects on Vegetation: Native, Crops, Forests by W.W.Heck, A.S. Heagle and D.S. Shriner. In: *Air Pollulion* (Ed.) A.S. Sterm. Vol. 6, pp. 247–350, Academic Press, New York.

Pfeffer, A. (1963). Insektenschadlinge an tannen in bereich der gas exha-lation-en. Z. Angew. *Entomol.*, 51: 203–207.

Pitcher, R.S. and D.G. McNamara (1972). The toxicity of low concentrations of silver and cuperic ion to three species of plant parasitic nematodes. *Nematologica,* 18: 385–390.

Prasad, D. and M.L. Chaudhary (1992). *Environmental Pollution: Air.* Venus Publishing House, New Dehli, pp. 255.

Prasad, D. (1991). Plant injury caused by air pollution. In: *The Fragile Environment,* (Ed.) I. Mohan. Ashish Publishing House, New Dehli, p. 169–182.

Prasad, D. (2001). Effect of pollution on sustainable crop production. *Proc. Nat. Symp. on Combating Pollutants, Accumulation in Ecosystem for Sustainable Agriculture.* Allahabad Agricultural Deemed University, p. 32–39.

Pratt, G.C. and S.V. Krupa (1979). In: *Proceedings of the 9*[th] *International Congress of Plant Protection, Washington, D.C.*, Aug. 5–11, 1979, Abst. 196.

Reinert R.A., D.T. Tingey and C.E. Koons (1971). The early growth of soybean influenced by ozone stress. *Agronomy,* 63: 148 (Abstract).

Reinert, R.A. and G.V. Gooding Jr. (1978). Effect of ozone and tobacco streak virus alone and in combination on *Nicotina tabacum. Phytopathology,* 68: 15–17.

Samoiloff, M.R. (1980). Action of chemical and physical agents on free-living. In: *Nematode as Biological Models* (Ed.) B.M. Zuckermann, Vol. 2, pp. 81–98. Academic Press, New York, London.

Samoiloff, M.R. and I. Bogaert (1984). The use of nematodes in marine ecotoxicology. In: *Ecotoxicological Testing for the Marine Environment,* (Eds.) G. Peroone, E. Jaspers, and Chr. Claus, part 1, pp. 407–426. State University of Ghent, Belgium.

Samoiloff, M.R., S. Schulz, Y. Jordan, K. Denich and E. Arnott (1980). A rapid simple long-term toxicity assay for aquatic contaminants using the nematode *Panagrellis redivivus. Canad. J. Fish Aquat. Sci.*, 37: 1167–1174.

Samoiloff, M.R., J. Bell, D.A. Birkhotz, G.R.B. Webster, E.G., Arnott, R. Pulak and A. Madrid (1983). Combined bioassay-chemical fractionation scheme for the determination and ranking of toxic chemicals in sediments. *Environ. Sci. Technol.*, 17: 329–334.

Sasser, J.N. (1979). Economic importance of *Meloidogyne* in tropical countries. In: *Root-knot Nematodes (Meloidogyne* spp.) *Systemics, Biology and Control*, (Eds.) F. Lamberti and C.E. Taylor, Academic Press, London, pp. 359–374.

Sasser, J.N. and D.W. Freckman (1987). A world perspective on Nematology: The role of the society. In: *Vistas on Nematology*, (Eds.) J.A. Veech and D.W. Dickson, Soc. Nematologists, Inc. Hyattsville, pp. 7–14.

Sasser, J.N. C.C. Carter and K. Hartman (1984). Standardization of host suitability studies and reporting of resistance to root-knot nematodes. *Coop. Pub. Dept. Plant Pathol., N.C. State Univ., and USAID, Raleigh*, pp. 7.

Saunders, P.J.W. (1966). The toxicity of sulphur dioxide to *Diplocarpon rosae* Wolf causing black spot of roses. *Ann. Appl. Biol.*, 58: 103–114.

Scheffer, T.C., G.G. Hedgcock (1955). Injury to north-western forest trees by sulfur dioxide from smelters. *U.S.D.A. Tech. Bull.* No. 1117, 49 pp.

Schoenbeck, Helfried, (1960). Beobachtungen zur frage des Einflusses industriellen immissionen aufdie Krankheits-bereitschaft der pflanze. *Ber. Landesanst. Bodennut-zungsschutz*, 1: 89: 98.

Schottler, (1972). Hydrochemische Untersuchung yon Sickerwässern unterhalb yon Abfallablagerungen, ihr Verhalten in Unt rgrund und Methodon Zur statistischen Darstellung. Thesis, Aachen.

Schraufnagel, F.R. (1962). Ridge and furrow irrigation for industrial waste disposal. *Sewage and Ind. Wastes*, 34: 1117–1132.

Shafer, S.R., R.I. Bruch and A.S. Heagle, (1985). Influence of acidic rain on *Phytophthora cinnamomi* and *Phytopathora* root-rot of blue lupine. *Phytopathology*, 75: 996–1003.

Shafer, S.R., S.R. Koenning and K.R. Barker (1992). Interactions of simulated acidic rain with root-knot or cyst nematodes on soybean. *Phytopathology*, 82: 962–970.

Sharma, N. and Trivedi, P.C. (1993). Studies on the biology of *Helerodera cajani* in three cultivars of *Sesamun indicum. J. Indian Bot. Soc.*, 73: 1–2.

Sharp, E.L. (1967). Atmospheric ions and germination of uredospores of *Puccinia striformis Science*, 156: 1359–1360.

Shew, B.B., R.A. Reinert and K.R. Barker (1982). Response of tomatoes to ozone, sulphurdioxide and infection by *Pratylenchus penetrans. Phytopathology*, 72: 822–826.

Shriner, D.S. (1978). Effects of simulated acidified rain on host-parasite interactions in plant diseases. *Phytopathology*. 68: 213–218.

Sierpinki, Z. (1967). Einfluss von industriellen lufterunreinigungen auf die populations dynamik einger primarer kieferschadlinege. *Proc.* 14th *Congress. Int. Union Forest, Res. Organ, (IUFRO Congress)*, 5: 518–513.

Sierpinski, Z. (1966). Znaczenie gospodarcze skosnika tuzinka (*Exoteleia dodecella L.*) na terenach uprzemyskowionych, *Sylwan*, 110: 23–31.

Simpkin, K.G. and G.C. Coles (1981). The use of *Caenorhabditis elegans* for anthelmintic screening. *J. Chem. Technol. Biotechnol.*, 31: 66–69.

Singh, R.P., S.K. Saxena and K. Kumari (1991). Effect of flyash on the growth of tomato plants, nematode population and available nutrients in soil. *Intern. J. Trop. Agric.*, 9(2): 88–92.

Singh, S.K. (1989). Studies on interaction of air pollutants and root-knot nematodes on some pulse crop. *Ph.D. Thesis*, A.M.U., Aligarh, pp.382.

Skye, F. (1968). Lichens and air pollution. *Acta Phytogeogr. Suec.*, 52: 1–123.

Smock, R.A. and R.D. Wastson (1941). Ozone in apple Refrig. *Eng.*, 42: 97–101.

Southey, J.F. (1986). Laboratory methods for work with plant and soil nematodes. *Min. Agri. Fish. Food, HMSO, London*, pp. 202.

Southey, J.F. and G.G. Samuel (1954). Potato root eelworm. I.A. review of the situation; 2 Research in progress. *Min. Agric. Fish.* N.A.A.S., pp. 12.

Spadling, D.H. (1966). Appearance and decay of strawberries peaches and lettuce treated with ozone. *USDA. MKT. Res. Rept.*, 756: 11.

Stapel, C. (1953). Plantasygd ommenes Ogskadedyrenes konamike, betydring Island brudget. *Tidsskr. Landokon Yr.*, 229–244.

Stark, R.W., P.R. Miller, F.W. Cobb Jr., D.L. Wood, J.R. Parmeter Jr. (1968). Photochemical oxidant injury and bark beetle infestation in injured trees. *Hilgardia*, 39: 121–126.

Sturhan, D. (1986). Influence of heavy metals and other elements on soil nematodes. *Revue Nematol.*, 9: 311.

Sturhan, D. (1987). Nematodes as potential indicators of heavy metals. Nematodes in natural system: A status report of a workshop held at the department of Nematology. Agric. Univ. Wageningen. The Netherlands, Dec. 16–18, pp. 41.

Taylor, A.L. (1967). *Introduction to Research on Plant Nematology*. Food and Agriculture, Organization of the United Nations, Rome, pp. 133.

Temple, P.H. and S. Bisessar (1979). Response of white bean to bacterial blight, ozone and antioxidant protection in the field. *Phytopathology*, 69: 101–103.

Temple, P.J. and S. Bisessar (1981). Uptake and toxicity of nickle and other metals in crops grown on soil contaminated by a nickle refinery. *J. Plant Hutr.*, 3: 473–482.

Templin, E. (1962). Zur populations dynamik einiger kiefernschadinsekten in raucheochaedigten bestacnden. *Wiss. Z. Tech. Universitect Dresden*, 11: 631–637.

Tietjen, J.B. and J.J. Lee (1984). The use of free living nematodes as a bioassay for estuarine sediments. *Mar. Environ. Res.*, 11: 233–251.

Treshow, M. (1965). Response of some pathogenic fungi to sodium fluoride. *Mycologia*, 57: 216–221.

Treshow, M.G. Dean and F.M. Harner (1967). Stimulation of tobacco mosaic virus-induced lesions of bean by flouride. *Phytopathology*, 57: 756–758.

Tyler, G. (1974). Heavy metal pollution and soil enzymatic activity. *Plant Soil*, 41: 303–311.

Vaituzis, A., J.D. Nelson, L.W. Wan and R.R. Colwell (1975). Effect of mercuric chloride on growth and morphology of selected strains of mercury resistant bacteria. *Appl. Microbial.*, 29: 275–286.

Van Berkum, J.A. and A.R. Seshadri (1970). Some important nematode problem in India. *10th Int. Nematol. Sym. of European Society of Nematologist*, Pascare, Holia.

Van Gundy, S.D. and I.J. Thomason (1962). Factors influencing storage life of *Hemicycliophora arenaria, Pratylenchus scribneri and Trichodorus christiei* (Abstr.). *Phytopathology*, 52: 366–367.

Van Kessel, W.H.M., R.W. Brocades Zaalberg and W. Seinen (1989). Testing environmental pollutants on soil organisms: A simple assay to investigate the toxicity of environmental pollutants on soil organisms, using $CdCl_2$ and nematodes. *Ecotoxicology and Environmental Safety*, 18: 181–190.

Vargo, R.H., E.J. Pell and S.H. Smith (1978). Induced resistance to ozone injury of soybean to tobacco ring spot virus. *Phytopathology*, 68: 715–719.

Vranken, G. and C. Help (1986). Toxicity of copper, mercury and lead to a marine nematode. *Mar. Pollut. Bull.*, 17: 453– 457.

Vranken, G., R. Vanderhaeghen and C. Help (1985). Toxicity of cadmium to free-living marine and *microphthalma, Monhystera disjuncta, pellioditis marina. Diseases of Aquatic Organism*, 1: 49–58

Weber, D.E., R.A. Reinert and K.R. Barker (1979). Ozone and Sulfur dioxide effects on reproduction and host-parasite relationships of selected plant-parasitic nematodes. *Phytopathology*, 69: 624–628.

Weidensaul, T.C. and S.L. Darling (1979). Effect of ozone and sulphur dioxide on the host pathogen relationship of scotch pine and *Scirrhia acicola. Phytopath.*, 69: 439–491.

Weinstein, L.H., D.C. McCune, A.L. Aluisio and P. Van Leuken (1975). The effect of sulphur dioxide on the incidence and severity of bean rust and early blight of tomato. *Environ. Pollut.*, 9(a): 145–155.

Weiss, B. and O. Larink (1991). Influence of sewage sludge and heavy metals on nematodes in an arable soil. *Biology and Fertility of Soils*, 12(1): 5–9.

Wentzel, K.F. (1965). Insekten-als Immissionsfalgeschadlinge. *Nature-wissenschaft*, 52: 113.

Wood, F.A. (1968). Source of plant pathogenic air pollutant. *Phytopathology*, 58: 1075–1084.

Yarwood, C.E., J.T. Middleton (1954). Smog injury and rust infection. *Plant Physiol.*, 29: 393–395.

Zullini, A. and E. Peretti (1986). Lead pollution and moss-inhabiting nematodes of an industrial area. *Water, Air and Soil Pollution*, 27: 403–410.

28

Ecofriendly Management of Nematode Problem of Vegetables and Pulses

Kumkum Dutta and M.G.Haidar
Department of Nematology, RAU, Pusa, Samastipur – 825 125, Bihar

Environmental pollution is an emerging threat and of great concern in today's context pertaining to its effect on the ecosystem. Intensification of agricultural production is generally achieved through a considerable increase in the use of agro-chemicals and mechanization. As pesticides are effective means of controlling injurious organism, agriculture has become largely dependent on chemical method of crop protection. Over the last decade, scientists, government officials and general public have come to realise that in order to make agriculture sustainable and to protect the environment, intensive farming should become less dependent on chemicals. The development of integrated system of crop production and protection and advances in agricultural sciences offer possibilities for a more rational and safe use of pesticides.

The large scale use of pesticides in current intensive agriculture impairs the quality of our environment, food and drinking water and pose threat to health of consumers and of those who apply pesticides. Secondly from an agricultural view point there is also an ample reason for reducing the use of pesticides because they are losing their efficacy rapidly due to increased resistance of pest organisms and accelerated degradation of pesticides by change in micro flora. As a result cost of farmers may rise and problems may occur in the export of plant material. Thirdly, import requirements made by other countries call for the production of agricultural products with maximum phytosanitary quality and minimum pesticide residue level. The environmental criteria will lead to a large scale elimination across the whole range of approved pesticides (Anon., 1991).

Phytonematodes, hidden enemies of crops, are real menace to successful and profitable cultivation of agricultural crops including pulses and vegetables. These nematodes cause 20–30 per cent yield loss in pulses and 20–50 per cent loss in vegetables. Among a large number of polyphagous nematodes, root-knot nematode (*Meloidogyne* spp.), reniform nematode (*Rotylenchulus reniformis*), cyst nematode (*Heterodera* spp.), stunt nematode (*Tylenchorhynchus* spp.) and lesion nematode (*Pratylenchus* spp.) are of major significance. The nematode problems become all the more complicated, accentuated and aggravated by soil borne wilt/rot fungi like *Fusarium*, *Rhizoctonia*, *Pythium* etc. In view of their economic importance the efforts for management of nematodes various methods have developed. The management practices attempt to keep the nematode population low at a safe level where they do not cause economic loss. Among various methods, use of nematicides is most effective and frequently used by the farmers.

Due to aspecific mode of action, nematicides act as broad spectrum biocides and thus they are risk for environment. Its frequent use results various harmful consequences on plants, human beings and ecosystem. Soil fumigants like DD (1,3 dichloropropane and 1,2-dichloropropane), EDB (1,2 dibromo ethane) and DBCP (1,2 dibromo-3-dichloropropane) contaminates ground water. Methyl bromide (MBr) leaves bromine residue in soil and edible products like lettuce, cowpea, green gram, black gram etc. There are chances of risk of occurrence of cancer, hereditary reproductive defects and neurotoxicity in humanbeings because of exposure to these nematicides. Accumulation of these toxic compounds in living organisms like fish and honey bee poison the food chain.

To overcome the harmful effects of pesticides on ecosystems and living organisms, focus of attention for managing nematode problems has shifted to ecofriendly approach. Ecofriendly management will restore the ecological balance of nature and conserve the rich fauna of parasitoids, predators, pollinators, earthworms and microbial biodiversity.

The most important and effective components of Ecofriendly Management can be identified as:

1. Physical method
2. Cultural method
3. Use of botanicals
4. Use of biocontrol agents
5. Use of Biotechnology
6. Novel agrochemical and chemicals inducers

Physical Method

Physical means of nematode control includes heat treatment of soil, solar drying, steam sterilization, hot water treatment and soil solarization. These aspects were reviewed by Jenkins (1960), Jones (1978), Mass (1987), Oostenbrink (1960), Southey (1965 & 1978) also reported the role of physical factors in the soil for regulating nematode population.

Plant parasitic nematodes killed by heat, desiccation, irradiation and high osmotic pressure are ecofriendly in nature. It is difficult to employ physical methods in killing nematodes in soil or planting material. It is virtually impossible when nematodes are actively growing in plants.

Some of the physical methods are described below:

Heat Treatment of Soil

It is one of the ancient practice for sterilizing soil against all microorganisms, harmful or beneficial. Alves and Campos (2003) studied the effect of soil warming of nematodes like *M. javanica* and *M. incognita* race-3 and found that antagonists were controlling nematodes more efficiently.

There are following steps of heat treatment of soil:

1. Soil is autoclaved at 30 psi for 30 minutes.
2. In greenhouse, steam in perforated pipes running through soil covered with plastic sheet.
3. Temperature should go upto 82°C throughout uniformly and at least for 30 minutes when bubbles release under the plastic sheet is completed.
4. High temperature for one or more days will kill nematodes, insects and fungi.

Steam and hot air treatment of soil is now employed in greenhouse for producing off season vegetables.

Solar Drying

Nematode eggs and juveniles are killed if exposed to sunlight for sun drying. Nematodes are killed in 2 days at 2.5 cm depth and 4 days at 5 cm depth. Frequent turning of soil may disinfect upto deeper layers (Van Gundy, 1972).

Hot Water Treatment of Plants

Hot water treatment is widely applied to particular planting stocks. The difficulty in this method lies in the fact that the thermal point of death for a nematode varies in relation to species and developmental stage of worm.

Hot water treatment is safe only when the difference is significant between the thermal point of death of the nematode and of the plant.

Southey and Jones (1963) applied hot water treatment (43.5°C for 2 hours) against stem nematode *Ditylenchus* spp. in onions and recommended that it should be done before winter storage of onions. This treatment is also successful in case of shallot.

The duration of treatment is 2 hours at a temperature of 43.5 to 44°C, 1 hour 30 min at 45°C and 1 hour at a temperature of 46°C. The germination of bulbs neither damaged nor suppressed by this treatment (Bruinsma and Seinhorst, 1954).

Soil Solarization

Soil solarization technique developed in Israel by Katan *et al.* (1976). It is a hydrothermal process for disinfection of soil (Stapleton and Devay, 1986). It is currently accomplished by incubating soil under transparent plastic film (primarily polyethylene or polyvinyl chloride PVC) during hot months with minimal cloud cover, thus producing a 'Greenhouse effect' which raises soil temperature to levels which are lethal or injurious to many plant pathogens and pests.

Mechanism of Soil Solarization

1. Accumulation of heat by transmission of short wave solar radiation but prevents loss of long wave solar radiation resulting in thermal inactivation at higher temperature, producing heat shock protein and irreversible heat injury.

2. High soil moisture by intermittent irrigation improves thermal conductivity and water vapour high latent heat.
3. Greater thermal sensitivity of hydrated rather than desiccated organism.
4. Antagonistic microorganism have lower thermal sensitivity and faster recolonization capacity enabling shift in favour of biocontrol.
5. Due to physico-chemical and microbiological process some toxic genes and minerals are released which are harmful to pathogens prolonged exposure to high temperature kills pathogens.

Effect of Soil Solarization

Chen and Katan (1980) reported that tomato seedling growth on extract of heated soil showed enhanced growth compared to extracts of untreated soil. Improvement in seedling growth following soil solarization in peach, walnut, radish, pepper and Chinese cabbage has been reported by several workers (Stapleton and Devay, 1982; Stapleton *et al.*, 1985). Reduction in nematode population on vegetables like tomato, cabbage, cauliflower, celery and egg plant using soil solarization had been reported by Anju Kamra and Gaur, 1993; Raymundo *et al.*, 1986; Mcsorley and Parrado, 1986; Kin and Hans, 1988. Similar results were also obtained in chickpea and Pigeonpea by Sharma and Nene (1990) and in soyabean by Kumar *et al.* (1993) respectively. Besides this, soil solarization also increases levels of ammonium (NH_3-H) and nitrate-nitrogen (NO_3-N) as well as other mineral nutrients such as K^+, P, Ca^{2+}, Mg^{2+}, Cl^- etc. (Black and Greb, 1962; Chen and Katan, 1980; Katan, 1987; Park, 1968; Stapleton and Devay, 1986; Stapleton *et al.*, 1990; Weir *et al.*, 1989).

Irradiation

X-ray exposure of 20000r (r=roentgen) of embryo nated eggs in female and γ-ray from Co have been found effective to control nematodes.

Mutagenic treatment with EMS and MH etc were found effective for inducing resistance in black gram (Routaray *et al.*, 2001).

Osmotic Pressure

Addition of salt, sugar, charcoal etc create osmotic stress on nematodes and can be used for controlling nematodes.

Cultural Method

One of the most effective ecofriendly management practice is the use of proper combination of cultural practices. The effectivity of cultural method depends upon the proper knowledge of life history, population dynamics, host range of the plant parasitic nematodes infesting crop.

Prevention of Spread

Nematode infested seeds and soil with nematodes like *Ditylenchus dipsaci* in onion and *Heterodera* spp. spread through rain, water and farm implements and cause severe disease on crop. It should be prevented by sanitation.

Adjustment of Date of Sowing

Soil temperature plays crucial role in the activities of plant parasitic nematodes, the time during which crop is planted is important (Ayoub, 1980). Most pathogenic nematodes are inactive at lower temperature. Manipulation of date of sowing of chickpea escaped root knot, lesion and raniform nematodes was reported by (Gaur *et al.*, 1979; Anon, 1999) and in cucumber reduced *Meloidogyne javanica* infection (Stephen *et al.*, 1988)

Fallowing

It is based on the principle starve the nematodes to death., Beneficial effect of fallow were reported to control *Meloidogyne* spp. (Brodie and Murphy, 1975; Widjaja and Widrich, 1974), Pratylenchus (Brodie and Murphy, 1975); *Rotylenchulus reniformis* (Brathwaite, 1974).

Flooding

Effective control of nematodes achieved through flooding (Brown, 1933; Hollis and Fielding, 1958; Van Gundy *et al.*, 1968). Reduction of nematode population may be due to high moisture level (Christie, 1959), less aeration in soil (Wallace, 1963), low pH and development of toxic substances. Alternate flooding and drying of soil is more effective against root-knot nematode (Rhodes, 1964; Newhall, 1965).

Use of Organic Amendment, Green Manure and Cover Crops

The use of organic amendments like organic manure, green manure, oil cakes etc as effective nematode control method is well documented by various workers. Linford and his associates (1938) first demonstrated the use of organic matter in reducing nematode damage.

Application of farm yard manure results reduction in population of root-knot nematode (Desai *et al.*, 1969) Pratylenchus (Morgan and Collins, 1964) and Tylenchulus penetrans (Mankau and Minteer, 1962). Organic amendments like neem, mustard or mahua cake @ 1.0 kg/m^2 reduced population of nematodes on egg plant. Similar results were observed on many vegetables (Ruello, 1983; Singh, 1965, Singh and Sitaramaiah, 1973).

Incorporation of green manure crops like Cowpea (*Vigna unguiculata*), Sesbania (*Sesbania aculeata*), Mung bean (*Vigna radiata*), Sunnhemp and cluster bean etc during ploughing in soil brings desirable physical, chemical and biological changes in soil. Crotolaria was found to be resistant against many plant parasitic nematodes. Incorporation of these crops increase the microbial population and nematophagous fungi might have reduced plant parasitic nematodes in amended soil (Mankau and Das, 1969; Tomerlin and Smart, 1969).

Singh and Sitaramaiah (1973) found that oilcakes of neem, castor, peanuts, linseed, mustard and mahua are capable of reducing *Meloidogyne* population in field plots when incorporated 3 weeks before planting. These oil cakes reduced population of *M.incognita, Helicotylenchus indicus, Tylenchorhynchus brassicae* and *Rotylenchulus reniformis* on mung bean and lentil (Tiyagi, 2001).

Nitrogenous, phosphorus and potassic fertilizers reported to reduce *Heterodera cajani* population on cowpea and other crops (Mukhopadhayaya *et al.*, 1972; Devi and Gupta, 1999).

Several legumes like cowpea, crotolaria, joint vetch, have potential use in nematode and nitrogen management can be used as cover crops for plant parasitic nematodes.

Deep Summer Ploughing

This practice reduces nematodes particularly in tropical and subtropical countries. It exposes nematodes to solar heat and dessication and accentuates the effect of fallowing. Three summer ploughing during June (av. atm. temp. ranging between 40 to 46°C) at Hissar, Haryana lead to 96.5 per cent reduction in nematode population *M.javanica* in tomato reported by Jain (1992).

Cropping Sequence

Nematode management through intercropping and crop rotation is based on the fact that some species of nematodes are able to feed and multiply only on host crop. Monoculture of vegetables particularly tomato, egg plant, chilli, okra, sponge gourd or rotation in all combinations causes manifold increase in population of *M.incognita* and other parasitic nematodes. Rotation of vegetables with graminaceous crops having antagonistic effect suppresses the growth of root-knot nematodes. The population of *Meloidogyne* spp. decreases in combination of eggplant–egg plant, wheat, eggplant-chilli- wheat, spinach-marigold-carrot, Red gourd-eggplant-mustard. Jain (2001) found cropping sequence of okra-wheat fallow-okra effective for controlling *M.javanica*. Intercropping of chickpea with sesame effective against *Meloidogyne* spp. (Tanda and Atwal, 1981). Similarly mixed intercropping of Pigeonpea with bajra sorghum and guar was more effective in management of *H.cajani* on Pigeonpea compared to crops like til, urd and green gram or even pigeonpea alone (Anon, 1997). Intercropping of sesame with Pigeonpea effectively reduces the population of *M.incognita, P.zeae, Hoplolaimus indicus,* Helicotylenchus and *Hemicriconemoides* except *Heterodera* cajani and *R. reniformes* (Haidar and Dutta, 2001) *Ditylenchus dipsaci* controlled in wheat-onion-maize-onion rotation (Khan *et al.*, 1987) and *M.incognita* in carrot-garlic rotation (Arya and Tyagi, 1980).

Use of Resistant Varieties

Growing of cultivars of pulses and vegetables resistant to nematodes is ecologically and economically viable method of nematode management.

Ali (1997) enlisted following cultivars of vegetables and pulses resistant to nematodes.

Table 28.1

Crop	*Nematode*	*Resistant Cultivars*
Brinjal	*Meloidogyne* spp.	Vijay, Black Beauty, Banaras Giant
Capsicum	*M. incognita*	NP 46A, G4, Mirch 1, Red Long CA(P)63
Tomato	*M. incognita, R. reniformis*	SL-120, Punjab NR-7, Hissar Lalit, Punjab NR-7, Hissar Lalit
Pea	*M. incognita M.javanica*	T44 DMR 7, KEP 130, HFD4, KFPD46
Lentil	*M. incognita M.javenica*	DPL-14 DPL-14, PL81-D, PL 81-340
Chickpea	*M. incognita M.javanica*	BGM 481, BGM 483, G 288341, BG 369, BGM 481, GL 88341, GMS 815
Mungbean	*M.javanica*	GM 85-2, ML 323
Uradbean	*M.javanica*	TPU3, WBU 105
Pigeonpea	*M. incognita M.javanica*	Pusa 23, GAUT 87-2 H82-1, 86-1, IPH 732, TH 9, UG 218
Cowpea	*R.reniformis*	Cowpea 1
Chilli	*R.reniformis*	Pusa Jawala

Use of Botanicals

Biopesticides of botanical origin, have become the focus of attention due to its ecofriendly properties like biodegradability, renewable nature and selective toxicity to target pests only.

Botanicals namely *Azadirachta indica, Pongamia glabra, Argemone mexicana, Datura metel, Eucalyptus globorus, Phyllanthus niruri, Madhuca indica, Shorea robusta, Asparagus sprengeri, Crotolaria juncea, Tagetes* spp., *Ocimum sanctum, Calotropis procera, Swertia chirata, Bauhinia variegata, Semecarpus anacardium, Eclipta alba, Cannabis sativa, Melia azadirach, Embelia ribes* and *Psidium guajava* have nematicidal properties were reported by various workers.

Neem plant parts like leaves, seed kernel, neem cake, aqueous extract and many neem product like Achook, Nimbicidin etc used as seed treatment and soil application in case of chickpea, pigeonpea, okra, tomato, chilli, pea, carrot were found effective against *Meloidogyne incognita* and *Rotylenchulus reniformis* (Ram and Gupta, 1980; Siddiqui and Alam, 1987; Mojumdar and Mishra, 1994).

In french bean, neem cake followed by karanj cake at 2 tonn/ha were found more effective than carbofuran and aldicarb @ 4 kg ai/ha in reducing reniform nematode population (Mishra and Padhi, 1985).

Singh and Sitaramaiah (1967) reported that the application of 5-10 per cent (w/w) of leaves of *Azadirachta indica, Melia azadirach, Cassia fistula, Crotolaria juncea* and *Sesbania aculeata* to infested soil reduces incidence of root-knot on tomato and okra. Mojumdar and Mishra (1992) found neem seed kernel as highly nematicidal when used as seed soaking and seed coating against *M.incognita* infecting mungbean and chickpea. Seed coating with neem seed kernel @ 20 per cent (w/w) provided 70 per cent reduction in root-knot galling.

Highest degree of root-knot reduction was observed in chilli plants grown with *Tagetes erecta* following in the order of enumeration by *Allium cepa, Allium sativum, Asparagus officinalis, Ipomoea palmata, Chrysanthemum indicum, Mirabilis jalapa, Datura metel* and *Ricinus communis* (Trivedi and Tyagi, 1984).

Few weeds like *Argemone mexicana, Lantana camara* and *Solanum surettense* have scope to be incorporated in INM schedule for management of *M. incognita* in okra and mung (Nath *et al.*, 1982; Prasad and Ram, 2001). Flowers, leaves and stem of *Tagetes* sp, *Calotropis* sp, Karanj in either aqueous extract or powder form suppresses the reproduction of nematodes and improves plant growth in vegetables and pulses.

Various concentration (0.1, 1 and 10 mg/ml) of aqueous methanolic and chloroform extracts of mangroove, viz *Avicennia* marina and Rhizophora mucronata reduced root-knot infection due to *M.javanica* in tomato seedlings (Mehdi *et al.*, 2001). Bare root dip treatment of tomato, chilli, egg plant, cabbage, cauliflower in leaf extract of lemon grass (*Cymbopogon flexuosus*) caused significant reduction in root penetration of second stage juveniles of *M.incognita* and their development (Prakash *et al.*, 2001).

Many of such botanicals have been exploited for their nematicidal value, but still many of them need yet to be exploited. Such botanicals are easy in application, environmentally safe and having no phytotoxic effect on crops.

Use of Biocontrol Agents

The application of natural enemies of plant parasitic nematodes for controlling nematode population is an essential component of ecofriendly management. This can be done by introducing or stimulating natural enemies of nematodes in the soil environment or through manipulation of the environment, hosts or antagonists or by mass introduction of one or more antagonists. Such micro organisms tend to be host specific and are able to maintain an equilibrium nematode density at or below the economic threshold level have greatest potential for use as biological control agents in nematode management.

Fungi

Among various natural enemies, wide range of fungi are nematophagous in nature and have developed very sophisticated strategies for either infecting or capturing these plant parasitic nematodes.

Nematophagous fungi mainly falls into two broad groups:

Endoparasitic Fungi

These fungi parasitize nematodes with small conidia or zoospores.

e.g., Haptoglossa zoospora, Harposporium anguillulea, Verticillium spp, Drechmeria coniospora etc.

Predators

Predator fungi capture nematodes using hyphal traps.

e.g., Arthrobotrys oligospora, Nematoctonus, Dactylella sp, *Dactylaria* spp. etc.

Wide range of opportunistic fungi parasitize eggs and cysts of nematodes. *Trichoderma harzianum, Trichoderma virens, Aspergillus niger, Paecilomyces lilacinus, Verticilium chlamydosporium* are found promising bio agents. *M. incognita* and *R. reniformis* in chickpea, tomato, coriander, okra (Haseeb *et al.*, 2003; Midha *et al.*, 2001; Nagesh and Hussani, 2001). *Paecilomyces lilacinus* is reported as an effective biocontrol agent against *Globodera pallida* and *M. incognita* (Franco *et al.*, 1981; Jatala *et al.*, 1981).

Commercial products of nematophagous fungi sold in the market under the name Royal 300, Royal 350, Trichonitrin etc. Management of *M.incognita* on tomato and eggplant using mycorrhizza, *Fusarium oxysporum* and *Verticillium chlamydosporium* was reported by (Diedhiou *et al.*, 2003 & Rao *et al.*, 2003). Association of *F. oxysporum* and *Aspergillus* niger with egg masse of *M.incognita* on egg plants reported by Goswami and Umarao (1995).

Mycorrhiza

Now mycorrhizza is not restricted to its use only as biofertilizer, its potential role in the biological control of plant parasitic nematodes is reported by many workers. Sikora (1979) found that prior presence of VAM fungi *Glomus mosseae* has resulted into an increase in plant resistance against *Meloidogyne* spp. The increased resistance may be due to vigorous root growth, enhancement of uptake of various mineral nutrients such as P, Ca, Cu, Mn, S and Zn (Hayman, 1982; Smith *et al.*, 1986), parasitism of eggs of nematodes, release of nematostatic compounds and competition with nematodes for food and space.

Sitaramaiah and Sikora (1982) reported *Glomus fasciculatum* suppresses the population of *Rotylenchulus reniformis* in tomato, Verma and Gupta (2002) worked on determination of efficacy of VAM (*G. fasciculatum*) at different phosphorus level (0, 30 and 60 kg/ha) for managing *M. incognita* on mungbean and reported more pronoureed reduction in nematode population at 0 level of P-application.

Diedhiou *et al.* (2003) reported preinoculation of tomato plants with *G. coronatum* or F 162 reduces nematode population and stimulates plant growth.

Bacteria

In case of bacteria, only less than 10 per cent of the total rhizosphere bacteria have potential of nematode control. They have added advantage over other organisms to be used as biological agent due to its easy multiplication on artificial media. But some obligate parasitic bacteria such as *Pasteuria penetrans* can be cultured an host plants only and root powder of the host plant is used for controlling nematodes.

Pasteuria penetrans found effective against root-knot nematodes in vegetables *viz.* tomato, Brinjal (Chand *et al.*, 2001; Mohanty and Mohapatra, 2002). It directly parasitize juvenile nematode, thus affects penetration and reproduction. Seed bacterization, soil drencing and bare root dip application with *Psudomonas fluoroscens, Bacillus subtilis, B. polymyxa* effectively controls plant parasitic nematodes was also reported by many workers. Gogoi and Neog (2003) reported *Pasteuria* treated plots significantly control root-knot nematode population on green gram, cowpea and french bean.

Predatory Nematodes

The nematodes of the groups Dorylaim, Nygolaim and Diplogasterid are predaceous in nature. Among them, mononchs may be proved efficient predators because of stronger predatory potential, high rate of predacity and high strike rate (Bilgrami, 1998).

Small (1979) found a significant reduction in the population of *G. rostochiensis* and *M. incognita* in presence of *Prionchulus punctatus*. The widespread and abundant presence of dorylaims and nygolaims makes them promising biological control agent of plant parasitic nematodes.

Application of infective stage juvenile suspension of *Steinernema feltiae* reduces *M.incognita* root galling, egg production, egg hatching on tomato seedling reported by (Lewis *et al.*, 2001).

Actinomycetes

A soil borne actinomycetes *Streptomyces avermitilis* produces a macrocyclic lactone Avermectin, is reported as novel nematicide (Ostlind *et al.*, 1981; Britle *et al.*, 1982). It can be used as soil treatment, seed treatment and seedling root dip treatment Cayrol *et al.* (1993) found root dip treatment of tomato in avermectin effective against *M.arenaria*. It is also compatible with other bioagents like *T. viride, P.fluoroscens* and also biodegrodable. Yield of avermectin treated plots of tomato is 30 per cent higher than methyl bromide treated plots (Xuhua *et al.*, 2000).

Besides them, blue green algae, amoeba and viruses are reported as natural enemies of plant parsitic nematodes. *Anabena oryzae, Nostoc calcicola* and *Spirulina* sp in combined treatment significantly reduces the number of galls and egg masses of *M. incognita* on cowpea (Youssef and Ali, 1998). Biofertilizers like Azolla and Azospirillum also found effective in managing *M.incognita* in Okra (Ramakrishnan *et al.*, 1997). An amoeba *Theratomyxa weberi* found predatory on plant parasitic nematodes (Sayre, 1973). Still more informations are required on various aspects of these biocontrol agents. The development and advancement in Genetics and Molecular biology is greatly improving the prospects of biological control.

Application of Molecular Biology and Biotechnology

Advances in Molecular biology and Biotechnology provides means to develop novel strategies for nematode management. It is used in classical breeding for plant resistance to nematodes, biological control of plant nematodes or insects with insects or insect parasitic nematodes or nematophagous fungi, taxonomy and use of bioregulators such as insect growth regulators and allelochemics for management of nematodes (Gaur and Sirohi, 1998).

Nematode diagnostics is essential for success of any nematode management programme even more when the control method is highly specific to the nematode species/pathotype/race. A number of new techniques for analyzing nucleic acids, proteins, carbohydrates and lipids can be helpful in the identification of pests, of those allozyme, monoclonal antibody and DNA based systems are most well developed for nematodes. Polyacrylamide gel electrophoresis (PAGE) has mainly focussed on cyst and root knot nematodes (Davies, 1997). PCR technique or RFLP analysis used as supplementary tools wherever necessary.

Nematode resistant transgenic plants can be designed by various approaches. The simple method is to introduce resistant genes effective against plant parasitic nematodes from wild species to commercial cultivars. The Mi gene from wild species of tomato *Lycopersicon peruvianum* into *L esculentum*, imparts resistance against *M. incognita*, *M. javanica* and *M. arenaria* but not to *M.hapla* (Guilbert and Mc Guire, 1956). Another gene from tomato (Hero) and a gene from potato (Gro 1) that confer resistance against all pathotypes of potato cyst nematode, *Globodera rostochiensis*, are being isolated (Ballvora *et al.*, 1995 & Ganel *et al.*, 1995).

The gene H1 that has been transferred from a wild species *Solanum tuberosum* sp. *andigena* to many commercial potato cultivars because if confers resistance to pathotypes Ro 1 and Ro 4 of *G.rostochiensis* (Pineda *et al.*, 1993). Resistance genes against soyabean cyst and sugarbeet are also being mapped with RFLP (Jung *et al.*, 1992; Vierling *et al.*, 1996). Transgenic plants produce monoclonal bodies that disrupts feeding, digestion, chemoreception, reproduction and other life processes of nematodes. Antinematode proteins (plantibodies, proteinase, inhibitors etc.) can be expressed constitutively in plants, suppresses protein digestion, resulting in slow starvation and delay in development and reproduction of pests. Trypsin inhibitor (CpTI) gene from cowpea (*Vigna unguiculata*) expressed in transgenic potato impedes reproduction of root-knot nematode and also interferes with development of *Globodera pallida*. (Vrain, 1997).

A lectin from garlic confer resistance against wide array of nematode pests. Similarly, protease from *Paecilomyces lilacinus* can be used to construct nematode resistant transgenic plants. Thus biotechnology is offering ecofriendly approach of nematode management in vegetables and pulses.

Novel Agrochemicals and Chemicals Inducers

Advances in Molecular biology and immunology provided new approach to discover novel agrochemicals, which are perceived as environmentally safe. Some encapsulated and slow releasing toxicants offer better promise as nematicides. Alginate bentonite CR slow release formulations offer efficient systems for reducing carbofuran leaching in clay soil (Prasad and Chawla, 2001). The newly developed controlled release phorate SX-10 drastically reduced the penetration, development of juveniles and reinvasion in brinjal cv. Pusa Kranti, CR formulation (SX-10) was found effective because the toxicant (a.i) was available in effective concentration in soil till 35th day and it is super ion over commercial formulations. (Thomas *et al.*, 2003).

Some chemicals stimulate the natural defense system of plants are known as chemical inducer. Pandey and Kalra (2001) reported soil drenching with chemical inducers i.e 4-chlorosalisylic acid, O-acetyl salicylic acid, 2-chloronicotinic acid drastically reduces root-knot severity in *Chamomilla recruita*. Now research in the field of management of nematodes is mainly focussed on developing novel ways to control nematodes. The use of chemical inducers, plant growth regulator and natural antinematode compounds. Few new compounds like furfural and others inhibiting sterine biosynthesis in plants have potential for future nematode management.

Potensized nematode cell extract of root-knot nematode (*M.incognita*) and reniform nematode (*R.reniformis*) induced in plants show resistance against *M. incognita* and *R. reniformis* (Rajendran *et al.*, 2003). It is due to presence of antigens in nematode cell extract induce defense response in plants.

The ecofriendly management is both conceptual and methodological. Integration of various management methods like physical, cultural, biological and various novel methods of nematode management in harmonious manner to keep all the economically important nematodes of vegetables and pulses below the economic threshold level with an objective of optimizing productivity, net return, stability and environmental quality.

Conclusion

It is evident that plant parasitic nematodes are causing significant yield losses in vegetables and pulses. The chemical method no doubt gives significant control over other methods however due to its harmful consequences on human being, environment and non-target organisms it is emerging as threat to ecosystem. Therefore, present day need is to use ecologically viable integrated nematode management strategies. The integrated application of physical, method, cultural methods, bioagents, botanicals and safer formulation of agrochemicals etc is proved as effective management tactics against plant parasitic nematodes in vegetables and pulses:

1. The first requirements in the design of pest management system for nematodes is the ability to identify species of nematodes. The identification techniques should be simple and permit much better estimation of species composition of specific population than do present morphological techniques.
2. Cropping system based management technology against polyspecific nematode communities should be practiced and more effective economically viable crop rotation practices should be developed.
3. Efforts should be diverted to explore the possibility of using bio control agents and botanicals under field condition. The production of biocontrol agents on commercial scale is also needed.
4. The use of information technology is likely to make certain impact on ecofriendly management system. The ease and speed of dissemination of information through internet will help researchers, extension, agencies and farmers. Use of organic amendments, bio-control agents, botanicals and host resistance should be encouraged.
5. Remote sensing techniques should be used to provide informations on the detection and quantification of damage to plants, assessment of distribution of main host plants, habitat of nematodes and surveillance of environmental factors favourable for pest and disease development.
6. Liberalization of trade between countries have given rise to the possibility of new pest introduction. Plant and animal quarantine has assumed considerable importance in this scenario. Hence government agencies must be more vigilant and formulate policy guidelines at national level regarding export and import of microbial agents.

REFERENCES

Ali, S.S. (1997). Status of nematode problems and research in India. In Diagnosis of key nematode pests of chickpea and pigeonpea and their management. In Proceedings of a Regional Training Course, 25–30 Nov., 1996, ICRISAT, Patancheru, pp: 74–82.

Alves, F.R. and V.P. Campos (2003). Effect of soil warming on the biological control of *M. javarica* and *M. incognita* race-3. Efeitos da temperatura sobre a atividade de fungae no controle biologico de *Meloidogyne javanica* e *M. incognita* race–3. *Ciencia e Agrotechnologia*, 27(1): 91–97.

Anju Kamra and H.S. Gaur (1993). Efficacy of soil solarization and formaldehyde application alone and in combination on plant parasitic nematodes and the growth of eggplant seedling (Abstr.). *Indian J. Nematol.*, 23: 6.

Anonymous (1991). Operational environmental criteria for the protection of soil and water against pesticides. Dutch Ministry of Housing, physical planning and the environment, July 1991. pp. 31.

Anonymous (1997). Project Report on effect of cropping practices in management of cyst nematode (*H. cajani*) problems of pigeonpea. Deptt. of Plant Pathology and Nematology, Allahabad Agricultural Institute, Allahabad, pp. 1–44.

Anonymous (1999). All India Coordinated Research Project on "Plant Parasitic Nematodes with Integrated Approach for their Control," *ICAR Biennial Report* (1997–99), *CCS, Haryana Agric. Univ.*, Hisar.

Arya, M. and Tyagi (1980). Biological control of root-knot incited by *M. incognita* 16 Dec., Jaipur, 20–30.

Ayoub, S.M. (1980). *Plant Nematology: An Agricultural Training Aid*. Nema aid Publ. Sacramento, C.A. p.195.

Ballvora, A., Hesselbach, J., Niewohner, J., Leister, D., Salamini, F. and Gebhardt, C. (1995). Marker enrichment and high resolution map of the segment of potato chromosome VII harbouring the nematode resistance gene Gro 1. *Mol. Gen. Genet.*, 249: 82–90.

Bilgrami, A.L. (1998). Predatory nematodes and protozoans as biopesticides of plant parasitic nematodes: In: *Plant Nematode Management: A Biocontrol Approach*, (Ed.) by P.C. Trivedi, pp 4–23.

Black, A.L., B.W. Greb (1962) Nitrate accumulation in soils covered with plastic mulch. *Agron. J.*, 54: 366.

Brathwaite, C.W.D. (1974). Effect of crop sequence and fallow on population of *R. reniformis* in fumigated and untreated soil. *Plant Dis. Reptr.* 58: 259–261.

Britle, A.J., A.E.Corps and D.J. Wright (1982). Biological effect of avermectins on nematodes. 16 *Inter. Symp. Nematol.* St. Andrews, Scotland (Abs.).

Brodie, B.B., W.S. Murphy (1975). Population dynamics of plant nematodes as affected by combination of fallow and cropping sequences. *J. Nematol.*, 7: 91–92.

Brown, L.N (1933). Flooding to control root-knot nematodes, *J. Agric. Res.*, 47: 883–888.

Bruinsma, F. and J.W. Seinhorst (1954). Warm water behandling vai sjalotten tegen aantasting door stengelaaltjes (*Ditylenehus dipsae* [Kuhn] Filipjev) Med. Direkteur Tuinbouw. No. 17 pp. 437–446.

Cayrol, J.C., Dijan and J.P. Frankowski (1993). Efficacy of avermectins B for control of *M. arenaria Fund Appl. Nematol.*, 16: 239–246.

Chand, R., J.S. Gill and B.D. Yadav (2001). Management of root–knot nematode using *Pasteuria penetrans* on the first and second subsequent tomato crop. (Abs.) In National Congress on Centenary of Nematology in India, Appraisal and Future Plans, 5–7 Dec.,2001, IARI, N. Delhi, pp: 141–142.

Chen, Y., J. Katan (1980). Effect of solar heating of soils by transparent polyethylene mulching on their chemical properties. *Soil Sci.*, 130: 271–277.

Christie, J.R. (1959). *Plant Nematodes: Their Bionomics and Control.* Jacksoniville, F.L.: the H. and WV.

Davies, K.G. (1997). Identification and quantification of plant parasitic nematodes using immunological techniques pp. 3–11. In Proceedings of a Regional Training Course, 25–30 Nov., 1996, ICRISAT, India (Ed. S.B. Sharma).

Desai, M.V., G.J. Patel, H.M. Patel, S.D. Naphade, (1969). Effect of some agronomic practices in the control of root–knot nematodes on tomato, All India Nematology Symp. N, Delhi, India.

Devi, S.L. and P. Gupta (1999). Effect of nitrogen, phosphorus and potash application against *H. cajani* on cowpea 29(2): 195–196.

Diedhiou, P.M., J. Hallmann, E.C. Oerke and H.W. Dehne (2003). Effect of arbuscular mycorrhizal fungi and a non–pathogenic *Fusarium oxysporum* on *M. incognita* infestation of tomato. *Mycorrhiza,* 13(4) 199–204.

Franco, J., P. Jatala, M. Bocangel (1981). Efficiency of *Paecilomyces lilacinus* as bicontrol agent of *Globodera pallida. J. Nemat.,* 13(4): 338–339.

Ganal, M.W., R. Simon, S. Bronmon Schenkel, M. Arndt, M.S. Phillips, S.D. Tanksley and A. Kumar (1995). Genetic mapping of a wide spectrum nematode resistance gene (*Hero*) against *Globodera rostochiensis* in tomato. *Mol. Plant Microbe Inter.,* 8: 886–891.

Gaur, H.S. and A. Sirohi (1998). Biotechnology in Plant Nematology. In: *Recent Advances in Plant Nematology,* (Ed.) P.C. Trivedi, CBS Publishers and Distributors, New Delhi, pp. 393–406.

Gaur, H.S., S.D. Mishra and U.C. Sud (1979). Effect of date of sowing on the rotation between the population density of root-knot nematode *M. incognita* and the growth of three varities of chickpea *Cicer arietinum Indian J. Nemat.,* 9: 152–159.

Gilbert, J.C. and D.C. Mc Guire (1956). Proc Am. Soc. Hort. Sci. 68: 437–442.

Gogoi, B.B. and P.P. Neog (2003) Performance of *Pasteuria penetrans* in a crop sequence. In 5th National Symposium on Biocontrol Agents for Sustainable Management of Pests. 18–20 Dec., 2003. G.B. Pant Univ., Pantnagar, pp.–184–185.

Goswami, B.K., Uma Rao (1995). Association of *Fusarium oxysporum* and *Aspergillus flavus* with the egg masses of *M. incognita* infecting egg plants (Abs.) In National Symposium on Nematode Management, IARI, N. Delhi, March, 24–26, 1995.

Haidar, M.G. and K. Dutta (2001). Effect of intercropping of sesame with pigeonpea on nematode population (Abs.) In National Congress on Centenary of Nematology in India, Appraisal and Future Plans, 5–7 Dec.,2001, IARI, N. Delhi, pp: 93–94.

Haseeb, A., P.K. Shukla, V. Kumar (2003). Management of Fusarium-Meloidogyne Disease Complex of Chilli by the application of biocontrol agents, organic amendments and pesticides under pot conditions. (Abs.) In 5th National Symposium on bio-control agents for Sustainable Management of Pest 18–20 Dec., 2003 GBPUAT,Pantnagar, pp: 192–193.

Hayman, D.S. (1982). The physiology of Vesicular arbuscular mycorrhizal symbiosis. *Canadian J. Bot.* 61: 944–963.

Hollis, J.P. and M.J. Fielding (1959). Population behaviour of plant parasitic nematodes in soil fumigation experiments *La Agric. Exp. Sta. Bull.,* 515 pp. 30.

J. Xuhua, Cao Voucherg, Yuan Huizhu, Wang Xuzhi, Zhang Jingsuo, Dong Wenguang and Zhao Yingrong (2000). Control of root–knot nematod disease in tomatose with avermectin Plant Protection 26(5): 46–47.

Jain, R.K. (1992). Nematode pests of vegetable crops. In: *Nematode Pests of Crop*, (Eds.) D.S. Bhatti and R.K. Walia, CBS Publishers and Distributors, New Delhi, pp. 77–97.

Jain, R.K. (2001). Influence of carbofuran on V A Mycorrhiza and *Meloidogyne incognita* interaction on cowpea. (Abs.) In National Congress on Centenary of Nematology in India, Appraisal and Future Plans, 5–7 Dec., 2001, IARI, New Delhi, p.163.

Jatala, P., R. Sales, R. Kaltenbach, M. Bocangel (1981). Multiple application and long term effect of *P. lilacinus* in controlling *M. incognita* under field condition. *J. Nemat.*, 13(4): 445.

Jenkins, W.R. (1960). Control of nematodes by physical methods. In: (Eds.) J.N. Sasser and W.R. Jenkins. Nematology Chapell Hill, North Carolina.

Jones, F.G.W. and R.A. Krempston (1978). In: *Plant Nematology*, (Ed.) J.F. Soc. The HMSO, London, pp.440.

Jung, C., R. Koch, F. Fisher, A. Brandes, G. Wricke and R.G. Herrmann (1992). DNA markers closely linked to nematode resistance genes in sugarbeet (*Beta vulgaris* L.). mapped using chromosome additions and translocations originally from wild beets of the *Procumbentes* section. *Mol. Gen. Genet.*, 232: 271–278.

Kali Ram and D.C. Gupta (1980). A note on the efficacy of fresh neem leaf extracts in the control of *M. javanica* infecting chickpea (*Cicer arietinum*). *Indian J. Nemat.*, 10: 96–98.

Katan, J. (1987). Soil solarization. In: *Innovative Approaches to Plant Disease Control*, (Ed.) I. Chet. John Wiley and Sons, New York, pp. 105–177.

Katan, J., A. Greenberger, H. Alon, A. Grinstein (1976). Solar heating by polyethylene mulching for the control of diseases caused by soil borne pathogens. *Phytopathology*, 66: 683–88.

Khan, A.S., A.K. Khanzada, F.M. Bilqees and M. Aslam (1987). Effect of different cropping sequences on population of *Ditylenehus dipsaci* (Khan, 1857). Filipjev, 1936, *Pakistan J. Zool.*, 19: 257–261.

Kim, J.I., S.C. Hans (1988). Effect of solarization for control of root-knot nematodes. *Korean J. Appl. Entomol.* 27: 1–5.

Kumar, B., N.T. Yaduraju, K.N. Ahuja and D.P. Prasad (1993). Effect of soil solarization on weeds and nematodes under tropical Indian conditions. *Weed Res.*, 33: 423–29.

Lewis, E. E., P.S. Grewal, S. Sardanelli (2001). Interaction between the *Steinernema feltiae–Xenorhabdus bovienii* insect pathogen complex and root–knot nematode *M. incognita*, *Biological Cont.*, 21(1): 55–62.

Linford, W.B., Y. Yap and J.M. Oliveira (1938). *Soil Sci.*, 45: 127–141.

Maas, P.W. Th (1987). Physical methods and quarantine. In: *Principles and Practices on Nematode Control of Crops*. Academic Press, Australia, 265–279.

Mankau, R. and S. Das (1969). The influence of chitin amendments on *M. incognita J. Nemat.*, 1: 15–16.

Mankau, R. and R.J. Minteer (1962). Reduction of soil population of the citrus nematode by the addition of organic materials. *Plant Dis Reptr.*, 46: 375–378.

McSorley, R. and J.L. Parrado (1986). Application of soil solarization to rock dale soils in a sub-tropical environment. *Nematropica*, 16: 125–140.

Mehdi, F.S., I.A. Siddqui, T. Zia, N.I. Ali (2001). Use of mangrove for the control of M. javanica in tomato, *Nematologia Mediterranea*, 29(2): 127–129, Bari, Italy.

Midha, R.L., P.C. Trivedi and B.D. Yadav (2001). *Paecilomyces* spp. as biocontrol agent against *M. incognita* on coriander (Abs.). In *National Congress on Centenary of Nematology in India: Appraisal and Future* Plans, 5–7 Dec.,2001, IARI, N. Delhi, pp: 146–147.

Mishra, R.P. and N.N. Padhi (1998). *Indian J. Nematol.*, 15: 270–271.

Mohanty, K.C. and S.N. Mahapatra (2002). Biological control of RKN infecting brinjal with *Pasteuria penetrans. Proceedings of National Symposium on biodiversity and management of nematodes in cropping system for sustainable agriculture* (11–13 Nov). 133–134.

Mojumdar, V. and S.D. Mishra (1992). Effect of seed soaking in aqueous extracts of neem seed on germination of mungbean and penetration of second stage juveniles of *Meloidogyne incognita. Ann. Agric. Res.*, 13(3): 279–299.

Mojumdar, V. and S.D. Mishra (1994). Effect of aqueous extracts of neem seed kernel and seed coat used as soil drenching against *M. incognita* infecting chickpea, *Curr. Nematol.*, 5(2): 173–175.

Morgan, M. and W.B. Collins (1964). The effect of organic treatments and crop rotation on soil populati... of *Pratylenchus penetrans* on strawberry culture, *Can. J. Plant Sci.*, 44: 272–275.

Mukhopadhya, M.C., M.R. Dalal and S.S. Khormi (1972). *Indian J. Nematol.*, 2:11–20.

Nagesh, M. and S.S. Hussaini (2001). Dose optimization of antagonistic fungi against root–knot and reniform nematodes in tomato under microplot conditions (Abs.). In *National Congress on Centenary of Nematology in India: Appraisal and Future* Plans, 5–7 Dec.,2001, IARI, N. Delhi, p. 146.

Nath, R., M.N. Khan, R.S. Kamalwanshi and R.P. Dwivedi (1982). Effect of *Argemone mexicana* on *M.javanica* in okra (*Abelmoschus esculentus*). *Indian J. Nematol.*, 12: 205–208.

Newshall, A.G. (1965). *Bot. Rev.*, 21: 189–250. In *Problems and Progress in Economic Phytonemtology,* (Ed.) D.S. Bhatti and R.K. Walia. HAU, Hisar, pp. 233–245.

Oostenbrink, M. (1960). *Tagetes patula* L. asvooruručht van. enkela land en tuinbouwgeswasson op zand endalgrond Meded, Land Opz Stn. *Gent.*, 25: 1065–1075.

Ostlind, D.A., S. Cifelli and R. Lang (1981). Insecticidal activity of the anti parasitic avermectins *Vet. Rec.*, 12: 105–168.

Pandey, R. and A. Kalra (2001). Chemical Inducers: A novel approach in root–knot nematode management in *Chamomilla recutita L.* In: *National Congress on Centenary of Nematology in India: Appraisal and Future Plan*, pp. 153–154.

Park, S.J. (1988). The role of polyethylene film mulching in enhancing phosphorus availibitlity in flue cured tobacco field. *Coresta Information Bulletin*, p. 88.

Pineda, O., M.W. Bonierbale, R.L. Plaisted, B.B. Brodie and S.D. Tanksley (1993). Identification of RFLP Markers linked to the H1 gene conferring resistance to the potato cyst nematode *G. rostochiensis Genome*, 36: 152–156.

Prakash, A., S.A. Tyagi and M.M. Alam (2001). Management of root-knot and reniform nematodes by root-dip in leaf extract of lemon grass (Abs.). In *National Congress on Centenary of Nematology in India: Appraisal and Future Plans*. 5–7 Dec., 2001, IARI, N. Delhi, p.103.

Prasad, D. and G. Chawla (2001). Future of nematicides in nematode management (Abs.). In: *National Congress on Centenary of Nematology in India: Appraisal and Future* Plans, 5–7 Dec., 2001, IARI, N. Delhi, pp: 60–61.

Prasad, D. and D. Ram (2001). Effect of nematicides and botanicals on growth of mung crop infected with *M. incognita* (Abs.). In *National Congress on Centenary of Nematology in India: Appraisal and Future Plans*, 5–7 Dec., 2001, IARI, N. Delhi, pp.111–112.

Rajendran, G., A. Santhi and K. Senthamizh (2003). Effect of potensized nematode induced cell extract against root-knot nematode *M. incognita* in tomato and reniform nematode, *R. reniformis* in turmeric, *Indian J. Nematol.*, Vol. 33(1): 67–69.

Ramakrishnan, S., C.R. Gunasekaran, S. Vadivelu (1997). Effect of bio-fertilizers Azolla and Azospirillium on root–knot nematode *M. incognita* and plant growth of okra. *Indian J. Nematol.*, 26(2): 127–130.

Rao, M.S., Dhananjay Naik, M. Shylaja (2003). Management of *M. incognita* on egg plant using a formulation of *Pochonia chlamydosporia* zare *et al.* (*Verticillium chlamydosporium*). *Pest Mgt. in Horticultural Ecosystem*, 9(1): 71–76.

Raymundo, S.A., J. Alcazar, R. Salas (1986). A technique for testing the efficacy of soil solarization in controlling root-knot nematodes at varying soil depth *J. Nematol.*, 18: 629.

Rhoades, H.L. (1964). Effect of *Crotolaria spectabilis* and *Sesbania exaltata* on plant nematode population on and subsequent yield of snapbean and cabbage. *Proc. Fla. State Host Soc.*, 77: 233–237.

Routaray, B.N., S.B. Dash, R.C. Mishra (2001). Effectiveness of mutagenic treatment in induction of resistance to *M. incognita* infection in black gram (Abs.). In National Congress on Centenary of Nematology in India: Appraisal and Future Plans (5–7 Dec., 2001), IARI, N. Delhi, pp. 154–155.

Ruello, J.S. (1983). Integrated control of *M. incognita* on tomato using organic amendments, marigolds and nematicide. *Plant Dis.*, 67: 671–673.

Sayre, R.M. (1973). *Theratomyxa weberi* amoeba predatory on plant parasitic nematode *J. Nematol* (5): 258–264.

Sharma, S.B., Y.L. Nene (1990). Effect of soil solarization on nematodes parasitic to chickpea and pigeonpea. *J. Nematol.*, 22: 658–664.

Siddiqui, M.A. and M.M. Alam (1988). Efficacy of seed dressing with the extract of neem, Persian lilac against *M. incognita* and *R. reniformis*, *Nematol, Medit.*, 15: 399–403.

Sikora, R.A. (1979). Predisposition to *Meloidogyne* infection by the endotropic mycorrhizal fungus *Glomus mosseae*. In *Root-knot Nematode Meloidogyne spp.: Systematic, Biology and Control*, (Eds.) F. Lamberti and C. Taylor, Academic Press, London, pp. 399–404.

Singh, R.S. (1965). Control of root-knot of tomato with organic soil amendments FAO *Plant Prot. Bull.*, 13: 35–37.

Singh, R.S. and K. Sitaramaiah (1967). Effect of decomposing green leaves, saw dust and urea on the incidence of root-knot of okra and tomato. *Indian Phytopath.*, 20: 349–355.

Singh, R.S. and K. Sitaramaiah (1973). Control of plant parasite nematodes with organic amendments of soil. Final Technical Report. Effect of organic amendments green manuring and inorganic fertilizers on root-knot of vegetable crops. *Exp. Stn.* and *Col. Agric.* Pantnagar, India. *Res. Bull.*, 6: 289.

Singh, R.S., K. Sitaramaiah (1973). Control of plant parasitic nematodes with organic amendments, green manuring and inorganic fertilizers on root–knot of vegetable crops *Exp. Stn.* and *Col. Agric.* Pantnagar, India, *Res. Bull.*, 6: 289.

Sitaramaiah, K. and R.A. Sikora (1982). Effect of mycorrhizza fungus *Glomus fasciculatum* on host parasite relationship of *Rotylenchulus reniformis* in tomato. *Nematologica*, 28: 412–419.

Small, R.W. (1979). A review of the prey of predatory soil nematodes. *Pedobiologia*, 30: 179–206.

Smith, G.S., R.S. Hussey and R.W. Roncadori (1986). Penetration and post infection development of *M. incognita* on cotton as affected by *Glomus intraradices* and phosphorous. *J. Nematol.*, 18: 429–435.

Southey, J.F. (1965). Physical method of control. In: *Plant Nematology*, (Ed.) J.F. Southey, GDI Min. Agr. Fish Food, HMSO, London, pp. 323–332.

Southey, J.F. (1978). Regulation controls. In: *Plant Nematology*, (Ed.) J.F. Southey, GDI Min. Agr. Fish Food, HMSO, London, pp. 326–332.

Southey, J.F. and A.G. Jones (1963). Hot water treatment of onion sets *Exper. Horticult.*, 9: 57–63.

Stapleton, J.J. and J.E. Devay (1982). Effect of soil solarization on population of selected soil borne microorganisms and growth of deciduous fruit tree seedlings. *Phytopathology*, 72: 323–326.

Stapleton, J.J. and J.E. Devay (1986). Soil Solarization: A non-chemical approach for management of plant pathogens and pests. *Crop Prot.*, 5: 190–98.

Stapleton, J.J. J.E. Devay, and B. Lear (1990). Simulated and field effects of ammonia based fertilizers and soil solarization on pathogen survival, soil fertility and crop growth. In: *Proceedings of the First International Conference on Soil Solarization*, Amman, Jordan, pp. 331–42.

Stapleton, J.J., J. Quick, J.E. De Vay (1985). Soil solarization: Effect of soil properties, crop fertilization and plant growth. *Soil Biol. Biochem.*, 17: 369–73.

Stephen, Z.A., A.H. Alwan and B.G. Antoon (1988). Effect of planting date on development of root–knot nematode *M. javanica*. Plant production and per cent of infection of tomato and egg plants. *Iraqi J. Agric. Sci. Zanco*, 6: 59–68.

Tanda, A.S. and A.S. Atwal (1989). Effect of seasame intercropping against root–knot nematode *M. incognita* in Okra, *Nematologica*, 34: 484–492.

Thomas, V., D. Prasad and J. Kumar (2003). Effect of controlled release formulation of Phorate on penetration and development of *M. incognita* infecting brinjal. (Abs.). In *5th National Symposium on Biocontrol Agents for Sustainable Management of Pests*, 18–20 Dec., 2003 GBPUAT, Pantnagar, p. 195.

Tomerlin, A.H.J. and G.C. Smart (1969). The influence of organic soil amendments on the nematode and other soil organisms *J. Nematol.*, 1: 29–30.

Trivedi, P.C. and B. Tyagi (1984). Effect of some plants on *M. incognita* population in chilli soil *J. Indian Bot. Soc.*, 63: 416–419.

Tyagi, S.A., A. Prakash and M.M. Alam (2001). Utilization of oilseed cakes for the management of plant parasitic nematodes and soil inhibiting fungi on lentil and Mungbean (Abs.). In National Congress on Centenary of Nematology in India, Appraisal and Future Plans, 5–7 Dec., 2001,IARI, New Delhi, p.153.

Van Gundy, S.D. (1972). In: *Pest Control Strategies for the Future*, pp. 317–329, National Acad. Sci., USA, pp. 376.

Van Gundy, S.D., F.D. Mc Elroy, A.F. Cooper and Lilt Stolzy (1968). Influence of soil temperature irrigation and aeration on *Hemicycliophora arenaria Soil Sci.*, 106: 270–274.

Verma, K.K. and D.C. Gupta (2002). Efficacy of VAM (*Glomus fasciculatum*). at various phosphorus levels for managing *M. incognita* in mungbean (*Vigna radiata*). In: *Proceedings of National Symposium on Biodiversity and Management of Nematodes in Cropping System for Sustainable Agriculture*, Jaipur, India 11–13. Nov., 2002 (Eds.) R.N. Singh, Pankaj, S.C. Dhawan, H.S. Gaur, IARI, N. Delhi, pp.161–163.

Vierling, R.A., J. Faghihi, V.R. Ferris and J.M. Ferris (1996). Association of RFLP Markers into loci conferring broad based resistance to the soyabean cyst nematode (*Heterodera glycines*). *Theor. Appl. Genet.*, 92: 83–86.

Vrain, T.C. (1997). Engineering resistance in crops parasitized by nematodes, In Diagnosis of key nematode pests of chickpea and pigeonpea and their management. In: *Proceedings of a Regional Training Course*, 25–30 Nov., 1996, ICRISAT, India (Ed. S.B. Sharma).

Wallace, H.R. (1963). *The Biology of Plant Parasitic Nematodes*. Edward Arnold Ltd. London, p. 280.

Weir, W.L., R.H. Garber, J.J. Sapleton, R.F. Gastelum, R.J. Wakeman, J.E. De Vay (1989). Control of potassium deficiency syndrome in cotton by soil solarization. *California Agri.*, 43: 26–28.

Widjaja, W.W. and W.A. Windrich (1974). Studies on nematological problems in horticulture pp: 393–402. In Research Reports, Agric. Cooperation Project, Indonesia The Netherlands, Sac 11–Technical contrib.. Jakarta.

Youssef, M.M.A. and M.S. Ali (1998). Management of *Meloidogyne incognita* infecting cowpea by using some native BGA *Anzeiger fer* Schadlingskunde, Pflanzenschutz, Umweltschutz, 71(1): 15–16.

29

Recent Taxonomic Status of Semi-endoparasitic Nematodes and their Management

Sudershan Ganguly* and M. Lal
Division of Nematology, Indian Agricultural Research Institute, New Delhi – 110 012
**E-mail: sg_nema@yahoo.com*

The plant parasitism is distributed in four distinct taxonomic groups *viz.*, orders Tylenchida, Aphelenchida, Dorylaimida (Longidoroidea) and Triplonchida (Trichodoroidea). Bulk of the plant parasitic nematode species are root parasites, while only a few species parasitise the above ground parts of plants. Most of the root parasites are *ectoparasites* as they do not enter the host and feed externally with the help of their spear. About 20 percent of the root parasites enter the plant tissues for feeding and develop inside are called *endoparasites*. A very small fraction of root parasites, feed on deeper tissues and their head and cervical region into the root tissue, are *semi-endoparasites*. The species of *Hoplolaimus, Tylenchorhynchus, Helicotylenchus, Scutellonema* and *Rotylenchus* are *migratory semi-endoparasites* because their larval and adult forms are active,vermiform and keep on changing their feeding sites. In contrast, Citrus nematode (*Tylenchulus semipenetrans*) and reniform nematode (*Rotylenchulus reniformis*) are two classical examples of *sedentary semi-endoparasites*. These nematodes, once establish their feeding site keep on feeding at the same place due to loss of mobility. Among all the species of semi-endoparasitic nematodes, reniform and citrus nematodes are the most agriculturally important nematodes, based upon their distribution and damage potential. This chapter deals with recent taxonomic status of sedentary semi-endoparasites of roots belonging to the order Tylenchida, and the management aspects of reniform and citrus nematodes.

Sedentary Semi-Endoparasitic Nematodes

The sedentary semi-endoparasitism is distributed in superfamily Hoplolaimoidea of Tylenchina and Tylenchuloidea of Criconematina with *Rotylenchulus* and allied genera in the former and *Tylenchulus* and allied genera in the latter. Among all the species of semi-endoparasites, reniform nematode and citrus nematode have been found to be economically important. Therefore, these nematodes have been studied extensively and much information has been generated on their vivid fundamental and applied aspects (Swarup *et al.*, 1989; Baghel, 1995).

Species Described/Recorded from India

There are five species of sedentary semi-endoparasites of roots, recorded from India. *Rotylenchulus reniformis* and *Tylenchulus semipenetrans* are widely distributed while others are confined to the isolated pockets. Only three species have been described from this country *viz. Trophotylenchulus piperis, T. andhriensis* and *T. waliai*. The details of 5 species known from India is given in Table 29.1.

Table 29.1: Semi-endoparasitic Nematode Species Recorded/Described from India

Nematode Species	*Distribution*	*Author*
Rotylenchulus reniformis	All the parts of India excepting temperate hilly regions	First recorded by Siddiqi and Basir (1959), and then several reports
R. parvus	Grass,Haryana	Bajaj and Bhatti,1987
Verutus mesoangustus	Vetiver Grass, Haryana	Bajaj and Walia, 1995
Tylenchulus semipenetrans	About 17 *Citrus* growing states of India	Siddiqi,1961; and then several reports
*Trophotylenchulus piperis**	*Piper nigrum* Kerala, India	Mohandas *et al.*,1985
*T.andhriensis**	A.P., India	Muthukrishnan and Shariff, 1985
*T. waliai**	*Cereus peruvianus*, Haryana	Bajaj and Bhatti, 1989

* Described from India.

Rotylenchulus and Allied Genera

The list of nominal species of *Rotylenchulus* and allied genera, along with their respective type host, type locality and authors, is given in Table 29.2.

Systematics and Taxonomic Position

Due to the lack of adequate knowledge about the evolutionary trends in plant parasitic nematodes, there have been controversies regarding the placement of these genera in the classification framework of Tylenchida. These genera with their distinct procorpus and metacorpus, overlapping oesophageal glands and didelphic gonad are placed under suborder Tylenchina and are differentiated from Criconematina. Their placement had been under subfamily Pratylenchinae (Thorne,1949), Nacobbinae (Chitwood and Chitwood,1950) Rotylenchulinae (Husain and Khan,1967). At the family rank, these nematodes have been variously accommodated in Tylenchidae (Thorne,1949) or Hoplolaimidae (Chitwood and Chitwood,1950) under Tylenchoidea or in Nacobbidae (Golden, 1971) under Heteroderoidea until family Rotylenchulidae was proposed by Husain and Khan (1967). This was recognised by Fortuner (1984) and Siddiqi (1986), but Maggenti et.al (1988) included these genera under Hoplolaimidae of Tylenchoidea,and suppressed Hoplolaimoidea, Heteroderoidea and several other higher categories, thus resulting in two suborders–Tylenchina and Aphelenchina, and only two superfamilies (Tylenchoidea and Criconematoidea) under Tylenchina, situation resembled almost that existed in 1971, or earlier. Khan and Ganguly (1990) proposed a moderate classification scheme of the order Tylenchida and discussed its origin and phylogenetic relationships. However, the most recent classification scheme for the order Tylenchida has been proposed by Siddiqi (2002), based upon which *Rotylenchulus* and allied genera are grouped as follows:

Table 29.2: Species of the Tylenchid Semi-endoparasitic Genera Along with their Type Hosts and Localities

	Species	*Type Host*	*Type Locality*	*Author*
I	***Rotylenchulus***			
	R. reniformis	Cowpea	Hawaii, USA	Linford and Oliveira, 1940
	R. parvus	Sugarcane	Mauritius	Williams (1960); Sher, 1961
	R. borealis	Grass	Amhem, Netherlands	Loof and Oostenbrink,1968
	R. clavicaudatus	*Strelitzia* sp.	S. Africa	Dasgupta *et al.*, 1968
	R. anamictus	*Acasia* Sp.	Merca, Somaliland	Dasgupta *et al.*, 1968
	R. leptus	Bamboo	Guelo, S. Rhodesia	Dasgupta *et al.*, 1968
	R. macrodoratus	Grapevines	Bari, Italy	Dasgupta *et al.*, 1968
	R. macrosoma	Olive	Hulda, Israel	Dasgupta *et al.*, 1968
	R. variabilis	Rumex	Inyanga, S. Rhodesia	Dasgupta *et al.*, 1968
	R. sacchari	Sugarcane	Natal, S. Africa	Van den Berg and Spaull, 1981
	R. brevitubulus	—	S. Africa	Van den Berg, 1990
II	***Senegalonema***			Germani *et al.*, 1984
	S. sorghi	Sorghum sp	Patar, Senegal	Germani *et al.*, 1984
III	**Verutus**			Esser, 1981
	V. volvigenitis	Button Weed	Florida,USA	Esser,1981
	V. californicus			Baldwin *et al.*, 1989
IV	**Bilobodera**			
	B. mesoangusta	*Mesoanthus sinensis*	Japan	Minagawa,1986 Sharma and Siddiqi,1992
V	**Acontylus**			Meagher,1968
	A. vipriensis	*Eucalyptus* Sp. and *Acacia* Sp.	Australia	Meagher1968

Order Tylenchida Thorne, 1949

Suborder Hoplolaimina Chizhov and Berezina, 1988

Superfamily Hoplolaimoidea Filipjev, 1934

Family Rotylenchulidae Hussain and Khan, 1967

1. Subfamily Rotylenchulinae Whitehead, 1958

Type genus: *Rotylenchulus* Linford and Oliveira, 1940

Other genus *Senegalonema* Germani, Luc and Baldwin, 1984

2. Subfamily Verutinae Esser, 1981

Type genus *Verutus* Esser, 1981

Other genus *Bilobodera* Sharma and Siddiqi, 1992

3. Subfamily Acontylinae Fotedar and Handoo, 1978

Type genus *Acontylus* Meagher, 1968.

Key to Subfamilies and Genera of Rotylenchulidae

1. Female and male with indistinct tail, young vermiform migratory stage absent, bursa absent. .... *Verutus* (Verutinae)

   Female and male with distinct tail, young vermiform migratory stage present, bursa present. 2

2. Oesophageal gland overlap mostly dorsal; one functional ovary; male head inflated knob like, juveniles almost straight .... *Acontylus* (Acontylinae)

   Oesophageal gland overlap mostly ventral; two functional ovary; male head not inflated or knob like, juveniles curved .... 3 (Rotylenchulinae)

3. DGO > half stylet length behind the spear base; young female with outstretched ovaries; bursa extending upto tail terminus .... *Senegalonema*

   DGO > half stylet length behind the spear base; young female with ovaries having double flexures; bursa not enclosing the tail tip .... *Rotylenchulus*.

Key to the Species of *Rotylenchulus* (Original)

1. Dorsal gland orifice 6–8 µm behind spear knobs; tail bifid .... *R. brevitubulus*

   Dorsal gland orifice >10 µm behind spear knobs; tail not bifid 2

2. Tail of young females and males broadly rounded .... *R. sacchari*

   Tail of young females and males conoid 3

3. Stylet 10–22 µm; knobs rounded 4

   Stylet 22–26 µm; knobs anchor shaped .... *R. macrodoratus*

4. h = 1–13; L < 0.45 mm 5

   h = 13–23; L > 0.45 mm 9

5. h = 1–3; L < 0.27 mm .... *R. parvus*

   h = 3–13; L > 0.27mm 6

6. V > 67 per cent 7

   V < 67 per cent 8

7. Stylet 12–14 µm; O = 107–141 per cent .... *R. anamictus*

   Stylet 14–22 µm; O = 63112 per cent .... *R. reniformis*

8. Stylet 12–14 µm; h = 3–7 .... *R. leptus*

   Stylet 14–16 µm; h = 8–13 .... *R. borealis*

9. V = 57–60; O = 75–85 per cent .... *R. clavicaudatus*

   V = 63–68; O = 139–188 per cent .... *R. macrosoma*

R. variabilis was synonimised with *R. borealis* by Germani (1978).

Reniform Nematode

Distribution

Reniform nematode was first discovered from cowpea fields in the island of Oahu, Hawaii islands, USA in1931, which was later described as *Rotylenchulus reniformis* by Linford and Oliveira, 1940. In India, it was first reported from roots of coffee by Siddiqi and Basir in 1959. Reniform nematode is generally confined to tropical and subtropical areas of the world and is reported to occur in more than 30 countries. In India the nematode has been reported from almost all the states excepting the temperate hilly regions of J and K and H.P.

Host Range

This nematode is known to infect about 115 plant species, mostly belonging to families Leguminaceae, Cucurbitaceae, Malvaceae and Solanaceae. Some of the crops on which noticeable damage have been recorded are: cotton, papaya, pineapple, cowpea, castor bean, banana, tea, sweet potato, coffee, pigeon pea, tomato, tobacco, okra, egg plant, beans, soybean and cantaloupe.

Symptoms

The gross symptomatology of reniform nematode infected plant has been studied for few crops only. The roots of infected plant exhibit " dirty root symptom" due to presence of soil particles and debris sticking to them. In the tender roots of tomato and pigeon pea, small necrotic spots are observed around the female head. A severe curling of young roots of many crops are observed in case of heavy infection. On cotton a few small and less dark green leaves are noticeable in nematode stressed plants and the plants remain stunted with less vigour. Severe infection often results in pruning of roots. Infected castor beans exhibit drying of leaf margins within one month of post infectional period. The auxillary buds in main stem develop into tiny leaves and later shed off, leaving the plants only with stem. The roots show moderate discolouration with the root-lets as dark brown and some times necrotic.

Management of Reniform Nematode

Chemical

The discovery of DD as a nematicide, led almost the entire pineapple industry in Hawaii to adopt this volatile soil fumigant against *R. reniformis*. Initially, the four halide soil fumigants, DD, EDB, DBCP and dichloropropene were widely used against this nematode for pineapple plantations in Hawaii. Due to the hazardous effects of these chemicals, non-volatile and contact chemicals including aldicarb, fensulfothion, carbofuran and other toxicants have been tested for control of this nematode attacking variety of crops like grapes, banana, tobacco, cotton, soybean, cantaloupe, sweet potato, sunflower and potato, which have been reviewed by Swarup *et al.*, 1989. The dosage used ranged from 2–16 Kg a.i./ha, depending on the plant species involved.

Foliar application of systemic pesticide oxamyl @ 600 and 2000 ppm on tomato and cotton, respectively could effectively manage this nematode (Rich and Bird, 1980), causing significant reduction in soil population and increase in yield. Seed treatment of okra and pulses with carbofuran or carbosulfan@ 1.25Kg a.i./ha is also effective against the entry of this nematode into roots during the early stage of plant growth.

Resistance

Use of resistant varieties is one of the most important component in the management of the nematode. Major efforts in this direction have been made on soybean and cultivar Forrest was found resistant wherein the resistance was governed by 2 pairs of genes with unequal effects (Harvillie *et al.*, 1985). Presently, resistant varieties/cultivars against reniform nematode, have been found in several crops *viz.*, coffee, cotton, cowpea, tomato, mustard, oats, onion, pigeonpea, potato, sorghum, soybean, sweet potato, pepper, etc., which has been compiled by Varaprasad, 1986.

Other Methods

Application of organic amendments either as green manure or in the form of cakes reduces the nematode population (Alam *et al.*, 1980; Khan et al., 1974). Two years crop rotation using green sorghum or resistant soybean (Pickett) in place of cotton was quite effective against this nematode (Gillman *et al.*, 1978). Hot water treatment is also effective in eliminating the nematode populations from roots (Martin, 1970). The mycorrhizal fungus, *Glomus fasciculatum* has been demonstrated to increase resistance of tomato plants against this nematode (Sitaramaiah and Sikora, 1982).

Tylenchulus and Allied Genera

The genus *Tylenchulus* was proposed by Cobb,1913 with its type species *T. semipenetrans* from citrus roots in Florida,USA. In 1957, Raski and Sher proposed *Sphaeronema* and in the same year Raski erected *Trophonema* and *Trophotylenchulus*. Certain species were described under the genus *Tylenchulus viz., T. mangenoti* (Luc,1957); *T.floridensis* (Raski,1957); *T.obscurus* (Colbran,1961), *T.clavicaudatus* (Colbran,1966) of which the last three are presently under *Trophotylenchulus* and *T. mangenoti* under *Ivotylenchulus* of Hashim,1984. There are now four species grouped under *Tylenchulus viz. T. semipenetrans,T. furcus* and two species *i.e., T.palustris* and *T. graminis* described by Inserra *et al.*, 1988.

The genus *Meloidoderita* Poghossian,1966 differs from all others genera of Tylenchida because its uterine walls transform to cystoid body to serve as protecting case for the retained eggs. In *Meloidoderita* and *Sphaeronema*, the post vulval region is almost absent due to which these genera differ from *Tylenchulus,Trophonema* and *Trophotylenchulus*. Siddiqi (1986) proposed genera *Goodeyella* and *Tumiota* for accommodating some species of *Sphaeronema*. The list of nominal species described under *Tylenchulus* and allied genera along with their respective type host, locality and authors is given in Table 29.3.

Table 29.3: Species of the Criconematid Sedentary Semi-endoparasitic Genera along with their Respective Type Host and Locality

	Species	*Type Host*	*Type Locality*	*Author*
I	***Tylenchulus***			Cobb,1913
	T. semipenetrans	Citrus	Brookville,USA	Cobb,1913
	T. furcus	Sugarcane, Grass	S. Africa	Van den Berg and Spaull,1982
	T. palustris	Pop ash (*Frarcinus caroliana*)	Florida, USA	Inserra *et al.*, 1988
	T. graminis	*Andropogon virginicus*	Florida, USA	Inserra *et al.*,1988
	Ivotylenchulus			Hashim,1984
	I. mangenoti	*Dorstenia embergeni*	Abidijan, Ivory Coast	Hashim,1984
	Trophotylenchulus *(Syn. Ivotylenchulus; Trophonema)*			Raski,1957
	T. floridensis	*Quercus falcatua*, *Magnolia* spp., *Diospyrus* spp.	Florida,USA	(Raski,1957) Maggenti, 1962

Contd...

Table 29.3–Contd...

	Species	*Type Host*	*Type Locality*	*Author*
	T. clavicaudatus	—	—	(Colbran,1966) Hashim,1984
	T. obscurus	—	Queensland, Australia	(Colbran,1961) Hashim,1984
	T. saltensis	*Olea europaea*	Jordan	Hashim,1984
	T. piperis	*Piper nigrum*	Kerala, India	Mohandas *et al.*,1985
	T. andhriensis	—	A.P., India	Muthukrishnan and Shariff, 1985
	T. waliai	*Cereus peruvinus*	Haryana, India	Bajaj and Bhatti,1989
	T. arenarius	*Juncus lesourii*	California, USA	Raski,1957
	T. asoensis	—	Japan	(Minagawa, 1983) Siddiqi, 1999
	T. okamotai	—	Japan	(Minagawa, 1983) Siddiqi, 1999
	T. arthenidis	—	—	Dolinski *et al.*, 1996
	T. colbrani	Mbalmayo forest	Cameroon	Siddiqi, 1999
	T. cunctus	Mbalmayo forest	Cameron	Siddiqi, 1999
	Boomerangia			Siddiqi, 1991
	B. benigna			Siddiqi, 1994
	B. bellula			Siddiqi, 1994
	Sphaeronema			Raski and Sher, 1952
	S. californicum	California laurel (*Umbellularia californica*)	California, USA	Raski and Sher, 1952
	S. comubiense	—	S. Africa	Ven den Berg and Spaull,1982
	S.rumicis	—	—	Kirjanova,1970
	S. sasseri	—	USA	Eisenback and Hartman, 1985
	S. alni			Turkina and Chizov, 1986
	S. salicis			Eroshenko, 1989
	Goodeyella			Siddiqi, 1986
	G. minutissima	Citrus	Pasar Minggu, Indonesia	(Goodey,1958) Siddiqi,1986
	Tumiota			Siddiqi,1986
	T. whittoni	Sweet Gum, Liquid ambar, *Styraciflua*	Gainesville, Florida, USA	(Sledge and Christie, 1962) Siddiqi,1986
V	***Meloidoderita***			Poghossian,1966
	M. kirjanovae	*Mentha longifolia*	Armanian SSR	Poghossian, 1966
	M. safrica	Sugarcane	S.Africa	Ven den Berg and Spaull, 1982
	M. polygoni	*Polygonum spp.*	USA	Golden and Handoo, 1984

Systematics and Taxonomic Placement

Due to the absence of gland overlap fused pro-metacorpus and true monodelphic condition, *Tylenchulus* and allied genera are grouped along with criconematids. Skarbilovich (1947) proposed the subfamily Tylenchulinae which was elevated to Tylenchulidae by Kirjanova(1955) and Raski

(1957), with two subfamilies Tylenchulinae and Sphaeronematinae, Geraert (1966) elevated two subfamilies to familial ranks and placed under Criconematoidea. Raski and Siddiqui (1975) erected superfamily Tylenchuloidea for these genera. Siddiqi (1980) proposed sub-order Criconematina which has been recognised by Khan(1986), Ganguly and Khan (1990) but Maggenti *et al.* (1987) did not recognise the suborder and placed all these genera under one subfamily Tylenchulinae under Tylenchulidae placed under Criconematoidea, a situation almost resembling that of two decade as in 1966. Keeping in view the present level of proposed earlier, we still hold the same opinion of retaining Tylenchuloidea under Criconematina and accept the broad framework of Siddiqi's classification of 1986 and 2002.

Systematics

Order: Tylenchida Thorne, 1949

Suborder: Criconematina Siddiqi, 1980

Superfamily: Tylenchuloidea Skarbilovich, 1947 (Raski and Siddiqi, 1975)

1. Type family: Tylenchulidae Skarbilovich, 1947 (Kirjanova, 1955)

Subfamily: Tylenchulinae Skarbilovich,1947

Type genus: *Tylenchulus* Cobb,1913

Other genera: *Boomerangia* Siddiqi, 1991

Trophotylenchulus Raski,1957

Synonyms: *Ivotylenchulus* Hashim,1984

Trophonema Raski,1957

Other families

2. Family: Sphaeronematidae Raski and Sher, 1952

Subfamily: Sphaeronematinae Raski and Sher, 1952

Type genus: *Sphaeronema* Raski and Sher, 1952

Other genera: *Goodeyella* Siddiqi, 1986

Tumiota Siddiqi, 1986

Subfamily *Meloidoderitinae* Kirjanova and Poghossian, 1973

3. Family: Paratylenchidae Thorne, 1947

Subfamily: Paratylenchinae Thorne, 1947

Type genus: *Paratylenchus*

Other genera: *Gracilacus* Raski, 1962

Cacopaurus Thorne, 1943

Subfamily: Tylenchocriconematinae Raski and Siddiqui, 1975

Type genus: *Tylenchocriconema* Raski and Siddiqui, 1975

The semi-endoparasitism is confined to only Tylenchulidae and Sphaeronematidae.

Key to Semi-endoparasitic Families of Tylenchuloidea

1. Adult female spherical or subspherical, lacking a post vulval region, excretory pore in oesophageal region .... Sphaeronematidae

   Adult female not spherical or subspherical, with a distinct post vulval region, Excretory pore much behind oesophageal region except (*Trophonema*) .... Tylenchulidae

Key to Genera of Tylenchulinae (After Siddiqi, 2002)

1. Excretory pore in female at 68–85 per cent of body length, surrounded by papilla like outgrowths .... *Tylenchulus*

   Excretory pore in female less than 61 per cent of body length, not surrounded by outgrowth 2

2. Female body arcuate, slightly swollen .... *Boomerangia*

   Female body spirally coiled; considerably swollen .... *Trophotylenchus*

Key to the Species of *Tylenchulus* (After Inserra *et al.*, 1994)

1. 2nd stage juveniles and adult females with distinct rectum and anus; mature females with swollen region 60 per cent or more 2

   2nd stage juveniles and adult females without distinct rectum and anus; Mature females with swollen portion 58 per cent or less 3

2. 2nd stage juveniles with furcate or bifid tail tip .... *T. furcus*

   2nd stage juveniles with tail tapering or variously conoid, not furcate .... *T. graminis*

3. Mature females with conoid post vulval section, Males with stylet knobs 1.6 µm wide or more; basal bulb > 8 µm wide; J2 tail cylindrical with bluntly rounded terminus .... *T. palustris*

   Mature female with digitate post vulval region with round terminus; male stylet knobs <1.2 µm wide; basal bulb <8 µm, juvenile tail tapering ..... *T. semipenetrans*

Family Sphaeronematidae (Raski and Sher,1952) Geraert 1966

Type subfamily: Sphaeronematinae Raski and Sher,1952

Other subfamily: Meloidoderitinae Kirjan. and Poghoss,1973

Key to Subfamilies and Genera of Sphaeronematidae

1. Mature female with a neck, uterine walls not forming a cystoid body for eggs ....Sphaeronematinae (2)

   Mature female without a neck; uterine walls may form a cystoid body for eggs .... *Meloidoderita* (Meloidoderitinae)

2. Mature female with vulva on a cone like protuberance of body .... *Sphaeronema*

   Mature female with vulva flushed with body surface 3

3. Females and juveniles with oesophageal glands in basal bulb; juveniles cephalic region with a truncate circum-oral elevatons .... *Goodeyella*

   Females and juveniles with oesophageal glands extending over intestine, juveniles cephalic region rounded .... *Tumiota*

Most of the species grouped under Criconematina are ectoparasites on roots of trees. They prefer undisturbed soil and feed at one site for long duration. The species of *Goodeyella* and *Sphaeronema* are the only endoparasites under this suborder. The sedentary semiendoparasitism is confined to only *Tylenchulus* and *Trophotylenchulus*..

Citrus Nematode

Distribution

Citrus neamtode was first recorded from citrus roots in 1912 in California, USA. Subsequently, it was found associated with unthrifty citrus orchards in other citrus growing area of USA, and also in Malta, Spain, Israel, India, Australia and South America. The neamatode is now recorded from almost all the citrus orchards of the world including India, and is an important limiting factor in citrus economy.

Host Range

Citrus nematode is known to parasitise 75 rutaceous species, 50 of which comprise of citrus species or citrus hybrids and 9 non-rutaceous species. Important commercial species like grapefruit (*Citrus paradisi*), sweet orange (*C. sinensis*), sour orange (*C. aurantium*) and lemon (*C. lemon*) are also susceptible.

Some unusual recorded non-rutaceous hosts are: *Olea eurpaea, Vitis vinifera, Syringia vulgaris, Andropogon rhizomatus, Calodendron copense, Eriobotrys japonica, Mikania batatifolia, Prunus communis, Diospyros lotus and Garcinia mangostana*..

Symptoms

Nematode infected trees suffer from " Slow decline or slow die back or Die back" disease. General decline of trees defoliation of shoots, smalling and yellowing of leaves and fruits, sign of starvation in leaves, root lesions, cortical sloughing of roots ultimately leading to root decay, are broad effects recorded in the trees infected by citrus nematode. Heavily infected trees decline within 3–5 years.

Management of Citrus Nematode

Physical Method

The bare root treatment art 45°C for 25 min. or 46.7°C for 10 min. can disinfect citrus seedlings from citrus neamtode (Bindra *et al.*, 1967).

Soil Treatments

Use of *neem, mahua*, ground nut and mustard oil cakes can effectively reduce citrus nematode population in the citrus rhizosphere.

Cultural Practices

Population of citrus namtode in soil and roots can be reduced by removing the older feeder roots before the growth flush. This practice is followed by application of FYM. This is a general cultural practice adopted by the farmers of Haryana and Punjab.

Plant Resistance

Use of resistant citrus root stock offer cost effective method of control of citrus nematode. Varietal screening studies against citrus nematodes revealed that trifoliage orange (*Poncirus trifoliate*). Carrizo and Troyer are highly resistant. Cleoptra mandarin from Morrocco, Smooth lemon Assam and Saoner were found moderately resistant. Rangpur lime, used as a root stock for Bold Red, Campbell Valencia and Hamlin scions, proved resistant to citrus neamtode Hybrid citrus root stocks developed by crossing Rangpur lime *(C. limona with P. trifoliate viz.*, CRH-3, CRH-5 and CRH- 41 were reported highly resistant (Reddy *et al.*, 1987).

Biological Control

Paecilomyces liacinus, endoparasitic fungus is an excellent biocontrol agent. This fungus can effectively manage citrus nematode population. *P. lilacinus* @ 4g in combination with carbofuran @ 1 kg gave maximum growth parameters and effectively reduced citrus nematode population in acid lime (Reddy *et al.*,1990). In the field experiment, *P.lilacinus* application proved effective in reducing citrus nematode population and in increasing fruit size (Jatala, 1986). Vesicualr arbuscular mycorrhizal fungus *Glomus mosseae*, symbiotic to citrus nematode on citrus, proved beneficial to crop. *G.mosseae* stimulates growth of *Citrus jambhiri* seedlings and its presence suppress the growth retarding effect of citrus nematode (Baghel *et al.*, 1990).

Plant Extracts

Application of certain plant parts and their extracts affect nematode population either by killing or paralysing them several *in vitro*, studies have been undertaken using different plant extracts. Castor (*Ricinus communis*, Oak (*Calotropis procera*) and Datura (*Datura stramonium*) chopped leaves applied @ 20 and 40g/kg soil in 9 Sq meter area, around citrus tree trunk, can result in reduction of nematode population and increase in yield.

Chemical Methods

Post- plant application of chemicals has been major mode of approach for the control of citrus nematode. In early seventies, DBCP, DD, Dimethoate, Metham Sodium and Dichlorofenthion were used for controlling this nematode in orchards. Due to their environmental and health hazards, theses were replaced by Organophosphates and Carbamates. Applications of Carbofuaran, Aldoxicarb and Oxamyl @ 1.1 Kg a.i./ha by low pressure drip irrigation system have proved to be beneficial to citrus trees (Vangundy *et al.*, 1984). Some of these chemicals can be used effectively as soil treatment also. Carbofuaran @ 2 Kg a.i./ha in combination with neem cake @ I Kg/plant could reduce 33 per cent nematode population and increased the yield by 33.8 per cent, with cost -benefit ratios varying from 1: 4 to 1: 7. (Baghel,1995).

REFERENCES

Baghel, P.P.S. (1995). Nematode problems in Citrus. In: *Nematode Pest Management: An Appraisal of Eco friendly Approaches*, (Eds.) G. Swarup, D.R. Dasgupta and J.S. Gill, Nematological Society of India, New Delhi, pp. 203–210.

Bajaj, H.K. and D.S. Bhatti (1989). *Trophotylenchulus waliai* n. sp. (Criconematina: Tylenchulidae) from Haryana India. *Indian J Nematol.* 19: 223–225.

Bajaj, H.K. and R.K. Walia (1995). *Verutus mesoangustus* Minagawa 1986 (Nematoda: Heteroderidae) a new record from India. *Indian J. Nematol.* 25: 48–51.

Chitwood, B.G. and N.B. Chitwood (1950). An Introduction to Nematology. Baltimore, USA. Monumental printing co. Revised edition, pp.213.

Cobb, N.A. (1913). New nematode genera found inhabiting fresh water and non-brackish soils. *J. Wash. Acad. Sci.*, 3: 432–444.

Colbran, R.C. (1966). Studies of plant and soil nematodes. 12. The eucalyptus cystoids nematode *Cryphodera eucalypti* n. g., n. sp. (Nematoda: Heteroderidae), a parasite of eucalypts in Queensland. *Qd. J. Agric. Anim. Sci.*, 23: 41–47.

Dasgupta, D.R. and Sheshadri, A.R. (1971). Reproduction, Hybridization and host adaptation in physiological races of the reniform nematode, *Rotylenchulus reniformis. Indian J. Nematol.*, 1: 128–144.

Dasgupta, D.R., D.J. Raski and S.A. Sher (1968). A revision of the genus *Rotylenchulus* Linford and Oliveira, 1940 (Nematoda: Tylenchidae). *Proc. Helminth. Soc. Wash.* 35: 169–192.

Esser, R.P. (1981). *Verutus volvingenitis* n.gen., n.sp. (Heteroderidae: Tylenchida) in Verutinae n. subf., a phytoparasitic nematode infesting buttonweed in Florida. *Proc. Helminth. Soc. Wash.* 48: 220–240.

Fortuner, R. (1984). Morphological variability in *Helicotylenchus* Steiner, 1945. 6: Value of the character used for specific identification. *Revue de Nematol.*, 7: 245–264.

Ganguly, Sudershan and E. Khan (1990). *Gracilpaurus corbetti* gen.n., sp.n. (Nematoda: Tylenchida: Paratylenchidae) from India with a key to the species of this genus. *Pakistan J. Nematol.*, 8: 65–71.

Geraert, E. (1966). The systematic position of the families Tylenchulidae and Criconematidae. *Nematologica*, 12: 362–368.

Germani, G., M. Luc and J.G. Baldwin (1984). A new Rotylenchulinae: *Senegalonema sorghi* n. gen., n. sp. (Nematoda: Tylenchida). *Revue de Nematol.*, 7: 49–56.

Gillman, D.F., Jones, J.E. and Birchfield, W. (1978). Cotton–soybean rotation for control of reniform nematodes. *La. Agric.*, 21: 10–11.

Golden, A.M. (1971). Classsification of the genera and higher categories of the order Tylenchida (Nematoda) In: *Plant Parasitic Nematodes. Volume 1: Morphology, Anatomy, Taxonomy and Ecology* (eds.) Zuckerman, B.M., W.F. Mai and R.A. Rhodes Academic Press Inc., New York and London, pp.191–232.

Golden, A.M. and Z.A. Handoo (1984). Description of *Meloidoderita polygoni* n.sp. (Nematoda: Meloidoderitidae) from USA and observations on *M. kirjanovae* from Israel and USSR. *J. Nematol.*, 16: 265–282.

Harvillie, D.G., Green, A. and Birchfield, W. (1985). Genetic resistance to reniform nematodes in soybeans. *Plant Dis. Reptr.*, 69: 587–589.

Hashim, Z. (1984). Description of *Trophotylenchulus saltensis* n. sp., with a comment on the status of *Trophotylenchulus* Raski, 1957 and a proposal for *Ivotylenchulus* n. gen. (Nematoda: Tylenchida). *Revue de Nematol.*, 6: 179–186.

Inserra, R.N., R.P. Esser and J.H. O'Bannon (1988). Identification of *Tylenchulus* species from Florida. *Florida Dep. Agri. Cons. Serv. Nematological Circular*, 153: 1–4.

Jatala, P. (1985). In: *An Advanced Treatise on Meloidogyne Vol. I: Biology and Control*, (Eds.) J.N. Sasser and C.C. Carter, North Car., S.V. Graphics, pp 303–308.

Khan, F.A. and F.M. Khan (1974). Studies on the reniform nematode, *Rotylenchulus reniformis*. I. Host range and population changes. *Indian J. Nematol.*, 3: 24–30.

Kirjanova, E.S. and E.E. Poghossian (1973). Redescription of *Melododerita kirjanovae* Poghossian, 1966 (Nematoda: Melododeteritidae, fam. n.). *Parazitologiya*, 7: 280–285.

Linford, M.B. and J.M. Oliveira (1940). *Rotylenchulus reniformis* nov. gen., nov. sp., a nematode parasite of roots. *Proc. Helmith. Soc. Wash.* 7: 35–42.

Loof, P.A.A. (1956). *Trophurus,* a new tylenchid genus (Nematoda). *Versl. Meded. Plziektenk. Dienst. Wageningen* 129: 191–195.

Loof, P.A.A. and M. Oostenbrink (1962). *Rotylenchulus borealis* n. sp., with a key to the species of *Rotylenchulus. Nematologica* 7: 83–90.

Luc, M. (1957). *Tylenchulus mangenoti* n. sp. (Nematoda: Tylenchulidae). *Nematologica.* 2: 239–334.

Maggenti, A.R., M.Luc, D.J.Raski, R. Fortuner and E. Geraert (1987). A reappraisal of Tylenchina (Nemata). 11. List of generic and supra-generic taxa, with their junior synonyms. *Revue de Nematol.* 11: 177–188.

Martin, W.J. (1970). Elimination of root knot and reniform nematodes and scurf infections from root lets of sweet potato plants by hot water treatment. *Plant disease Reporter,* 54: 1056–1058.

Meagher, J.W. (1968). *Acontylus vipriensis* n.g., n. sp. (Nematoda: Hoplolaimidae) parasitic on *Eucalyptus* sp. in Australia. *Nematologica.* 14: 94–100.

Minagawa, N. (1986). Taxonomic studies on Criconematidae (Nematoda: Tylenchida) of Japan I. Genera *Neolobocriconema, Paralobocriconema* n.gen. and *Macrocroconema* n. gen. *Bull. Nat. Inst. Agro–Environ. Sci. Ibaraki, Japan.* 1: 95–126.

Poghossian, E.E. (1966). A new nematode genus and species of the family Heteroderidae from the American SSR (Nematoda). *Dokl. Akad. Nauk Armyan,* SSR 42: 117–123.

Raski, D.J. (1957). *Trophotylenchulus* and *Trophonema,* two new genera of Tylenchulidae n. fam. (Nematoda). *Nematologica,* 2: 85–90.

Raski, D.J. and S.A. Sher (1952). *Sphaeronema californicum,* nov. gen. Nov. spec.,(Criconematidae: Sphaeronematinae, nov. subfam.) an endoparasite of the roots of certain plants. *Proc. Helminth. Soc. Wash.* 19: 77–80.

Raski. D.J. and I.A. Siddiqui (1975). *Tylenchocriconema alleni* n. g., n. sp. from Guatemala (Tylenchocriconematidae n. fam.; Tylenchocriconematoidea n. superfam.; Nematoda). *J.Nematol.* 7: 247–251.

Rich, J,R. and Bird, G.W. (1980). Inhibition of *Rotylenchulus reniformis* penetrations of tomato and cotton roots with foliar application oxamyl. *J. Nematol.,* 5: 221–224.

Siddiqi, M. R. (1961). Studies on *Tylenchorhynchus* spp. (Nematoda: Tylenchida) from India. *Z. Parasitkde.* 21: 46–64.

Siddiqi, M. R. (1999). *Tylenchorhynchus colbrani* sp. n. and *T. cunctus* sp. n. (Criconematina: Tylenchulidae) from Mbalmayo forest in Cameroon. *International J. Nematol.* 8: 179–184.

Siddiqi, M.R. (1994). Nematodes of tropical rain forest 3– Three new genera and five new species of Tylenchs. *Afro–Asian J. Nematol.* 4: 22–31.

Siddiqi, M.R. (2000). *Tylenchida: Parasites of Plants and Insects.* II edition, CAB International, UK, pp. 831.

Siddiqi, M.R. and M.A. Basir (1959). On some plant parasitic nematodes occurring in south India with the description of two new species of the genus, *Tylenchorhynchus.* Proc. 46th Indian Science Congress., Part IV, pp 35.

Sitaramiaha, K. and R.A. Sikora (1982). Effect of the mycorrhizal fungus *Glomus fasciculatus* on the host parasite relationship of *Rotylenchulus reniformis* in tomato. *Nematologica*, 28: 412–419.

Skarbilovich, T.S. (1947). Revision of the systematics of the nematodes family Anguillulinidae Baylis and Daubney, 1926.*Dokl. Akad. Nauk*, SSR 57: 307–308.

Steiner, G. (1949). Plant nematodes the grower should know. *Proc. Soil Sci. Soc. Fla.*, 4: 72–117.

Swarup, G., D.R. Dasgupta and P.K. Koshy (1989). *Plant Diseases*. Anmol Publications, New Delhi, pp. 395.

Thorne, G. (1949). On the classification of the *Tylenchida*, new order (Nematoda, Phasmidia). *Proc. Helminth. Soc. Wash.* 16: 37–73.

Vangundy, S.D. and S.Garabedian (1984). Application of nematicides through drip irrigation systems. *Meded de Facult.* Landbou. Rajksuniversiteit, Gent, 49: 629–634.

Varaprasad, K.S. (1986). *Rotylenchulus reniformis* Linford and Oliveira, 1940. A comparative account of systematics, biology and management. In: *Plant Parasitic Nematodes of India: Problems and Progress*, (Eds.) G. Swarup and D.R. Dasgupta, pp. 196–200, IARI Publication.

30

Induced Plant Resistance

Varghese P. Thomas, Berin Pathrose** and D. Prasad**

**Division of Nematology, Indian Agricultural Research Institute, New Delhi – 110 012*

***Division of Entomology, Indian Agricultural Research Institute, New Delhi – 110 012*

What does the plant do to protect itself from various types of pathogens? Plants do not have an immune system in the sense that we think of for ourselves. However, they don't just sit around passively waiting for the next pathogen to attack. Instead, plants are equipped with a variety of mechanisms to fend off such attackers. Plants, like animals, also have a natural defense system against disease. Plants have their own networks of defense against plant pathogens that include a vast array of proteins and other organic molecules produced prior to infection or during pathogen attack. Plants can also acquire enhanced resistance to pathogens, In other words the plant acquires a Systemic Resistance. Typically the host defense mechanisms are divided into two groups:

Constitutive Defenses

Defenses common to all healthy plants. These defenses provide general protection against invasion, colonization, infection, and infectious disease caused by pathogens. The constitutive defenses have also been referred to as "natural" or "innate" resistance, since they are inherent to a specific host, but these terms are better reserved for certain types of constitutive defense.

Inducible Defenses

Defense mechanisms that must be induced or turned on by host exposure to a pathogen (as during an infection). Unlike the constitutive defenses, they are not immediately ready to come into play until after the host is appropriately exposed to the parasite. The inducible defenses involve the *immune responses* to a pathogen causing an infection. The inducible defenses are generally quite specifically directed against an invading pathogen but constitutive defenses are not so specific, and are directed toward general strategic defense.

In this chapter we will be dealing with the infection induced resistance factors in plants. Induced plant resistance involves various biochemical interactions occurring between host and the pathogen. The host resistance is due to a combination of physical and chemical barriers that are induced only after infection. A wide range of host responses includes antimicrobial compounds (phytoalexins), proteins, callose-deposition, lignification, proteinase inhibitors and hydrolytic enzymes such as glucanases and chitinases.

Inducible plant responses to herbivory have gone from being viewed as anomalous cases 30 years ago to perhaps a ubiquitous feature today. Karban and Myers (1989) suggested breaking plant responses to herbivory or infection by pathogens into three classes:

Inducible Responses

Changes in the physical or chemical nature of a plant in response to attack.

Inducible Resistance

Changes in the physical or chemical nature of a plant in response to attack that can be demonstrated to reduce the intensity of the current attack or prevent future attack.

Inducible Defense

Changes in the physical or chemical nature of a plant in response to attack that can be demonstrated to reduce the intensity of the current attack or prevent future attack, and have been demonstrated to enhance plant fitness.

Inducible plant responses are sometimes classified by their speed and extent within the plant. Based on speed of response they are classified as:

Delayed Inducible Resistance (DIR)

Influences herbivores in the next generation or year, often associated with passive plant responses to nutrient loss during defoliation

Rapidly Inducible Resistance (RIR)

Influences current herbivores or pathogens during their lifetime.

Based on the extent of response they are classified as:

Localized

Restricted to areas around herbivore or pathogen attack, may be due to a hypersensitive response of localized cell death to prevent pathogen spread.

Systemic Acquired Resistance (SAR)

The entire plant responds to herbivore or pathogen attack.

Many types of plant chemicals may be involved in inducible responses. Examples include certain types of tannins, alkaloids, glucosinolates, proteinase inhibitors, terpenes, and volatile compounds that may be attractive to predators and parasitoids. Compounds that act against pathogens like bacteria and fungi are termed phytoalexins.

There may be many different individual messengers involved in the systemic response to herbivores or pathogens, but the best-studied are in four major categories:

Proteins

The best studied of these is systemin, a protein discovered to be critical in the systemic production of proteinase inhibitors in tomato and other plants.

Salicylic Acid Derivatives

These are related to the compound in aspirin, which was originally derived from willows. Most, if not all plants produce related compounds at low levels, and possess a biosynthetic pathway (appropriately called the salicylic acid (SA) pathway) that interconvert a series of related compounds. Several of these have hormonal activity that assists in the systemic production of phytoalexins. Perhaps the best studied of these is methyl salicylate.

Jasmonic Acid Derivatives

These are compounds related to fatty acids that are inter-converted by a biosynthetic pathway called the Jasmonic acid (JA) pathway. Some of these molecules are related to cell membrane lipids that may be released as a result of damage or infection. Perhaps the best studied of these is methyl Jasmonate.

Nitric Oxide

Nitric oxide is a gaseous free radical, which has emerged as a major signaling molecule of ancient origin and ubiquitous importance. Being a reactive free radical it has a relatively short half-life, in the order of a few seconds. Thus, the range of its effects is limited to the cell in which it is generated, or to cells in the near neighbourhood.

Carbohydrates and Related Compounds

These are derived from plant or pathogen cell walls and probably have a more localized effect. In fact, they may serve as elicitors, or compounds that stimulate SAR. Elicitors may also be present in herbivore saliva. The role of carbohydrate related compounds in SAR is probably mediated via the SA or JA pathways and their signaling molecules.

Current evidence suggests that the SA and JA pathways may be interrelated by a phenomenon called *biochemical cross talk* (Figure 30.1). This introduces additional complexity into the SAR response by potentially allowing for simultaneous activation by a single elicitor, or interference between the JA and SA pathways. To distinguish systemic SA-dependent defences resulting from pathogen pretreatments (or pretreatments with SA or SA like compounds) from other systemic responses that operate without SA, Kloepper and recently Van Loon proposed to term the former reactions systemic acquired resistance (SAR) and the latter reactions induced systemic resistance (ISR). This second term encompasses all systemic defence reactions (Kloepper, *et al.*, 1992; Van Loon, 1997).

Induced plant "responses" to attack consists of any active or passive change in the plant following herbivory or infection. Induced plant "resistance" refers to reduced preference, performance, or pathogenicity of the attacker on induced plants compared to controls. One of the first products most plants produce when their defense system is stimulated is SA. These molecules then move around the plants system stimulating the production of other natural defense products. Such induced responses that increase plant resistance are thought to be adaptive in that they can increase the fitness of plants in environments with herbivory, but may reduce fitness in the absence of herbivory. One benefit of induced responses over constitutive strategies is that the plant-defense strategy is not on all of the

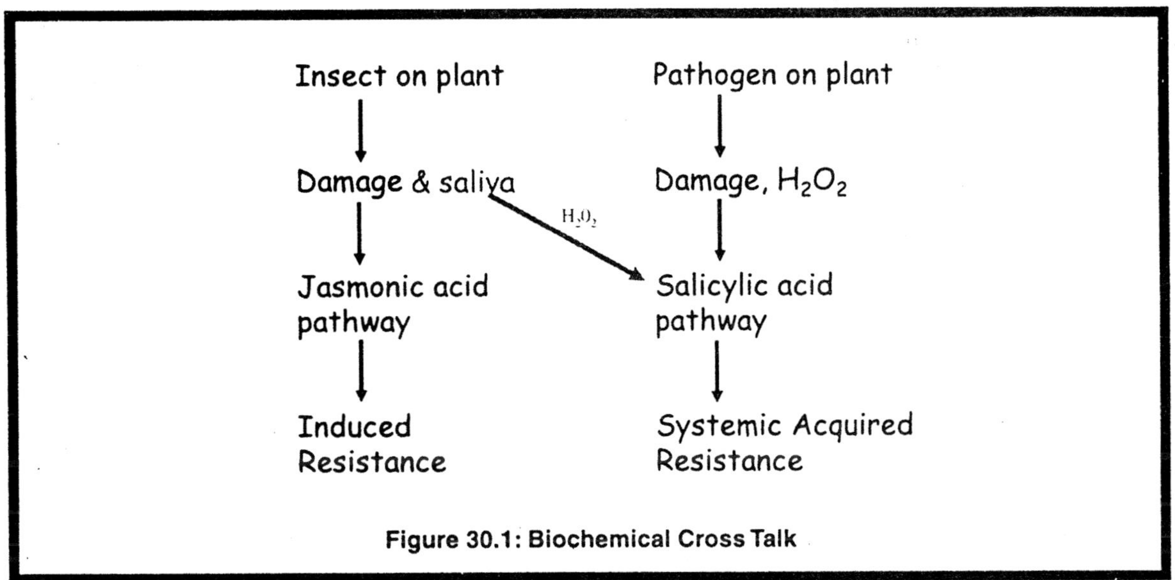

Figure 30.1: Biochemical Cross Talk

time. Pathogen infection and wound can also result in the expression of genes associated with disease resistance. The expression of *Xa1*, a gene conferring resistance to bacterial blight, caused by *Xanthomonas oryzae* pv. *oryzae*, is induced by pathogen infection and wound (Yoshimura *et al.*, 1998).

Plant intrinsic responses that can be engineered to attain a wider, more durable resistance include the Hypersensitive Response (HR) and SAR. HR is a complex, early defense response that causes necrosis and cell death to restrict the growth of a pathogen. Resistance (R) genes detect the pathogen and change the membrane potential and ion permeability of the plasma membrane. In phase one of the response, the R genes trigger an increase in extra cellular pH and K^+ (Orlandi *et al.*, 1992), while eliciting an influx of calcium and hydrogen ions into the cell. The outward K^+ and inward Ca^{2+} and H^+ ion flux are dependent and trigger the HR, resulting in cell death and formation of local lesions, which contain antimicrobial compounds. In phase two cells undergoing the HR produce reactive oxygen species (ROS; oxidative burst), including super oxide anions, hydrogen peroxide, and hydroxyl radicals (Baker *et al.*, 1993). Lipid per oxidation and lipid damage may be partially responsible for some of these cell changes and probably affect membrane function. Phenolics and phytoalexins, such as glyceollin (in soybean), and other compounds are synthesized in cells surrounding the lesion. Callose, lignin, HGRP is deposited and pathogen related (PR) proteins are induced and include beta 1, 3-glucanase and chitinases.

The HR precedes the secondary resistance response, the *SAR* and is characterized by necrotic lesions around the infection site; biochemical changes include generation of active oxygen species, cell death, overproduction of lignin-related materials, and the induction of certain (PR) proteins.

Several enzymes may be involved in generation of Reactive Oxygen Species (ROS). Copper amine oxidase is found in the apoplast and is loosely associated with the cell wall. It catalyzes the oxidative deamination of polyamines, especially putrescine, releasing hydrogen peroxide and ammonia (Koyanagi *et al.* 2000). In pea it is a copper-containing, extra cellular, homodimer glycoprotein. Other enzymes thought to play a role in ROS production includes xanthine oxidase; NADPH oxidase, oxalate oxidase, peroxidases, amine oxidases (copper containing and flavin containing).

Systemic Acquired Resistance (SAR)

Systemic acquired resistance implies the production by the plant of one or several translocated signals that are involved in the activation of resistance mechanisms in uninfected parts. Thus, a first infection predisposes the plant to resist further attacks. This latter aspect makes it comparable to immunization in animals and humans. As a result the entire plant is more resistant to a secondary infection. SAR is long lasting and often confers broad-based resistance to different pathogens. Inducible defenses play a major role in conferring disease resistance against plant pathogens, and their effects on phytophagous insects can include increased toxicity, delay of larval development, or increased attack by insect parasitoids. Both biological and chemical inducers can activate the resistance response. Biological inducers include avirulent pathogens that usually cause necrotic lesions, whereas chemical inducers include salicylic acid, BTH (benzo-(1, 2, 3)-thiadiazole-7-carbothioic acid *S*-methyl ester) and 2,6 dichloroisonicotinic acid. Salicylic acid is a signaling compound involved in transmission of the defense response throughout the plant to produce SAR. Development of SAR has been closely correlated with the expression of a set of nine gene families called SAR genes. Five of these gene families encode the PR proteins, a set of extracellular polypeptides associated with pathogen infection. Onset of SAR correlates with the coordinate induction of nine classes of mRNAs. Representative proteins of SAR include antifungal chitinases, beta-1, 3-glucanases, PR-1 and PR-5.

Plant growth promoting rhizobacteria (PGPR) can also induce systemic acquired resistance that operates independently of SA. Plant growth promoting rhizobacteria (PGPRs), belonging mainly to the group of fluorescent *Pseudomonas* spp, have been known to control plant diseases by suppressing pathogen development through antibiotic effects or by competing for iron through siderophores. Investigations into the mechanisms of resistance induction by PGPRs revealed that these bacteria also confer systemic protection against pathogenic organisms by inducing defensive capabilities in the plant. Induced systemic resistance (ISR) due to *P. fluorescens* has been demonstrated in several species including carnation, radish, arabidopsis, cucumber, and tobacco (Mauch-Mani and Metraux, 1998). Pieterse, *et al.*, 1998; postulated that rhizobacteria-mediated ISR follows a novel signaling pathway in which components from the jasmonate and ethylene response are engaged successively to trigger a defense reaction and is independent of salicylic acid accumulation and PR gene expression.

Induced resistance may be a particularly useful biological-control strategy because it reduces herbivorous pest populations directly and indirectly by causing a shift toward predation in naturally occurring omnivores. In the presence of prey (mite eggs), adult and larval *Frankliniella occidentalis* consumed nearly half the amount of plant tissue as did thrips without prey. In the absence of prey, systemically induced plant responses following herbivory reduced the amount of feeding by thrips compared to that on uninduced control plants. In the presence of prey, induced responses caused a shift in the feeding preferences of thrips: thrips consumed half the amount of plant material and twice the number of prey on induced plants as did thrips on uninduced control plants (Agrawal *et al.*, 1999).

Induced responses in some cases can be a double-edged sword, increasing herbivory by some herbivores under certain conditions, while reducing herbivory by other herbivores (Agrawal and Sherriffs, 2001). In wild radish responses induced by damage of *Pieris rapae* attracted oviposition by *P. rapae* and increased damage by flea beetles, *Phylotreta* spp; both of which are specialist herbivores. Specialist herbivores frequently detoxify or sequester plant defense compounds and are attracted to secondary metabolites from their hosts (glucosinolates in this case). But these are deterrents for non-adapted insects.

Oxidative Burst

Elicitor molecules from bacteria and fungi trigger a plant defense response called the *oxidative burst*, which involves the production of reactive oxygen species (ROS), super oxide and hydrogen peroxide. It is possible that a plant resistance (R) gene may detect the presence of the pathogen and trigger the production of ROS through a signaling process that may include an increase in cytosolic Ca^{2+}, H^{+}, and aromatic monoamines such as catecholamines, phenylethylamine, tyramine, dopamine and other structurally related compounds. Two models involving plasma membrane NADH oxidase, cell wall peroxidase, and copper amine oxidase have been proposed to account for ROS production. It is likely that peroxisomes are also involved, because they appear to produce large quantities of ROS.

One of the most effective defenses in plants is mediated by *R* genes that are able to detect specific pathogen races through recognition of pathogen-encoded Avirulence (Avr) proteins. *R*-gene-mediated resistance (also termed gene-for-gene resistance is commonly, although not invariably, associated with rapid necrosis of plant cells at the site of invasion (HR) and this results in the efficient containment of pathogen. A large number of plant *R* genes specifying resistance to bacterial, fungal, viral or nematode pathogens and aphids have now been cloned (Feys and Parker, 2000).

Identification of pathogen-derived molecules (elicitors) helps in better understanding of the basic cellular and molecular mechanisms that control the outcome of a plant-pathogen interaction, which determine early recognition of the microorganism by a given plant. The elicitors analyzed to date vary widely in their chemical nature but are all characterized by their ability to trigger defense responses when applied to a plant, thus mimicking pathogen attack. Elicitins, a family of low molecular weight proteins, are secreted by all analyzed *Phytophthora* species when cultured in vitro. Upon application to tobacco, these proteins elicit necrosis formation and histological disturbances in the plant, which are similar to those observed during HR in non-host interactions between *Phytophthora* species and tobacco (Keller *et al.*, 1996).

Defense-related Proteins

By changing their physiological conditions, higher plants protect themselves from various stresses such as pathogen attacks and a series of proteins actively synthesized with this reaction is called *defense-related proteins*. In particular, protective plant proteins specifically induced in pathological or related situations have been intensively studied from the agricultural interest and called *pathogenesis related proteins* (PR's). PR-1 and PR-5 have activity against oomycetes. In a sense, PR proteins constitute a point where the various response networks intersect by reacting with different inducers such as salicylic acid, jasmonic acid, systemin, and ethylene. PR proteins have been classified into 12 major groups or families. Some of them show antifungal activity. The functions of most PR proteins remain a mystery but some of them are known to be 1, 3-glucanases (PR-2), chitinases (PR-3) or fungal membrane permeabilizers (PR-5). Enzymes indispensable for phytoalexin biosynthesis, isoflavone reductases, for example, are also a kind of defense-related proteins. Transgenic tobacco plants constitutively expressing a *PR-1a* gene are resistant to the fungal pathogens *Phytophthora parasitica* and *Peronospora tabacina*. Because of these associations with pathogen resistance, PR gene expression is used as a molecular marker for SAR.

On the other hand, many of the reserve proteins accumulated in seeds and fruits take the constitutive defense function against microbial pathogens and invertebrate pests in addition to their storage function. These inducible or constitutive defense mechanisms of higher plants are relatively conserved in the course of evolution. Accordingly, most plants produce or accumulate similar proteins under certain situations irrespective of their morphological differences.

In theory, the constitutive expression of PR proteins, either singly or combined, might confer decreased susceptibility to a specific group of pathogens. The sequence similarities, serologic or immunologic relationships, and enzymatic properties are the basis of their classification. Remarkably, the defensive proteins are drawing the much attention of plant breeders. Because defense-related proteins usually provide a plant with resistance to stresses, varieties that are apt to intensively induce such proteins are agriculturally valuable. Less toxic substances that cause crops to express defensive proteins are being investigated as a new type of agrochemical. Moreover, some defense-related proteins are going to be constantly produced in genetically modified plants. However, several of them have been proved to be latex allergens and cross-reactive allergens in fruits, vegetables, and pollen.Even though defense-related proteins can be useful agriculturally, their allergenicity should be evaluated carefully.

Pathways

Jasmonic Acid

Jasmonic acid (JA) is a growth inhibitor, senescence-promoting substance. Ammoniates, such as jasmonic acid, are cyclopentanone derivatives synthesized from linolenic acid via the octadecanoic pathway. JA is structurally related to prostaglandins, which are autacoidal hormones found in mammals. JA is synthesized from alpha-linolenic acid, which is a C18 poly-unsaturated fatty acid. It is synthesized from plant membranes by enzymes similar to lipase. Key JA biosynthetic enzymes have chloroplast transit peptides that directing their import into the chloroplast. Key enzymes include lipoxygenase, allene oxide synthase, and allene oxide cyclase (Figure 30.2).

JA can be conjugated to amino acids, *i.e.*, leucine, valine, isoleucine and the sugar, ß-glucoside using UDP-glucose. JA and methyl jasmonate are major JAs in plants. JA induces vegetative storage proteins, osmotin, thionin (antifungal) and defensin. It induces the phytoalexin-related enzymes chalcone synthase, PAL, and hydroxymethylglutaryl-COA reductase. It also induces protease inhibitors to control insects. JA and ethylene induce PR-3, PR-4, PDF1.2, chitinases (CHI-B), and hevein-like protein. 12-oxo-phytodienoic acid is also a physiological signal for defense.

The plant produces Jasmonic acid after caterpillar damage and results in increased production of compounds involved in resistance. Thaler, 1999; studied the effect of exogenous application of JA in tomato. Foliar jasmonic acid application increased the levels of polyphenol oxidase, an oxidative enzyme implicated in resistance against several insect herbivores. Induced plants received 60 per cent less leaf damage by *Helicoverpa zea* than did control plants. Induced plants produced fewer flowers than control plants, but this did not translate into differences in yield between treatments. In addition, there was no difference in yield between induced and control plants under natural and experimentally reduced herbivore levels.

Salicylic Acid

Salicylic acid induces flowering, inhibits K^+ and P^+ uptake, and inhibits ethylene synthesis. It is probably antagonistic to jasmonic acid. SA triggers the SAR and may move as a long-distance component of a signal transduction pathway. SA induces the pathogenesis related proteins PR-1, PR-2, PR-5, beta-1, 3-glucanase, and glutathione S-transferase. SA participates in SAR and in the localized hypersensitive response. It induces alternative pathway respiration. Two alternative pathways for SA synthesis have been proposed. One pathway would convert chorismate via isochorismate synthase and pyruvate lyase to SA. In the other pathway phenylalanine is converted by PAL and BA-2-hydroxylase in to SA (Figure 30.3). Other reactions using SA and derivatives use UDP-glucose:SA

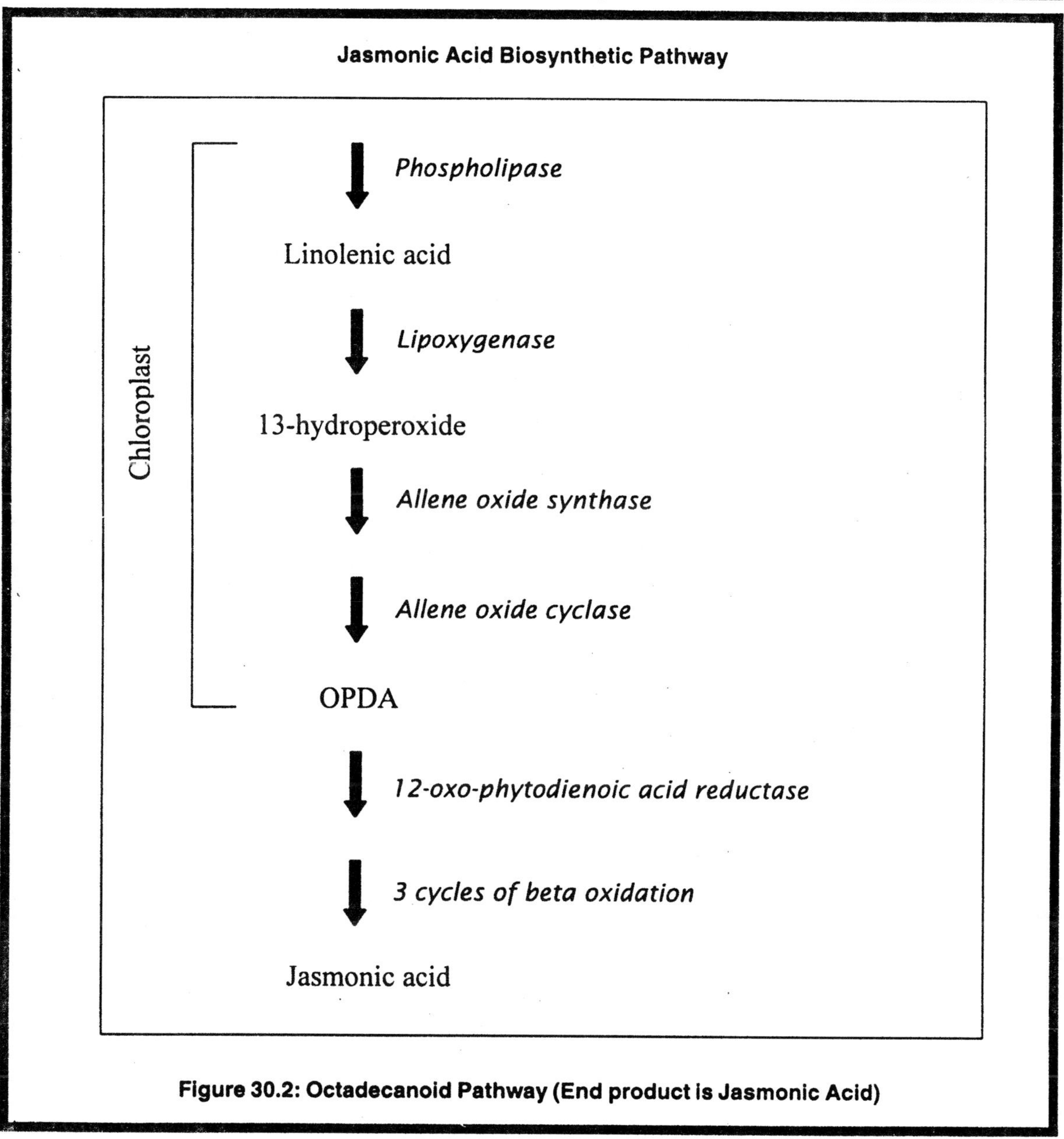

Figure 30.2: Octadecanoid Pathway (End product is Jasmonic Acid)

glucosyltransferase, and SA-ß-glucosidase for interconversion of SA derivatives. SA elevates H_2O_2 levels and binds and inhibits catalase, which degrades H_2O_2, which, in turn, induces PR-1. SA also inhibits ascorbate peroxidase, which also scavenges H_2O_2, but does not inhibit peroxidases involved in lignification and cell wall cross-linking. SA inhibits ACC synthase, reducing ethylene production and aconitase.

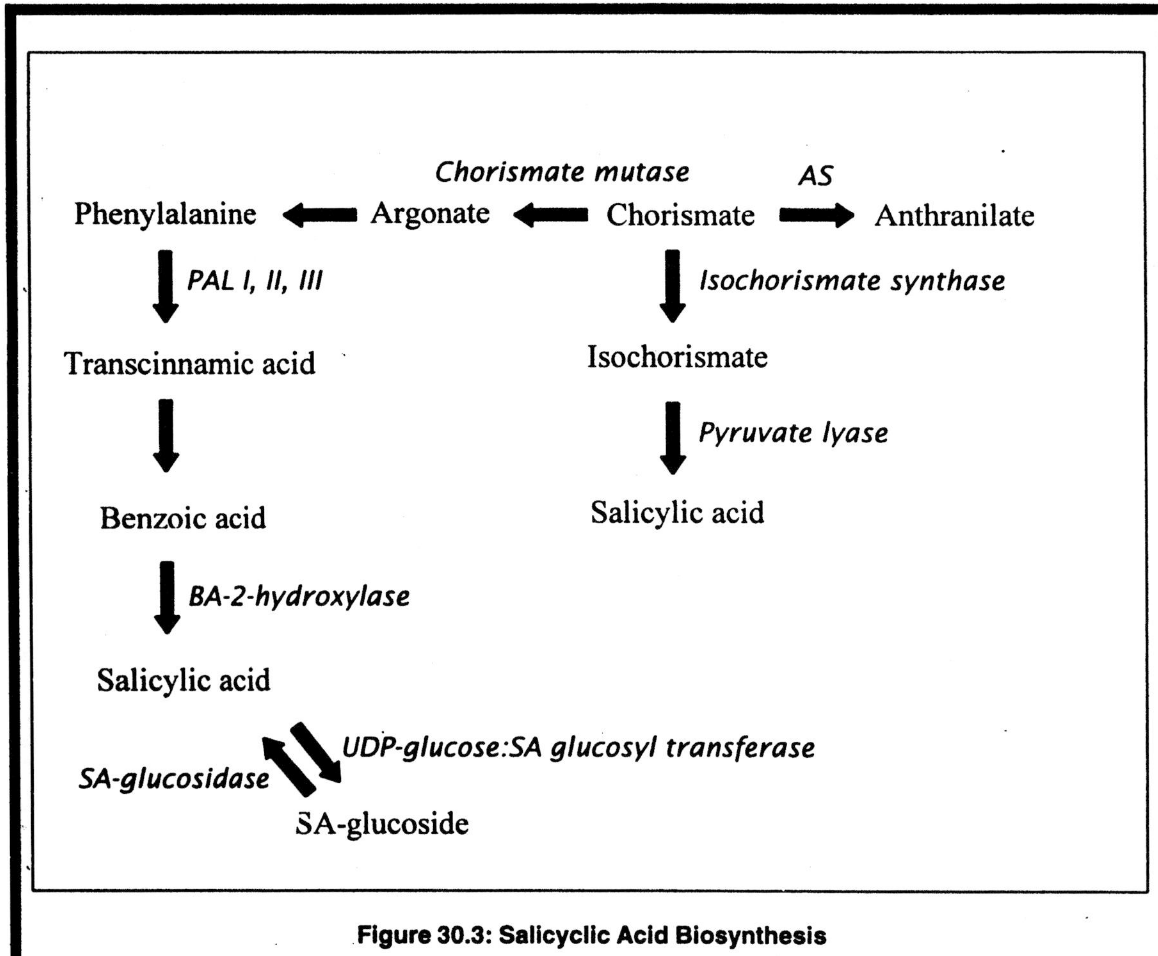

Figure 30.3: Salicyclic Acid Biosynthesis

Ethylene

Ethylene is a plant hormone synthesized from the amino acid, methionine, a member of the aspartate family of amino acids. Ethylene is involved in fruit ripening, abscission, senescence, epinasty, dormancy, leaf expansion, flower induction, hypertrophy, and other physiological responses and may play a role in defense against pests and pathogens (Figure 30.4).

Nitric Oxide

Nitric oxide has been previously shown to activate mammalian defense responses. Only recently has NO's involvement in the plant defense response to pathogens been addressed. NO is sparingly soluble in water, with the solubility increasing in the presence of ferrous salts. Therefore, it is able to move by diffusion in aqueous parts of the cell, such as cytoplasm, but also move freely through the lipid phase of the membranes. Once produced, it can move from one cell to another or within a cell. There is evidence of NO production in plants from nitric oxide synthase-like enzymes (NOS), nitrate

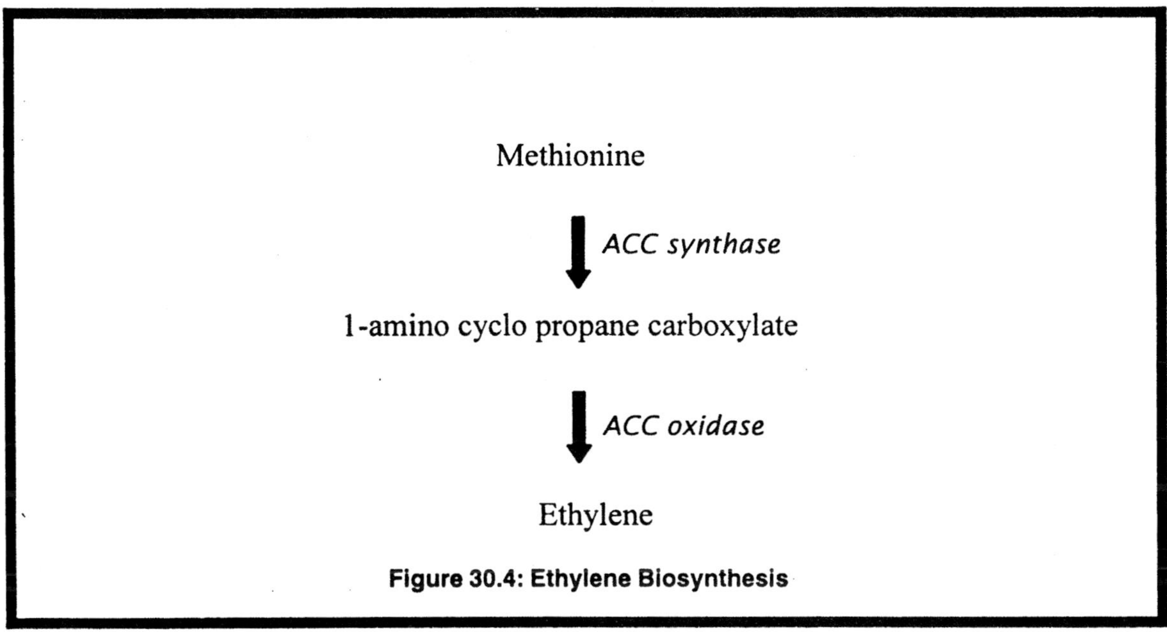

Figure 30.4: Ethylene Biosynthesis

reductase, or nonenzymatic sources (Figure 30.5). NO is involved in the induction of HR. Associated with HR is an oxidative burst, in which there is greatly increased ROS generation, programmed cell death (PCD), and the activation of signaling pathways driving the expression of various defense-related genes. Treatment of soybean cultures with avirulent (HR-inducing) but not virulent (HR-noninducing) *Pseudomonas syringae* pv. *glycinea*, induced rapid NO synthesis. Moreover, NO donors induced cell death as well as expression of PAL and chalcone synthase (*CHS*) genes. Pathogen induced NO produced via NOS, interacts with H_2O_2 to mediate HR. NO also induced the synthesis of SA and expression of defense-related gene *PR-1*.

NO-induced cell death also possesses the characteristics of PCD, such as chromatin condensation, requirement for gene expression, and activation of a caspase like cascade. NO by itself does not induce PCD, but NO: superoxide ratio determines PCD. If superoxide levels are greater than those of NO, then NO reacts with superoxide to form peroxynitrite, which does not result in PCD. However, if more NO than superoxide occurs, then NO reacts with H_2O_2 (arising from dismutation of superoxide) to induce cell death.

Conclusion

Plants defend themselves against various pathogens both actively (induced resistance) and passively (constitutive resistance).The level of resistance obtained in a host-pathogen combination depends on the sum total of combinations of the various resistance mechanisms in both constitutive and inducible responses.

Plant induced resistance is an approach to control plant diseases through induction and enhancement of plants own defense mechanisms which would not involve the application of toxic compounds. Plant activates a highly coordinated resistance response upon recognition of a potential pathogen or its products (elicitors). An induced resistance to an infection includes a cascade of induced

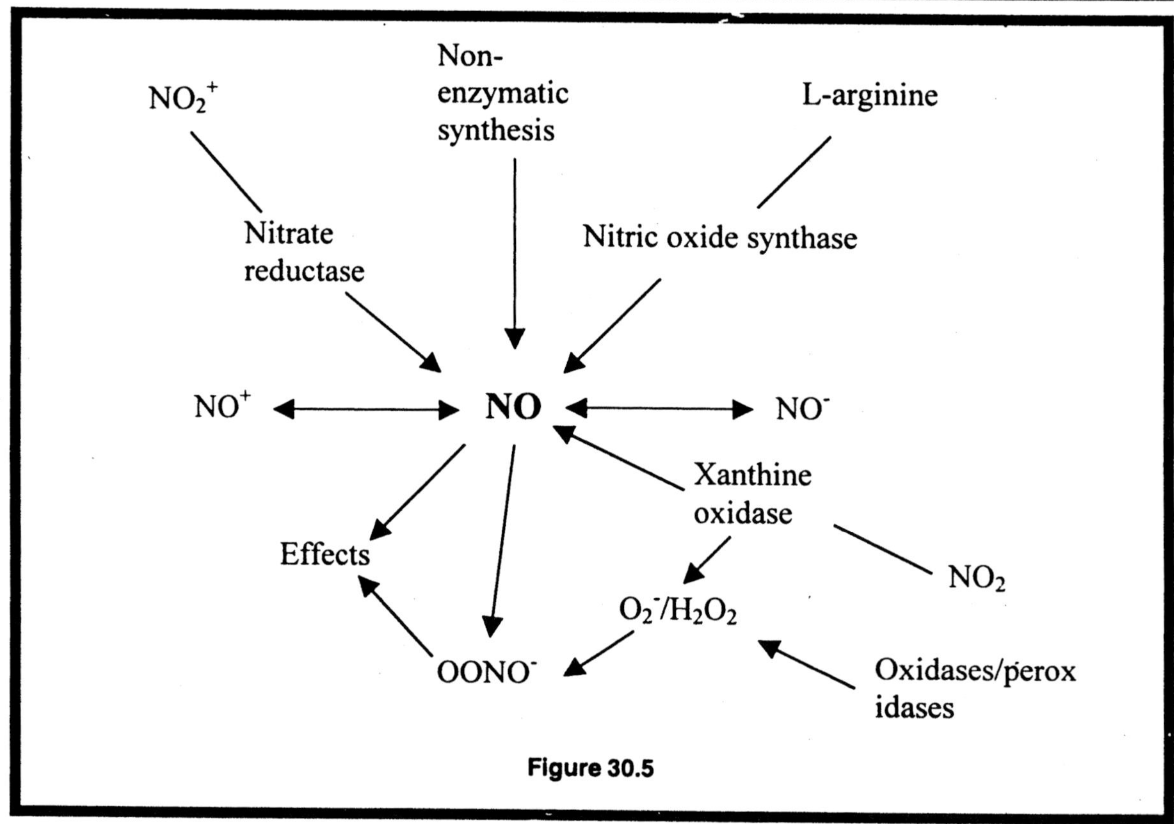

Figure 30.5

host response. These include phytoalexins, proteins (PR's) and physical barrier to penetration. The signal transduction pathway contributes significantly towards host resistance. Salicylic acid is reported as an endogenous signal in inducing defense response in host system. The abiotic factors such as salicylic acid and biotic factors such as non-pathogenic organisms could used to enhance the host-plant resistance.

Recombinant DNA technology and genetic engineering has made it possible to recombine genes irrespective of their source of origin. Together with Recombinant technology and the current knowledge on signal transduction pathway will allow us with the opportunity to exploit the potential of resistant crops. Still further research is needed to understand the innate/intrinsic factors which are responsible for the plants enhanced resistance upon encountering some pathogens for which we need to study the various ecological and biochemical interactions occurring between host and the pathogen.

REFERENCES

Agrawal, A.A. and M.F. Sherriffs (2001). *Ann. Entomol. Soc. Am.*, 94(1): 71–75.

Agrawal, A.A., C. Kobayashi and J.S. Thaler (1999). *Ecology*, 80(2): 518–523.

Baker.C. J., N. Mock and E.W. Glazener Orlandi (1993). *Physiol. Molec. Plant Path.*, 43: 81–94.

Bent, A.F. and I.C. Yu (1999). *Advances in Agronomy*. 66: 251–298.

Broglie, K., I. Chet, M. Holliday, R. Cressman, P. Biddle, S. Knowlton, C. J. Mauvais and R. Broglie (1991). *Science*, 254: 1194–1197.

Feys, B.J. and J.E. Parker (2000). *Trends in Genetics*. 16(10): 449–445.

Honèe, G. (1999). *Eur. J. Plant Path.*, 105:319–326.

Jongedick, E., H. Tigelaar, J.S.C. Van Roekel, S. A. Bres-Vloermans, P.J.M. Van den Elzen, B.J.C. Cornelissen, and L.S. Melchers (1995). *Euphytica*, 85:173–180.

Keen, N.T. (1999). *Advances in Botanical Research*, 30: 292–328.

Keller, H., J. Blein, P. Bonnet and Pierre Ricci (1996). *Plant Physiol.*, 110: 365–376.

Kloepper, J., S. Tuzun and J. Kuc (1992). *Biocontrol Science and Technology*, 2: 349–351.

Koyangi, Matsumara K., S. Kuroda and K. Tanizawa (2000). *Biosci. Biotechnol. Biochem.*, 64: 717–722.

Liu, D., K.G. Raghothama, P.M. Hasegawa and R. A. Bressan (1994). *Proc. Natl. Acad. Sci. USA*, 91: 1888–1892.

Mauch-mani, B. and J. Metraux (1998). *Annals of Botany*, 82: 535–540.

Orlandi, E.W., S.W. Hutchison and C.J. Baker (1992). *Physiol. Molec. Plant Path.*, 40: 173–180.

Pieterse, C.M.J., S.C.M. Van Wees, E. Hoffland, J.A. Van Pelt, L.C. Van Loon (1996). *Plant Cell*, 8: 1225–1237.

Strittmatter. G., J. Janssens, C. Opsomer, and J. Botterman (1995). *Bio/Technology*, 13: 1085–1088.

Rakshit, S. (2000). In: *Crop Pest and Disease Management: Challenges for the Millennium*, (Eds.) D. Prasad and S.N. Puri, pp. 260–266.

Tabei, Y., S. Kitade, Y. Nishizawa, N. Kikuchi, T. Kayano, T. Hibi and K. Akutsu (1998). *Plant Cell Reports*, 17: 159–164.

Thaler, J.S. (1999). *Environ. Entomol.*, 28(1): 30–37.

Van Loon, L.C., W.S. Pierpoint, T. Boller, V. Conejero (1999). *Physiol. Molec. Plant Path.*, 58: 85–97.

Van Loon, L.C. (1997). *European J. Plant Path.*, 103: 753–765.

Yoshimura, S., U. Yamanouchi, Y. Katayose, S. Toki, Z. Wang, I. Kono, N. Kurata, M. Yano, N. Iwata and Takuji Sasaki (1998). *Proc. Natl. Acad. Sci., USA*, 17; 95(4): 1663–1668.

31

Precision Farming Technologies in Nematode Management

A.K. Ganguly, R.N. Sahoo** and Umarao**

**Division of Nematology, Indian Agricultural Research Institute, New Delhi – 110 012*

***Division of Agricultural Physics, Indian Agricultural Research Institute, New Delhi – 110 012*

The net deficits in the total food grain requirements (demand over supply) in our country are expected to be about 1.3 million tonnes in 2005, which may swell to about 24 million tonnes by the year 2020. The Green Revolution, which is now fading, had strong positive and some negative impacts. Increase in yield in farmer's field has been achieved through Green Revolution at considerable expense of its resource base and largely by means of i excessive and indiscriminate use of external inputs such as irrigation, seeds, fertilizers, pesticides, *etc.* But, intensively managed cropping systems in India seem to be unsustainable as productivity levels have reached a plateau or declined with a decreasing response to farm inputs and a widening gap between the potential and realized yields and deteriorated the environmental quality. Agriculture is presently not seen as a stimulating and economically rewarding activity in India. In the coming decades, the farm families of India will have to produce more food grains and other agricultural commodities under conditions of shrinking per capita land and water resources and expanding biotic and abiotic stresses. A better upkeep of key functions of ecosystems through adoption of environment-friendly technologies and holistic management of farm resources is, therefore, essential for agricultural sustainability. Precision farming is one of such modern eco-friendly technologies which offer opportunities to optimize yields and profits and reduce pressure on natural resources, thereby leading to total quality management under heterogeneous and complex

Indian farming systems. This technology is based upon blending the best of farmer's local field knowledge with developments in geographic information systems (GIS), global positioning systems (GPS), remote sensing and the information technology.

Farming systems are continuing to change in response to economic, technological and social trends. Farming practices are becoming questioned not only by farmers but the public at large. Concerns are profitability and environmental impact. Farmers look to adopt new technology. In Canada, the trend over the past two decades has been to reduce tillage systems (direct seeding) and increase use of fertilizers and herbicides. The cost savings of reduced tillage does not outweigh the added cost of crop chemicals so profit margins have remained narrow. Our labour costs are high and our natural resources finite so we must utilize technology to maintain competitiveness in a world market. Developments in sensor-controller technology, computers and positioning systems now bring new opportunities for farm management (Goddard *et al.* 1995).

Precision Farming and its Components

The term 'Precision farming means carefully tailoring soil and crop management to fit the different conditions found in each field. It is sometimes called 'Prescription farming', site-specific farming' or 'variable rate technology farming'. Precision farming is the term used to describe the goal of increased efficiency in the management of agriculture. It is a developing technology that modifies existing techniques and incorporates new ones to produce a new set of tools for the manager to use. Inevitably it integrates a significant amount of computing and electronics but higher levels of control require a more sophisticated system approach.

Therefore, Precision Farming is not simply the ability to apply treatments that are varied at local level, but must be considered as the ability to precisely monitor and assess the agricultural enterprise at a local and farm level and to have sufficient understanding of the processes involved to be able to apply the inputs in such a way as to be able to achieve a particular goal. This is not necessarily maximum yield but may be to maximize financial advantage while operating within environmental constraints.

Furthermore, the techniques of Precision Farming should be regarded as indivisible from the concept of sustainable land management. In sustainable development, the requirement is for non-negative changes in the stock of natural resources and in waste capacity of the receiving environment (Pearce *et al.*, 1988). This definition is helpful in Precision Farming as it contains the concept of stewardship of the land for future generations, preserving potential for multi-use, and the importance of off-site as well as on-site effects. Sustainable land management implies more than maintaining crop yields.

This technology is based upon blending the best of farmer's local field knowledge with developments in Geographic information systems (GIS), Global positioning systems (GPS), remote sensing (RS) and the information technology. The success of the precision farming system relies on the integration of these technologies into a single system that can be operated at farm level with sustainable effort.

Geographic Information System

GIS is a computerized data storage and retrieval system, which can be used to manage and analyze spatial data relating crop productivity and agronomic factors. It can integrate all types of information and interface with other decision support tools. GIS can display analyzed information in maps that allow (a) better understanding of interactions among yield, fertility, pests, weeds and other factors, and (b) decision-making based on such spatial relationships. Many types of GIS software with varying functionality and price are now available. A comprehensive farm GIS contains base maps such as topography, soil type, N, P, K and other nutrient levels, soil moisture, pH, etc. Data on crop rotations, tillage, nutrient and pesticide applications, yields, etc. can also be stored. GIS is useful to create fertility, weed and pest intensity maps, which can then be used for making maps that show recommended application rates of nutrients or pesticides. Figure 31.1 shows role of GIS in Precision Agriculture.

Remote Sensing

It is the process of retrieving information on an object without having physical contact with it. Role of Remote Sensing technologies becomes more crucial in order to address the spatial variabilities of soils and crops across the various scales of mapping. The remote sensing technology inputs also capture the vulnerability and dynamism of agricultural systems. The developments in space-borne imaging sensors, particularly their spatial, spectral and temporal resolutions are well characterized to capture these features. While high spatial resolution images enable mapping and monitoring the structural attributes of agro-ecosystems, high spectral resolution or hyper-spectral imaging addresses their functionalities. The high temporal resolution captures the dynamisms of agro-ecosystems.

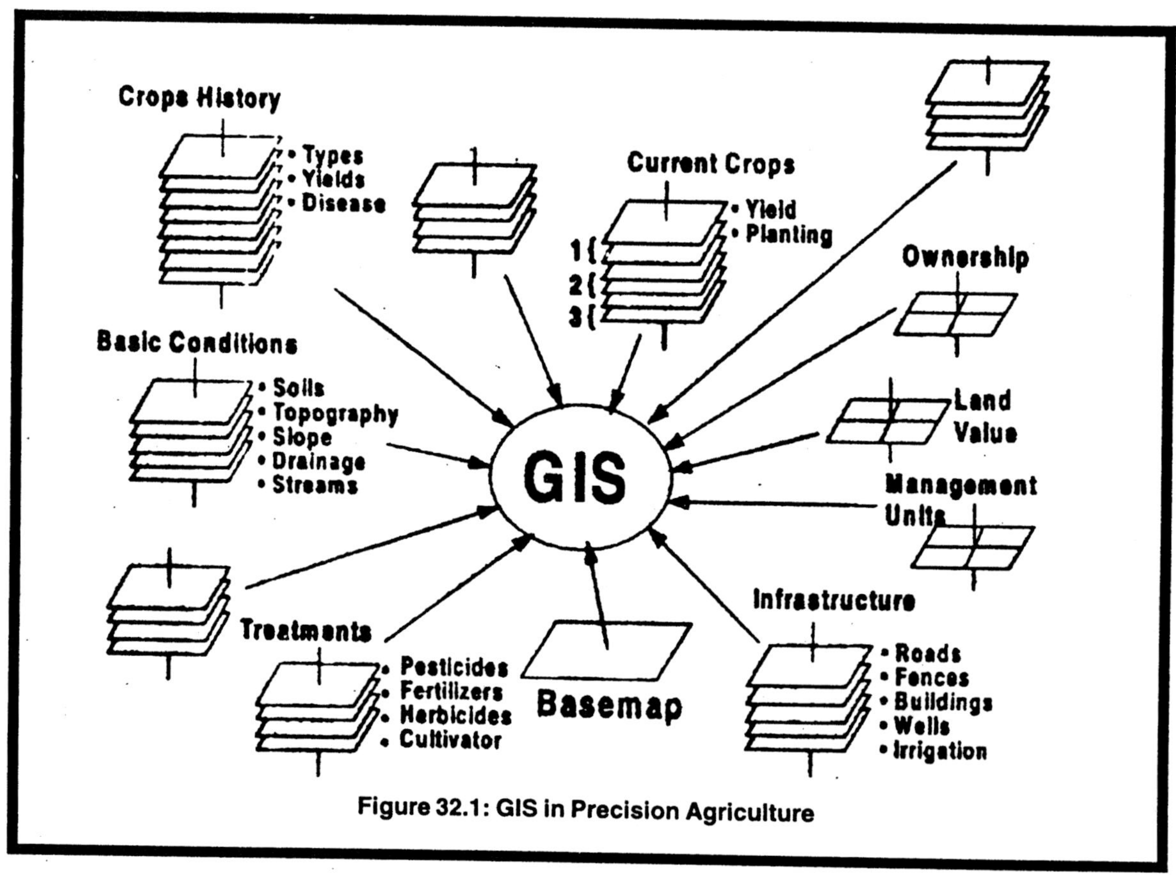

Figure 32.1: GIS in Precision Agriculture

Global Positioning System

The development of the publicly available global positioning system (GPS) new doors in opportunities for spatial data. This is a passive positioning system from constellation of 24 orbiting radio-navigation satellites. They provide continuous position data provided the receivers have a line of site access (or nearly so) to the satellites. Positioning can be two or three dimensional in real-world coordinates.

The GPS can be used in two modes; single receiver mode and differential mode using two receivers. In the single mode one receiver collects the timing information and processes it into position. This system is the cheapest and easiest but its accuracy suffers due to the introduced positional errors. In the differential mode, one receiver is mounted in a stationary position, usually at the farm office, while the other is on the tractor or combine. The stationary receiver appears to move due to the introduced randomised positional error. This error signal is then transmitted to the mobile receiver and deducted from the incoming signal, thus reducing the overall positional errors. These errors can be identified from a stationary receiver, but conversely, accuracy decreases as mobility increases.

There is an equivalent system called GLONASS that was developed by the former Soviet Union but is now under the control of the CIS Space Command and will consist of 24 satellites in the full constellation. European users will have a much better coverage and accuracy with GLONASS than GPS as Selective Availability cannot be applied to GLONASS (Lachner, 1994).

More recently farmers have gained access to site specific technology though GPS. GPS makes use of a series of military satellites that identify the location of farm equipment within a meter of an actual site in the field. The information like:

1. Locations of soil samples and comparing the laboratory results with a soil map
2. Requirement fertilizer and pesticides to fit soil properties (clay and organic matter content) and soil conditions (relief and drainage)
3. tillage adjustments according to various conditions across the field and
4. monitoring and recording yield data as one goes across the field befit the farmer to enhance the overall.

The availability of GPS approaches to farming will allow all field-based variables to be tied together. This tool has proven to be the unifying connection among field variables such as weeds, crop yield, soil moisture, and remote sensing data, *et al.* Through the enhanced understanding of the agronomic system the farmer can make management decisions within a field to both enhance economic return and to minimize environmental problems, namely, site-specific applications of fertilizer, herbicides and insecticides.

The potential of remote sensing products are currently being analyzed to determine how well yield can be explained by remote sensing data collected during the growing season, how well weeds can be mapped, and how other crop variables can be explained. If field scouting efforts can reduced by one-half or more with use of remote sensing products, farmers and crop consultants can be more effective and do a better job of managing those agronomic variables which can be altered. The relationships between spatial resolutions of IKONOS data (4m) and IRS ID LISS-III (23.5 m) will be examined through their prediction of crop yields. If 23m data are sufficient and they are significant, prediction of crop yields during the growing season can be accomplished for individuals or large areas. This information will help producers make better decisions concerning storage, harvesting schedules and marketing. If nematode and weed mapping can be accurately accomplished with remote sensing products, nematocide or herbicide application can be applied selectively within a field. This will save product and should provide a positive economic and environmental return.

Yield Mapping

Yield mapping and soil sampling tend to be the first stage in implementing PF. Yield maps are produced by processing data from an adapted combine that has a vehicle positioning system integrated with a yield recording system. Massey Ferguson were the first company to produce a commercial yield mapping combine. This combine has a Differential Global Positioning System (DGPS) fitted to it that can be identified by the GPS receiver on the roof of the cab and the differential aerial above the engine. The output from the combine is a data file that recorded every 1.2 seconds the position of the combine in Longitude and Latitude, with the yield at that point. This data set can then be processed by various geo-statistical techniques (usually involving Krigging) into a yield map.

Variable Rate Technology

The variable rate technology is the most advanced component of PF technologies, provides "on-the-fly" delivery of field inputs. A GPS receiver is mounted on a truck so that a field location can easily be recognized. An in-vehicle computer, which contains the input recommendation maps, controls the distribution valves to provide a suitable input mix by comparing to the positional information received from the GPS receiver. Current commercial VRT systems are either map-based or sensor-based (NRC, 1997). The map- based VRT systems require a GPS/DGPS geo-referenced location and a command unit that stores a plan of the desired application rates for each field location. The sensor-based VRT systems do not require a geo-referenced location but include a dynamic control unit, which specifies application through real time analysis of soil and/or crop sensor measurements for each field location. New VRT systems like the manure applicator being developed at Purdue University may soon enable precise application of manure in cropping systems. There are two methods of VRT such as:

The first method, *Map-based*, includes the following steps: grid sampling a field, performing laboratory analyses of the soil samples, generating a site-specific map of the properties and finally using this map to control a variable-rate applicator. During the sampling and application steps, a positioning system, usually DGPS (Differential Global Positioning System) is used to identify the current location in the field.

The second method, *Sensor-based*, utilizes real-time sensors and feedback control to measure the desired properties on-the-go, usually soil properties or crop characteristics, and immediately use this signal to control the variable-rate applicator.

The use of remote sensing, GIS and GPS for crop monitoring, condition assessment and yield modeling has already been well established. Crop simulation models (CSMs) provide potential production under the different scenario of constraints, including weather, soils, crops, cultural practices etc. The conjunctive use of VTR, remote sensing, GIS, GPS and CSMs provides technological framework for Precision Agriculture.

Precision Farming and Nematode Management

Phytonematodes and their Importance

Plant parasitic nematodes pose a threat to agricultural crops throughout the world like other pests and diseases particularly in tropical and sub-tropical regions. For centuries man's essential crop plants have been affected by these microscopic organisms hidden enemy in the soil that feed on the roots, buds, stems, crowns, leaves, and developing seed. The degree of damage to a particular crop and cultivars depends on type of nematode species, level of soil infestation and environment. These deleterious effects on plant growth result in low yields and poor quality. Crops such as banana, citrus, coconut, coffee, cotton, fruits, peanut, pineapple, potato, rice, soybean, tobacco, and vegetables are affected.

Nematodes vary greatly in shape and size. For example, a very small marine nematode, *Greeffiella minutum*, which is only 82 µm in length may be compared to another species, *Placentonema gigantissima*, which lives in the placenta of the sperm whale and may reach the length of 8 m (Poinar, 1983). In contrast, most species of plant- parasitic nematodes are 1 to 2 mm in length, but may vary in shape from filiform to fusiform, pear-shaped, lemon-shaped or kidney-shaped.

These microscopic organisms are the most abundant. multicellular organisms on earth, are found in all habitats and ecosystems of the biosphere. They occur in soil, decaying organic matter, all forms of plant life and most animals, including domesticated and wild species (Norton,1978). Billions of nematodes may live in a hectare of farm land, or mud at ocean bottoms, as well as in forest floors. Marine, fresh water, and soil nematodes make up the majority of these organisms. To date, some 15000 species of nematodes have been described, and the total number of species may exceed 500000 (Poinar, 1983). Although many nematodes parasitize plants or animals, others are beneficial as they contribute to nutrient cycling or serve as biological control agents of plant pests.

Economic damage caused by phytonematodes ranges from slight, perhaps less than 1 per cent to total loss. The average annual loss when all crops are considered is believed to be 10 per cent or more. The crop losses due to nematodes are related to their population density and crop growth conditions. International estimates have indicated crop losses amounting to 8.8 per cent in developed countries and 14.6 per cent in the developing countries, which were priced at US$ 100 billion in 1988. Recent projections have estimated that in India nematodes may be causing about 5 per cent loss in oilseeds, 8 per cent in cereals and pulses, 10 per cent in sugarcane and fruit crops, 12 per cent loss in vegetables, etc. The price of the crops lost has been estimated to be about Rs.242 billion annually (Seshadri and Gaur 1999).

The symptoms of nematodes on plants may be observed on shoots and/or roots. The above ground symptoms expressed as a result of nematode feeding on roots usually non-specific. The affected plant may exhibit leaf chlorosis, reduced vigour in ability to tolerate adverse condition, wilting under dry condition. This symptom could be due to lack of micronutrient in soil/soil pH/cation exchange capacity or affected by other soil microorganisms. Symptoms of nutrient deficiency, such as yellowing of foliage, result even after there is an adequate supply of nutrients in the root zone. It is known that the impact of nematodes on plant growth is worse when sol nutrients are less favourable to plant growth. Symptoms of nutrient deficiency, such as yellowing of foliage, result even after there is an adequate supply of nutrients in the root zone.

For example, in below ground root lesions, root pruning, root galling and cessation of root growth are common symptoms of nematode feeding. Roots damaged in this manner cannot efficiently use the moisture and nutrients available in the soil. Furthermore, roots so weakened and damaged root become susceptible to many types of bacteria, fungi, and other organisms, which invade the roots and accelerate root decay. This type of damage usually does not draw immediate attention, and the foliage of the crop may appear to be healthy. Yields are affected, however, and losses resulting from this kind of damage, because of its wide occurrence, constitute a sizeable portion of the overall damage incited by nematodes. The primary reason that the presence of plant parasitic nematode is often not realized before yield losses are incurred is because it may take several years for the nematode to reach population densities that result in visible above ground symptoms. At lower population densities, aggregated patches of stunted plants may be visible in the field. These patches are round to oval in shape and are not easily observed by scouts walking the fields, yet crop is being adversely affected. At higher nematode population densities, patches of various sizes within fields are chlorotic and severely stunted. The stunting of plants and the slower rate of plant growth both result in reductions in healthy (green) leaf area index (GLAI). Differences in GLAI in fields are extremely difficult to visually detect and labor-intensive to quantify at ground level (Nutter *et al.*, 1999). However, nematode infested patches should be clearly obvious using remote sensing and geographic information systems (GIS) technologies that can accurately and precisely measure and map variations in GLAI as well as yield, protein, and oil concentrations over time and space (Adcock *et al.*, 1990; Guan and Nutter, 2000; Nutter, 1989; Nutter *et al.*, 2000).

Efficient management of plant parasitic nematodes require carefully integrated combination of several practices such as crop rotation (cropping sequence), use of resistant varieties/trap crop/ antagonistic crop, judicious use of nematocide and biological control.

Need for Precision Management

The yield of the crop is governed by number yield limiting factors such as fertility, weather, crop type, crop husbandry, weeds, pests and disease management as shown in the figure below. Among these, there are some on which very little human control is there and others can be controlled for increasing the yield. One of those manageable factors is pests and diseases. The suit of Precision, farming technologies would help in precise monitoring of pests, their effect on crop growth and possible remedial measures. This would result in more effective management, reduced pesticides, pesticide usage, and less contamination of groundwater with agricultural chemicals such as nematocide, nitrogen and fertilizers. It is a developing technology that modifies the existing techniques and incorporate new areas to produce a new set of tools for the manager to use. The advantages associated with this technologies are:

1. It helps in increasing the ability to identify, diagnose, to understand crop and field problems.
2. Improve the equipment efficiency operation through better scheduling.
3. Varietal choice against pest and pathogen.
4. Provide environmental stewardship by accurate and precise application of chemicals and fertilizers to reduce the potential for leaching and run off precise application of nematocide.
5. It helps to introduce refugia system of growing plants.

Factors to be considered in Precision Nematode Management

One of the basic component of precision agriculture is that the presence, distribution, and intensity of any yield-reducing factor (or factors) within fields must be known. Growers often ignore that the nematode is the limiting factor. The lack of awareness of infestation by growers results in the building up the nematode population densities that eventually came catastrophic crop losers. One obvious reason could be due to not routinely checked for the presence of nematodes because of costs of sampling fields. Development of a means for remote details of nematode infestation via aerial images would reduce the costs associated without scouting of field for the pest. To obtain such knowledge, one must collect soil samples within fields using a systematic grid of quadrants and then have these samples analyzed to determine the presence and population density of the nematode. Unfortunately, collection and processing of soil samples are both labor intensive and cost prohibitive.

Nematodes are basically distributed uniformly, randomly, or in clusters. The most common pattern of spatial distribution of nematodes is the cluster, resulting in the well known "pocket effect". Clusters are often not well defined and may be any arrangement in which nematodes may seem to be grouped together with intervening spaces separating them from other groups. Clustering of nematodes has caused unsolved problems in sampling. A species or several species may cluster in one place. Spatial clusters of nematodes result from spatial patterns of the present or previous hosts, morphology and distribution of the roots, almost imperceptible environmental changes, land management practices, fortuitous events of climate, inherent reproductive capacities, migratory ability, and other biological aspects of the nematode.

Severe crop damage caused by a nematode can sometimes be fairly uniform in a field, but more frequently the nematode causes ragged and uneven stands or obliterates a crop in uniform to irregular spots. This pattern is largely a reflection of nematode population size. Differences in vegetation are major factors in many situations, and usually with vegetational changes in non-cultivated areas there are concomitant, edaphic changes that may be just as important. Major differences in nematode distribution may occur in areas separated by only a few inches to a few feet in cultivated fields. Species of *Meloidogyne* and *Xiphinema americanum* had different spatial arrangements in vineyards, the latter being distributed mainly within the row and in the upper soil layers, while the former was more widely distributed throughout the profile and occurred frequently between as well as within the rows (Ferris and McKenry, 1974). Thus, there is something more than root distribution that can govern nematode occurrence. Changes in soil properties that seem small to people may loom large for the nematode.

Obtaining information on pest population dynamics and interactions among pests over a landscape can enhance models of pest introduction, dissemination, and population changes that occur during rotational cropping systems. This new knowledge regarding pest population dynamics should contribute to implementation of new cultural practices and alteration of current management strategies, such as site-specific management.

To accurately estimate pest clustering across a field landscape, the level of spatial structure and temporal stability needs to be accurately quantified.

Benefit of Technologies in Nematode Management

Highly accurate GPS information will allow researchers to return to exacting the same spot in a given field enabling long term monitoring of nematode densities in research plots. As fields are surveyed, growers will be able to create details of soil type drainage, nutrient status and nematode infestation levels. Farm equipment linked to the survey database and real time GPS approaches will allow precision placement of fertilizers and nematicides as the weeding traverse the field.

Detailed information on the spatial distribution of nematode in relation to various environmental parameters will be digitally compared/collated using GIS. Such compilation of information will eventually allow historically assessment of nematode disease prevalence, variability and dissemination.

Remote sensing offers a potential new method to diagnose correctly and quantify the actual causes of plant stress. The benefit of remote sensing will be helpful in demonstrating the uneven distribution nematode within a field and nematode dissemination. This will further illustrate the merits of various sampling strategies. Aerial and satellite images can be used to monitor vegetation status and have potential use in assessing geographic prevalence of nematode disease.

Remote sensing and GPS are tools to detect changes and locating ground positions, and GIS is key to quantify the variabilities in soil texture and landscape information. Indeed, these technologies are being widely applied in various agricultural and non-agricultural systems, especially in managing soil-driven yield-limiting problems (Cassel *et al.*, 2000; Pierce and Sadler, 1997). As the ecological and economic awareness of wise pesticide use and demands in conservation tillage increase, the potential of applying precision technologies to manage biotic crop yield-limiting factors on a site-specific basis also increases (Nelson *et al.*, 1999; Strickland *et al.*, 1998). Exploiting the potential and sustainability of site-specific management (SSM), however, depends on identifying the right spatio-temporal conditions of SSM that, in turn, are dependent on applying appropriate diagnostic sampling strategies (Pierce and Sadler, 1997), establishing cause- and-effect relationships of yield-limiting factor (s), and asking strategic questions before tactical questions.

Site-specific nematode management offers the possibility of using less pesticide overall or treating nematode-infested areas with nematicide based on spatial parameters. The environmental argument for this approach is supported by a powerful economic argument, as nematicides are by far the most expensive input that has the potential to be applied in nematode management. The high cost of nematicides relative to other potentially variable inputs for potato production, and the fact that some growers use two types of nematicide on a single crop, combine to make powerful economic and environmental arguments for a proper assessment of site-specific application of nematicides to be made (Evans *et al.*, 2002). The high cost involved mean that the potential savings are large, this, in turn, means that sufficiently intense sampling to make reliable maps is economically viable. Failure to detect latent potato cyst nematode infestations that are less than the detection threshold at the time of sampling is inevitable. Such infestations would cause no yield loss but, with potential multiplication rates in excess of 100-fold, could cause future management problems. The best current advice in terms of site-specific application of nematicides is to treat "hot spots" of infestation with the more expensive fumigants and treat the whole field with a granular nematicide to prevent excessive multiplication of the nematode.

The precision farming technologies provide numerous research opportunities to investigate the plant pathogen population dynamics and discover interactions among plant pathogens and other field parameters. Information and applications of these technologies also make excellent educational demonstrations for producers to visualize the effect of management and cultural practices on pest populations. This research identified the influence of management strategies, cultural practices, and field location on the introduction of progression of early, blight and root lesion nematode across a production field landscape (Morgan *et al.*, 2002). Based on this information, management strategies may be modified to minimize pesticide application areas and maximize economic return. Likewise site-specific strategies for pest management may be useful in changing grower practices from uniformly applied pesticide treatment to timely and targeted treatments with highly effective and environmentally compatible materials., Chemical control is most potent method of nematode management since it is quick in action, effective against most nematodes, can be applied as and when required and visible efficacy on crop improvement can well be recognized. Precise and judicious use of nematicides are being used by nursery bed and seedling and root dip treatment in to raise healthy and uninfected transplant crop and spot and row application in field.

Sampling Methodology to Establish Cause and Effect Relationship for Precise Management

It is necessary to apply the right sampling strategy that would map the yield limiting factor in case of site specific management. Because the current sampling strategies and cost effectiveness limitation, it is not uncommon to see a management decision for biotic and abiotic yield limiting factor based on the problem (s) from composite samples of large areas. By compositing samples spatial structure variability is a key component of site specific management is eliminated. Many sampling schemes (SS) are applied to detect yield-limiting problems like plant-parasitic nematodes that may have similar spatial structures. If one were to test three sampling schemes in these fields, the outcome would vary widely. Most current management strategies, samples are collected along zigzag line across the whole field and the sample is composited and sub samples analysed to identify the problem (McSorley, 1998). This is the most economical way to find out whether or not problems exist in the field. The limitation is that this sampling scheme does not identify the spatial structure of yield limiting factors. A yield-limiting factor in any given field could have small and large clusters that vary widely in density and spatial structure. This limiting scheme may give limited information of the location of the yield-limiting problem if GPS is applied.

A key to effective decision making in precision agriculture is the ability to obtain remotely sensed images that clearly depict the spatial and temproal pattern of plant stress at a resolution that allows the farm producer and (or) crop consutlant to cost effectively deploy a crop management tactic (Nutter *et al.*, 2002). Aerial images can replace ground-based radiometers to efficiently assess the health and productivity of soybean fields.

The ultimate goal is to develop methodologies to provide timely assessments concerning the health and productivity for regional cropping areas as well as for individual fields. The combination of remotely sensed data ftom different remote sensing platforms (ground, aircraft, satellite), coupled with the use of GIS and Global Positioning System (GPS) technologies and sophisticated data analysis tools will benefit precision agriculture by providing growers timely techniques to diagnose and quantify the causes of plant stress, reliable, and cost effective disease/crop management information. There are several factors need to be considered for efficient sampling:

1. Sampling procedure
2. Collection of samples (fallow, summer/winter at the time of growing crop)
3. Spatial structure and temporal changes of yield limiting factor. Temporal movements are important while considering insect and airborne pathogen, but less significant in case of plant parasitic nematode since mobility in their own is limited. Nematode population dynamics peak is important when management strategy need to be applied.
4. Value of the crop
5. Economics of sampling.

Examples of Precision Nematode Management

The greatest constraint to potato production in the United Kingdom (UK) is damage by the potato cyst nematodes (PCN) *Globodera pallida* and *G. rostochiensis*. Management of PCN depends heavily on nematicides, which are costly. The inverse relationship between population density before planting and multiplication rate of PCN makes it difficult to devise reliable spatial nematicide application procedures, specially when the pre-planting population density is just less than the decision threshold. Also, the spatial dependence found suggests that the coarse sampling grids used commercially are likely to produce misleading distribution maps. This concern to protect the environment, coupled with the recent narrow gross margins in potato production in the UK, have stimulated the investigation of site-specific application of nematicides for controlling potato cyst nematodes (PCN). Global Positioning System (GPS) technologies have made it possible for modulated treatment with nematicide to be accurately targeted (Haydock and Evans 1995), and commercial packages have followed (Anonymous, 1997).

Evans *et al.* (2002) investigated the possibility of site specific application of nematicides for controlling potato cyst nematode and the following observation have been made The efficient management practices depend on accurate distribution maps based on number of sample analysis. But extra data cost more and there is limit to the frequency at which fields can be sampled. Sampling several fields at sample spacing of 20 m and geostatistical analysis has shown that:

1. Nematode density is usually correlated on a range of 40 to 60 m. The general pattern seen from the 20 m spaced sampling is still visible at 40 m but the patterns are increasingly different as the distance between samples increases. The pattern generated from 100 m spaced samples could be a poor reflection of the true distribution of PCN in the field.
2. An analysis of potential cost benefit ratios shows that sampling costs would be recovered even for a sample spacing as close as 20 m, provided that the resulting map permitted the decision not to treat 42 per cent of the field with granule nematicide (cost is 360 pounds/hectare), 27 per cent with a fumigant (cost of fumigant 550 pounds/hectare) or 16 per cent of the grower were considering joint application of granular and fumigant nematicides. Spatial applications are most easily made if decisions over whether to apply nematicide are made in one hectare blocks, but the sophisticated positioning equipment now available makes it possible to apply nematicide to very specific position within a field.
3. When a distribution map of potato cyst nematode was prepared before planting the patches of PCN density were observed in 34 per cent, 14 per cent and 8 per cent of the field. Whatever the strategy is adopted for taking individual samples, there is always chance of missing the natural patchiness of PCN the largest single patch that can be missed is one that fits nearly between sampling station. For example, sample station spaced 20 m apart on a square grid could miss a patch of up to 28 m diameter.
4. Sampling at pre-cropping and post harvest also vary the distribution maps.
5. Sometimes the population density may be small either because probably a patch is still developing or the density has yet to exceed the detection threshold or became a previously detectable density has declined to one that is no longer detectable without a host crop.
6. The most practical approach to site specific application of nematicide is to apply the material to discrete patches, the more squared off these area, the more meaningful/manageable the application. This was the approach adopted in field where the pre-planting population at densities likely to cause any effect on the crop was largely confined to a particular spot of the field. This area was treated with DD in the autumn before planting and the whole field was treated with oxamyl immediately before planting. The post harvest map shows that the population density reached in the more heavily infested but fumigated area was no greater than in the rest of the field and that spatial application of the fumigant was therefore successful in minimizing long-term problem.

This precise management tools have been prescribed for nematode control in high-intensive crop production such as *Radopholus similis* on banana and root-knot nematodes on potato in the northwestern United States. Approaches that focus on a harvest index to locate environmental stress should be able to relate nematode kinds and numbers to poor yield and other stress factors. This approach is now being used in some banana operations in which fruit is harvested in small subunits and yield data are recorded and analyzed by computer. Poor yielding sections can be examined for nematode densities and other potential problems. This approach was also developed in Costa Rica on banana plantation (Stoorvogel *et al.*, 1999; Stoorvogel *et al.*, 2001). On the banana plantations, management relies on manual labour (except for aerial spraying of fungicides). Therefore, high-tech gadgets to be placed on machinery were not suitable. The spatial variation in growing conditions were mapped through a traditional soil survey. The temporal and spatial variation in crop production were monitored through a yield mapping system that was based on automatic weighing and coding of banana bunches as they entered the processing facilities. When certain areas with a given soil type produced below the average value for this particular soil type, on-site investigations were made to find the reason which can be many. Nematodes may occur, soil structure may be degraded, drainage may not work, etc. Corrective measures are taken for these particular areas to raise the yields. This has proved to be an effective screening tool to differentiate management. This is an example to indicate high tech gadgets are not always necessary to introduce precision farming.

Many growers do not realize that their fields are infested with the nematode until soybean cyst nematode (SCN) *H. glycines* population densities reach population densities that result in clearly visible above ground symptoms. Even when soybean producers are aware that their fields are infested with SCN, they usually do not know the extent of SCN infestation and amount of yield loss being caused by this pest. The time of onset of plant yellowing is an important diagnostic feature. For example, symptoms caused by iron deficiency chlorosis usually appear in early June, within a few weeks of planting, whereas yellowing due to *H. glycines* typically appears in July or August, a month or two after planting. Consequently, the temporal pattern of remotely sensed images obtained over time may provide a useful means to accurately differentiate these two stresses. Thus, remote sensing coupled with GIS technologies offers new tools that may help in the early detection of *H. glycines* due to the distinctive spatial pattern (aggregations) of plant stress caused by the nematode. The use of multiple, narrow wavelength bands (measured or extracted from images) also might be used to define characteristic *H. glycines* injury spectral signatures. Nutter *et al.* (2002) developed GIS field map using the GIS software programme ArcView, and observed soybean cyst nematode *Heterodera glycines* densities ranged from zero to >10,000/100 cm^3 soil, soybean yield ranged from 175.9 to 111.989g/quadrat (586.3 to 3796.3 kg/ha, soyabean oil concentration ranged from 15.9 to 18.6 per cent and protein concentration ranged from 29.0 per cent to 37.0 per cent. Percentage reflectance measurements explained up to 60 per cent of the variation in initial SCN population densities within soybean quadrats and up to 91 per cent of the variation in soybean yield. Aerial image and satellite image intensities explained up to 80 per cent and 47 per cent of the variation in soybean yield, respectively. Percentage reflectance data also explained 36 per cent and 54 per cent of the variation in oil and protein concentrations of the harvested soybeans, respectively. These results indicate that remote sensing coupled with GIS technologies may provide new tools to detect and quantify SCN population densities and their impacts on the quantity and quality of soybean yield.

Remote sensing and GIS technologies could soon become invaluable tools for alerting soybean growers when and where to check their fields for the presence of *H. glycines* such remote sensing also may prove valuable in monitoring the performance of SCN-resistant soybean varieties for loss of effectiveness due to directional selection of SCN populations that can successfully reproduce on resistant soybean varieties. Currently, assessments of plant health and productivity are conducted using field scouting, ground-based soil sampling, and costly SCN testing methods. These scouting, sampling and testing methods are extremely time and labour intensive, particularly when the crop area to be assessed is large.

Morgan *et al.* (2002) determined spatial distribution. correlation between spatial distribution, soil parameters crop yield and temporal stability of root lesion nematode in three years crop rotation using GPS and GIS. The population of dynamics of *Alternaria solani, Verticillium dahliae,* and *Pratylenchus penetrans* were investigated in commercial potato fields. A 0.5-ha diamond grid-sampling scheme was georeferenced, and all disease ratings and nematode samples were taken at these grid points. Per cent disease severity was rated weekly, and *P. penetrans* densities were quantified 4 weeks after potato emergence. Spatial statistics and interpolation methods were used to identify the spatial distribution and population dynamics of each pathogen. Interpolated maps and aerial imagery identified *A. solani* intra-season progression across the fields as the potato crop matured. Late-season nitrogen application reduced *A. solani* severity. The spatial distributions of *dahliae* and *P. penetrans* were spatially correlated. It was observed that root lesion nematode densities were typically the larger near tillage implement entry points into the fields. Higher densities may be resulted from pathogen introduction to the fields after the most recent soil fumigation or improper application of soil fumigant at or near the field entry points. Densities of nematode were significantly lower in the fumigated portion off field than the non-fumigated portion. This knowledge on population dynamic should contribute to implementation of new cultural practices and provides opportunities to reduce pesticides quantities that are traditionally used in broadcast application.

Melakeberhan (2002) stressed the significance of spatio temporal characteristics of various distribution pattern having single or double limiting factors. The degree of accurately identify the spatial structure of yield limiting factor increase with the number of samples and how systematically the samples are collected. For identifying the spatial structure of the yield limiting factor most intensive and the most accurate sampling plan are to be undertaken but it is most expensive one. The type of sampling plan depends on value of the crop. In addition to spatial structure SSM requires determination of temporal changes of yield limiting factor. It is important for insects and airborne pathogen but less important for plant parasitic nature since their mobility on their own is limited.

Conclusion

These technologies allow better decision-making in the management of crop yield-limiting biotic and abiotic factors and their interactions on a site-specific (SSM) basis in a wide range of production systems. Characterizing the nature of the problem(s) and public education are among the challenges that scientists, producers, and industry face when adapting PA technologies. To apply SSM, spatio-temporal characteristics of the problem(s) need to be determined and variations within a field demonstrated. Spatio- temporal characteristics of a given pathogen or pest problem may be known but may not be the only or primary cause of the problem. Hence, exact cause-and-effect relationships, need to be established by incorporating GIS, GPS, and RS-generated data as well as possible interactions, exploiting the potential of Precision technologies in sustainable ways. Integrating remote sensing technologies into day to day management of cropping systems (precision agriculture) offers tremendous new opportunities farmers to cost effectively narrow the yield gaps that currently constrain agricultural cropping systems worldwide. Yield gaps are defined as the potential yields that growers could achieve for specific fields if plant stress were alleviated, minus the yields they actually harvest (Nutter *et al.*, 1993; Nutter and Littrell, 1996). Remote sensing, GIS, and GPS technologies have the potential to revolutionize the way in which farmers manage their crops (Anderson *et al.*, 1999); however, key conceptual and quantitative links concerning the relationships between remote sensing data and biological systems are often lacking. To fully maximize the management/yield benefits that might be achieved by integrating GIS, GPS, and remote sensing technologies into disease management programs, it is first necessary to concentrate on:

1. Methodology to differentiate accurately vegetation types (crops, natural vegetation, forests, etc.).
2. To recognize when plants (fields) are stressed.
3. Accurately determine reason for causing plant stress.
4. To quantify the degree of plant stress within fields.
5. To quantify and relate remote sensing assessments for plant stress with ground truth (visual) assessments for plant health and crop/plant productivity, and
6. To develop crop stress thresholds for plant populations that alert the farmer (in a timely manner) when available crop stress mitigation measures should be deployed to optimize the farmer's net return on investment.

Finally whatever tactics are used, the overall strategy must be planned to fit one of two possible objectives:

1. To employ short-term tactical management simply to prevent yield loss in the current crop, or
2. To aim for longer-term strategic management that will ensure that plant nematode populations remain static or even decline over the rotation.

In making these decisions, it must be recognized that every field will have a unique pattern of plant nematode distribution, and that population will behave in a unique manner when crops are grown. If site-specific treatments are to be employed, then information on distribution of nematodes and density must be collected at a scale appropriate to the decisions that might be made.

REFERENCES

Adcock, T.E., F.W. Nutter Jr. and P.A. Banks (1990). Measuring herbicide injury to soybeans (*Glycine max*) using a radiometer. *Weed Sci.*, 38: 625–627.

Anderson, J.E., R.L. Fischer and S.R. Deloach (1999). Remote sensing and precision agriculture: Ready for harvest or still maturing. *Photogrammetric Engineering Remote Sensing*, 10: 1118–1123.

Bouma, J. and J.J. Stoorvogel (2002). Precision agriculture as a generic strategy for different technology levels. Extended Summaries Vol. 2: 2nd International Agronomy Congress, November 26–30, 2002, New Delhi.

Cassel, D.K., O. Wendroth and D.R. Neilson (2000). Assessing spatial variability in an agricultural experiment station field: Opportunities arising from spatial dependence. *Agronomy J.*, 92: 706–714.

Evans, Ken, Webster, M. Richard Paul D. Halford, Barker, D. Anthony and Michael D. Russell (2002). Site-specific management of nematodes: Pitfalls and practicalities. *J. Nemat.*, 34(3): 194–199.

Ferris, H. and M.V. McKenry (1974). Seasonal fluctuation in the special distribution of Nematode population in a California vineyard. *J. Nemat.*, 6: 203–210.

Guan, J. and F.W. Nutter Jr. (2001). Factors affecting the quality and quantity of sunlight reflected from alfalfa canopies. *Plant Disease*, 865–874.

Lechner, W. (1994). Navstar GPS and GLONASS. In: *Computers and Electronics Agriculture*, Special Issue; GPS in Agriculture, Elsevier Science Publishers, pp. 139.

McSorley, R. (1998). Population dynamics. In: *Plant and Nematode Interactions*, (Eds.) K.R. Barker, G.A. Pederson, and G.L. Windham, Agronomy Monograph No. 36. Madison, WI, American Society of Agronomy, Crop Science Society of America, and Soil Science Society of America Publishers, pp. 109–134.

Melakeberhan, H. (2002). Embracing the emerging precision agriculture technologies for site-specific management of yield limiting factors. *J. Nemato.*, 34(3): 185–188.

Morgan, G.D., A.E. MacGuidwin, J. Zhu and L.K. Binning (2002). Population dynamics and distribution of root lesion nematode (*Pratylenchus penetrans*) over a 3-year potato crop rotation. *Agronomy J.* (In Press).

Nelson, M.R., T.V. Orum and R. Jaime-Garcia (1999). Applications of geographic information systems and geostatistics in plant disease epidemiology and management. *Plant Disease*, 83: 308–319.

Norton, D.C. (1978). *Ecology of Plant Parasitic Nematodes*. A Wiley-interscience Publications, pp. 268.

Nutter, F.W. Jr. (1989). Detection and measurement of plant disease gradients in peanut with a multispectral radiometer. *Phytopathology*, 79: 958–963.

Nutter, F.W. Jr. and R.E. Gaunt (1996). Recent developments in methods for assessing disease losses in forage/pasture crops. In: *Pasture and Forage Crop Pathology*, (Ed.) Madison, W.I., American Society of Agronomy, Crop Science Society of America, and Soil Science Society of America Publishers, pp. 93–118.

Nutter, F.W. Jr. and R.H. Littrell (1996). Relationships between defoliation, canopy reflectance, and pod yield in the peanut-late spot pathosystem. *Crop Protection*, 15: 135–142.

Nutter, F.W. Jr., J. Guan and R.J. Percifield (2000). Assessing crop health in time and space by measuring spectral reflectance of solar radiation from plant canopies. In: 2nd *International Conference on Geospatial Information in Agriculture and Forestry*, ERIM, International, Inc., Ann Arbor, MI, pp. 165–176.

Nutter, F.W. Jr., R.J. Guan, R.J. Percifield, B. Menelas, T.R. Ronburg, S. Bidney, J.P. Basart, C.S. Chong and M. Cook (1999). Use of remote sensing to quantify the impact of plant disease epidemics on plant biomass and productivity. 9th *Annual Space Grant Conference, 5th November*, University of Northern Iowa, Cedar Falls, IA.

Nutter, F.W. Jr., P.S. Teng and M.H. Royer (1993). Terms and concepts for yield, crop loss, and disease thresholds. *Plant Disease*, 77: 211–215.

Nutter, F.W. Jr., G.I. Tylka, J. Guan, A.J.D. Moreira, C.C. Marett, T.R. Rosburg, J.P. Basart and C.S. Chong (2002). Use of remote sensing to detect soybean cyst nematode-induced plant stress. *J. Nemat.*, 34(3): 222–231.

Parker, W.E. (1998). Does mapping have a role in potato cyst nematode (*Globodera rostochiensis* and *G. Pallida*) management strategies? *Aspects of Applied Biology*, 52: 367–374.

Pearce, D., E. Barbier and A. Markanda (1988). Sustainable development and cost-benefit analysis. *London Environmental Economics Centre, UK*, 88(1): 38.

Pierce, F.J. and E.J. Sadler (1997). The state of site-specific management for agriculture. Madison, WI: American Society of Agronomy Miscellaneous Publication. American Society of Agronomy, Crop Science Society of America, and Soil Science Society of America Publishers.

Poiner, G.O. Jr. (1983). *The Natural History of Nematodes*. Prentice Hall Inc., Englewood, N.J.

Seshadri, A.R. and H.S. Gaur (1999). Integrated Nematode Management Approaches for Sustainable Agriculture. Changing scenario in farming practices: Policies and management Edtd. S. Kushwaha farm book series, pp. 286–219.

Stoorvogel, J.J., L. Kooistra and J. Bouma (1999). Spatial and temporal variation.in nematocide leaching, management implications for a Costa Rican banana plantation. In: *Assessment of Non-Point Source Pollution in the Vadose Zone*, (Eds.) D.L. Corwin, K. Loague and T.R. Ellswroth, Geophysical Monograph Volume 108, AGU, Washington D.C., pp. 281–289.

Strickland, R.M., D.R. Ess and S.D. Parsons (1998). Precision farming and precision pest management, the power of new crop production technologies. *J. Nemat.*, 30: 431–435.

32

Acarines as Biocontrol Agents of Plant Parasitic Nematodes

Rachna Gulati*, Kumkum Walia** and Ranjana Rani

**Department of Entomology, CCS Haryana Agricultural University, Hisar – 125 004 (India)*

***Department of Nematology, CCS Haryana Agricultural University, Hisar – 125 004 (India)*

The community of soil arthropods is one of the most species rich communities in terrestrial ecosystem. The coexistence leads to competition for exploitation of available food and habitat resources among the organisms, which affects the productivity. To increase the productivity, it is vital to understand the life processes and activities of various biota occupying the soil ecosystem especially soil acarines. In addition to their role in nutrient recycling, their predatory behaviour is of significance since it helps in natural regulation of some harmful organisms. Soil acarines are instrumental in keeping the populations of plant parasitic nematodes under check in some of the undisturbed ecosystems. An attempt has been made to put together the available literature on predator- prey relationship between predacious soil acarines and plant parasitic nematodes.

Acarines belonging to all the four sub-orders *viz.*, Astigmata, Cryptostigmata, Mesostigmata and Prostigmata have different feeding habits. Some are parasites of higher animals, some feed on plants, others are pests of store products. Some beneficial acarines are predatory on insects, acarines and even soil inhabiting nematodes. Acarines also play a significant role in mineralisation and decomposition processes of organic matter in the soil. The feeding organs in acarines are chelicerae and pedipalps. Pedipalps hold the prey and chelicerae are used in cutting, chewing or devouring the prey as a whole. Depending on the type of food, the morphology of chelicerae too varies (Buryn and Brandl, 1992; Walter and Ikonen, 1989). Granivorous mites *viz., Tyrophagus putrescentiae* have small and stout chelicerae with few teeth on *digitus mobilis* and *digitus fixus*, arthropod feeders possess large sized chelicerae with large number of teeth while nematode feeders have small sized chelicerae with large number of teeth adapted for holding worm- like (nematodes) prey.

Linford and Oliveira (1938) observed six unidentified mite species feeding on nematodes. However, no attention was paid to these organisms for almost two decades. The literature pertaining to nematophagy in mites is accumulating slowly but steadily over the past few years and some recent reviews are available on the subject (Small, 1988 & Walia, 1994; 2003). The distribution of nematode feeding acarines, their trophic and reproductive behaviours have been described under different sub-orders of the class Acarina.

Astigmata

Astigmatic acarines are comparatively less in soil ecosystem and are predominant inhabitants of store products and animals. Muraoka and Ishibashi (1976) could identify only one out of 41 soil inhabiting acarines as astigmatid (*Caloglyphus* sp.) when fed upon the second stage juveniles of two nematode species, *Heterodera glycines* and *Meloidogyne incognita*. Only three genera *i.e., Tyrophagus, Sancassania* (=*Caloglyphus*) and *Rhizoglyphus* have been reported with predacious activity (Sturhan and Hampel, 1977; Walter *et al.*, 1986; Sell, 1988a; Walia and Mathur, 1995). *T. putrescentiae, T. similis* and *T. zachvatkini* could be cultured on a nematode diet too in addition to their regular diet (baker's yeast) and they preferred nematodes over yeast or alga (Walter *et al.*,1986 & Walter,1987b;1988a). *R. echinopus* consumed a variety of nematodes *viz., Aphelenchoides ritzemabosi, Ditylenchus dipsaci, D. destructor, Heterodera avenae, Longidorus elongatus* and *Xiphinema index*. *Tyrophagus putrescentiae* predated upon root-knot nematode and large nematodes *e.g., Longidorus* sp. (Walia and Mathur, 1995; Bilgrami, 1994; 1997a&b; Ranjana, 2002). The mite also consumed infective stages of entomopathogenic nematodes, *Heterorhabditis heliothidis* and *Steinernema feltiae* (Epsky *et al.*, 1988). *Sancassania ultima* (=*Caloglyphus*) devoured a number of plant parasitic nematodes namely *Globodera rostochiensis, H.schachtii, Paratylenchus, Pratylenchus, Tylenchorhynchus* sp. and root- knot nematode (Sell, 1988b).

Food preference studies on two astigmatic mites revealed the difference in feeding behaviour of *Schwiebea rockettii* and *Histiostoma bakeri*. The former fed on a variety of lower organisms including dead nematodes and could be cultured on a plant parasitic nematode, *Radopholus citrophilus* whereas *H. bakeri* only fed on decaying matter (Walter and Kaplan, 1990 a).

Cryptostigmata

The oribatid mites popularly also known as ' Beetle mites' are heavily sclerotised and slow moving and so are not well adapted for predatory mode of life. These mites are abundantly available in organic matter rich surface soil and litter. Their role in decomposition and mineralisation of organic matter has been widely studied and it has been established that mites do play a very significant role in increasing the soil fertility. There are limited reports of predation by oribatid mites on nematodes. Rockett and Woodring (1966) observed an oribatid mite species, *Pergalumna omniphagus* predating upon free-living nematode, *Pelodera lambdiensis*. This was a significant finding for these primarily detritus feeders. The work of Muraoka and Ishibashi (1976) added a list of 22 cryptostigmatid mite species to our knowledge of nematophagous mites. Rockett (1980) studied the phenomenon of nematode predation by 14 genera of oribatid mites. *Pergalumna* voraciously fed upon *Chiloplacus* sp. and *Rhabditis* sp., the two free living nematode genera. Specimens of *Galumna* sp., *Fuscozetes* sp. and *Nothrus* sp. fed less frequently whereas *Ceratozetes* sp. only chewed and did not devour the nematodes. Rest of the mite genera were not nematophagous. Further observations on feeding behaviour indicated that the mite neither searched nor preferred nematodes and the time required for a tritonymph or adult pergalumnid to grasp and completely devour an average sized nematode (± 500μm) varied considerably from 15 seconds to 25 minutes. As many as three nematodes could be consumed successively.

Mesostigmata

The most of the nematode predacious acarines belong to this group and are quite abundant and frequent geographically in the soil ecosystem. Some of the nematode specialists *viz., Alliphis halleri* (Buhlmann, 1984) also belong to this sub-order. Some mesostigmatid mites prefer nematodes to their regular diet *e.g., Macrocheles muscaedomesticae* nymphal stages (Rodriguez *et al.*, 1962). Others though feed on nematodes yet prefer their regular diet for reproduction and multiplication (Afifi and Van der Geest, 1984). Fungus feeding and free-living nematodes, *Aphelenchus avenae, Aphelenchoides composticola, Panagrellus redivivus, Rhabditis elongata* and *R. teres* also provide a rich food resource for *M. muscaedomesticae* (Ito, 1971). All the stages of small nematodes and immature stages of large nematodes were predated upon effectively by *Rhodacarellus silesiacus, R. calcarulatus* and *Rhodacarus roseus* (Karg, 1961; 1971). Many other mite species (*Lasioseius penicilliger, L. scapulatus, R. roseus* and 18 other species) have been reported to feed upon nematodes, *Tylenchorhynchus dubius, Paratylenchus* sp., *Pratylenchus* sp., *Rotylenchus* sp., *Cephalobus* sp., *H. glycines*, root-knot nematode, *Meloidogyne* spp., *A. avenae, Acrobeloides* sp., *Caenorhabditis* sp., *Seinura* spp. etc. (Sharma, 1971; Van de Bund, 1972; Muraoka and Ishibashi, 1976; Imbriani and Mankau, 1983; Inserra and Davis, 1983; Ishibashi *et al.*, 1987). Thirteen nematode species belonging to different trophic groups were fed upon by *H. calcuttaensis* in *in vitro* studies (Bilgrami, 1994b). In greenhouse cultures of *Tylenchulus semipenetrans* and root- knot nematode at Florida, USA, the densities of nematophagous mites often exceeded 1000 mites per liter (Walter *et al.*, 1993).

The food and feeding behaviour of mites depends on the type of mouth parts (esp. chelicerae), size of the prey (smaller consumed more than larger ones) and size of the predator (large predators survive on both nematode diet and arthropod diet). Nematodes sometimes provide a high nutrient rich diet to nematophagous acarines, which develop faster and reproduce more under such conditions. In *M. muscaedomesticae* 23.11 ±5.58 progeny per female per day is produced in presence of both, nematodes and house fly eggs (Rodriguez *et al.*, 1962). *Gamasellodes vermivorax* completes its life cycle on rhabditid nematode just in 10 days at 24 °C and population increases four fold after 16 days depending upon nematode densities in microcosms (Walter, 1987a). *Contortylenchus brevicomi*, a nematode found in galleries of bark beetles is essential for the development of *Dendrolaelaps neodisetus*, which completes its life cycle in 6 to 13 days at 25 °C on the nematode diet (Kinn, 1984). *G. vermivorax, Macrocheles* sp. and *Cosmolaelaps vacua* take lesser time to develop on a nematode diet than on an arthropod diet and lay eggs only in presence of nematode diet (Walter *et al.*, 1987). *Hypoaspis calcuttaensis* too reproduced on nematode diet and not on fungus (Walia and Mathur, 1998). *Lasioseius thermophilus*, a mite found in mushroom compost feeds on the eggs of *D. myceliophagus*. The mite population increased from 10 and 50 to 603 and 1611, respectively in the presence of nematodes whereas nematode population declined to 18,550 (50 mites inoculated) from 504,000 in control (absence of mites). Cayrol (1970) suggests its possible use as a biocontrol agent against *D. myceliophagus* in mushroom beds.

Prostigmata

The prostigmatid mites belonging to the traditional preda Stigmaeidae, Caligonellidae, Raphignathidae, Cryptognathi in nematodes (Walter *et al.*, 1988). Among prostigmatids, th important nematode predators (Walter 1988 a, b). Nematode difficult to recognize in gut contents owing to lack of skelet small to locate easily. *Alycus roseus* is a nematode specialist efficient nematode predator consumed other food too. The infe nematodes, *H. heliothidis* and *S. feltiae* are reported to be preda *al.*, 1988) and do hamper the populations of these nematod develop faster on a nematode diet than an arthropod diet ye compared to nematode diet (Taha *et al.*, 1988). Another cunaxi greenhouse cultures of root-knot nematode in Florida could c juveniles of the nematode and not the adults and egg masses

Potential of Acarines as Biocontrol Agents

The acarines predating upon nematodes are euedaphic forms with rapid generation time and high reproductive out put. They do possess certain survival strategies too. However, the acarines have not been exploited as efficient biocontrol agents. The main reason being the inability to culture them in bulk due to lack of complete knowledge about the biology and ecology of these efficient predators. Efforts to explore their potential as biocontrol agents are limited to pot experiments only.

In Egypt, the association of soil mites, *Rhizoglyphus echinopus* and *T. putrescentiae* with root-knot nematode, *Meloidogyne javanica* on *Vicia faba* seedlings was studied in pots (Salem and Abo- Korah, 1981). Out of the two mites, only *T. putrescentiae* multiplied and the population of the nematode decreased in the presence of the mite. The mites helped in mitigating the suppressive effect of nematode on plant growth.

Under certain field ecosystems the mesostigmatic mite, *Hypoaspis aculeifer* could limit the populations of *M. chitwoodi, M. hapla* and *Heterodera schachtii* (Inserra and Davis, 1983). Addition of organic manures, leaf litter, intercropping in field crops favours the development and multiplication of the mites which in turn helps in regulating the population of plant parasitic nematodes (Karg, 1983).

The precise mite-nematode relationship operating in soil is not fully understood. Why the population of the prey (*Cephalobus* sp.) increased at the end of the experiment in presence of the predator (*Caloglyphus* sp) could not be explained by Muraoka and Ishibashi (1976). Such observations need to be explored extensively to harness the full potential of the mites.

In greenhouse trials, the citrus nematode, *Tylenchulus semipenetrans* could be restricted by *Macrocheles muscaedomesticae* successfully (Osman *et al.*, 1988). The mesostigmatic mites have been judged as better predators in soil ecosystem than astigmatids, which are rare in soil (Walia and Mathur, 1996). These also hold potential to be used for the nematode pests of mushroom. The preliminary experiments with *H. calcuttaensis* against *Aphelenchoides composticola* (Walia and Mathur, 1994) and *Lasioseius thermophilus* against *Ditylenchus myceliophagus* are encouraging (Cayrol, 1970) and can be explored further. *H. calcuttaensis, Lasioseius athiasae* and *Protogamasellus discorus* have been found to reduce soil population and root galls caused by root-knot nematodes. The mite population increased 8 times after 45 days in the presence of nematodes (Walia and Mathur, 1996; Amin *et al.*, 1999).

Cryptostigmatic and prostigmatic mites though reported with predacious activity have not been exploited as biocontrol agents. Some of the prostigmatic mites are 'nematode specialists' *i.e.*, feed only on nematodes and have better chances of establishing in presence of high nematode populations than any other group.

Predatory acarines play very different roles in insect-nematode interrelationship in woody trees. The mites may feed on *Bursaphelenchus xylophilus*, the nematode involved in pine-wilt disease of pines in Japan (Tamura and Enda, 1980 a, b) or the nymphal stages feed on nematodes which increase the population of adults instrumental in lowering the population of bark beetles (*Ips* sp.) by predation (Kinn, 1984) or the mite may hamper the control of insects by feeding on entomogenous nematodes (Kinn, 1983; Epsky *et al.*, 1988).

The predacious mites are important in breakdown of litter and decomposition process. The bacteriophagous nematodes are predated upon by the mites, which indirectly enhance the process of decomposition in presence of decomposing bacteria, which are otherwise grazed upon by the nematodes in the absence of predatory acarines (Santos and Whitford, 1981; Santos *et al.*, 1981; Elkin and Whitford, 1982; Laakso *et al.*, 1999).

The predatory soil acarines are distributed globally and are quite abundant too. Predation has been established in all the four sub orders of the class Acarina. However, these studies are mostly restricted to trophic and reproductive behaviour and have been carried out by either soil zoologists or ecologists. A few mesostigmatids have been employed in pot experiments to varying degree of success. Short life cycle, mobility, parthenogenetic and high reproductive out put gives them an advantage for a potential biocontrol agent. However, there are disadvantages too. Firstly, they are large in size, mostly omnivorous and more abundant in the upper layers of the soil. Some members of Mesostigmata and Prostigmata need particular attention since these are truly subterranean forms (*e.g.*, fam. Rhodacaridae), narrow bodied, best suited for their pursuit of prey in narrow pore spaces. But, the culturing of these mites has not been successful till now (Walter and Kaplan, 1990).

The more practical approach to use predatory acarines as biocontrol agents will be the implementation of cultural methods which increase the organic matter and hence the acarines population. The perennial crops or horticultural crops can be the ideal ecosystem where such a strategy can be successful. Studies on bioecology of mite and nematode interaction are required so that the stage at which, disturbing the ecological balance would retard the nematode population can be worked out. There is a need to carry out more pot and field experiments to harness these potential biocontrol agents against the nematodes.

Conclusion

Soil acarines play a significant role in increasing the soil productivity and in regulation of some harmful organisms including plant parasitic nematodes. Only three genera of Astigmatids *viz.*, *Tyrophagus*, *Sancassania* (= *Caloglyphus*) and *Rhizoglyphus* have been reported with predacious activity. The cryptostigmatids primarily detritus feeders are also not very good predators of nematodes. A large number of genera in Mesostigmata and a few in Prostigmata are voracious predators. Some are nematode specialists, prefer nematode diet to regular diet and have short life cycles with high reproductive output on diet supplemented with nematodes. Plant parasitic nematodes have been successfully controlled by predatory acarines in pot experiments. However, field application has not been possible due to lack of mass cultures. More studies are required on bio-ecological interactions of this predator-prey relationship.

REFERENCES

Afifi, A. M. and L. P. S. Van der Geest (1984). Notes on the development and biology of the predacious soil mite *Cosmolaelaps claviger* (Berlese, 1883) (Gamasida: Laelapidae). In: Acarology VI Vol. 1, (Eds.) D.A. Griffiths and C.E. Bowman. Ellis Horwood Ltd., Chichester, England, pp. 585–590.

Amin A.W., M.H. Mowafe and S. A. Fatma (1999). Effect of predaceous mesostigmatid mites in the control of *Meloidogyne javanica*, root-knot nematode on kidney bean. *Pakistan Journal of Nematology* 17: 91–96.

Bilgrami, A.L. (1994). Predatory behaviour of a nematode feeding mite *Tyrophagus putrescentiae* (Sarcoptiformes: Acaridae) *Fundam. appl. Nematol.* 17: 293–297.

Bilgrami, A.L. (1997a). Prey catching and feeding mechanisms of nematode feeding mites, *Tyrophagus putrescentiae* and *Hypoaspis calcuttaensis*. *Annals of plant protection Sciences* 5: 90–93.

Bilgrami, A.L. (1997b). Evaluation of the predation abilities of the mtie *Hypoaspis calcuttaensis*, predaceous on plant and soil nematodes. *Fundam. appl. Nematol.* 20: 96–98.

Buhlmann, A. (1984). Influence of agricultural practices on the population of *Alliphis halleri* (G. and R. Canestrini, 1881) (Acari: Gamasina). In: Acarology VI, Vol. 2, (Eds.) D..A. Griffiths and C.E. Bowman. Ellis Horwood Ltd., Chichester, England, pp. 901–909.

Buryn, R. and R. Brandl (1992). Are the morphometrics of chelicerae correlated with diet in mesostigmatid mites (Acari)? *Exp. Appl. Acarol.* 14: 67–82.

Cayrol, J. C. (1970). Action des autres composants de la biocenose du champignon de couche sur le nematode mycophage, *Ditylenchus myceliophagus* J. B. Goodey, 1958, et etude de son anabiose: forme de survie en conditions defavourables, *Rev. Ecol. Biol. Sol* 7: 409–440.

Elkins, N. Z. and W. G. Whitford (1982). The role of microarthropods and nematodes in decomposition in a semi arid ecosystem. *Oecologia*, 55: 303–310.

Epsky, N.D., D.E. Walter and J.L. Capinera (1988). Potential role of nematophagous microarthropods as biotic mortality factors of entomogenous nematodes (Rhabditida; Steinernematidae, Heterorhabditidae). *J. Eco. Ento.*, 81: 821–825.

Imbriani, J.L. and R. Mankau (1983). Studies on *Lasioseius scapulatus*, a mesostigmatid mite predaceous on nematodes. *J. Nematol.* 15: 523–528.

Inserra, R. N. and D. W. Davis (1983). *Hypoaspis nr. aculeifer* : a mite predacious on root- knot and cyst nematodes. *J. Nematol.* 15: 324–325.

Ishibashi, N., F Young, M. Nakashima, C. Abiru and N. Haraguchi (1987). Effect of application of DD -136 on silk-worm, *Bombyx mori*, predatory insect, *Agriosphodorus dohrni*, parasitoid, *Trichomalus apanteloctenus*, soil mites, and their non target soil arthropods, with brief ntoes on feeding behaviour and predatory pressure of soil mites, tardigrades, and predatory nematodes on DD- 136 nematodes. In: *Recent Advances in Biological Control of Insect Pests by Entomogenous Nematodes in Japan*, (Ed.) N. Ishibashi. Ministry of Education, Culture and Science, Sago, Japan, pp. 158–164.

Ito, Y. (1971). Predation by manure inhabiting mesostigmatids (Acarina: Mesostigmata) in some free-living nematodes. *Appl. Ent. Zool.*, 6: 51–56.

Karg, W. (1961). Okologische Untasuchungen von edaphischen gamasiden (Acari: Parasitiformes). *Pedobiologia* 1: 58–98.

Karg, W. (1971). Acari (Acarina), Milben, Unterordnung Anactinochaeta (Parasitiformes). Die freilebenden Gamasina (Gamasides), Raubmilben. In: Die Tierwelt Deutschlands und der angrenzenden Meeresteile, 59. *Teil Gustav Fischer Verlag, Jena*, 475.

Karg, W. (1983). Verbreitung und Bedeutung von raubmilben der cohors Gamasina als Antagonisten von Nematoden. *Pedobiologia* 25: 419–432.

Kinn, D. N. (1983). The life cycle of *Proctolaelaps dendroctoni* Lindquist and Hunter (Acari: Ascidae); a mite associated with pine bark beetles. *Internat. J. Acarol.* 9: 205

Kinn, D.N. (1984). Life cycle of *Dendrolaelaps neodisetus* (Mesostigmata: Digamasellidae), a nematophagous mite associated with pine bark beetles (Coleoptera: Scolytidae). *Environ. Entomol.* 13: 1141–1144.

Laakso, J. and H. Setala (1999). Population and ecosystem– level effects of predation on microbial–feeding nematodes. *Oecologia* 120: 279–286.

Linford, M.B. and J.M. Oliveira (1938). Potential agents of biological control of plant parasitic nematodes. *Phytopathology* 28: 14.

Muraoka, M. and N. Ishibashi (1976). Nematode feeding mites and their feeding behaviour. *Appl. Ent. Zool.* 11: 1–7.

Osman, G.Y., A.M. Zaki, F.M. Salem and E.T.L. Darwish (1988). Biological control study on *Tylenchulus semipenetrans* (Cobb) (Nematoda) by certain soil mesofauna. *Anzeiger fur Schadlingskunde Pflanzenschutz, Umweltschutz*, 61: 116–118.

Ranjana Rani (2002). Studies on the Nematophagy in acarines fauna and their feeding potential on seed gall nematode. M.Sc. Thesis submitted to CCS Haryana Agricultural University, Hisar, pp. 71.

Rockett, C.L. (1980). Nematode predation by oribatid mites (Acari: Oribatida). *Internat. J. Acarol.*, 6: 219–224.

Rockett, C.L. and J.P. Woodring (1966). Oribatid mites as predators of soil nematodes. *Ann. Ento. Soc. Amer.* 59: 669–671.

Rodriguez, J.G., C.F. Wade and C.N. Wells (1962). Nematodes as a natural food for *Macrocheles muscaedomesticae* (Acarina; Macrochelidae) a predator of the house fly eggs. *Ann. Ento. Soc. Amer.* 55: 507–511.

Salem, F.M. and S.M. Abo–Korah (1981). Inter– relationship between plant parasitic nematodes and soil acari IV. Interaction between *Meloidogyne javanica* and certain acarid species in relation to horse– bean nematode infection. *Menoufeia Journal of Agric. Research* 4: 345–353.

Santos, P.F. and W.G. Whitford (1981). The effects of microarthropods on litter decomposition in a Chihuahuan desert ecosystem. *Ecology*, 62: 654–663.

Santos, P.F., J. Phillips and W.G. Whitford (1981). The role of mites and nematodes in early stages of buried litter decomposition in a desert. *Ecology*, 62: 664–669.

Sell, P. (1988a). *Caloglyphus* sp. (Acarina: Acaridae), an effective nematophagous mite on root–knot nematodes (Meloidogyne spp.). *Nematologica*, 34: 246.

Sell, P. (1988b).Milben zur biologischen bekampfung von wurzelgallen–Nematoden (Meloidogyne spp.). *Med. Fac. Landbouww. Rijksuniv. Gent* 52: 859–865.

Sharma, R. D. (1971). Studies on the plant parasitic nematode *Tylenchorhynchus dubius. Meded. Landbouwhogeschool, Wageningen*. 71: 98–104.

Small, R.W. (1988). Invertebrate Predators. In: *Diseases of Nematodes*, Vol II, (Eds.) G.O. Poinar Jr. and H.B Jansson. CRC Press Inc., Boca Raton, Florida, pp. 73–92.

Sturhan, D. and G. Hampel (1977). (Plant parasitic nematodes as prey of the bulb mite, *Rhizoglyphus echinopus* (Acarina: Tyroglyphidae). *Anzieger fur Schadlingskunde, Pflanzenschutz Umweltschutz*. 50:115–118.

Taha, H.A., M.E.E., El–Naggar, M.M. Abou–El–Nagaga and S.M. Soliman (1988). Effect of different prey species on the development and fecundity of the predacious mite, *Neocunaxoides andrei* Baber and Hoff (Acari: Cunaxidae). *Agricultural Research Review* 66: 129–135.

Tamura, H. and N.Enda (1980a). Life histories of three species of nematode feeding mesostigmatid mites associated with the pine sawyer beetle *Monochamus alternatus. J. Jpn. For. Soc* 62: 301–307.

Tamura, H. and N. Enda (1980b). Mesostigmatid mites associated with Japanese pine sawyer beetle. *Jpn. J. Appl. Ent. Zool.* 24: 54.

Van de Bund, C. F. (1972). (Some observations on predatory action of mites on nematodes). *Zesz. Probl. Postepow Nauk Roln.* 129: 103–110.

Walia, K.K. (1994) Nematophagous Mites – a review. In: *Nematode Pest Management in Crops*, (Eds.) D.S. Bhatti and R.K. Walia, CBS Publ. and Distr. New Delhi, pp. 257–269.

Walia, K. K. (2003). Bioecological interactions between nematophagous mites and plant parasitic nematodes In: *Mites, their identification and management*, (Eds.) P.R. Yadav, R. Chauhan, B.N. Putatunda and B.S. Chillar, Deptt. Of Entomology, CCS Haryana Agricultural University, Hisar. Pp. 190–194.

Walia, K.K. and S. Mathur (1994a). Acarine fauna of arable soils and their screening for nematophagy. *Indian J. Nematol.* 24: 69–79.

Walia, K.K. and S. Mathur (1994b). Predatory potential of two nematophagous mites on fungivorous nematode, *Aphelenchoides composticola in vitro. Indian J. Nematol.* 24: 243–245.

Walia, K.K. and S. Mathur (1995). Predatory behaviour of two nematophagous mites, *Tyrophagus putrescentiae* and *Hypoaspis calcuttaensis*, on root–knot nematode, *Meloidogyne javanica. Nematol. medit*. 23: 255–261.

Walia, K.K. and S. Mathur (1996). Potential of two nematophagous mites in suppressing root– knot nematode, *Meloidogyne javanica* on Okra. *Indian J. Nematol* 26: 172–175.

Walia, K. K. and S. Mathur (1998). Reproductive behaviour of two nematophagous mites on nematodes and alternate diet. *Annals of Plant Protection Sciences* 6: 178–181.

Walter, D.E. (1987a). Life history, trophic behaviour and description of *Gamasellodes vermivorax* new species (Mesostigmata: Ascidae), A predator of nematodes and arthropods in semi–arid grasslands. *Can. J. Zool.* 65: 1689–1695.

Walter, D.E. (1987b). Trophic behaviour of 'Mycophagous' microarthropods. *Ecology* 68: 226–229.

Walter, D.E. (1988a). Nematophagy by soil arthropods from the short grass steppe, Chihuahuan desert and rocky mountains of the central United States. *Agric.Ecosys. Environ.* 24: 307–316.

Walter, D.E. (1988b). Predation and mycophagy by endeostigmatid mites (Acari: Prostigmata). *Exp. Appl. Acarol.* 4: 159–166.

Walter, D.E., R.A. Hudgens and D.W. Freckman (1986). Consumption of nematodes by fungivorous mites, *Tyrophagus* spp. (Acarina: Astigmata: Acaridae). *Oecologia* 70: 357–361.

Walter, D.E., H.W Hunt and E.T. Elliott (1987). The influence of prey type on the development and reproduction of some predatory soil mites. *Pedobiologia* 30: 419–424.

Walter, D.E., H.W. Hunt and E.T. Elliott (1988). Guilds or functional groups? An analysis of predatory arthropods in a short grass steppe soil. *Pedobiologia* 31: 247–260.

Walter, D.E. and F.K Ikonen. (1989). Species, guilds and functional groups: taxonomy and behaviour in nematophagous arthropods. *J. Nematol.* 21: 315–327.

Walter, D.E. and D.T. Kaplan (1990a). Feeding observations on two astigmatid mites, Schwiebea rocktii (Acaridae) and *Histiostoma bakeri* (Histiostomatidae) associated with citrus feeder roots. *Pedobiologia* 34: 281–286.

Walter, D.E. and D.T. Kaplan (1990b). A guild of thelytokous mites (Acari: Mesostigmata) associated with citrus feeder roots in Florida. *Environ. Entomol.* 19: 1338–1343

Walter, D.E. and D.T. Kaplan (1991). Observations on *Coleoscirus simplex* (Acarina: Prostigmata), a predatory mite that colonizes greenhouse cultures of root- knot nematode (*Meloidogyne* sp.) and a review of feeding behaviour in Cunaxidae. *Exp. Appl. Acarol.* 12: 47–59.

Walter, D.E, D.T. Kaplan and E.L. Davis (1993). Colonization of greenhouse cultures by nematophagous mites and fungi. *Supp. to J. Nematol.* 25: 789–794.

33

Biocontrol Agents for Management of Phytonematodes

D. Prasad* and Veena**

**Division of Nematology, Indian Agricultural Research Institute, New Delhi – 110 012*

***Department of Zoology, M.M.H. College, Ghaziabad – 201 001, U.P.*

Plant parasitic nematodes form an important niche in agro-ecosystems. They feed on plant organs as ectoparasites, semi-endoparasites and endoparasites and cause reduced plant productivity. They totally depend on their respective hosts for sustenance and survival. They have developed requisite morphological and biochemical adaptations through course of evolution for their feeding habits and mode of survival during host-free periods. Endoparasitic nematodes with or without sexual dimorphism constitute the most destructive category of plant parasitic nematodes. They cause diverse damages in agro-ecosystems and yield losses suffered may be enormous. According to an estimate plant parasitic nematodes cause losses of over US $ 78 billion in agricultural production on worldwide basis. The share of developing countries in this loss is greater than developed countries. Yield loss estimates available in India for a number of crops are not very precise and adequate but available information indicate that they are important factor in low productivity of a number of crops in different regions of the country.

Development of two soil fumigants DD and EDB in mid-1940's which showed, excellent nematode killing properties was the beginning of the efforts for management of plant parasitic nematodes. Development and use of nematicides increased substantially in the following years. Granules, volatile liquids, gases, emulsifiable concentrates etc., were manufactured and the nematicides were accordingly classified as soil fumigants and non-fumigants (systemic and non-systemic). They were used for injection in soil, drenching, seed treatment, root dip and foliar applications. Most of the nematicides were effective and provided adequate control and a general agreement amongst the nematologists persisted that nematode problems in agriculture can be solved by application of nematicides and the growers became enthusiastic and used over-doses in order to maximize profits. In 1977, dependence on nematicides received a serious set-back. Many of the employees at a chemical plant at Lathrop, California, USA who handled DBCP, a soil fumigant, were found to be either azoospermic or oligozoospermic. Low sperm counts were found among formulators, custom applicators, farmers and salesmen. The United. States Environmental Protection Agency banned production and sale of this nematicide. Several regulatory measures followed as most of the nematicides were found to be the cause of environmental concerns and human health risks. Nematicides cause soil pollution and adversely affect useful micro flora and fauna which interact with pathogens at the soil root interface. They leave residues in crop plants and soils which present risks to human health and are hazards to grazing livestock. Nematicides pollute water and deteriorate water quality. Most of them have been detected in groundwater. This has changed entire scenario of management efforts for plant parasitic nematodes. Non-chemical measures are preferred to chemical measures. Biological or biocontrol of plant parasitic nematodes has become most significant and intensive area of research. Biological control is defined as the action of parasities, predators or pathogens in maintaining another organism's population density at a lower average than would occur in their absence (Jatala 1986). Several organisms, such as fungi, bacteria, viruses, predatory nematodes, insects, mites and some invertebrates have been found to parasitize or prey nematodes (Stirilling, 1991), and among these some have shown as a potential biocontrol agents against nematodes. The most widely studied organisms are fungi, bacteria and predatory nematodes than others.

Fungi Affecting Plant-parasitic Nematodes

The microorganisms that parasitize or prey on nematodes, fungi hold an important position and some of them have shown great potential as biocontrol agents (Jatala, 1986; Stiriling, 1991). Fungi may contribute up to 80 per cent of the total microbial biomass in many soils. In nature, fungi continuously destroy nematodes in virtually all soils because of their constant association with nematodes in the rhizosphere. Some 70 genera and 160: species of fungi have been found associated with nematodes but only few of them are successful biocontrol agents.

The fungi involved in the biological control of nematodes are grouped into:

1. Endoparasitic fungi
2. Predacious fungi
3. Opportunistic fungi

Endoparasitic Fungi

Endoparasites do not produce extensive mycelium but exist in the environment as conidia, which infect nematodes by either adhering to the surface of the nematodes or being ingested. The conidia rapidly germinate and invade the entire nematode, with assimilate hyphae absorbing all the body contents. The life cycle is completed by the fungus breaking through the body wall of the host to form reproductive structures such as conidiophores which support conidia, or evacuation tubes from which zoospores are released (Figure 33.1). Two zoosporic fungi, *Catenaria auxiliaris* and *Nematophthora gynophila*, have a role in regulating population of *Heterodera* spp. Some other endoparasires which have some role in biocontrol are *Acrostaiagmus*, *Harposporium*, *Myzocytium*, and *Nematoctonus* etc.

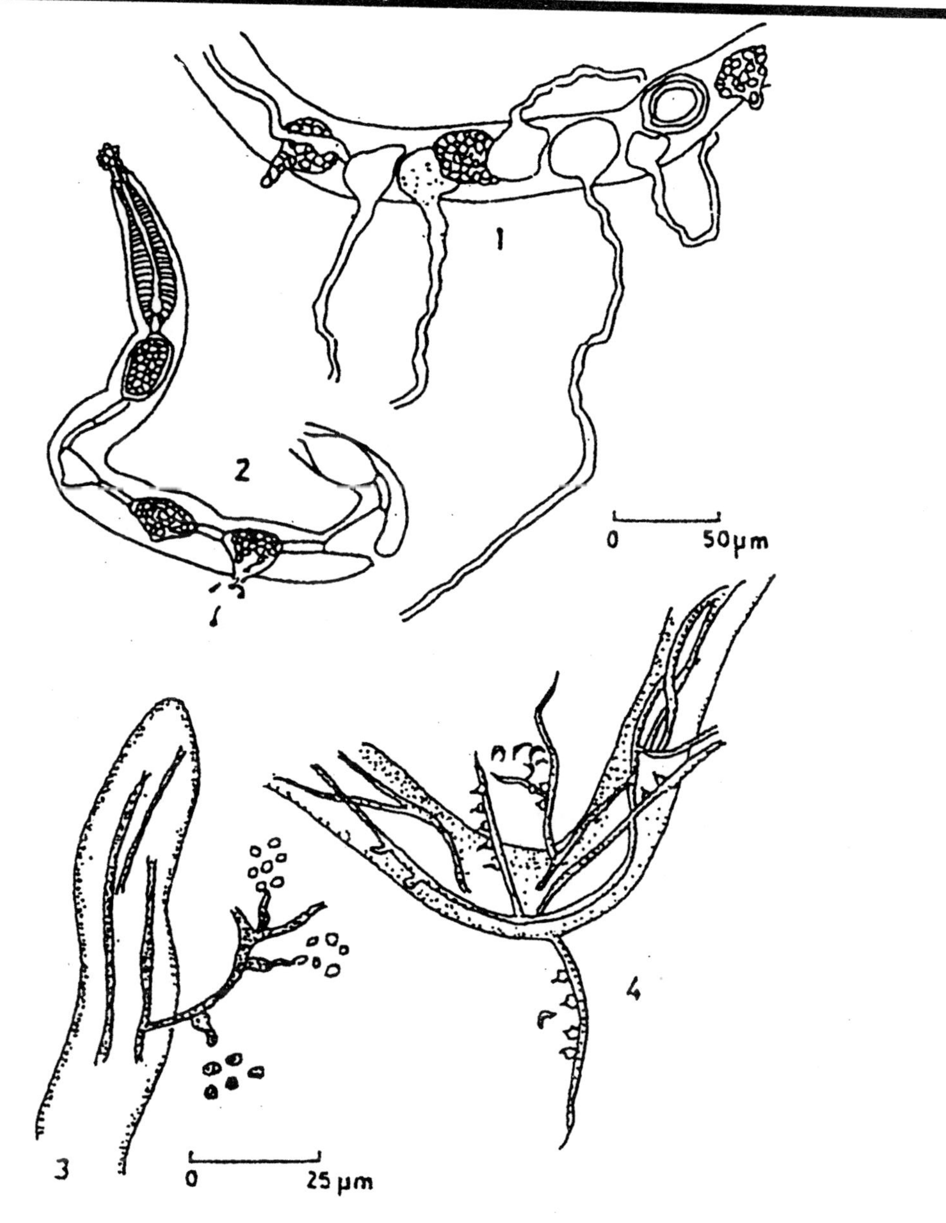

Figure 33.1: Endoparasites

(1) Nematode body with infection hypha of *Myzocytium humicola,* empty sporangia having long, contortuous evacuation tubes and an oospore; (2) Nematode infected at the month with zoospores of *Catenaria anguillulae.* The whole body full of sporagina and two of them filled with zoospores; (3) Conidiophorous hyphae of *Acrastalagmus obovatus* with sterigmata and conidia; (4) Conidiophores or *Harpasporium illliputanum* emerging from an infected nematode, with phialides and conidia.

Predacious Fungi

There are more than 50 species of predacious fungi which capture and kill nematodes in soils. With age, these fungi exhibit reduction in nematode-trapping efficiency. For effective control, it is necessary that their limited period of activity coincides with the period of nematode invasion of crop roots. Peak fungal activity is f reached only 12–15 days after introduction. Predatory fungi produce extensive hyphal network in the environment and produce trapping devices at intervals along each hypha. The traps capture nematodes either mechanically or by adhesion, allowing fungus to penetrate the nematode rapidly and digest the body contents. Predatory fungi form six different types of traps, such as unmodified adhesive hyphae (*Stylopage* spp.), three-dimensional adhesive nets (*Arthrobotrys* spp), two dimensional adhesive nets (*Arthrobotrys* spp), adhesive knobs (*Dactylella* spp.), constricting rings (*A. dactyloides* and *Dactylaria brochopaga*) and non-constricting rings (*Dactylaria candida*) to capture the nematodes (Figure 33.2 and 32.3). Despite the poor performance of some predacious fungi and biocontrol agents there are some examples of their success. Few developed a commercially prepared isolate of *Arthrobotrys* sp. to protect mushrooms against against *Ditylenchus myceliophagous* and this gave about 40 per cent control when used with compost. Another commercial preparation of an Arthrobotrys isolate gave good protection against root-knot nematodes on tomato. These results give better hope for their use in biocontrol of nematodes.

Opportunistic Fungi

Opportunistic fungi can colonize nematode reproductive and have the ability to affect them. Nematodes belonging to Heteroderid group (*Heterodera* and *Meloidogyne*) and at the sedentary stages of their life-cycle are vulnerable to attack by these fungi either within the *roots* or when exposed on the root surface or in the soil. Once in contact with cysts and egg masses, these fungi grow rapidly and eventually parasitize all eggs that are in early embryonic developmental stages. It is generally believed that these fungi have better role in biocontrol of nematodes than the two earliest groups discussed. Although large number of opportunistic fungi are known, *Paecilomyces lilacinus* and *Verticillium chlamydosporium* have been studied by number of workers (Tables 33.1 and 33.2).

Table 33.1: Effect of *Paecilomyces lilacinus* on Nematode Reproduction and Plant Growth

Nematode	*Crop*	*Effect P. lilacinus on Nematode*	*References*
Meloidogyne javanica	Cucumber	Reduction of nematode population	Stephan *et al.*, 1991
M. incognita *Rotylenchulus reniformis*	Water melon	Reduction of nematode population	Vincente *et al.*, 1991
M. javanica	Okra	Fungal improved plant top growth	Walia *et al.*, 1991
M. javanica	Chickpea	Fungus 2 g/pot one week and before nematode inoculation was effective	Zaki & Maqbool, 1991
M. incognita acrita	*Luffa aegyplica*	Reduction in root galling	Khan *et al.*, 1992
M. incognita race 3	Chickpea	Reduction in nematode multiplication	Siddiqui & Mahmood, 1992
M. incognita	Chilli	Reduction in galls, egg/egg mass and nematode population	Pandey and Trivedi, 1992
M. incognita	Brinjal	Reduction in nematode	Trivedi, 1992
M. incognita *R. reniformis*	Pepper	Fungus improve plant growth and fruit-yield	Vincente and Acosta, 1992
R. renifomis	Pigeonpea	Fungus found-superior than green manuring in reducing nematode multiplication	Mahmood and Siddiqui, 1993

Contd...

Table 33.1–Contd...

Nematode	*Crop*	*Effect P. lilacinus on Nematode*	*References*
M. incognita race 3	Chickpea	Fungus found superior than *Bacillus subtilis* in reducing nematode multiplication	Siddiqui and Mahmood, 1993
M. javanica	Tomato and okra	Significant reduction in nematode galling	Parveen *et al.*, 1993
M. incognita	Tomato	Reduction in nematode population	Eknayake and Jayasundra, 1994
M. javanica	Tomato	Fungus 4g/kg soil was effective	Zaki, 1994
M. incognita	Tomato	Fungus found superior than *Bacillus subtilis* in reducing nematode multiplication	Gautam *et al.*, 1995
Helerodera cajani	Pigeonpea	*P. lilacinus* was less effective than *Verticillium chlamydosporium* in reducing nematode population	Siddiqui and Mahmood, 1995
M. javanica	Chickpea	*P. lilacinus* along with *V. chlamydosporium* and *Trichoderma harzianum* suppressed nematode	Siddiqui *et al.*, 1996
M. javanica	Tomato	Organic amendments increased the parasitism of the fungus	Khan and Saxena, 1997
M. incognita	Okra	Fungus combined with VAM gave maximum nematode control	Sharma and Trivedi, 1997

Table 33.2: Effect or *Verticillium chlamydosporium* on Nematode Reproduction

Nematode	*Effect of V. chlamydosporium on Nematode*	*References*
Meloidogyne spp.	2000 propagules/g of the fungus reduced nematode multiplication	De Leij and Kerry, 1990
Heterodera schachtii	75 per cent control of nematodes	Crump, 1991
M. arenaria	Reducrion nematode multiplication	De Leij and Kerry, 1991
H. schachtii *Globodera pallida*	75 and 76 per cent reduction in eggs of first generation of both nematodes respectively	Crump and Irving, 1992
M. incognita *M. javanica, M. arenaria*	All nematodes infected with the fungus	De Leij *et al.*, 1992
H. schachtii	Fungal parasitism had influence on nematode reproduction	Muller, 1992
M. incognita, *M. hapla*	Nematode control was greater in peaty sand than in other soil types	De Leij *et al.*, 1993
M. incognita	Aerated soil layers favoured better nematode reduction	De Leij *et al.*, 1993
H. trifolil	23 per cent of young white females parasitized	Hay and Skipp, 1993
H. cajani	*V. chlamydosporium* was more effective than *P. lilacinus* in reducing nematode multiplication	Sidiqui and Mahmood, 1995
H. cajani	*V. chlamydosporium* with *Trichoderma harzianum* and *Glomus mosseae* caused highest reduction of nematode	Sidiqui and Mahmood, 1996
M. javanica	Cow manure and rice bran favoured fungal growth and reduced nematode multiplication	Machado and Campos, 1997
M. exiqua	Fungus cultured in wheat grain suppressed nematode eggs	Campos and Campos, 1997
H. avenae	8g/kg of fungus inoculum found effective against nematodes	Bharadwaj *et al.*, 1997

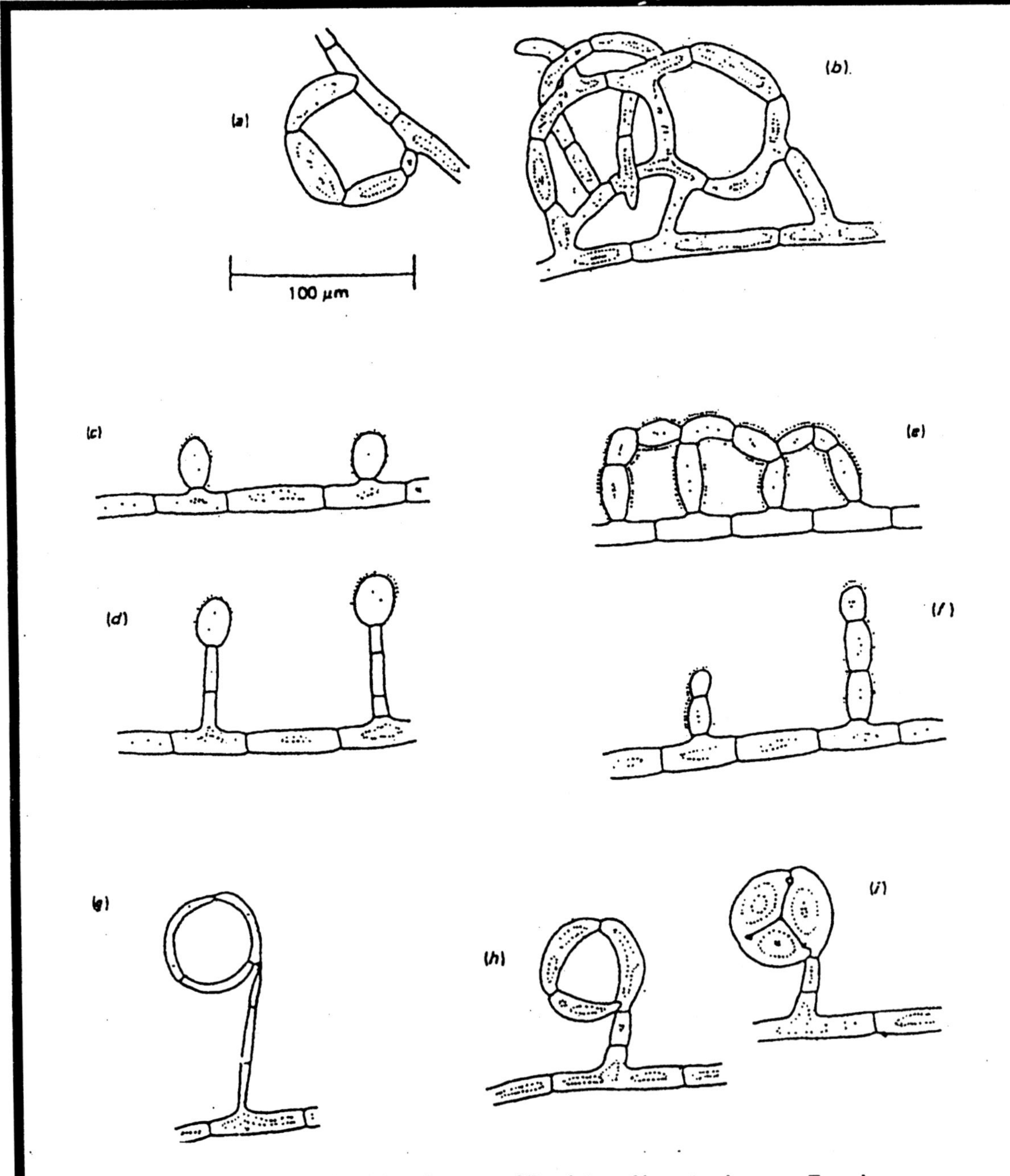

Figure 33.2: Trapping Organs of Predatory Nematophagous Fungi

Adhesive nets forming simple (*a*) and complex (*b*) three-dimensional traps; sessile (*c*) and stalked (*d*) adhesive knobs; and adhesive branches (*f*) often forming simple two-dimensions adhesive networks (*e*) non-constricting rings (g) and constricting rings open (*h*) and closed (*i*).

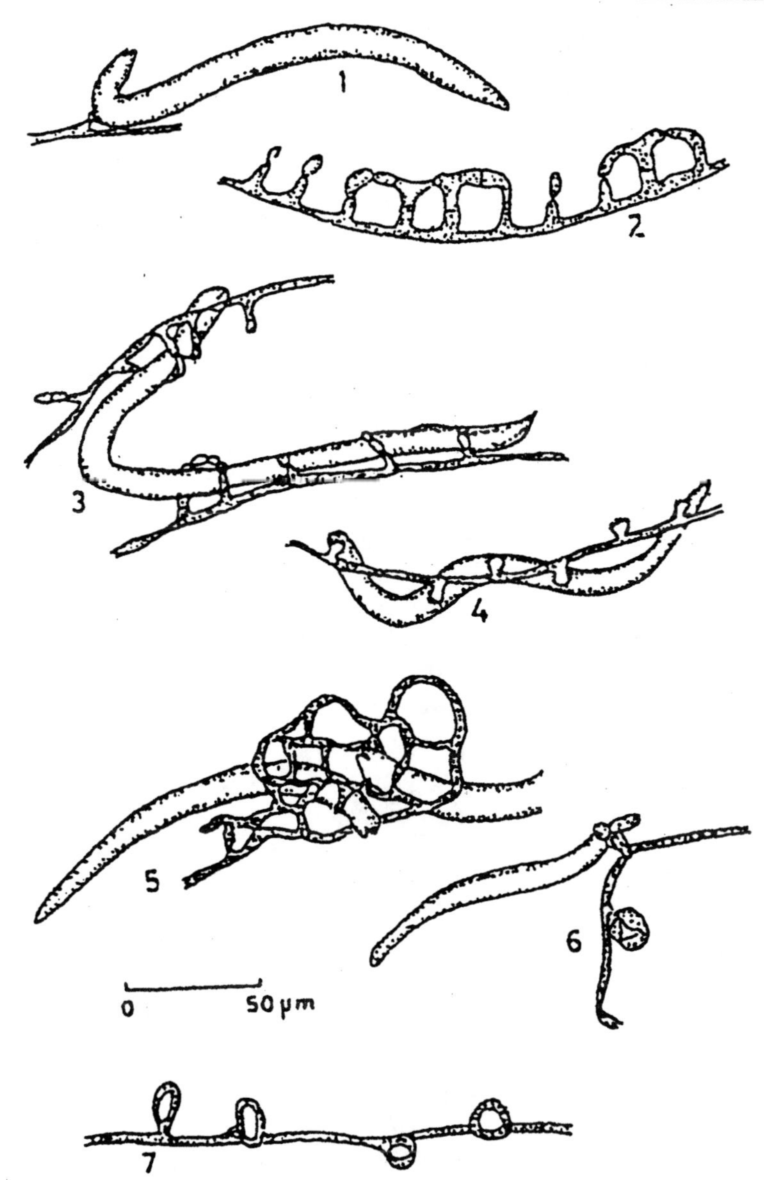

Figure 33.3: Adhesive Hyphae, Branches, Knobs, Net and Constricting Rings

(1) Nematode attached to the adhesive hypha of *Stylopage hadra* by means of globular protuberance at the point of contact. (2) Hyphae or *Monacrasporium gephyraphagum* producing adhesive columnar outgrowths connecting with each other to form scalariform network. (3) Nematodes attached to adhesive branches or *M. cionopagum.* (4) Nematode captured by adhesive knobs of *M. parivcollis.* (5) Nematode entangled in adhesive network of *Arthrobotrys musiformis.* (6) Nematode captured at head region in constricting ring of *A. brochopaga.* (7) Open 3 celled constricting rings of *A. brochopaga.*

Bacteria Affecting Plant-parasitic Nematodes

Pasteuria spp.

Pasteuria penetrans (Thorne) Sayre and Starr is an obligate parasite of some plant parasitic nematodes. It was first described by Thorne (1940) as the microsporidian *Duboscqia penetrans,* but was renamed *Bacillus penetrans* by Mankau (1975), who demonstrated that it was prokaryotic. However, the name *B. penetrans* was not officially accepted, and Sayre and Starr (1985) concluded that the organism was better placed in the genus *Pasteuria,* although they agreed with other authors that it had certain affinities with the Actinomycetales. *Pasteuria* spp. have been reported from over 135 different species of nematodes belonging to some genera, and from 37 countries on five continents (Sayre and Starr, 1988). At this time, however, only three species, *P. penetrans, P. thornei* Starr and Sayre, and *P. nishizawae* Sayte, Wergin, Schmidt and Starr, that parasitize nematodes, have been descried based on host range and morphological differences (Sayre *et al.,* 1991).

Nematodes become infected in soils when they contact the endospores, which adhere to their cuticles. Most studies of the life cycle of *P. penetrans* have been made in *Meloidogyne* spp., but similar developmental stages have been found in other nematodes. Infected *Meloidogyne* second stage juveniles enter roots and begin feeding before the spores germinate. A germ tube penetrates the cuticle and gives rise to a vegetative microcolony that fragments become filled with spores and egg production is prevented (Figure 33.4). Adult females and males may also be parasitized, but it is not clear whether the bacterium is able to complete its life cycle after such late infection. *Pasteuria penatrans* is very virulent and has reduced *Meloidogyne* populations in pots by 99 per cent in only 3 weeks (Birchfield

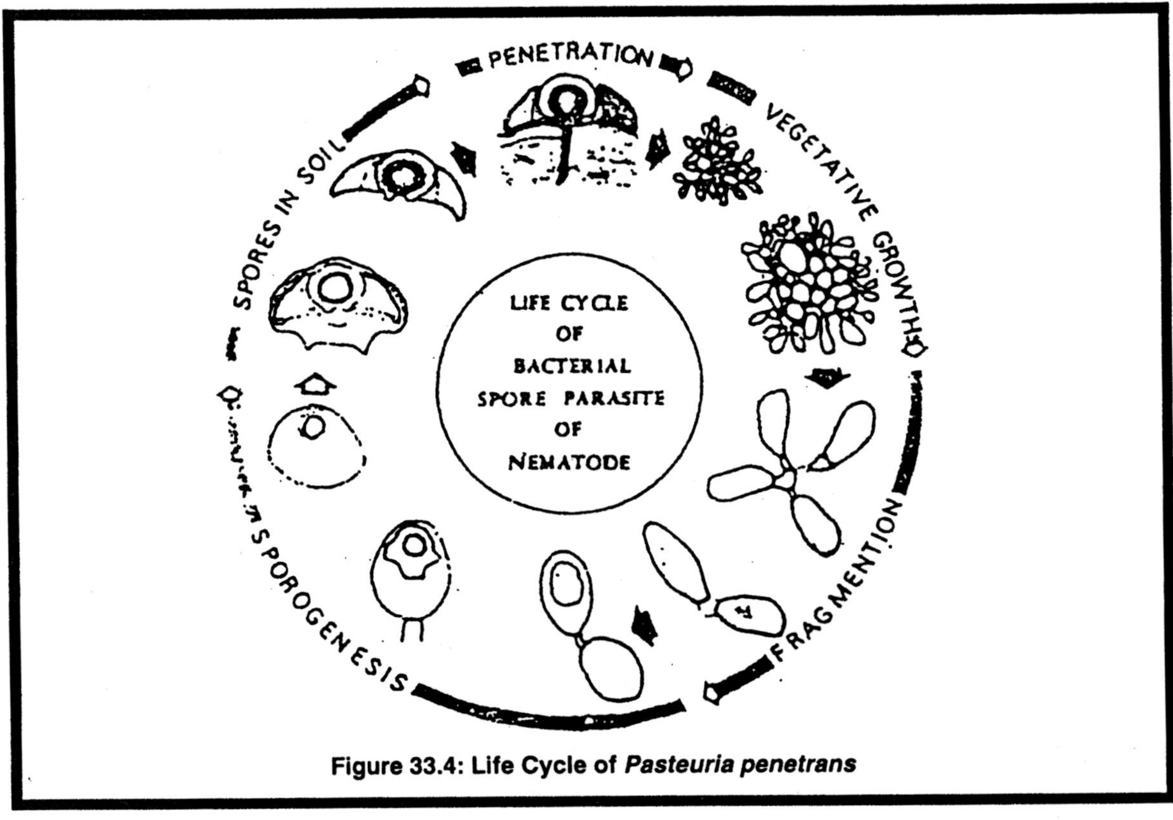

Figure 33.4: Life Cycle of *Pasteuria penetrans*

and Antonopouls, 1976). *Pasteuria panetrans* can survive several years in air dried soil apparently without loss of viability (Mankau and Prasad, 1977).

Pasteuria penatrans has not been cultured *in vitro* because of its obligate nature. Stiriling and Wachtel (1980) devised an *in vivo* method of producing *P. penetrans* spores: juveniles infected in water were added to soil around the roots of tomato plants, which were harvested after 7–8 weeks and the roots washed, air dried and milled into a powder that contained female nematodes filled with spores.

P. penatrans appears to have considerable potential as a biological control agent, but commercial exploitation is prevented by a lack of methods for large scale production. *P. penatrans* has most potential as a agent for inundation application where small quantities of soil need to be treated and where the organism can be thoroughly mixed, such treatments would probably be too expensive for field crops, even though, once established, control is likely to be long-lasting.

Rhizosphere Bacteria

Rhizosphere bacteria, mainly fluorescent *Pseudomonas* (Oostendorp and Sikora, 1989; Spiegel *et al.*, 1991) and certain others like *Bacillus subtilis* and *B. cereus* (Oka *et al.*, 1993), *B. sphaericus* (Recke and Sikora, 1992), *Arthrobacter* (Kloepper *et al.*, 1988) *Serratia* (Kloepper *et al.*, 1988) and *Agrobacterium* (Racke and Sikora, 1992) are antagonisric to nematodes. These bacteria colonize roots and express their antagonism to nematodes through one or more of the following mechanisms (Schippers *et al.*, 1987; Becker *et al.*, 1988, Oostendorp and Sikora, 1990; Westcott and Klupfel, 1993; Keuken- and Sikora, 1995; and Hoffman and Sikora, 1995).

Among the rhizobacteria, *P. fluorescens* has been investigated for its antagonism to nematodes more extensively than others and found effective against *Meloidogyne* spp. (Table 33.3).

Table 33.3: Interaction Rhizosphere Bacteria with Nematodes

Nematode	*Bacteria*	*Mechanism of Action*
M. incognita *M. arenaria* *R. reniformis*	*B. subtilis*	Repellent metabolites
M. javanica	*P. chitonolytica*	Blockage of lectin-carbohydrate recognition sites and production of toxic metabolites
M. incagnira	*Pseudomonas* spp. *B. subtills*	Nematicidal action against juveniles
Criconemella xenoplax	*P. aureofaciens*	Inhibition of egg hatch
Heteradera schachtii *Globodera pallida*	*P. fluorescens* > *A. radiobactor* >	Alteration of root exudates which, affect hatching and mortality of juveniles

Other Nematode Antagonistic Soil Bacteria

Soil bacteria like *Clostridium butyricum*, *Disulfovibrio desulficans*, *Chromobacterium* sp. and *Bacillus thuringiensis* var. *thuringiensis* producing butyric acid, hydrogen sulphide, cynamide and exotoxins respectively, have been demonstrated to be antagonistic to nematodes. Among these *B. thuringiensis* is.considered a potential biocontrol agent. *In vitro* studies established that the exotoxin produced by the bacterium during growth and sporulation had high nematicidal activity against juveniles and eggs of *Meloidogyne* spp. Delta endotoxins of *B. thuringiensis* var. *israeliensis* (strain *Bti*-H-14) and *B.*

thuringiensis var. *kurstaki* (*Btk*-HD-1), applied to soil showed an increase or decrease in oviposition and juvenile emergence from cysts of *Heterodera glycines* Field application of *B. thuringiensis* to tomato and pepper in Peurto Rico reduced galling in roots due to *M. incognita* and soil population of *Rotylenchulus reniformis* and increased yield significantly.

Predatory Nematodes

Predacious nematodes are the most studied of the predators, but even so, little is known of their effects on nematode populations. Predatory nematodes are found in four main taxonomic groups (Mononehids, Aphelenchids, Dorylaimids and Diplogasterids), with each group having quite different feeding mechanisms and food preferences. The modification of large buccal cavity equipped with large dorsal tooth or with small ventral teeth enable the nematode to swallow the prey or suck the body contents.

Methods of Isolating Some Antagonists of Nematodes from Soil

Isolation of Nematophagous Fungi

Drechsler's Soil Sprinkle Technique (Gray, 1987)

Sprinkle 0.5–Ig soil

↓

Water agar or weak cornmeal agar plate

↓

Inoculate 5000 nematodes

↓

Observe weekly for 6–8 weeks

Predatory fungi can be detected by the presence of trapped nematodes and by their distinctive conidiophores. Pure cultures can be established by transferring spores to a suitable medium (*e.g.*, PDA, Cornmeal agar etc.).

Baermann Funnel Technique (Barron, 1977)

Nematodes extracted from decanting and sieving method

↓

Nematodes spread on 2 per cent agar

↓

Examine endozoic fungi for 7–10 days

↓

Isolation and culturing in suitable medium (PDA, cornmeal agar, etc.)

Detecting Antagonist of Specific Nematodes

Nematodes extracted from decanting and sieving method should discarded and replaced with 500 target nematodes stained with chrysodin

↓

Incubation in soil

↓

Nematodes extraction

↓

Spread over on water agar and
Stained nematodes observed for parasitism for 10 days

Differential Centrifugation Technique (Nicolay and Sikora, 1988)

Centrifuge soil-water suspension for 2 min.

↓

Supernatant decant and again centrifuge at 4400 rpm for 1 hr

↓

Pellet mixed with few drops of water and spread over 2 per cent agar

↓

Nematode inoculation

↓

Observed for 7–10 days

↓

Isolation and culturing

Isolation of Fungal Parasites of Eggs and Saccate Females

Spread eggs/females over 2 per cent agar

↓

Examine for 24–48 hrs

↓

Transfer eggs/females showing fungal proliferation to PDA/Cornmeal agar with antibacterial antibiotic

Selective Media for *Paecilomyces lilacinus*

Medium of Mitchell *et al.* (1987)

Serial dilution of soil and 1 ml of an appropriate dilution of the soil-water suspension is pipetted into an empty sterile petri dish. The cooled selective medium is added and incubated at 25–27°C for 7–10 days. The medium contains,

NaCl .... 10 g

Pentachloronitrobenzene .... 50 mg

Benomyl .... 0.5 mg

PDA .... 39 g

Deionized water .... 1 lit. and after autoclaving add Chlortetracycline .... hydrochloride 50 mg and Tergitol 1 ml

PSM Medium (Cabanillas and Barker, 1989)

Distilled water .... 1 lit.

Streptomycin sulphate .... 50 mg

Chlortetracycline hydrochloride .... 50 mg

Dichloran .... 50 mg

Oxgall .... 250 mg

PDA .... 39 g

Spread known volume of soil-water suspension over the medium and incubate at 25–27°C for 7–10 days

Rose Bengal-chitin-iprodione Medium (Gaspard *et al.*, 1990)

Colloidal chitin .... 2 g

Rose Bengal .... 20 mg

Streptomycin sulphate .... 50 mg

K_2HPO_4 .... 1 g

$MgSO_4.7H_2O$ .... 0.5 g

Iprodione .... 50 mg

Agar 20g

Trace element 10 ml (1 gm each of $FeSO_4.7H_2O$, $MnSO_4$ and $ZnSO_4$ dissolved in 1 lit. to which 2 drops of conc. H_2ZSO_4 added)

Distilled Water .... 1 lit.

Soil is diluted serially with water and a 0.5 ml aliquot is placed on the medium and incubated for 7 days at 25°C.

Semi-selective Medium for *Verticillium chlamydosporium* (De Leij and Kerry, 1991)

Distilled water .... 1 lit.

Corn meal agar .... 17 g

Rose bengal .... 75 mg
NaCl .... 17.5 g
Streptomycin sulphate .... 50 mg
Aureomycin .... 50 mg
Chloramphenicol .... 50 mg
TritonX–100 .... 3 ml
Carbendazim .... 37.5 mg
Thiabendazole .... 37.5 mg

Soils are processed using standard serial dilution tech. and a 0.2 ml aliquot of the appropriate dilution is plated on the medium and incubated at 18–25°C and colonies observed after 5–15 days.

Rose Bengal-chintin + Benomyl Medium (Gaspard *et al.*, 1990) for *V. chlamydosporium*

The medium and incubation condition are the same as for the *P. lilacinus* medium except that iprodione is replaced with benomyl.

Identification of Nematophagous Fungi

It is not possible to identify nematophagous fungi by the condia alone, the trapping mechanism is also an important diagonistic character. Most of the isolation techniques used for the group rely on the fungus making itself known by capturing or infecting organs, and the dimensions of the conidia and conidiaphores are often different from those examined on baited plates. Because of this, the character used to identify these fungi are those found an nematode-infected cultures. An early key by Dallfus (1946) was extensively revised by Cooke and Godfrey (1964), and while this contains details of all the commonly found species, only 97 species are included, so another 50–60 species are not covered.

Major Steps in the Identification of Nematophagous Fungi by Cooke and Godfrey

Endoparasites with assimilative hyphae (conidia-forming species)

Endoparasites with vegetative or hyphal bodies (mainly species producing Zoospores)

Predators

Hyphae with clamp connexions

- present (*Nematoctonus*-adhesive cells formed)
- absent

Conidia borne on

- sterigmata (*Meria*)
- phialides

Conidia

- arcute, slightly curved, straight generally on subspherical phialides (*Harposporium*-ingested conidia)
- oblong, ellipsoidal, spherical generally on flask-shaped phialides (*Verticillium*, *Acrostalagmus*, *Cephalosporium*-adhesive conidia)

Unmodified hyphae (aseptate, septate)

Hyphal traps formed (aseptate, septate)

Adhesive branches

Adhesive knobs

conidiphore usually branched near apex, knobs always branched

conidiophore usually simple, knob sessile or branched

Non-constricting rings

Constricting rings

conidia borne in terminal cluster

conidia borne singly

Adhesive nets

I septa

> I septa

Conidia

not truly capitate

truly capitate

loosely capitate

Single conidium

Cluster of conidia.

34

Paecilomyces lilacinus: Fungal Bioagent for Controlling Phytonematodes

Archana Mittal and D. Prasad

Division of Nematology, Indian Agricultural Research Institute, New Delhi – 110 012

Paecilomyces lilacinus is cosmopolitan in distribution and occurs with greatest frequency in areas with a warm climate, particularly in the tropics and subtropics. Jatala *et al.* (1979) was the first to report *P. lilacinus* as a parasite of eggs of *M. incognita var. acrita* on potato roots in Peru. *P. lilacinus* has been frequently isolated from eggs of *M. arenaria* from peanuts (Godoy *et al.*, 1983) and from eggs of *M. incognita* and cysts of *H. glycines* from soybeans all in Alabama (Gintis *et al.*, 1982, Morgan Jones *et al.*, 1984). Kerry *et al.* (1982) reported the association of *P.lilacinus* with root knot nematodes in Europe. It was also isolated from the diseased eggs of *H. avenae* samples collected from Sweden. Stirling (1991) found that *P.lilacinus* and *V. clamydosporium* were the only fungal egg parasites recovered from root-knot nematodes in a survey of tropical and subtropical soils in Australia. Goswami and Rao (1997) isolated this fungus from *M. incognita* egg masses. Mizobutsi *et al.* (1999) observed *P.lilacinus* infecting the eggs of *H. glycines* in Brazil in soybean fields. Yang *et al.* (2000) isolated the fungus from nematode affected areas in Fuijian province, China.

Among soil hyphomycetes consistently associated with sedentary plant parasitic nematodes, *P. lilacinus* (Thom) Samson is known to be potential egg parasite of many economically important nematode pests of crops. There are certain attributes of *P. lilacinus* for which it is advantageous to use as biocontrol agent are:

1. It readily produces abundant inocula in the form of long chain of conidia.
2. It is a good competitor in most agricultural soils particularly in warmer regions. In addition, it is primarily saprophyte and able to compete for use a wide range of common organic substrates in soil.
3. It is a facultative parasite and can be easily mass cultured.
4. It is a chitin degrader and has strong proteolytic activity. These enzymatic activities are important because egg shell of *M. incognita* comprises of protein and chitin.
5. It has a strong competitive saprophytic ability because of its antagonistic activity against bacteria and fungi.

Classification

It belongs to kingdom–Mycota; Division–Eumycota; Sub-division–Deuteromycotina; Class–Hyphomycetes; Order–Turberculariales; Family–Aspergillaceae (Cannayane and Sivakumar, 2002). This was classified in the genus *Penicillium* Link, as *P. lilacinus* Thom, the species was reclassified in *Paecilomyces* Bainier by Samson (1970) on the basis of conidiophore morphology.

Morphology

The fungus has well developed septate and profusely branched hyaline mycelium. The septa have a central pore through which the cell contents migrate freely from one cell to another cell. The most important means of reproduction is formation of conidia and terminal chlamydospores. The conidiophores ramify in grouped branches or irregularly. The one celled conidia are separated from the phialides in the form of chains. Most of the conidia are prismatic and some are oval or short elliptical. The cell wall is made of chitin. The conidiospore is single celled and transparent.

Biological Mode of Action

This fungus has been reported to parasitize eggs of many sedentary endoparasitic nematodes. The infection process begins with growth of the fungal hyphae in the gelatinous matrix; eventually the eggs of *Meloidogyne* or *Tylenchulus* species are engulfed by the mycelial network, which becomes prostrate and spirals over the smooth egg surface. The proliferated hyphal branches penetrate the eggs. In cysts of *Globodera* species the fungus penetrates the vulva or the broken and exposed neck region. After entering the cysts the fungus grows saprophytically on the mucilaginous body content surrounding the eggs during or before its parasitism of the eggs. In all cases, eggs in the early embryonic developmental stages prior to the gastrulation process are more vulnerable to infection. Once the hyphae is in contact with the egg a series of ultrastructural changes occurs in the egg, shell due to the effect of exogenous metabolites and to the chitinolytic activities of the fungus. Morgan Jones *et al.*, 1984 studied the ultra structural variation of *P.lilacinus* infected *Meloidogyne arenaria* eggs and juveniles. When an egg became occupied by fungal hyphae, three changes in the egg shell structure occurred. The vitelline layer split into three constituent plasma membrane like brands which appeared to be unevenly thickened along their length. The two inner bands were more electron dense than the outer band. A large number of more or less isodiametric regularly spaced vacuoles appeared in the chitin layer, around each of which a very narrow electron dense zone was evident. The inner limits of the chitin layer became indistinct and the lipid layer largely disappeared.

Egg Parasitism

It was observed that *P. lilacinus* was an effective egg parasite of *M. incognita acrita* and *Globodera pallida* (Jatala *et al.*, 1979). This fungus has the greatest potential for application as a biocontrol agent in subtropical and tropical agricultural soils (Morgan-Jones *et al.*, 1984). Its association with nematode eggs was first reported by Lysek (1976). This fungus frequently infects nematode eggs and occasionally females of *M. incognita acrita*. The penetration of eggs and embryo destruction occurred within five days. Jatala *et al.* (1980) found that sixty-eight per cent of *M. incognita* egg mass infection and more than 50 per cent destruction of eggs.

The effect of protease and chitinase enzymes produced by *P.lilacinus* were observed on *M. hapla* eggs and 60 per cent undifferentiated eggs vacuolated and deformed were recorded. The eggs containing larvae emerged increased in number but less mobile. (Fitters *et al.*, 1992) Freire and bridge (1985) found that *P.lilacinus* can parasitize the eggs, females and juveniles of *M. incognita*. Jimenez and Gallo (1986) reported that this fungus infected eggs and females occasionally.

Croshier *et al.* (1985) recorded 76.25 per cent egg infection of *M. javanica* under in vivo condition. Sharma and Trivedi (1989) reported that *P.lilacinus* penetrated the eggs of *M.incognita* nematode and fed upon the internal contents leaving empty shells. The infected eggs were observed swollen than uninfected eggs. Oduor and Waudo (1995) found to have equal egg parasitism on the eggs of *M. incognita, M. arenaria* and *M. javanica*. Carnerio and Gomes (1993) found differences in in vitro pathogenicity among twelve *P. lilacinus* isolates to the eggs of *M. javanica* but all were egg parasitic.

Similarly, Lin *et al.* (1993) observed 54 per cent parasitization of *M. incognita* eggs under laboratory condition and 41.2 per cent on tomato plants inoculated with 8g *P.lilacinus* inoculum. Santos *et al.* (1992) observed isolate variation between *P.lilacinus* isolate No.4 and 50 for egg parasitism of *M. incognita* race 3 which showed 80 per cent and 3 per cent respectively. The egg parasitization caused by *P.lilacinus* differed in in vitro and in vivo conditions. Under in vitro condition the fungus parasitized 50 per cent eggs in egg mass of *M. incognita* but only 14.95 on the black pepper seedlings treated with fungus and nematode. (Freire and Bridge, 1985)

Khan and Saxena (1996) showed that *P.lilacinus* significantly controlled *M. incognita* followed by *M. arenaria* and *Rotylenchulus reniformis*. The egg masses were more vulnerable to fungal infection compared to eggs and females.

Effect of culture filtrate of *P. lilacinus*

Arai *et al.* (1973) isolated peptide antibiotics lencinostatin, lilacinin from *Penicillium lilacinum. P. lilacinus* exhibits chitinase (Okafor, 1967; Gintis, *et al.*, 1983) and proteolytic activity (Endreeva *et al.*, 1972). It also produces a peptidal antibiotic P168 which exhibited wide range of toxicity on fungi, yeast and gram-positive bacteria (Isogai *et al.*, 1980a, 1980b, 1981).The embryonated eggs treated with culture filtrates showed morphological changes in eggs and within2-4 days, development of eggs stopped, embryo disintegrated and vacuole formed (Fitters *et al.*, 1993). The death of nature eggs, colonized internally was due to fungal metabolites (Carneiro and Gomes, 1993). Khan *et al.* (1988) observed inhibition of hatching of *M. incognita* juveniles at different concentrations of fungal culture filtrates and complete inhibition was recorded at 50 per cent and 100 per cent concentrations. The filtrates suppressed the root galling of tomato plants caused by *M. incognita*.

Cayrol *et al.* (1989) reported nematicidal properties of culture filtrates of *P. lilacinus*. The toxin production were different on different media and their effect differed on various nematode species. The mechanism of toxic activity was considered to be neurotropic. Acetic acid was identified from culture filtrates of *P. lilacinus* and it affected the movement of nematode juveniles (Djian *et al.*, 1991).

Caroppo *et al.* (1990) found ovicidal activity of *P. lilacinus, P. maraquandi, P.variotii* and *P. coleus* against *M. inocgnita*. The culture broth of these fungi reduced 78 to 93 per cent egg hatching under excised root culture of tomato. Siddiqui and Hussain (1991) tested that culture filtrate of *P. lilacinus* was quite effective against *M. incognita* and *Macrophomina phaseolina* while Zaki (1994a) reported that culture filtrate of *P. lilacinus* had immobilizing effect on *M. javanica* and inhibited hatching of eggs. Khan and Khan (1992) reported that concentration of culture filtrates of *P.lilacinus* had direct relation to the inhibition of hatching and percentage mortality of M. incognita. Chen-Sen Yu *et al.* (1996) observed reduced hatching ranging between 21 per cent and 56 per cent was observed in *Heterodera glycines* by P.lilacinus. Oduor and Waudo (1995) also reported that *P.lilacinus, P. herbarum* and *Fusarium oxysporum-1* had suppressive effect on hatching of *Meloidogyne* spp. but hatched juveniles were not parasitized.

Sun-ManHong *et al.* (2002) studied the effect of *Paecilomyces lilacinus* on egg hatching and juvenile mortality of *Heterodera glycines*. The results showed that M-14 filtrates and its hyphal homogenate inhibited the nematode egg hatching more intensively than the IPC filtrate did when immersed in M-14 filtrate or its half-dilution solution, the juveniles became paralyzed. Only few juveniles could be recovered in distilled water when treated with M-14 filtrate for 24h. The mortality of the juveniles increased over time. These results demonstrates that M-14 metabolites had strong effects against the soybean cyst nematode.

Goswami and Mittal (2002) studied the effect of some fungal bioagents on root-knot nematode, *Meloidogyne incognita* infecting brinjal. In vitro studies showed that *Aspergillus niger* was highly toxic followed by *Aspergillus fumigatus* exhibiting 85 per cent mortality at 1 : 10 dilution while *P.lilacinus* and *Geotrichum candidum* showed least mortality. *A.niger* and *A. fumigatus* showed maximum inhibition of hatching and *G. candidum* the least. However, remarkably high egg parasitization was observed by *P.lilacinus* (75 per cent) and *G. candidum* (72 per cent). In vitro studies shows that *A. niger* and *P. lilacinus* along with mustard cake gave maximum reduction in root-knot nematode population both in soil and root.

Survival of *P. lilacinus* in Soil

Godoy *et al.* (1983) recovered *P.lilacinus* with high frequency from the inoculated soils at the end of the experiment. This showed its competitive and colonizing ability in soil. Carnerio and Cayrol (1991) investigated a commercial product (containing

10^{11} spores/g) of *P.lilacinus* against *M. arenaria* on tomato. The fungus significantly reduced the second and third generation of nematode. It was found most effective at spore density of 10^6/g soil. However, fungal density in soil rapidly decreased below the acceptable level of control.

Cabanillas *et al.* (1989a) observed that *P.lilacinus* viability was high in wheat grain and diatomaceous earth granule. The fungus provided 25 per cent increase in tomato yield and 25 per cent gall suppression in *M. incognita* infested field. It had also carried over protection to tomato against this nematode and provided three times higher fruit yield with 25 per cent galls suppression in the second year. The number of colony forming unit (CFU) per gm of soil increased 10 folds than the initial population inoculated at planting. Hewlett *et al.* (1988) evaluated *P.lilacinus* against *M. javanica* on tobacco and found that the fungus could not control the nematode though the average fungal propagules in soil was high (1.6-2.3 × 10^4/ g) for two years period. The highest fungal propagules was observed at 15 cm depth at the harvest and then declined below this level.

Management of Phytonematodes by *P. lilacinus*

Davide and Zorilla (1985) while investigating the biocontrol potential of *P. lilacinus* against *M. incognita* on okra found the fungus to be quite effective and economically better than nematicides.

According to Shahzad and Ghaffar (1987) carbofuran @ 1 kg a.i./ha was less effective than *P. lilacinus* against *M. incognita*. This fungus also effectively controlled nematodes on okra and mung in subsequent seasons but nematicide was no longer effective. Similarly, in the fungus treated plots, higher yields were recorded than carbofuran @ 2.5 kg a.i./ha. The efficacy of *P. lilacinus* had been comparable to that of most commonly used nematicides (Jatala, 1985). The fungus, *P. lilacinus* reduced the soil population of *M. incognita* by 66 per cent to 77 per cent whereas isazophos reduced it by 89 per cent (David and Zorilla, 1986).

Experimental results from the infested fields in different countries have indicated that *P. lilacinus* effectively controls *M. incognita* and even better than number of commonly used nematicides (Candanedo *et al.*, 1983; Jatala, 1983). The successful field results of *P. lilacinus* for biocontrol of *Tylenchhulus semipenetrans* on oranges in Peru have been reported. The fungus reduced *T. semipenetrans* population to relatively lower levels than nematicides tested. There was significant increase in root development, plant growth and fruit diameter with the inoculation of *P. lilacinus* (Jatala, 1983).

Dube and Smart (1987) obtained higher crop yields and reduced *M. incognita* population when *P. lilacinus* and *Pateuria penetrans* were applied together in field microlpots. Maheshwari amd Mani (1988) also found that simultaneous inoculation of *P. lilacinus* and *Pateuria penetrans* were more effective and capable of significant reduction of soil population of *M. javanica*. The effect of single and multiple application of *P. lilacinus* against *M. incognita* under field trials revealed that only one inoculation was sufficient to establish it and to get substantial degree of nematode control (Jatala *et al.*, 1981). However, Cabanillas and Barker (1989) indicated that effective biocontrol for *M. incognita* required more than one application and at proper time. The application of *P. lilacinus* prior to nematode inoculation resulted in reduced number of *M. incognita* juvenile population and low gall indices were recorded. (Romann and Rodriguez-Marcano, 1985).

The introduction of *P. lilacinus* 10 days before and at planting of tomato resulted in greatest protection against *M. incognita* and mid-season inoculation also provided higher percentage of eggs mass infection (Cabanillas and Barker, 1989). Khan and Esfahani (1990) found that *P. lilacinus* was more effective for controlling *M. javanica* infecting tomato when fungus was added prior to nematode or simultaneous inoculation. A great number of nematode eggs were infested, juvenile development inhibited, eggs devoid of juveniles and only filled with fungal mycelium. The developed juveniles were also found attacked and mycelial growth over their bodies were observed.

Tandingan and Assunction (1990) recorded reduced soil population of *M. incognita* and number of galls, egg mass on ramie roots with the introduction of *P. lilacinus*. Walia *et al.* (1991) reported that application of *P. lilacinus* through wheat bran carrier was better than carbofuran @ 1 kg a.i./ha in gall reduction on okra plants infecte with *M. javanica*. Vicente *et al.* (1991) evaluated *P. lilacinus* for controlling *M. incognita* and *Rotylenchulus reniformis* in infested fields of watermelon. The application of this fungus two weeks before planting was better than carbofuran or fenamiphos for controlling nematodes.

Pandey and Trivedi (1992) observed that *P. lilacinus* significantly reduced number of galls in *Capsicum annuum*, eggs per egg mass, final soil population and hatching of *M. incognita* eggs. Vicente and Acosta (1992) conducted field trial to compare the effect of *P. lilacinus* and carbofuran against *M. incognita* and *R. reniformis* on pepper. The inoculation of fungus one week before planting and application of carbofuran were equally effective for significant nematode reduction and obtained heavier fruit yield of pepper.

Sosamma *et al.* (1994) reported that application of *P. lilacinus* 25 days before *Radopholus similis* inoculation in *Piper betle* was effective in reducing nematode damage. The neem cake extracts (5 per cent and 10 per cent) were found very useful carrier for *P. lilacinus* and it effectively controlled *M. incognita* on brinjal (Rao and Reddy, 1994).

Walter and Barker (1994) investigated the efficacy of *P. lilacinus* both in greenhouse and field condition against *R. reniformis* infecting tomato. The results indicated that the fungus had detrimental effect on *R. reniformis* population development. This fungus was also effective in reducing the population of *Heterodera cajani* race A (Rana and Dalal, 1994), The higher growth and yield of mungbean was obtained in the fungus inoculated plants.

Davide and Zorilla (1995b) studied efficacy of *P. lilacinus* in farmtrial against many plant parasitic nematodes infecting potato. The fungus was found better than carbofuran @ 4.5 kg/ha. The application of fungus as tuber-dip treatment in spore suspension for 10 min and mycelial suspension at planting proved better than soil-mix application. The higher yield of potato and healthier crop was obtained in *P. lilacinus* treated plots. Mittal *et. al.*, (1995) found enhanced suppression of *M. incognita* while applying *P. lilacinus* with chitin. Geneeralao and Davie (1995) proved that *P. lilacinus* was more effective and comparable with nemacur (fenamiphos 10G at 200 ppm) against *R. similis* infecting banana. Similarly, David and Zorilla (1995a) found that *P. lilacinus* was comparable with ethoprophos and carbofuran but not with fenamiphos.

Noe and Sasser (1995) studied the efficacy of *P. lilacinus* for controlling *M. incognita* under field conditions and higher vegetable yield was obtained in fungus treated plots. The cumulative yield data of tomato and okra showed increasing differences between fungus treated and control plots with the progress of the season. The lower juvenile population was also found and the fungus provided same level of nematode suppression as the nematicide. Oduor et. al., (1996) observed better plant performance and low gall index with the application of *P. lilacinus* in combination with aldicarb or nematicidal plants.

Khanna (2000) studied the biomanagement of *Meloidogyne incognita* by *Paecilomyces lilacinus* on egg plant. An inoculum level of 3.6 g fungus per 600 g soil was estimated as the optimum level for keeping the root-knot gall index under check and enhancing growth of aubergine. The best plant growth and minimum root gall index was achieved when fungus was inoculated 15 days earlier than nematode inoculation but statistically at par with the treatment in which inoculations of fungus and nematode were done simultaneously.

Nagesh *et al.* (1997) studied the integrated management of *Meloidogyne incognita* on tuberose using *Paecilomyces lilacinus* in combination with plant extracts. Combination of *P.lilacinus* (at 2×10^4 spores/ml) with 5 per cent neem leaf extract resulted in significantly higher fresh plant weight and flower yield. Root gall index was least under *P.lilacinus* plus neem leaf extract (5 per cent) combination followed by *P.lilacinus* plus castor leaf extract (5 per cent) treatment. Although per cent egg and egg mass parasitization by *P.lilacinus* was higher when integrated with the leaf extracts, neem leaf extract improved the parasitization by *P.lilacinus* more than that with castor leaf extract.

It was also found that the addition of *P.lilacinus* @ 4g (on rice grain husk medium) per kg soil with or without 2 per cent (w/w) neem or mustard oil cake extract was effective in reducing *M. incognita* on tomato. The efficacy of *P.lilacinus* or oilcakes in controlling *M. incognita* declined sharply on the succeeding brinjal crop. (Saikia *et al.*, 1999). Cannayane and Sivakumar (2001) found that tomato plants whose roots were treated with *P.lilacinus* spores (1×10^6, 4×10^6 and 8×10^6 spores/ml) showed reduced gall formation (32.7, 33.7 and 59.2 per cent reduction). Similarly, the egg parasitism rate and the number of juvenile nematodes in the soil decreased with the increase in the number of fungal spores.

P.lilacinus was found to be effective against root-knot nematode in betelvine. At 4 g/kg soil inoculum level of the fungus, the plant growth and yield of *Piper betle* were increased, and the root-knot infestations decreased. The fungus and Furadan 3 G were found to be compatible (Hazarika *et al.*, 2000).

Conclusion

The success of biocontrol agents in the management of plant diseases and nematodes are gaining importance worldwide. This opportunistic fungus that infects the eggs and females are found abundantly in the rhizosphere. It is strongly proteolytic and chitinolytic. Apart from being nematicidal and nematostatic it has antagonistic activity against many pathogenic fungi and bacteria. Some isolates have been reported to cause infection in animals and human beings. There is a great need to standardize the dosage, time and method of application of this fungus. In future our research on this fungus should be focused on the role of soil environment on their action and efficiency. It is also found that the effect is more when this fungus is used in combination along with the toxic fungus like *Trichoderma spp.* or *Aspergillus spp.* in the INM programme. This is an environmentally friendly alternative to chemical pesticides. More focus on mode of action of this fungus against plant parasitic nematodes will be useful in exploiting its biocontrol potential.

REFERENCES

Arai, M., Mikami, Y., Fukushima, K., Utsumi, T. and Yazawa, K., (1973). A new antibiotic, leucinostatin, derived from *Penicillium lilacinum*. *J. Antibiot.* 26: 157-161.

Bansal, R. K., Walia, R. K. and Bhatti, D. S. (1988). Evaluation of some agro-industrial wastes for mass production of nematode parasitic fungus, *Penicillium lilacinum*. *Nematol. Medit.* 16: 135-136.

Bilai, V.I., Pidoplichko, N. M. and Dymovich, V. A. (1964). Antibacterial properties of *Pencillium* species from the rhizosphere of agricultural plants. *Mykrobiol. Zh.* 26: 31-37.

Bonants, P. J. M., Fitters, P. F. L., Belder, E-den, waalwijk, C., Henfling, J. W. D. M. and Den, Belder-E. (1995). A basic serine protease from *Paecilomyces lilacinus* with biological activity against *Meloidogyne hapla* eggs. *Microbiology Reading* 141: 775-784.

Brian, P. W. and Hemming, H. G. (1947). Production of antifungal and antibacterial substances by fungi, preliminary examination of 166 strains of Fungi Imperfect. *J. Gen. Microbiol.* 1: 158-167.

Cabanillas, E. and Barker, K. R. (1989). Impact of *Paecilomyces lilacinus* inoculllum level and application time on control of *Meloidogyne incognita* on tomato. *J. Nematol.* 21 : 115-120.

Cabanillas, E., Barker, K. R. and Daykin, M. E. (1988). Histology of the interactions of *Paecilomyces lilacinus* with *Meloidogyne incognita* on tomato. *J. Nematol.* 20 : 362-365.

Cabanillas, E., Barker, K. R. and Nelson, L. A. (1989a). Survival of *Paecilomyces lilacinus* in selected carriers and related effects on *Meloidogyne incognita* on tomato. *J. Nematol.* 21 : 121-130.

Cabanillas, E., Barker, K. R. and Nelson, I. A. (1989b). Growth of isolates of *Paecilomyces lilacinus* and their efficacy in biocontrol of *Meloidogyne incognita* on tomato. *J. Nematol.* 21 : 164-172.

Candanedo, E.., Lara, J., Jatala, P. and Gonzales, F. (1983). Control biologico del nematodo *Meloidogyne incognita* con el hongo *Paecilomyces lilacinus* Abstr. XXIII Meeting, APS Caribbean Division.

Cannayane, I. and C. V. Sivakumar (2001). Nematode egg–parasitic fungus 1. *Paecilomyces lilacinus*–A review. *Agricultural Reviews*, 22 (2): 79-86.

Carneiro, R. M. D. G. and Cayrol, J. C. (1991). Relationship between inoculum density of the nematophagous *Paecilomyces lilacinus* and control of *Meloidogyne arenaria* on tomato. *Revue Nematol.* 14 : 629-634.

Carneiro, R. M. D. G. and Gomes, C. B. (1993). Methodology and pathogenicity tests, effect of *Paecilomyces lilacinus* and *P. fumosoroseus* isolates to eggs of *Meloidogyne javanica*. *Nematologiaa Brasileris*. 17 : 66-75.

Caroppo, S., Perito, B. and Pelagatti, O. (1990). In vitro evaluation of nematicide activity by several fungi against *Meloidogyne incognita* eggs. *Redia*. 73 : 451-462.

Cayrol, J. C., Djian, C. and Pijarowski, L. (1989). Study of the nematicidal properties of the culture filtrate of the nematophagous fungus, *Paecilomyces lilacinus Revue Nematol.* 12 : 331-336.

Croshier, R., Montecinos, G., Jimenez, M. and Gallo, P. (1985). Effectiveness of *Paecilomyces lilacinus* Thom in the control of the root-knot nematode *Meloidogyne javanica* (Trueb, 1885) Chitwood, 1949. *Int. Nematol. Network Newsl.* 2 : 3.

Davide, R. G. and Batino, E. (1985). Biological control of roo-knot nematodes on cotton through the use of fungi *Paecilomyces lilacinus* (Thom) Samson and *Gliocladium roseum* Bainler as seed treatment. *Phil. Agric.* 68: 159-167.

Davide, R. G. and Zorilla, R. A. (1986). Evaluation of a fungus *Paecilomyces lilacinus* for the biological control of root-knot nematodes *Meloidogyne incognita* on okra compared with a nematicide isazofos. *Int. Nematol. Network Newsl.* 3: 32-33.

Davide, R. G. and Zorilla, R. A. (1995a). Evaluation of *Paecilomyces lilacinus* and some nematicides for the control of potato cyst nematode *Globodera rostochiensis*. *Biocontrol* 1 (3): 69-76.

Davide, R. G. and Zorilla, R. A. (1995b). Farm trials by potato growers on the use of *Paecilomyces lilacinus* fungus for the biological control of potato cyst nematodes and other species. *Biocontrol* 1 (4): 63-75.

Djian, C., Pijarowski, L.,Poncheet, M. Arpin, N. and Favre-bonvin, J. (1991). Acetic acid : a selective nematicidal metabolite from culture filtrates of *Paecilomyces lilacinus* (Thom) Samson and *Trichoderma longibrachiatum* Rifai. *Nematologica*. 37: 101-112.

Dube, B. and Smart Jr., G. C. (1987). Biological control of *Meloidogyne incognita* by *Paecilomyces lilacinus* and *Pasteuria penetrans*. *J. Nematol.* 19: 222-227.

Eguiguren-Carrion, R. (1995). Control of *Meloidogyne incognita* and *Nacobbus* sp. with *Paecilomyces lilacinus* in the greenhouse and effect of nematicides on the fungus. *Biocontrol*. 4: 41-49.

Endreeva, N. A., Ushakova, V. I. and Egorov, N. S. (1972). Study of proteolytic enzymes of different strains of *Paecilomyces lilacinus* Thom in connection with their fibrinolytic activity. *Microbiological* 41: 364-368.

Fioretto, A. M. C. and Villacorta, A. (1989). Thermal requirements for the development of the nematogentic fungi *Paecilomyces lilacinus*. *Pesquisa Agropecuria Brasileira* 24: 975-978.

Fitters, P. F. L., Belder, E-den and Den, Belder E. (1992). Effect of fungal products on nematodes. *Gewasbescherming* 23(1): 16.

Fitter, P. F. L., Belder, E-den and Den, Belder E. (1993). A time-lapse technique to study the effect of fungal products on embryogensis of nematode eggs. Mededelingen Van de Faculteit Laanbouww-etenschappen, Universiteit-Gent. 58: 751-756.

Freire, F. C. O. and Bridge, J. (1985). Parasitism of eggs, females and juveniles of *Meloidogyne incognita* by *Paecilomyces lilacinus* and *Verticilllium chlamydosporium*. *Fitopatologia Brasileira* 10: 577-596.

Generalao, L. C. and Davide, R. G. (1995). Evaluation of biological control efficiency of three fungi grown in different substrates against *Radopholus similis* on banana. *Biocontrol* 1: 35-43.

Gintis, B. O., Morgan-Jones, G. and Rodriguez-Kabana, R. (1983). Fungi associated with several developmental stages of *Heterodera glycines* from an Alabama soybean field soil. *Nematropica* 113: 201-213.

Godoy, G. R., Rodriguez-Kabana and Morgan-Jones, G. (1983). Fungal parasites of *Meloidogyne arenaria* egg in an Alabama soil. Amycological survey and greenhouse studies. *Nematropica* 13: 201-213.

Goswami, B. K and Mittal, A. (2002). Effect of some fungal bioagents on root-knot nematode, *Meloidogyne incognita* infecting brinjal. *Pakistan Journal of Nematology*, 20: 1, 5-59.

Goswami, B. K. and Uma Rao. (1997). Studies on different isolates of *Paecilomyces lilacinus* collected from different agroclimatic regions in India. *Indian Journal of Nematology*, 27(2): 238-240.

Hazarika, K., Dutta, P.K., Saikia, L and Deka, S. C. (2000). Biological control of root-knot nematode in betelvine using parasitic fungus, *Paecilomyces lilacinus*. *Crop Research*. 19: 2, 338-342.

Hewlett, T. E., Dickson, D. W., Mitchell, D. J. Kannwischer-Mitchell, M. E. and Mitchell, M. E-Kannwischer (1988). Evaluation of *Paecilomyces lilacinus* as a biocontrol agent of *Meloidogyne javanica* on tobacco. *J. Nematol*. 20: 578-584.

Isogai, A., Suzuki, A., Higashikawa, S., Kuyama, S. and Tamura, S. (1980a.) Constituents of a peptidal antibiotic P 168 produced by *Paecilomyces lilacinus* (Thom)Samson, *Agr. Biol. Chem*. 44: 3029-3031.

Isogai, A., Suzuki, A., Higashikawa, S. and Tamura, S. (1980b). Structure of peptidal antibiotic P 168 produced by *Paecilomyces lilacinus*. *Agr. Biol. Chem*. 44: 3033-3036.

Isogai, A., Suzuki, A., Higashikawa, S., Kuyama, S. and Tamura, S. (1981). Isolation and biological activity of a peptidal antibiotic P 168. *Agr. Biol. Chem*. 45: 1023-1024.

Jatala, P. (1983). Potential and prospects of biological control of nematodes. Abstr. XXIII meeting APS Caribbean Division, 43.

Jàtala, P. (1983). Biological control of nematodes. In An Advances Treatise on *Meloidogyne*. Vol. 1. Biology and Control. (Eds. Sasser, J. N. and Carter, C. C.) N.C.S. University Graphics, Raleigh, pp. 303-308.

Jatala, P. (1983). Biological control of plant parasitic nematodes *Ann. Rev. Phytopathol*. 24: 453-489.

Jatala, P., Kaltenbach, R. and Bocangel, M. (1979). Biological control of *Meloidogyne incognita* and *Globodera pallida* on potatoes. *J. Nematol*. 11. 303.

Jatala, P., Kaltenbach, R., Bocangel, M., Devaux, A. J. and Campos, R. (1980). Field application of *Paecilomyces lilacinus* for controlling *Meloidogyne incognita* on potatoes. *J. Nematol*. 12: 226-227.

Jatala, P., Salas, R., Kaltenbach, R. and Bocangel, M. (1981). Multiple application and long-term effect of *Paecilomyces lilacinus* in controlling *Meloidogyne incognita* under field applications. *J. Nematol*. 13: 445.

Jimenez, R. M. and Gallo, D. D. (1986). New contribution on effectiveness of *Paecilomyces lilacinus* (Thom) Samson in the control of root-knot nematodes (*Meloidogyne* spp.) IDESIA 10: 29-33.

Khan, M. W., Ashraf, S. S. and Haider, S. R. (1988). An analysis of biocontrol efficacy of *Paecilomyces lilacinus* in management of root-knot nematodes. Int. Conf. Res. Plant Sciences and its Relevance to the Future. Delhi. India p. 67.

Khan, M. W. and Esfahani, M. N. (1990). Efficacy of *Paecilomyces lilacinus* for controlling *Meloidogyne javanica* on tomato in greenhouse in India. *Pakistan J. Nematol.* 8: 95-100.

Khan, S. T. and Khan, T. A. (1992). Effect of culture filtrates of soil fungi on the hatching and mortality of root-knot nematode (*Meloidogyne incognita*) *Curr. Nematol.* 3: 53-60.

Khan, T. A. and Saxena, S. K. (1996).Comparative efficacy of *Paecilomyces lilacinus* in the control of *Meloidogyne* spp. and *Rotylenchulus reniformis* on tomato. *Pakistan J. Nematol.* 14: 111-116.

Khanna, A. S. (2000). Biomanagement of *Meloidogyne incognita* (Kofoid & white) Chitwood by *Paecilomyces lilacinus* (Thom.) Samson on egg plant. *Pest management and Economic Zoology* 8:2, 133-136.

Lin, M. S., Ma, M. A., Shen, S. W. and Mao, Q. F. (1993). Culture of the fungus *Paecilomyces lilacinus* and its use in the control of the tomato root-knot nematode, *Meloidogyne incognita*. *Chinese J. Biological Control* 9: 116-118.

Liu, X. Z., Liu, W. M. and Zhang, D. S. (1995). Biological characteristics of *Paecilomyces lilacinus* associated with soybean cyst nematodes. *Chinese J. Biological Control.* 11: 70-74.

Lysek, H. (1976). Auto dehelminthization of soil in lowland deciduous forests. Universitatis Paalackianae Olomucensis Facultatis Medicae. 41: 73-106.

Maheshwari, T. U. and Mani, A. (1988). Combined efficacy of *Pasteuria penetrans* and *Paecilomyces lilacinus* on the biocontrol of *Meloidogyne javanica* on tomato. *Int. Nematol. Network Newl.* 5: 10-11.

Mikami, Y., Yazawa, K., Fukushima, K., Arai, T., Udagawa, S. and Samson, R. A. (1989). Paeciltoxin production in clinical or terrestrial isolates of *Paecilomyces lilacinus* strains. *Mycopathologia.* 108: 195-199.

Mittal, N., Saxena, G. and Mukherji, K. G. (1995). Integrated control of root-knot diseases in three crop plants using chitin and *Paecilomyces lilacinus. Crop Protection* 14: 647-651.

Mizobusti, E. H., S. Ferraz, R. C. F. Ribeiro and M. Menezes. (1989). Isolation of fungi from *Heterodera glycines* eggs collected in different soyabean–producing areas of Brazil. *Nematologia Brasileira* 23(2): 69-75.

Morgan-Jones, G., White, J. F. and Rodriguez-Kabana, R. (1984a). Fungal parasites of *Meloidogyne incognita* in an Alabana soybean field soil. *Nematropica* 14: 93-96.

Nagesh, M., Reddy, P.P and Rao, M. S. (1997). Integrated management of *Meloidogyne incognita* on tuberose using *Paecilomyces lilacinus* in combination with plant extracts. *Nematologia Mediterranea* 25: 1, 3-7.

Saikia, M. K., Puzari, K. C and Saikia, U. N. (1999). An integrated approach to control root-knot nematode, *Meloidgyne incognita* on vegetables. *Annals of Biology* 15: 1, 81-84.

Noe, J. P. and Sasser, J. N. (1995). Evaluation of *Paecilomyces lilacinus* as an agent for reducing yield losses due to *Meloidogyne incognita. Biocontrol* 1: 57-67.

Oduor, O. P., Sikora, R. A., Waudo, S. W. and Schuster, R. P. (1996). Effects of aldicarb and mixed cropping with *Datura stramonium, Ricinus communis* and *Tagetes minuta* on the control and integrated management of *Meloidogyne javanica. Nematologica* 42: 127-130.

Oduor, O. P. and Waudo, S. W. (1995). Effects of fungi on hatching and parasitism of eggs, juveniles and females of root-knot nematode. *Pakistan. J. Nematol.* 13: 119-127.

Okafor, N. (1967). Decomposition of chitin by microorganisms isolated from temperate and tropical soil. *Nova Hedwigia* 13: 209-226.

Pandey, R. and Trivedi, P. C. (1992). Biological control of *Meloidogyne incognita* by *Paecilomyces lilacinus* in *Capsicum annuum*. *Indian Phytopath.* 45: 134-135.

Rana, B. P. and Dalal, M. R. (1994). Effect of different inoculum levels of *Paecilomyces lilacinus* in controlling *Heterodera cajani* race A on mungbean. *Ann. Pl. Protect. Sci* 2: 43-45.

Rao, M. S. and Reddy, P. P. (1994). A method for conveying *Paecilomyces lilacinus* to soil for management of root-knot nematodes on egg-plant. *Nematol. Medit.* 22: 265-267.

Roman, J. and Rodriguez-Marcano, A. (1985). Effect of the fungus *Paecilomyces lilacinus* on the larval population and root-knot formation of *Meloidogyne incognita*. *J. Agric. Univ. Puerto Rico* 69: 159-167.

Samson, R. A. (1974). *Paecilomyces* and some allied Hyphomycetes. *Stud. Mycol.* 6: 1-119.

Santos. M. A., Ferraz, S. and Muchovej, J. J.(1992). Evaluation of 20 species of fungi from Brazil for biocontrol of *Meloidogyne incognita* race 3. *Nematropica* 22: 183-192.

Shahzad, S. and Ghaffar, A. (1987). Field application of *Paecilomyces lilacinus* and furadam for the control of root-knot disease of okra and mung. *Int. Nematol. Network Newsl.* 4: 33-34.

Sharma, A. and Trivedi, P. C. (1989). Influence of inoculum levels of fungus *Paecilomyces lilacinus* (Thom) Samson on the biocontrol of root-knot nematode. *Meloidogyne incognita* (Chitwood) *Int. Nematol. Network Newsl.* 6: 27-29.

Siddiqui, Z. A. and Hussain, S. I. (1991). Control of *Meloidogyne incognita* and *Macrophomina phaseolina* on chickpea by fungal filtrates. *Pakistan J. Nematol.* 9: 131-137.

Sosamma, V. K., Geetha, S. M. and Koshy, P. K. (1994). Effect of the fungus, *Paecilomyces lilacinus* on the burrowing nematode, *Radopholus similis* infesting betelvine. *Indian J. Nematol.* 24: 50-53.

Stirling, B. R. (1991). Biological Control of Plant Parasitic Nematodes. C. A. B. International, U. K. pp. 196.

Sun-ManHong., Liu-XingZhong., Jin-Zhibo., Sun-MH., Liu-Xz and Jin-Zb. (2002). Effects of *Paecilomyces lilacinus* on egg hatching and juvenile mortality of *Heterodera glycines*. *Acta Phytophylacia Sinica* 29: 1, 57-61.

Tandingan, J. C. and Asuncion, E. V. (1990). Biological control of root-knot nematodes in ramie using *Paecilomyces lilacinus* (Thom) Samson. USM-Research and Development. 1: 18-24.

Vicente, N. E. and Acosta, N. (1992). Biological and chemical control of nematodes in *Capsicum annuum*. *J. Agric. Univ. Puerto Rico.* 76: 171-176.

Vicente, N. E., Sanchez, L. A. and Acosta, N. (1991). Effect of granular nematicides and the fungus *Paecilomyces lilacinus* in nematode control in watermelon. *J. Agric. Univ. Puerto Rico.* 75: 307-309.

Villanueva, L. M. and Davide, R. G. (1984). Influence of pH, temperature, light and agar media on the growth and sporulation of a nemtophagous fungus, *Paecilomyces lilacinus* (Thom) Samson. *Phil. Agric.* 67: 223-231.

Walia, R. K. and Bansal, R. K. (1992). Factors governing efficacy of nematode oviparasitic fungus, *Paecilomyces lilacinus* : effects of fungus inoculum level. *Afro-Asian Nematol. Network.* 1: 9-11.

Walia, R. K., Bansal, R. K. and Bhatti, D. S. (1991). Effect of *Paecilomyces lilacinus* application time and method in controlling *Meloidogyne javanica* on okra. *Nematol. Medit.* 19: 247-249.

Walter, S. A. and Barker, K. R. (1994). Efficacy of *Paecilomyces lilacinus* in suppressing *Rotylenchulus reniformis* on tomato. *J. Nematol.* 26: 600-605.

Yang Xiu Juan; He Yu Xian, Chen Fu Ru and Zhenglian. (2000). Isolation and selection of egg mass fungi of *Meloidogyne* sp. in Fujian Province. *Fujiann Journal of Agricultural Sciences,* 15(1): 12-15.

Zaki, F. A. (1994a). Effect of culture filtrates of *Paecilomyces lilacinus* on *Meloidogyne javanica. Nematol. Medit.* 22: 41-43.

35

Regulatory and Phytosanitary Measures for Nematode Management

Rajan and Arjun Lal

National Bureau of Plant Genetic Resources, New Delhi – 110 012

Plant introduction in the form of seeds, cuttings, bulbs, rhizomes, etc. has played and shall play a pivotal role in the development of agricultural and allied resources in the country. The prehistoric plant introduction activity led to domestication of plants in areas now recognized as centers of origin of crops. However, benefits derived from these introductions must be measured against the associated risks and their future impacts on agriculture production. It is well-established fact that when man moves plants, seeds and even goods from country to country, there is a distinct possibility that he may move pest or pathogen from its native habitat to a new location. No one can predict accurately, how the introduced pest would perform in its new environment and location in the event of its establishment in new location. There are examples, that in the absence of its natural enemies, competitors and availability of susceptible host plant, the pest (nematode) in its new environment has more serious effects than in its place of origin.

The threat of the accidental introduction of a major pest species, which could potentially cause massive problems for agriculture, forestry or public health, has encouraged governments to take steps to reduce the likelihood of such introductions occurring. Often this is achieved through the enforcement of what are generally referred to as quarantine regulations. Quarantine is one of the important tools for preventing the large scale spread of plant diseases. Although it can be expensive, under the present scenario of world trade and research needs, role of plant quarantine has become vital for the safe movement of seed and planting material. The various aspects of plant quarantine in relation to nematodes are discussed hereunder.

Need for Exchange of Plant Genetic Resources (PGR)

Plant germplasm exchange activity started with the introduction of new crops and economic plants. As plant science activities developed and plant scientists put forth new targets for breaking the yield barriers, it was increasingly felt that the basic requirement for a plant breeder is to have extensive gene pool/germplasm of diverse origin. Exchange of PGR is essential for following purposes:

1. For continued improvement in the yield of existing crops through various plant breeding methods;
2. For developing crop cultivars which are resistant to various biotic agents and abiotic stress conditions;
3. For introducing genotypes having potentiality for higher oil, protein, certain alkaloids and quality traits.
4. For introducing plant species which has ability to grow under harsh weather and soil conditions or a new source of food, fuel or fiber.
5. For conducting multi -location trials under the agreement with various international institutes, and with in the country under All India Crop Improvement Programs (AICIP).

During 1976-1998, about two million samples of different agri-horticultural crops meant for research were imported into India through National Bureau of Plant Genetic Resources (ICAR), New Delhi. Besides the introduction of plant material from other countries, a large amount of indigenous plant material is being collected every year from different parts of the country. For example the Division of Plant Exploration and Collection (NBPGR) undertook/organized/coordinated hundreds of explorations and collected thousands of accessions. This material is evaluated by growing at various places. An unaccounted number of plant materials and soil are being moved from one field to another and from one state to another state by farmers, extension workers, private and government nurseries without any inspection for associated pests and pathogens.

Early Records of Nematode Spread

Although mankind has been benefited by this exchange, undoubtedly several species of plant parasitic nematodes associated with the seed or planting material have been introduced to clean areas (Thorne, 1961; Bingefors, 1967; Southey and Aitkenhead 1972; Sanwal and Mathur, 1976; Lal and Mathur, 1988). Species that were previously capable of dispersal only over short distances by natural means can now be transferred inadvertently from country to country or from one geographical region to another by means of fast international transport links. For instance potato cyst nematodes (*Globodera rostochiensis* and *G. pallida*) were introduced into India (Nilgiri hills) with imported potato tubers from UK; soybean cyst nematode (*Heterodera glycines*) and pine wood nematode (*Bursaphelenchus xylophilus*) got introduced into Japan from USA; stem and bulb nematode (*Ditylenchus dipsaci*) has spread to most European countries with the exchange of infected *Medicago sativa* seeds. Not only species but also biotypes or races can be introduced with imported plant material. Indigenous collection of germplasm and their transfer to different centers for evaluation or multiplication without proper quarantine is helping in spread of some nematodes like *Radopholus similis* with banana and plantation crops; *Aphelenchoides besseyi* with paddy and foxtail millet seeds.

History of Plant Quarantine

In ancient time, quarantine was used for a widow who was permitted by law to remain in her deceased husband's house for a period of 40 days–after which she was required to pay rent to his heirs for the privilege of remaining longer. During the 13th century, ships entering Venetian ports were required to wait offshore 40 days before docking to prevent seamen or passengers possibly infected with plague or other diseases from entering the city. This period was considered adequate to permit latent cases of disease in the ship's crew to develop and detected before anyone was allowed to land. Quarantine gained importance/attention following the catastrophes which took place in Europe-Irish Blight of potato (1830), Powdery Mildew of grapevines (1847), *Phylloxera* of grapevines (1861), Downymildew of grapes (1875) and back rot of grapes (1888). The first plant quarantine law was promulgated in France in 1860 to suppress and prevent spread of the common barberry, the alternate host plant of wheat rust. Germany, France, Australia and USA were the first few countries to establish plant quarantine. California Pest Act was passed in 1881 to prevent the spread of grape *Phylloxera*. Plant Quarantine work in US may be said to have started in 1891 when California set up a seaport inspection at San Pedro, the first anywhere in the world. In 1914, first effort was made towards International Agreement on Plant Protection in Rome. This was followed by an International Convention of Plant Protection by over 50 countries in 1919 and certain agreements regarding the issue and acceptance of Phytosanitary certificate were finalized. In 1951, FAO sponsored an International Plant Protection Convention and under the article 3, of the convention the Plant Protection Agreement for South East Asia and Pacific Region was sponsored in 1956 and India became a party to this agreement. Indian Government agreed to adopt legislative measures specified in the convention for the purpose of securing common and effective action to prevent the introduction and spread of plant pests.

The Government of India had taken legislative steps as far back as 1914 for preventing the introduction of pests into India. An Act known as Destructive Insects and Pests Act, 1914 (DIP Act, 1914) was passed by the then Governor General of India in Council on the 3rd February, 1914. Plant Protection and Quarantine Organization (now known as Directorate of Plant Protection Quarantine and Storage) was established in May 1946.

Plant Quarantine: Definition and Principles

The Quarantine is a restriction imposed by duly constituted authorities, whereby the production, movement or existence of plants, plant products, animals, animal products or any other article or material or the normal activity of the person is brought under regulation, in order to prevent the introduction or further spread of a pest or pathogen. Quarantine is the use of legal or regulatory means to prevent material, in this case plant material from moving in an unrestricted fashion into or out of a specific area or jurisdiction. As used today, the term quarantine does not necessarily imply a holding period, although such a holding period may be required in certain cases. If movement of material is completely prevented the quarantine is referred to as embargo. More often, restrictions requiring certification of materials are imposed upon the importer or exporter of the materials. To make quarantine to be effective there must be laws and there must be mechanisms for their enforcement.

There are different reasons for imposing quarantines. Probably the most clear cut situation occurs when a plant pest has never been found in one region or jurisdiction and there is a desire to import plant material of the same type into that area. In this case it is relatively easy to justify a strong quarantine on the importation of plant materials. Further there are many types and levels of certifications that may be required by a jurisdiction for movement of plant material.

The objective of plant quarantine for nematodes is to prevent the entry of new nematodes and to prevent the further spread of those species which have gained entry (or are indigenous to the country) but have a limited geographical distribution in the country, thereby reducing or avoiding losses that would otherwise occur through the damage done by the pest (nematode) or through continuing cost of control measures. On the operational level, quarantine activities can be divided into two categories *viz.*, preventing introduction of exotic species (termed as international quarantine or external quarantine) and preventing further spread of species which have a limited geographical distribution (termed as domestic quarantine or internal quarantine).

A quarantine pest is defined in the International Plant Protection Convention as "pest of potential national economic importance to the country endangered thereby and not yet present there or present but not widely distributed and being actively controlled".

In terminology of the convention a pest is defined as "any form of plant or animal life, or any pathogenic agent, injurious or potentially injurious to plants or plant products". Therefore a pest may be fungus, bacterium, virus, weed, nematode, insect, mite or other harmful animal.

Features of DIP Act

1. It authorizes the Central Government to prohibit or regulate the import into India or any part there of or any specific place therein, of any article or class of articles.
2. It authorizes the officers of the customs at every port to operate.
3. It authorizes the Central Government to prohibit or regulate the export from a state or the transport from one state to another state in India of any plant materials, diseases or insects, likely to cause infection or infestation. It also authorizes the control of transport and carriage and gives power to prescribe the nature of documents to accompany such plants and plant material and articles.
4. It authorizes the State Government to make rules for the detection, inspection, disinfestations or destruction of any insect or class of insects or any article or class of articles, in respect of which the Central Government has issued notification. It also authorizes the state Government for regulating the powers and duties of the officers whom it may appoint on his behalf.
5. It provides penalty for persons who knowingly contrivers the rules and regulations issued under the Act.
6. It also protects the plant quarantine persons from any suit or prosecution or other legal proceedings for any thing done in the good faith or intended to be done under the Act.

Specific Requirements

Under the IPPC, member countries have reached international agreement on "specific requirements" which are the measures required from exporting country by importing country. These requirements are formulated in relation to the quarantine pest status. These requirements may include crop inspections, area freedom, disinfection treatments, prohibition, ease of detection during routine inspection, survival etc. These requirements are stringent for A 1 pests than for A 2 pests.

When plant germplasm is regulated by quarantine, two means of nematode management are utilized: exclusion and eradication. Under exclusion, the entry of a nematode is prevented by inspection and treatment, or the host is banned or otherwise restricted. Under eradication, infected or infested plants or plant parts are removed at the ports of entry, or the plant material is subjected to treatment or therapy. The tolerance level is zero for any exotic nematode species intercepted from imported plant quarantine.

Domestic Plant Quarantine

Domestic quarantine or internal quarantine is aimed to prevent the spread of introduced exotic species or indigenous key nematode pests to clean (pest free) areas within the country. India is a large country having vast geographical area, diversified agro climatic zones and varied cropping patterns. It is almost essential to strengthen the domestic quarantine regulatory measures so as to contain the spread of introduced pest and undertake effective timely eradication programs.

A study of the fauna and flora of any country would reveal that some of the pests and pathogens found are alien to that country. In-spite of the best precautions taken and stringent measures of quarantine applied, even the most advanced countries have not been able to prevent the entry of pests. In such cases man has taken the advantage of domestic quarantine. When a foreign pest is noticed, immediately the area is cordoned off and measures are taken to kill or eradicate the nucleus of the infestation and stop the further spread of pest. In India DPPQ and S takes necessary steps to regulate the inter- state movement of plants and planting materials in order to prevent the further spread of destructive pests which have already entered the country.

Some economically important plant parasitic nematodes are known to occur in some localized areas and are spreading to new areas with the movement of nematode infested seed or planting material. Due to farmers ignorance and absence of domestic quarantine, the burrowing nematode (*Radopholus similis*) has spread to most banana growing areas in the country. Also the seed borne nematode *Aphelenchoides besseyi* causing white tip disease in rice has spread to most rice growing areas by the movement of infested seed and its use for sowing.

The legislative measures to prevent the introduction and spread of destructive pests of crops are operative through the DIP Act, 1914. The domestic quarantine regulations are operative by the Central Government through powers vested under section 4A, B and D and Section 5 authorizes the State Governments to enact similar regulations and Section 5A provides for penalties.

The domestic quarantine regulation was issued by the Central Government in 1944 against fluted scale (*Icerya purchasi*) and San Jose scale (*Quadraspidiotus perniciosus*) in 1953. To prevent the spread of cyst nematodes of potato (*Globodera pallida* and *G. rostochiensis*) from Nilgiri hills of Tamil Nadu to other parts of India, the Central Government issued a notification in 1983. Other notifications issued by the Central Govt. to prevent the spread of introduced pests are given to Table 35.1.

Table 35.1: Domestic Quarantine Regulations Issued Against Introduced Pests in India

Sl.No.	*Pest*	*Year of Notification*	*Main host*	*Prevalent in Sstates*
1.	Fluted scale (*Icerya purchasi*)	1944	Apple	Himachal, J&K, Uttar Pradesh
2.	San Jose scale (*Quadraspidiotus perniciosus*)	1953	Apple	Himachal, J&K, Uttar Pradesh
3.	Codling moth (*Carpocapsa pomonella*)	1977	Apple and walnut	Ladakh (J&K)
4.	Coffee berry borer (*Hypothenemus hampei*)	1992	Coffee	Karnataka, Kerala and Tamil Nadu
5.	Bunchy top virus	1951	Banana	Assam, Kerala, Orissa, Tamil Nadu and W. Bengal
6.	Banana mosaic virus	1961	Banana	Maharashtra, Gujarat
7.	Potato wart disease (*Synchytrium endobiotiecum*)	1959, 1997	Potato	West Bengal,* Darjeeling
8.	Potato cyst nematodes (*Globodera pallida, G. rostochiensis*)	1983	Potato	Nilgiri hills (TN)
	Apple scab (*Venturia inacqualis*)	1977, 1978	Apple	Himachal Pradesh, J&K

* Also prevalent in Nepal and Bhutan

Limitations and Constraints of Domestic Quarantine

1. Lack of basic information on the occurrence and distribution of major key nematode pests in the country. In other words nematode distribution maps are lacking for most of the key nematode pests.
2. Absence of concerted action and enforcement of internal quarantine regulations by the state governments. India must develop organized services of plant quarantine at state level parallel to Australia and USA.
3. Lack of interstate border quarantine check-posts at rail and road lines greatly added to the free movement of planting material across the states.
4. Lack of close cooperation and effective coordination between state governments and center for timely notification of introduced pests, organizing pest detection surveys for delineating the affected areas and immediate launching of eradication campaigns in affected areas.
5. Lack of public support and quarantine awareness.
6. Lack of international support for developing rapid diagnostic tools/kits for quick detection/identification of exotic pests at the field level.
7. Lack of rigorous seed/stock certification or nursery inspection programs to make available the pest free seed/planting material for farmers.

Post-entry and Intermediate Plant Quarantine

Post-entry and Intermediate Quarantine services refer to growing, inspection and screening of plant propagating material in isolation, employing techniques that will (to the extent possible) preclude the entry of locally existing pests, prevent the escape of those pests which may accompany introduced planting material and result in the introduction of plant–propagating material guaranteed to be free from pests not known to occur within the importing country. The environment in which the plants are grown may be controlled partly on completely depending on existing season climate and plant growth requirements or it may be an isolated area (island) away from mainland *i.e.*, separated by natural barrier.

Post- entry quarantine facilities (environment controlled glasshouses or screen houses) are located in the campus of plant quarantine station or germplasm introducing organization of the importing country. Whereas intermediate plant quarantine station is located in a region where the plant propagating material under quarantine does not occur, so that any escaped pest (nematode) cannot find a favorable host for its establishment. Each plant species should be carefully reviewed before it is assigned to either post entry or intermediate quarantine.

International Quarantine Regulations: Fundamental Requirements

The establishment of plant quarantine measures should be based on certain fundamental prerequisites, which are as follows:

1. The measure should be based on sound biological principles.
2. They should not be used for the furtherance/barrier of the trade. The quarantine must be derived from adequate law and authority. Quarantine laws should be reviewed from time to time and modified on the basis of facts available.
3. Cooperation from both on international scale and from the public is required.

4. Quarantine measures can be effective only if those responsible for them are well informed.
5. Quarantine measures are only one of the facets of domestic pest management programs. A careful integration is needed to achieve the maximum effect.

Plant Quarantine: Legal Aspects

The legal basis for plant quarantine in plant introductions found in legislation's issued by national and state (or provincial) government is in the form of enabling legislation or rules and regulations. Standards or guidelines for promulgating these regulations are often suggested by the international organizations or intergovernmental agreements or conventions. These organizations include:

1. The Plant Protection Committee for South East Asia and the Pacific Region
2. Near East Plant Protection Commission
3. Caribbean Plant Protection Commission
4. European and Mediterranean Plant Protection Organization
5. Organismo Internacional Regional de Sanidad Agropecuaria
6. Comite Interamericano de Proteccion Agropecuria
7. Organizacion Bolivariana de Sanidad Agropecuaria
8. Inter-African Phytosanitary Council
9. North American Plant Protection Organization.

The organizations either are organized within the framework of the Food and Agricultural Organization (FAO) or cooperate within the framework of FAO. They suggest a minimum set of regulations that member governments should include in their regulations for the protection of the country and the region. The concept of protecting a bio-geographical area and not individual countries is a prerequisite to the success of quarantine activity.

In India, import of seeds, plants, planting materials and their products is regulated by the Rules and Regulations contained in the DIP Act, 1914. The Act has been amended or corrected a number of times, since then, the latest one has come in the form of Plant Quarantine Order, 2004. Under the revised act, plant species imported into India shall be accompanied with the official Phytosanitary Certificate containing additional declaration that the material is free from nematode pests specified against it (Table 35.2). As a general rule, modification of a quarantine regulation should be considered whenever it is necessary in the interest of the country.

In order to protect the agricultural and forest plants of our country, the Government of India has taken legislative steps as far as 1914 and established the Directorate of Plant Protection, Quarantine and Storage to implement the various regulations framed under the Order. This Directorate in turn, has established Plant Quarantine and Fumigation Stations at major seaports; airports and land frontier check posts. Indian custom is entrusted with implementation of many regulations under the Order. The custom officials are authorized to check plants/plant materials imported as accompanied baggage by international passengers at airports/seaports and land frontier check posts. In the declaration form, a statement "Plant and propagating materials are subjected to inspection and treatment on arrival" is included in the Disembarkation card to be submitted by all passengers. Plant material imported as accompanied baggage/cargo (air, sea or land) consignments as parcels are passed on to the Plant Quarantine Officers for detailed inspection and treatment. The quarantine inspection of the plant material imported for the purpose of research and consumption are dealt separately.

Table 35.2: Special Conditions for the Import of Plants, Fruits and Seeds with Respect to Nematodes under the Revised Regulation Order, 1989

Sl.No.	Plant Species	Nematode Species*	Special Condition for Import
1.	All species of *Allium* (Onion, garlic, leek, shallot etc)	Stem and bulb nematode (*Ditylenchus dipsaci*)	Nil
2.	Coconut seeds and plants (all species of *Cocos*)	Red ring nematode (*Rhadinaphelenchus cocophilus*)	1. Consignment can only be imported by Director CPCRI, Kasargod, Kerala 2. Imported material shall be grown in individual pots under post-entry quarantine isolation
3.	Potato (all species of *Solanum*)	Cyst nematodes (*Globodera pallida* and *G.rostochiensis*)	1. Consignment can only be imported by Director, CPRI, Shimla 2. Imported material shall be grown in Post Entry quarantine.
4.	Berseem (all species of *Trifolium*)	Stem and bulb nematode (*Ditylenchus dipsaci*)	Nil
5.	Beans (all species of *Vigna*)	Cyst nematode (*Heterodera schachtii*)	Nil
6.	Beet root	Beet cyst nematode (*Heterodera glycines*)	Nil
7.	Broad bean (*Vicia faba*)	Soybean cyst nematode (*Heterodera glycines*)	Nil
8.	Chrysanthemum	Foliar nematodes *Aphelenchoides fragariae, A. ritzemabosi* stem and bulb nematode (*Ditylenchus dipsaci*)	Nil
9.	Narcissus Daffodil etc.	Stem and bulb nematode (*Ditylenchus dipsaci*)	Nil
10.	Oat	Stem and bulb nematode (*Ditylenchus dipsaci*)	Nil
11.	Pea	Pea cyst nematode	Nil

* Nematode species which require additional declaration in Phytosanitary Certificate from country of origin.

Phytosanitary Certification (PC)

Phytosanitary certificate is a legal document issued by the government authorized officials of the country from where the plant material is being explored. Plant Quarantine Regulations of many countries have a number of measures in common.

1. Embargoes–specified prohibitions against import of plant species *viz.*, Import of skinned cotton seeds and Mexican jumping beans.
2. Grant exceptions to prohibitions for special purposes.
3. Require import permits from competent authority.
4. Require phytosanitary certificates and/or certificates of origin.
5. Stipulate inspection upon arrival and/or before dispatch.
6. Prescribe treatments to eliminate risks.
7. Prescribe isolation after entry, or other safeguards.

Theoretically, exclusion of the plant material on which a nematode is likely to enter is the most effective means of preventing establishment provided the nematode has no other means of spread. However, such prohibitions exclude trade and indispensable exchange of germplasm to increase genetic diversity of crops. In most cases exceptions to prohibitions may be achieved at least by quarantine after entry, in containment facilities like plant introduction stations where the material is regularly examined by specialists. Soil accompanying the plant material is undoubtedly a serious quarantine risk and is prohibited in most quarantine regulations. The importation of rooted plants, bulbs, rhizomes and tubers with adhering soil must always be regarded as major risk with regards to nematodes.

For most plant material, quarantine regulations stipulate entry requirements. The PC is generally required together with a description of the consignment including its origin. The IPPC (1951) International Plant Protection Convention stipulated that contracting Governments shall issue certificates conforming to FAO model which states:

"This is to certify that the plants, parts of plants or plant products described below, or representative sample of them were thoroughly examined on (date) by (name) an authorized officer of Plant Protection Service of (country) and were found to the best of my knowledge to be substantially free from injurious diseases and pests; and the consignment is believed to conform with the current phytosanitary regulations of the importing country both as stated in the additional declaration hereon and otherwise".

Additional declarations may refer to special security treatments (*e.g.*, fumigation) and freedom from soil and specified pests. The extent to which nematode risk is lowered by the phytosanitary certificate depends on efficiency of detection methods employed, expertise in accurate species identification and size of sample taken for inspection. Nevertheless, the PC is not meant to substitute for inspection upon arrival. Reliability of inspection can be increased by extending the period of observation before exportation or after entry in containment facilities. In case of nematodes, except few developed countries, phytosanitary certificates are issued without any nematologists. Hence, special emphasis is given for inspection and detection of nematodes in imported consignments received with/without PC, by nematologists.

Regulations on World Trade

The movement of articles in international trade is subject to a wide spectrum of rules, regulations, policies and guidelines implemented by government agencies of various nations. The principal objective of plant quarantine activities is the exclusion of exotic, quarantine significant plant pests and pathogens along man made pathways. International trade can be considered a two way street; the exports of one country are the imports of another country.

The regulatory activities and international trade are closely interwoven with pest risk analysis serving as an interface. The IPPC standards for Sanitary and Phytosanitary (SPS) Agreement are in place that provides the biological justification for regulatory activities, including the entry status of imported agricultural commodities. In India and more than 105 other countries, the legal basis for regulatory activities related to agricultural exports is the IPPC of 1951, as amended in 1952 and administered by FAO of United Nations. Under the SPS portion, FAO has been given the responsibility to develop international standards for PRA. Member countries, including India must be able to biologically justify their regulations and quarantine operations, particularly when they interfere with international trade.

Under the SPS Agreement member countries need to rationalize their phytosanitary risk management measures to ensure that they:

1. Are in harmony with international standards (Article 3)
2. Are scientifically justified (Article 5)
3. Are proportionate to the risk involved (Article 5.4)
4. Do not discriminate between members (Article 2.3)
5. Are similar or identical to those applied in their territory and to other members (Article 2.3)

Pest Risk Analysis

According to International Plant Protection Convention, "phytosanitary measures" against pest must be based on sound "Pest Risk Analysis" (PRA). In other words the risk to the importing country from the introduction of an exotic pest must be adequately/scientifically assessed by the process of PRA. Essentially, this consists of a verification of the quarantine pest status with regard to its geographical distribution and regulatory measures adopted followed by assessment of potential economic importance and potential for introduction, establishment and spread. This is followed by careful analysis of the phytosanitary measures which can be applied.

Approaches to risk assessments of exotic pests have been worked out and reviewed by several workers (Baker, 1981; Kahn, 1982, 1999; Khan, 1999 and Hopper, 1996). It is the biological uncertainty that dictates the need for development of the PRA. Although it is not possible to identify which non-indigenous organisms will create problems, it assumes that some will. PRA focuses on those organisms for which biological information is available. By developing detailed assessments for known pests under new environmental situations effectively mitigate measures to eliminate unknown organisms.

An exotic species of nematode will establish only when a susceptible host is present, the species is virulent and capable of reproducing and the environment is favorable. Risk assessment of exotic pests has been reviewed by several authors (Silvestro and Tacconi, 1988; Bianco and Talame, 1998; Phillips *et al.*, 1994).

Appropriate Level of Protection

The SPS Agreement, under WTO recognizes and maintains the right of countries to determine and set an "appropriate level of protection" (ALP) for all pest or disease threats. The SPS Agreement maintains the right of countries to determine their own/ALP. In setting the ALP, the Agreement requires countries to avoid arbitrary or unjustifiable distinctions in the levels of protection it considers to be appropriate in different situations, if such distinctions result in discrimination or a disguised restriction on international trade.

A conservative entry status might be adopted even for pests of low risk because the benefits from safe importation exceed the potential control costs of an inadvertent introduction. Alternatively a high risk and a liberal entry status might be employed by governments importing grain during periods of famine or food shortage. As in case onion shortage in recent past, when bulk of it was imported under OGL (Open General Licence) scheme. Internationally, regulations usually specify risk categories that safeguard with imported plant germplasm. If the pest is categorized as prohibited, the risks of its introduction are great, and the existing safeguards to prevent entry are considered inadequate.

Nematodes: As Quarantine Pests

Nematodes exist in large numbers in all sorts of habitats, still there are several of these which are not yet reported in many countries and if present, have a limited distribution. They feed on nearly every kind of plant and they are difficult to control or eradicate. The following are known to occur on plants, based upon their mode of spread/dissemination and quarantine strategies:

1. Seed gall nematodes, which may be carried with the seed.
2. True seed borne nematodes–in the seed/nuts in the husk.
3. The foliar and stem nematodes, which may be introduced with living, plant material, cut flowers etc.
4. The endo-parasitic nematodes–spread through rooted cuttings, plants, soil, under ground parts etc.
5. The cyst forming nematodes–are easily transported independently of their hosts particularly in soil.
6. All above–along with soil and packing material.

Categorization of Quarantine Pests

In order to contain the number of quarantine pests and facilitate the international trade, various plant protection organizations in consultation with member countries, have divided quarantine pests in to two categories: A 1 pests (which are not present in any part of the region); and A 2 pests (which are present in some areas of the region or country). All member countries are urged to take measures against A 1 pests whether or not they are of direct concern to them in order to protect the region as a whole, while the countries are free to take any action against A 2 pests which concern them. EPPO has categorized *Globodera pallida, G.rostochiensis, Heterodera glycines, Ditylenchus destructor, D. dipsaci, Bursaphelenchus xylophilus, Aphelenchoides besseyi* as A1 pests.

Nematode Dissemination

Nematodes are small microscopic animals (average body length less than 1 mm) and lack organs for movement. By themselves nematodes can move about 20- 30 cm in a year under optimum conditions. For their spread from one field to another or to distant places they require help of other agencies like wind, water or animals (including man). Amongst the pests which can be transported with the movement of infested soil or plant material, plant parasitic nematodes form an important group. For their long distance movement no other agency has contributed more than the man. With the modern fast means of transport, nematodes associated with planting material could reach to distant places with in a few hours which helps the nematodes to retain their energy required for feeding plants and get established in new areas. Some important bio-ecological factors responsible for dissemination and establishment of plant parasitic nematodes of quarantine significance are listed below:

In Transit Phase

1. Endo-or ecto parasitic nature of nematode in host tissue
2. Developmental stage of life cycle involved in transit
3. Amount of inoculum and frequency of transport
4. Type and condition of carrier (soil/root/seeds/leaves etc.)
5. Sensitivity to desiccation.

In Establishment Phase

1. Number and relative susceptibility of the host
2. Reproduce at a tremendous rate

3. Amount of inoculums and frequency and transport
4. Temperature requirements
5. Moisture requirements and availability
6. Mode of reproduction and fecundity
7. Survival mechanism
8. Capacity of adaptation to new conditions.

In Developmental and Economic Importance Phase

1. Host range, economic importance and intensity of host plant cropping
2. Reproductive potential in prevailing environmental conditions
3. Pathogenicity and interaction with other organisms
4. Land management practices used.

Nematode Detection Techniques

Nematode detection techniques used in regulatory nematology are same as used for general nematology work but they are used more carefully and with greater precision. The technique to be used depends upon part of plant to be examined for presence of nematodes. The most commonly used methods are as follows:

1. Visual examination of seeds or plant material for the presence of contaminants (soil clods, plant debris, seed galls etc.), shriveled or abnormal/deformed seeds.
2. Soaking the seeds in water for 24–48 hours and thereafter removing the seed coat to free nematodes lodged inside *viz. Aphelenchoides besseyi, Ditylenchus dipsaci.*
3. Visual examination of vegetatively propagated plant material for nematode symptoms on the roots such as galls, stubby roots, discoloration of roots etc. Nematodes are recovered from infected roots by dissecting needles or root incubation techniques.
4. Nematodes from small soil clods are recovered by soaking them and direct examination under stereoscopic binocular microscope and picking individual cyst or through decantation and sieving techniques, when the quantity is sufficient..

Above methods can be modified as per the sample size and requirements.

Nematodes Intercepted from Imported Plant Material

Imported seeds and planting material of agri-horticultural crops meant for research are routed through NBPGR, New Delhi/Hyderabad. During the period, 1980-2003, fifty five thousand samples were found infested or contaminated with plant parasitic nematodes. Detailed examination of imported samples by quarantine nematologists has yielded over 300 interceptions of nematodes comprising of 56 species belonging to 27 genera. Some economically important nematode species of quarantine significance, intercepted from imported seeds and planting material is listed in Table 35.3. Most of these interceptions have been made from consignments accompanied with official Phytosanitary Certificates. Thus quarantine examination has prevented introduction of several exotic nematodes in the country. However, it must be admitted that light infestations in large sized samples might have been overlooked and evaded detection (Latha *et al.*, 1999; Mani, 1998; Kehlenbeck, 1998; Hockland,

2000; Schiessendoppler, 1996). Hence regular surveillance of released material during crop growth period is necessary for early detection of escaped nematodes, if any.

Table 35.3: Some important Nematode Species Intercepted from Imported Seed/Planting Material at NBPGR

Nematode Species	*Seed/Plant Material*	*Country of Origin*
Anguina tritici	Wheat (soil clods)	Turkey, UK, Italy
Aphelenchoides besseyi	Paddy, *Digitaria smutsii*, *Oryza glabirima*, *Stylosanthes hamata*	Philippines, USA, Nigeria, Brazil, Madagascar, Burma, Indonesia, Australia, Costa Rica.
Aphelenchoides species	*Arachis* sp.; *Juglans* sp.; *Pyrethrum* sp.; Coconut; Red Palm; Aster	Brazil, USA, USSR, Syria, Canada, Holland, Australia, Indonasia
Bursaphelenchus species	Oil Palm (nuts)	Indonesia
Ditylenchus angustus	*Oryza sativa* (seeds)	Madagascar
D. destructor	Potato (tubers)	USA
D. dipsaci	*Alfaalfa*, garlic (bulbs) flower bulbs	USA, France, Holland
Heterodera goettingiana	Pea (soil clods)	UK
H. schachtii	Sugarbeet (soil clods)	UK
Rhadinaphelenchus cocophilus	Coconut (husk)	Guyana

* Species not yet reported from India

Nematodes of Quarantine Significance: An Analysis

An exotic species of nematode will establish only when a susceptible host is present, the species is virulent and capable of reproducing and the environment is favorable. Risk assessment of exotic pests has been reviewed by several authors. They all draw attention to the indispensable need for reliable information on:

1. The vulnerability of economicly important cultivars of hosts in the threatened areas
2. The ecological requirements of parasites to become an economic pest
3. The chances of introduction of exotic parasites or races of parasites.

Vulnerability of important cultivars of the host crop to an exotic nematode could be assessed by either taking the domestic cultivars, breeding lines and germplasm to regions where the pest is endemic or epidemic; or importing a virulent and aggressive population of parasite and study them on domestic material under controlled conditions in containment facilities. Both the approaches have their own advantages and disadvantages. However, it is always better to send the germplasm rather than import the parasite for screening the domestic germplasm.

Information on the ecological requirements of a parasite to become an economic pest is obtained from continued observations in infested areas, supplemented with experimental data. The probability of the pest becoming established is based on the volume of vector material imported; the hitch hiking potential of the parasite and the ease of establishment of the parasite on arrival. The economic impact is based on following factors:

1. The amount of host grown or area of cultivation
2. The ecological range of the pest *viz.* the range of host
3. The loss in yield when normal controls are used
4. The added cost to control the new pest
5. The added cost of increasing the growing area to compensate the losses caused by the new pest.

On the basis of above analysis, *Hetrodera zeae* was ranked 15; *H.avenae* 16; *H.latipons* 17 and *Globodera rostochiensis* 48 among 1333 exotic pest risk analysed. Plant parasitic nematodes of quarantine importance to India are listed in Table 35.4.

Table 35.4: Plant Parasitic Nematodes of Quarantine Importance in India

Exotic Species (Not known to occur in India)	*Introduced/Indigenous Species (Known to occur in India)*
Anguina agrostis	*Anguina tritici*
A. graminis	*Aphelenchoides besseyi*
Aphelenchoides arachidis	*A. fragariae*
Belonolaimus longicaudatus	*A. ritzemabosi*
Bursaphelenchus xylophilus	*Ditylenchus angustus*
Ditylenchus destructor	*Globodera pallida*
D. dipsaci	*G. rostochiensis*
Heterodera glycines	*Heterodea avenae*
H. goettingiana	*H. cajani*
H. schachtii	*H. zeae*
*H. trifolli**	*H. zeae*
Meloidogyne naasi	*Hirschmanniella oryzae*
Naccobus aberrans	*Pratylenchus brachyurus*
Radopholus citrophilus	*P. coffeae*
Rhainaphelenchus cocophilus	*P. vulnus*
Trichodorus virulifrus	*Radopholus sirnilis*
Xiphinema diversicaudatum	*Rotylenchulus reniformis*
	Xihinema americanum

* Reported from India but its occurrence is doubtful

Limitations in Early Detection of Introduced Nematodes

There are certain limitations in early detection and subsequent containment/eradication of introduced nematode species in a region. These are:

1. Absence of reliable symptoms in new soil infestations. Visual inspection for field symptoms may not give any indication of nematode infestation in the crop.
2. Once introduced in a locality/area, nematodes are easily carried in viable condition to neighboring fields through movement of infested soil or planting material. Man has no control over natural agencies like floods, cyclones, birds and animals.

3. Nematode extraction methods have limited reliability especially when nematode number is very low. According to one estimate if there are 37 million cysts per hectare sampled, statistically no cyst would be observed in five samples out of every 100.
4. Crop rotation has little value since most nematode cysts can persist in soil for a long time. Other nematodes have a very wide host range.
5. Several plant and weeds may serve as reservoirs for introduced pests.
6. Majority of key nematode pests being endoparasites don't come in contact with nematicides or natural enemies.
7. Eradication in fields/soil, by chemicals is not economically feasible and creates environmental pollution hazards.
8. Development of resistant varieties against nematodes is a time consuming process and resistance is not stable in many cases.

Establishment of Nematode Regulatory Laboratory

For establishment of an efficient quarantine laboratory for nematodes, the basic equipment and physical needs, a regulatory Nematologist (with strong foundation in nematode taxonomy and morphology and expertise in nematode identification), one laboratory technician (with training in nematology) a compound microscope and soil sample washing room with efficient drainage system.

Latest Developments

With a view to provide the best planting materials available in the world to Indian farmers Government of India has been working very hard. While liberalizing the procedures for seed and other planting materials to achieve the desired goals, the policy also states that absolutely no compromise on the requirements of plant quarantine procedures will be made to prevent entry into India of exotic pests, diseases and weeds detrimental to Indian agriculture. The Government later brought forward a comprehensive 'Plants Fruits and Seeds (Regulation of Import into India) Order in 1984 which was again revised in 1989. The latest Plant Quarantine (Regulation of Import into India) Order 2004, effective from 1 January 2004, demands preparation of PRAs for all materials to be imported except for those mentioned in Schedule IV, V and VI which specifies the lists of material either prohibited or restricted for import with additional declarations.

Though all consignments need to be examined at the port of export and a certificate in this regard must be produced along with the consignment at the port of entry; except for cut flowers, garlands, bouquets, dry fruits/nuts etc. weighing not more than two kilograms, with PC is not required (Chapter II, Article 21). The Article is not taking into consideration foliar nematodes which could accompany flowers, buds and leaves.

Suggestions to Strengthen Nematode Quarantine

Although the related actions were initiated in 1914, Plant Quarantine activities are still inadequate (both in terms of legal and functional aspects) to meet the demands of large countries like India. The situation for nematode quarantine is far from satisfactory (Kornobis, 1998). In many cases quarantine measures are not sufficient to stop an epidemic. Often this is because measures are enacted too late and are associated with a pathogen that can spread very quickly or cross long distances (Szonyegi,1999). Some suggestions are given hereunder to improve the existing plant quarantine system in India with particular reference to nematodes.

1. Plant Quarantine Order, 2004 needs further amendments to include some more exotic nematode species of quarantine importance, specially the ones which can disseminate with wood/packing material.
2. Efforts should be made through a "National Project" to obtain basic information on exact geographical distribution of major key nematode pests and based on this information, distribution maps should be prepared for knowing pest free areas.
3. Developing infrastructure such as glasshouse, insect proof net houses and post entry quarantine isolated nurseries in geographically isolated areas to grow the imported planting material suspected to carry pests. It is always safe to supply the seeds/cuttings obtained from planting material grown in isolation and inspected for pest and diseases during the entire growth period. Such facilities should be developed in different agro climatic zones of the country to meet the specific growth requirements of different crops.
4. For the containment of the indigenous pests, seeds/planting material registration and certification by the government should be implemented and enforced under the supervision of trained personnel. No nursery stock and seed material infected with nematodes should be allowed to grow in the field.
5. Implementation of "Emergency Projects" with the objective to early detection of serious pests and preventing their establishment and spread. Such project should have close coordination between germplasm exchange agency and the plant protection scientists to keep the track of introduction material. Such a survey is immediately required for occurrence of pine wood nematode (*Bursaphelenchus xylophilus*) in North East states and foot hills of the Himalaya in India.
6. Seed and planting material collected in India by the exploration scientists must be sent for quarantine examination before being grown for evaluation in the fields or storage in gene banks or distribution to different centers.
7. Exchange of plant genetic resources should be in the form of aseptic culture of plantlets.
8. Bud woods rather than rooted cuttings/root stocks should be exchanged.
9. To lower down the risk of associated nematodes with plant material, PC may contain additional declaration of special treatments to meet the quarantine requirements of the importing country.
10. Publicity about plant quarantine and its importance in preventing spread of nematodes must be disseminated by full sized billboards, public service announcements, press releases, leaflets, farmers/growers meetings etc. There should be a separate wing for publicity of plant quarantine within the directorate of Plant protection, Quarantine and Storage.
11. Development of salvaging techniques, race identification and other aspects related to quarantine may be worked upon.
12. Regulatory measures should be enforced strictly as our national economy is still so much dependent upon agriculture that we stand to lose significantly through the introduction of unwanted pests and pathogens, including nematodes.
13. To promote export, pest free areas need to be identified.

International Quarantine: Obligations

Several international agreements to which India is a signatory are relevant–as these agreements are influencing quarantine policies, obligations and international trade.

WTO and its Subsidiary Agreements

The WTO's role and scope is defined, of which two annexes have particular relevance to quarantine–the Agreement on the Application of Sanitary and Phytosanitary Measures (SPS Agreement) and the Agreement on Technical Barriers to Trade (TBT Agreement).

IPPC

The IPPC was adopted by FAO conference in 1951. It was amended by the FAO conference in 1979 and the amendments came into force in 1991. The convention has as one of its main objectives securing common and effective action to prevent the spread and introduction of pests and disease of plants and plant products and to promote measures for their control. It describes the principles of plant quarantine and relevant actions to be taken by national governments in the implementation of plant quarantine.

The Convention on Biological Diversity (CBD)

The Convention came into force on 29 December 1993. Two provisions in Article 8 of the convention are particularly relevant to quarantine issues:

1. Establish or maintain means to regulate, manage or control the risks associated with the use and release of living modified organisms resulting from biotechnology which are likely to have adverse environmental impacts that could affect the conservation and sustainable use of biological diversity and
2. Prevent the introduction or control and eradicate those alien species which threaten ecosystems habitats and species.

Future Thrust Areas

1. The Department of Biotechnology (DBT) has responsibility for regulating GMO's. A national containment facility for testing, environment safety of bulk consignments is urgently required.
2. Unlike chemical pollutants, once genetic pollution due to GM crops and their wild weedy relatives is introduced, it cannot be cleaned up by simply allowing it to delay but it will be passed on to all future generations.
3. The establishment of WTO has given new impetus to the development of plant quarantine standards under the IPPC. The SPS Agreement encourages governments to harmonize or base their national measures on the international standards, guidelines and recommendations developed in other international organizations.
4. In order to strengthen quarantine services and pest management, exchange of technical experts with practical experience should be encouraged.
5. Linkages are needed to train officials who can execute phytosanitary and quarantine issues and establish local mechanisms and guidelines for commercialization of transgenic crops.
6. GATT Article III states the principle of non discrimination, demanding equal treatment of domestic and imported goods. It implies that a country cannot use import tariffs, levies, bans or quotas as a measure to safeguard the competitiveness of a domestic industry.
7. To carry out extensive surveys of diseases/nematodes of national and international importance in crops of high export potential. Thereby locating nematode/pest free areas for export of crops.
8. To generate a comprehensive epidemiological data on important diseases in order to be able to develop Pest Risk Analysis (PRA) and identify pest free areas as per the WTO norms.

9. To develop detection and salvaging techniques for destructive nematodes those are applicable to testing samples of bulk consignments in batches.
10. To evaluate the ecological and quarantine risks in import of transgenics and beneficial biocontrol agents, on local nematode population.
11. To evolve the national regulatory mechanisms of plant protection and biosafety in order to meet the international obligations.
12. There is a need to reorganize the National Plant Quarantine System (NPQS) as a whole.

Conclusions

Quarantine is one of the important tools for preventing the large-scale spread of plant pests. Although it can be expensive and difficult to implement, recent streamlining of international, national and state regulations provides hope to avoid repetition of some of the historical quarantine failures that resulted in major epidemics. Rapid and targeted action against imported pathogens should be the goal of future quarantine efforts. Exclusion practiced on a small scale includes the use of pathogen-free propagation stock. To be successful, growers must obtain and utilize this material at the beginning of each new production cycle. Seed certification programs are necessary to ensure that crops grown from seed are vigorous and free from pathogens. Nematodes associated with seed can be found on seed surfaces mixed with debris, associated with seed coats, within the seeds and just under the seed coat. Eradication strategies usually involve the destruction of nematodes, infected host plants and alternate hosts. It is known that general eradication measures are usually not effective against nematodes once in soil.

From the foregoing account it can be concluded that Plant quarantine has an important role to play in the modern agriculture to prevent the introduction of exotic pest and to prevent the further spread of serious indigenous pests within the country. However, an efficient quarantine system has yet to be developed in view of the changing pest strategies. Public awareness and cooperation are very essential to achieve the objectives of plant quarantine.

It is subjective to decide whether or not quarantine procedures are effective in keeping out exotic pests. The argument comes from the fact that it is not possible to measure how many pests would have entered a new region had quarantine procedures not been in operation. Pests including nematodes are not uniformly distributed through out country/world because of quarantine and some of the nematodes may not survive even if introduced knowingly. It in unlikely that any nation would be willing to withdraw quarantine procedures just to test the hypothesis.

REFERENCES

Baker, C.R.B. (1981). Assessing risks from exotic pests. *EPPO Bull.*, 11: 145–150.

Bianco, M and M. Talame (1998). Nematodes: Quarantine and certification of grapes. *Informatore Fitopatologico*, 48(7/8): 17–19.

Bingefoprs, S. (1967). International dispersal of nematodes. *Netherlands J. Plant Path.*, 73 (suppl.1): 44–60.

Hockland, S. (2000). Significant interceptions and other new or usual plant parasitic nematodes recorded in England 1996–1999. *Nematology*, 2(7): 762.

Hopper, B.E.Ed. (1996). NAPPO Compendium of Phytosanitary Terms, NAPPO, Nepean, Ontario, Canada.

Kahn, R.P. (1982) The host as vector: Exclusion as a control. In: *Pathogens: Vectors and Plant Disease,* (Eds.) K.F. Harris and K. Maramorosch, Academic Press, New York.

Kahn, R.P. (1991) Exclusion as a plant disease control strategy. *Ann. Rev. Phytopathology,* 29:219.

Kehlenbeck, H. (1998). Costs and benefits of the impact of new EEC regulations concerning plant health. Part 2: Benefits analysis and Conclusions. *Nachrichtenblatt des Deutschewn Pflanzenschutzdienstes,* 50 (9): 217–224.

Khan, R.M. (1999). Distribution of Radopholus similes in India, its spread in new regions and an analysis of the nematfauna of banana crop phosysatem. *Nematol. Medit.,* 27: 239–245.

Lal, A and V.K. Mathur (1988). Seed health testing for plant parasitic nematodes. In: *Proceedings of Seed Science and Technology,* (Eds.) Yadava and Changi Ram, HAU Press, Hisar, pp. 318–320.

Latha, S., D.V.N. Babu, N. Sathyanayana, O.R. Reddy and R. Sharma (1999). Plant parasitic nematodes intercepted from imported ornamental plants. *Indian Phytopathology,* 52(3): 283–284.

Mani, A. (1998). Plant parasitic nematodes intercepted from imported plant and organic materials. *Nematologia Medit.,* 26(1): 93–95.

Phillips, D., M. Chandrasheker and W.P. Roberts (1994). Pest risk analysis and its implications for pest and disease exclusions from Australia. *Australian Pl. Pathol.,* 23(3): 97–105.

Sanwal, K.C. and V.K. Mathur (1976). Interception of Plant Parasitic Nematodes from imported Seeds. *Indian J. Nemat.,* 5: 251–255.

Schiessendoppler, E. (1996). EU guidelines and diagnosis of quarantine organisms in seed potatoes. *Vereinigung Osterreichischer Pflanzenzuchter,* 47: 129–130.

Silvestro, D.Di. and R. Tacconi (1988). Nematode: passport and certification of fruit crops. *Infrmatore Fitopatologico,* 48(7/8): 25–29.

Southey, J.F. and P. Aitkenhead (1972). Soil borne pests and pathogens associated with imported plants with special reference to nematodes. *OEPP/EPPO Bull.,* (7): 49–59.

Thorne, G. (1961). *Principles of Nematology.* McGraw Hill Book Co., New York, 552 pp.

Part IV

AGROCHEMICALS & WEED SCIENCE

36

Decontamination of Pesticide Residues

Sumitra Arora and Madhuban Gopal***

**National Center for Integrated Pest Mangement, Indian Agricultural Research Institute, Pusa Campus, New Delhi – 110 012*

***Division of Agricultural Chemicals, Indian Agricultural Research Institute, New Delhi – 110 012*

The pesticides and allied agrochemicals have become an integral component of modern agriculture and public health, but their excessive and injudicious use has raised several environmental issues. The extensive application of pesticides creates an environmental risk due to the total pesticide residues, identified in air, water, soil, foods and human tissues (Dikshith, 1991). Besides contaminating agricultural commodities, their transport may increase their level beyond acceptable limits and this way they may have an impact on the delicate equilibrium of our environment. Some pesticides (particularly organochlorine compounds) were banned after evidence of their toxicity, persistence and bioaccumulation in the environment and their ability to concentrate in the food chain. Organophosphate compounds are considered to degrade either biologically or chemically in plants, animals and soils, more rapidly depending on their formulation, method of application, climate and soil types and conditions (Khan, 1980) but are much more toxic than other insecticides. Municipal and industrial treatment plants, effluents, discharge from factories, accidental spillage during transportation or storage, deliberate pesticide dumping and leakage from waste disposal sites also contribute substantially to pollution.

The concept of poison or toxicant implies interaction between a living organism and a chemical substance, which cause pathological changes to appear in it. Acute poisoning of an organism by a pesticide occurs when the pesticide acts at once, whereas chronic poisoning of an organism is the result of the repeated action of relatively small amounts of a pesticide and manifests itself in slowly developing the malfunctioning of a normal vital activity. The entry of a poisonous substance into an organism calls forth the response in the form of protective reactions that restrict the toxic action of the poison. These reactions include excretion of the substance (as such or in modified form) from the organism, deposition of it in the tissues, decomposition of the poison into simpler substances with their excretion or inclusion into the general process of metabolism. Therefore it is necessary to have awareness as to how to detoxify the pesticides and reduce harmful effects of these pesticides on the environment by decontamination of various components of the environment or commodities with which they might have come in contact and in which they persist in undesirable amounts.

Detoxification

Detoxification and decontamination are two different terms. Detoxification is breaking or modification of the pesticide in order to reduce its toxicity or to reduce the hazard due to residues. While the first term is usually employed before pesticide disposal, decontamination is a process for removal or segregation (Roy, 2003) of toxicant for example from an agricultural commodity by say washing.

The methods for detoxification have been divided into three categories (Gopal, 2003 and Gopal *et al.*, 2000):

Physical

1. Incineration
2. Deep well injection
3. Ground burial

Biological

1. Land fill
2. Soil incorporation
3. Activated sludge
4. Enzymatic treatment

Chemical

1. Acid and base hydrolysis
2. Oxidation and reduction
3. Fixation
4. Solvent extraction
5. Adsorption
6. Chlorinolysis

Several basic types of reactions which occur in an organism are hydrolysis, oxidation, reduction, dehydrochlorination, isomerization and conjugation. As a result, either less toxic (detoxication) or more poisonous (activation) products may be formed. In the long run, metabolism results in more hydrophilic (polar) substances. These are catalyzed by enzymes often in the presence of a hydrogen donor. In this article, microorganism mediated or enzymatic methods of detoxification have not been covered, only chemical methods are discussed.

Great importance has been attached to chemical reactions, which reduce the toxicity of the parent compound. Oxidation reactions are useful for decomposing stable pesticides *e.g.*, those possessing aromatic rings or halogen derivatives of hydrocarbons. Hydroxylation, of aromatic ring in various pesticides, makes the molecule more polar, reduces its toxicity and makes it more amenable for decomposition. Reduction of nitro group and dehydrochlorination can also reduce the toxicity and have lower physiological activity *e.g.*, amino derivative of parathion–methyl.

The degradation products of the xenobiotics form conjugates within the plant or animal body, which are as a rule less toxic, more polar and mobile. Among such reactions acetylation, formation of sulfates, conjugation with amino acids, glucose and glutathione, and O- and S-methylation are known. The conjugation process involves both the pesticides themselves and the products of their metabolism namely, alcohols, phenols, carboxylic acid, amines, thiols, heterocyclic compounds. Taking examples of these processes, it is known that derivatives of dithiocarbamic acid when inside plants form conjugates with glucose and 2,4-D forms conjugates with amino acids. Atrazine conjugates with glutathione in an animal's organism and it leads to detachment of chlorine and loss of toxicity. Detoxification of phosalone in the animals takes place by S-methylation of this heterocyclic moiety.

Most chemical degradation studies pertaining to acid or base hydrolysis, oxidation, reduction or irradiation are reported for simple disposal methods for use by the pesticide consumers. Other methods such as fixation, wet scrubbing, chlorinolysis, neutralization, precipitation, ion exchange and solvent extraction have been examined for treating industrial pesticide production plant effluents and other industrial scale wastes.

Hydrolysis (Acid or Base)

Most pesticides are lipophilic and are not excreted in an unchanged form. Due to hydrolysis, lipophilic substances transform into hydrophilic compounds and the reaction products can penetrate only slightly through the membranes to the vitally important centres and are excreted from the organism at a faster rate. Though, toxicity in general decreases in such processes, exceptions are known. For example, acetyldiptrex get hydrolyzed under the influence of enzymes to more toxic trichlorfon.

Acid or base hydrolysis as an effective method for pesticide detoxification and disposal is restricted primarily to the more toxic organophosphate and phenyl carbamate insecticides (oral LD_{50}<200 mg/Kg). The substituted phenols and degradation products derived from organophosphates and carbamate insecticides are generally less toxic (LD_{50}>300 mg/Kg) than the parent molecule (Handa *et al.*, 1999). However, phenyl ureas, acid anilides and other hydrolysis products can have toxicities equivalent to or greater than the parent pesticides.

$$R_2P(=O)\text{-}O\text{-}R \xrightarrow{\text{acid or base hydrolysis}} R_2P(=O)\text{-}OH + R\text{-}OH$$

Phenylamide pesticides are a group of closely related compounds, namely, phenylcarbamates, phenyl ureas and anilides. El-Dib and Aly (1976) investigated hydrolysis of phenylamide in aqueous solution. The observed reactivity of phenylamides and their susceptibility for nucleophilic attack by hydroxyl ions are mainly linked with the polarity of the carbonyl carbon atom. The rate of hydrolysis can be improved by using additives to enhance the water solubility of the pesticide. Faust and Gomma (1972) reported that ethanol (25 per cent) as additive created a three fold increase in the rate of parathion hydrolysis. Muhlmann and Schrader (1957) had earlier reported that detergents, emulsifiers and increasing temperatures would also increase reaction rates.

Oxidation and Reduction

Compounds, which resist hydrolysis, prefer oxidation as the main route of their metabolism in organisms. For example beta-oxidation mechanism in living organisms, consists of stepwise decomposition of fragments of the hydrocarbon chain with an even number of carbon atoms to acetic acid. Phenoxybutyric acid is oxidized to more phytotoxic phenoxyacetic acid in plants. The oxidation of side bonds of cyclic and heterocyclic compounds occurs when direct hydroxylation of the ring is hampered. More polar and less toxic products thus formed accelerate ring decomposition processes.

The use of chlorine, potassium permanganate, potassium persulfate, hydrogen peroxide, lithium, sodium or liquid ammonia for oxidation or reduction of pesticides has been examined for use in pesticide disposal and detoxification procedures. Kennedy *et al.* (1969) degraded 15 of 19 examined pesticides using a liquid ammonia metallic sodium solution, while treatment with biphenyl, hydrogen peroxide or metallic liquid ammonia was not successful. Here commercial formulations of pesticides were subjected to selected concentration of five different chemicals, to partially degrade or to decompose the compounds. The compounds selected were–atrazine, malathion, nemagon, carbaryl, dieldrin, trifluralin, bromacil, picloram, diuron, dicamba, DNBP and 2,4-D. Solutions of pesticide formulations were treated with three different concentrations of hydrogen peroxide, nitric acid, sulphuric acid, sodium hydroxide or ammonium hydroxide. After addition of chemicals, the mixtures were thoroughly shaken, allowed to stand for 5–10 minutes, shaken again and allowed to react at room temperature for 24–36 hours. The solvent layer was drawn off, evaporated to approximately 1 ml and analyzed by TLC or by Infra red spectroscopy (IR). The aqueous layer was extracted three times with either chloroform-benzene or n-hexane and subjected to IR.

There was no significant effect by hydrogen peroxide. Nitric acid induced changes in carbaryl and atrazine. The 16 N acid treatment produced nitrobenzene from carbaryl and 2-OH compounds from atrazine, which is believed to have resulted from the acid hydrolysis of 2-chloro-S-triazine. Sulfuric acid effected a definite change in the structure of bromacil. Sodium hydroxide (NaOH) caused only slight changes in the structure of dieldrin while malathion was broken down sufficiently to yield inorganic phosphate.

Dicamba exhibited no change when treated with NaOH, yet 2,4-D was changed to its sodium salt. DNBP was altered structurally by NaOH, but the spectra was still unresolved. Picloram was decarboxylated with NaOH and chlorine present was replaced by hydroxyl group. Structural changes were also produced in bromacil when treated with NaOH. Ammonium hydroxide treatment of carbaryl produced 1-naphthol but affected only a color change in trifluralin. Mauk and Prengle (1976) reported that combined O_3 and UV light treatment greatly improved the chemical oxidation of refractory organic compounds in wastewaters. Bolzacchini *et al.* (1994) studied oxidative pathway for degradation of simazine, atrazine and terbutylazine herbicides both with cerium ammonium nitrate and ozone at acidic pH.

Chemical Industrial Method

The detoxification methods, which the firms producing or formulating pesticides, generally utilize involves:

1. Optimization of the chemical reactions involved in the production of the pesticides, not only increases yield of the product but also decreases the demand on the air and water pollution control system.
2. The chemicals with low water solubility can be extracted using an organic solvent.
3. Neutralization by addition of concentrated solution of acid or base or by precipitation and coagulation.
4. Wet scrubbing of gaseous pesticide containing effluents. The effluents are either sent through trapping or detoxifying solution and then through an activated charcoal filter or they are scrubbed in towers, where scrubbing liquid, generally water, is sprayed concurrently to the upward flowing gas.
5. Gruber (1975), Hager (1976) and Harris *et al.* (1991) reported activated charcoal adsorption as an effective method for removal of organic material and pesticides from industrial and municipal wastewaters.
6. Chemical and Engineering News (1974) informed use of Amberlite polymeric adsorbent of phenols in particular for absorption.

Fixation and Chlorinolysis

The chemical fixation of toxic wastes involves the reaction of at least two reactants with the hazardous waste molecule to form a chemically and mechanically stable, cross-linked, three dimensional matrix. This immobilized waste is fairly resistant to leaching and the fixed material can be deposited in a landfill (Newman, 1976).

In chlorinolysis, organochlorine wastes are converted into tetrachloro methane (CCl). The chlorine gas reacts with HCH to produce CCl. This process operates at 500–600°C temperature and at high pressure (300 psi) and requires Ni alloy reactor liner. Envir. Sci. Tech. (1974) and Krekeler and Weber (1972) reported that CCl_4 produced from herbicide orange (mixture of 2,4-D and 2,4,5-T) as free of TCDD (2,3,7,8-tetra chlorodibenzo-p-dioxin), which is a toxic contaminant in the herbicide mixture. Low *et al.* (1991) demonstrated that titanium dioxide mediated photocatalytic oxidation of organic compounds can be an alternative to conventional methods for the removal of organic pollutants in water. Semiconductor such as titanium dioxide and zinc oxide possess a characteristic band structure in which a filled valence bond is separated by an energy gap from a vacant conduction band.

When titanium dioxide particles are irradiated with light of band gap energy, an electron hole pair is created *i.e.*, photo excited electrons, e- and holes h+ are produced with the promotion of the electrons into the conduction band. Some of the electrons and holes migrate to the surface (oxide surface) of the particle, before recombination occurs. In the aqueous phase, the illuminated surface is widely regarded as a producer of hydroxyl radicals. *e.g.*,

$$h^{+} + OH^{-} \longrightarrow \cdot OH$$

These and other oxidizing products attack oxidizable contaminants.

Formation of Radicals from Illumination of TiO_2 in Water

$$TiO_2 \xrightarrow{hv} h^+ + e^-$$
$$h^+ + e^- \longrightarrow \Delta$$
$$e^- + O_2 \longrightarrow \cdot O_2^-$$
$$\cdot O_2 + H^+ \longrightarrow HO_2\cdot$$
$$HO_2 + O_2^- \longrightarrow HO_2^- + O_2$$
$$HO_2^- + h^+ \longrightarrow HO_2\cdot$$
$$h^+ + 2H_2O \longrightarrow \cdot OH + H_3O^+$$
$$h^+ + OH^- \longrightarrow \cdot OH$$

Both positive holes and hydroxyl radicals have been proposed as the oxidizing radical species responsible for initial attack on organic solutes. The study was done on a related series of primary, secondary and tertiary amines and other N- and S- containing compounds. The compounds studied were–n-phenylamine, piperidine, pyridine (pyrichlor, picloram, trichlopyr etc.), phenylalanine, desipramine, thioridazine, penicillamine, isosorbide dinitrate, 4-nitro-catechol, 2,4-dinitro phenol (DNOC, DINOSEB, DINOTERB), cyclophosphamide, 5-fluorouracil, atrazine, ethylenediamine tetraacetic acid (disodium) and tetra butyl ammonium phosphate. All compounds give rise to both NH_4^+ and NO_3^- ions, but in different relative concentrations. It was observed that NO_2 groups gave predominantly NO_3^- and primary amines gave predominantly NH_4^+ ions.

The reaction is initiated by the addition of a ·OH radical to form the 3-hydro-3-OH-pyridine radical, followed by rapid addition of O_2 to give 2,3-dihydro-2-peroxy-3-OH-pyridine radical, with the eventual opening of the ring to form N-(formylimino)-2-butanol. The latter compound finally decomposed in water to produce the dialdehyde and formamide.

Dennis and Kobylinski (1983) reported a treatment system based on re-circulation of pesticide contaminated wastewater through a bed of granular activated carbon, which was found to be of particular use to the small waste generators. The carbolator is light in weight and portable consisting of an 18-gallon reinforced epoxy tank with an O-ring seal cover. Two porous polypropylene bags, which hold the granular carbon, are placed within the tank. Pesticide contaminated water is pumped continuously from a waste holding tank into the bottom of the carbolator and upward through the carbon bed. The water or effluent is returned to the tank. On site analysis of the waste by TLC showed the presence of malation, dimethoate and baygon with traces of diazinon and dursban. Samples were taken at various intervals. The system was allowed to operate unattended overnight, and no problem was encountered. After 24 hours of continuous operation, no pesticides were detected in the water by TLC method. This carbon filtration system offers an economical and dependable means for small generators of pesticide-laden wastewater.

Maguire (1990) subjected deltamethrin [(S)-α-CN-S-phenoxy benzyl (1R, 3R)-cis-2,2-dimethyl-3-(2,2-dibromovinyl) cyclopropane carboxylate] in natural water in the dark to cis/trans isomerization, yielding the α-S, 1S isomer 2'-deltamethrin, which is inactive against mice and insects.

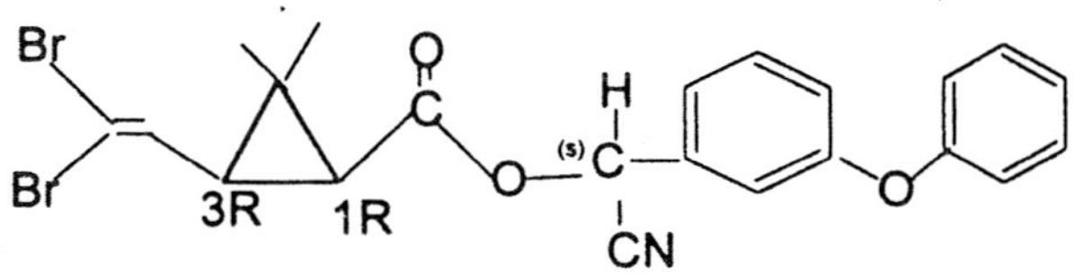

The cis 1R, 3R configuration about the cyclopropane ring and the S-configuration for the cyano group at the benzylic C-atom are essential for its high toxicity. The major degraded products were:

Cis-3-(2,2-dibromovinyl)-2,2-dimethylcyclopropane carboxylic acid [DBCA] and 3-phenoxy acetic acid.

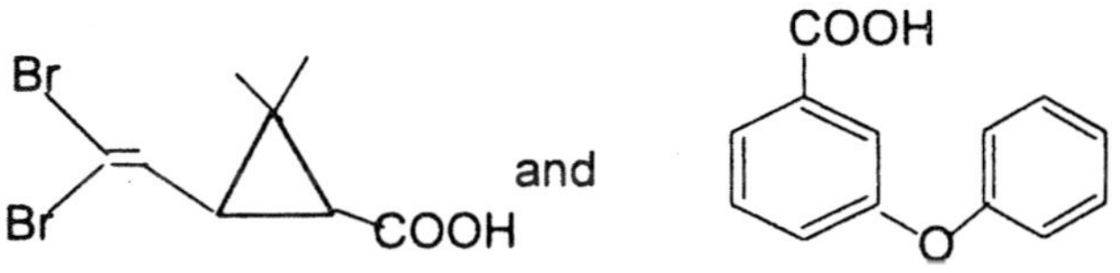

It was concluded from the work that (2+2′)-deltamethrin was the only product formed when the isomerization of parent deltamethrin occurred in dark in water. Moreover 2′, 3 and 4 isomers were the only isomers formed from sunlight irradiation of 1-deltamethrin in thin films or in hexane solutions. The use of riverbeds can prove to be a potential option for wastewater renovation in arid or semi arid regions (Ojha *et al.*, 2002) in future. Use of lagoons can be an alternative for domestic wastewater reclamation which provides elimination of organic pollution (Driche *et al.*, 2002).

Koshioka *et al.* (1986) investigated photodegradation of an organophosphorus insecticide, propaphos {dipropyl-4-(methylthio)-phenyl phosphate} under xenon lamp irradiation. Propaphos absorbs UV light in the range 190-320 nm. Photodegradation was done in two solvents-methanol and distilled water. In both solutions, the photodegradation of propaphos was more significant at lower wavelengths. A variety of reactions were involved in photodegradation of propaphos, but major route of degradation was oxidation at the S-atom. Four compounds were identified as photodegradation products of propaphos by GC-MS.

Miles (1991) investigated degradation of three insecticidal species of Temik-aldicarb sulfoxide and aldicarb sulfone in chlorinated water. The degradation rates of pesticides in chlorinated water in pH range 6-9 are several orders of magnitude faster than simple base hydrolysis, which is the primary detoxification mechanism in water. It is clear from Figure 36.1 that I-rapidly oxidized to II in chlorinated water. III- quickly and quantitatively formed N-Cl derivative (IX), which decomposed to the acid (XI) and dichloromethylamine (X). These results suggest that Cl_2 treatment of drinking water contaminated with these pesticide residues will result in significantly lower residues, but the effects of most of these species on human health is unknown.

Pignatello (1992) degraded the herbicides 2,4-D and 2,4,5-T in acidic aerated solutions of H_2O_2 and Fe^{2+} or Fe^{3+}. Both Fe^{2+} or Fe^{3+} react with H_2O_2. Mechanism involves hydroxyl radical intermediates, which can attack organic compounds. This classical reaction of Fe^{2+} with H_2O_2 known as Fenton reaction generates OH in the rate limiting step. The ·OH may be scavenged by reaction with another Fe^{2+} or react with an organic compound.

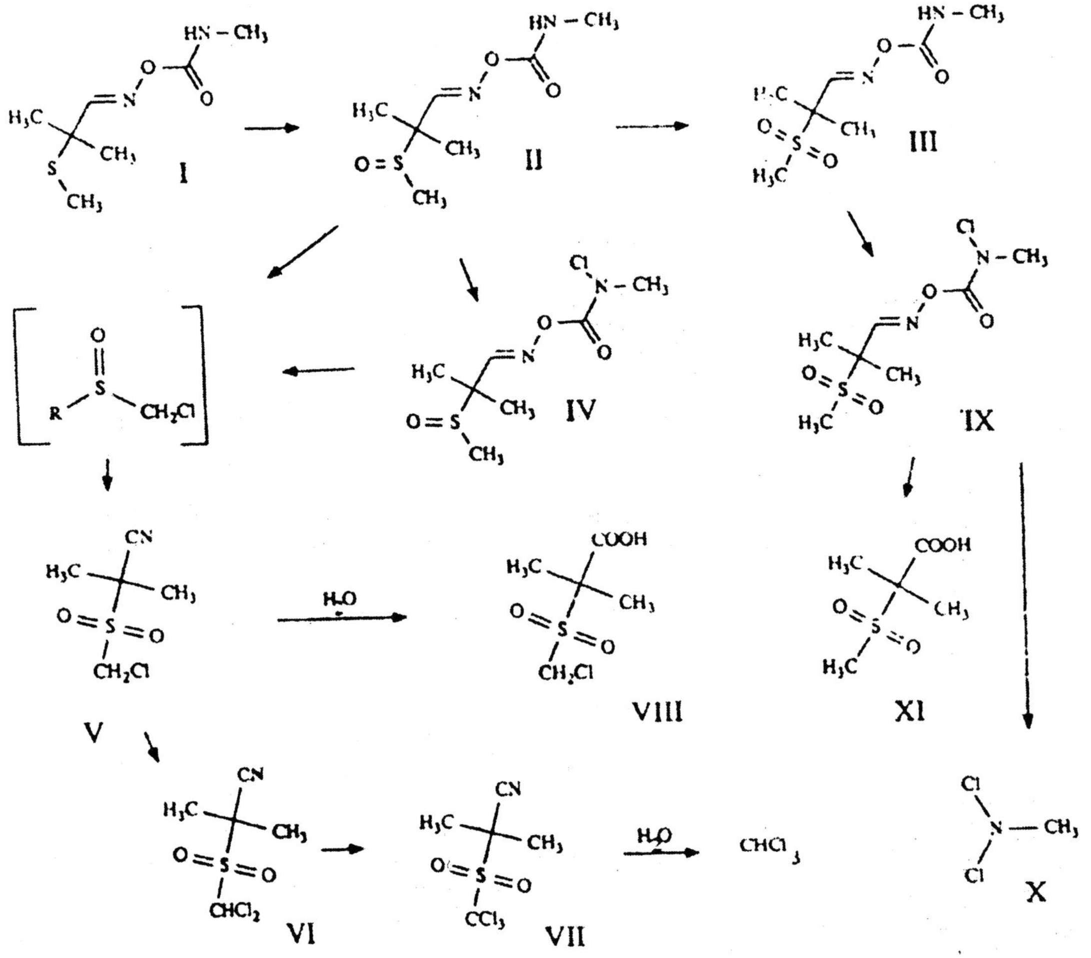

Figure 36.1: Proposed Reaction Pathways for I-III in Chlorinated Water. Reactions other than chlorination are noted.

$Fe^{2+} + H_2O_2 \longrightarrow Fe^{3+} + HO^-$ (rate limiting step)

$\cdot OH + Fe^{2+} \longrightarrow HO^- + Fe^{3+}$

Fe^{3+} categorically decomposes H_2O_2 to O_2 and H_2O.

$Fe^{3+} + H_2O_2 \longrightarrow Fe\text{-}OOH^{2+} + H^+$

$Fe\text{-}OOH^{2+} \longrightarrow HO\cdot_2 + Fe^{2+}$

$Fe^{2+} + HO\cdot_2 \longrightarrow Fe^{3+} + HO^-_2$

$Fe^{3+} + HO\cdot_2 \longrightarrow Fe^{2+} + O_2 + H^+$

$OH\cdot + H_2O_2 \longrightarrow H_2O + HO^\cdot_2$ Scavenging step

OH reacts rapidly and non-selectively with most organic compounds by hydrogen abstraction and addition to C-C unsaturated bonds. Compared to.OH, $HO_2^{\cdot}$ is much less reactive and its conjugate base $O_2^{\cdot}$ is practically unreactive as a free radical. Carbon centered radicals generated by hydroxyl radical attack, may react with O_2 (if present) to give organo peroxy radical (ROO·), which can decompose to form $HO_2^{\cdot}$ or ultimately non radical oxygenated products.

Treatment of herbicide with Fe^{3+}/H_2O_2 leads to complete dechlorination of the herbicides in less than one hour and extensive conversion to CO_2. The principle oxidant is assumed to be.OH.

$$\text{(2,4-dichlorophenoxyacetic acid: } OCH_2COOH, Cl, Cl) + 7H_2O_2 + 4O_2 \xrightarrow{Fe^{3+}} 2HCl + 8CO_2 + 9H_2O$$

Kenley (1985) investigated decomposition kinetics of four OP (organophosphorous) esters in aqueous sodium perborate. Sodium perborate greatly accelerated the organophosphate ester decomposition and the reactions are best described by perhydroxyl ion (which is in equilibrium with perborate) undergoing nucleophilic attack on phosphorous. The reactions involved were:

$$PB \rightleftharpoons HO^- + B(OH)_3$$

$$H_2O_2 \rightleftharpoons H^+ + HO_2^-$$

$$H_2O + B(OH)_3 \rightleftharpoons B(OH)_4^- + H^+$$

$$HO_2^- + OP \longrightarrow PNP + \text{products}$$

$$HO^- + OP \longrightarrow PNP + \text{products}$$

$$H_2O + OP \longrightarrow PNP + \text{products (PNP–para nitro phenolate)}$$

It was concluded that sodium perborate serves as a buffered source of H, O and perhydroxyl anion (HO). The fraction perborate dissociated to perhydroxyl depends on pH of the reaction medium and the concentration of borate present. Increasing pH and borate concentration decreased the perhydroxyl ion concentration.

Saha and Sarkar (1991) provided a chemical detoxification process to remediate soils contaminated with halogenated pesticides, BNAs and the like. The process uses zinc and a mild acid such as acetic acid to generate nascent or atomic hydrogen, which replaces aliphatic and aromatic halogen atoms in toxic organics in the presence of heat and a catalyst. The halogenated toxic organics are converted to nontoxic hydrodehalogenated compounds. Suitable catalysts include nickel, cobalt, molybdenum, yttrium, and iridium. Handa *et al.* (1990) suggested that decontamination method for organochlorine pesticides often involves the reductive removal of chlorine from the toxicant. Nickel boride generated in situ by the reaction of a soluble nickel salt with sodium borohydride, degraded lindane producing 94–100 per cent chloride ions

$$\text{Lindane} \xrightarrow[CH_3OH]{NiB_3/NaBH_4} \text{benzene} + \text{cyclohexene} + \text{cyclohexane} \quad Cl$$

Decontamination

Decontamination means reduction of contamination of edible or other commodity from pesticide residues without spoiling the commodity. Easiest example is that of washing the fruits. In decontamination, the pesticide is present as a residue in a complex biological matrix, so the residue is to be inactivated with as little change to bulk of the sample as possible and the decontamination procedures must also be economically feasible. Due to the diversity in chemical properties of pesticides and their formulated products and to the quantity and composition of pesticide wastes which must be treated, no single treatment system can be universally applied.

Chemical pesticide decontamination methods generally are restricted to hydrolysis procedures. The pesticide residues present in plant parts used as feed and fodder get concentrated in the animal tissues. When this material (if not destroyed) is consumed as food by human or animals, it may cause different forms of harmful effects. In literature, in most of the cases, the commodity is decontaminated by washing, cooking, baking, boiling and frying etc. Dixit *et al.* (1984) reported various decontamination methods for cowpea. The insecticide residues on mango fruits, resulting from plant protection sprays, were reduced to 66-68 per cent for dimethoate and fenthion as against 21–27 per cent for fenvalerate and cypermethrin simply by washing (Awasthi, 1993). Peeling off the fruit pericarp was found to dislodge 100 per cent of residues in all cases. Patel *et al.* (2001) reported considerable reduction of lindane residues by washing and washing + cooking of aubergine and okra fruits. Jacob and Verma. (1989, 1990) also described some home processing methods in reducing the residues of malathion on cauliflower, and methamidophos on cauliflower and okra.

In order to decontaminate the farm machinery from lindane, a detergent solution (Kompleks) containing sodium perborate 3 per cent, soda ash 13 per cent, sodium salts of synthetic fatty acids 10 per cent, triethylamine (TEA) 5 per cent and surfactant 2 per cent, water 67 per cent were recommended.

Kompleks reagent

A 3 per cent boiling solution of Kompleks has been found to decompose 80 per cent lindane in 4 hours and the products were described as a low toxicity mixture of cyclohexenes. Sprayer contamination through previous use of sulfonylurea herbicides caused growth inhibition and red leaf discoloration of sugarbeet and rape. Neururer and Womastek used a 0.03 per cent aqueous solution of liquid ammonia, which proved an effective cleansing agent for sulfonylurea-contaminated equipment and unlike sodium hypochlorite, previously recommended for this purpose (Gopal, 1988), it is not an environmental threat.

Surface residues of fresh fruits and vegetables can be removed by peeling. The decontamination is enhanced by washing with water, detergent and water charcoal suspension, which is further enhanced by blanching with hot water or steam. During processing of tomatoes into juice, a high temperature process usually accomplished with the aid of the alkali resulting in higher levels of dithiocarbamate residues as compared to whole tomatoes, due to degradation of the compound to ethylenethiourea (ETU) which is a potent tumorigen and teratogen. Surface residues of mancozeb, ETM and ETU on fresh produce can be decontaminated by an alkaline water wash containing hypochlorite. This can further be washed with sodium sulfite for removing any hypochlorite remaining.

Nelson *et al.* (1992) investigated the efficacy of laundering procedures for the removal of pesticide residues from contaminated protective clothing fabrics. Compounds selected for the tests represented 6 chemical classes and included: carbaryl, methyl parathion, fonofos, terbufos, deltamethrin, cyfluthrin, cypermethrin, alachlor, atrazine, tri-allate and trifluralin. The use of prewash treatment reduced pesticide residues regardless of fibre type. Differences in pesticide residue level attributable to fibre type were not consistent.

Sedlak *et al.* (1991) reported that application of quicklime (CaO) to soils at a poly chlorobiphenyl (PCB) contaminated site resulted in significant decrease in PCB concentrations. The hydration of quicklime is a highly exothermic reaction and relatively high temperatures can be reached on quicklime amended soils.

$$CaO + H_2O \longrightarrow Ca(OH)_2$$

The alkalinity of $Ca(OH)_2$ was exploited by Sinha and Gopal (2002) in detoxification of eggplant treated with β-cyfluthrin. They explained that synthetic pyrethroids are esters, so this edible alkali could reduce the residues of this ester containing insecticide. They reported that simple culinary processes are quite efficient in reducing the pesticide residues from eggplant fruits. The lime-water soaking was found better for decontamination of eggplant fruits than simple washing with tap water (Figure 36.2). This is due to the alkaline nature of the calcium hydroxide, which helps in degrading the ester linkage of β–cyfluthrin. This facilitates in removal of the pesticide from the eggplant fruits as the level of pesticide residues in eggplant fruits was found to be less than 0.1 ppm after lime-water

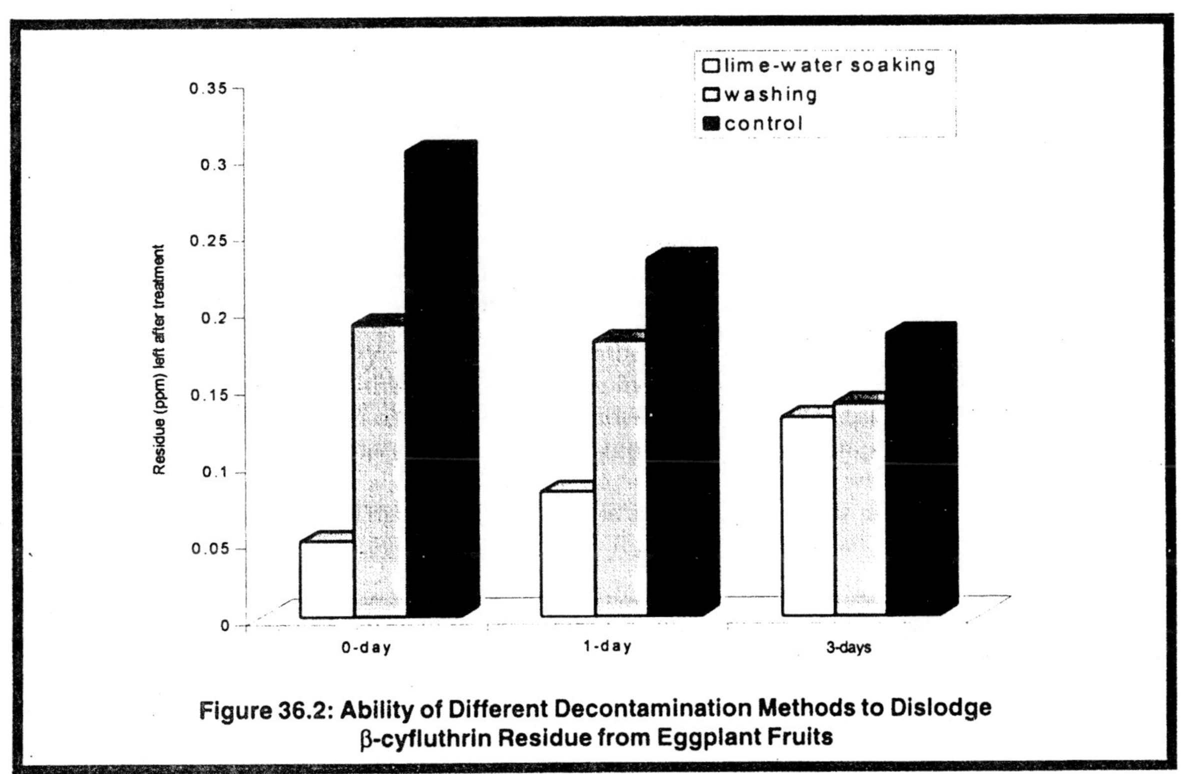

Figure 36.2: Ability of Different Decontamination Methods to Dislodge β-cyfluthrin Residue from Eggplant Fruits

soaking. This process was however effective only for the 0 day and 1 day samples. In the later stages, these treatments were found not very effective, as the insecticide might have penetrated in the skin of eggplant fruit.

Simple washing reduced the residues of fenvalerate and monocrotophos by 23–25 per cent and 44–55 per cent, respectively, in fresh samples of vegetables. The effectiveness of the process varied from 0–5 days and per cent reduction was reduced with time. Due to inherent nature of water-soluble monocrotophos and fat-soluble fenvalerate, the reduction by simple washing with water differed (Gopal, 2003).

Garg (1988) reviewed removal of residues of pyrethroid insecticides from various food crop plants by processes like washing, peeling, cooking and chemical decontamination. Zabik *et al.* (2000) studied the effect of processing on pesticides in apple fruit, which was shown to be very effective in reducing the levels of pesticides. Producing single-strength apple juice reduced azinphos-methyl, chlorpyrifos, esfenvalerate, and methomyl residues by 97.6, 100, 97.8, and 78.1 per cent, respectively. Production of apple sauce reduced all four compounds by 95 per cent. Azinphos-methyl, chlorpyrifos, esfenvalerate, and methomyl residues were reduced in apple slices by 94.1, 85.7, 98.6, and 94.7 per cent, respectively.

Wani *et al.* (2000) detected pesticide residues in adipose tissue, meat, liver and kidney samples of 40 goats, slaughtered at the Bareilly abattoir in North India. Cooking caused a reduction in DDT, HCH and their metabolites/isomers with concomitant increase in the level of DDD residue. Broiling was the most effective method for reducing pesticide residues in meat but pressure-cooking and microwave cooking were also effective.

Monocrotophos (0.05 and 0.1 per cent) and carbaryl (0.15 and 0.3 per cent) were sprayed on tomato crops grown in Hyderabad, India, at 3 stages of development, specifically, 25 d after transplanting, at 50 per cent flowering and 50 per cent fruit formation. Gopichand *et al.* (1998) reported initial deposits of monocrotophos in fruits as 2.45 and 3.48 ppm at low and high doses, respectively. Treatment with 2 per cent tamarind removed 32.26 and 43.18 per cent of the residues treated with 0.05 and 0.1 per cent monocrotophos, respectively, when tamarind was applied on the same day as monocrotophos. Applying the tamarind treatment 5 d after monocrotophos resulted in a 38.45 and 29.41 per cent reduction in monocrotophos residues for the two treatments, respectively. Treatment with 2 per cent tamarind removed 37.95 per cent (0.15 per cent treatment) and 36.04 per cent (0.3 per cent treatment) of the carbaryl residues when applied on the same day as carbaryl. When tamarind was applied 5 d after the carbaryl application the residue levels were reduced by 16.53 per cent (0.15 per cent treatment) and 39.39 per cent (0.3 per cent treatment).

Burgos *et al.* (2002) studied decontamination of aflatoxin B (AFB1)-contaminated maize, which is required if maize is to be suitable for alternative use, by an ammoniation-fermentation integrated process. This process could be used for the production of fuel ethanol from aflatoxin-containing maize. Different concentrations (0.25, 0.50, 1.00, 1.5 and 2.0 per cent w/w) of ammonium persulphate were tested in the detoxification of AFB1-contaminated maize during fermentation. In order to increase the decontamination of corn, 0.5 and 1.0 per cent (w/w) azodicarbonamide, benzoyl peroxide and hydrogen peroxide were tested. Peroxides were added at 3 different stages of the fermentation process: liquefaction, saccharification and fermentation. Levels of AFB1 and ethanol were determined after each fermentation process. Treated maize was tested for mutagenic potential using the Ames test with TA100 tester strain and pure AFB1 as positive control. Addition of 2.0 per cent (w/w) ammonium persulphate caused the highest level of decontamination without affecting ethanol production. Addition of peroxides did not significantly ($P < 0.05$) increase ethanol production or significantly

(P < 0.05) improve the decontamination process. The best processes for decontamination of maize and for ethanol production included the addition of 2.0 per cent (w/w) ammonium persulphate for both and of 1.0 and 0.5 per cent (w/w) benzoyl peroxide respectively. All treated maize samples showed no mutagenic potential.

Kumar *et al.* (2000) studied dissipation and decontamination of cypermethrin (300 g a.i. ha^{-1}) in chilli fruits in field of Andhra Pradesh, India, in 1997–98. Dipping of green chillies in 2 per cent salt solution for 10 min followed by a water wash proved to be effective, facilitating the removal of 90.56 and 66.93 per cent residues correspondingly at 0 and 5 d after spraying.

Both laboratory-scale and full-scale studies by Troxler *et al.* (1993) have shown that thermal desorption is an effective technology for treating soils contaminated with a variety of organochlorine pesticides. Soil decontamination results are presented from six full-scale and nine laboratory, bench and pilot-scale thermal desorption treatment studies. It was reported that:

1. The removal efficiency of pesticides from contaminated soils was affected by soil treatment temperature, total treatment time and the treatment time;
2. Thermal desorption achieved residual pesticide concentration of 0.1–1.0 mg/kg at soil treatment temperature of 500–1000 °F and treatment times of 8–40 min;
3. Pesticide removal efficiencies of >99 per cent can be achieved;
4 DDT decomposes to DDE at high temperature resulting in apparent low removal efficiency for DDE at soil treatment temperature < 84 C (550 °F);
5. The initial soil water content affects the residence time required to heat the feed material and the target treatment temp.;
6. Soil particle size distribution affects mass transfer rates and adsorption of pesticides to soil, with soils high in clay or silt being more difficult to treat than sandy soils; and
7. Adsorption of pesticides to soil organic matter affects their removal efficiency.

A soil washing/cloud point extraction technique was used for the decontamination of soil polluted with DDT (Evdokimov *et al.*, 1998). Soil samples spiked with DDT were washed with 3 per cent and 5 per cent nonionic detergent solutions, consisting of a mixture of Igepal ICO-630 and Triton X-114. The results showed that a 2-h washing period with a 5 per cent detergent solution, and a 3-h period with a 3 per cent detergent solution, removed practically all DDT from the soil. Extraction efficiencies depended on the surfactant concentration and the volume of washing liquid. The cloud point of the extraction medium was 34°C, and clouding experiments succeeded in extracting >93 per cent of the DDT in the washing solution into the surfactant (coacervate) phase. The method holds promise for large-scale treatment of contaminated soils. To decontaminate soil from pesticide a technique of soil solarization is reported by Gopal *et al.* (2000). The technique was found to be selective for faster degradation of pendimethalin.

Waliszewski *et al.* (2000) subjected samples of cow's milk to heat treatment, by heating the milk to boiling point or leaving it to boil for up to 10 minutes. The determined organochlorine pesticides (HCB, alpha, beta, gamma-HCH isomers, pp' DDT, op' DDT and pp' DDE) in the fat of raw milk, heated milk and boiled milk indicated a significant increase of levels in the heated milk and a decrease in the boiled milk. The phenomenon was attributed to the specific linking of pesticides to the lipophilic components of milk and by denaturation of these linkages by the heat treatments.

For cleanup of pesticide rinse water waste, Mullins *et al.* (1998) used a field equipment consisting of batch sorption of pesticides onto an organic sorbent followed by separation and finally degradation of sorbed pesticide in bioreactors during solid state fermentation. Investigations involved the use of organic sorbents and other amendments, such as used vegetable oil soil, in an effort to first retard pesticide movement and subsequently to facilitate biodegradation.

Rodriguez *et al.* (1993) studied montmorillonites modified with tetradecyltrimethylammonium and trimethylphenyl ammonium and the hydroxy-Al polymer as sorbents for the insecticides parathion and paraoxon in aqueous media. Sorptive characteristics of modified clays depended on their organic carbon content and the hydrophobic nature of the pesticide. In contrast tetradecyltrimethylammonium-montmorillonite appeared to be an effective sorbent for removing hydrophobic pesticides from water and this fact could be applied for decontamination of sediments and waters.

Banerjee and Martin (1997) have developed a large-scale pesticide decontamination plant which utilizes ozone treatment and granulated carbon. In this, four-chambered tank is attached to an ozone supply. Raw wastewater is poured in the tank and the impurities are oxidized by ozone. Then partially treated water is passed through a column of activated charcoal. The detractors of the process complain that the water is not fit for drinking as it contains ozone, metabolites of pesticides, carbon particles and its taste is not satisfactory.

The detoxification methods have been divided mainly into three categories, namely physical, chemical and biological methods. The chemical method includes hydrolysis and oxidation-reduction reactions. Besides these it also includes fixation and chlorinolysis for chemical industrial method. The decontamination methods are generally washing, cooking, boiling and peeling but involvement of chemical is on the rise. Simple washing is effective only for the decontamination of fresh samples, for aged samples, special methods are still needed.

It is therefore suggested not to use old synthetic pesticides; especially at fruiting stage instead prefer botanical and biopesticides. Make residue analysis a compulsory part of field efficacy or IPM trials and bold to reject contaminated commodity. If residues are above MRL, methods of decontamination and detoxification should be used. The pesticides have a shelf life and obsolete or old pesticides have to be detoxified before they are dumped in environment. Therefore in order to ensure the safe use of pesticides in food commodities, it should be degraded to some non-hazardous compounds on/inside edible commodities. For registration of pesticides, Government agencies are considering making of decontamination methods as a requirement.

REFERENCES

Awasthi, M.D. (1993). Decontamination of insecticide residues on mango by washing and peeling. *J. Food Sci. Tech.*, Mysore, 30 (2): 132–133.

Banerjee, R. and M. Martin (1997). Yamuna the river of death. In: *Homicide by Pesticide*, (Ed.) A.Agarwal, State of the Environment Series 4, Chapter 3, p 80–84.

Bolzacchini, E., A. Brambilla, M. Orlandi, S. Polesello and B. Rindone (1994). Oxidative pathways in the degradation of triazine herbicides: a mechanistic approach. *Water Sci. Tech.*, 30 (7): 129–136.

Burgos Hernandez, A., R.L. Price, K. Jorgensen-Kornman, R. Lopez-Garcia, H. Njapau and D.L. Park (2002). Decontamination of aflatoxin B1-contaminated corn by ammonium persulphate during fermentation. *J. Sci. Food Agril.*, 82 (5): 546–552.

Chemical and Engineering News (1974). A way to remove and recover phenol from waste streams. 52(47): 16.

Dennis, W.H. and E.A. Kobylinski (1983). Pesticide-laden wastewater treatment for small waste generators. *J. Env. Sci. Hlth., B Pesticides, Food Contaminants and Agricultural Wastes.* 18 (3): 317–331.

Dikshith (1991). *Toxicology of Pesticides in Animal.* CRC Press Boca Raton.

Dixit, A.K., M.D. Awasthi and S.K. Handa (1984). Decontamination of insecticide residues from cowpea. *Pesticides,* 10: 43–44.

Driche, M., D. Abdessemed and G. Nezzal (2002). Domestic wastewater tertiary treatment by natural lagoon for reuse. In: *Environmental and Groundwater Pollution,* (Eds.) Sherif Singh and Al-Rashed Balkema Publishers, Tokyo.

El-Dib, M.A. and O.A. Aly (1976). Persistence of some phenylamide pesticides in the aquatic environment. *Water Res.,* 10: 1047–1050.

Environmental Science Technology (1974). *Emerging Technology of Chlorinolysis,* 8: 18.

Evdokimov, E., Wandruszka-R von and R. Von-Wandruszka (1998). Decontamination of DDT-polluted soil by soil washing/cloud point extraction. *Analytical Letters,* 31(13): 2289–2298.

Faust, S.D. and H.M. Gomaa (1972). Chemical hydrolysis of some organic phosphorus and carbamate pesticides in aquatic environment. *Environ. Letters,* 3: 171.

Garg, P.K. (1988). Effect of processing on the removal of pyrethroid residues. *Pesticides,* 22(12): 21–22.

Gopal, M. (1988). Keetnashi ke dushprabhav kaise kam karein. *Vigyan Pragati,* 7: 308–309.

Gopal, M., I. Mukherjee, D. Prasad and N.T. Yaduraju (2000). Soil solarization: Technique for decontamination of an organophosphorus pesticide from soil and nematode control. *Bull. Environ. Contam. Toxicol.,* 64: 40–46.

Gopal, M., I. Mukherjee and A.K. Dikshit (2000). Pesticide decontamination strategies for the future in Short term training Course on *Recent Developments in Agrochemical Research to meet the challenges of the 21st century.* Winter School 6th–15th November, Division of Agricultural Chemicals, IARI, New Delhi, pp. 168–177.

Gopal, M. (2003). Measures to reduce pesticide residues in Cotton, Rice and Vegetables. In: *Training Course Material on IPM in Cotton, Rice and Vegetables* (Ed.) D.K. Garg. National Centre for Integrated Pest Management, New Delhi.

Gopichand, K., T.R. Babu, K.N. Reddy, D.D.R. Reddy and T.B. Gour (1998). Dissipation and decontamination of monocrotophos and carbaryl residues in tomato. *J. Res. ANGRAU,* 26(2): 13–16.

Gruber, G.I. (1975). Assessment of industrial hazardous waste practices, organic chemicals, pesticides and various explosive industries. Distributed by Nat. Tech. Inf. Service PB 251–307, Various paging. Springfield.

Hager, D.G. (1976). Waste water treatment via activated carbon. *Chem Engg. Prog.,* 72: 57.

Handa, S.K., N.P. Agnihotri and G. Kulshreshtha (1999). *Pesticide Residues Significance, Management and Analysis.* Pesticide Residue Analysis Research Periodicals and Book Publishing House, Texas, USA, pp. 165–183.

Harris, D.A., K.S. Johnson and J.M.E. Ogilvy (1991). A system for the treatment of waste water from agrochemical production and field use. *Proceedings of the Brighton Crop Protection Conference, Weeds.* 2: 715–722.

Jacob, S. and S. Verma (1989). Persistence and decontamination of malathion on cauliflower (*Brassica oleracea var. botrytis L.*). *Indian J. Ent.*, 1990, 51(4): 377–383.

Jacob, S. and S. Verma (1990). Decontamination of cauliflower and okra treated with methamidophos. *Plant Protection Bulletin*–Faridabad. 42 (1–2): 17–19.

Kenley, R.A., G.C. Lee and J.S. Winterle (1985). Organophosphorus ester reactions with aqueous sodium perborate. *J. Org. Chem.*, 50: 40–44.

Kennedy, M.V., B.J. Stojanovic and F.L. Shuman (1969). Chemical and thermal methods for disposal of pesticides. *Residue Rev.*, 29: 89.

Khan, S.U. (1980). *Pesticides in the Soil Environment*. Elsevier Sci. Publishing Co., Amsterdam, The Netherlands, p. 161.

Koshioka, M., J. Kanazawa and T. Murai (1986). Photodegradation of an insecticide, dipropyl 4-(methyl-thio)phenyl phosphate, by xenon lamp. *J. Pest. Sci.*, 11(4): 557–562.

Krekeler, H. and H. Weber (1972). Chlorinolysis of aliphatic and aromatic compounds to carbon tetra chloride. Presented 164th Nat. Meeting Amer. Soc. Sep. 1972.

Kumar, K.P., D.J. Reddy, K.N. Reddy, T.R. Babu and V.V. Narendranath (2000). Dissipation of cypermethrin residues in chilli. *Pest. Res. J.*, 12(1): 130–132.

Low Gary, K.C., S.R. McEvoy and R.W. Matthews (1991). Formation of nitrate and ammonium ions in titanium dioxide mediated photolytic degradation of organic compounds containing nitrogen atoms. *Environ. Sci. Technol.*, 25: 460–467.

Maguire, R.J. (1990). Chemical and photochemical isomerization of deltamethrin. *J. Agril. Food Chem.*, 38(7): 1613–1617.

Mauk, C.E. and H.W. Prengle (1976). Ozone with ultra violet light provides improved chemical oxidation of refractory organics. *Poll. Engg.*, 1: 42.

Miles, C.J. (1991). Degradation of aldicarb, aldicarb sulfoxide, and aldicarb sulfone in chlorinated water. *Env. Sci. Tech.*, 25(10): 1774–1779.

Muhlmann, R. and G. Schrader (1957). Hydrolysis der insektizide phosphorsaureester. *Z. Naturforsch*, 126: 196.

Mullins, D.E., S.E. Gabbert, J.E. Leland, R.W. Young, G.H. Hetzel and D.F. Berry (1998). Organic sorption/ biodegradation of pesticides. *Rev. Toxicol.*, 2: 195–201.

Nelson, C., J. Laughlin, C. Kim, K. Pigakis, M. Raheel and L. Scholten (1992). Laundering as decontamination of apparel fabrics: residues of pesticides from six chemical classes. *Arch. Env. Cont. Toxicol.*, 23(1): 85–90.

Neururer, H. and R. Womastek (1992). Sulfonylureas: new cleaning possibility. *PSP Pflanzenschutz Praxis.* 1: 6.

Newman, J.H. (1976). Fixing hazardous wastes. *Processing*, 22(6): 41.

Ojha, C.S.P., V.P. Singh and P. Nema (2002). Wastewater renovation using a river bed in a semi–arid zone. In: *Environmental and Ground Water Pollution*, (Eds.) Singh Sherif and Al-Rashed, Balkema Publishers, Tokyo.

Patel, B.A., P.G. Shah, M.F. Raj, B.K. Patel and J.A. Patel (2001). Dissipation of lindane in/on brinjal and okra fruits. *Pesticide Res. J.*, 13(1): 58–61.

Pignatello, J.J. (1992). Dark and photoassisted Fe^{3+}–catalyzed degradation of chlorophenoxy herbicides by hydrogen peroxide. *Env. Sci. Tech.*, 26(5): 944–951.

Rodriguez Gonzalo, E, M.J. Sanchez Martin and M. Sanchez Camazano (1993). Adsorption of parathion and paraoxon by modified montmorillonites. *Toxicol. Env. Chem.*, 37 (3–4): 157–163.

Roy, A.K. (2003). Mycological problems of crude herbal drugs– Overview and challenges. *Indian Phytopath.*, 56 (1): 1–13.

Saha, A.K. and A. Sarkar (1991). Chemical detoxification process for treatment of soils contaminated with halogenated organic compounds. United States Patent. US 5,051,030, 7 pp.; Issued Sep. 24, 1991. Applied Jun. 5, 1990. Assigned to Roy F. Weston, Inc., West Chester, PA, USA.

Sedlak, D.L., K.E. Dean, D.E. Armstrong and A.W. Andren (1991). Interaction of quicklime and polychlorobiphenyl–contaminated solids. *Env. Sci. Tech.*, 25 (11): 1936–1940.

Sinha, S. and M. Gopal (2002). Evaluating the safety of β-cyfluthrin insecticide for usage in eggplant crop. *Bull. Environ. Contam. Toxicol.*, 68 (3): 400–405.

Troxler, W.L., S.K. Goh and L.W.R. Dicks (1993). Treatment of pesticide contaminated soils with thermal desorption technologies. *Air and Waste*, 43(12): 1610–1619.

Waliszewski, S.M., A.A. Aguirre, R.M. Infanzon and J. Siliceo (2000). Variation of organochlorine pesticide levels in cow's milk during heat treatment. *Revista Internacional de Contaminacion Ambiental*, 16(2): 61–66.

Wani, S.A., B.N. Kowala, V.K. Rao (2000). Effect of cooking on the pesticide residues in goat tissues. *Indian J. Small Ruminants*, 6(1): 23–28.

Zabik, M.J., M.F.A. El-Hadidi, J.N. Cash, M.E. Zabik and A.L. Jones (2000). Reduction of azinphos-methyl, chlorpyrifos, esfenvalerate, and methomyl residues in processed apples. *J. Agril. Food Chem.*, 48(9): 4199–4203.

37

Microbial Metabolites: Future Ecofriendly Agrochemicals

Rajib Karmakar and Gita Kulshrestha

Division of Agricultural Chemicals, Indian Agricultural Research Institute, New Delhi – 110 012, India

Global agrochemical industry, since 1990s, has been evolving rapidly under the impact of a spate of mergers and acquisitions among the agrochemical companies in the top league. The leading players in the international market are engaged in grand alliances to mop up resources for the capital intensive activities of development, registration and marketing of new crop protection molecules and agro-biotechnological research for producing transgenic plant of different economically important agricultural crops for the world market.

Biopesticides have been gaining increased attention and interest among those concerned with developing environmentally friendly, safe and Integrated Crop Management (ICM)-compatible approaches and tactics for pest management. Biopesticides is a term that encompasses many aspects of pest control such as microbial (viral, bacterial and fungal) organisms, entomophagus nematodes, plant-derived pesticides (botanicals), secondary metabolites from microorganisms.(antibiotics), insect pheromones, genes used to transform crop to express resistance to insect, pests, fungal and viral attacks.

The increasing exploitation of secondary metabolite to synthesize new ecofriendly.agrochemicals either directly or as a template is getting popularized day-by-day in the arena of IPM phase. The path

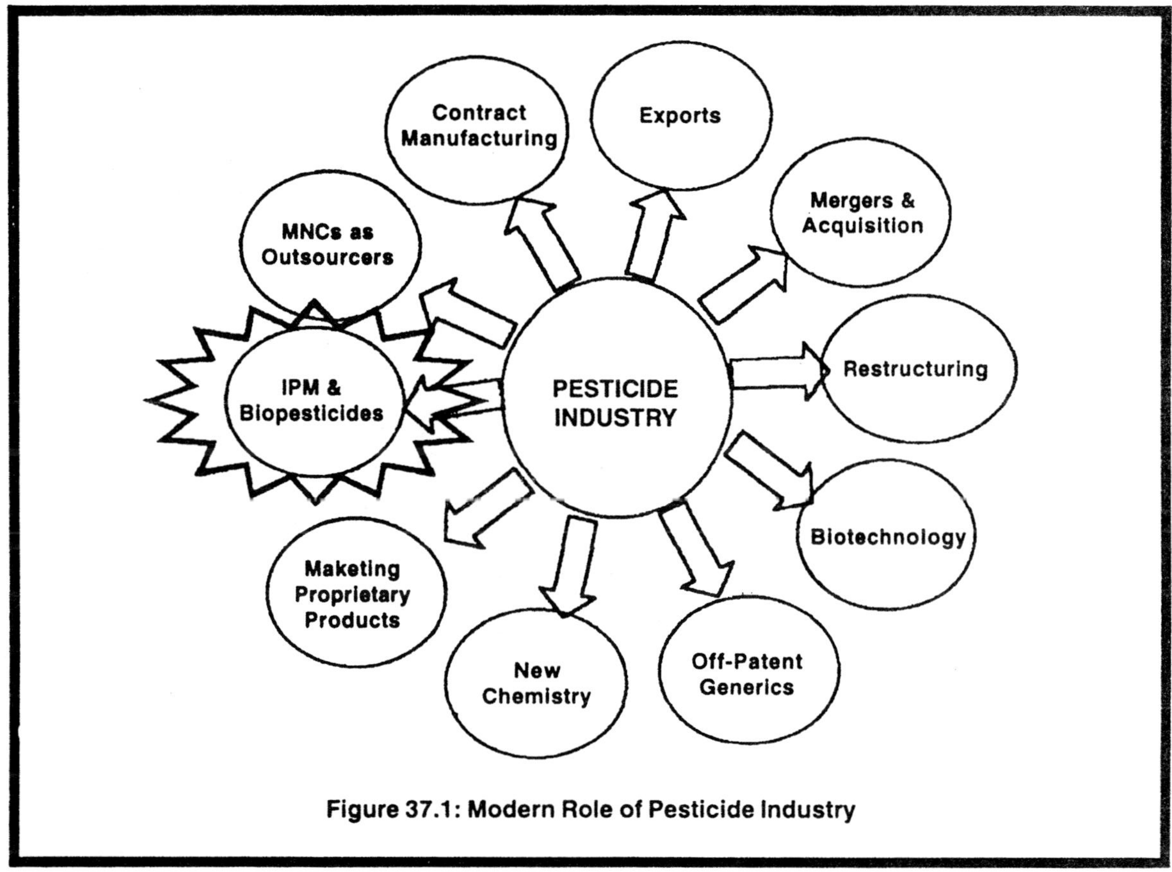

Figure 37.1: Modern Role of Pesticide Industry

to successful commercialization and large-scale production of microbial metabolites has proved long and difficult and yet, in the face of these challenges research and development of microbial metabolites have not only continued but also accelerated.

This chapter presents a review on the use and efficacy of microbial secondary metabolites that have potential as natural herbicides, insecticides, acaricides, nematicides etc., either directly or as templates for bio-rational ecofriendly agrochemicals. Their merits as alternatives to synthetic chemicals and biological control agents have been highlighted for a holistic approach in Integrated Pest Management (IPM).

Insecticides

Spinosyns and Spinosoids

Natural products can be an excellent source of novel chemistries. Among the recent insect control agents to reach the market place is the natural product, spinosad (Tracer®, Spin-Tor®, Success®), a naturally occurring mixture of spinosyn A (primary component) and spinosyn D. The source of activity was found to be a new species of actinomycete, *Saccharopolyspora spinosa* Mertz and Yao. More than 20 spinosyns and 800 spinosoids (semi- synthetic analogues) have been isolated or synthesized respectively.

Structure Activity Relationship (SAR)

In the light of the excellent insecticidal properties of spinosyn A and spinosad, as well as their novel mode of action, a programme was undertaken, initially by Lilly Research Laboratories (LRL), and then by Dow Agrosciences, to explore spinosyn chemistry furher so as to both expand the spectrum and increase insecticidal potency (Sparks *et al.*, 2001). This programme consisted of preparing semi-synthetic derivatives/analogues of the spinosyns, termed spinosoids (Crouse *et al.*, 2001; Sparks *et al.*, 2000) as well as the continued isolation and identification of new spinosyns (naturally occurring analogues of spinosyn A) (Sparks *et al.*, 1999).

Spinosyn A is highly active against neonate larvae of the tobacco budworm *Heliothis virescens* (Table 37.1) as are spinosyn B, C, D and Q. As demonstrated by spinosyn E, F, S and Y, a reduction in the size of alkyl group at C 16 or C 21 (Figure 37.2 and Table 37.1) reduces *Heliothis virescens* activity.

Table 37.1: Chemical Structures and LC_{50} values of Various Products

Sl.No.	*Compound*	*R1*	*R2*	*R21*	*R16*	*R6*	*R2'*	*R3'*	*R4'*	*LC_{50}*
	Spinosyns									
1.	Spinosyn A	Me	Me	Et	Me	H	OMe	OMe	OMe	0.3
2.	Spinosyn B	H	–	–	–	–	–	–	–	0.4
3.	Spinosyn C	H	H	–	–	–	–	–	–	0.8
4.	Spinosyn D	–	–	–	–	Me	–	–	–	0.8
5.	Spinosyn E	–	–	Me	–	–	–	–	–	4.6
6.	Spinosyn F	–	–	–	H	–	–	–	–	4.5
7.	Spinosyn H	–	–	–	–	–	OH	–	–	5.7
8.	Spinosyn J	–	–	–	–	–	–	OH	–	> 80
9.	Spinosyn K	–	–	–	–	–	–	–	OH	3.5
10.	Spinosyn L	–	–	–	–	Me	–	OH	–	26
11.	Spinosyn M	H	–	–	–	–	–	OH	–	22.6
12.	Spinosyn N	H	–	–	–	Me	–	OH	–	40
13.	Spinosyn O	–	–	–	–	Me	–	–	OH	1.4
14.	Spinosyn P	–	–	–	–	–	–	OH	OH	> 64
15.	Spinosyn Q	–	–	–	–	Me	OH	–	–	0.5
16.	Spinosyn R	H	–	–	–	–	OH	–	–	14.5
17.	Spinosyn S	–	–	Me	–	–	OH	–	–	53
18.	Spinosyn T	–	–	–	–	–	OH	OH	–	> 64
19.	Spmosyn U	–	–	–	–	–	OH	–	OH	22
20.	Spinosyn V	–	–	–	–	Me	OH	–	OH	17
21.	Spinosyn W	–	–	–	–	Me	–	OH	OH	> 64
22.	Spinosyn Y	–	–	Me	–	–	–	–	OH	20

Contd...

Table 37.1–Contd...

Sl.No.	Compound	R1	R2	R21	R16	R6	R2'	R3'	R4'	LC_{50}
	Spinosoids									
23.	N-demethyl D	H	–	–	–	Me	–	–	–	5.6
24.	N-demethyl K	H	–	–	–	–	-	–	OH	9.9
25.	N, N-didemethyl K	H	H	–	–	–	–	–	OH	7.4
26.	N-demethyl P	H	–	–	–	–	–	OH	OH	> 64
27.	2'-H A	–	–	–	–	–	H	–	–	0.23
28.	2'-H D	–	–	–	–	Me	H	–	–	0.23
29.	2'-O-ethyl A	–	–	–	–	–	OEt	–	–	0.30
30.	3'-H A	–	–	–	–	–	–	H	–	0.36
31.	3'-O-ethyl A	–	–	–	–	–	–	OEt	–	0.03
32.	3'-O-n-propyl A	–	–	–	–	–	–	OnPr	–	0.05
33	3'-O-n-butyl A	–	–	–	–	–	–	OnBu	–	0.38
34.	3'-O-allyl A	–	–	–	–	–	–	Oallyl	–	0.06
35.	3'-O-CH_2CF_3 A	–	–	–	–	–	–	OCH_2CF_3	–	0.43
36.	4'-H A	–	–	–	–	–	–	–	H	0.41
37.	4'-O-ethyl A	–	–	–	–	–	–	–	OEt	0.24
38.	2',3' and 4'-tri-O-ethyl A	–	–	–	–	–	OEt	OEt	OEt	0.02
39.	**Cypermethrin**	–	–	–	–	–	–	–	–	0.18

With exception of spinosyn Q, loss of a methyl group from the rhamnose moiety at the 2′-position (spinosyn H and its analogues Q, R, S), 3′-position (spinosyn J and its analogues L, M, N) or 4′-position (spinosyn K and its analogues O, Y) all result in decrease in activity towards *H. virescens*

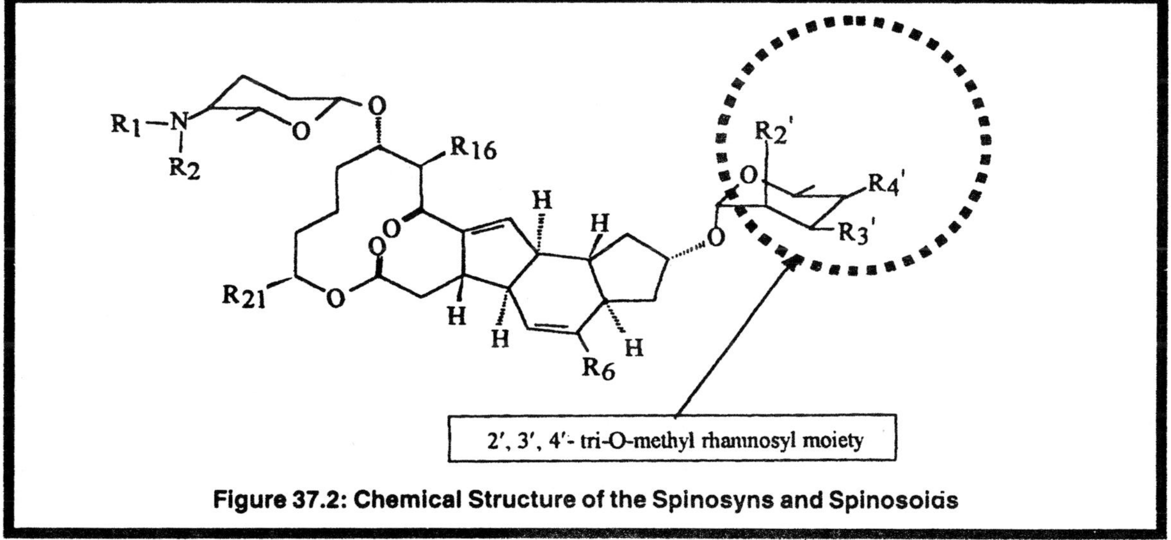

Figure 37.2: Chemical Structure of the Spinosyns and Spinosoids

larvae when compared to spinosyn A. The effect of losing of this methyl group is greatest at the 3'-position, where activity decreases in some cases by two orders of magnitude (Table 37.1). The di-demethyl rhamnosyl spinosyns (*e.g.*, spinosyns P, U, V and W) are only weekly active at best, with those lacking a 3'-methyl group being nearly inactive (Table 37.1.). As noted previously these data clearly demonstrate that relatively small changes in the spinosyns structure are often associated with large changes in biological activity. Although spinosyn A is slightly more active than spinosyn D in general spinosyns possessing a methyl group at C6 (spinosyn D related analogues) tend to be more active and less effected by changes in the rest 6 of the molecule than corresponding analogues possessing only a H at C6.

Role of Artificial Neural Network (ANN) to Spinosyn SAR

ANN is usually software-based systems that are modeled on concept derived from the mammalian nervous system. ANN-based QSAR studies of the spinosyn/spinosoids (Sparks *et al.*, 2000) suggested several possible approaches to improve activity, one of which involved extending the alkyl chain of the 2', 3' and 4'-O-methyl groups. Subsequent synthesis and testing of 4'-tri-O-ethyl analogue of spinosyn A clearly demonstrated a marked improvement in activity against *H. virescens* larvae. Further ANN analysis and synthesis pinpointed the 3'-position as the most critical for modification in the rhamnose moiety (Sparks *et al.*, 2000). As noted in the ANN analysis, modifications to the 3'-position of the rhamnose moiety have the greatest effect on activity towards *H. virescens* larvae (Table 37.1). Extending the alkyl chain from methyl to ethyl at the 2' and 4'positions (Table 37.1, compounds 29 and 37) produces no improvement in the activity compared with spiposyn A. In contrast, the 3'-O-ethyl analogue of spinosyn A (Table 37.1, compound 31) is about 10 folds more active than spinosyn A (Table 37.1, compound 1). Further extension of the 3'-P-ethyl chain to propyl, allyl or butyl results in no improvement in activity compared with the 3'-O-ethyl derivative (Table 37.1), and with the chain length reaching four carbons (butyl), there is a distinct decline in activity. Likewise, the presence of fluorine atoms on the 3'-O-alkyl group (3'-O-CH_2CF_3) reduces activity (Table 37.1). Thus for the 3'-position of the rhamnose moiety, there appears to be an alkyl chain length optimum at around 2–3 carbons.

Mode of Action

The spinosyns cause hyperexcitation, and ultimately disruption of the insect central nervous system. Electrophysiological evidence indicates that the spinosyns can alter nicotinic currents in neuronal cell bodies from the central nervous system of the American cockroach (*Periplanata americana* L.). This nicotinic effect is correlated with toxicity to neonate *H. virescens* larva. In addition to altering nicotinic function, electrophysiological evidence also indicates that spinosyns can disrupt the function of γ-amino butyric acid (GABA) receptors of small neurons from the central nervous system of *P. americana*. Insecticidal spinosyns and spinosoids disrupt GABA receptor function in these neurons, while spinosyns lacking insecticide activity do not. The initiation of a relatively small amplitude chloride conductance by the insecticidal spinosyns coincides with this loss of GABA receptor responses, and may underlie the loss of receptor function (Gerald Watson, Dow AgroSciences, pers comm). However, spinosyns and spinosoids have not been found to affect the binding of either nicotinic or GABA receptor radiolegands (nailah orr, Dow AgroSciences, pers comm), suggesting that the spinosyns donot interact directly with no nicotinic or GABA receptor binding sites. These results suggest that the spinosyns can affect nicotinic/GABA receptors through an undetermined mechanism that may underline the insecticidal properties of the spinosyns and their semi-synthetic analogues (spinosoids).

Natural Cyclopeptides from Fungi

Difficulties in synthesis make natural cyclopeptides challenging targets for chemists. Two natural toxic cyclopeptide series produced by pathogenic fungi: tentoxin [cyclo-(N-MeAla-Leu-N-MeΔ Phe-Gly)] and the dextruxins [cyclo-(Pro-Ile-N-MeAla-β-Ala-HA)]. The total synthesis of these two bioactive series were optimized and several analogues were designed and synthesized to establish SAR. Such systemic investigations can determine the crucial features responsible for the activity of the natural compound and help the design of more powerful and more selective products.

Dextruxins were first isolated by Kodaira (1961) from the entomopathogenic fungus *Metarhizium anisopliae*. Tentoxins were produced by phytopathogenic fungus of the same family, *Alternaria alternata*. The structures of destruxin and tentoxin analogues were shown below.

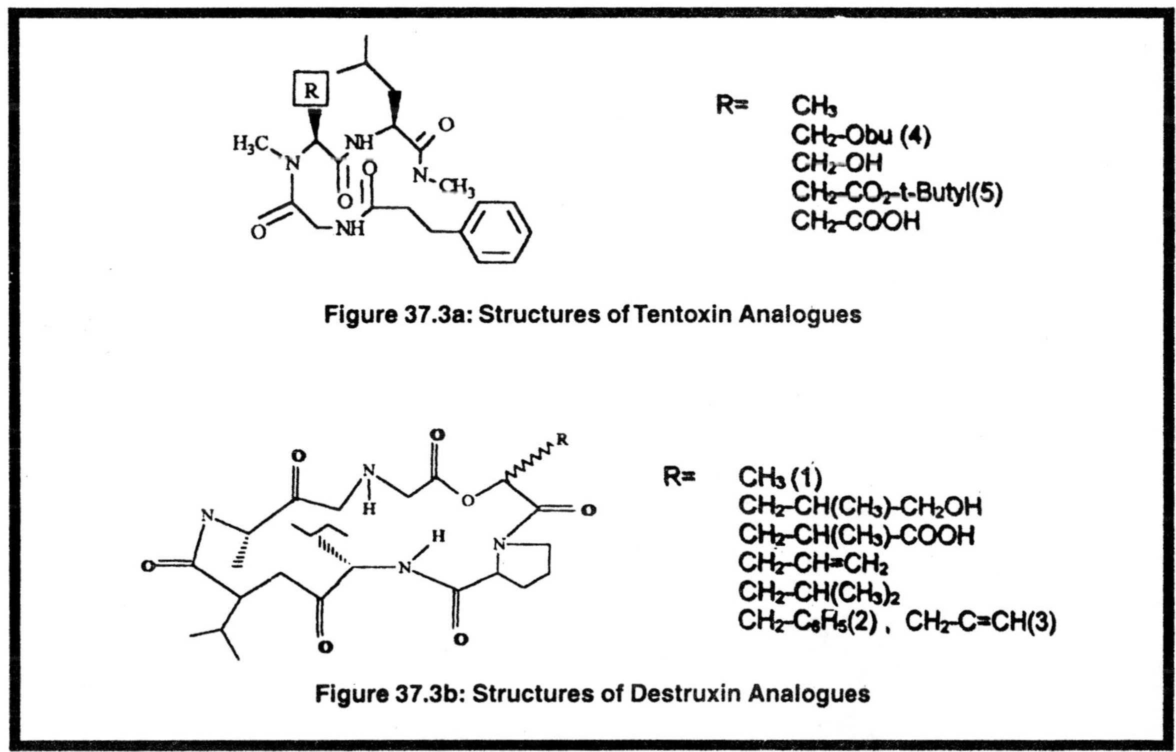

Figure 37.3a: Structures of Tentoxin Analogues

Figure 37.3b: Structures of Destruxin Analogues

Bacillus thuringiensis

B. thuringiensis (known as 'Bt') is an insecticidal bacterium, marketed worldwide for control of many important plant pests mainly caterpillars of the Lepidoptera (butterflies and moths) but also for control of mosquito larvae, and simuliid blackflies that vector river blindness in Africa. Bt products represent about 1 per cent of the total 'agrochemical' market (fungicides, herbicides and insecticides) across the world. The commercial *Bt* products are powders containing a mixture of dried spores and toxin crystals. They are applied to leaves or other environments where the insect larvae feed.

Recent studies on the delta-endotoxin structure show that it has three domains, Domain I is a bundle of 7 alpha helices, some or all of which can insert into the gut cell membrane, creating a pore through which ions can pass freely. Domain II consists of three antiparallel beta-sheets, similar to the antigen-binding regions of immunoglobulins. Domain III is a tightly packed beta-sandwich which is thought to protect the exposed end (C- terminus) of the active toxin, preventing further cleavage by gut proteases. Interestingly, the diphtheria toxin (of another bacterium) has an essentially similar structure to the Bt toxin.

B. thuringiensis strains produce two types of toxins. The main types are the Cry (crystal) toxins, encoded by different cry genes, and this is how different types of Bt are classified. The second types are the Cyt (cytolytic) toxins, which can augment the Cry toxins, enhancing the effecti veness of insect control. Over 50 of the genes that encode the Cry toxins have now been sequenced and enable the toxins to be assigned to more than 15 groups on the basis of sequence similarities.

Table 37.2: Bioactivity of Different Crystal Shapes of Cry Genes

Gene	*Crystal Shape*	*Insect Activity*
Cry I [several subgroups: A (a), A (b), A (c), B, C, D, E, F, G]	Bipyramidal	Lepidoptera larvae
Cry II [subgroups A, B, C]	Cuboidal	Lepidoptera and diptera larvae
Cry III [subgroups A, B, C]	Flat/irregular	Coleoptera
Cry IV [subgroups A, B, C, D]	Bipyramidal	Diptera

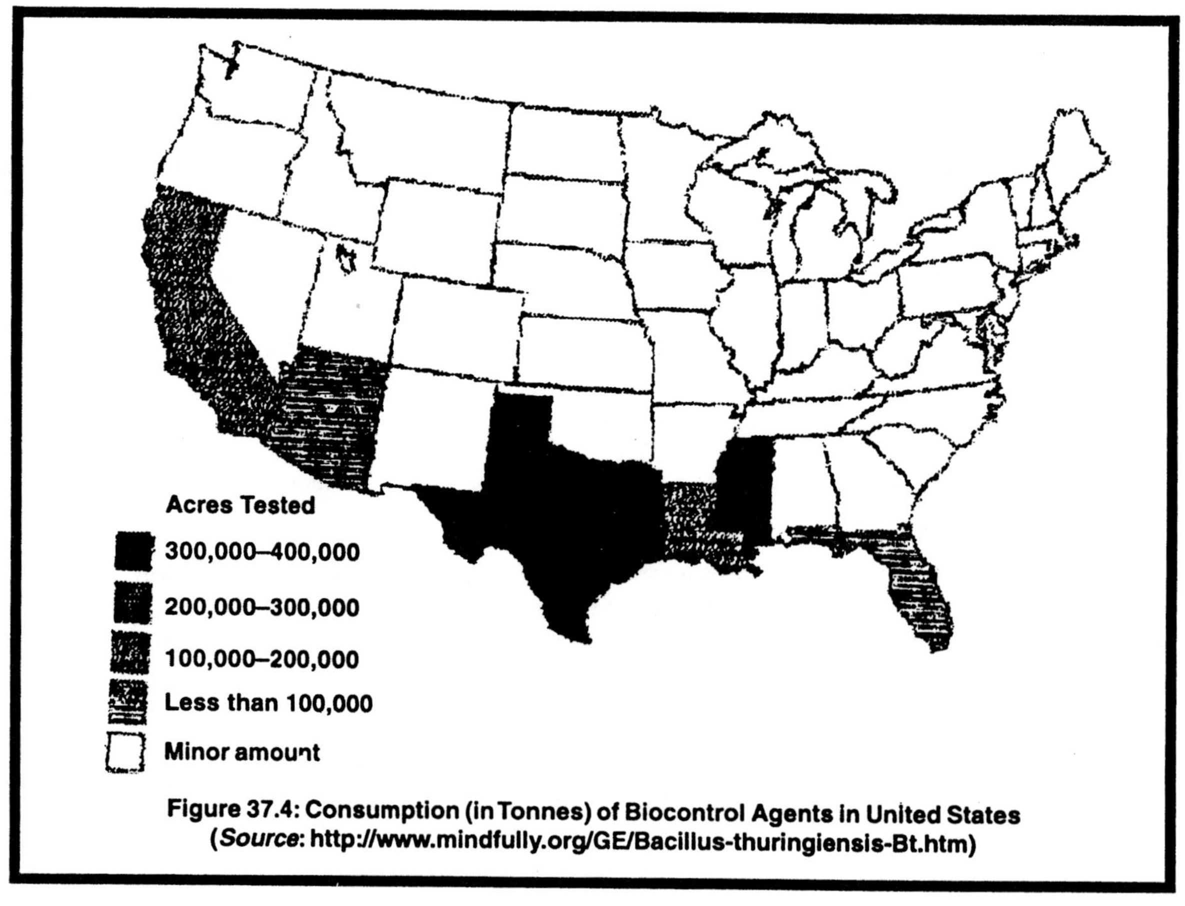

Figure 37.4: Consumption (in Tonnes) of Biocontrol Agents in United States (*Source*: http://www.mindfully.org/GE/Bacillus-thuringiensis-Bt.htm)

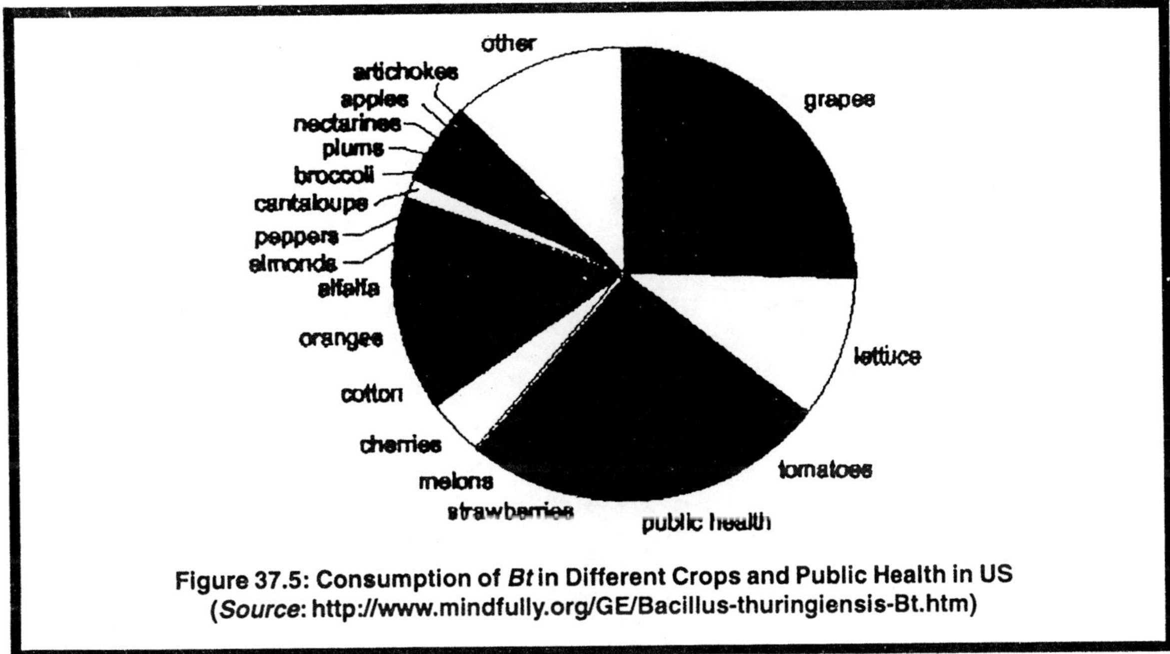

Figure 37.5: Consumption of *Bt* in Different Crops and Public Health in US (*Source*: http://www.mindfully.org/GE/Bacillus-thuringiensis-Bt.htm)

Fungicides and Bactericides

Strobilurin

Strobilurins are one of the most important classes of agricultural fungicides. Their invention was inspired by a group of fungicidally active natural products. The outstanding benefits they deliver are currently being utilized in a wide range of crops throughout the world. First launched in 1996, the strobilurins now include the world's biggest selling fungicide, azoxystrobin. Now six Strobilurin active ingredients are commercially available for agricultural use.

The discovery of Strobilurin fungicides was inspired by a natural fungicidal derivatives of β-methoxyacrylic acid, the simplest of which are Strobilurin A; Oudemansin A, and Myxothiazol A. These natural products are produced by a wide range of Basidiomycete, wood-rotting fungi, for example *Oudemansiella mucida* (Schrad ex Fr) Hoehn and *Strobilurus tenacellus* (Pers ex Fr) Singer and, in the case of myxothiazol A (Figure 37.6), the gliding bacterium *Myxococcus fulvus* (Bartlett *et al.*, 2002).

Table 37.3: Companies Marketing Strobilurins and Related Fungicides

Fungicides	*Company*	*Announced*	*First Sale*
Azoxystrobin	Syngenta	1992	1996
Kresoxim-methyl	BASF	1992	1996
Metominostrobin	Shionogi	1993	1999
Trifloxystrobin	Bayer	1998	1999
Picoxystrobin	Syngenta	2000	2002
Pyraclostrobin	BASF	2000	2002
Famoxadone	DuPont	1996	1997
Fenamidone	Aventis	1998	2001

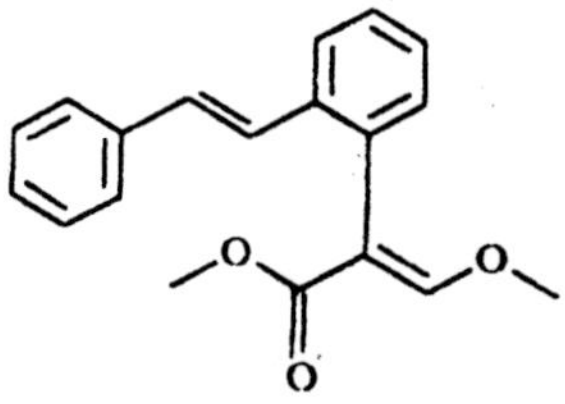

Strobilurin A

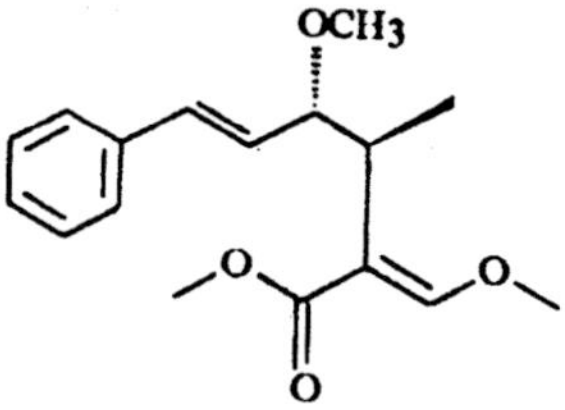

Oudesmanin A

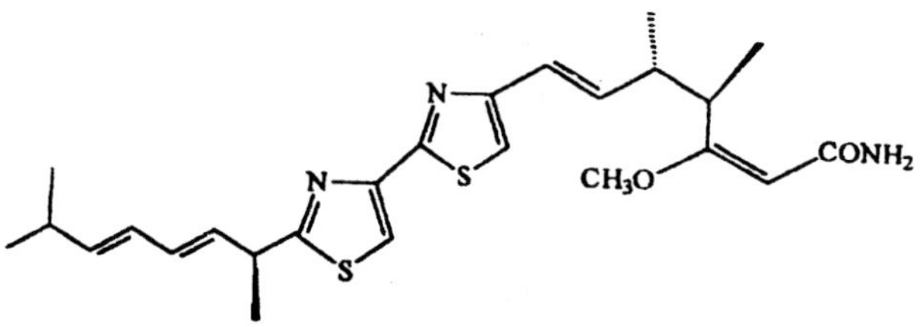

Myxothiazole A

Figure 37.6: Chemical Structures of Various Biofungicides

The natural products are unsuitable for use as agricultural fungicides in their own right. However a knowledge of their structures and physical properties provided the starting point for independent programmes of research within Syngenta and BASF which led, after several years' work, to the discovery and almost simultaneous launch of the first of the synthetic strobilurins. The two companies have each published detailed accounts of this work. For Syngenta, this programme led to the discovery of azoxystrobin, selected from about 1400 synthesized compounds (Godwin *et al.*, 1992; Clough *et al.*, 1998). For BASF, the result was kresoxim-methyl (Ammermann *et al.*, 1992; Sauter *et al.*, 1999).

Milestone in the discovery of azoxystrobin are shown in Figure 37.7. Strobilurin A, although active against fungi growing on agar, does not control fungi growing on plants in the glasshouse

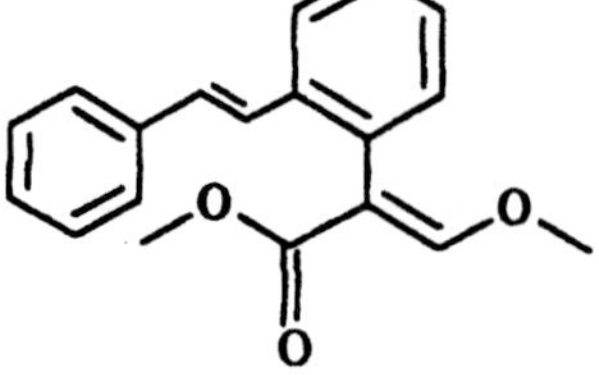

Stilbene- less volatile and much more stable in light than the natural products

1

Figure 37.7–Contd...

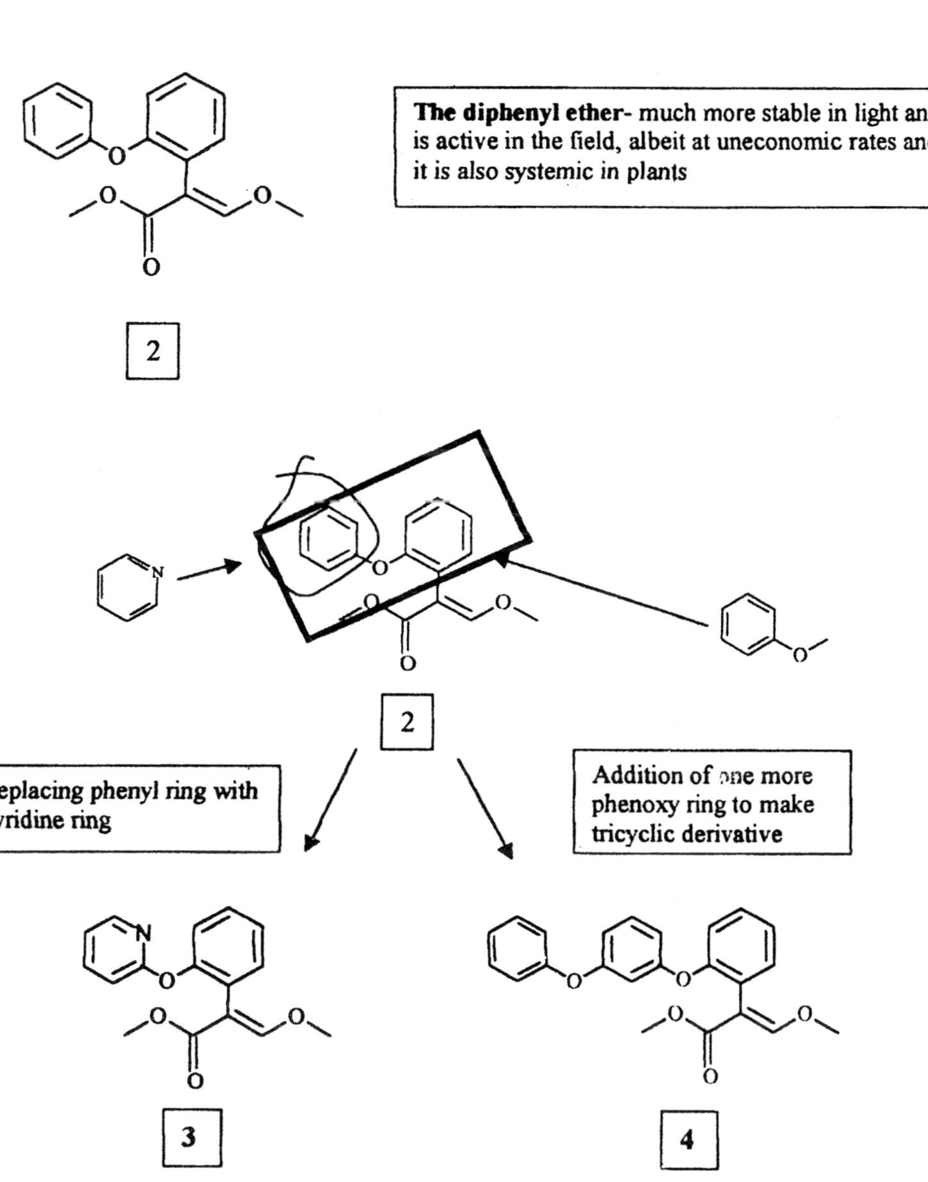

The pyridine retains the activity of its isostere [2] and, as expected, is highly mobile (probably too mobile) in plants.

The tricyclic derivative has improved levels of fungicidal activity, but is too lipophilic to determine systemic movement

Figure 37.7–Contd...

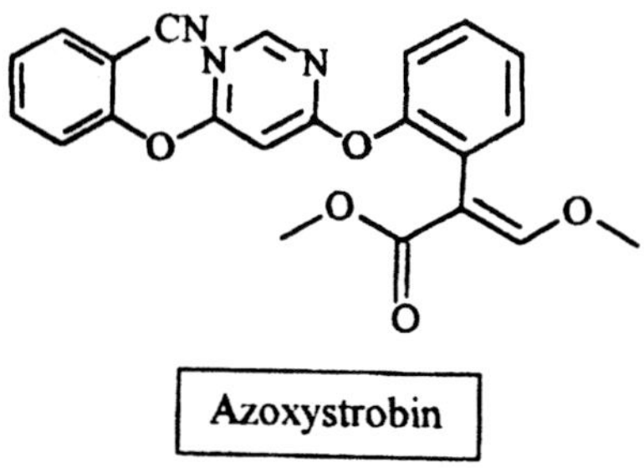

Figure 37.7: Structure and History of Discovery of Azoxystrobin

because of its photochemical instability and volatility. However, it was discovered that the (Z)-olefinic bond of strobilurin A could be replaced by an O-disubstituted benzene ring to give the stilbene [Figure 37.7(1)], which is less volatile and much more stable in light than the natural products. The stilbene is active in the glasshouse but still degrades too quickly in light to express high activity in the field. The diphenyl ether [Figure 37.7(2)] is much more stable in light and is active in the field, albeit at uneconomic ràtes. Furthermore (2) is systemic in plants, an important property of many modern fungicides which improves field performances by redistribution of the (2) within plant tissue after spraying. Modification to the compound led to the analouges [Figure 37.7(3)] and [Figure 37.7(4)]. The pyridine (3) retains the activity of its isostere (2) and, as expected, is highly mobile (probably too mobile) in plants. The tricyclic derivative (4) has improved levels of fungicidal activity, but is too lipophilic to determine systemic movement. A combination of these ideas which led from (2) to (3) and (4), together with a huge amount of experimentation, led finally to the discovery of azoxystrobin. The synthetic scheme of preparation of azoxystrobin is shown in Figure 37.8.

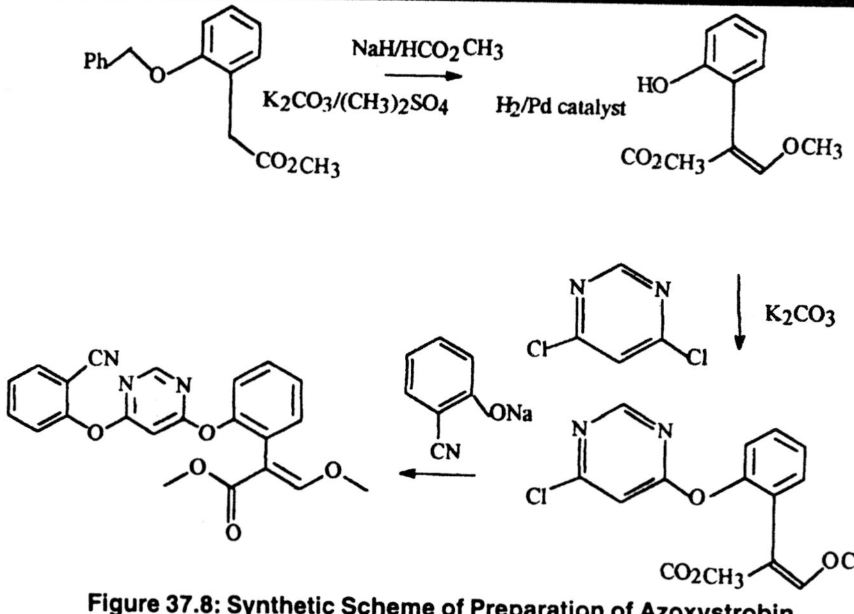

Figure 37.8: Synthetic Scheme of Preparation of Azoxystrobin

Milestone in the discovery of kresoxim-methyl are shown in Figure 37.9. Again, the stilbene [Figure 37.7(1)] was prepared as stabilized analogue of strobilurin A. The reduced analogue (5) and the isosteric benzyl phenyl ether (6) were then prepared and found to be fungicidal,.and the frame work of (6) is present in kresoxim-methyl.

The most widely used route to compounds containing methyl-β-methoxyacrylate toxophore involves a Claisen condensation between a methyl phenyl acetate and methyl formate in the presence of a base such as sodium hydride, followed by O-methylation of the resulting methyl β-hydroxyacrylate. The sequence is highly streoselective, and leads to the desired (E)-isomer almost exclusively. The wittig reaction shown in Figure 37.10 has also been used to construct the methyl-β-methoxyacrylate group, but in this approach an (E/Z)-mixture of isomers is usually obtained (Beautement *et al.*, 1991).

The leading approach to the methyl methoxy imino acetate toxophore is shown in Figure 37.11. Methyl benzoyl formates are readily converted into compounds containing the methyl methoxyiminoacetate toxophore by treatment with methylamine. (Hayase *et al.*, 1995; Hayase *et al.*, 1989).

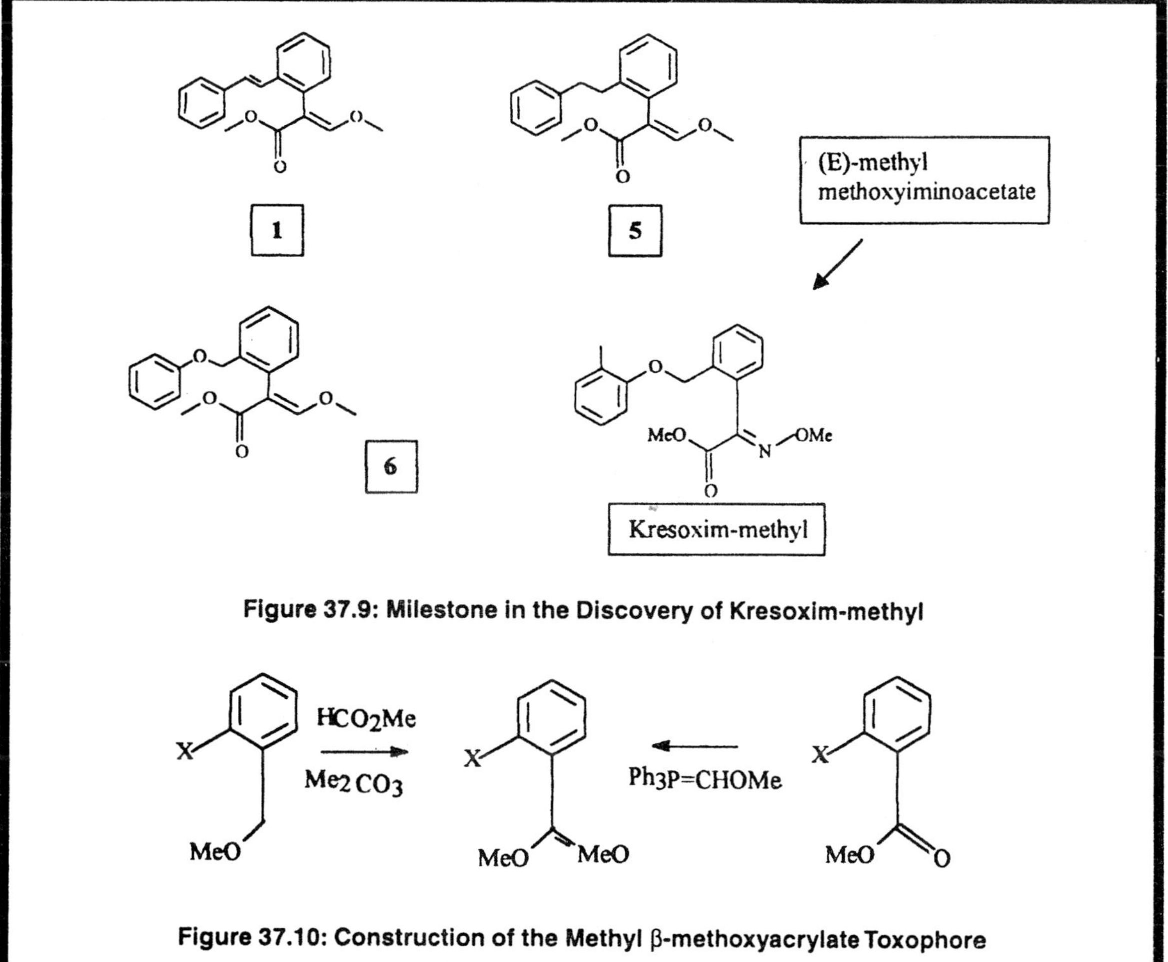

Figure 37.9: Milestone in the Discovery of Kresoxim-methyl

Figure 37.10: Construction of the Methyl β-methoxyacrylate Toxophore

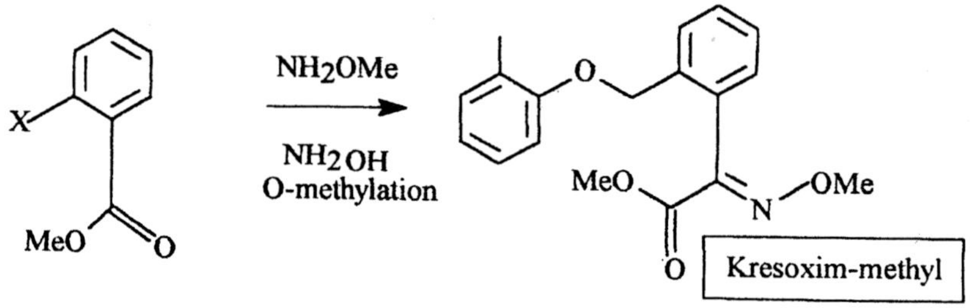

Figure 37.11: Construction of the Methyl Methoxyiminoacetate

Trichoderma sp. Metabolites as Fungicide Lead

Problems with fungicides and soil fumigants such as methyl bromide have let researchers to seek better methods to protect crops against soil borne pathogens. Biocontrol with antagonistic microbes such as the fungus *Trichoderma* is one area of research. *Trichoderma* has been used for years for biocontrol of pathogens.

Trichoderma (*Gliocladium*) *virens* is one of the most widely studied species of fungal biocontrol agents. It has proven effective as a suppressant of several soil-borne pathogens (G.L. Papavizas, 1985), but *T. virens* has received little attention as a biocontrol agent for wood attacking fungi. However, T.L. Highley and L. Ferge (1995) found that a commercially formulated *T. virens* biofungicide, Gliogard-GL-21 (W.R. Grace and Co., CT) 3, was an effective pretreatment in the protection of wood against decay by all the white-rot and brown-rot fungi examined in laboratory tests.

Antagonistic strains belonging to *Trichoderma* genus are able to produce various secondary metabolites which can play a role in the mechanism of their biological activity. Particularly, they produce antifungal secondary metabolites, which by spreading through soil cracks could ensure to the antagonists an ecological advantage in the competition for the soil colonization against the pathogens (http://www.item. ba.cnr.it/biopesti.htm).

Many of the metabolites produced by *Trichoderma* sp. serve as a lead for fungicides. These are shown in Figure 37.12.

Fungicide from *Pseudomonas* sp.

Soil borne bacteria which interact with plant roots and protect them against pathogenic microorganisms are commonly called plant growth-promoting rhizobacteria (Kloepper and Schrot, 1978). Pseudomonus are common member of the plant growth-promoting rhizobacterial microflora (Thomashow, 1996). The ability of Pseudomonads to suppress soil borne fungal pathogens depends in part of their ability to produce antifungal secondary metabolites (Vincent et al., 1991). Among the antifungal compounds synthesized by the *Pseudomonads*, 2,4-diacetyl phloroglucinol (DAPG) is a key secondary metabolite associated with the biological control of a range of plant pathogens worldwide (Defago, 1993). The chemical structures of pyrrolnitrin and pyoluteorin, derived have been shown in Figure 37.13.

Fusarium wilt diseases, caused by the fungus *Fusarium oxysporum*, lead to significant yield losses of crops. One strategy to control fusarium wilt is the use of antagonistic, root- colonizing *Pseudomonas* sp. It has been demonstrated that different strains of these bacteria suppress disease by different

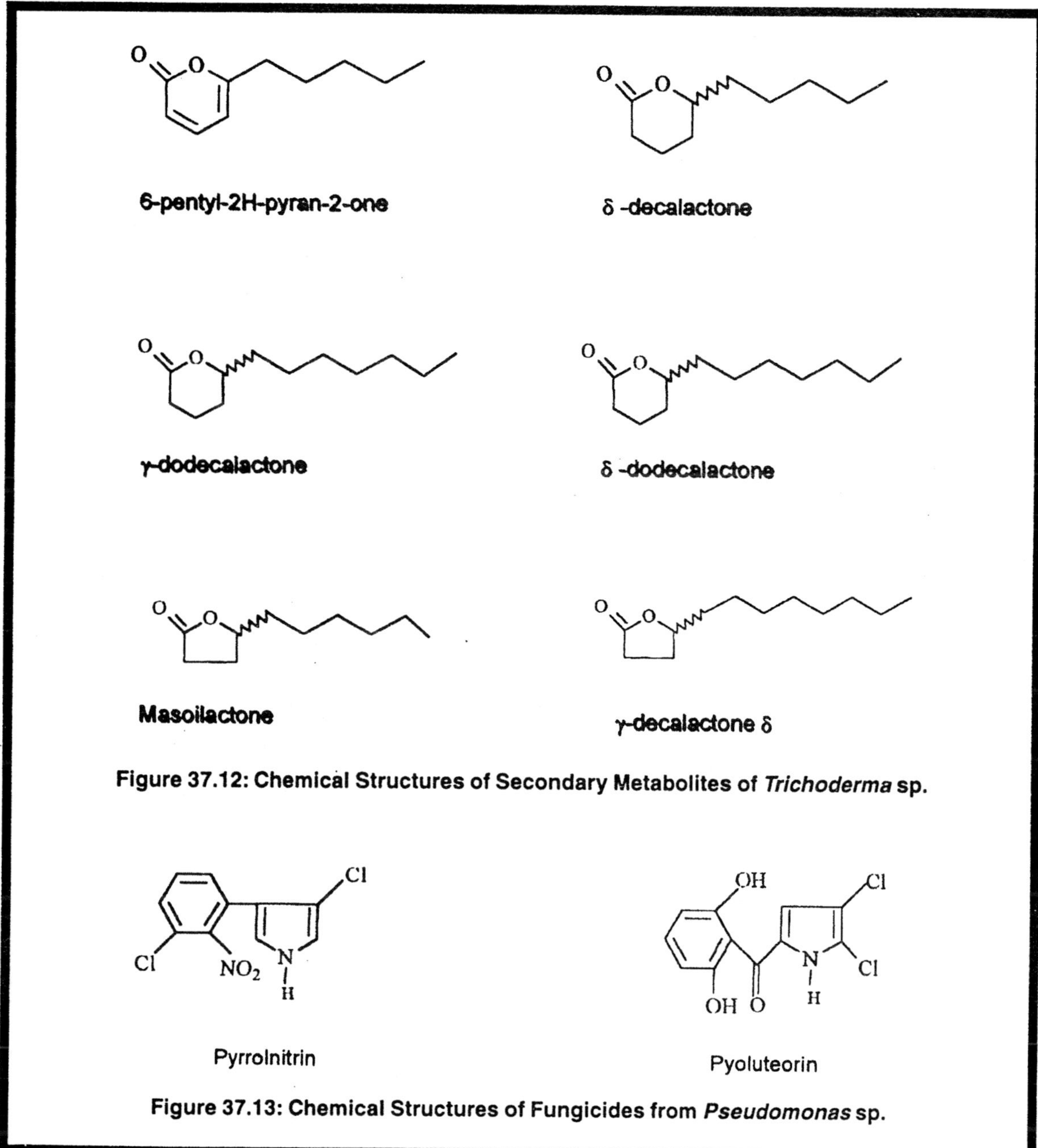

Figure 37.12: Chemical Structures of Secondary Metabolites of *Trichoderma* sp.

Figure 37.13: Chemical Structures of Fungicides from *Pseudomonas* sp.

mechanisms. Therefore, application of a mixture of these biocontrol strains, and thus of several suppressive mechanisms, may represent a viable control strategy. A prerequisite for biocontrol by combinations of biocontrol agents can be the compatibility of the co-inoculated microorganisms (Marjan de Boer *et al.*, 1999) Figure 37.14 describes the mechanism.

Pseudomonads are producing a variety of compounds that can serve as a good lead for the synthetic fungicides. The different secondary metabolites that are produced by Pseudomonads are described in Figure 37.15 (Copping and Menn, 2000).

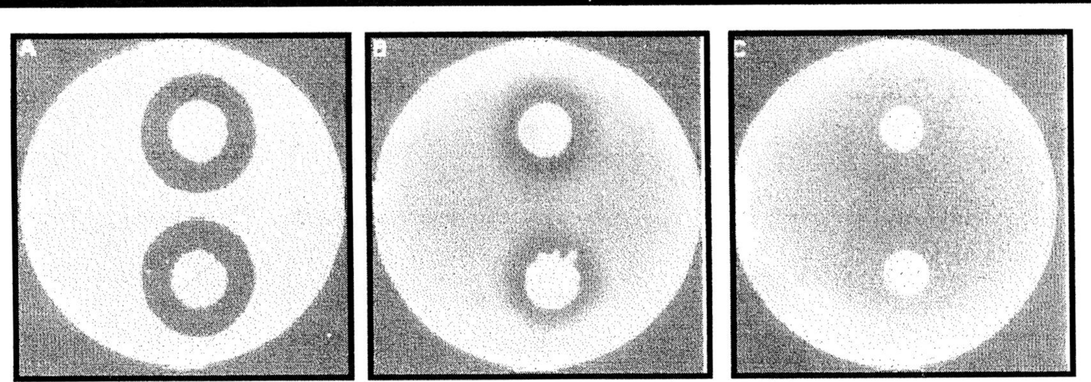

Figure 37.14: *In vitro* Antagonism Between Fluorescent *Pseudomonas* spp. Strains.

The strains were spot-inoculated by pipetting 2 droplets of 5 µl of a suspension (10^7 cfu/ml) on KB agar plates. After 48h of growth at 27°C a suspension of the target strain (10^7 cfu/ml) was atomized over the spot-inoculated plates. After an additional 24h several interactions were observed: A; the atomized strain was inhibited strongly by the spot- inoculated strain resulting in a clear zone of strain inhibition. B; the sprayed strain, was slightly inhibited resulting in a zone of reduced growth or C; the sprayed strain was not inhibited by the spot-inoculated strain.

Toxins from *Fusarium* sp.

Two types of compounds nonaketide and heptaketide were isolated from two week old Sorghum culture of *F. oxysporum* f *sp. Vasinfectum* ATCC isolate 24908 (Bell *et al.*, 2003). The structures are shown in Figure 37.16.

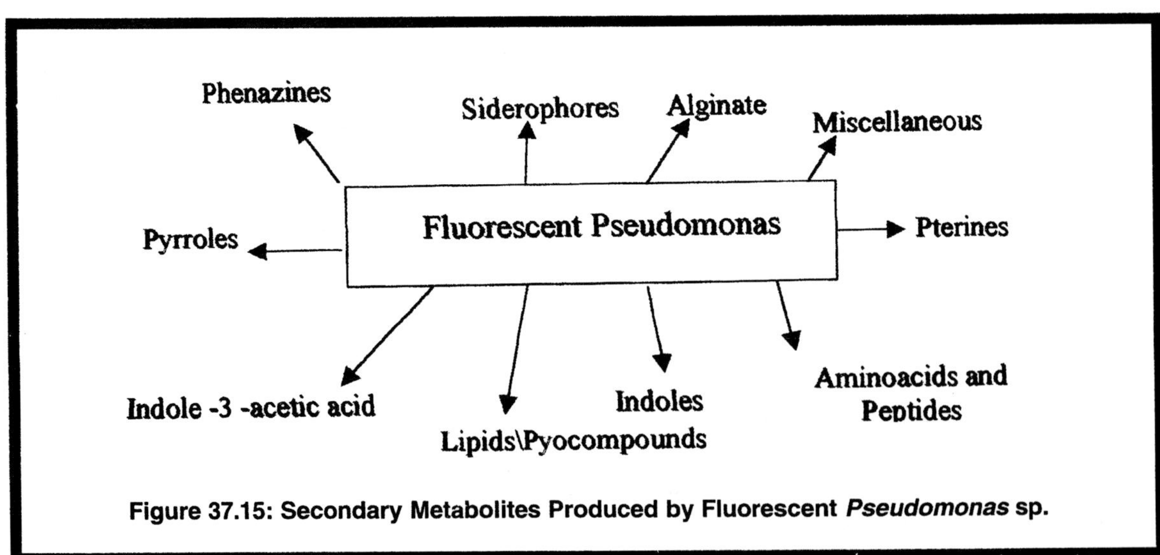

Figure 37.15: Secondary Metabolites Produced by Fluorescent *Pseudomonas* sp.

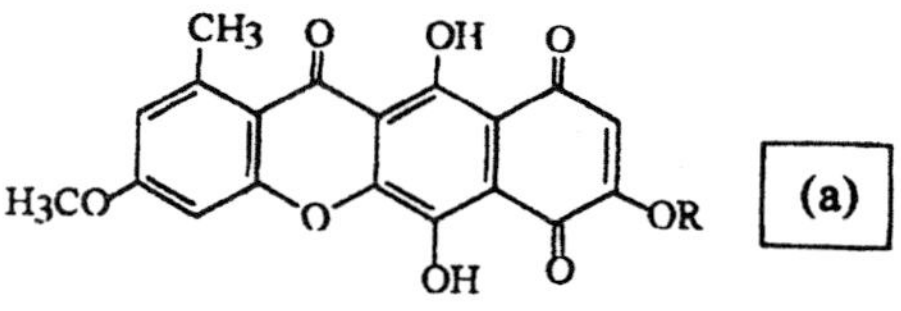

Bikaverin: R= $-CH_3$

Norbikaverin: R= -H

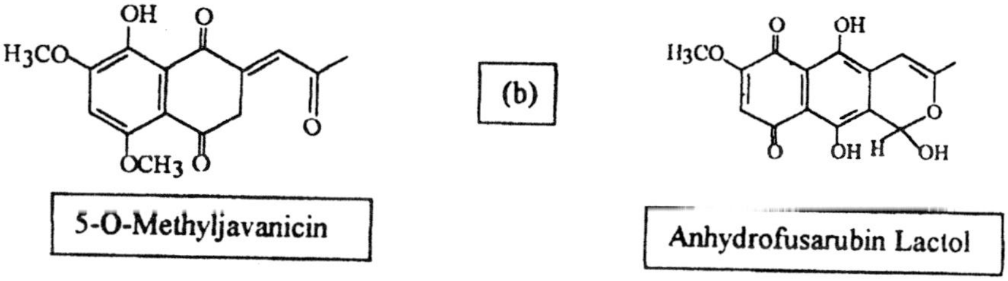

Figure 37.16: Structures of (a) Nonaketide and (b) Hepaketide Toxins from *F. oxysporum* f sp. *Vasinfectum*

Herbicides

Streptomyces sp.

Herbicidal activity of the compounds excreted/secreted by the *Streptomyces* sp. Actinomycetes group came into the limelight with the discovery of bialaphos from *Streptomyces hygroscopica* (Saxena and Pandey, 2001). In plant, it is converted to phosphinothricin by the hydrolytic cleavage of two alanyl groups and irreversibly inhibits GS (Omura *et al.*, 1984)

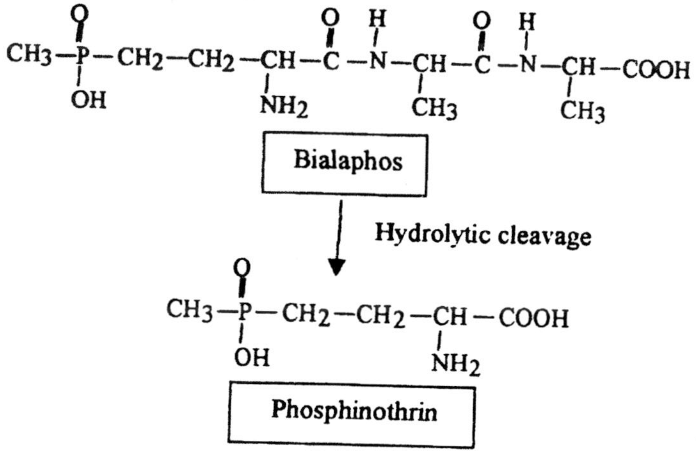

Figure 37.17: Biochemical Transformation of Bialophos in Plant

Anisomycin from *Streptomyces* sp. was the first microbial product used for the development of the synthetic herbicide methoxyphenone (3,3-dimethyl-4-methoxy benzophenone) (Yamada *et al.*, 1974). Herbimycin is a new benzaquinoid ansamycin antibiotic found to be an efficient herbicidal agent. It is produced by *Streptomyces* sp. AM 3672.

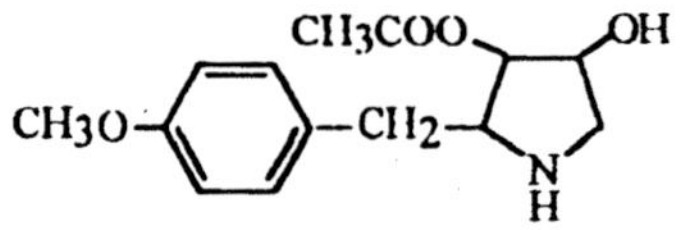

Anisomycin

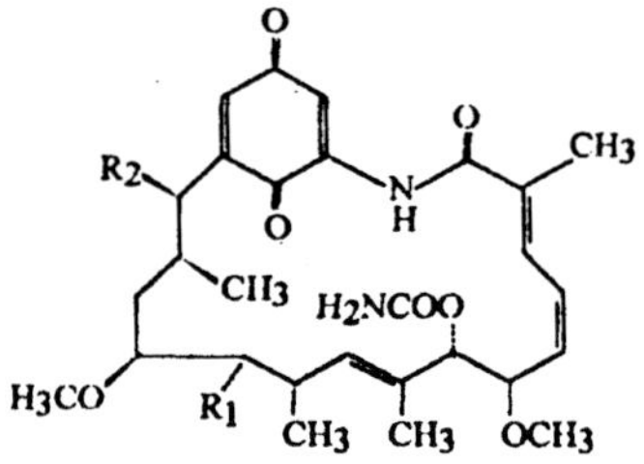

Herbimycin

Figure 37.18: Chemical Structures of Anisomysin and Herbimycin

Herbicide Leads from Fungi

Plant pathogen especially fungi are capable of inducing symptoms of disease in their respective hosts by virtue of the phytotoxin they produce (Strobel *et al.*, 1992). Over the past decades, toxin production of a few fungal pathogens causing diseases on some important weedy plants has been examined by some groups (Cutler, 1988). These have been categorized as host-specific and non host-specific This is described in Figure 37.19.

Conclusion

Microbial secondary metabolites offer the best source of novel compounds with bioactivity against various pests. Tremendous efforts have been underway in chemical characterization of these bioactive molecules. Most of the information available is not useful in evaluating their potential as pesticides. In only a few cases in which such compounds have been discovered to be effective has a mechanism of action been determined. In the few cases in which a molecular target site has been established, it has generally been one that has not yet exploited. The potential for future discovery is enormous. With the rapidly increasing incidents of pests becoming resistant to herbicides, new mechanism of action will be a highly desirable trait in agrochemicals of future.

Some microbial products such as bialaphos and glufosinate (Phosphinothricin) provide ample evidence of the potential of already discovered compounds for development and marketing. These serve as example leads for further increasing the interest in microbial products as pesticides. Modification of the structures of microbial toxins will improve their ability as pesticides.

There are, however, certain obstacle to success with such a strategy including the structural complexity of many microbial toxins, difficulty in working with very small amounts of materials, and the cost of rediscovering new compounds.

HERIBICIDE LEADS FROM *FUNGI*

HOST SPECIFIC	NON-HOST SPECIFIC
Some compounds are so specific that they are only toxic to certain cultivars	Non-host-specific phytotoxins have a broader range of activity and applicability
✓ Maculosin (from *Alternaria alternata* and Host specific to *Cantaeurea maculosa*) ✓ Bipolaroxin (from *Bipolaris cyanodontis*)	✓ Cornexitin from *Paecilomyces variotii* ✓ Tentoxin

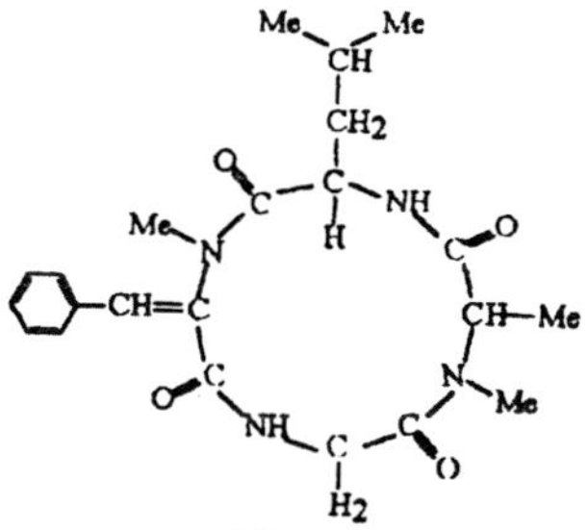

Tentoxin

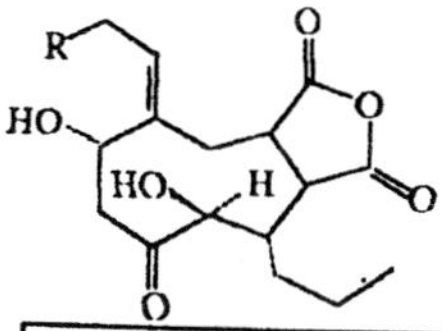

Cornexistin (R=H)

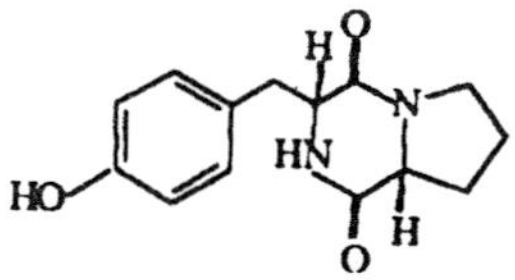

Maculosin

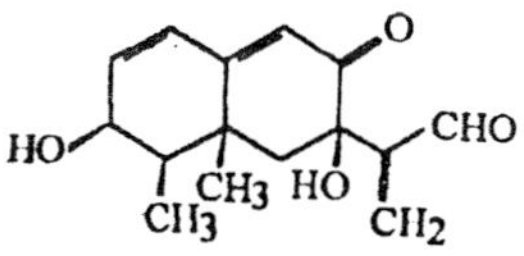

Bipolaroxin

Figure 37.19: Chemical Structures of Host Specific and Non-host Specific Bioherbicides

REFERENCES

Ammermann, E., G. Lorenz, K. Schelberger, B. Wenderoth, H. Sauter, C. Rentzea (1992). BAS 490 F: A broad spectrum fungicide with a new mode of action. In: *Proc. Brighton Crop Peotect. Conf. Pests and Diseases*, BCPC, Farnham, Surrey, UK, pp. 403–410.

Bartlett, D.W., J.M. Clough, J.R. Godwin, A.A. Hall, M. Hamer and B. Parr-Dobrzanski (2002). The Strobilurin fungicides. *Pest Mgt. Sci.*, 58: 649–662.

Beautement, K., J.M. Clough, P.J. de Fraine and C.R.A. Godfrey (1991). Fungicidal β-methoxyacrylates: from natural products to novel synthetic agricultural fungicides. *Pest. Sci.*, 31: 499–519.

Bell, A.A., H.H. Wheeler, J. Liu, R.D. Stipanovic, L.S. Puekhaber and H. Orta (2003). United State Department of Agriculture-Agricultural Research Service studies on polyketide toxins of *Fusarium oxysporum* f sp. *vasinfectum:* Potential target for disease control. *Pest Mgt. Sci.*, 59: 736–747.

Clough, J.M. and C.R.A. Godfrey (1998). The strobilurin fungicides, in Fungicidal activity, chemical and biological approaches to plant protection. (Eds.) D.H. Hutson and J. Miyamoto, Wiley series in Agrochemicals and Plant protection, John Wiley and Sons, pp. 109–148.

Copping, L.G. and J.J. Menn (2000). Biopesticides: A review of their action, application and efficacy. *Pest Mgt. Sci.*, 56: 651–676.

Crouse, G.D., T.C. Sparks, J. Schoonover, J.M. Gifford, T. Bruce, T.V. Worden and J.G. Martynow (2001). Recent advances in the chemistry of the spinosyns. *Pest Mgt. Sci.*, 57: 177–185.

Curtler, H.G. (1988). Perspective on discovery of microbial phytotoxins with herbicidal activity. *Weed Tech.*, 2: 525.

Defago, G. (1993). 2,4-diacetyl phloroglucinol, a promosing compound in bio control. *Plant Patholo.*, 42: 311–312.

Godwin, J.R., V.M. Anthony, J.M. Clough and C.R.A. Godfrey (1992). ICIA5504: A novel, broad spectrum, systemic β-methoxyacrylate fungicide. In: *Proc. Brighton Crop Peotect Conf. Pests and Diseases*, BCPC, Farnham, Surry, UK, pp. 435–442.

Hayase, Y., T. Kataoka, M. Masuko, M. Niikawa, H. Takenaka, M. Ichinari, T. Takahashi, Y. Hyashi and R. Takeda (1995). Phenoxyphenyl alkoxyiminoacetamide: New broad-spectrum fungicides. In: *Synthesis and Chemistry of Agrochemicals* IV, (Eds.) D.R. Baker, J.G. Fenyes and G.S. Basarab. ACS Symposium Series No. 584, American Chemical Society, Washington DC, USA, 30: 343–353.

Hayase, Y., T. Kataoka, H. Takenaka, M. Ichinari, M. Masuko, T. Takahashi and N.S. Tanimoto (1989). Preparation of substituted phenyl (alkoxyimino) acetamides and their use as fungicides. *European Patent Appl.*, pp. 398–692.

Highley, T.L. and L. Ferge (1995). Antagonism of Gliocladium virens against wood decay fungi. Internat. Res. Group on Wood Pres., Doc. No. IRG/WP95–10102.

http://www.item.ba.cnr.it/biopesti.htm.

http://www.mindfully.org/GE/Bacillus-thuringiensis-Bt.htm.

Kloepper, J. and M. Schrot (1978). Plant growth promoting rhizobacteria on radishes. *Proc. Int. Conf. Plant. Pathog. Bacateriol.*, 2: 879–882.

Kodaira, Y. (1961). Biochemical studies on the muscardine fungi in the silkwork, *Bombix moro. J. Fac. Textile Sericult., Shinshu Univ. Ser.*, E. 29: 1–68.

Marjan de Boer, Ientse van der Sluis, Leendert C. van Loon and Peter A.H.M., Bakker (1999). Combining fluorescent *Pseudomonas* spp. strains to enhance, suppression of *fusarium* wilt of radish. *European J. Pl. Path.*, 105: 201–210.

Mertz, F.P. and R.C. Yao (1990) *Saccharopolyspora spinosa* sp. nov isolated from soil collected in a sugar mill rum still. *Internat J. Syst Bacteriol.*, 40: 34–39.

Omura, R.S., M. Murata, H. Hanaki, K. Hiotozawa, R. Oiwa, H. Tanaka (1984). Phosalacine: A new herbicidal antibiotic containing phosphinothricin. Fermentation, isolation, biological activity and mechanism of action. *J. Antibiot.*, 37: 829.

Papavizas, G. (1985). Trichoderma gliocladium: Biology, ecology and potential for biocontrol. *Ann. Rev. Phytopathol.*, 24: 23–54.

Sauter, H., W. Steglich and T. Anke (1999). Strobilurins: Evolution of a new class of active substances. *Angew. Chem. Internat. Edn.*, 38: 1328–1349.

Saxena, S. and A.K. Pandey (2001). Microbial metabolites as ecofriendly agrochemicals for the next millennium. *Appl. Microb. Biot.*, 55: 395–403.

Sparks, T.C., P.B. Anzeveno, J.G. Martynow, J.M. Gifford, M.B. Hertlein, T.V. Worden and H.A. Kirst (2000). The application of artificial neural network to the identification of new spinosoids with improved biological activity towards larvae of *Heliothis virescens*. *Pestic. Biochen. Physiol.*, 67: 187–197.

Sparks, T.C., G.D. Crouse and G. Durst (2001). Natural products as insecticides: The biology, biochemistry and quantitative structure-activity relationships of spinosyns and spinosoids. *Pest Mgt. Sci.*, 57: 896–905.

Sparks, T.C., G. Durst and T.V. Worden (2000). Structure activity relationship of the spinosyns. *Proc. 2000 Beltwide Cotton Production Conf.*, National Cotton Council, Memphis, TN, USA, pp. 1225–1229.

Sparks, T.C., G.D. Thompson, H.A. Kirst, M.B. Hertlein, J.S. Mynderse, J.R. Turner and T.V. Worden (1999). Fermentation derived insect control agents: The Spinosyns. In: *Biopesticide Use and Delivery*, (Eds.) F. Hall and J.J. Menn, Humana, Totowa, NJ., USA, pp. 171–188.

Strobel, G., F. Sugawara, J. Hersenhom (1992). Pathogens and their products affecting weedy plants. *Phytoparasitica*, 20: 307.

Thomashow, L.S. (1996). Biological control of plant root pathogens. *Curr. Opin. Biotechnol.*, 7: 343–347.

Vincent, M., L. Harrison, J. Brackin, P. Kovacevich, P. Mukherji, N.M. Weller and E. Pierson (1991). Genetic analysis of the antifungal activity of a soil-borne *Pseudomonas aureofaciens* strain. *Appl. Environ. Microbiol.*, 57: 2928–2934.

Yamada, O., Y. Kaise, F. Futatsunga, S. Ishida, K. Ito, H. Yamamato, J. Munakata (1974). Studies on the plant growth regulating activity of anisomycin and toycamycin. *Agric. Biol. Chem.*, 36: 2013–2015.

38

Current Researches on Development of New Organophosphorus Pesticides

Deb Prasad Ray, D. Prasad**, Gita Kulshrestha* and R.L.Gupta**

**Division of Agricultural Chemicals, Indian Agricultural Research Institute, New Delhi – 110 012, India*

***Division of Nematology, Indian Agricultural Research Institute, New Delhi – 110 012, India*

Pesticides have played a pivotal role in bringing about a successful green revolution in India. India is the third largest consumer of pesticides in the world and highest among the South Asian countries. During the last four decades, the consumption of pesticides in India has increased several hundred folds, from 154 MT in 1953-54 to 80,000 MT in 1994-95 (Agnihotri, 2000). However, thereafter consumption of pesticides steadily declined to the present level 54,135 MT (based on 1999-2000) (Table 38.1). The decline was primarily because of ban or restriction on use of organochlorine pesticides such as HCH, DDT, aldrin etc. in agriculture.

Table 38.1: Production and Consumption of Technical Grade Pesticides in Agriculture in India

Years	*Production (MT)*	*Consumption (MT)*
1994–95	90,758	80,000
1995–96	96,880	73,652
1996–97	94,350	66,677
1997–98	84,154	60,143
1998–99	88,751	57,240
1999–2000	93,710	54,135

Source: Agnihotri (2000) *Pestic Res. J.*, 12(1): 150–155

Upto 1995-96, the major group of chemicals used in agriculture was insecticides (80 per cent) followed by fungicides (10 per cent), herbicides (7 per cent) and others (3 per cent). Thereafter, percentage of consumption of pesticides declined with simultaneous increase in the percentage consumption of herbicides and fungicides. In 1999-2000, the consumption of insecticides was 60 per cent, fungicides 21 per cent, herbicides 14 per cent and others 5 per cent. During this period, type of insecticides used also changed. The percentage of organochlorines decreased from 40 to 14.5 per cent, carbamates from 15 to 4.5 per cent and pyrethroides from 15 to 4.5 per cent, but there was a significant increase in percentage of organophosphates from 30 to 74 per cent. The consumption pattern of insecticide in India is presented in Table 38.2.

Table 38.2: Current consumption pattern of commercial insecticides

Class	*Insecticide Market (per cent)*
Organophosphate	74.0
Organochlorines	14.5
Synthetic pyrethroids	5.0
Carbamates	4.5
Natural	2.0

Source: *Pestic. Res. J.* (2000), 12(1): 150-155.

Organophosphorus (OP) pesticides are very much effective to control the agricultural pests. Out of about 900 pesticides in use globally, more than 250 belong to this group. Although OPs are known to be best insecticides and acaricides, they are equally effective as fungicides, nematicides, herbicides, plant growth regulators, insect chemosterilants and rodenticides (Eto, 1974; Fest and Schmidt, 1982). The spectrum of activity depends on the design of molecule around the phosphorus atom. Besides a very few trivalent P-compounds, majority of the OP pesticides are derivatives of P (V) oxyacids carrying a phosphoryl group or thiophosphoryl group.

Phosphoryl group

Thiophosphoryl group

The above property of OP compounds has been well exploited in the development of new OP pesticides with improve bioefficacy, safety and diverse biological activities. Here, an attempt has been made to review the status of current research on development of new OP pesticides.

Recent OP Scenario in India

The use of OP pesticides in the India started as early as 1960s. In the beginning the pesticides were imported. Cynamid (India) Ltd. was the first company to erect a plant in 1963 to manufacture malathion. After this breakthrough, phorate, fenitrothion, parathion and several other insecticides are also started to manufacture. India is now producing more than 40,000 t comprising 32 organophosphorus pesticides (Table 38.3).

Malathion, aluminium phosphide, zinc phosphide and glyphosate are currently being exported to different countries. The production of most of the OP pesticides was at first in the hands of multinational companies (MNCs). However, in the mid 70s, Government of India decided to identify some organisation which could develop indigenous technologies for pesticides needed in the country. As a result, some laboratories of Council of Scientific and Industrial Research (CSIR) and the Indian Agricultural Research Institute (IARI), New Delhi were identified for the purpose.

Table 38.3: Production capacity of technical Organophosphorus pesticides (metric tonnes)

Pesticide	*Production in*		
	1998–99	*1999–00*	*2000–01*
Acephate	4200	4200	4200
Aluminium phosphide	955	1800	2000
Anilophos	650	650	600
Chlorpyriphos	4478	4300	4000
Dichlorvos (DDVP)	1715	1700	1700
Diazinon	–	–	–
Dimethoate	1570	1480	1480
Edifenphos	130	180	150
Ethion	2550	2800	3000
Ethephon (PGR)	–	–	–
Fenitrothion	–	–	–
Fenthion	–	–	–
Formothion	–	–	–
Fesetyl-Al	39	50	41
Glyphosate	560	800	950
Iprobenfos	165	175	180
Kitazin	–	–	–
Malathion	1100	1000	900
Methyl parathion	920	850	800
Metasystox	720	600	540
Monocrotophos	7488	6000	6000
Phenthoate	–	–	–
Phorate	4810	4800	5000
Phosalone	175	137	147
Phosphamidon	2285	2000	1000
Primiphos methyl	–	–	–
Profenophos	800	800	1000
Propetamphos	–	–	–
Quinalphos	4609	3500	3500
Temephos	21	–	–
Triazophos	529	650	700
Zinc Phosphide	250	300	330

Source: Mathur and Tannan (1999) *Pesticide Information*, Annual issue, 2001.

The technologies developed so far in the field of organophosphorus pesticides are given in Table 38.4. Rallis India Ltd., Gharda Chemicals, Excel Industries Ltd, Montari Industries Ltd. And other industries are also actively involved in the developing indigenous technologies for OP pesticides.

Table 38.4: Indigenous Organophosphorus Pesticide Developed

Institute/Laboratory	*Name of Chemical*	*Released to*
Indian Agricultural Research Institute, New Delhi	Edifenphos	M/s Dharamsi Morarji Chemicals Company, Bombay
	Acephate	M/s Ventech Pesticides Pvt. Ltd., Hyderabad
Indian Institute of Chemical Technology, Hyderabad	Monocrotophos Diazinon Chlorophyriphos DDVP	Most processes have been released to Industry
Regional Research Lab, Jorhat	Quinalphos Phosphamidon Chlorfenvinfos Phosmet Glyphosate Fenthion	Most processes have been released to Industry
National Chemical Lab, Pune	Dimethoate Ethephon Phenthoate Abate Fenitrothion	Most processes have been released to Industry

Existing Classes of OP Pesticides

The existing organophosphorus can be classified into phosphorothioates, phosphonothioates, phosphates, phosphonates and phosphoroamidates etc. on the basis of their chemical nature. Their application ranges from human medicines (*e.g.*, metrifonate for Schistosomiasis and for Alzheimer's disease), to animal medicines (*e.g.*, Diazinon for sheep scab), disease vector control (*e.g.*, malathion for mosquitoes), and economic pest control (*e.g.*, chlorpyriphos for termites). (Ray, 1998). All OP pesticides have a common inhibitory action on acetyl choline esterase, which is primary target mode of action in both pests and man. The present pesticides of this class are belong to the different sub- classes as given in Table 38.5.

All the organophosphorus pesticides used in India have been developed in the abroad. Only a handful of indigenous technology have been developed in India. Unfortunately pesticides become obsolete very soon due to some unavoidable factors. The widespread use of pesticides in an attempt to increase crop production generate a series of toxicological and environmental problems, in particular the appearance of resistant species of pests, toxic effects to man and animals and the persistence of pesticides in the environment. Therefore, one has to be always in search for the newer products which are better in comparison. Although the use of organophosphorus pesticide has been increased in the world as well as in India during the past few decades, yet a vast lacuna has been created in the field of OP research. Very limited development work is going on in Japan and India and in few other countries.

Table 38.5: Different Subclasses of OP Pesticides

Sl.No.	Sub-classes	General Structure
1.	Phosphorothioates	$(RO)_2P(=O)-SR$ $(RO)_2P(=S)-OR$
2.	Phosphonothioates	$(RO)_2P(=S)-R$
3.	Phosphates	$(RO)_2P(=O)-OR$
4.	Phosphonate	$(RO)_2P(=O)-R$
5.	Phosphoroamidate	$(RO)_2P(=O)-NH_2$
6.	Phosphorodithioate	$(RO)_2P(=S)-SR$
7.	Phosphinate	$R_2P(=O)-OR$
8.	Phosphinothioate	$R_2P(=S)-OR$

This is may be due to the concern about the potential toxic effects of OPs. A number of pressure groups are actively campaigning for the banning of all OP pesticides, particularly in the high profile application of sheep dipping (Ray, 1998). Whilst most scientists would recognize that a blanket ban across a whole class of chemicals is irrational, not all people are aware of just how wide range of structures toxicity and applications there is for these pesticides. Therefore, there is a need both for continued research in this area and also proper monitoring of particular uses.

Current Research on Development of New OP Pesticides

During the last few decades, there is enthusiastic growth in the application of OPs in the farmers field. At the same time generation of new molecule is objectionable. Although, OP related work is going on throughout the world, but the main emphasis is given on the environmental safety and biochemical research. Only a handful of organization are engaged to develop the organophosphorus pesticides. An indepth screening of literature revealed that mainly there are three area of research related to organophosphorus pesticides.

1. Development of molecule with potential insecticidal, fungicidal, acaricidal, herbicidal and plant growth regulatory properties.
2. Stereospecific synthesis of newer molecule
3. Biochemical aspects

Development of New OP Molecules

Insecticides

The existing well known OP insecticides like phorate (LD_{50} 1.6 mg kg–^{1}), parathion (LD_{50} 3 mg kg–^{1}) and tetraethylpyroophosphate (LD_{50} 1.12 mgkg–^{1}) are highly toxic to the higher animals. In the later stage more safer pesticides were developed such as diazinon (LD_{50} 108 mg kg–^{1}), acephate (LD_{50} 945 mg Kg–^{1}) and malathion (LD_{50} 1200 mg kg–^{1}). Except a few enolphosphates such as DDVP and monocrotophos, almost all the earlier OPs were Thiono (P=S) type, which are in general more stable and lipophilic and penetrate the insect body more easily than corresponding oxon (P=O). Recently, a few assymetric S-propyl O-alkyl O-aryl phosphorothiolate and thiolothionate type compounds *viz.* profenphos and prothiophos have been identified which are found effective against some OP resistant insects. Profenophos has been registered in India for use in cotton. Among the two stereoisomers of profenophos, R (-) isomers is found to be an effective insecticide

A number of proinsecticides such as acephate and butonate (Figure 38.2) which shows higher selective toxicity have been developed to replace the toxic insecticides methamidophos and DDVP, respectively.

A number of selective OP insecticides like fenitrothion, fenthion and chlorthion having a methyl group or Cl atom at meta position of phenyl ring, have been developed based on the difference in mammalian and insect anticholinesterase activity. The selectivity attributed to the difference in the distance between the binding and esteritic sites of two AchEs.

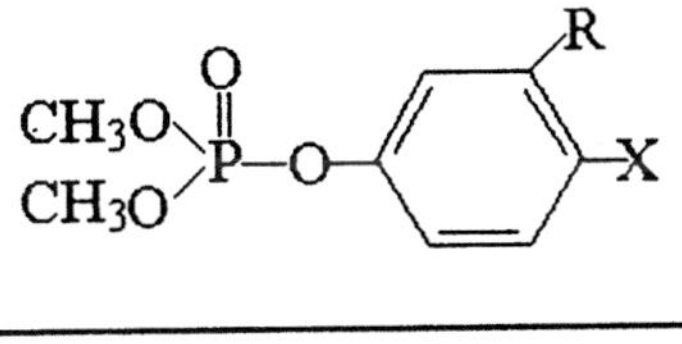

$$(CH_3O)_2P(=O)\text{–O–}C_6H_3(R)(X)$$

	R	*X*
Fenitrothion	Me	NO_2
Fenithion	Me	SMe
Chlorthion	Cl	NO_2

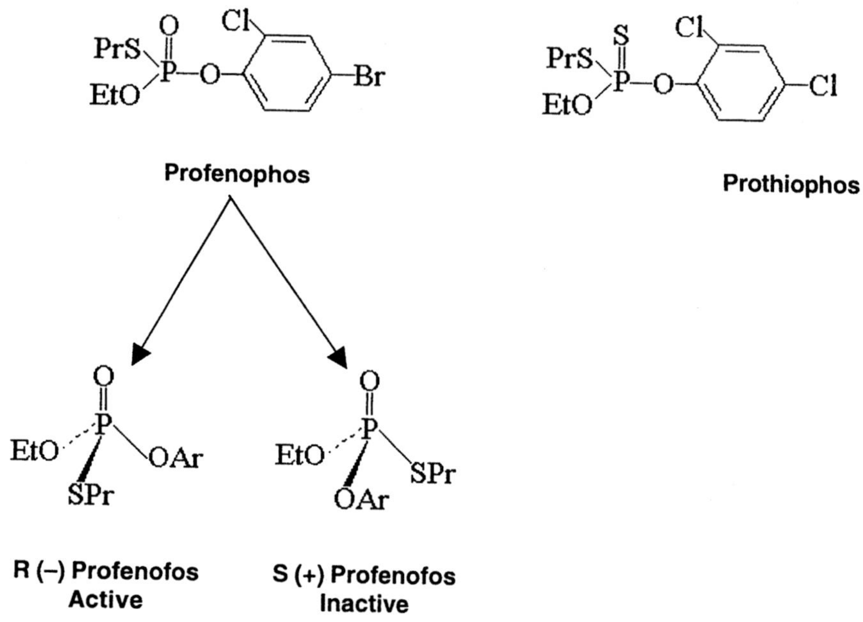

Profenophos

Prothiophos

R (–) Profenofos
Active

S (+) Profenofos
Inactive

Figure 38.1: Assymetric Phosphorothiolate and Thiolothionate Type of Insecticides

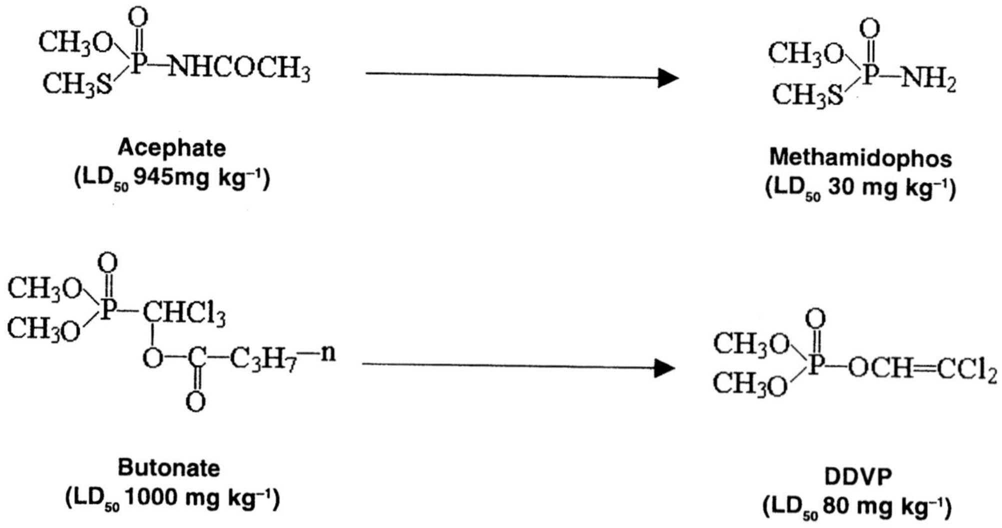

Acephate
(LD_{50} 945mg kg^{-1})

Methamidophos
(LD_{50} 30 mg kg^{-1})

Butonate
(LD_{50} 1000 mg kg^{-1})

DDVP
(LD_{50} 80 mg kg^{-1})

Figure 38.2: Organophosphorus Pro-insecticides

In this structure Cl or Me at m-position are more effective to insect and safe to mammals.

The six membered cyclic phosphorus compounds such as salithion possess high stability where as five membered cyclic phosphates are less stable due to the ring strain and thus ineffective to insects. Introducing an electron donor in the ring such as "N" enhance the stability of the molecule which confers better stability. This led to development of five membered cyclic phosphoroamidates based on amino acids £ leucine 4-(isobutyl, 2-methoxy-1,3,2- oxazaphospholidine-2-sulfides (Hirashima *et al.*, 1991).

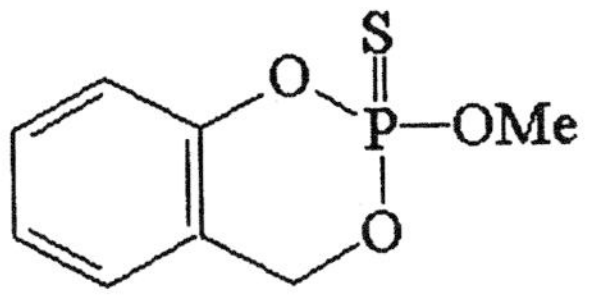

Salithion (LD_{50} = 0.05 mg kg–^{1})

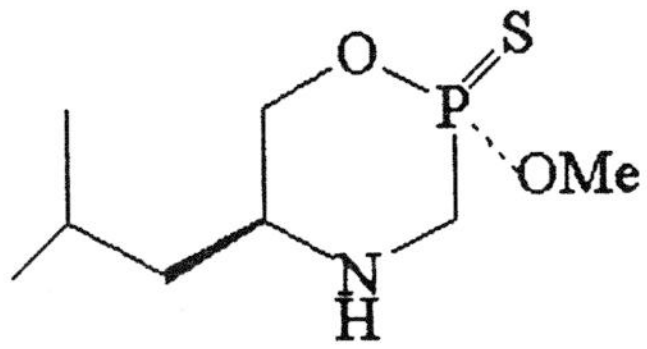

BMOS

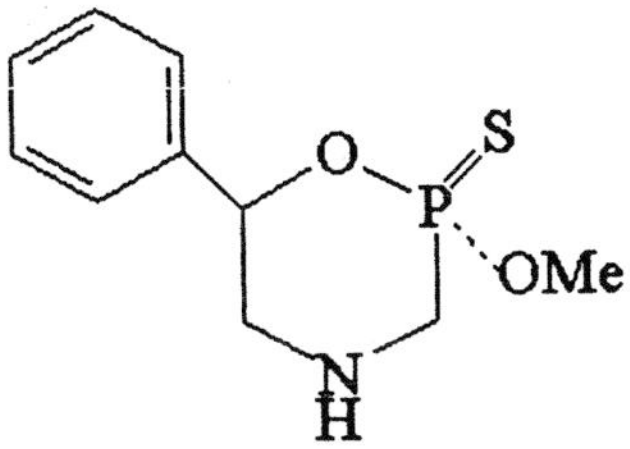

PMOS

Figure 38.3: Insecticidal Cyclic Phosphorus Compounds

Another 5-phenyl analog of 2-methoxy–1,3,2- oxazaphospholidine 2-sulfides (5-PMOS) derived from 2-amino-1-phenylethanol (APE) which is an octapamine analogoue. It was found to be stronger insecticide than BMOS (Wu *et al.*, 1988). These two compounds showed an insecticidal activity to the housefly (*Musca domestica*) and the flour beetle (*Tribolium castaneum*) by inhibiting acetylcholine esterase (Ach) at lethal concentrations (Hirashima *et al.*, 1989).

The same group described the synthesis of 1,3,2- thiazaphospholidines and their biological activities including the insecticidal activity against *M. domestica*, larval growth inhibitory activity against *T. castaneum* and the effect on adenylate cyclase activity of *P. americana* ventral nerve chords.

In the course of investigation about the methylphosphonothioic dichloride (I). Yoshikawa (2000) found that I reacted with alkyl N-alkylcarbamates (II) in the presence of triethylamine to afford the corresponding N-alkoxycarbonyl–N- alkylmethylphosphonamidothioic chloride (III). These were reacted with appropriate phenols which were suitable for leaving groups of OP insecticides. O-aryl N-alkoxycarbonyl–N-alkylmethylphosphonamidothionates (IV) were obtained in about 70 per cent yield.

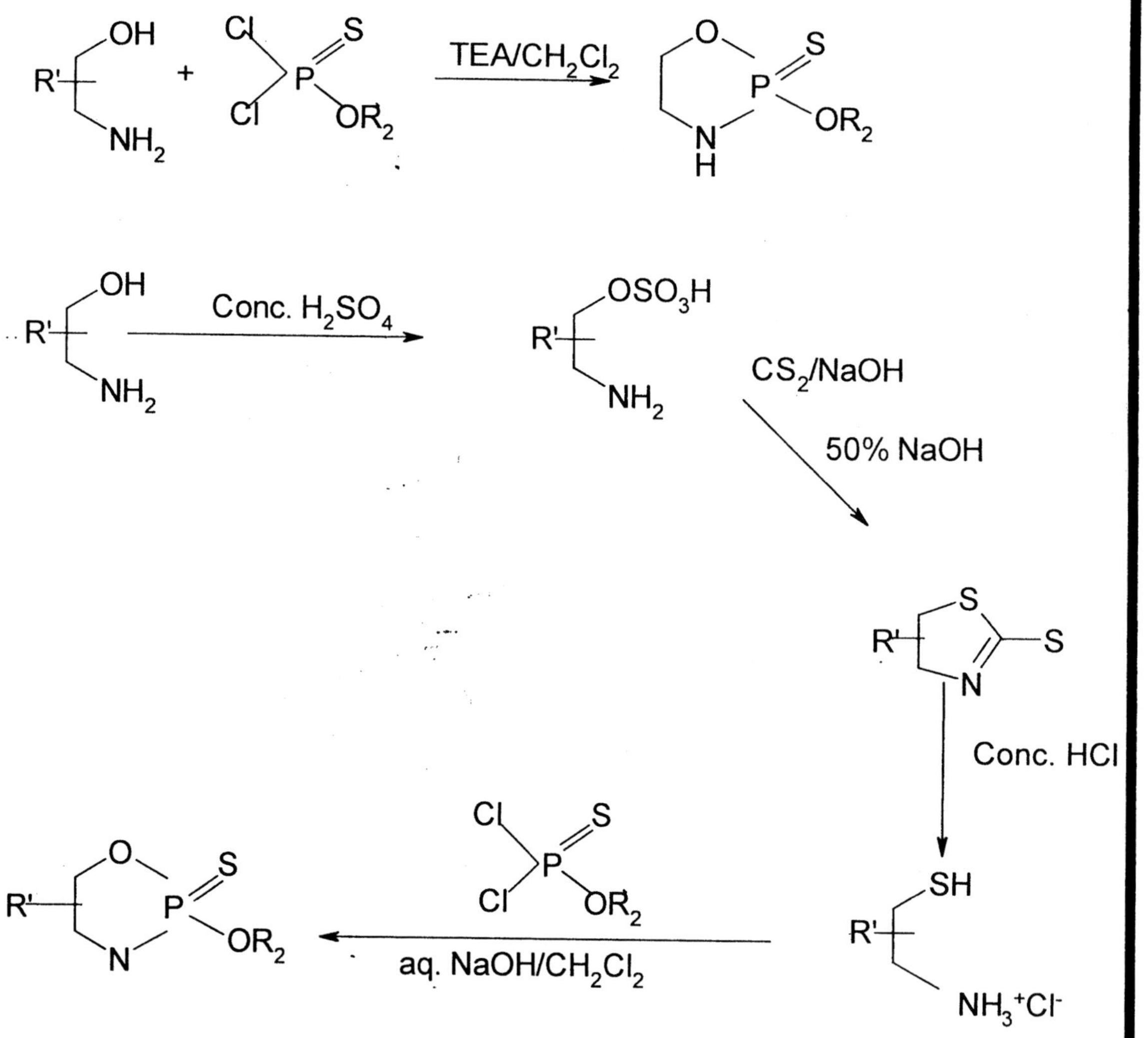

Figure 38.4: Synthetic Routes of 2-alkoxy-1,3,2-oxaza Phospholidine and Thiazaphospholidine-2-sulphides

The insecticidal activity of the above product was examined against *Delphacodes striatellus* and *Spodoptera litura*. The n-propyl group R^2-substituent and isopropyl group for R^1 was apparently suitable for insecticidal activity.

Development of Fungicides

The prominent OP pesticides developed in the past are edifenphos (Hinosan), Kitazin-P etc. which are used against blast diseases and are of thiolate type. The recently developed fungicides are structurally different and have different spectrum of activity. These includes tolclofosmethyl,

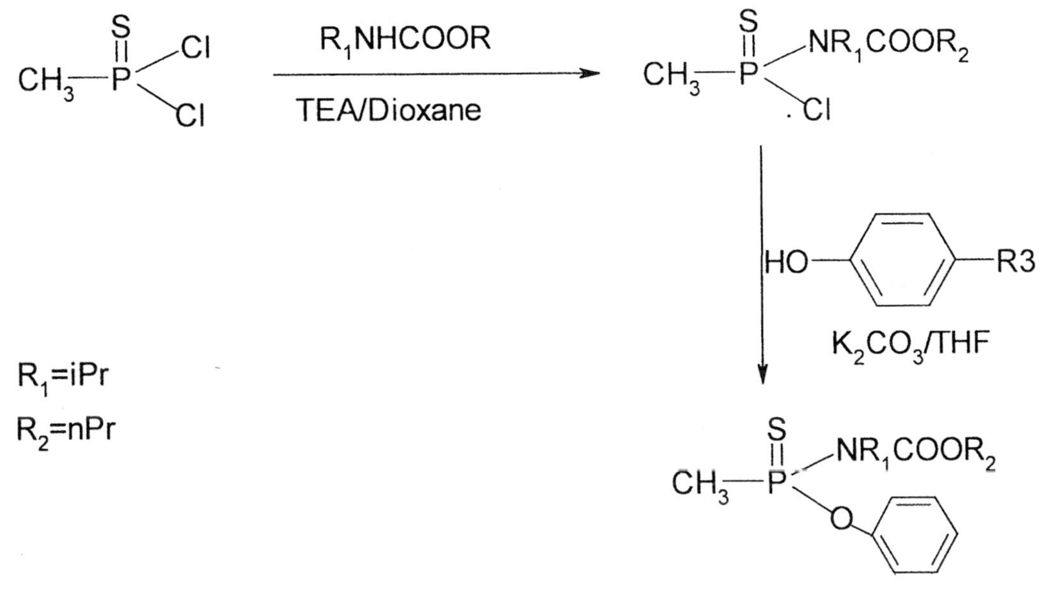

Figure 38.5: Synthesis of O-aryl N-alkoxycarbonyl-N-alkylmethylphospho-namidothionates

pyrazophos; amide type, like phosbutyl, triamiphos; phthalimide derivatives like ditalimphos; phosphonic acid derivatives like NINA, phosetyl-Al and phosphonomycin; a phosphate, phosdiphen as well as some azole type systemic fungicides, namely triazolyl methyl (di-p-fluorophenyl) phosphine sulfide and phosphinate analogue, ethyl hydroxy (pyridine-3-yl) methyl-4-fluorophenyl phosphinate.

Since the inception of the Division of Agricultural Chemicals, IARI, New Delhi research on development of new OP pesticides has been undertaken and a large number of new OP compounds of diverse structure has been synthesized, evaluated and reported for various pesticidal activities.

Bedi and Roy (1978) and Roy *et al.* (1980) prepared diaryl dichloromethyl/ethyl phosphonates and evaluated against rice blast (*P. oryzae*). The nitro- group substitution at 4 para position of benzene ring gave the most active compound in this series.

Chloroalkyl Diaryl Phosphonates

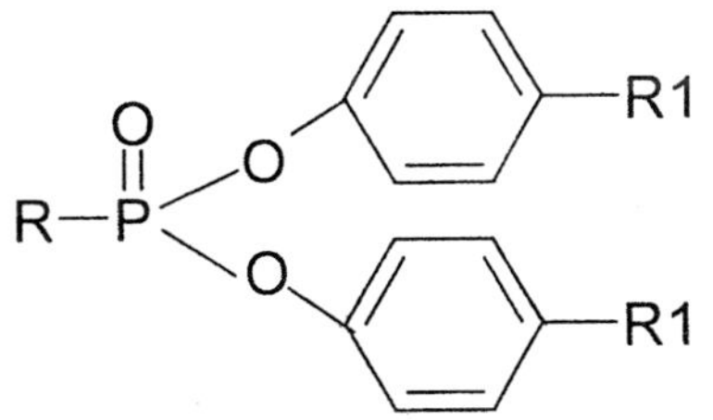

R	*R₁*	*ED₅₀ ppm (P. oryzae)*
$CHCl_2$	H	3.3
$CHCl_2$	4-NO_2	1.1
$CHCl_2$	2,4,5-Cl_3	3
CH_3CCl_2	2,4,5–Cl_3	20
CH_3 CCl_2	Cl_5	30
$ClCH_2$ CH_2	2,4,5- Cl_3	6
Cl CH_2 CH_2	Cl_5	4

Roy and Bedi (1982) synthesized a series of O,O-diaryl-S-methyl phosphorothiolates and reported their fungicidal activity against *P. oryzae*, *H. oryzae*, *A.alternata*, *R. bataticola* and *P. aphanidermatum*. In this series, the most active compound against most of the test fungi was O,O-di (2,4,5, trichlorophenyl) S-methyl phosphorothiolate.

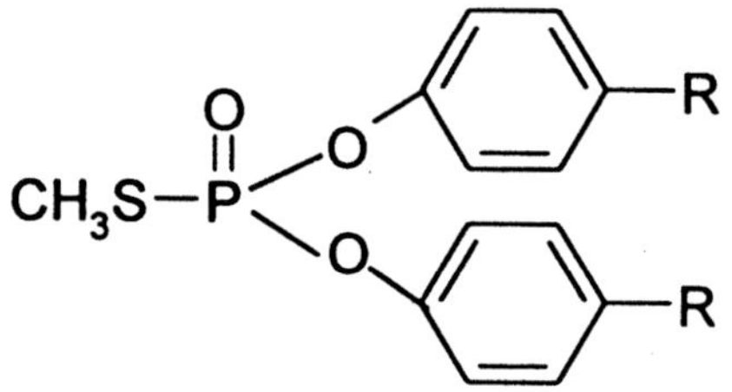

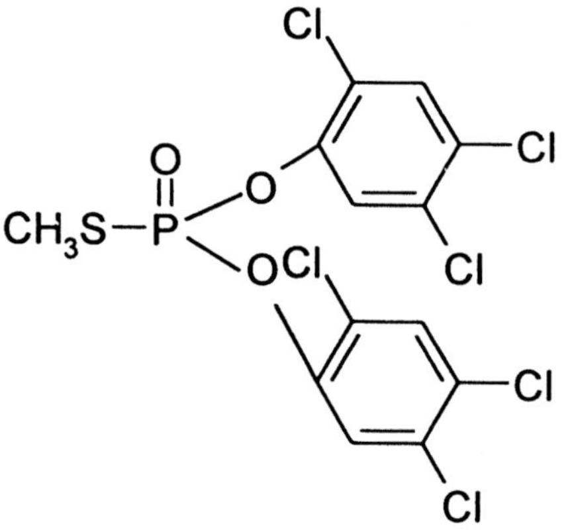

The synthesis and fungicidal activity of another similar series of phosphorothiolates; O,O-diaryl S-ethyl phosphorothiolates against *P. oryzae*, *H. oryzae*, *A.alternata*, *R. bataticola* and *P. aphanidermatum* was reported by Gupta and Roy (1984).

S-alkyl Diaryl Phosphorothiolates

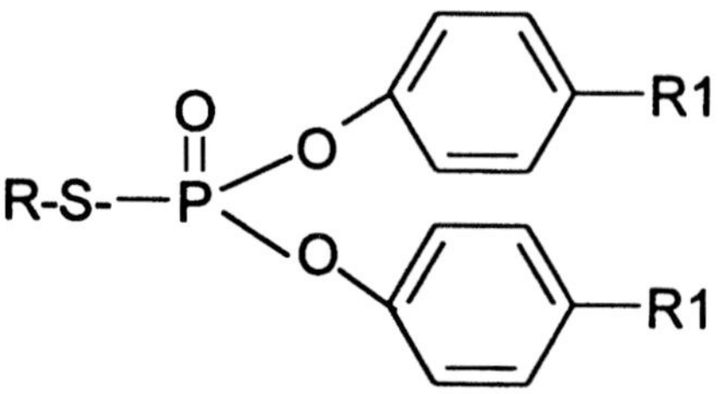

R	R_1	ED_{50} ppm (*P. oryzae*)	Reference
CH_3	2,4,5-Cl_3	5	Roy and Bedi (1982)
CH_3	Cl_5	27	— do —
C_2H_5	2,4,5-Cl_3	13	Gupta & Roy (1984)
C_2H_5	Cl_5	60	— do —

In this series, the high activity against all test fungi was exhibited by O,O-di (2,4,5-trichlorophenyl) S-ethyl phosphorothiolate.

Development of Phosphorothiolate Fungicides and Related Esters

The well known OP antiblast agents such as ediphenphos, IBP are thiolate type. Newer OP fungicides formulated in the recent past are structurally different and have different spectrum of fungicidal activity.

Sl.No.	Compounds	References
1.	O,O-diaryl-S-methyl phosphorothiolates	Roy & Bedi (1982)
2.	O,O-diaryl-S-ethyl phosphorothiolates	Gupta & Roy (1984)
3.	O,O-dialkyl-O-aryl phosphorothiolates	Sasaki *et al.* (1984)
4.	O,O- dialkyl-S-isopropyl phosphorothiolates	Gupta *et al.* (1988)
5.	O-alkyl S,S-diaryl phosphorothioates	Gupta & Roy (1990)
6.	S-alkyl-S,S-diaryl phosphorothioates	Gupta & Roy (1991)
7.	S-methyl S,S-diphenyl phosphorothiolates	Gupta & Roy (1993)
8.	1,3,2-oxazaphospholidine-2-sulfides	Taira *et al.* (1994)
9.	Monoterphenoid derivative of phosphorothionates and thiophosphonates	Tawata *et al.* (1996)
10.	Heterocyclic organothiophosphates	Jaiswal *et al.* (1998)
11.	O,O-diaryl O-n-propyl phosphorothionates	Patra & Gupta (2000)
12.	O,O-diaryl-O-methyl phosphorothioates	Gupta (2000)

Phosphates and Phosphites

1.	O,O,O-triaryl phosphite	Vasu (1982) Vasu and Roy (1985)
2.	O,O-diaryl-O-2-chloroethyl phosphates	Roy & Taneja (1990)
3.	O-2-chloroethyl-O-methyl phosphate	Singh *et al.* (1996)
4.	O-aryl-O-2-chloroethyl-O-ethyl phosphates	Singh *et al.* (1998)

Some Recent Development in OP Fungicides

The insecticidal and fungicidal efficacy of S-aryl thiophosphoroamidates against many kind of crop insects and diseases are well known. In their continuous screening to enhance the fungicidal potentialities of chemicals in similar groups, Ishikawa and Morita (1993) found another class of fungicidal S, N-diarylphosphoroamidates and a new synthetic method relevant to the derivatives.

Because of the structural novelty, they investigated the SAR of this type of chemicals to develop potential fungicides, and found that S,N-diaryl thiophosphoroamidates had high activity against rice blast (*Pyricularia oryzae*) and rice sheath blight (*Rhizoctonia solani*) without being phytotoxic to the host plant.

Synthesis of the Compound

PhS–P(=O)(Cl)Cl + X–C₆H₄–NHR → PhS–P(=O)(NR–C₆H₄–X)(Cl)

MeOH

PhS–P(=O)(NR–C₆H₄–X)(OMe)

Diphenylphosphonic Amide Derivatives

To elicit pesticide potentialities of chemicals in similar groups, Ishikawa *et al.* (1993) again found a new class of fungicidal diphenyl phosphinic amide derivatives.

The diphenylphosphinic amides are structurally unrelated to any other class of fungicides. Because of the structural novelty they investigated the SAR of this type of chemicals to develop potential fungicides, and found the compounds displayed promising activity against cucumber downy mildew (*Pseudoperonospora cubensis*) without being phytotoxic to the host plants.

Thirty four diphenylphosphinic and diphenyl thiophosphinic amides they synthesized by alkylation of corresponding (thio) phosphinic amides.

$$Ph_2-P(=S)-N=P(R)-Ph_2$$

Benzyl, cyanomethyl, alkoxy carbonyl alkyl groups at position R were found to be most suitable.

1,3,2-Oxazaphospholidine-2-sulfides

Alpinia speciosa is one of the Zingiberaceae species whose leaves emit a pleasure fragrance. In the produced substances from *A. speciosa*, there are volatile and nonvolatile components. Volatile components such as geraniol and isothymol in the essential oil possess high antimicrobial activity against plant pathogenic bacteria.

Taira *et al.* (1994) synthesized 23 kinds of 1,3,2-Oxazaphospholidine-2-sulfides by reacting (S)-2-aminoalcohols, PSCl3 and the components of essential oil; geraniol, isothymol, thymol, citronellol and eugenol.

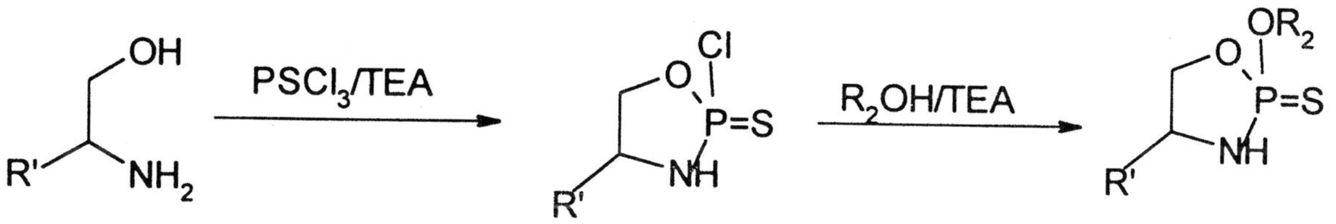

PSCl$_3$/TEA → R$_2$OH/TEA →

R= alkyl or aryl

R-OH= essential oil

Among the prepared compounds (4S)–2-citronelloxy-4-isopropyl- 1,3,2-oxaza phospholidine 2-sulfide and (4S)–2-geraniloxy–4-isopropyl-1,3,2-oxazaphospholidine- 2-sulfide had strongest antimicrobial activity against *Pythium* sp and *Corticium* sp., respectively. L-valine derivatives showed stronger activity than the other L-amino acids derivatives.

Thiophosphorylated Monoterpenoids and Related Compounds

Tawata *et al.* (1996) took the same plant *Alpinia speciosa* due to the antifungal activity of its essential oil constituents such as isothymol, thymol and eugenol. They changed these volatile constituents into nonvolatile phosphorothionates by reaction with thiophosphoric agents using triethyl amine or aq. NaOH as a base. They synthesized 28 kinds of thiophosphorus esters.

During volatile testing, phenethyl alcohol and isothymol were almost volatilized in 9 and 14 days, respectively and eugenol was volatilized 89.2 per cent in 14 days. Dimethyl isothymol phosphorodthionate, dimethyl eugenyl phosphorothionate, dimethyl phenethyl phosphorothionate and diethyl phenethyl phosphorothionate and derived from each essential oil were scarcely volatilized and their volatility were reported to be 4.0, 2.3, 12.6 and 7.6 per cent in 14 days, respectively. Among the synthesized compounds, dimethyl phenethyl phosphorothionate showed the strongest antifungal activity, 39.6 and 56.6 per cent inhibition at 100 ppm against *Pythium* sp and *Corticium rolfsii*, respectively.

$R1-P(=S)Cl_2 \longrightarrow R1-P(=S)(OR_3)Cl$

R_3OH/ 6N NaOH

$R1-P(=S)(OR_3)OH + R1-P(=S)(OR_3)OR_3$

$R1-P(=S)(R2)Cl \longrightarrow R1-P(=S)(R2)OR_3$

R_1 & R_2 = MeO, EtO, Ph

R_3OH = Essential oil

Efficacy of OP Derivatives Against Fungal Pathogens of Sugarcane

Jaiswal *et al.* (1998) prepared several novel OP derivatives by the reactions of O,O-diethyl chlorophosphate, thiophosphate with three important series of heterocyclic compounds, *viz.* bis-(mercaptotriazoles), bis (mercapto oxadiazoles and bis (mercaptothiadiazoles). The derivatives have been characterised on the basis of analyses and spectral data. They evaluated the fungicidal activity of these derivatives against *Colletotrichum falcatum, Fusarium oxysporium,* and *Curvularia pallescens.* They also correlated the screening result with the structural features of the tested compounds. The greater potency has been observed with dithiophosphates compared to phosphates, with OP derivatives containing bis (mercaptotriazole) rings compared to other heterocyclic rings and with spacers between the rings.

6-alkyl-2-alkylamino-4H-1,3,2-benzodioxaphosphorin-2-sulfides

Six membered cyclic phosphates, phoshoramidates and phosphoramidothionates having an aryloxy group or bulky alkyloxy group at position 2 showed less insecticidal activity than that of a small alkyloxy group. Eto *et al.* (1963) reported the introduction of any electron withdrawing or releasing group in the ring decreased the activity of salithion. Nevertheless, salithion may be changed into a non-toxic compound by introducing an alkyl group on its benzene ring. Salithion showed a negligible activity against both *Pythium* and *Corticium*. Tawata *et al.* (1998) presented the correlation between the chemical structure of 6-alkyl-2-alkylamino-4H- 1,3,2- benzodioxaphosphorin-2-sulfides and antifungal activity against pathogenic fungi.

Among the compounds evaluated 6-n-propyl-2-n-proylamino- 4H-1,3,2-benzodioxaphosphorin-2-sulfide showed the strongest activity of 66 per cent inhibition at 10 ppm against *Pythium* sp. Other compounds of these series exhibited more than 40 per cent inhibition and were highly active than the referential fungicide iprobenfos. The activity was varied by both the substituent of R^1 on the benzene ring at position 6 and R^2 on the nitrogen atom at position 2 in the relative order of (R^2 = n propyl)> isoproyl > n-butyl isobutyl >5 (sec-butyl) when R^1 was n-propyl, isopropyl and n-butyl groups. Compounds having a n-propyl group as the substituent R^1 exhibited relatively stronger inhibition than those of other compounds with isopropyl, n-butyl and S-butyl groups without remarkable influences of the substituent R^2 on the nitrogen atom.

Synthesis of 6-alkyl-2-alkyl amino- 4H-132- benzodioxaphosphorin-2-sulfides

(Tawata *et al.*, 1998)

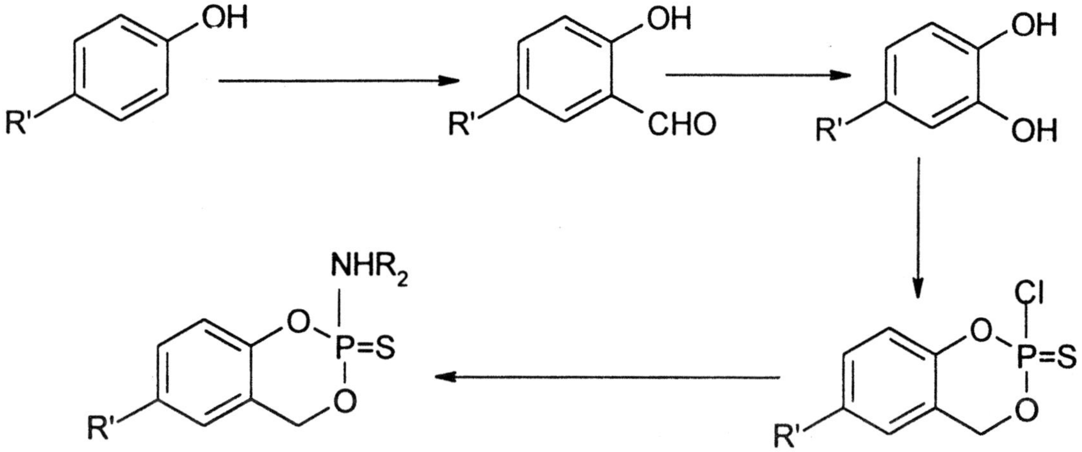

R' and R_2 are alkyl groups

Against *Corticium rolfsii* some of these compounds exhibited more than 50 per cent inhibition at 10 ppm. Among these compounds 6-n-propyl-2-isopropylamino-4H-1,3,2-benzodioxaphosphorin 2-sulfide showed strongest activity of 60 per cent inhibition at 10 ppm and n-propylamino derivative which was the most active compound against *Pythium* sp had 63 per cent inhibition. The activity against C. *rolfsii* was also influenced by the substituent R^1 on the benzene ring and the relative order of growth inhibition against C *rolfsii.* was (R^1=n-propyl) > isopropyl > n-butyl. Unfortunately, the activity of these compounds against C.*rolfsii* was a little bit lower than that of iprobenfos.

O,O-diaryl-O-2-chloroethylphosphorothionates

In a recent study, chattapadhyay and Gupta (2001) synthesised 20 O,O-diaryl-O-2-chloroethylphosphorothionates having different phenyl substitution and characterised

These by physico-chemical techniques. The compounds were evaluated for their fungicidal activity against two phytopathogenic fungi *viz. Rhizoctonia solani* and *Sclerotium rolfsii.*

Synthesis

PCl_3 + S —($AlCl_3$/S)→ $PSCl_3$ + Cl-CH_2-CH_2-OH → Cl-CH_2-CH_2-O-P(=S)(Cl)-Cl

+ 2 (R-substituted phenol, OH) + TEA → Cl-CH_2-CH_2-O-P(=S)(O-phenyl)(O-phenyl)

The most active compounds were identified from these series were O,O-di (pentachlorophenyl) O-2-chloroethyl phosphorothionate (ED_{50} 419 ppm) and O,O-di (2-methylphenyl O-2-chloroethyl phosphorothionate (ED_{50} 426 ppm) against *R. solani* and *S. rolfsii*, respectively. The QSAR study was done using *in vitro* fungitoxicity data and various physico-chemical parameters for hydrophobic, electronic and steric properties of the phenyl ring by multiple regression analysis. The QSAR model for the fungicidal activity of O,O-diaryl O-2-chloroethyl phosphorothionates against *R. solani* revealed that the presence of electron donating groups in the benzene ring and a high value of , hydrophobic parameter irrespective of being negative or positive imparted high fungicidal activity to the compounds. The shape of the para substituent as determined by high and low value of STERIMOL length [ΣL (p)] and width [ΣB4 (p)] parameters, respectively and also low value of hydrophobic parameter RM are favourable for fungicidal activity of compounds in this series of against *S.rolfsii.*

Development of OP Nematicides

The OP nematicides like dichlofenthion, fensulfothion, triazophos etc. are all phosphorothionate type except fenamiphos, a phosphoramidate. A few recently commercialised OP nematicides namely heterophos, ethoprophos and rugby[R] are of phosphorothiolate type. Some new phosphoramidate and diamidates nematicides are fosthietan and diamidophos.

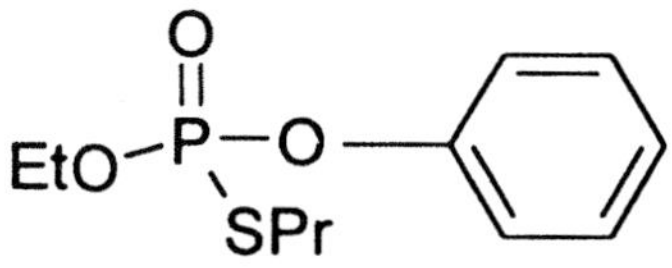

Heterophos

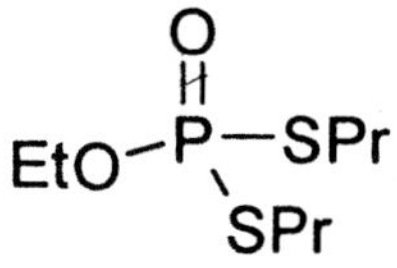

Ethoprophos

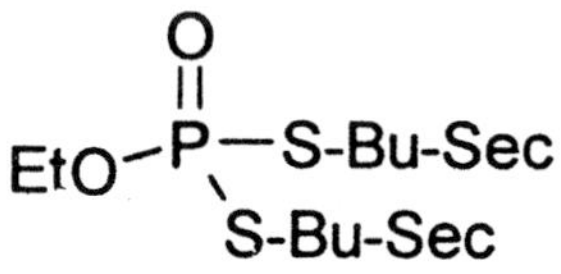

Rugby[R]

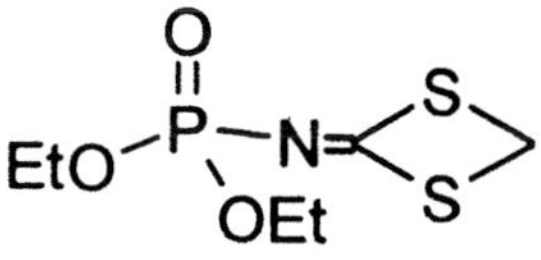

Fosthietan

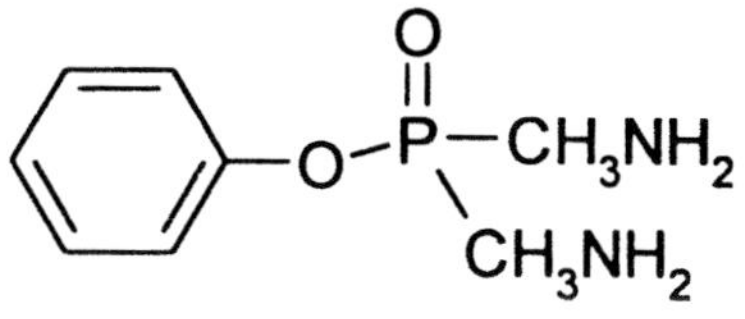

Diamidofos

New Potential OP Nematicides

1. Phosphorothionate (Gupta *et al.*, 1994)

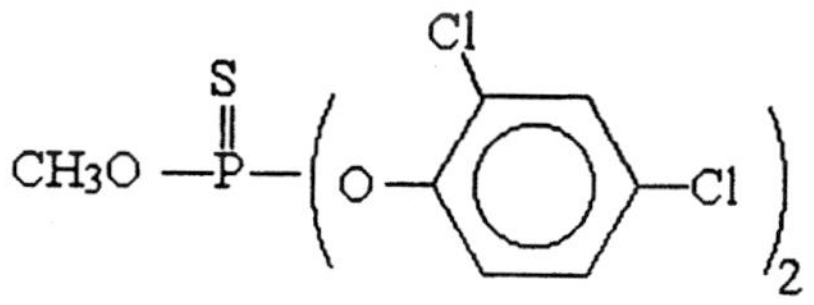

2. Phosphorothiolate (Gupta *et al.*, 1984, 86, 89)

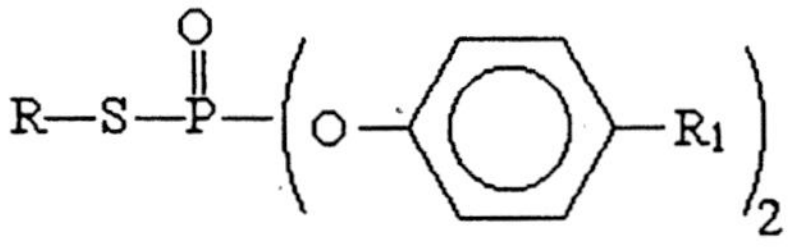

R	R_1	*Activity*
Me	2,4,5 Cl_3	99.6 at 125ppm
Et	2,4 $(CH_3)_2$	63.4 at 62.5ppm
i-Pr	2,3,5 $(CH_3)_3$	LC_{50} 26ppm

3. Phosphorotrithiolate (Gupta *et al.*, 1993)

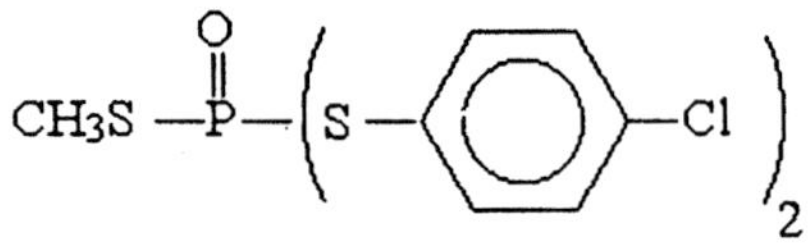

LC_{50} 36ppm

4. Phosphonate (Roy *et al.*, 1991)

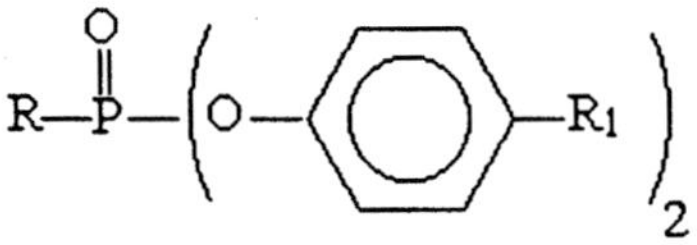

R	R_1	*Activity*
Cl_2CH	2,4,5-Cl_3	52 at 62.5ppm
CH_3CCl_2	2,4,5-Cl_3	100 at 125ppm
Cl CH_2 CH_2	2,4-Cl_2	100 at 125ppm

5. Phosophonoamidates (Khazanchi *et al.*, 1984)

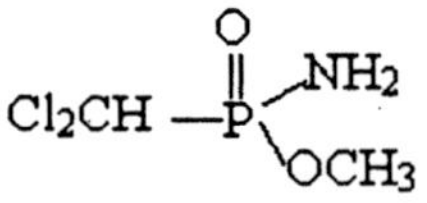

87 at 62.5 ppm

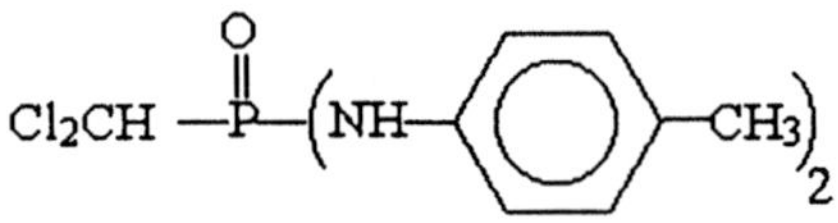

86 at 62.5 ppm

6. Phosphoramidates (Khazanchi *et al.*, 1988)

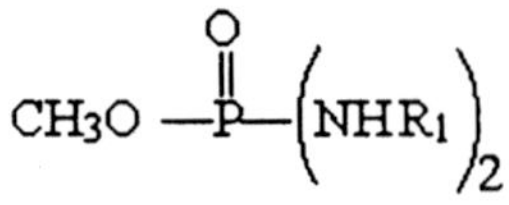

R_1= n-Pr ; i-Pr 96-98 at 62.5 ppm

7. Phosphates (Singh *et al.*, 1998)

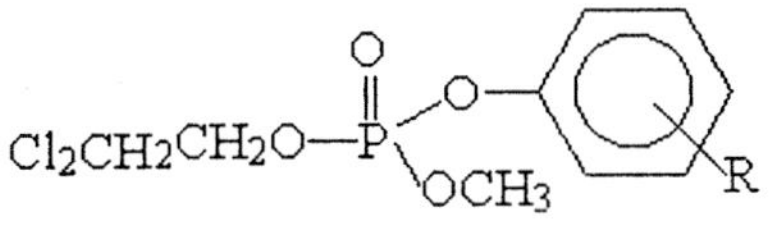

$Cl_2CH_2CH_2O-P(=O)(OCH_3)-O-C_6H_4R$

Most active compound from this series was reported to be O-2,4,6-trichlorophenyl-O-2-chloroethyl-O-methyl phosphate.

New Nematicides

Sl.No.	*OP Compounds*	LC_{50} *Value (ppm)*
1.	Diaryl S-ethyl phosphorothiolates	53
2.	O,O, Dimethyl phosphoramidates and Methyl phosphoric diamides	32
3.	O,O, diaryl O-2 chloroethyl phosphates	72
4.	Bisaryl 2-chloroethyl phosphonates	28
5.	2-Chloroethyl O-aryl O-methyl phosphonates	18
6.	S-alkyl S, S diaryl phosphorotrithioates	35
7.	O-alkyl S, S diaryl phosphorotrithionates	44
8.	O-aryl-O-ethyl-2-chloroethyl phosphonates	20
9.	O, O-bis (aryl) Isobutyl phosphonates	26
10.	O-aryl O-2 Chloroethyl O-methyl phosphates	29
11.	O, O-diaryl O-methl phosphothionates	55

An ecofriendly molecule was developed and tested against root knot and reniform nematodes *in vitro* and *in vivo* and forwarded for "PATENT" as new nematicide to ICAR.

Development of OP Herbicides

Organophosphorus compounds are distinguished in the field of pesticide chemistry for their versatility in biological action. At the early stages of development they were known as good insecticides, but their herbicidal activity was discovered much later. Amongst various groups of OP compounds, phosphonic acid derivatives, *viz.* phosphonamidates, find distinct place for their phytotoxic and herbicidal properties on weeds as well as on crop plants. A few known herbicides of this class, *e.g.*, glyphosate and DMPA, have already been commercialised. However, phosphonamidates need more exploration as herbicides. The research in this area was first initiated by Khazanchi and Roy (1983), who reported moderate to good herbicidal activity of a series of O-methyl dichloromethyl phosphonamidates and diamides.

$Cl_2CH-P(=O)(OMe)-NHPr\text{-}i$ $\qquad$ $Cl_2CH-P(=O)(NH_2)-NH_2$

Phosphonamidates/ diamides

Further, a series of monoamidates and diamidates of 1,1-dichloroethyl phosphonic acid were prepared and screened for their herbicidal and phytotoxic activities alongwith quantitative structure activity relationship (Chakravarty, 1992). The same workers Khazanchi and Roy (1986) reported some phosphoramidates and diamides for herbicidal activity against *Phalaris minor*.

Compounds	*Effective aginst*
$(CH_3O)_2$- P – NH C_6H_5	*Phalaris minor* Premergence
$(CH_3O)_2$- P – NH C_6H_4-CH_3	50 per cent inhibition at 100 ppm
$(CH_3O)_2$- P – NH C_6H_4-CH_3	Post emergence
$(CH_3O)_2$- P – NH C_6H_5N	Complete kill below 150 ppm

In 1998, Das and Roy reported fourteen new O-phenyl / 4-chlorophenyl-N-alkyl/aryl-2-chloroethyl phosphonamides and their phytotoxicity towards monocotyledonous (wheat) and dicotyledonous (mustard) plants. They also reported the herbicidal activity of these compounds against wild oat.

Synthesis:

Scheme I

$$PCl_3 + AlCl_3 + ClCH_2CH_2Cl \longrightarrow ClCH_2CH_2\text{-}PCl_3^+ \, AlCl_4^-$$

$$\xrightarrow[-10^\circ C]{CH_2Cl_2/H_2O} ClCH_2CH_2\text{-}P(=O)Cl_2$$

Scheme II

$$ClCH_2CH_2\text{-}P(=O)Cl_2 \longrightarrow ClCH_2CH_2\text{-}P(=O)(Cl)\text{-}O\text{-}C_6H_4\text{-}R$$

$$\xrightarrow{RNH_2/TEA} ClCH_2CH_2\text{-}P(=O)(NHR)\text{-}O\text{-}C_6H_4\text{-}R$$

Among all of these compounds O-phenyl-N-propyl-2-chloroethyl phosphonamidate was found to be the most phytotoxic (preemergence) and also possessed herbicidal properties. There was a reduction in phytotoxicity with increase in the number of carbon atoms in the alkyl chain. Dicotyledonous plants were, however, more susceptible to foliage application than were moncotyledonous plants. Wheat showed maximum resistance to most of the compounds. N-propyl 4-chlorophenyl-3-chloroethyl phosphonamidate was found to be least active molecule in this series.

Chemosterilant OP's

1. TEPA

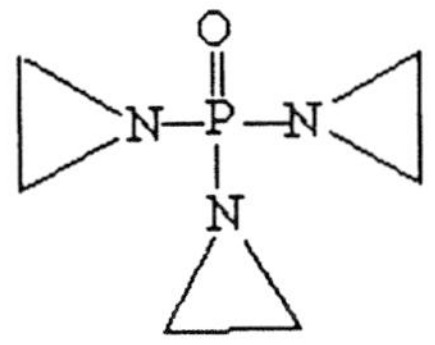

2. HEMPA

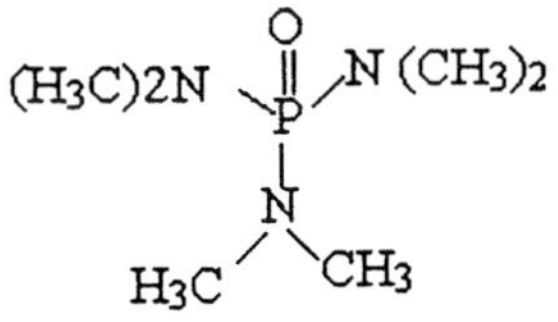

$(H_3C)_2N$ — P(=O) — $N(CH_3)_2$; $N(CH_3)_2$

3. Apholate

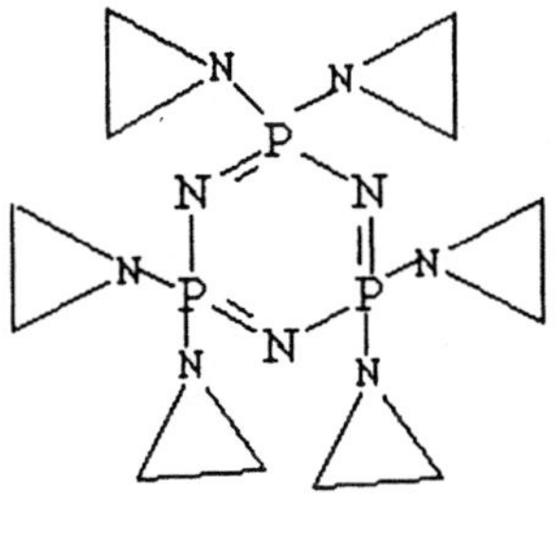

Rodenticide OP's:

1. Gophacide

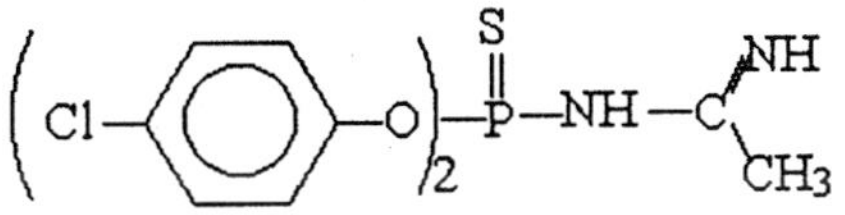

$(Cl-C_6H_4-O)_2P(=S)-NH-C(=NH)CH_3$

A selective rodenticide

OP's used as Plant Growth Regulators

1. 2,4-DEP — $(Cl-C_6H_3(Cl)-OCH_2CH_2O)_3P$

2. Ethepon (Ethrel) — $(HO)_2P(=O)-CH_2CH_2Cl$

3. Chlorphonium — $Cl-C_6H_3(Cl)-CH_2P^+(Bu)_3Cl^-$

4. Glyphosine — $HO-C(=O)-CH_2N(CH_2-P(=O)(OH)_2)_2$

Heterocyclic OP pesticides

Although a variety of OP comounds have been reported to show unique bioactivities as herbicidal, insecticidal, and fungicidal ones, however, there have not been enough studies on the phosphinoyl derivatives where the N-atom of the heterocycle is linked to the phosphorus atom.

In the synthetic studies for the purpose of finding the candidate compound as agro-chemicals, Haga *et al.* (1964) showed the following compound as excellent insecticidal and acaricidal

R1-pyridyl-NH-P(=X)(SR$_2$)(OR$_3$)   X= O,S R_2,R_3= Alkyl gı

Among these derivatives, 3-chloro-5-trifluoromethyl pyridine derivative showed the highest activities.

CF_3-(3-Cl-pyridyl)-NH-P(=O)(S-nPr)(OEt)

However, the above compound possessed one drawback is its high mammalian toxicity [(LD_{50} mg/kg (mouse)]. Since, in this compound 3-Chloro-5-trifluoromethyl pyridine acts as strongly electron withdrawing group, the e-density of the N-atom in the P-N-bond is estimated to decrease.

So, in the next step Koyanagi *et al.* (1997) began to synthesize N-phosphinoyl heterocylces, where an e-deficient N- atom in the heterocycle is directly bonded o phosphorus atom.

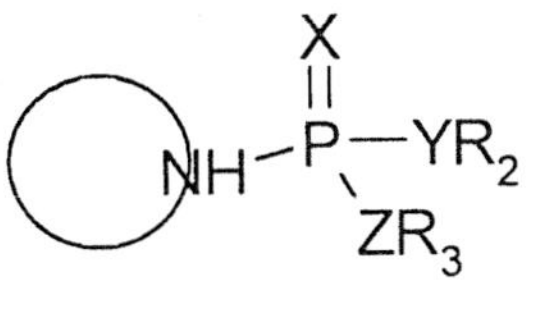

X= O,S
Y,Z= O,S.NH
R2, R3= Alkyl gı

Synthesis

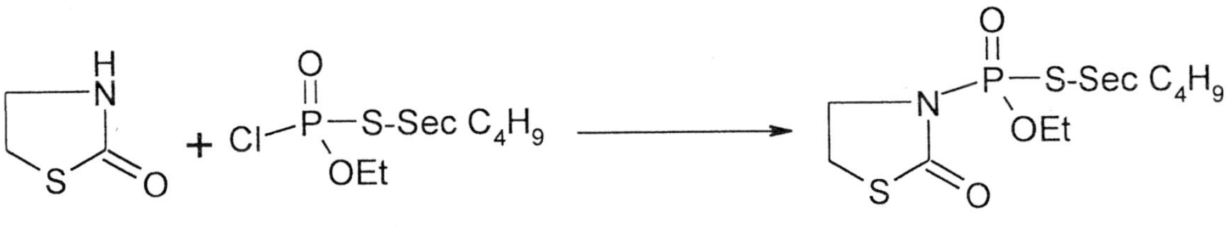

$$\text{thiazolidin-2-one (NH)} + Cl\text{-}P(=O)(OEt)\text{-S-Sec } C_4H_9 \longrightarrow \text{N-}P(=O)(OEt)\text{-S-Sec } C_4H_9$$

Biological Activities

Acaricidal Activities

Few compounds of these series shown to be very good acaricidal in nature. The acaricidal activity of these compounds were tested against two spotted spidermite (*Tetramychus utricea*) having the resistance to dicofol and conventional OP insecticides.

Aphicidal Activities

The aphicidal activities were obtained against green peach aphid (*Myzus persicae*). The aphicidal activities were graded from A to E and presented in the Table .

Nematicidal Activities

The nematicidal activitiers were obtained against southern root knot nematode (*Meloidogyne incognita*). Soil contaminated with the nematodes was put in each pot. Test solution was poured into the pot and the treated soil was uniformly mixed. Four tomato seedlings were transplanted in the pot. Assessment was made 3 weeks after transplanting. The nematicidal activities were graded A to E in the Table .

Heterocyclic Phosphonates and their Antibacterial and Antifungal Activities

There are many heterocyclic compounds exhibit versatile pharmacological activities such as antibacterial, antifungal, anthelmentic antiarlergic, antineoplastic, local analgesic, antihistaminic, vasodilator, hypotensif and spasmolytic activity. Many phosphonic acids and their derivatives have been also known to exhibit important biological properties including antibiotic, antileukemic, herbicides, plant growth regulators and insecticidal activity, depending on the nature of substituent on the phosphonic group. Therefore, it seemed of interest to synthesize the compounds that contains two pharmacological important parts i) heterocyclic and ii) phosphonate group and investigate their antibacterial and antifungal activities. Mete *et al.* (1999) provided details relating to the synthesis, characterisation of some heterocyclic compounds and their antimicrobial properties against gram positive, *Enterococcus faecalis* (ATCC 29212) and *Staphytococcus aureus* (ATCC 29213) and gram negative *Escherichia coli* (ATCC 25922), *Pseudomonas aeroginosa* (ATCC 27853) standard strains and yeast like fungi, *Candida albicans* and *Cabduda tropicalis*

Synthesis

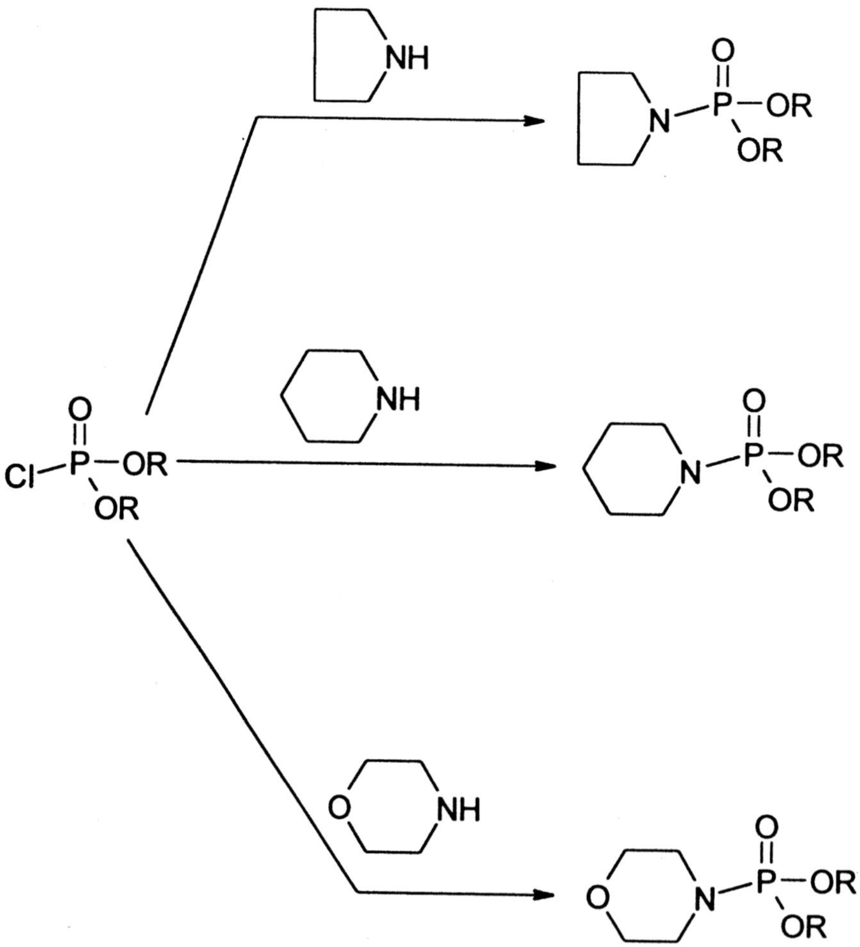

Among the tested compounds 12, 13, 16, 17, 18, 22 and 23 are found effective to inhibit the growth of *Candida albicans* and *Candida tropicals* at the MICs between 100-800 mg/ml. The compounds tested here generally do not exhibit considerable antibacterial activity at the concentration studied (100-800 mg/mL), except 18 which exhibits antibacterial activity against gram positive bacteria at the MIC of 400 µg/ mL.

Stereospecific Synthesis of Newer Molecule

Synthesis of Optically Active O-aryl-O-ethyl-phenyl Phosphonothionates

Although many OP insecticides with a chiral phosphorus atom are in use today, it is well known that their insecticidal activity depends on the chirality of the P-atom. For example(R) p (+) EPN is more active than corresponding (S) p (-) EPN (Ohkawa,1982). The synthesis of chiral isomers of O-aryl O_ethyl phenylphosphonothionates by the conventional methods has been illustrated below:

$$R1-P(=S)(OH)(OR_2) \longrightarrow Cl-P(=S)(R1)(OR_2) \longrightarrow R1-P(=S)(OR_2)-O-C_6H_4-X$$

The resolved phosphonothioic acid is converted to the corresponding phosphochloridothionate by reacting with phosphorus pentachloride or oxalyl chloride. This reaction occurs with inversion of the central P atom. Reaction of the chloridothionates with phenols gives corresponding thionate esters with additional inversion of at the P-atom. The enatiomers of EPN (Ohkawa *et al.* (1977), cyanofenphos (Ohkawa *et al.*, 1977) and leptophos (Allahyyari *et al.*, 1980) have been prepared by the conventional method.

As this pathway involves two reaction steps of displacement at the phosphorus atom with inversion source recemisation inevitably occurs, particularly in the chlorination process. Keeping this in mind Yoshikawa (1999) planned to synthesize such optically active phenyl phosphonothionates by the mixed anhydride (MA) method which has commonly been used for peptide synthesis. O-ethyl phenyl-phosphonothioic acid wasused as the starting material. Ethyl chloroformate, isobutyl chloroformate, isobutyryl chloride and pivaloyl chloride commonly used for peptide synthesis gave the corresponding anhydrides in good yields, but they did not afford objective O-aryl-O-ethyl phenyl phosphonothionates by reacting with appropriate phenols. The reaction of O,O-diethyl phosphorochloridate with O-ethylphenyl phosphonothioic acid in the presence of triethylamine afforded the corresponding mixed anhydride in about 90 per cent yield. The reaction of the anhydride with appropriate phenols gave the objective O-aryl ethyl O-ethyl phenyl phosphonothionates in about 35 per cent yields. These results may be explained by the higher reactivity of the P-atom at the thiophosphonyl center than that at the phosphoryl center. Yoshikawa (1990) applied this method to synthesize optically active O-aryl O-ethyl phenylphosphorothionate

Optically active O-ethyl phenyl phosphonothioic acids which were resolved by fractional recrystallization as a brucine salt were converted to the corresponding mixed anhydrides in about 85 per cent yields. Conversion from anhydride to the corresponding O-aryl-o-ethyl phenylphosphonothionates was under milder conditions than those used for recemate synthesis. When the reaction was pursued at 30oC, no racemization was apparen in the synthesis of 8a, 8b, 9a and 9b.The absolute configuration of the central P-atom was determined according to the (S) p(-) cyanofenphos whose configuration has been defined by an X-ray diffraction analysis (Ohkawa, 1980). The yield in the last step of the reaction seemed too low for a preparative synthesis. However, about 60 per cent of O-ethyl phenylphosphonothioic acid could be recovered with little loss of the optical purity. In addition, both the enantiomeric O-aryl O-ethyl phenylphosphonothionate could be obtained

(R)p(+) acid

(S)p(-) Acid

(S)p(-) anhydride

(R)p(+) anhydride

(R)p(+) Enantiomer

(S)p(-) Enantiomer

from a single enantiomer of the acid by using a combination of conventional method and MA method. These results indicate that this method would be useful for investigating phosphorus Stereochemistry and Biochemistry.

Biochemical Aspects

The Discovery of a Novel Site of Action for Herbicidal Bisphosphonates

Following renewed interest in a class of herbicides containing a 1,1-bisphosphonate moiety, Cromartie *et al.* (1999) initiated studies to identify the molecular target of these compounds. N- (2- (3-methylpyridyl) amino methylene disphosphonic acid (compound I), a representative of example of this group, was found not to inhibit soluble plant pyrophosphatase or H^+ translocating pyrophosphatase in the vacuolar membrane. Fatty acid biosynthesis from acetate or from pyruvate was also not supressed by this herbicide. Germination of plant seeds was not inhibited by the bisphosphonate, but root growth in dark was strongly inhibited without a significant effect on respiration. The bisphosphonate had little effect on maleic acid biosynthesis. However, the biosynthesis of carotenoids from isopentyl pyrophosphate in daffodil (*Narcissus* sp.) chromoplasts was found to be strongly inhibited by compound I at 2.5 mm. Further, study showed isopentyl pyrophosphate dimethylallyl pyrophosphate isomer was not blocked by the compound but that farnezyl pyrophosphate synthase was very strongly inhibited by compound I with IC_{50} of 23 nM. It is proposed that biophosphonates such as compound I are herbicidal as a result of the inhibition of pyrophosphate synthase in the *prenyl biosynthetic pathway* and that this enzyme represents a new target site to the development of herbicides.

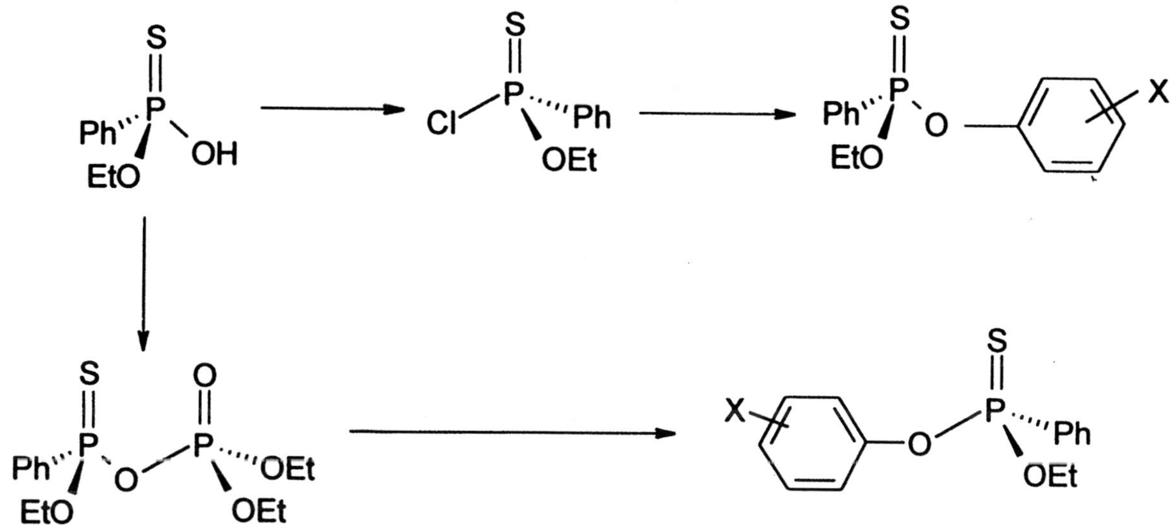

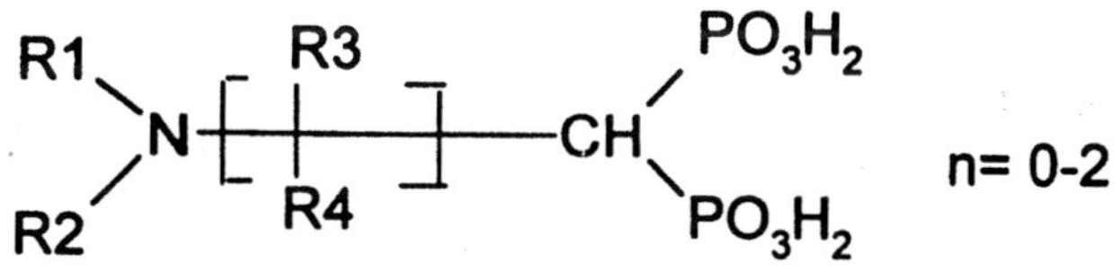

n= 0-2

Compound I

A Novel Bioactivation Mechanism of Phosphoramidate Insecticides

In order to elucidate the activation mechanism of phosphoramidate insecticides: Miyamoto *et al.* (1999) chosen 2,4-dichlorophenyl methyl, N-alkylphosphoramidates (alkyl: methyl, ethyl, propyl, isopropyl, and butyl), as model compounds and compared with isofenphos oxon. Although they showed a potent insecticidal activity against susceptible housefly except for butyl homologue, their in vitro acetylchlonesterase (AchE) inhibitory activity was very poor. N-desalkyl compound has been known as an activated form. When bovine erythrocyte AchE inhibition of N-isopropyl homologue

Mechanisms

isomerase

CH_2Opp CH_2Opp

IPP

Dimetyl allyl pyrophosphat€

GPP-Synthase

CH_2Opp

Geranyl-pp

+ IPP

FPP-synthase

Cis-phytoene

pytoene synthase

+ IPP

Farnesyl-pp → Geranyl geranyl-pp

synthase

and isofenphos oxon were carried out in rat liver microsomal system, they gave a strong inhibitory activity under the oxidative condition with NADPH, but in the presence of SKF-525A or piperonyl butoxide as a mixed function oxidase (MFO) inhibitor, isofenphos oxon lost all the inhibitory activity. On the other hand, N-isopropyl homologue as a flavin-containing monooxygenase (FMO) inhibitor.

Comparison of these data with insecticidal activity suggested the involvement of another bioactivation mechanism. Peracid oxidation as a model of biological oxidation resulted in the production of des-N-alkyl compound (HO-PO) in addition to N-desalkyl compound (H_2N-PO), which suggested the structure of the activated form as an oxidized form of phosphoramide moiety of the starting compound. Three structures were considered from a computational chemistry point of view. N-Oxide structure (RNH(O)-PO) is isomerized to O-aminophiosphate structure (RNHO-PO) depending on the alkyl substituent on the nitrogen atom, and the latter is the intermediate to give des-N-alkyl compound, which has no AchE inhibitory action. The formation of O-aminophosphate structure from the adduct of the starting compound with peracid was analyzed using model compounds by the theory of Frontier orbital. Also, O-aminophosphate structure might be produced via phosphorus oxyoxonate structure (RNH-P(OO). The possibility of hydroxylamine structure (RN(OH)-PO) to give N-oxide or O-aminophosphate structure was small if any in this model phosphoramidate compounds. In any way, O-aminophosphate structure was small if any in this model phosphoramidate compounds. In any way, O-aminophosphate structure is most probable as a novel activated form mainly formed by FMO or peracid.

Mechanism

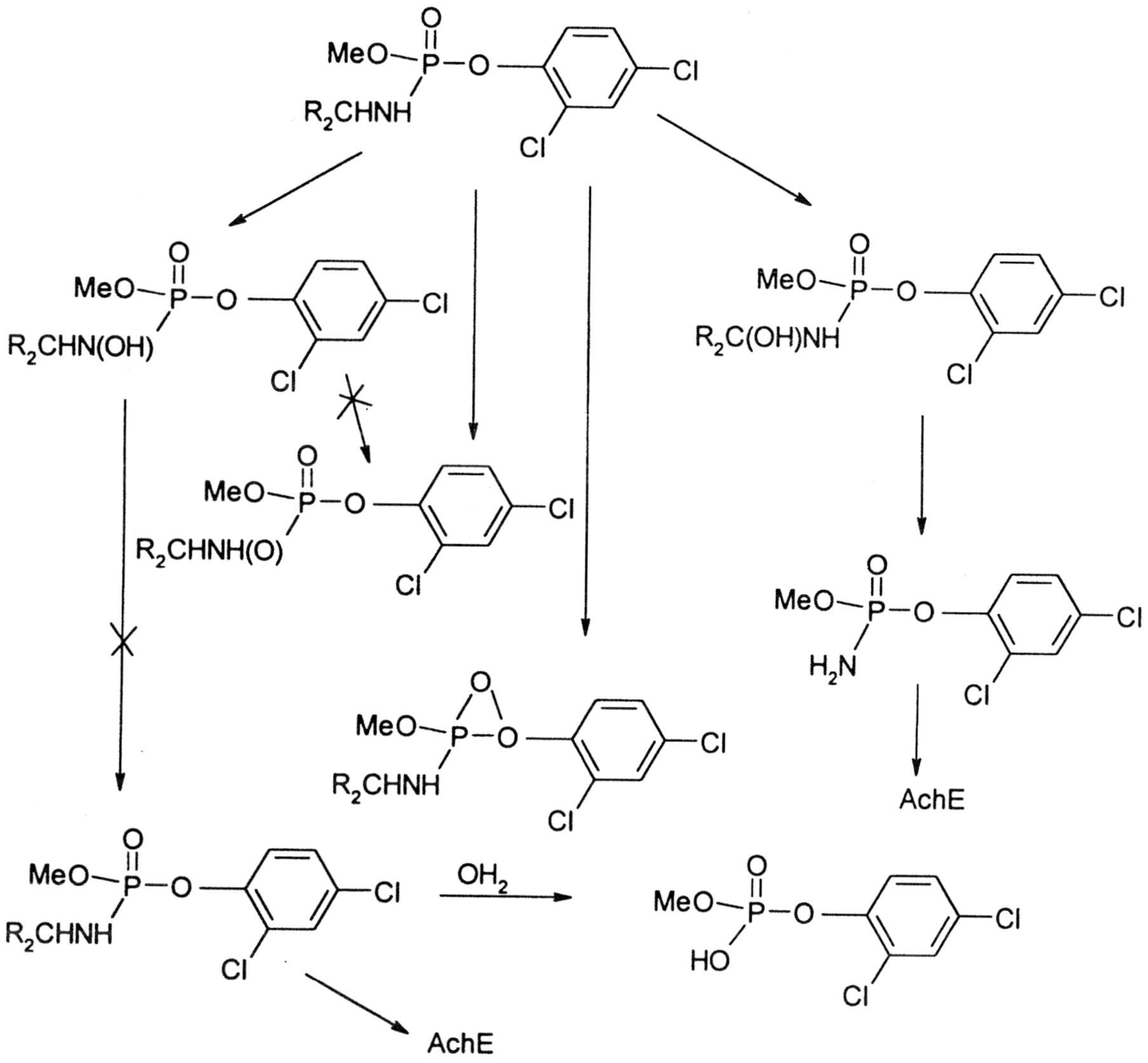

MeO—P(=O)(R$_2$CHNH)—O—C$_6$H$_3$Cl$_2$

R$_2$CHN(OH)　R$_2$CHNH(O)　R$_2$C(OH)NH　H$_2$N　R$_2$CHNH　HO

OH$_2$

AchE

AchE

Conclusion

Organophosphorus compounds are the leading pesticides which are most widely used in the farmers field. Despite the risk of acute poisoning from OP compounds they are still preferred over other pesticides because of their contact, systemic and broad spectrum activity and short residual life. The newer of pesticides are more selective in action and have relatively low mammalian toxicity. The environmental protection agency (EPA) has published its plan to screen up to 87,000 synthetic chemicals in the environment for their ability to mimic the feminising effect of the female hormone estrogen in which many OP pesticides also included. Such endocrine disruptors have been shown to result in development and reproductive abnormalities in wildlife, and are suspected by some scientists of causing adverse effect in humans, including birth defects, breast cancer, prostrate cancer and infertility. However, the above mentioned risks are negligible when the recommended safety precautions are observed.

Although, the risks associated with low level exposure over a long period (*i.e.*, chronic toxicity) are less well understood, there is little or no evidence of lasting harm. However, we have much to learn about OP pesticides, and there is still a need both for continued research in this area and also for particular uses to be properly monitored.

Moreover, there is an urgent need to synthesise the pesticides which are used in low dosage and environment friendly. For that collaborative work between botanical and synthetic workers are very much needed.

Finally, the discovery of novel site of action of newly synthesized compounds and their QSAR study may prove immense help in design and development of new and useful OP pesticides.

REFERENCES

Agnihotri, N.P. (2000). Pesticide consumption in Agriculture in India: An update. *Pestic. Res. J.*, 12(1): 150–155.

Allahyyari. R., J.G. Hoolingshaus, R.L. Rapp, R.A. Jacobson and T.R. Fukuto (1980). Resolution absolute configuration and acute and delayed neurotoxicity of the chiral isomers of o–aryl o–methyl phenylphonothionate analogous related to Letophos. *J. Agric. Food Chem.*, 28: 594–599.

Bedi, S. and N.K. Roy (1978a). Preparation and evaluation of diaryl dichloromethyl phosphonates against rice blast. *Indian J. Agric. Sci.*, 48: 248–251.

Bedi, S. and N.K. Roy (1978b). Residues of diaryl dichloromethyl phosphonates in/on rice plant and their chemical stability. *Indian J. Agric. Sci.*, 48: 701–704.

Chakravarty, S. (1992). Synthesis and herbicidal activity of amides 1,1– dichloroethyl phosphonic acids and their structural activity relationships. *Ph D. Thesis*, P.G. School, IARI, New Delhi, pp. 247.

Chattapadhyay, T.K. and R.L. Gupta (2001). Synthesis, fungitoxicity and QSAR study of some new phosphorothionates. *Tropical Agricultural Research*, 13: 249–259.

Cromartie, T.H., K.J. Fisher and J.N. Grossman (1999). The discovery of Novel site of action of herbicidal bisphosphonates. *Pestic. Biochem. Physiol.*, 63: 114–126.

Das, S.K. and N.K. Roy (1998). Preparation of some novel phosphonamidates, their toxicity and herbicidal properties. *Pestic. Sci.*, 52: 263–267.

Eto, M., Y. Kinoshita, T. Kato and Y. Oshima (1963). *Nature*, 200: 171.

Eto, M. (1974). Phosphonates. In: *Organophosphorus Insecticides: Organic and Biological Chemistry*. CRC Press Inc, Cleveland, pp. 75–78.

Fest, C. and K.J.O. Schmidt (1982). *The Chemistry of Organophosphorus Pesticides*. Springer Verlag, New York, pp. 1–360.

Gupta, R.L. (2000). Quantitative Strucure Activity Relationships for fungitoxicity of O, diaryl O–methyl phosphorothionates. *Ann. Pl. Protec. Sci.*, 8: 223–229.

Gupta, R.L. and N,K. Roy (1984). Synthesis and fungitoxicity of O,O–diaryl S–ethyl phosphorothiolates. *Pestic. Sci.*, 15: 553–556

Gupta, R.L. and N,K. Roy (1990). Synthesis and quantitative structure activity relationship for antifungal activity of O–alkyl S,S–diaryl phosphorodithioates. *Indian J. Chem.*, 29B: 870–875.

Gupta, R.L. and N,K. Roy (1991). Synthesis and anti–rice blast activity of S–alkyl S,S–diaryl phosphorotrithioates. *Indian J. Chem.*, 30B: 320–323.

Gupta, R.L. and N,K. Roy (1993a). Fungitoxicity and quantitative structure activity relationship of S–alkyl S,S–diaryl phosphorotrithioates. *Pestic. Res. J.*, 5(1): 16–22.

Gupta, R.L. and N,K. Roy and Chandrahas (1988). Quantitative structure activity relationships study of fungicidal O,O–diaryl S–ethyl phosphorothioates. *Pestic. Sci.*, 32: 139–144.

Haga T, Toki T, Koyanagi T, Yoshida K and Imai O (1964). Ishihara Sangyo Kaisha Ltd Japan Kokai Tokkyo Koho JP 59–42396.

Haga T (1992). "Rational approaches to structure activity and ecotoxicology of Agrochemicals" Eds: W Draber &T Fujita, CRC Press, Florida pp. 103–119.

Hirashima A, Ishaaya I, Ueno R, Oyama K and Eto M (1989). *Pestic Biochem Physiol* 34: 205.

Hirashima A, Yoshii Y, Harada, Y, He HW and Eto M (1991). Synthesis of biological activitiers of substituted 2–alkoxy 1,3,2–thiazaphospholidine–2–sulfides. *J Pestic Sci* 16: 499–507.

Ishikawa H and Morita K (1993). New synthesis and fungicidal efficacy of S,N–diaryl thiophosphoramidates. *J Pestic Sci* 18: 197–199.

Ishikawa H, Kido T, Umeda T and Ohyama H (1993). Fungicidal activity of new diphenyl phosphinic amide derivatives *J Pestic Sci* 18: 125–126.

Jaiswal AK, Rao GP, Pandey OP and Sen Gupta SK (1998). Efficacy of organophosphorus derivatives against fungal pathogens of sugarcane. *J Agric Food Chem* 46: 1609–1613.

Khazanchi R and Roy NK (1983). Synthesis and herbicidal activity of O–methyl p – (dichloromethyl). phosphonamidates and diamides. *Agric Biol Chem* 47: 331–335.

Khazanchi R and Roy NK (1986). Synthesis and phytotoxic properties of O, O– dimethyl phosphoroamidates and O–methyl phosphonic diamides. *Indian J Agric Sci* 56(10): 726–731.

Koyanagi T, Okada H. Imai O, Toki T and Haga T (1997). Synthesis and pesticidal activities of N–phosphinoyl heterocycles. *J Pestic Sci* 22: 187–192.

Mathur SC and Tannan SK (1999). Future of Indian Pesticide industry in next millennium. *Pesticide Information* XXIV (4): 923.

Mete A, Sener S, Kucukbay H, Gunal S and Darmaz R (1999). Snythesis of some heterocyclic phosphonates and their antibacterial and antifungal activities. *Indian Journal of Chemistry* 38 (B): 197–200.

Miyamoto T, Kasagami T, Asai M and Yamamoto I (1999). A Novel bioactivation mechanism of phosphoramidate insecticides. *Pestic Biochem Physiol* 63: 151–16.

Ohkawa, H (1982). Stereoselecivity of organophosphorus insecticides In:"Insecticide Mode of Action" Ed. JR Coats, Academic Press, New York pp. 163–168.

Ohkawa H, Mikami N, Okuno Y and Miyamoto J (197). Stereospecificity in toxicity of the optical isomers of EPN *J Bul Environ Contam Toxicol* 18: 534–537.

Ohkawa H, Mikami N, Miyamoto J (1977). Stereoselectivity in metabolism of the stereoisomers of cyanofenphos (o,p–cyanophonyl o–ethyl phenylphosphonothiate in rats and liver microisomers. *Agric Biol Chem* 41: 369–376.

Patra D and Gupta RL (2000). Quantitative structure activity relationships for the fungicidal activity of O,O– diaryl O– n– propyl phosphorothionates. *Pestic Res J* 12(2): 267–270.

Pesticide Information (2001). Annual Issue, pp. 68–76.

Prasad D (2000). Integrated nematode pests management strategies: A futuristic approach (Lead talk). *Nat. Symp. on Important Plant Diseases of North-eastern India and their Management*. Dec. 6–7. Org. N.D.U.A&T., Faizabad, p. 8–9.

Ray D (1998). How safe are organophosphorus pesticides? *Pesticide Outlook*, October Issue, p. 7–10.

Roy NK, Laljee B and Bedi S (1986). Synthesis and fungicidal activity of diaryl 1,1–dichloroethyl phosphonates. *Agric Biol Chem* 44: 2995–2997.

Roy NK and Bedi S (1982). A novel transformation to S–methyl phosphorodichloridothiolate and the fungitoxicity of diaryl S–methyl phosphorothiolates. *Agric Biol Chem* 46 (7): 1935–1937.

Roy NK and Taneja HK (1990). Synthesis and antifungus activity of O–methyl O–aryl 2–chloroethyl phosphonates and bisaryl 2–chloroethyl phosphonates *J Indian Chem Soc* 69: 42–43.

Sasaki M, Ooishi T, Kato T, Takayama C and Mukai K (1984). Synthesis and antifungal activity of O,O–dialkyl O–aryl phosphorothioates and related compounds. *J Pestic Sci* 9: 737–744.

Singh N (1998). Synthesis, fungitoxicity and Quantitative structure activity relationships of O–aryl O–2– choloroethyl phosphates against *Sclerotium rolfsii. Pestic Res J* 37B(3): 257–261.

Singh N, Gupta, RL and Roy NK (1996). Synthesis and Quantitative structure activity relationships of aryl O– 2– chloroethyl methyl phosphate fungicides. *Indian J Chem* 35B(7): 697–702.

Singh N and Prasad D (2000). Nematotoxicity and SAR of aryl 2–chloroethyl phosphates against root knot nematode *M. incognita. Pestic Res J* 12(1): 48–53.

Taira S, Tawata S, Kobamoto N, Toyama S and Yasuda M (1994). Synthesis and fungicidal activity of new 1,3,2–oxazaphospholidine 2–sulfides. *J Pestic Sci* 19: 299–304.

Tawata S, Taira S, Kobamoto N, Ishihara M and Toyama S (1996). Synthesis and fungicidal activity of new thiophosphorylated monoterpenoids and related compounds. *J Pestic Sci* 21: 141–146.

Tawata S, Taira S, Kikizu H, Kobamoto N, Ishihara M and Toyama S (1998). Synthesis and fungicidal activity of new 6–alkyl–2–alkylamino–4H–1,3,2–benzodioxaphosphorin 2–sulfides. *J Pestic Sci* 23: 137–140.

Vasu K (1982). Investigation of some organophosphorous compounds having fungicidal activity. Ph.D. Thesis, P.G. School, IARI, New Delhi.

Vasu K and Roy NK (1983). Synthesis and fungitoxicity of diaryl methyl phosphonates *Agric Biol Chem* 47: 2657.

Vasu K and Roy NK (1985). Laboratory evaluation of triaryl phosphites as potential fungicides. *Indian J Agric Chem* 18(1): 105–107.

Wu SY, Hirashima A, Takeya R and Eto M (1988). *Agric Biol Chem* 52: 2911.

Yoshikawa, H. (1999). New synthesis of optically active O–aryl O–ethyl phenyl phosphonothionates. *Biosci. Biotechnol. Biochem.*, 63(2): 424–426.

Yoshikawa, H., T. Shono and M. Eto (1984). A facile synthesis and insecticidal activity of optically active phenylphosphonothiolates. *J. Pestic. Sci.*, 9: 455–462 .

Yoshikawa, H. (1999). New synthesis of optically active O–aryl O–ethyl phenylphosphonothionates. *Biosci. Biotechnol. Biochem.*, 63 (2): 424–426.

Yoshikawa, H. (2000). Synthesis and insecticidal activity of O-aryl N-alkoxycarbonyl N-alkyl-methylphosphonamidothionates. *J. Pestic. Sci.*, 25: 392–394.

39

Weed Management for Sustainable Agricultural Production

Sahadeva Singh

Assistant Commissioner (Crops), Ministry of Agriculture, Department of Agriculture and Cooperation, Krishi Bhawan, New Delhi – 110 001

Increased population will demand more food production and this can only be achieved by increasing crop yields. While human populations have increased steadily in the developing countries, developments in technology have lagged behind and the honored crop production practices of the past have been preserved intact. The ideal crop production system is one that combines good land, crop and pest management to enable the farmer to:

1. Obtain moderate to high crop yields on a sustainable basis
2. Protect crop from pests, and
3. Use of land intensively and economically without soil degradation and irreversible loss of soil organic matter, structure and composition.

Above all in the context of improved production technologies for maximizing crop yields, the weed menace has to be curbed in time otherwise the very purpose of adoption of improved technologies will be defeated.

The present chapter intends to give an idea of the importance of weed management and how crop production is improved with technologies for better weed control.

Integrated Pest Management

Since mid 60s IPM has been promoted by FAO worldwide as a preferred strategy for pest control. IPM is defined as a pest management system that utilizes all suitable techniques and methods in as compatible a manner as possible and maintains the pest populations at levels below those causing economic injury.

There is general agreement that IPM must be science-based and economically viable for farmers. Prevention is an important requirement as well pest control during the crop cycle. Preventive approach obliges to evaluate regularly pest density and risk of entry of new pests into the fields. Common control strategies are:

Cultural Methods

Cultural methods include crop rotation, resistant cultivars, use cover crops, inter-cropping, mechanical control, and efficient water management and fertilizer application.

Biological Control

Biological control either classical method introducing new exotic organisms evaluated previous to their introduction, and augmentative method through rearing and release of existing beneficial natural enemies.

Chemical Control

Rational and judicious use of pesticides.

Other Methods

Use of soil solarization, and other physical methods, which are still under investigation.

IPM approach has had major impact in insect and mite control, with some other positive results of disease control and very little effect on weeds. The reason of this behaviour with weeds is due to:

1. Little understanding of problems caused by weeds, particularly in developing countries.
2. Few or no ecological elements of weeds included in IPM programmes.
3. A pronounced tendency to use of herbicides as the sole control option in developing countries or large crop areas, or to rely on hand–weeding in poor or subsistence sites.

In contrast to other pest management disciplines, devising integrated weed management strategies that address a diversity of weed species with a diversity of life history traits is difficult. In the contrast of IPM particular importance must be given to the improvement of weed management, which is still a minor component of the majority of available programmes on IPM. Auld (1994) classifies the impact if weeds on t any agricultural system as follows:

1. Reduced agricultural output (quantity and/or quality).
2. The costs involved in a given level of control of weeds within the existing farming system.
3. A change in the farming system to a new system (*e.g.*, new crop) brought out mainly because of particular weeds.
4. External costs caused by weeds spreading across farm boundaries.

The importance of weed control increases worldwide. While other pests are under control in various crops, weeds continue to be limiting factor to crop yields. Weeds directly affect crop growth and development by competing effectively for light, nutrients, and water; releasing substances from their roots and leaves which are toxic to various crop plants; and serving as hosts of many pest organisms which pose serious risks to crop production. Weeds also seriously affect inland water bodies and related activities (irrigation, fishery, and energy supply from hydroelectric stations).

Weeds cause important social problems. Small scale farmers in developing countries spend more than 40 per cent of their labour time weeding, and this includes mainly female and child labour of the farmer's family.

Weed management is a systematic approach to minimize weed impacts and optimize land use, and it is an approach that combines prevention and control (Aldrich, 1984). It has been defined as "an environmentally sound system of farming using all available knowledge and tools to produce crops free of economically damaging, competitive vegetation" (Fishers *et al.*, 1985).

Weed management emphasizes minimizing the effects of, but not eliminating, weed populations. The central objective of an effective weed management programme will be to manipulate the crop-weed relationship so that growth of the crop is favored over that of weeds.

Components of Weed Management

The inclusion of weed component in IPM compels to study various elements of weed ecobiology and interference to really achieve long term effects of weed control. Sustainable weed management should foresee control of weeds during the periods of weed competition but also should guarantee reduction of viable weed seed bank in soil.

Basic component of weed management include:

1. Regular monitoring of weed infestation, which may include detection of herbicide resistant weeds in areas regularly treated with herbicides.
2. Biology and ecology of the prevalent weed species.
3. Plant interference and critical period of crop-weed competition.
4. Technically effective, economically viable and environmentally safe control methods and integration of several weed control techniques including selective use of herbicides.

Monitoring of Weed Infestation

A fundamental basis for sound weed management is to know the species present and their level of infestation. The composition of weed communities varies greatly and is closely linked to cropping history. Regular evaluation of weed infestation in agricultural fields is advisable, particularly in areas currently treated with herbicides. It is essential to distinguish perennial or parasitic weeds, which may not respond to the standard traditional weeding practices, while more precise determination, even of annual weed species can be vital for the optimum selection of any control method including herbicides. The persistence of most weed species is contingent upon the presence of a buried seed bank and knowledge of size of seed bank and flux of seeds with in it, in relation to land cultivation practices, is essential to predict weed infestation in medium term at test (Forcella and Burnside, 1993).

It has been considered the best predictor of future weed problems is knowledge of previous weed problems. By keeping long-term records of weed problems in replicated untreated check areas within each field, growers should be able to not only anticipate severity of weed infestations, but recognize early changes in species composition of weed composition.

Weed Ecobiology

Weed management means control of weed population and to this end it is important to have elements of population equilibria and life stages for regulating population size. Weed populations at a given site are very dynamic. Population density often changes as a result of selection pressures such as herbicide use, tillage, cultivation, grazing, and cropping sequence changes. The intensive use of herbicides over the past 30 years has introduced a selection pressure that has altered the species composition of arable sites probably more than any other single production activity. Changes in rotational and tillage practices may greatly impact species composition and population density. The persistence of annual species is generally favored by tillage that continually restores the seedbed to a condition favorable for germination and establishment as opposed to sites that are not tilled annually. In contrast, no tillage may favor perennial species that are perpetuated by underground storage organs such as rhizomes, tubers, corns, and/or rootstocks.

Ecological studies of weeds and weed management are required for diverse objectives such as:

1. Auto-ecological comparison of weed species.
2. Measuring the impact of new technologies on the dynamics of weed flora.
3. Quantifying weed community structures and analyzing its causes, among other studies.

In practical terms, it is necessary to have information of the behaviour of prevailing weeds. Data on germination of diaspores, phases when plants start its fruit setting and for emissions of new vegetative organs incase of perennial species, factors affecting their growth and development, production of reproductive organs. In tropical and subtropical areas specially in hot climatic condition, several weed flushes occur during the crop cycle, its knowledge is essential to design better control and prevention of weed seed build up.

Critical Periods of Competition or Interference

Weeds severely interfere with the growth and development of crops. Interference comprises competition with crops for light, nutrients and water, and allelopathy, which Is the effect of substances root exuded cr leaf leached by a plant on growth and development of other plant(s).

For studying competition, it is common to study the critical period of weed competition, which has been defined as the period during which weeds must be controlled to prevent losses (Sattin *et al.*, 1998).

The knowledge of critical periods of weed competition may help to determine when to implement control measures in order to minimize yield losses for many crops (Zimdahl, 1995). The crop yield level obtained by weeding during this short span should provide crop yield sufficiently close to that obtained by full crop season free from weeds. A fundamental principle of plant competition is that early occupants on a soil tend to exclude the later ones. This principle finds application in weed control. Many weeds are prolific producers of seeds which remain viable or dormant in soil and germinate when they get suitable atmosphere. The seeds that germinate either before or along with emergence of crop plants are the most successful competitor. The duration of weed infestation and time of weed removal have a significant influence on crop growth and economic yield. Variations in critical period of required weed free period depends on many factors, such as the relative growth of the crop and weeds, the density of planting, the variety grown, the time of moisture and nutrient stress, etc. The critical period of weed competition with different crop have been given in Table 39.1.

Table 39.1: Critical Period of Crop-weed Competition and Range of Yield Loss Due to Weeds in Different Crops

Crop	*Critical Period of Crop-weed Competition*	*Yield Loss Due to Weed (%)*	*Reference*
Rice (direct sown)	10–45 DAS	42–84	Raja Ram Mohan Rao *et al.* (1974)
Rice (transplanted)	4–6 weeks after transplanting	12–51	Shetty and Gill (1974) and DRR (1990)
Wheat	Up to 30–45 DAS	15–25	Gill and Brar (1972) and Solanki *et al.* (1980)
Maize	Up to 30–45 DAS	29.5–74	Pandey and Mishra and Krishna Murthy *et al.* (1981)
Sorghum	Up to 40 DAS	6.2–42	Badhe *et al.* (1984)
Finger millet	15–30 DAS	11–56	Prakash *et al.* (1982) and Mani *et al.* (1968)
Sugarcane	Up to 90 days after planting	15–72	Bhogat and Bhan (1988)
Sugarbeet	20–80 DAS	70.3	Mani *et al.* (1968)
Greengram and blackgram	4–6 weeks after sowing	Up to 50	Vats and Sidhu (1976)
Gram	60–80 DAS	30–70	Bisen and Tiwari (1983)
Peas	15–45 DAS	20–50	Sandhu *et al.* (1978)
Soybean	30–40 DAS	40–60	Krishnamurthy *et al.* (1981) and Maurya *et al.* (1990)
Groundnut	Up to 60 DAS	18–52	Krishna Gowda *et al.*, (1977)
Mustard	20–40 DAS	30	Hill *et al.* (1984)
Sesame	14–35 DAS	17–41	Kondap *et al.* (1983) and Rao and Narayana Rao (1985)
Cotton	15–60 DAS	47.5	Ali and Sankaran (1984) and Mani *et al.* (1968)
Jute	35–42 DAS	56.87	Tosh (1982), Saraswat and Bibhas Ray (1988)
Potato	20–40 DAS	20–60	Mani *et al.* (1968)
Onion	Up to 40 DAS	48–85	Chetty and Sankaran (1988) and Bhallia (1978)

Control Strategies

Weed must be controlled within the critical period of weed crop competition. The critical period of crop-weed competition under Indian conditions coincides with other labour intensive agricultural operations. Mechanically weeding is not possible when the soil is too wet to work. In fields heavily infested with weeds, herbicides can reduce the need for hand weeding and, therefore, more area can be covered by using selective herbicides. It would, however, be wrong to assume that herbicide use is always beneficial. Improved varieties, high quality seeds, adequate soil preparations, sowing time and method, seed rate, timely application of fertilizers at recommended rates, all produce vigrous crop which compete well and dominate the weeds. The integration of appropriate crop management practices can assist herbicides for long term gains. Improved growing technologies and cropping patterns have accounted for variability of different types of weeds.

The main weed control strategies can be classified as follows:

1. Tillage systems (minimum till, zero till and conventional tillage).

2. Cultural: Crop rotation, use of competitive and/or allelopathic cultivars, living or dead cover crops, water management, fertilizer application, mulching, inter-cropping.
3. Physical: Inter-row cultivation and hand-weeding.
4. Biological control releasing insects or other animals, using beneficial pathogens.
5. Chemical control, through the use of herbicides.
6. Other control methods *e.g.*, soil solarization.

Tillage Systems

Tillage systems are essential for land preparation, and weed infestation may dictate the type of tillage to be used. Tillage based on ploughing and harrowing during land preparation has been a conventional method of weed control and still very much used for growing roots and tubers, and vegetables. Minimum or zero tillage systems are becoming more used for the production of direct seeded grains. These tillage systems allow leaving residues and stubble from preceding crop covering the soil, which reduce weed infestation to some extent. Tillage system's alone may reduce the stand of certain species. In conventional agriculture no till or zero till may succeed reducing weed infestation in general if combined properly with other control measures, *e.g.*, crop rotation, using of cover crops, increased crop seed densities and rational use of herbicides.

Cultural Weed Control

Crop Rotation

The key for good cropping system is to have good crop rotation schemes, which help reducing pest problems. Rotation with morphologically different crops and with different growing requirements may help to break the cycle of adapted weeds. Cropping sequence has enormous influence on variations of weed populations as a direct result of implementing different weed management practices associated with crops involved in rotation, Incorporation of densely seeded and competitive plants as preceding crops, such as cereals and legumes reduce significantly weed density from one season to another.

Competitive and/or Allelopathic Crops

One of the ways to combat weeds is giving necessary competitive ability to crops at early stage of growth. Therefore, those plants able to compete with weeds and to suppress them are the best in the context of integrated weed management. Some studies have been carried out of the competitive ability of rice cultivars with weeds *Echinochloa colona, E. phyllopogon, E. oryzoides, Brachiaria brizantha,* or *Brachiaria decumbens* (Fischers and Gibson, 2001), which showed that it is possible to have modern high-yielding, semi-dwarf cultivars able to compete with weeds efficiently. Competitive rice cultivars, such as semi-dwarf cv.M-f 2002, effectively suppressed the infestation of *Echinochloa oryzoides* and *E. phyllopogon* in California and may help reduce herbicide dependency and decrease selective pressure for resistance(Gibson *et al.*, 2001).

Plants in general are able to produce and exude substances suppressing the growth of other plants in their vicinity. Several crops-cultivars may exhibit a high level of allelopathic effect over certain weed species. Results have shown that there is large variation in allelopathy among rice cultivars, being some of them able to suppress both mono-and dicot weed species (Olofsdotter, 2001). Similar results have been obtained in South Korea (Lin *et al.*, 2000), which represents a potential for improved weed management. Wheat varieties differ in allelopathic potential against weeds, indicating that selection of allelopathic varieties might be useful strategy in integrated weed management.

Allelopathic Cover Crops

Allelopathic cover crops are one of the most promising application for weed control. Fujii (2001) found that leguminous cover crops such as hairy vetch. (*Vicia villosa*) and velvetbean (*Mucuna pruriens*), grass cover crops, such as oat (*Avena sativa*) and rye (*Secale cereale*), certain cultivars of wheat (*Triticum aestivum*) and barley (*Hordeum vulgare*) were promising. Cover crops fit well into many different cropping systems, during fallow or non-crop season. Legume cover crops reduce the need for the application of N fertilizers, but indirectly may reduce the incidence of some pests, including weeds, with subsequent reduction of pesticide inputs. So promising the use of cover crops is a contribution for increasing sustainability of crop production.

Fertilizer Application

Application of phosphatic fertilizers without nitrogen stimulates the growth of broad leaf weeds. Contrary to it, nitrogen fertilizers may stimulate the growth of grassy weeds. Appropriate combination of nutrients and herbicides may change the crop-weed completion in favour of crop. The efficiency of herbicides has been found more in fertilized plots than in unfertilized plots. Fertilizer dose can be effectively integrated with herbicides by lowering the dose of fertilizer.

Water Management

Surface water for irrigation can increase weed problems by applying water contaminated with weed seeds. Surface irrigation water is considered a significant source of weed seeds on irrigated farmlands. It has been considered a potential source of adding weed seeds to the soil from surface sources of irrigation water. The problem can be greatly reduced by controlling weeds along irrigation laterals and by removing weed seed from irrigation water with appropriate screening devices.

Mulching and Inter-cropped

Mulches are good tools of weed management. They normally exclude light and serve as physical barrier to emergence, in addition, they also affect soil temperature and moisture. Non living mulches can be divided into either natural, such as plant leaves, stalks, straw, other residues, compost and dry soil, or synthetic materials like plastic polyethylene widely used in pineapple and strawberry production. Natural mulches add organic matter and nutrients to the soil with their decomposition, and are inexpensive. Natural mulches exert good control of annual weeds but a weak one over perennials.

A cover crop, inter-cropped or inter-planted with an annual or perennial crop, is called living mulch, which is useful to suppress weed infestation, to prevent soil erosion, enhancing soil fertility and improving water infiltration. Cover crops also exclude light with their shading limiting weed emergence.

Physical Weed Control

Inter-row Cultivation or Weeding During Crop Cycle

To control weeds with cultivation or manual weeding in the inter-rows is a traditional practice. Its use depend on economic feasibility and other requirements. Organic agriculture has to use this method in several crops avoiding using herbicides. Inter-row cultivation can increase the competitive ability of crops to reduce weeds. However, the magnitude of reduction in weed and growth depends largely on nature of intercrops, their relative proportion in the mixture and the spatial management of plants.

Biological Weed Control

Weed control through specific natural enemies is a highly sustainable way of controlling troublesome weeds. One the most important progress in this area is the control of floating aquatic weeds using various specific insects. There are several weeds introduced in various countries and the best way to control them is using classical approach, *i.e.*, introducing specific natural enemies from area of origin of weeds. A list of organisms used effectively for biological control of some weeds is given in Table 39.2.

Table 39.2: List of Organisms Used Effectively for Biological Control of Some Weeds

Weed	*Biological Control Agent*
Eichhornia crassipes solms	Weevils *Neochetina bruchi* Hustache, *N. eichhorniae* Warner Moth *Sameodes albigutalis*
Hydrilla verticillata	Grass carp *Ctenopharyngodon idella* Tuber Weevil *Bagous affinis* Leaf mining Fly *Hydrellia pakistanae*
Pistia stratiotes Linn.	Weevil *Neohydronomus affinis*
Salvinia molesta	Weevil *Cyrtobagous salviniae*
Alternanthera philoxeroides Griseb	Flee Beetle *Agasicles hygrophila* Stem borer *Vogtia malloi*
Parthenium hysterophorus Linn.	Fungus *Puccinia abrupta* var. partheniicola
Euphorbia esula	Flea beetles *Apthona* spp.
Chromolaena odorata	Moth *Pareuchaetes pseudoinsulata*
Lantana camera Linn.	Lacebug *Teleonemia scrupulosa*
Cordia curassavica	Beetle *Metrogaleruca obscura*
Orobanche spp.	Stem Feeding Fly *Phytomyza orobanchia*

Biological control involves the deliberate use of organisms such as insects or fungi to control weeds, while biological management also involves the use of crop competition, allelopathy, resistant varieties, natural chemical agents, and other approaches. Biological appraoches to weed management are part of an overall cropping system, consisting of many interacting components,.designed to reduce the impacts of weeds on potential crop productivity. Biological management is more complex than chemical weed control because it requires long-term planning, multiple tactics and manipulation of cropping systems.

Chemical Weed Control

Herbicides were introduced into the agricultural practice more than 55 years ago. At that time 2,4-D was the only chemical available and just for the control of annual weeds, mainly broad leaf weeds in cereals. By mid 50 new selective herbicides started to be used in several crops and five decades after the advent of 2,4-D, list of herbicides grew enough. A list of important herbicides used in agricultural crops is given in Table 39.3.

Table 39.3: List of Important Herbicides Used in Agricultural Crops

Herbicide Group	*Mode of Action*	*Herbicides*
Limidazolinones, sulfonylureas, triazolopyrmidines, pyrimidinyl benzoates, Sulfonylaminocarbony-ltriazolinones	Inhibition of acetolactate synthase ALS	Chlorsulfuron, imazaquin, flutmesulam, bispyribac-Na-flucarbazone-Na
Aryloxyphenoxypropionates, Cyclohexanediones	Inhibition of acetyl CoA carboxylase (ACCase)	Diclofop, fenoxapropethyl, sethoxydim
S-Triazines, trazinones, triazolinone ureas, uracils, amides	Inhibition of photosynthesis at photosystem II	Atrazine, diuron, amicarbazone, propanil, metribuzin, bromacil, desmediphan, pyrazon
Nitriles	Inhibition of photosynthesis at photosystem II	Bromoxynil, ioxynil
Bipyridiliums	Photosystem-I-electron diversion	Paraquat, diquat
Dinitroanilines	Microtubule assembly inhibition	Trifluralin, nitralin, dinitramine, pendimethalin
Chloroacetamides, oxyacetamides and acetamides	Inhibition of cell division	Butachlor, napropamide
Synthetic auxins	Synthetic auxins	2,4-0, MCPA, MCPB
Thiocarbamates	Inhibition of lipid synthesis-not ACCase inhibition	Triailate, EPTC, butylate
Glycines	Inhibition of EPSP synthase	Glyphosate, sulfosate
PPO Inhibitors (Diphenylethers, Phenylpyrazoles, N-phenylph-thalimides, Oxadiazoles)	Inhibition of protoporphyrinogen oxidase (PPO)	Oxyfluorfen, fluazolate, oxadiazon

In several crop situations, mixtures of two or three herbicides have been used with results better than the application of single herbicide. A number of factors must be considered in determining the proper herbicides to integrate into a weed management system, including product and application costs, tillage system, crop rotation sequence, specific crop-weed associations, environmental conditions, and other general crop management practices. The recent advent of highly active, selective, and broad-spectrum postemergence herbicides has increased the total number of options available for chemical weed management in several major crops. Combinations of soil-applied and/or postemergence herbicides will often provide the broadest spectrum of weed control and help prevent establishment of dominant weed species complexes. There is no single herbicide that will selectively control all weed species, so efficient weed management with chemicals requires that herbicide selection be based on the weed species present in the target area. In crop production areas where weed species combination is diverse, an integrated approach to weed management may prove most effective and include cultural and mechanical weed management practices in addition to herbicides use.

The problem of herbicide resistance shows that weed control cannot be solved just using herbicides alone. Herbicides rationally applied are effective tools combined with other control methods (crop rotation, tillage, cover crops and others), but chemical control should be considered as another tool and not the only one. With the use of herbicides weed monitoring becomes important to detect shift of weed population and possible herbicide-resistance weeds. To prevent herbicide resistance it is essential to make available to all interested available case studies indicating cropping system in place, number of years and applications for the resistance to evolve. It is also necessary to avoid continued use of herbicides having the same mode of action unless it is integrated with other weed control practices, and in probable metabolic resistance, use herbicides which degrade by different metabolic means. To devise effective herbicide resistance management strategies it is important to have good understanding of the causes, nature, genetics, mechanisms, and solution for herbicide-resistance weeds (Heap and LeBaron, 2001).

However, the recent commercialization of crop cultivars containing transgenes for herbicide resistance(HRC) in different part of world, poses a potential blessing to agriculture as well as potential threat to the well being of the agro-ecosystem (Rubin, 1998). There are several concerns about the consequences of development and deployment of transgenic herbicide-resistance (HR), among them:

1. The potential transfer of genes from herbicide resistant crops (HRC) to wild relatives thus creating super weeds.
2. Possibility of HRC volunteers to become weeds in subsequent crops.

The above problems compels to avoid the repeated use of one herbicide, which may cause a switch in the weed flora towards less sensitive species. Therefore, alternate weed control strategies will have to be adopted to prevent problems. Particular attention should be given to the use of HRCs in centres of crop origin, where the risks of gene transfer are higher. Special care should be taken to ensure that native germplasm is not jeopardized by the transfer of herbicide resistance genes.

Integrated Weed Management

Herbicide were mainly seen as substitute to cultural and mechanical methods of weed control. A range of herbicides was generally compared to the traditional weeding practice. The need for developing an integrated approach of weed control was recognized about 25 years back. However, the situation for the control of weeds has seriously deteriorated due to persistence application of the same chemical herbicides and routinely repeated cultivation practices which has caused widespread emergence of weeds that are resistance to all currently used herbicides. The. methods applicable in one area may not be applicable elsewhere because farmers may lack necessary resources (labour and capital) to afford the implementation of new methods. However, due to the implications of weed interference in the whole agricultural process improved crop production *i.e.*, higher yields and quality, will only be reachable when weed problems will be minimized and related expenses. Therefore, social and environmental concern as well as desire to improve weed control efficiency and agricultural production, have led to increase emphasis on integrating several control methods.

Preventive Measures in Context with Free Trade

Several weeds are alien plants in many countries and regions of the world, where they cause serious problems. Increased world trade, international travel and world demographic growth bring serious problems of introduced plants, some of them highly invasive. Active free market is one of the major causes of introduced weeds into new territories. Some of the species such as *Parthenium hysterophorus* and *Phalaris minor* have been introduced in India in grains coming from USA. Some other weeds such as *Chromolaena odorata* which was brought from the Caribbean and introduced as cover crop in plantations of Africa and Asia, becoming later an invasive weed, which seriously affect crop production and plant biodiversity. The perennial grass *Sorghum halepense* from Mediterranean area was introduced in North America as a pasture and at present causing problems in several crops as a highly competitive weed. More recently, there are evidences that certain species of Prosopis introduced in arid areas of Africa and the Middle East are invading large fertile areas of the countries.

The invasiveness and economical damage of introduced weeds should be minimized. To this end it is required to conduct some activities to prevent these adverse phenomena, such as:

1. Risk assessment of possible adaptation of exotic plants
2. Provide knowledge and tools to quarantine officers/inspectors about exotic weed species potentially able to be introduced and methods for detecting them.

In each country, quarantine should have a list of weeds to prevent their entry. The list should be prepared based on the list of weeds absent and potentially able to establish themselves in the country. Quarantine officers/inspectors should also have methodologies for the detection of weed species in imported agricultural products. Prevention is an important activity in the context of integrated weed management to avoid future problems and losses caused of exotic weeds. Some of the exotic weeds introduced in different regions of world is given in Table 39.4.

Table 39.4: List of Important Weeds Introduced in Various Regions of World

Species	*Area of Introduction*
Sorghum halepense	Western Hemisphere
Rottboellia cochinchinensis	Western Hemisphere
Cyperus rotundus	Western Hemisphere, Africa
Mimosa pigra	Africa, Asia and Oceania
Chromolaena odorata	Africa, Asia and Oceania
Parthenium hysterophorus	Africa, Asia and Oceania
Euphorbia esula	North America
Lantana camera	Africa, Asia and Oceania
Eichhornia crassipes	Central America, Caribbean, Africa
Hydrilla verticillata	North America
Prosopis spp.	Middle East and Africa
Striga asiatica	North America
Orobanche ramose	Pacific Coast of Mexico and Cuba

Future Strategies for Weed Management

Modem agriculture or its improvement will largely depend on the methods used for weed control. Research and development and feed back from scientists and farmers and extension workers is regularly provide information and validate weed control methods for effective weed management. The following action points are suggested for future improvement in weed management:

1. Ecobiology of prevailing weed species, including weed seed bank, effect of biotic and abiotic factors on the growth and productivity of the species.
2. Weed interference to determine critical periods of competition.
3. Tillage systems and their short and long-term effects on weed flora.
4. Use of cover crops and mulches.
5. Identification of competitive and/or allelopathic crops for better competition with weeds.
6. Inter-cropping for suppressing weeds.

7. Preventive methods of weed control, including regulatory measures and weed risk assessment.
8. Biological control of prevailing weeds.
9. Monitoring of herbicide resistance.
10. Hazard assessment of ecological impact of the use of HRCs.
11. Prevention of the disposal of organic wastes in water bodies.
12. Reactivate weed control whenever possible as a key weed control method.

REFERENCES

Aldrich, R.J. (1984). *Weed-crop Ecology: Principles in Weed Management*. Breton Publishers, North Scituate, Massachusetts.

Ali, A.M. and S. Sankaran (1984). Crop-Weed competition studies in winter irrigated *gossypium hirsutum* (var.) MCU-9 (Abstr.). In: *Proceedings of the Annual Conference of Indian Society of Weed Science*, p. 61.

Auld, B. (1994). Economic criteria for implementation of weed management. In: *Weed Management for Developing Countries*, (ed.) R. Labrada, J.C. Caseley and C. Parker. FAO Plant Production and Protection Paper No. 120 pp. 237–246.

Badhe, S.B., R.V. Nalamwar and P. Khadekar (1984). A study of crop-weed competition in hybrid sorghum (Abstr.). In: *Proceedings of the Annual Conference of Indian Society of Weed Science*, p. 21.

Bhallia, P.L. (1978). Weed competition, crop losses and chemical weed control on onion–A review. *Pestology*, 2(1): 3–39.

Bhogat, B.S. and V. Bhan (1918). Influencing of time of weed removal on emergence of weeds in sugarcane (Abstr.). In: *Proceedings of the Indian Society of Weed Science*, p. 50.

Bisen, C.R. and J. Tiwari (1983). Chemical weed control in gram. *Indian J. Weed Sci.*, 15(2): 223–225.

Chetty, K.N. and S. Sankaran (1988). Crop weed competition studies in onion (Abstr.). In: *Proceedings of the Biennial Conference of Indian Society of Weed Science*, p. 58.

Directorate of Rice Research, Hyderabad (1990). *Annual Report, 1988–89*, Directorate of Rice Research, Hyderabad, p. 37.

Fischer, A.J. and K.D. Gibson (2001). Cultivars competitvos como herramienta para el Manejo Integrado de Malezas. Resumenes XV Cngreso ALAM, Maracaibo, Venezuela, pp. 26–30.

Fisher, B.B., E.A. Yeary and J.E. Macroft (1985). *Vegetation Management Systems in Principles of Weed Control in California*, Thomson Publishers, Fresno, California, pp. 213–228.

Forceila, F. and O.C. Burnside (1993). Pest Management: Weeds. In: *Sustainable Agriculture Systems*, (eds.) J.L. Hatfield D.L. Karlen, Lewis Publishers, Boca Raton, pp.157–197.

Fujii, Y. (2001). Screening and future exploitation of allelopathic plants as alternative herbicides with special reference to hairy vetch. *J. Crop Prod.*, 4(2): 257–275.

Gibson, K.D., J.E. Hill, T.C. Foin, B.P. Caton and A.J. Ficher (2001). Water-seeded rice cultivars differ in ability to interfere with water-grass. *Agronomy J.*, 93(2): 326–332.

Gill, H.S. and L.S. Brar (1972). Investigation on the chemical control of Phalaris minor in wheat. *J. Res.*, Punjab Agricultural University, 9: 287–290.

Heap, I. and H. LeBaron (2001). Introduction and overview of resistance. In: *Herbicide Resistance and World Grains*, (Ed.) S.B. Powles and D.L. Shaner, CRC Press Inc., USA, pp. 1–22.

Hill, H.S., K.S. Sandhu, S.P. Mehra and T. Singh (1984). Efficacy of some herbicides for control of weeds in Indian mustard. *Indian J. Sci.*, 16: 171–175.

Kondap, S.M., M.R. Reddy and K. Ramakrishna (1980). Studies on yield reduction due to *Cyperus rotundus* in different crops (Abstr.). In: *Proceedings of the Conference of Indian Society of Weed Science*, p. 25.

Kondap, S.M., Y.V.N. Murthy and P. Venugopal (1983) More seeds through weed control. *Indian Farming*, 33(8): 25–29.

Krishna Gowda, K.J., S. Purushothan and K. Krishnamurthy (1977). Effect of weed free environment during specific growth periods in groundnut. In: *Proceedings of Indian Society of Weed Science*, Andhra Pradesh Agricultural University Weed Science Conference, held at Hyderabad, pp. 203.

Krishnamurthy, K.B., V.C. Reddy and K. Kenchaiah (1981). Critical stages for weed competition in soybean groundnut and maize. In: *Proceedings of the 8th Asian Pacific Weed Science Society Conference*, (2): 123–127.

Lin, W.X., K.U. Kim and D.H. Shin (2000). Rice allelopathic potential and its modes of action on barnyardgrass (Echinochloa crus-galli). *Allelopathy J.*, 7: 2, 215–224.

Maurya, B.M., Gogulnar and J.P. Tiwari (1999). Herbicidal weed control efficiency and nutrient removal by weeds in soybean. Indian *J. Weed Sci.*, 22(3&4): 51–56.

Olofsdotter, M. (2001). Rice: A step toward to use allelopathy. *Agronomy J.*, 93: 1, 3–8.

Pandey, J. and S.S. Mishra (1982). Effect of *Cannabis sativa* L. on yield of rabi maize (*Zea mays* L.) (Abstr.). In: *Proceedings of the Conference of the Indian Society of Weed Science*, p. 20.

Prakash, V., S.P. Singh and J.P. Tandon (1982). Critical period of crop-weed competition in ragi (*E. coracana*) and barn yard millet (*E. frumentacea* L.) (Abstr.). In: *Proceedings of the Annual Conference of Indian Society of Weed Science*, p. 19.

Rao, R.R.M. and S.G. Singh and V.M. Bhan (1974). Best weeds to be part of your rice yield. *Indian Farming Digest*, 7(6): 19–22.

Rao, A.S. and K.N. Rao (1985). Performance of some herbicides on weed control in sesamum. *J. Oilseeds Res.*, 2(1): 117–119.

Rubin, B. (1998). Herbicide use and transgenic crops resistant to herbicides. In: *Report of FAO Expert Consultation on Weed Ecology and Management*, FAO, Rome Italy, September 22–24, 1997, Plant Production and Protection Division, pp. 73–78.

Sandhu, K., S. Singh, J.S. Kolar and A.P. Singh (1997). Weed control in peas. *J. Res.*, Punjab Agricultural University, 15: 236–240.

Saraswat, V.N. and B. Ray (1988). A decade of weed control in jute: A review. *Indian J. Weed Sci.*, 17(4): 54–60.

Satin, M., A. Berti and G. Zamin (1998). Influence of weed time of emergence and removal on crop. In: *Report of FAO Expert Consultation on Weed Ecology and Management*, FAO, Rome, Italy, September 22–24, 1997, Plant Production and Protection Division, pp. 53–64.

Shetty, S.V.R. and H.S. Gill (1974). Critical period of crop-weed competition in rice. *Indian J. of Weed Sci.*, 6(2): 101–107.

Solanki, S.S., C.R. Bisen and J.P. Tiwari (1980). Critical duration of weed competition in dwarf wheat var. Sonalika (Abstr.). In: *Proceedings of Indian Society of Weed Science*, Orissa University of Agriculture and Technology, Weed Science Conference held at Bhubaneshwar.

Tosh, G. (1982). Crop-weed competition studies in low land jute (Abstr.). In: *Proceedings of Conference of Indian Society of Weed Science*, pp. 16.

Vats, O.P. and M. Sidhu (1976) Critical period of crop-weed competition in mung (*Vigna radiata*). *Indian J. of Weed Sci.*, 8(1): 64–69.

Zimdahl, R.J. (1995). Weed science in sustainable agriculture. *American J. Altern. Agri.*, 10: 138–142.

40

Plant Protection Management through Seeds

Razia Khatoon Zaidi
Botany Section, Women's College, Aligarh Muslim University, Aligarh – 202 002, U.P.

In modern age, agriculture commodity has attained industrial importance, in other words, agriculture alone can keep the wheels of industry moving. Thus, the seed which is the source of origin of agriculture cannot be neglected. Good seed is not only meaningful end of the hunger of the world but a symbol of better way of life and foundation of an age of plenty, peace and freedom (Freeman, 1961).

A variety of micro-organisms like, bacteria, fungi and viruses tend to invade seeds and render them diseased. Among the seed invaders, the fungi constitute a major group. For more than 150 species of it are already known to be actively engaged in biodeterioration of various seeds and grains (Christensen and Kaufmann, 1965). Periodic failure of crops raised from even quality seeds due to diseases and production of abnormal seedlings and dead seeds in germination test have drawn the attention of research workers, administrators and seed producers. Considerable work has been done in advanced countries after the Second World-War for boosting up their agricultural production. The increasing demand of the healthy seeds in producing the disease free crops raised a new branch for phytopathology known as seed pathology.

Heavy losses have been observed to be caused by seed-borne pathogens in various crops, seed rots, seedling rot, *i.e.*, pre- and post-emergence losses, diseases at various stages of plant growth like leaf spots, stem rots, wilt, blights, root rots and fruit rots, etc., influence the crop stand and ultimate yield. The expectation of a cultivar are shattered if the seed used is not of a high quality and pathogen free.

In case of wheat, losses smut and bunts are the most important seed-borne diseases. In India, 30–40 per cent losses have been observed. *Alternaria triticina* Prasad and Prabhu caused leaf blight which resulted in 60 per cent losses in some fields. *Septoria nodorum* Berk. is prevalent in North India. In paddy *Drechslera oryzae* Breda de Haan. Subram and Jain., *Pyricularia oryzae* Cav. and *Xanthomonas oryzae* Uyeda and Ishiyama Dawson are important seed-borne pathogens. Bengal Famine of 1942–43 was due to *D. oryzae* when the losses were between 50–90 per cent. Losses of 6 to 60 per cent have been reported due to *X. oryzae*. In 1967, bacterial blight was in epiphytotic form in certain rice growing areas of Madhya Pradesh. White tip nematode (*Aphelenchoides oryzae* Yakvo) has been reported to cause losses of 50 per cent in Uttar Pradesh. In maize *Diplodia zeae* (Schw.) Lev., *Drechslera maydis* (Nisikado) Subram. and Jain, and *Fusarium moniliforme* Scheldon are important. Several smuts in sorghum are seed-borne. A number of leaf spot pathogens are also seed-borne like *Colletotrichum graminicola* (Ces.) Wilson, *Curvularia lunata* (Wakker) Boedijn. *Myrothecium* sp. etc. Ergot disease is important in pearl millet. In legumes, a number of fungi like *Macrophomina phaseoli* (Maublane) Ashby, *Colletotrichum lindemuthianum* (Sacc. and Magn.) Br. and Sav., *Ascochyta* sp. cause important diseases. Similarly, seeds of vegetables, Crucifers, Solanaceous crops, oilseed crops, forage crops, fiber crops, etc., carry pathogens which are detrimental to the crop. A number of destructive pathogens are associated with vegetatively propagated plant parts. In India, wart disease and golden nematode of potato were introduced through tuber and are now restricted in the areas of Darjeeling and Nilgri hills respectively. A number of viruses are seed-borne pathogens which result in heavy losses.

Infection of seeds results in seed rot, discolouration, reduced size, distortion, shrivelledness etc. Seed-borne infections generally reduce germination capability of seeds, deteriorate seed quality and reduce the market value. In rice, *D. oryzae* infection decreased the starch and sugar contents. In black gram protein contents were reduced due to *M. phaseoli* infection in seed.

Brief History of Seed Pathology

The earliest reference to the association of cause of plant diseases with seed is found in the works of Remnant as early as 1637 and of Tull 1773. Studies of Hell Wig (1699), with *Claviceps purpurea* (Fr.) Tul. on rye, Micheli (1723), with *Orobanche minor* L. on beans, Needham (1743, 1745), with *Anguina tritici* (Steinbuch) Filipjev on wheat and Tillet (1755), with *Tilletia* sp. on wheat are probably the earliest records which gave a factual basis to the growing suspicions of farmers that some disease agent is carried by seed. Frank (1883), demonstrated the seed-borne nature of C. *lindemuthianum* causing anthracnose of bean.

In Saxony, first official seed testing station was established in order to determine germination capacity and purity fo seeds (Malone and Muskett, 1964) although as early as 1916, Switzerland passed a seed law which required that at least two men in country where clover seed was sold were to serve as inspectors for seed offered for sale (Porter, 1949). In Europe, seed testing got importance after 1869 and a number of small testing and control stations were established (Brown, 1939). Association of fungi with germinating seed was soon observed by workers engaged in seed testing. In 1886, Bassey in Iowa, U.S.A. published the results of germination test with notes on the fungi found. Smith (1879 & 1901), in Great Britain, published illustrated notes on fungi detected on germinating farm seeds.

Seed health as an international problem attracted the attention of scientists as early as 1905 when the first meeting of international analyst was held in connection with Botanical Congress in Wien. Association of official seed analysts of North America was organized in 1908. Leningard Plant

Protection Station came into existence in 1920 and Dorogin started working on seed-borne fungi of crops in Russia. The first seed testing station was set up at Wageningen in Netherlands and the first seed pathology laboratory was established in 1918. Dorogin recommended testing of seeds for health as early as 1923. In 1938, Doyer published a handbook consisting of methods for detection of seed-borne organisms, their classified list and control measures. Muskett and Malone described ulster method for the study of seed-borne organisms of flax seed in 1941. Studies by Russel and Ledingham (1941), Leukel and Martin (1950), Gordon (1944), Greanly and Machacek (1946) on seed health and seed-borne fungi provided information regarding development of seed pathology in early days.

Seed Mycoflora Legumes

Among agricultural crops legumes occupy an important position because of their high nutritional value and because they are among the popular pulse as well as vegetable and forage crops. An attempt has been made in these pages to discuss the various aspects of phytopathological investigations carried out by different seed pathologists from time to time.

Phytopathological tests performed on normal and off coloured seeds of *Vigna mungo* by Dharamlingan and Ranakrishna (1978) revealed that 100 per cent of the off coloured seeds were infected by *Rhizoctonia solani*. Fungus cause wilting of susceptible varieties of chickpea, the most common fungi among the seed borne fungi are *Achrophialophora fusispora* and *Fusarium oxysporum ciceri* (Purkayastha and Chakarvarty, 1981).

As early as 1968 De Tempe made an analysis of the laboratory testing requirements of pea seeds infected with *Ascochyta*. Before Tempe some other workers like Popava (1961) studied ascochytosis of peas. Black spot of pea seeds, hollow heart of peas, influence of storage on deterioration of stored pea seeds, phytopathogenic and saprophytic fungi associated with forage legumes (Crosier 1961; Field and King 1962; Tetravanikova *et al.*, 1963; Leech 1965). Wharton and Eusor (1970) performed a presumptive test to detect *Ascochyta* infection in pea seeds. Gupta (1970), Mathur, Neergaard (1972) performed experiments and explained different aspects of seed borne fungi of rice, wheat, blackgram and soybean. *Aspergillus rubber* was found to deteriorate stored pea seeds. Storage conditions play an important role in seed quality losses (Onuorah and Herman *et al.*, 1973), *Pythium* species cause seed decay and stem rot of *Vigna sinensis* and *Pisum sativum*. Perry (1973) showed that susceptibility is negatively correlated with the proportion of seedling emergence.

Being an important crop, extensive literature is there regarding the seed pathogen of soybean, Gupta (1974) and Keeling (1974) studied soybean seed fungi, seed rot, and relation of seed exudates to host susceptibility. Seed borne micro-organisms of soybean decrease the percentage of germination. It depends on the concentration and age of inoculum (Foor *et al.*, 1976). Fungi associated with *Phaseolus mungo* and *Vigna mungo* seeds by (Suhag and Suryanarayana 1977, Saxena and Sinha 1978). Spilker *et al.* (1981) effect of humidity temperature fertility and cultivar on the reduction of soybean seed quality. Thermotherapy of soyabean seeds can check the seed-borne fungi. Hydrogen peroxide seed disinfection treatment proved successful (Helton and Dilbeck, Dawood, Thomas and Sinclair 1982). In the same year D. Ercole and sportelli reduce *Fusarium* infection with sodium hypochlorite on *Cicer arietinum*. Vyas and Nene (1984) improved seed germination of *Cicer* by seed treatment methods. Dleep and Madhu (2001) give a brief note on seed infection and transmission of *Fusarium oxysporum* in pigeonpea. Screening of blackgram and greengram germplasm against *Heterodera cajani* (Jagan *et al.* 2004) chemical control of powdery mildew of peas was carried out by Pandey (2004). Avadesh Kumar *et al.* (2004) carried out pathogenicity tests and showed tomato variety VFN-8 is resistant against *Fusarium oxysporum*. There is a need of the development of resistant varieties of important crop plants including legumes.

Some Aspects of Seed health Testing

As intensive agricultural development is taking place in India with a view to accelerate the food production for feeding the ever increasing population. All available resources are being mobilized to set up our production and the farmers are being advised to take up scientific farming. Increased productivity can be achieved by using cultivars of high yielding varieties and avoiding crop failure. This involved the better quality seeds in terms of germination, purity and health by the farmers. Seed health is therefore important for ensuring good plant stand and plant health.

A number of workers like Christensen (1957), Neergaard and Saad (1962), Suhag and Suryanarayan (1977), Bhattachary (1984), Khare (1996), Hooda and Srivastava (1998) have indicated that no single method is adequate enough in the field of seed pathology for detecting all the fungi associated with seeds. Some steps are:

Inspection of Dry and Pre-soaked Seeds

Inspection of dry and pre-soaked seeds is important to reveal the presence of adherent fungi. Certain seeds are deformed, damaged, discoloured, contaminated with mycelial fragments (Konmendahl *et al.*, 1957). Damaged seeds act as avenues for the entry of pathogens. Such seeds are much more susceptible to mould, saprophytes and pathogens than the normal seed both during storage and under field conditions.

Seeds of legumes examined after soaking have been found infected with conidia of *Alternaria, Drechslera, Curvularia, Fusarium, Aschochyta, Trichothecium roseum* (Wharton and Eusor, 1970; De Tempe 1973; Siddaramarian *et al.*, 1981).

Isolation of Seed Mycoflora

Isolation of seed mycoflora shows differences in quantity and quality. M.N. Khare (1996) described different methods to test the seeds for associated fungi. During seed health testing common blotter method and agar plate methods were used by Parveen and Prakash (1983) for isolation of fungal flora.

According to Mishra and Kanaujia (1973) and Lutey (1963) the variation in the type of fungi in different varieties may be in part due to variation in agricultural operations, climatological conditions, host specification and physico-chemical nature of seeds, and these factors influence fungal invasion.

Histopathology

Histopathology is an important aspect of the seed health testing, to ascertain the exact location of fungi in different seed parts.

Maceration techniques and histopathological studies were carried out by different workers to locate the site of seed infection and transmission of the infection. Schneider *et al.* (1974) isolated *Colletotrichum truncatum* within the seed coat of surface sterilized soybean seeds. Similar studies were made by Ilyas *et al.* (1975) and they observed *Diaporthe phaseolorum* var. *sojne* and *Cercospora kikuchii* in the seed coat, cotyledons and embryos of surface sterilized seeds of soybean. *Peronospora mushurica* (Naum) Syd. was detected in soybean seeds (Pathak *et al.*, 1978).

Presence of fungal mycelium in different parts of seeds was reported by Hyde (1950), Christensen (1951), Jain and Patel (1969), Agarwal and Jain (1978), Fulco (1979). Saxena and Sinha (1978) conducted histopathological investigations of the field fungi associated with the seeds of *Vigna* var. *radiate* and *V. mungo* from U.P. and Haryana. Fungal spores of *Botrytricinaria, Chaetomium indicum, Colletotrichum lendimuthianum, Fusarium equiseti* were found adherent on surface. Spores of *Aspergillus niger, A. flavus* and *Alternaria alternata* exist as extra embryonal infection of seeds. The spores get entry through the pores.

Wan Zainus *et al.* (1980) showed that *Phoma sorghina* and *D. phaseolorum* spores grow within seed coat, cotyledonary tissues and embryos. Hill and West (1982) found fungal penetration of soybean seeds through pores.

Pathogenicity

Pathogenicity of fungal flora and adverse effect of culture filtrates were studied both in laboratory as well as in field conditions. Inhibition of seed germination may partly be due to some pathogenic metabolites secreted by fungi in the culture medium (Curtis, 1958; Lugankas, 1962) Forgacs (1962), Conducer *et al.* (1963), Spensly (1963), Diener and Davis (1966) reported that Aflatoxin produced by *A. flavus* damage the seeds to appreciable degree. These findings are in conformity with Dwidi and singh (1973), Kanaujia (1974), Abraham (1978), Kamal and Verma (1979), Vishnunavat and Shukla (1981). Singh and Gupta (1984) found that culture filtrates of *Alternaria alternate, Aspergillus niger, Fusarium moniliforne* are quite effective in reducing the germination percentage of seeds and inhibition in root and shoot elongation increased as the culture grew older. Adverse effect of culture filtrates were observed in cowpea and grow seeds and reduction in germination percentage was observed (Zaidi 1999). Pathogenicity test of tomato against *M. incognita* and *Fusarium oxysporum* were carried out by Avadesh Kumar *et al.* (2004) in Bioved Research Center, Allahabad.

Plant Protection

Plant protection includes the experiments conducted to test the efficacy of different fungicides, nematicides, antibiotic and biocides against seed mycoflora.

For centuries seed treatment has been in use. About 1860 salt brine was accidently discovered to be effective against bunt of wheat. The modern era of fungicides and seed treatment started with introduction of organic mercury compounds and since then these chemicals are famous seed dressing compounds (Zote, Mayee and Kraft 1982). Karwasra and Singh (1982) successfully controlled the mycoflora of cluster bean by seed treatment. *Rhizoctonia solani* was effectively controlled by carboxin, carbendazin, Benomyl and quintozene when used as seed treatments, foliar spray and as soil drench. Carboxin was most effective (Venkatasubbariah *et al.*, 1983).

According to Vyas and Nene (1984) greater losses in *Cicer arietinum* seed germination occurred with increase in duration of storage. This was accelerated by storing the seed at high relative humidity. Treatment the seeds at higher rate of thirun resulted in considerably lesser loss in germination. However thiram treatment above the rate of 0.30 per cent did not result in any additive effect.

A.K. Jain (1998) efficacy and economics of some fungicides for the control of grain smut in barren yard. Millet control of seed borne fungi by antibiotics has also been observed by Rahalkas and Neergaard (1969), Thirumlachar (1967), Dharamvir and Raychaudry (1968 & 1971), Lo (1977). Streptomycin has been found most effective as it eliminated majority of the fungi in *vigna anguiculata* and *Cicer arietinum* seeds (Zaidi and Prakash 1983). Nematicides could be used for controlling fungi Van Berkun and Hoestra (1979).

Biocontrol Management

The continued search for better and ecofriendly methods of combating fungal pathogens has in recent past moved away from the traditional empirical search for new synthetic chemicals. The growing realization that plant and their products have the potential in providing materials deadly to fungi, has prompted a renewed look at these natural biocides. Further, the idea that there can be an alternative to chemicals for management of some plant diseases have recently gained attention by several workers. According to Shahi, Singh and Singh (1998) Margosa oil and cake extract were tested to observe its potential as a biocide against collar rot pathogen, *Sclerotium rolfsii* of chickpea and their effect on seed germination of the host. Both oil and cake extract inhibited the effect of the pathogen to some degree without affecting the seed germination of chickpea.

Khan *et al.* (1974) reported two bitter principals of neem *viz.*, Nibidin and Thionimone were highly toxic to plant parasitic nematode and fungi. Alam *et al.* (1975) observed that all concentrations of root exudates of margosa were inhibitory to larval hatch of *Meloidogyne incognita*. Neem leaf extract and neemgum has found quite useful in checking the seed-borne fungi of cowpea and gram. Number of fungi isolated from seeds treated with these plant products were 9 and 8 respectively as compared to 21 in control (Zaidi and Prakash, 1983). Chavan (1983) reported alkanes from neem leaves to be insecticidal.

Latex of *Calotropis procera* has been found quite effective in controlling the seed mycoflora of gram. Pretreatment of seeds with latex not only reduce the fungal flora but also increase the percentage germination of the seeds (Zaidi 1997). Parveen (1983) has also reported the reduction in fungal flora of *Triticale* and *Penisetum* by latex of *Calotropis procera*.

Varietal reaction and pathogenicity test of tomato against *M. incognita* and *Fusarium oxysporum* was studied by Avadesh *et al.* (2004). Biological control of post harvest citrus disease using yeast by Neeta Sharma *et al.* (2004). Control of bacterial blight with a number of plant extracts from *Eucalyptus*, garlic, *Curcuma* (Neetu Rani *et al.*, 2004). Comparative efficacy of different biocontrol agents gainst *M. incognita* and *Fusarium oxysporum* on Brinjal were tested by Akhtar Haseeb *et al.* (2004) at A.M.U. Aligarh and they observed an enhancing effect on plant fresh weight and these treatments were quite effective against the pathogen.

Conclusion

It is evident from the foregoing discussion that the seeds of important crops harbour a definite external and internal seed mycoflora, which reduce the seedling stand in the field, should invariably be tested for seed health before planting so as to avoid the introduction of pathogens in new areas and be treated with pesticides to attain maximum yield of crops. But the emphasis should be on the use of biopesticides and plant product which are ecofriendly which can minimize the pollutant load of the environment (air, water and soil). These biopesticides could be employed an encouraging results have been obtained in reducing the seed mycoflora which in a way could be ranked equal to that obtained by chemical pesticides. These biopesticides are easy, economical, indigenous and safe methods of plant protection management.

REFERENCES

Abraham, T.A. (1978). *Indian J. Mycol. and Plant Path.*, 31(3): 402–403.

Agarwal, N.K. and B.L. Jain (1978). *Indian J. Mycol. and Plant Path.*, 8(1): 59.

Alam, M.M., A. Masood and S.I. Husain (1975). *Indian J. Exp. Biol.*, 13(4): 412–414.

Avadesh, K.S., R.C. Pandey, B.K. Dwivedi and S.P. Singh (2004). *Proc. 6th Indian Agricultural Scientist and Farmer's Congress*, Allahabad.

Bessey, C.E. (1886). Quoted by Malone and Musket (1964).

Bhattacharya, K.K., K. Sengupta, B.N. Chatterji (1984). *Seed Res.*, 12(1): 40–44.

Brown, E. (1939). Development of seed testing. U.S. Deptt. of Agric. Bur. Ind. Div. Crops and Disease Minn. Siro.

Chavan, S.R. (1983). *Second International Neem Conference 1983* held in Germany.

Christensen, C.M. and H. Henry Kaufman (1965). *Annual Rev. Phytopath.*, 3: 69–84.

Conducer, R.C., K. Sargeant and R. Yeo (1963). *Biotech. Bioling*, 5: 185–192.

Crosier, W.F. (1961). *Phytopath.*, 25(7): 827–829.

Curtis, R.W. (1958). *Science*, 128: 661–682.

Dawood, M.K.M. (1982). *Acta Mycologia*, 18(1): 83–112.

D'Ercole, N. and M. Sportelli (1982). *Informatore Fitopathologico*, 32(6): 51–54.

Dleep and Madhu (2001). *J. Indian Bot. Soc.*, 80: 99–102.

De Tempe, J. (1968). *Proc. Int. Seed Test. Ass.*, 33: 583.

De Tempe, J. (1973). *Seed Sci. and Technol.*, 1: 845.

Dharamlingan, C. and V. Rama Krishna (1978). *Seed Res.*, 6: 101–109.

Dharamvir and S.P. Raychaudhry (1968). *Hindustan Antibiotics Bull.*, 10: 319–321.

Dharamvir and S.P. Raychaudhry (1971). *Plant Prot. Bull., New Delhi*, 23: 1–3.

Diener, U.L. and N.D. Davis (1966). *Phytopath.*, 56: 1390–1393.

Dorogin, G.N. (1923). *Rev. Appl. Mycol.*, 3: 24.

Doyer, L.C. (1938). *Int. Seed Test Ass.* N. Veerman and Zonen Wageningen, Holland, 58.

Dwivedi, R.S. and S.P. Singh (1973). *Acta Bot. Indica* 1: 41–44.

Fields, R.W. and T.H. King (1962). *Phytopath.*, 52(4): 336–339.

Forgaces, J. (1962). *Foodstuffs*, 34: 124–134.

Freeman, O.L. (1961). Cited from the year book of Agriculture on Seed. U.S.D.A. Washington, D.C.

Fulco, W., S. Da Soares M.H.G. and Dc. O. Carmargo M.R. (1977). *Agronome Sulviogradense*, 13(2): 309–318.

Gordon, W.L. (1944). *Can. J. Res.*, 22: 282.

Gupta, O.M. and Kotashanta (1970). *Indian J. Mycol. and Plant Path.*, 8: 62.

Gupta, P.C., D.A. Millar and C.H. Mittle (1974). *Ind. J. Agric. Sci.*, 43(6): 617–618.

Harman, G.E., C.E. Heit, Pfleger and S.W. Braveman (1973). *Pl. Dis. Reptr.*, 57: 592.

Haseeb, A., S. Abuzar and P.K. Shukla (2004). *Proc. of 6th Indian Agricultural Scientists and Farmer's Congress*, Bioved Research, Allahabad.

Helton, A.W. and R. Dilbeck (1982). *Plant Dis.*, 66(9): 784–787.

Hill, H.J. and H.S. West (1982). *Crop Science*, 22(3): 602–605.

Hooda, K.S. and M.D. Srivastva (1998). *Indian J. Plant Path.*, 16(122): 64–66.

Hyde, M.B. (1950). *Ann. Appl. Biol.*, 37: 179–186.

Ilyas, M.B., O.D. Dhigra, M.A. Ellis and J.B. Sinclair (1975). *Plant Disease Rep.*, 59(1): 17–19.

Jain, J.P. and P.N. Patel (1969). Seed mycflora of 'guar', their role in emergency and vigour of seedlings and efficacy of fungicides. *Indian Phytopath.*, 29(2): 245–250.

Jain, A.K. (1998). *Indian J. Plant Path.*, 16(1&2): 57–59.

Kanujia, R.S. (1974). *Acta Bot. Indica*, 2: 129–135.

Karwasra, S.S. and M. Singh (1982). *Indian Phytopath.*, 35(3): 501–502.

Keeling, B.L. (1974). *Phytopath.*, 64(11): 1445–1447.

Khan, M.W., M.M. Alam, A.M. Khan and S.K. Saxena (1974). *Acta Botanica Indica*, 2(2): 120–128.

Khare, M.N., R.P. Mishra, S.M. Kumar and J.N. Chand (1972). *Indian Phytopath.*, 26(1): 69–75.

Kraft, J.M. (1982). *Plant Disease*, 66(9): 798–800.

Lal, J., U. Kumar and S.D. Mishra (2004). *Proc. 6th Indian Agricultural Scientists and Farmer's Congress*, Allahabad.

Lo, S.L. (1977). *Plant Prot. Bull.*, Taiwan, 19(1): 55–60.

Malone, I.P. and A.E. Muskett (1964). *Proc. Int. Seed Test. Ass.*, 29: 1–383.

Mathur, S.B. and P. Neergaard (1972). *Proc. Int. Seed Test Ass.*, 37: 723.

Muskett, A.E. and J.P. Malone (1941). *Annl. Appl. Boil.*, 28: 8–13.

Needham, T. (1743). *Chil. Trans. Roy. Soc. London*, 42: 534.

Needham, T. (1945). New microscopical discoveries (F. Needhum, Halborn, England, 126).

Neergaard, P. and A. Saad (1962). *Indian Phytopath.*, 15(2): 85–111.

Ounorah, P.E. (1973). *Plant and Soil*, 39(1): 187–191.

Parveen, Q. (1983). Studies on seed mycoflora of *Triticale* and *Pennisetum*. Ph.D. Thesis, Aligarh Muslim University, Aligarh.

Pathak, V.K., S.B. Mathur and P. Neergaard (1978). *Eppo Bull.*, 8: 21.

Perry, D.A. *Transactions of the British Mycological Society*, 61(1): 135–144.

Popova, M.I.V. (1961). *Trud. Vseros, Nanch issled Inst. Sakhar. Svekly I sakhara* 1(2): 47–54,

Porter, R.H. (1949). *Bot. Rev.*, 15: 221–344.

Purkayastha, R.P. and B.N. Chakraborty (1981). *Plant Disease*, 65(4): 362.

Rahalkar, P.W. and P. Neergaard (1969). *Hindustan Antibiotics Bull.*, 11(3): 163–165.

Rani, N., P. Kumar and S. Rathi (2004). *Proc. 6th Indian Agricultural Scientists and Farmer's Congress*, Allahabad.

Russel, R.C. and R.K. Ledingham (1941). *Sci. Argr.* 21: 761.

Saxena, R.M. and S. Sinha (1978). *Indian J. Mycol. and Plant Path.*, 8: 61.

Schneider, R.W., O.D. Dhingra, J.F. Nicholson and J.B. Sinclair (1974). *Phytopath.*, 64: 154–155.

Shahi, R.K., R.B. Singh and N.B. Singh (1998). *Indian J. Plant Path.*, 16(1&2): 50–54.

Sharma, N., V.K. Verma, P. Avasthi (2004). *Proc. 6th Indian Agricultural Scientists and Farmer's Congress*, Allahabad.

Shortt, B.J., J.B. Sinclair, C.G. Helm, M.R. Jeffords and M. Kgan (1982). *Phytopath.*, 72(6): 615.

Siddarmaiah, A.L., S.A. Desai and R.K. Hegde (1981). *Curr. Res.*, 10(8): 131–132.

Singh, P.N. and K. Gupta (1984). *Seed Res.*, 12(1): 123–126.

Sinha, O.K. and M.N. Khare (1977). *Indian Phytopath.*, 30: 337.

Smigh, A.L. (1879–1901). *Trans. Brit. Mycol. Soc.*, 1: 182.

Spensley, P.C. (1963). *Endeavour*. 22: 75–79.

Spilker, D.A., A.F.R. Schmittuenner and C.W. Ellett (1981). *Phytopath.*, 71(10): 1027.

Suhag, L.S. (1978). *Indian J. Mycol. Plant Path.*, 3(1): 40–43.

Teterevni Kova-Babayen, D.N. Mme (1963). *Inzu. Akad. Nauk. Armyam. S.S.R. Soil. Sci.*, 16: 9.

Thirumalachar, J.M. (1968). *Adv. Appl. Microbiol.*, 10: 313–337.

Tull, J. (1733). *Proc. II. Bot. Conf. II*, 2, 14.

Van Berkum, J.A. and H. Hoestra (1979). Amsterdam, Oxford, New York.

Venkatasubbaiah, P., H. Shokarashetty and K.M. Safeeullah (1983). *Indian Phytopath.*, 36(2): 279.

Verma, R.K. and S.C. Vyas (1977). *Pesticides*, 11: 20.

Vishunavat, K. and P. Shukla (1981). *Indian Phytopath.*, 34(4): 512–514.

Wan Zainum, N. and D.G. Parbery (1980). *Science and Tech.*, 8(2): 193–201.

Zaidi, R.K. and D. Prakash (1981). *J. Sci. Res.*, B3: 191–193.

Zaidi, R.K. and D. Prakash (1982). *J. Sci. Res.*, 4(2): 77–79.

Zaidi, R.K. and D. Prakash (1983). *Ph.D. Thesis*, A.M.U. Aligarh.

Zaidi, R.K. (1997). *New Agriculturist*, 8(2): 183–185.

Zote K.K. and C.D. Mayee (1982). *Pesticides*, 16(4): 10–12.